Social Sciences

Life Sciences

General Interest

Finite Mathematics

for the Managerial, Life, and Social Sciences

SIXTH EDITION

S. T. Tan
Stonehill College

Australia • Canada • Mexico • Singapore • Spain • United Kingdom • United States

Sponsoring Editor: *Curt Hinrichs*
Marketing Team: *Karin Sandberg, Beth Kroenke, Laura Hubrich*
Editorial Assistant: *JoAnne von Zastrow, Emily Davidson*
Production Coordinator: *Marjorie Z. Sanders*
Production Service: *Cecile Joyner, The Cooper Company*
Interior Design: *Delgado Design, Inc.*
Cover Design: *Lisa Henry*
Cover Illustration: *Judith Harkness*
Photo Editor: *Terry Powell*
Manuscript Editor: *Betty Duncan*
Print Buyer: *Vena Dyer*
Typesetting: *Bi-Comp/The PRD Group*
Cover Printing: *Phoenix Color Corp.*
Printing and Binding: *World Color Corp., Versailles*

Photo Credits: p. xvi, The Photographer's Window; Pablo Corral V/Corbis; Danny Lehman/Corbis; **p. xvii,** The Photographer's Window; Bob Rowan/Progressive Image/ Corbis; **p. 3,** The Photographer's Window; **p. 77,** PhotoDisc; **p. 181,** PhotoDisc; **p. 219,** Tony Freeman/Photo Edit; **p. 283,** PhotoDisc; **p. 335,** PhotoDisc; **p. 383,** PhotoDisc; **p. 455,** The Photographer's Window; and **p. 531,** The Photographer's Window.

For more information, contact:

BROOKS/COLE PUBLISHING COMPANY
511 Forest Lodge Road
Pacific Grove, CA 93950 USA
www.brookscole.com

Printed in the United States of America

10 9 8 7 6 5 4 3 2

Library of Congress Cataloging-in-Publication Data

Tan, Soo Tang.
Finite mathematics for the managerial, life, and social sciences / S. T. Tan.—6th ed.
p. cm.
Rev. ed. of: Applied finite mathematics. 5th ed. © 1997.
Includes index.
ISBN 0-534-36960-X
1. Mathematics. I. Tan, Soo Tang. Applied finite mathematics. II. Title.

QA39.2.T34 2000 99-047847
510—dc21

To Pat, Bill, and Michael

Contents

* Sections marked with an asterisk are not prerequisites for later material.

PREFACE

Finite Mathematics for the Managerial, Life, and Social Sciences, Sixth Edition, is directed toward students in these fields. The only prerequisite for understanding this book, which treats the standard topics in finite mathematics, is 1 to 2 years, or the equivalent, of high school algebra. The objective of the text is twofold: (1) to provide students with background in the quantitative techniques necessary to better understand and appreciate the courses normally taken in undergraduate training and (2) to lay the foundation for more advanced courses, such as statistics and operations research. We hope to accomplish this by striking a careful balance between theory and applications. This edition incorporates many valuable comments and suggestions by users of the fifth edition and reviewers of the sixth edition.

FEATURES

The following list includes some of the many important features of the book:

■ **Coverage of Topics** Since the book contains more than ample material for a one-semester or two-quarter course, the instructor may be flexible in choosing the topics most suitable for his or her course. The following chart on chapter dependency is provided to help the instructor design a course that is most suitable for the intended audience.

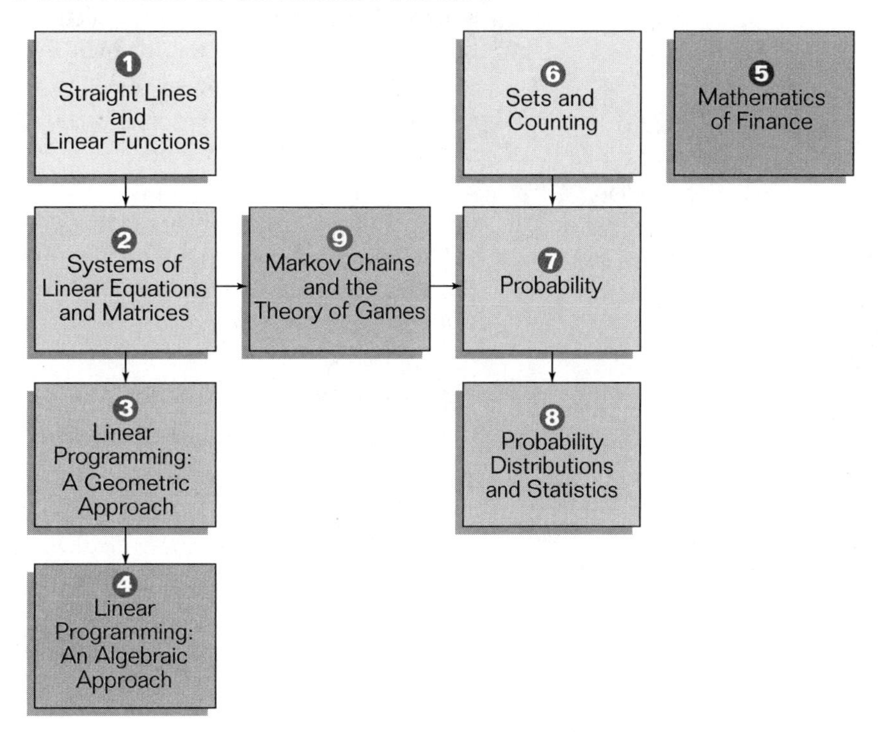

■ **Approach** The problem-solving approach is stressed throughout the book. Numerous examples and solved problems are used to amplify each new concept or result in order to facilitate students' comprehension of the material. Graphs and pictures are used extensively to help students visualize the concepts and ideas being presented.

■ **Level of Presentation** Our approach is intuitive, and we state the results informally. However, we have taken special care to ensure that this approach does not compromise the mathematical content and accuracy.

■ **Applications** The text is application oriented. Many interesting, relevant, and up-to-date applications are drawn from the fields of business, economics, social and behavioral sciences, life sciences, physical sciences, and other fields of general interest. Some of these applications have their source in newspapers, weekly periodicals, and other magazines. Applications are found in the illustrative examples in the main body of the text as well as in the exercise sets.

■ **Sources** We have included sources for those applications that are based on real-life data.

■ **Exercises** Each section of the text is accompanied by an extensive set of exercises containing an ample set of problems of a routine, computational nature that will help students master new techniques. The routine problems are followed by an extensive set of application-oriented problems that test students' mastery of the topics. Included among these problems are numerous graphing calculator exercises. Each chapter of the text also contains a set of review exercises. Answers to all odd-numbered exercises appear in the back of the book.

■ **Self-Check Exercises** Every section has self-check exercises, with solutions, to help students monitor their own progress.

■ **Portfolios** These interviews are designed to convey to the student the real-world experiences of professionals who have a background in mathematics and use it in their professions.

■ **Group Discussion Questions** These are optional questions, appearing throughout the main body of the text, that can be discussed in class or assigned as homework. These questions generally require more thought and effort than the usual exercises. Complete solutions to these exercises are given in the *Complete Solutions Manual.*

TECHNOLOGY

Exploring with Technology Questions

These optional questions appear throughout the main body of the text and serve to enhance the student's understanding of the concepts and theory presented. Complete solutions to these exercises are given in the *Complete Solutions Manual.*

Using Technology Subsections

These pages contain optional material and are placed at the end of the sections for which their use is appropriate. The subsections are written in the traditional example–exercise format, with answers given at the back of the book. They may be used in the classroom if desired or as material for self-study by the student.

As many up-to-date and relevant applications have been introduced in these subsections, they provide students with an opportunity to interpret results in a real-life setting.

Student Resources on the Web

Students and instructors will now have access to these additional materials at the Brooks/Cole World Wide Web site:
http://www.brookscole.com/math/authors/tans

- Review material and practice chapter quizzes and tests
- Group projects and extended problems for each chapter
- Instructions, including keystrokes, for the procedures referenced in the text for specific calculators (TI-82, TI-83, TI-85, TI-86, and other popular models)
- Modified *Using Technology* sections for CAS systems, including the command statements for Mathematica, Maple, and other popular systems
- New spreadsheet examples as described below

New in the Sixth Edition

- Exercises that emphasize the understanding of concepts and theory have been added to most sections. These new exercises are usually near the end of the exercise sets and take the form of true or false questions.
- More real-life applications, with sources, have been added; for example, the narrowing gender gap, net-connected computers in Europe, the social ladder, on-line banking, and the growth of portable phone services. Examples of other new applications are 401K retirement plans, restaurant health-code violations, the probability of success of a kidney transplant, small-town revival, and refinancing a home.
- A graphical illustration in 3-dimensional space of the solution of a problem using the Simplex Method is given in Chapter 4. This affords students an opportunity to visualize the steps associated with the iterations of the Simplex Method in solving the problem. (See Figures 4.4 and 4.5 on pages 233 and 234.)
- More Group Discussion questions and Exploring with Technology questions have been added. Also, more exercises have been added to the Using Technology sets.
- In Sections 9.1–9.3 (Markov Chains), the distribution vectors have been written as column vectors instead of row vectors. This change was made so that the matrix multiplication in these sections would conform with that given in Chapter 3.
- Coverage of the equation of a circle has been added to Chapter 1.

- New spreadsheet examples and exercises for selected sections that may be solved using the Microsoft® Excel program are now given at the Brooks/Cole Web site: http://www.brookscole.com/math/authors/tans/
- Instructions for the TI-83 and TI-86 calculators have been added to the Web site.
- New supplements include a text-specific graphing calculator manual and a standalone Microsoft® Excel Manual.

SUPPLEMENTS

- *Student's Solutions Manual,* available to both students and instructors, includes the solutions to odd-numbered exercises. ISBN 0-534-37001-2
- *Complete Solutions Manual* includes solutions to all exercises. ISBN 0-534-37002-0
- *Printed Test Bank with Chapter Tests,* free to adopters of the book, contains sample tests for each chapter. ISBN 0-534-37009-8
- *Thomson Learning Testing Tools* is a fully integrated suite of test creation, delivery, and reporting tools for exams, quizzes, and tutorials.
- *Graphing Calculator Supplement,* by Ryan & Hester, both of Texas A&M University, is available to both students and instructors. The manual develops selected examples and exercises and also includes additional problems for reinforcement. It is specifically written for use with the TI line of programmable graphics calculators. ISBN 0-534-37403-4
- *Finite Mathematics with Microsoft® Excel,* by Chester Piascik, Bryant College, illustrates key topics in finite mathematics through the use of Microsoft Excel. Explanations of Excel instructions and formulas reinforce underlying mathematical concepts. The author encourages students to be active learners, asking them to verbalize and verify the mathematical concepts behind the spreadsheet results. ISBN 0-534-37057-8

ACKNOWLEDGMENTS

I wish to express my personal appreciation to each of the following reviewers, whose many suggestions have helped make a much improved book.

Daniel D. Anderson
University of Iowa

Ronald D. Baker
University of Delaware

Ronald Barnes
University of Houston—Downtown

Frank E. Bennett
Mount Saint Vincent University

Teresa L. Bittner
Canada College

Michael Button
San Diego City College

Frederick J. Carter
St. Mary's University

Charles E. Cleaver
The Citadel

Leslie S. Cobar
University of New Orleans

Matthew P. Coleman
Fairfield University

William Coppage
Wright State University

Jerry Davis
Johnson State College

Michael W. Ecker
Pennsylvania State University, Wilkes-Barre Campus

Bruce Edwards
University of Florida—Gainesville

Robert B. Eicken
Illinois Central College

Charles S. Frady
Georgia State University

Howard Frisinger
Colorado State University

Larry Gerstein
University of California—Santa Barbara

John Haverhals
Bradley University

Sharon S. Hewlett
University of New Orleans

Patricia Hickey
Baylor University

Harry C. Hutchins
Southern Illinois University

Frank Jenkins
John Carroll University

Bruce Johnson
University of Victoria

David E. Joyce
Clark University

Martin Kotler
Pace University

Paul E. Long
University of Arkansas

Larry Luck
Anoka-Ramsey Community College

Sandra Wray McAfee
University of Michigan

Gary MacGillivray
University of Victoria

Gary A. Martin
University of Massachusetts—Dartmouth

Norman R. Martin
Northern Arizona University

Ruth Mikkelson
University of Wisconsin—Stout

Maurice Monahan
South Dakota State University

John A. Muzzey
Lyndon State College

James D. Nelson
Western Michigan University

Ralph J. Neuhaus
University of Idaho

Lloyd Olson
North Dakota State University

Wesley Orser
Clark College

Lavon B. Page
North Carolina State University

James Perkins
Piedmont Virginia Community College

Richard D. Porter
Northeastern University

Sandra Pryor Clarkson
Hunter College—SUNY

Richard Quindley
Bridgewater State College

C. Rao
University of Wisconsin

Robert H. Rodine
Northern Illinois University

Thomas N. Roe
South Dakota State University

Arnold Schroeder
Long Beach City College

Donald R. Sherbert
University of Illinois

Ron Smit
University of Portland

Lowell Stultz
Texas Township Campus

Lawrence V. Welch
Western Illinois University

A special thanks also goes to Arthur J. Rosenthal, Salem State College, and Bruce R. Johnson, University of Victoria. I also wish to thank Lea Rosenberry for her assistance in preparing the Test Banks for this edition. My thanks also go to the editorial, production, and marketing staffs of Brooks/Cole: Curt Hinrichs, Emily Davidson, Marjorie Sanders, Vernon Boes, Karin Sandberg, Beth Kroenke, and Samantha Cabaluna for their thoughtful contributions and patient assistance and cooperation during the development and production of this book. Finally, I wish to thank Cecile Joyner of The Cooper Company and Betty Duncan, for doing an excellent job ensuring the accuracy and readability of this sixth edition, and Delgado Design for the design of the interior of the book.

S. T. Tan

APPLICATIONS

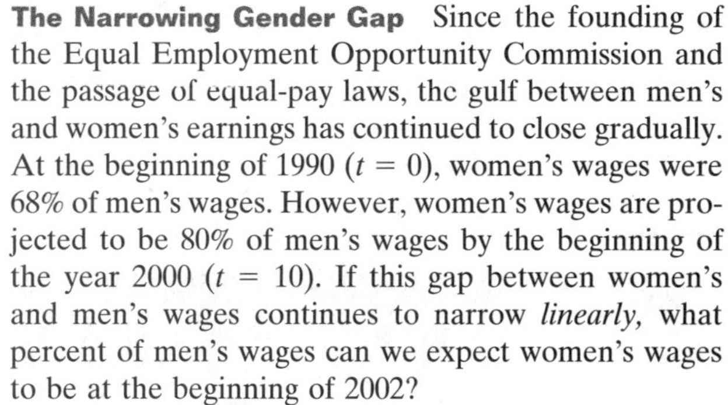

The Narrowing Gender Gap Since the founding of the Equal Employment Opportunity Commission and the passage of equal-pay laws, the gulf between men's and women's earnings has continued to close gradually. At the beginning of 1990 ($t = 0$), women's wages were 68% of men's wages. However, women's wages are projected to be 80% of men's wages by the beginning of the year 2000 ($t = 10$). If this gap between women's and men's wages continues to narrow *linearly,* what percent of men's wages can we expect women's wages to be at the beginning of 2002?

Source: Journal of Economic Perspectives

Television Advertising As part of a campaign to promote its annual clearance sale, the Excelsior Company decided to buy television advertising time on Station KAOS. Excelsior's television advertising budget is \$102,000. Morning time costs \$3000 per minute, afternoon time costs \$1000 per minute, and evening (prime) time costs \$12,000 per minute. Because of previous commitments, KAOS cannot offer Excelsior more than 6 minutes of prime time or more than a total of 25 minutes of advertising time over the two weeks in which the commercials are to be run. KAOS estimates that morning commercials are seen by 200,000 people, afternoon commercials are seen by 100,000 people, and evening commercials are seen by 600,000 people. How much morning, afternoon, and evening advertising time should Excelsior buy to maximize exposure of its commercials?

Restaurant Health-Code Violations Suppose that 30 percent of the restaurants in a certain town are in violation of the health code. If a health inspector randomly selects five of the restaurants for inspection, what is the probability that

a. none of the restaurants are in violation of the health code?
b. just one of the restaurants is in violation of the health code?
c. at least two of the restaurants are in violation of the health code?

The Social Ladder The following table summarizes the results of a poll conducted with 1154 adults by the *New York Times/CBS News.*

Annual Household Income ($)	Percent of Respondents Within That Income Range	Percent of Respondents Who Call Themselves Rich	Middle Class	Poor
Less than 15,000	11.2	0	24	76
15,000–29,999	18.6	3	60	37
30,000–49,999	24.5	0	86	14
50,000–74,999	21.9	2	90	8
75,000 and higher	23.8	5	91	4

a. What is the probability that a respondent chosen at random calls himself or herself middle class?
b. If a randomly chosen respondent calls himself or herself middle class, what is the probability that the annual household income of the individual is between $30,000 and $49,999, inclusive?
c. If a randomly chosen respondent calls himself or herself middle class, what is the probability that the individual's income is less than or equal to $29,000 or greater than or equal to $50,000?

Source: New York Times/CBS News, Wall Street Journal Almanac

Refinancing a Home The Martinez's are planning to refinance their home. The outstanding balance on their original loan is $150,000. Their finance company has offered them two options:
Option A is a fixed-rate mortgage at an interest rate of 7.5% per year compounded monthly, payable over a 30-year period in 360 equal monthly installments.
Option B is a fixed-rate mortgage at an interest rate of 7.25% per year compounded monthly, payable over a 15-year period in 180 equal monthly installments. (Assume that there are no additional finance charges.)
a. Find the monthly payment required to amortize each of these loans over the life of the loan.
b. How much interest would Mr. and Mrs. Martinez save if they chose the 15-year mortgage instead of the 30-year mortgage?

Finite Mathematics

for the Managerial, Life, and Social Sciences

SIXTH EDITION

1 STRAIGHT LINES AND LINEAR FUNCTIONS

This chapter introduces the Cartesian coordinate system, a system that allows us to represent points in the plane in terms of ordered pairs of real numbers. This in turn enables us to compute the distance between two points algebraically. We also study straight lines. *Linear functions*, whose graphs are straight lines, can be used to describe many relationships between two quantities. These relationships can be found in fields of study as diverse as business, economics, the social sciences, physics, and medicine. In addition, we see how some practical problems can be solved by finding the point(s) of intersection of two straight lines. Finally, we learn how to find an algebraic representation of the straight line that "best" fits a set of data points that are scattered about a straight line.

Which process should the company use? The Robertson Controls Company must decide between two manufacturing processes for its Model C electronic thermostats. In Example 4, page 51, you will see how to determine which process will be more profitable.

1.1 The Cartesian Coordinate System

The Cartesian Coordinate System

The real number system is made up of the set of real numbers together with the usual operations of addition, subtraction, multiplication, and division. We assume that you are familiar with the rules governing these algebraic operations (see Appendix B).

Real numbers may be represented geometrically by points on a line. Such a line is called the **real number,** or **coordinate, line.** We can construct the real number line as follows: Arbitrarily select a point on a straight line to represent the number zero. This point is called the **origin.** If the line is horizontal, then choose a point at a convenient distance to the right of the origin to represent the number 1. This determines the scale for the number line. Each positive real number x lies x units to the right of zero, and each negative real number $-x$ lies x units to the left of zero.

In this manner, a one-to-one correspondence is set up between the set of real numbers and the set of points on the number line, with all the positive numbers lying to the right of the origin and all the negative numbers lying to the left of the origin (Figure 1.1).

FIGURE 1.1
The real number line

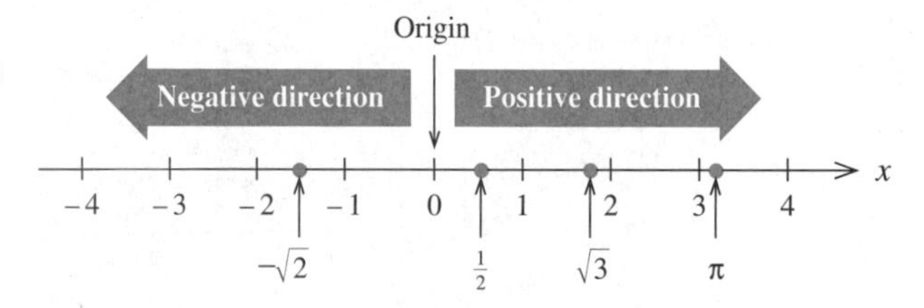

In a similar manner, we can represent points in a plane (a two-dimensional space) by using the **Cartesian coordinate system,** which we construct as follows: Take two perpendicular lines, one of which is normally chosen to be horizontal. These lines intersect at a point O, called the **origin** (Figure 1.2). The horizontal line is called the **x-axis,** and the vertical line is called the **y-axis.** A number scale is set up along the x-axis, with the positive numbers lying to the right of the origin and the negative numbers lying to the left of it. Similarly, a number scale is set up along the y-axis, with the positive numbers lying above the origin and the negative numbers lying below it.

FIGURE 1.2
The Cartesian coordinate system

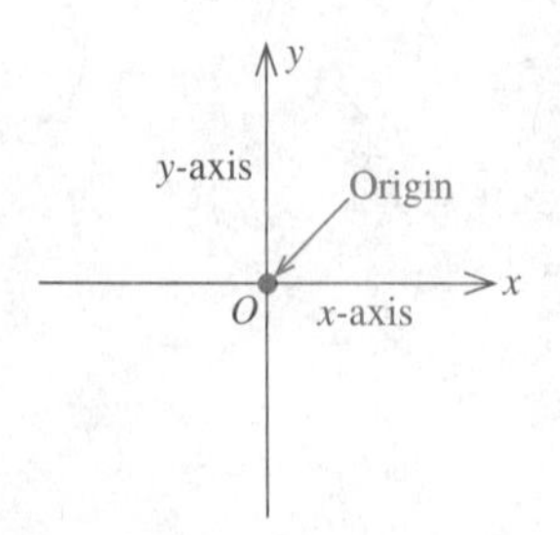

REMARK The number scales on the two axes need not be the same. Indeed, in many applications different quantities are represented by x and y. For example, x may represent the number of typewriters sold and y the total revenue resulting from the sales. In such cases it is often desirable to choose different number scales to represent the different quantities. Note, however, that the zeros of both number scales coincide at the origin of the two-dimensional coordinate system. ■■■

FIGURE 1.3
An ordered pair in the coordinate plane

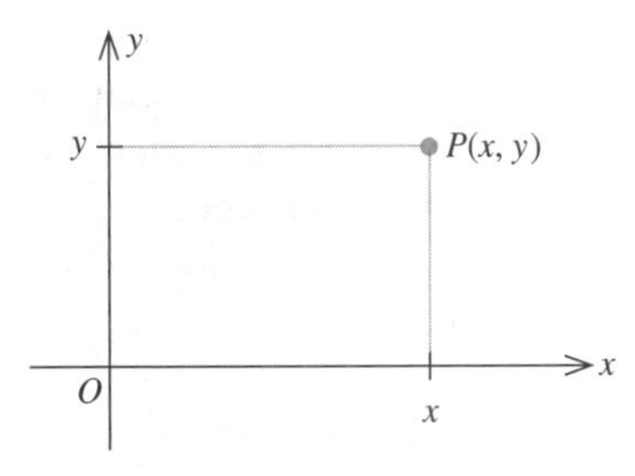

We can represent a point in the plane uniquely in this coordinate system by an **ordered pair** of numbers—that is, a pair (x, y), where x is the first number and y the second. To see this, let P be any point in the plane (Figure 1.3). Draw perpendiculars from P to the x-axis and y-axis, respectively. Then the number x is precisely the number that corresponds to the point on the x-axis at which the perpendicular through P hits the x-axis. Similarly, y is the number that corresponds to the point on the y-axis at which the perpendicular through P crosses the y-axis.

Conversely, given an ordered pair (x, y), with x as the first number and y the second, a point P in the plane is uniquely determined as follows: Locate the point on the x-axis represented by the number x and draw a line through that point parallel to the y-axis. Next, locate the point on the y-axis represented by the number y and draw a line through that point parallel to the x-axis. The point of intersection of these two lines is the point P (Figure 1.3).

In the ordered pair (x, y), x is called the **abscissa,** or **x-coordinate,** y is called the **ordinate,** or **y-coordinate,** and x and y together are referred to as the **coordinates** of the point P. The point P with x-coordinate equal to a and y-coordinate equal to b is often written $P(a, b)$.

The points $A(2, 3)$, $B(-2, 3)$, $C(-2, -3)$, $D(2, -3)$, $E(3, 2)$, $F(4, 0)$, and $G(0, -5)$ are plotted in Figure 1.4.

REMARK In general, $(x, y) \neq (y, x)$. This is illustrated by the points A and E in Figure 1.4. ■■■

The axes divide the plane into four quadrants. Quadrant I consists of the points P with coordinates x and y, denoted by $P(x, y)$, satisfying $x > 0$ and $y > 0$; Quadrant II, the points $P(x, y)$, where $x < 0$ and $y > 0$; Quadrant III, the points $P(x, y)$, where $x < 0$ and $y < 0$; and Quadrant IV, the points $P(x, y)$, where $x > 0$ and $y < 0$ (Figure 1.5).

FIGURE 1.4
Several points in the coordinate plane

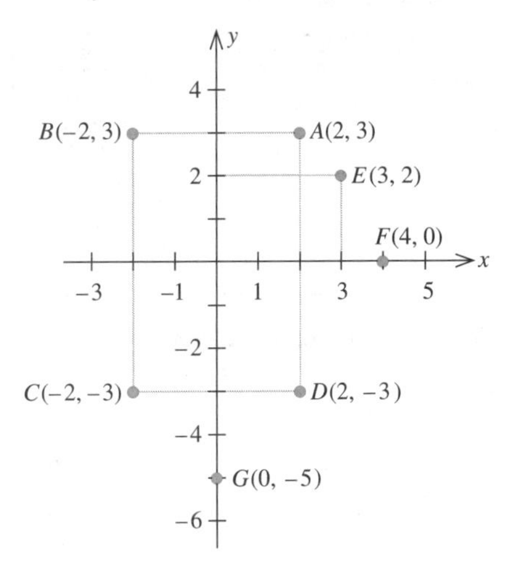

FIGURE 1.5
The four quadrants in the coordinate plane

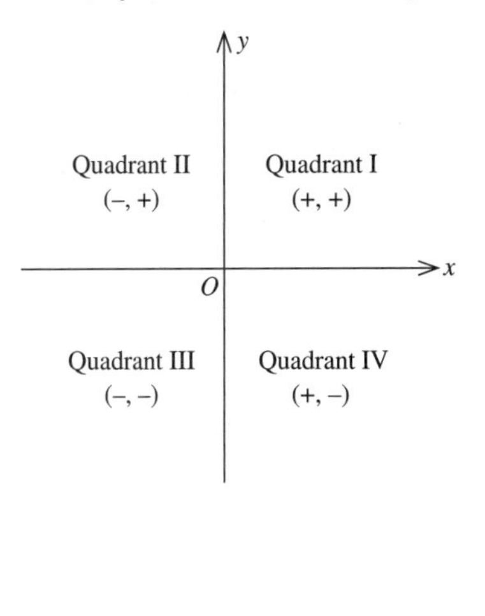

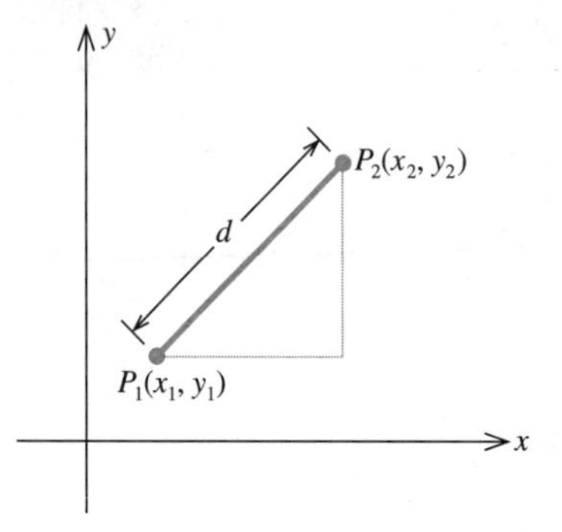

FIGURE 1.6
The distance between two points in the coordinate plane

The Distance Formula

One immediate benefit that arises from using the Cartesian coordinate system is that the distance between any two points in the plane may be expressed solely in terms of the coordinates of the points. Suppose, for example, (x_1, y_1) and (x_2, y_2) are any two points in the plane (Figure 1.6). Then the distance d between these two points is, by the Pythagorean theorem,

$$d = \sqrt{(x_2 - x_1)^2 + (y_2 - y_1)^2}$$

For a proof of this result see Exercise 45, page 12.

Distance Formula

The distance d between two points $P_1(x_1, y_1)$ and $P_2(x_2, y_2)$ in the plane is given by

$$d = \sqrt{(x_2 - x_1)^2 + (y_2 - y_1)^2} \qquad (1)$$

In what follows, we give several applications of the distance formula.

EXAMPLE 1 Find the distance between the points $(-4, 3)$ and $(2, 6)$.

SOLUTION ✓ Let $P_1(-4, 3)$ and $P_2(2, 6)$ be points in the plane. Then, we have

$$x_1 = -4 \quad \text{and} \quad y_1 = 3$$
$$x_2 = 2 \qquad\qquad y_2 = 6$$

Using Formula (1), we have

$$\begin{aligned} d &= \sqrt{[2 - (-4)]^2 + (6 - 3)^2} \\ &= \sqrt{6^2 + 3^2} \\ &= \sqrt{45} \\ &= 3\sqrt{5} \end{aligned}$$

■■■■

Group Discussion
Refer to Example 1. Suppose we label the point $(2, 6)$ as P_1 and the point $(-4, 3)$ as P_2. (a) Show that the distance d between the two points is the same as that obtained earlier. (b) Prove that, in general, the distance d in Formula (1) is independent of the way we label the two points.

APPLICATION

EXAMPLE 2

In Figure 1.7 S represents the position of a power relay station located on a coastal highway, and M shows the location of a marine biology experimental station on a nearby island. A cable is to be laid connecting the relay station with the experimental station. If the cost of running the cable on land is \$2 per running foot and the cost of running the cable under water is \$6 per running foot, find the total cost for laying the cable.

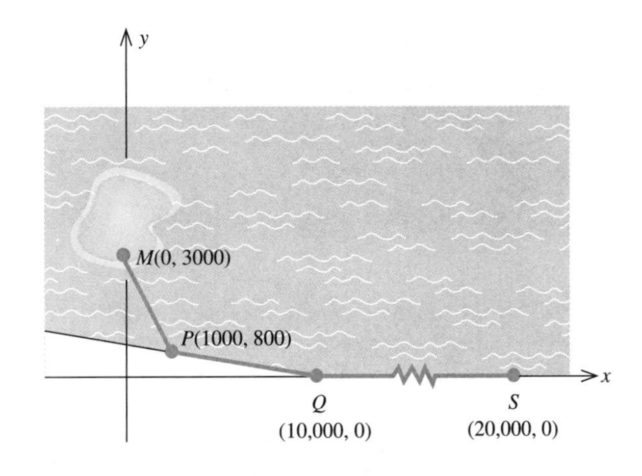

FIGURE 1.7
The cable will connect the relay station S to the experimental station M.

SOLUTION ✓

The length of cable required on land is given by the distance from P to Q plus the distance from Q to S—that is,

$$\begin{aligned}
&\sqrt{(10{,}000 - 1000)^2 + (0 - 800)^2} + \sqrt{(20{,}000 - 10{,}000)^2 + (0 - 0)^2} \\
&\quad = \sqrt{9000^2 + 800^2} + 10{,}000 \\
&\quad = \sqrt{81{,}640{,}000} + 10{,}000 \\
&\quad \approx 19{,}035.49
\end{aligned}$$

or approximately 19,035.49 feet. Next, we see that the length of cable required underwater is given by

$$\begin{aligned}
\sqrt{(0 - 1000)^2 + (3000 - 800)^2} &= \sqrt{1000^2 + 2200^2} \\
&= \sqrt{5{,}840{,}000} \\
&\approx 2416.61
\end{aligned}$$

or approximately 2416.61 feet. Therefore, the total cost for laying the cable is

$$2(19{,}035.49) + 6(2416.61) = 52{,}570.64$$

or approximately \$52,571. ■■■■

EXAMPLE 3

Let $P(x, y)$ denote a point lying on the circle with radius r and center $C(h, k)$ (Figure 1.8). Find a relationship between x and y.

FIGURE 1.8
A circle with radius r and center $C(h, k)$

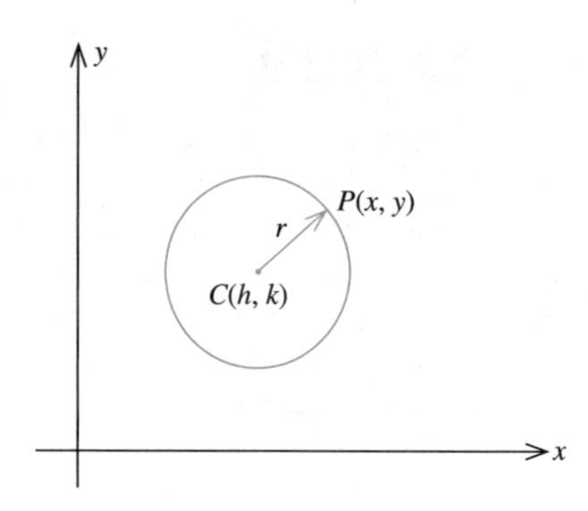

SOLUTION ✓ By the definition of a circle, the distance between $C(h, k)$ and $P(x, y)$ is r. Using Formula (1), we have

$$\sqrt{(x-h)^2 + (y-k)^2} = r$$

which, upon squaring both sides, gives the equation

$$(x-h)^2 + (y-k)^2 = r^2$$

that must be satisfied by the variables x and y. ■■■■

A summary of the result obtained in Example 3 follows.

Equation of a Circle

An equation of the circle with center $C(h, k)$ and radius r is given by

$$(x-h)^2 + (y-k)^2 = r^2 \qquad \textbf{(2)}$$

EXAMPLE 4 Find an equation of the circle with (a) radius 2 and center $(-1, 3)$ and (b) radius 3 and center located at the origin.

SOLUTION ✓ **a.** We use Formula (2) with $r = 2$, $h = -1$, and $k = 3$, obtaining

$$[x-(-1)]^2 + (y-3)^2 = 2^2$$
$$(x+1)^2 + (y-3)^2 = 4$$

(Figure 1.9a).

FIGURE 1.9

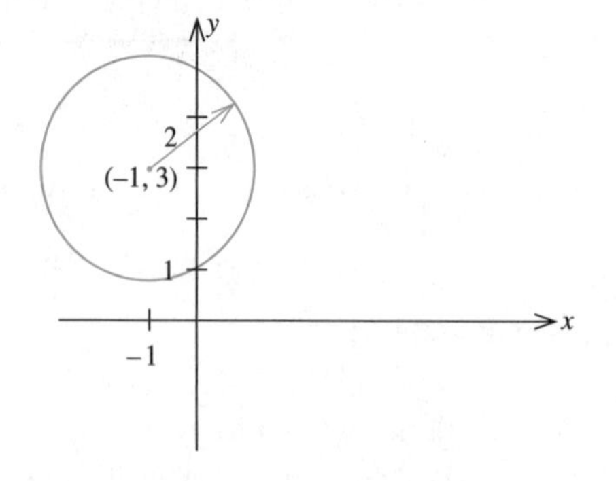

(a) The circle with radius 2 and center $(-1, 3)$

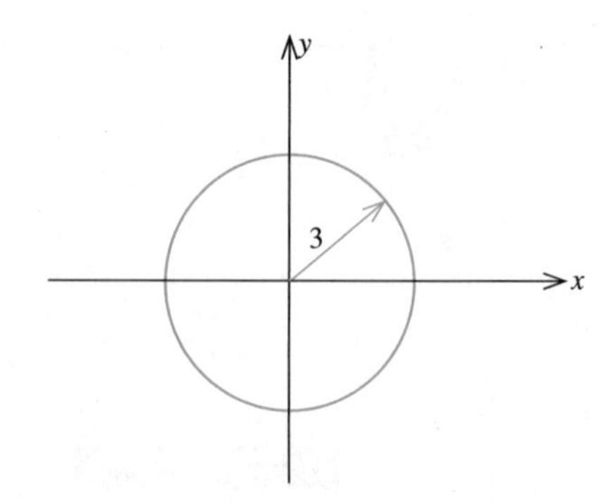

(b) The circle with radius 3 and center $(0, 0)$

b. Using Formula (2) with $r = 3$, $h = k = 0$, we obtain

$$x^2 + y^2 = 3^2$$
$$x^2 + y^2 = 9$$

(Figure 1.9b).

Group Discussion

1. Use the distance formula to help you describe the set of points in the xy-plane satisfying each of the following inequalities.

a. $(x - h)^2 + (y - k)^2 \leq r^2$ **c.** $(x - h)^2 + (y - k)^2 \geq r^2$

b. $(x - h)^2 + (y - k)^2 < r^2$ **d.** $(x - h)^2 + (y - k)^2 > r^2$

2. Consider the equation $x^2 + y^2 = 4$.

a. Show that $y = \pm\sqrt{4 - x^2}$.

b. Describe the set of points (x, y) in the xy-plane satisfying the equation

(i) $y = \sqrt{4 - x^2}$ (ii) $y = -\sqrt{4 - x^2}$

Self-Check Exercises 1.1

1. a. Plot the points $A(4, -2)$, $B(2, 3)$, and $C(-3, 1)$.

b. Find the distance between the points A and B, between B and C, and between A and C.

c. Use the Pythagorean theorem to show that the triangle with vertices A, B, and C is a right triangle.

2. The accompanying figure shows the location of cities A, B, and C. Suppose a pilot wishes to fly from city A to city C but must make a mandatory stopover in city B. If the single-engine light plane has a range of 650 mi, can the pilot make the trip without refueling in city B?

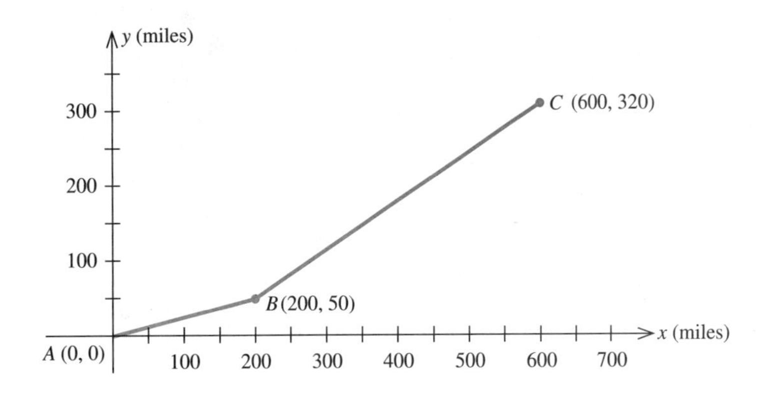

Solutions to Self-Check Exercises 1.1 can be found on page 12.

1.1 Exercises

In Exercises 1–6, refer to the accompanying figure and determine the coordinates of the given point and the quadrant in which it is located.

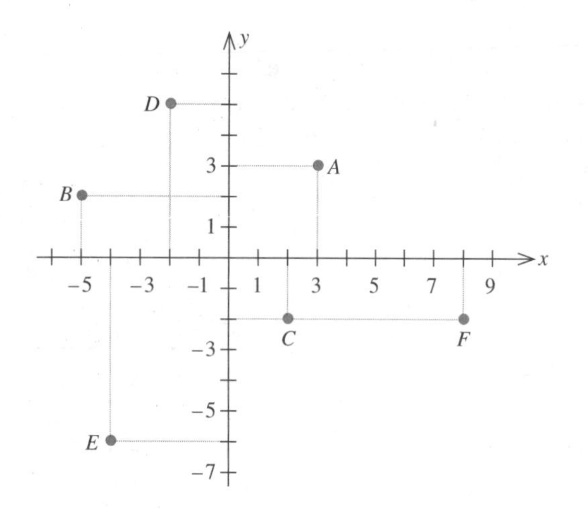

1. A 2. B 3. C
4. D 5. E 6. F

In Exercises 7–12, refer to the accompanying figure.

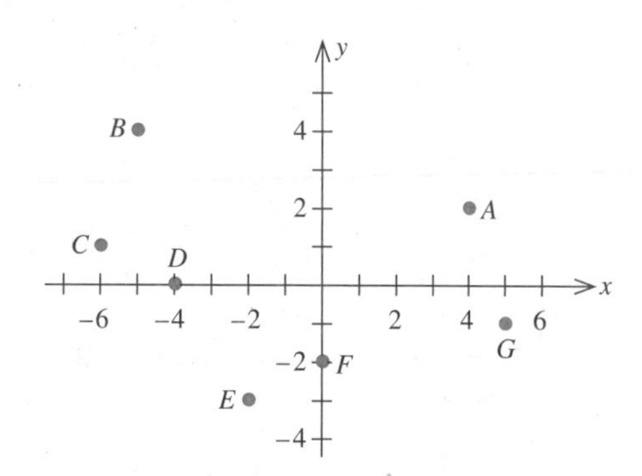

7. Which point has coordinates (4, 2)?

8. What are the coordinates of point B?

9. Which points have negative y-coordinates?

10. Which point has a negative x-coordinate and a negative y-coordinate?

11. Which point has an x-coordinate that is equal to zero?

12. Which point has a y-coordinate that is equal to zero?

In Exercises 13–20, sketch a set of coordinate axes and plot the given point.

13. $(-2, 5)$ 14. $(1, 3)$

15. $(3, -1)$ 16. $(3, -4)$

17. $(8, -7/2)$ 18. $(-5/2, 3/2)$

19. $(4.5, -4.5)$ 20. $(1.2, -3.4)$

In Exercises 21–24, find the distance between the given points.

21. $(1, 3)$ and $(4, 7)$ 22. $(1, 0)$ and $(4, 4)$

23. $(-1, 3)$ and $(4, 9)$ 24. $(-2, 1)$ and $(10, 6)$

25. Find the coordinates of the points that are 10 units away from the origin and have a y-coordinate equal to -6.

26. Find the coordinates of the points that are 5 units away from the origin and have an x-coordinate equal to 3.

27. Show that the points $(3, 4)$, $(-3, 7)$, $(-6, 1)$, and $(0, -2)$ form the vertices of a square.

28. Show that the triangle with vertices $(-5, 2)$, $(-2, 5)$, and $(5, -2)$ is a right triangle.

In Exercises 29–34, find an equation of the circle that satisfies the given conditions.

29. Radius 5 and center $(2, -3)$

30. Radius 3 and center $(-2, -4)$

31. Radius 5 and center at the origin

32. Center at the origin and passes through $(2, 3)$

33. Center $(2, -3)$ and passes through $(5, 2)$

34. Center $(-a, a)$ and radius $2a$

A calculator is recommended for Exercises 35–40.

35. **Distance Traveled** A grand tour of four cities begins at city A and makes successive stops at cities B, C, and D before returning to city A. If the cities are located as

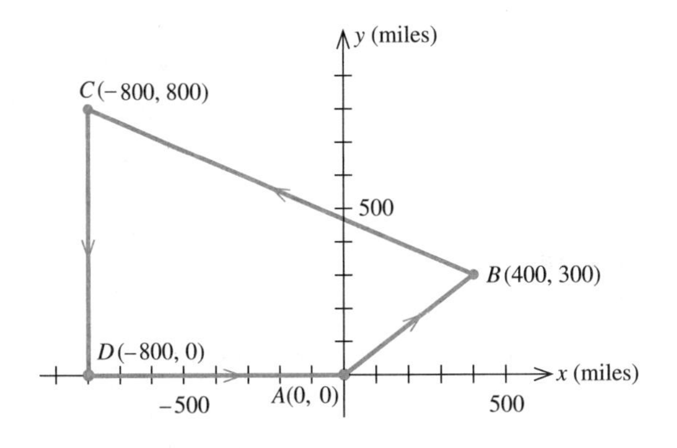

shown in the accompanying figure, find the total distance covered on the tour.

36. **Delivery Charges** A furniture store offers free setup and delivery services to all points within a 25-mi radius of its warehouse distribution center. If you live 20 mi east and 14 mi south of the warehouse, will you incur a delivery charge? Justify your answer.

37. **Travel Time** Towns A, B, C, and D are located as shown in the accompanying figure. Two highways link town A to town D. Route 1 runs from town A to town D via town B, and route 2 runs from town A to town D via town C. If a salesman wishes to drive from town A to town D and traffic conditions are such that he could expect to average the same speed on either route, which highway should he take in order to arrive in the shortest time?

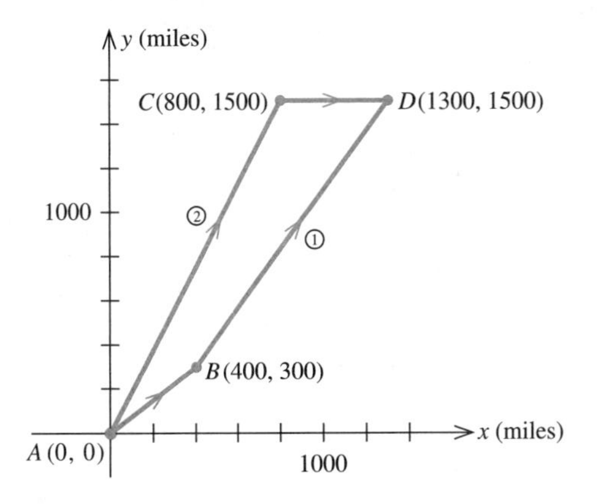

38. **Minimizing Shipping Costs** Refer to the figure for Exercise 37. Suppose a fleet of 100 automobiles are to be shipped from an assembly plant in town A to town D. They may be shipped either by freight train along Route 1 at a cost of 11 cents/mile/automobile or by truck along Route 2 at a cost of $10\frac{1}{2}$ cents/mile/automobile. Which means of transportation minimizes the shipping cost? What is the net savings?

39. **Consumer Decisions** Will Barclay wishes to determine which antenna he should purchase for his home. The TV store has supplied him with the following information:

Range in miles			
VHF	**UHF**	**Model**	**Price**
30	20	A	\$40
45	35	B	50
60	40	C	60
75	55	D	70

Will wishes to receive Channel 17 (VHF), which is located 25 mi east and 35 mi north of his home, and Channel 38 (UHF), which is located 20 mi south and 32 mi west of his home. Which model will allow him to receive both channels at the least cost? (Assume that the terrain between Will's home and both broadcasting stations is flat.)

40. **Calculating the Cost of Laying Cable** In the accompanying diagram, S represents the position of a power relay station located on a coastal highway, and M shows the location of a marine biology experimental station on a nearby island. A cable is to be laid connecting the relay station with the experimental station. If the cost of running the cable on land is \$2/running foot and the cost of running cable underwater is \$6/running foot, find an expression in terms of x that gives the total cost of laying the cable. Use this expression to find the total cost when $x = 900$. When $x = 1000$.

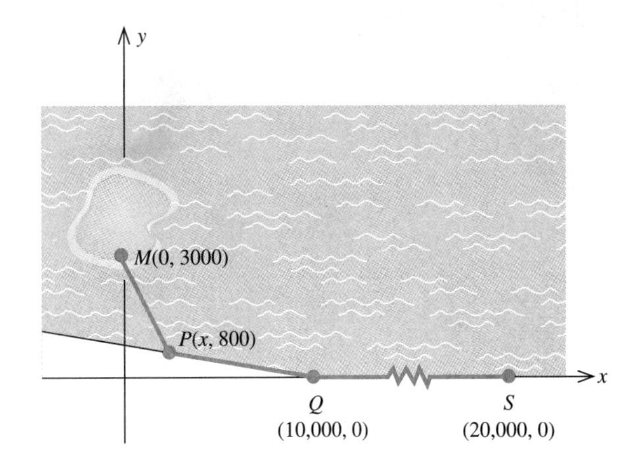

41. Two ships leave port at the same time. Ship A sails north at a speed of 20 mph while ship B sails east at a speed of 30 mph.

a. Find an expression in terms of the time t (in hours) giving the distance between the two ships.
b. Using the expression obtained in part (a), find the distance between the two ships 2 hours after leaving port.

In Exercises 42 and 43, determine whether the statement is true or false. If it is true, explain why it is true. If it is false, give an example to show why it is false.

42. If the distance between the points $P_1(a, b)$ and $P_2(c, d)$ is D, then the distance between the points $P_1(a, b)$ and $P_3(kc, kd)$ $(k \neq 0)$ is given by $|k|D$.

43. The circle with equation $kx^2 + ky^2 = a^2$ lies inside the circle with equation $x^2 + y^2 = a^2$, provided $k > 1$.

44. In the Cartesian coordinate system, the two axes are perpendicular to each other. Consider a coordinate system in which the x- and y-axis are noncollinear (that is, the axes do not lie along a straight line) and are not perpendicular to each other (see the accompanying figure).
a. Describe how a point is represented in this coordinate system by an ordered pair (x, y) of real numbers. Conversely, show how an ordered pair (x, y) of real numbers uniquely determines a point in the plane.
b. Suppose you want to find a formula for the distance between two points, $P_1(x_1, y_1)$ and $P_2(x_2, y_2)$, in the plane. What advantage does the Cartesian coordinate system have over the coordinate system under consideration? Comment on your answer.

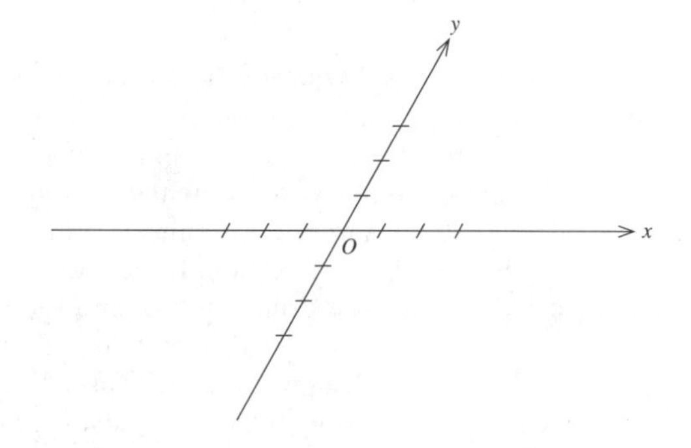

45. Let (x_1, y_1) and (x_2, y_2) be two points lying in the xy-plane. Show that the distance between the two points is given by

$$d = \sqrt{(x_2 - x_1)^2 + (y_2 - y_1)^2}$$

Hint: Refer to the accompanying figure and use the Pythagorean theorem.

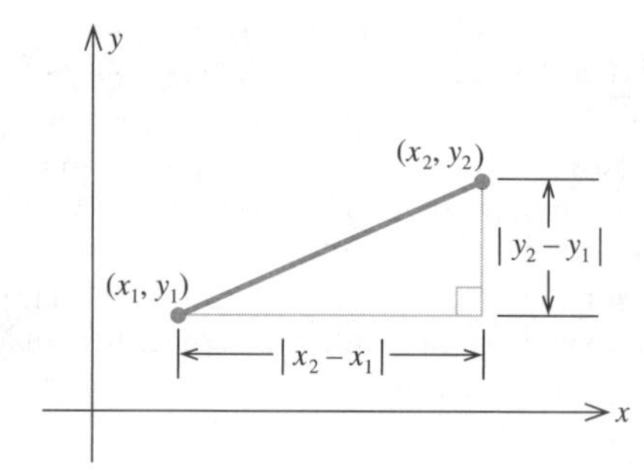

SOLUTIONS TO SELF-CHECK EXERCISES 1.1

1. a. The points are plotted in the accompanying figure.

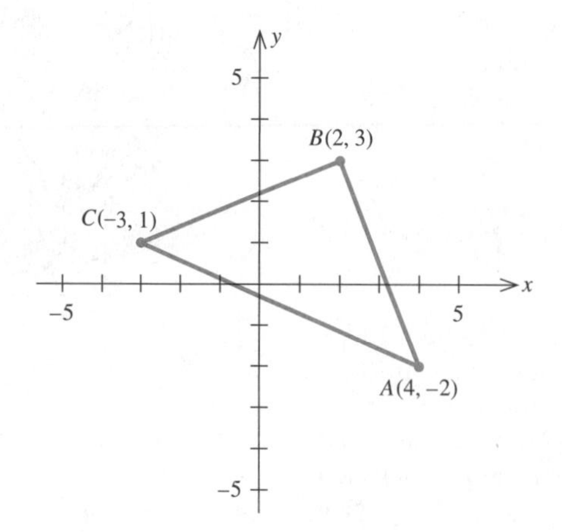

b. The distance between A and B is

$$\begin{aligned} d(A, B) &= \sqrt{(2-4)^2 + [3-(-2)]^2} \\ &= \sqrt{(-2)^2 + 5^2} = \sqrt{4+25} = \sqrt{29} \end{aligned}$$

The distance between B and C is

$$\begin{aligned} d(B, C) &= \sqrt{(-3-2)^2 + (1-3)^2} \\ &= \sqrt{(-5)^2 + (-2)^2} = \sqrt{25+4} = \sqrt{29} \end{aligned}$$

The distance between A and C is

$$\begin{aligned} d(A, C) &= \sqrt{(-3-4)^2 + [1-(-2)]^2} \\ &= \sqrt{(-7)^2 + 3^2} = \sqrt{49+9} = \sqrt{58} \end{aligned}$$

c. We will show that

$$[d(A, C)]^2 = [d(A, B)]^2 + [d(B, C)]^2$$

From part (b), we see that $[d(A, B)]^2 = 29$, $[d(B, C)]^2 = 29$, and $[d(A, C)]^2 = 58$, and the desired result follows.

2. The distance between city A and city B is

$$d(A, B) = \sqrt{200^2 + 50^2} \approx 206$$

or 206 miles. The distance between city B and city C is

$$\begin{aligned} d(B, C) &= \sqrt{[600-200]^2 + [320-50]^2} \\ &= \sqrt{400^2 + 270^2} \approx 483 \end{aligned}$$

or 483 miles. Therefore, the total distance the pilot would have to cover is 689 miles, so she must refuel in city B.

1.2 Straight Lines

In computing income tax, business firms are allowed by law to depreciate certain assets such as buildings, machines, furniture, automobiles, and so on, over a period of time. *Linear depreciation,* or the *straight-line method,* is often used for this purpose. The graph of the straight line shown in Figure 1.10 describes the book value V of a computer that has an initial value of \$100,000 and that is being depreciated linearly over 5 years with a scrap value of \$30,000. Note that only the solid portion of the straight line is of interest here.

FIGURE 1.10
Linear depreciation of an asset

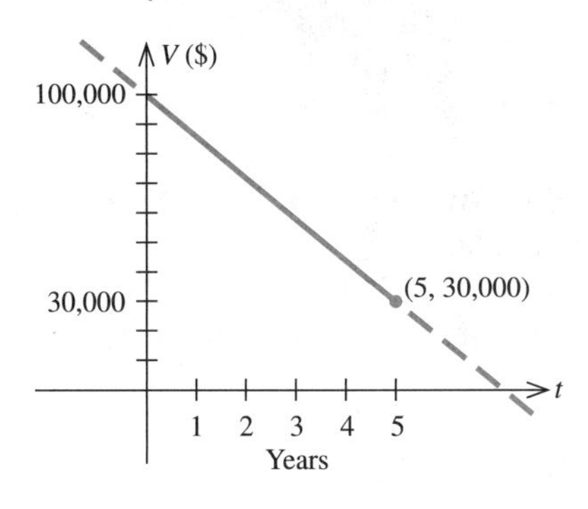

The book value of the computer at the end of year t, where t lies between 0 and 5, can be read directly from the graph. But there is one shortcoming in this approach: The result depends on how accurately you draw and read the graph. A better and more accurate method is based on finding an *algebraic* representation of the depreciation line. (We will continue our discussion of the linear depreciation problem in Section 1.3.)

To see how a straight line in the xy-plane may be described algebraically, we need to first recall certain properties of straight lines.

Slope of a Line

Let L denote the unique straight line that passes through the two distinct points (x_1, y_1) and (x_2, y_2). If $x_1 = x_2$, then L is a vertical line, and the slope is undefined (Figure 1.11).

FIGURE 1.11
m is undefined.

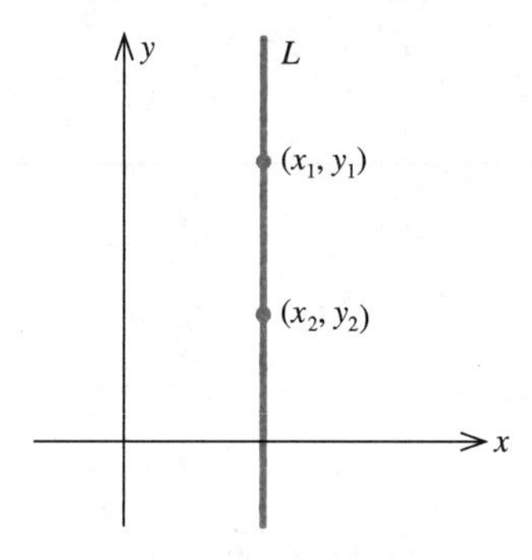

If $x_1 \neq x_2$, we define the slope of L as follows.

Slope of a Nonvertical Line

If (x_1, y_1) and (x_2, y_2) are any two distinct points on a nonvertical line L, then the slope m of L is given by

$$m = \frac{\Delta y}{\Delta x} = \frac{y_2 - y_1}{x_2 - x_1} \tag{3}$$

(Figure 1.12).

FIGURE 1.12

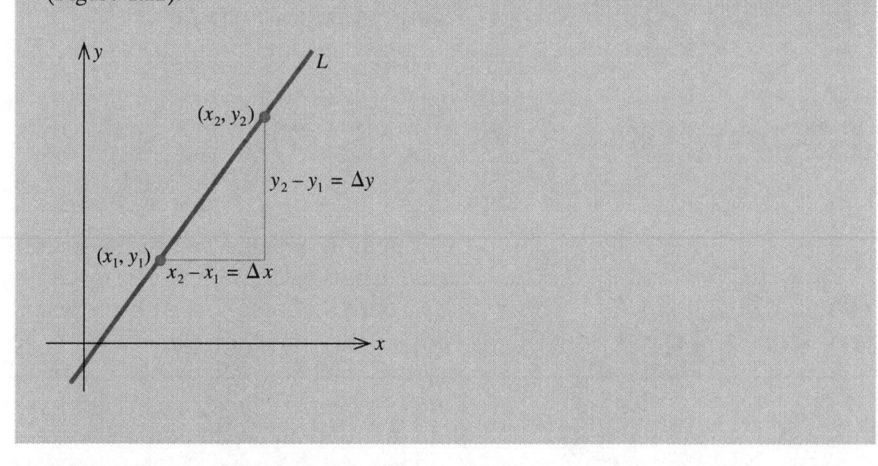

Since the slope of a nonvertical line is a ratio of two real numbers, the slope of a straight line is a constant whenever it is defined.

The number $\Delta y = y_2 - y_1$ (Δy is read "delta y") is a measure of the vertical change in y, and $\Delta x = x_2 - x_1$ is a measure of the horizontal change in x as shown in Figure 1.12. From this figure we can see that the slope m of a straight line L is a measure of the *rate of change of y with respect to x.* Furthermore, our earlier observation concerning the slope of a nonvertical straight line tells us that this rate of change is constant.

FIGURE 1.13

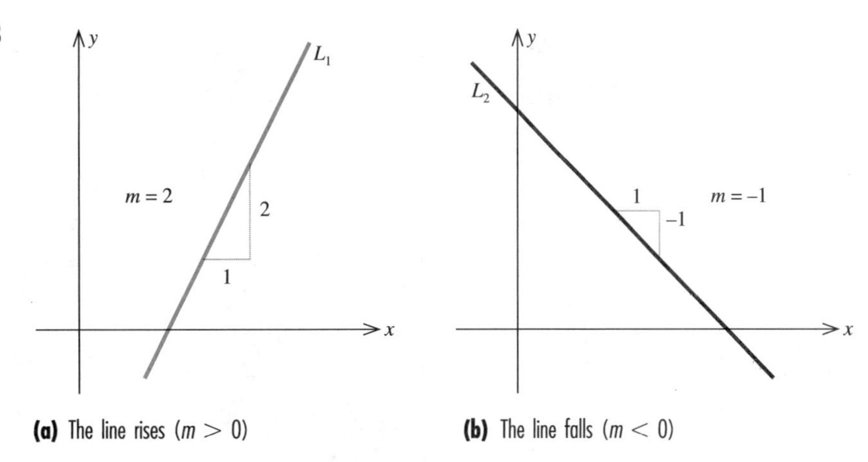

(a) The line rises ($m > 0$) **(b)** The line falls ($m < 0$)

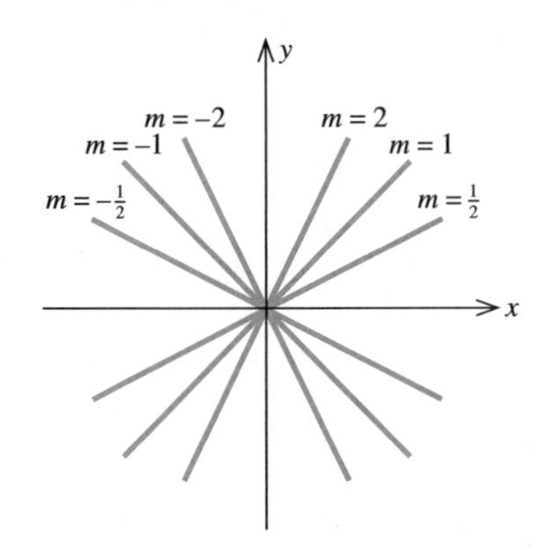

FIGURE 1.14
A family of straight lines

Figure 1.13a shows a straight line L_1 with slope 2. Observe that L_1 has the property that a 1-unit increase in x results in a 2-unit increase in y. To see this, let $\Delta x = 1$ in Equation (3) so that $m = \Delta y$. Since $m = 2$, we conclude that $\Delta y = 2$. Similarly, Figure 1.13b shows a line L_2 with slope -1. Observe that a straight line with positive slope slants upward from left to right (y increases as x increases), whereas a line with negative slope slants downward from left to right (y decreases as x increases). Finally, Figure 1.14 shows a family of straight lines passing through the origin with indicated slopes.

> **Group Discussion**
> Show that the slope of a nonvertical line is independent of the two distinct points used to compute it.
> **Hint:** Suppose we pick two other distinct points, $P_3(x_3, y_3)$ and $P_4(x_4, y_4)$ lying on L. Draw a picture and use similar triangles to demonstrate that using P_3 and P_4 gives the same values as those obtained using P_1 and P_2.

EXAMPLE 1 Sketch the straight line that passes through the point $(-2, 5)$ and has slope $-4/3$.

SOLUTION ✓ First, plot the point $(-2, 5)$ (Figure 1.15). Next, recall that a slope of $-4/3$ indicates that an increase of 1 unit in the x-direction produces a *decrease* of 4/3 units in the y-direction, or equivalently, a 3-unit increase in the x-direction produces a 3(4/3), or 4-unit, decrease in the y-direction. Using this information, we plot the point $(1, 1)$ and draw the line through the two points.

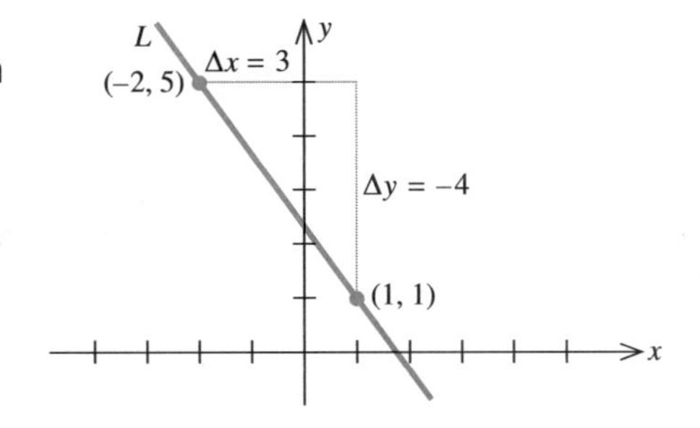

FIGURE 1.15
L has slope $-4/3$ and passes through $(-2, 5)$.

■■■■

EXAMPLE 2 Find the slope m of the line that passes through the points $(-1, 1)$ and $(5, 3)$.

SOLUTION ✓ Choose (x_1, y_1) to be the point $(-1, 1)$ and (x_2, y_2) to be the point $(5, 3)$. Then, with $x_1 = -1$, $y_1 = 1$, $x_2 = 5$, and $y_2 = 3$, we find, using Equation (3),

$$m = \frac{y_2 - y_1}{x_2 - x_1} = \frac{3 - 1}{5 - (-1)} = \frac{2}{6} = \frac{1}{3}$$

(Figure 1.16). Try to verify that the result obtained would have been the same had we chosen the point $(-1, 1)$ to be (x_2, y_2) and the point $(5, 3)$ to be (x_1, y_1).

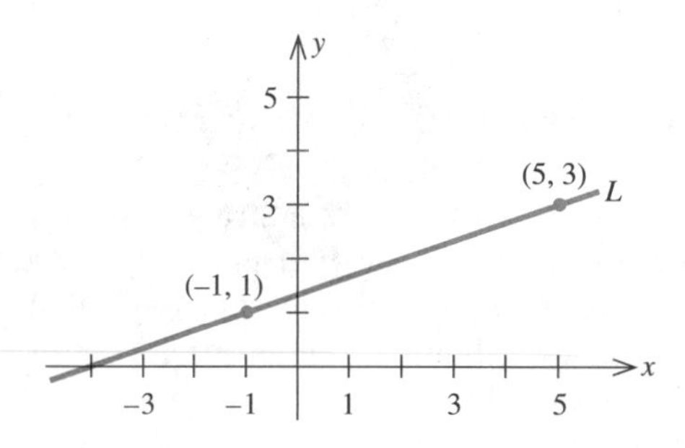

FIGURE 1.16
L passes through $(5, 3)$ and $(-1, 1)$.

■■■■

EXAMPLE 3 Find the slope of the line that passes through the points $(-2, 5)$ and $(3, 5)$.

SOLUTION ✓ The slope of the required line is given by

$$m = \frac{5 - 5}{3 - (-2)} = \frac{0}{5} = 0$$

(Figure 1.17).

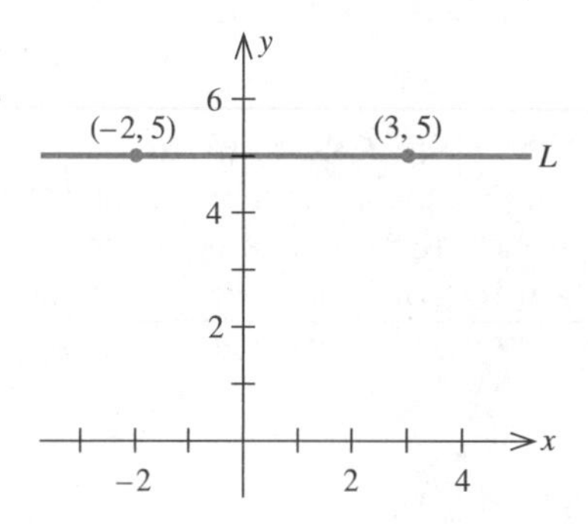

FIGURE 1.17
The slope of the horizontal line L is zero.

■■■■

REMARK In general, the slope of a horizontal line is zero. ■■■

We can use the slope of a straight line to determine whether a line is parallel to another line.

Parallel Lines

> Two distinct lines are **parallel** if and only if their slopes are equal or their slopes are undefined.

EXAMPLE 4 Let L_1 be a line that passes through the points $(-2, 9)$ and $(1, 3)$ and let L_2 be the line that passes through the points $(-4, 10)$ and $(3, -4)$. Determine whether L_1 and L_2 are parallel.

SOLUTION ✓ The slope m_1 of L_1 is given by

$$m_1 = \frac{3 - 9}{1 - (-2)} = -2$$

The slope m_2 of L_2 is given by

$$m_2 = \frac{-4 - 10}{3 - (-4)} = -2$$

Since $m_1 = m_2$, the lines L_1 and L_2 are in fact parallel (Figure 1.18). ■■■■

FIGURE 1.18
L_1 and L_2 have the same slope and hence are parallel.

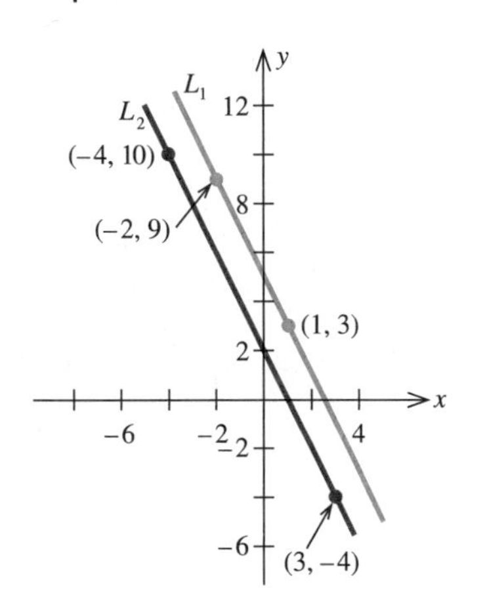

EQUATIONS OF LINES

We now show that every straight line lying in the xy-plane may be represented by an equation involving the variables x and y. One immediate benefit of this is that problems involving straight lines may be solved algebraically.

Let L be a straight line parallel to the y-axis (perpendicular to the x-axis) (Figure 1.19). Then L crosses the x-axis at some point $(a, 0)$ with the x-coordinate given by $x = a$, where a is some real number. Any other point on L has the form $(a, \bar{y})$, where $\bar{y}$ is an appropriate number. Therefore, the vertical line L is described by the sole condition

$$x = a$$

and this is accordingly an equation of L. For example, the equation $x = -2$ represents a vertical line 2 units to the left of the y-axis, and the equation $x = 3$ represents a vertical line 3 units to the right of the y-axis (Figure 1.20).

FIGURE 1.19
The vertical line $x = a$

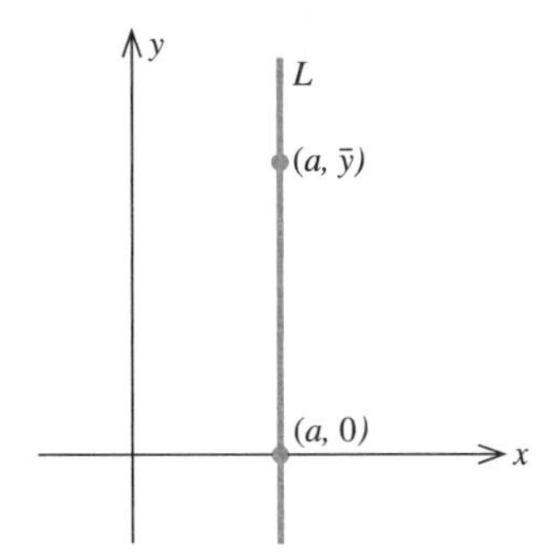

FIGURE 1.20
The vertical lines $x = -2$ and $x = 3$

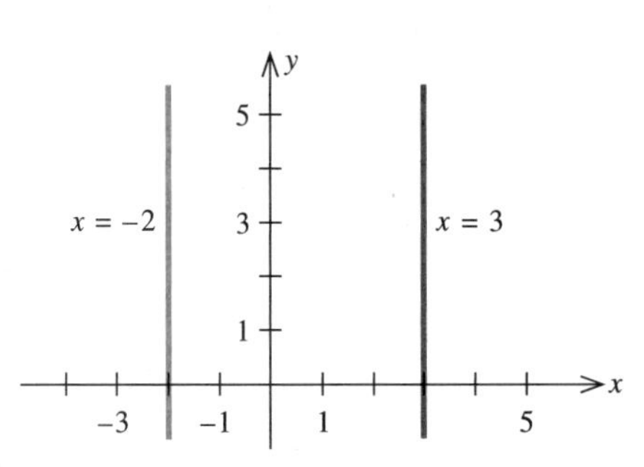

Next, suppose L is a nonvertical line so that it has a well-defined slope m. Suppose (x_1, y_1) is a fixed point lying on L and (x, y) is a variable point

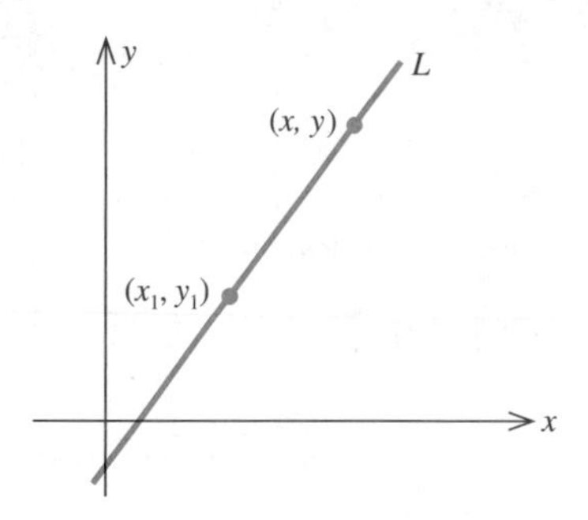

FIGURE 1.21
L passes through (x_1, y_1) and has slope m.

on L distinct from (x_1, y_1) (Figure 1.21). Using Equation (3) with the point $(x_2, y_2) = (x, y)$, we find that the slope of L is given by

$$m = \frac{y - y_1}{x - x_1}$$

Upon multiplying both sides of the equation by $x - x_1$, we obtain Equation (4).

Point-Slope Form

> An equation of the line that has slope m and passes through the point (x_1, y_1) is given by
>
> $$y - y_1 = m(x - x_1) \qquad \textbf{(4)}$$

Equation (4) is called the point-slope form of the equation of a line since it utilizes a given point (x_1, y_1) on a line and the slope m of the line.

EXAMPLE 5 Find an equation of the line that passes through the point (1, 3) and has slope 2.

SOLUTION ✓ Using the point-slope form of the equation of a line with the point (1, 3) and $m = 2$, we obtain

$$y - 3 = 2(x - 1) \qquad [(y - y_1) = m(x - x_1)]$$

which, when simplified, becomes

$$2x - y + 1 = 0$$

(Figure 1.22).

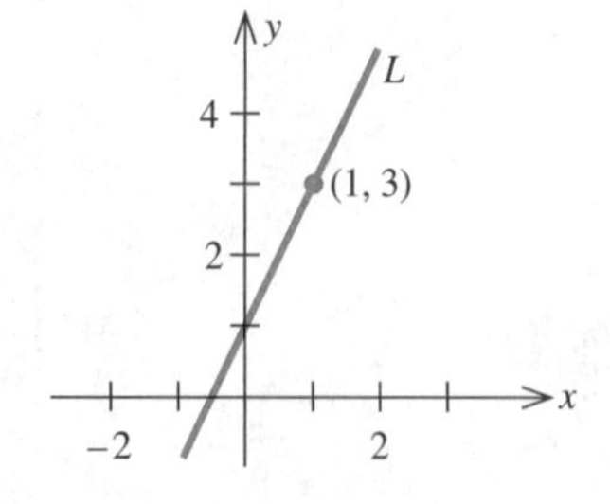

FIGURE 1.22
L passes through (1, 3) and has slope 2.

■■■■

EXAMPLE 6 Find an equation of the line that passes through the points $(-3, 2)$ and $(4, -1)$.

SOLUTION ✓ The slope of the line is given by

$$m = \frac{-1 - 2}{4 - (-3)} = -\frac{3}{7}$$

Using the point-slope form of the equation of a line with the point $(4, -1)$ and the slope $m = -3/7$, we have

$$\begin{aligned} y + 1 &= -\frac{3}{7}(x - 4) \qquad [(y - y_1) = m(x - x_1)] \\ 7y + 7 &= -3x + 12 \\ 3x + 7y - 5 &= 0 \end{aligned}$$

(Figure 1.23).

FIGURE 1.23
L passes through (−3, 2) and (4, −1).

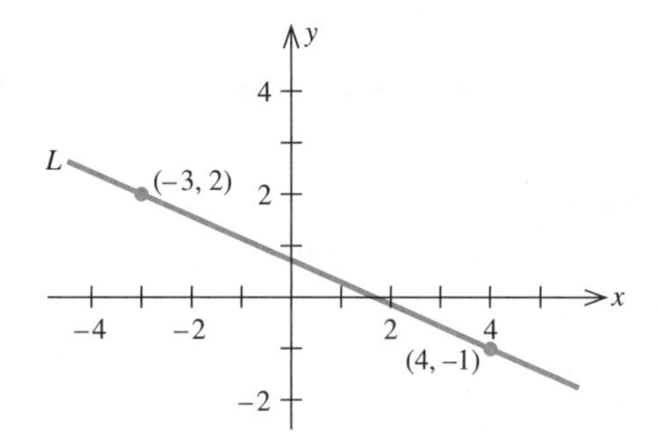

■■■■

We can use the slope of a straight line to determine whether a line is perpendicular to another line.

Perpendicular Lines

> If L_1 and L_2 are two distinct nonvertical lines that have slopes m_1 and m_2, respectively, then L_1 is **perpendicular** to L_2 (written $L_1 \perp L_2$) if and only if
>
> $$m_1 = -\frac{1}{m_2}$$

If the line L_1 is vertical (so that its slope is undefined), then L_1 is perpendicular to another line, L_2, if and only if L_2 is horizontal (so that its slope is zero). For a proof of these results, see Exercise 84, page 29.

EXAMPLE 7 Find an equation of the line that passes through the point $(3, 1)$ and is perpendicular to the line of Example 5.

SOLUTION ✓ Since the slope of the line in Example 5 is 2, the slope of the required line is given by $m = -1/2$, the negative reciprocal of 2. Using the point-slope form of the equation of a line, we obtain

$$y - 1 = -\frac{1}{2}(x - 3) \qquad [(y - y_1) = m(x - x_1)]$$

$$2y - 2 = -x + 3$$

$$x + 2y - 5 = 0$$

(Figure 1.24). ■■■■

A straight line L that is neither horizontal nor vertical cuts the x-axis and the y-axis at, say, points $(a, 0)$ and $(0, b)$, respectively (Figure 1.25). The numbers a and b are called the **x-intercept** and **y-intercept,** respectively, of L.

Now, let L be a line with slope m and y-intercept b. Using Equation (4), the point-slope form of the equation of a line, with the point given by $(0, b)$ and slope m, we have

$$y - b = m(x - 0)$$

$$y = mx + b$$

This is called the **slope-intercept form** of a line.

FIGURE 1.24
L_2 is perpendicular to L_1 and passes through (3, 1).

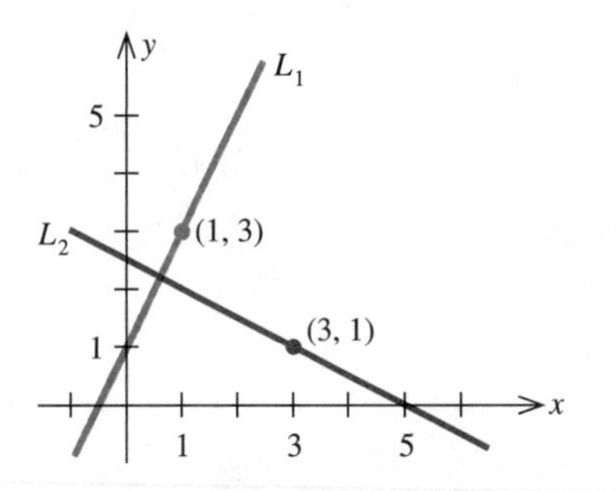

FIGURE 1.25
The line L has x-intercept a and y-intercept b.

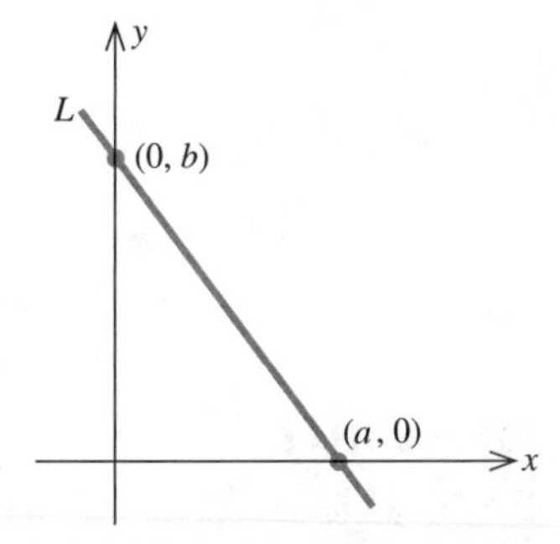

Exploring with Technology

1. Use a graphing utility to plot the straight lines L_1 and L_2 with equations $2x + y - 5 = 0$ and $41x + 20y - 11 = 0$ on the same set of axes using the standard viewing rectangle.
   a. Can you tell if the lines L_1 and L_2 are parallel to each other?
   b. Verify your observations by computing the slopes of L_1 and L_2 algebraically.
2. Use a graphing utility to plot the straight lines L_1 and L_2 with equations $x + 2y - 5 = 0$ and $5x - y + 5 = 0$ on the same set of axes using the standard viewing rectangle.
   a. Can you tell if the lines L_1 and L_2 are perpendicular to each other?
   b. Verify your observation by computing the slopes of L_1 and L_2 algebraically.

Slope-Intercept Form

The equation of the line that has slope m and intersects the y-axis at the point $(0, b)$ is given by

$$y = mx + b \tag{5}$$

EXAMPLE 8 Find an equation of the line that has slope 3 and y-intercept -4.

SOLUTION ✔ Using Equation (5) with $m = 3$ and $b = -4$, we obtain the required equation:

$$y = 3x - 4$$

■■■■

EXAMPLE 9 Determine the slope and y-intercept of the line whose equation is $3x - 4y = 8$.

SOLUTION ✔ Rewrite the given equation in the slope-intercept form and obtain

$$y = \frac{3}{4}x - 2$$

Comparing this result with Equation (5), we find $m = 3/4$ and $b = -2$, and we conclude that the slope and y-intercept of the given line are 3/4 and -2, respectively.

■■■■

Exploring with Technology

1. Use a graphing utility to plot the straight lines with equations $y = -2x + 3$, $y = -x + 3$, $y = x + 3$, and $y = 2.5x + 3$ on the same set of axes using the standard viewing rectangle. What effect does changing the coefficient m of x in the equation $y = mx + b$ have on its graph?
2. Use a graphing utility to plot the straight lines with equations $y = 2x - 2$, $y = 2x - 1$, $y = 2x$, $y = 2x + 1$, and $y = 2x + 4$ on the same set of axes using the standard viewing rectangle. What effect does changing the constant b in the equation $y = mx + b$ have on its graph?
3. Describe in words the effect of changing both m and b in the equation $y = mx + b$.

APPLICATIONS

EXAMPLE 10 The sales manager of a local sporting goods store plotted sales versus time for the last 5 years and found the points to lie approximately along a straight line (Figure 1.26). By using the points corresponding to the first and fifth years, find an equation of the *trend line*. What sales figure can be predicted for the sixth year?

FIGURE 1.26
Sales of a sporting goods store

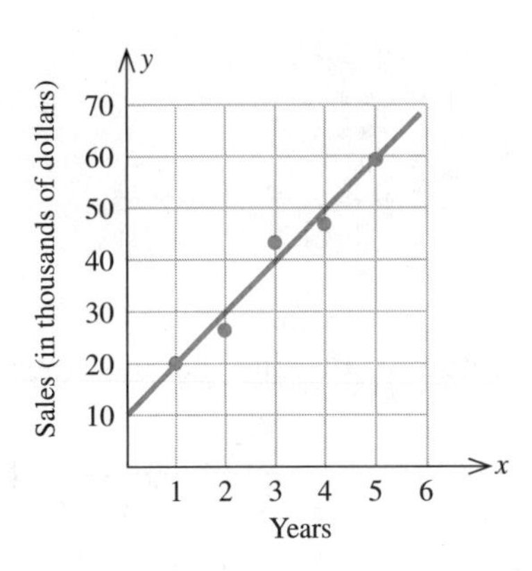

SOLUTION ✓

Using Equation (3) with the points (1, 20) and (5, 60), we find that the slope of the required line is given by

$$m = \frac{60 - 20}{5 - 1} = 10$$

Next, using the point-slope form of the equation of a line with the point (1, 20) and $m = 10$, we obtain

$$y - 20 = 10(x - 1)$$
$$y = 10x + 10$$

as the required equation.

The sales figure for the sixth year is obtained by letting $x = 6$ in the last equation, giving

$$y = 70$$

or \$70,000.

EXAMPLE 11

Suppose an art object purchased for \$50,000 is expected to appreciate in value at a constant rate of \$5000 per year for the next 5 years. Use Equation (5) to write an equation predicting the value of the art object in the next several years. What will be its value 3 years from the purchase date?

SOLUTION ✓

Let x denote the time (in years) that has elapsed since the purchase date and let y denote the object's value (in dollars). Then, $y = 50{,}000$ when $x = 0$. Furthermore, the slope of the required equation is given by $m = 5000$ since each unit increase in x (1 year) implies an increase of 5000 units (dollars) in y. Using Equation (5) with $m = 5000$ and $b = 50{,}000$, we obtain

$$y = 5000x + 50{,}000$$

Three years from the purchase date, the value of the object will be given by

$$y = 5000(3) + 50{,}000$$

or \$65,000.

Group Discussion

Refer to Example 11. Can the equation predicting the value of the art object be used to predict long-term growth?

GENERAL EQUATION OF A LINE

We have considered several forms of the equation of a straight line in the plane. These different forms of the equation are equivalent to each other. In fact, each is a special case of the following equation.

General Form of a Linear Equation

> The equation
>
> $$Ax + By + C = 0 \tag{6}$$
>
> where A, B, and C are constants and A and B are not both zero, is called the general form of a linear equation in the variables x and y.

We now state (without proof) an important result concerning the algebraic representation of straight lines in the plane.

> An equation of a straight line is a linear equation; conversely, every linear equation represents a straight line.

This result justifies the use of the adjective *linear* in describing Equation (6).

EXAMPLE 12 Sketch the straight line represented by the equation

$$3x - 4y - 12 = 0$$

SOLUTION ✔ Since every straight line is uniquely determined by two distinct points, we need find only two such points through which the line passes in order to sketch it. For convenience, let's compute the points at which the line crosses the x- and y-axes. Setting $y = 0$, we find $x = 4$, so the line crosses the x-axis at the point (4, 0). Setting $x = 0$ gives $y = -3$, so the line crosses the y-axis at the point (0, −3). A sketch of the line appears in Figure 1.27.

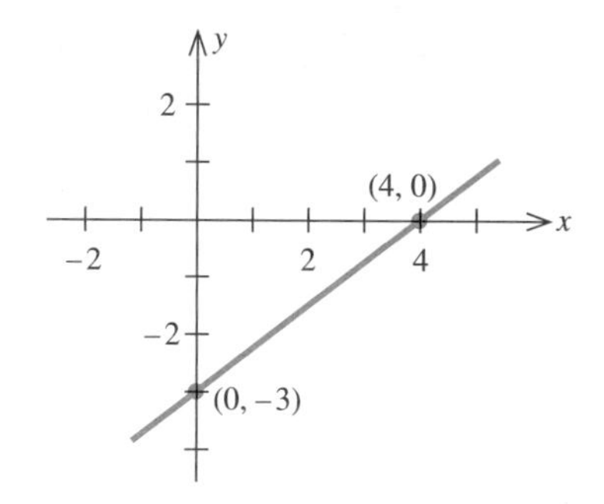

FIGURE 1.27 The straight line $3x - 4y = 12$

■■■■

Following is a summary of the common forms of the equations of straight lines discussed in this section.

Equations of Straight Lines

Vertical line:	$x = a$
Horizontal line:	$y = b$
Point-slope form:	$y - y_1 = m(x - x_1)$
Slope-intercept form:	$y = mx + b$
General form:	$Ax + By + C = 0$

SELF-CHECK EXERCISES 1.2

1. Determine the number a so that the line passing through the points $(a, 2)$ and $(3, 6)$ is parallel to a line with slope 4.
2. Find an equation of the line that passes through the point $(3, -1)$ and is perpendicular to a line with slope $-1/2$.
3. Does the point $(3, -3)$ lie on the line with equation $2x - 3y - 12 = 0$? Sketch the graph of the line.
4. The percentage of people over age 65 who have high school diplomas is summarized in the following table:

Year, x	1960	1965	1970	1975	1980	1985	1990
Percentage with Diplomas, y	20	25	30	36	42	47	52

Source: U.S. Department of Commerce

a. Plot the percentage of people over age 65 who have high school diplomas (y) versus the year (x).
b. Draw the straight line L through the points (1960, 20) and (1990, 52).
c. Find an equation of the line L.
d. Assume the trend continues and estimate the percentage of people over age 65 who will have high school diplomas by the year 2005.

Solutions to Self-Check Exercises 1.2 can be found on page 29.

1.2 Exercises

In Exercises 1–4, find the slope of the line shown in each figure.

1.

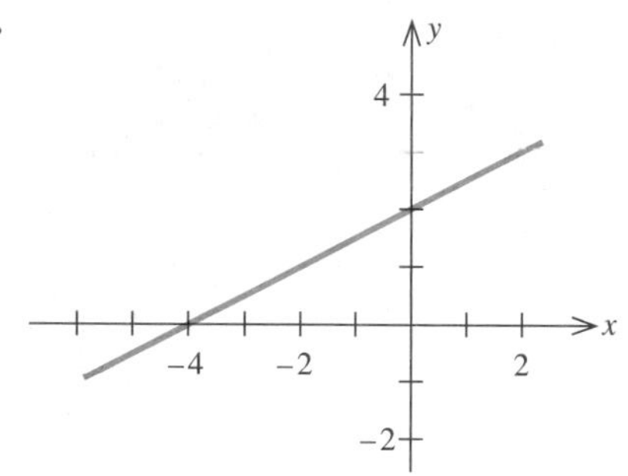

2.

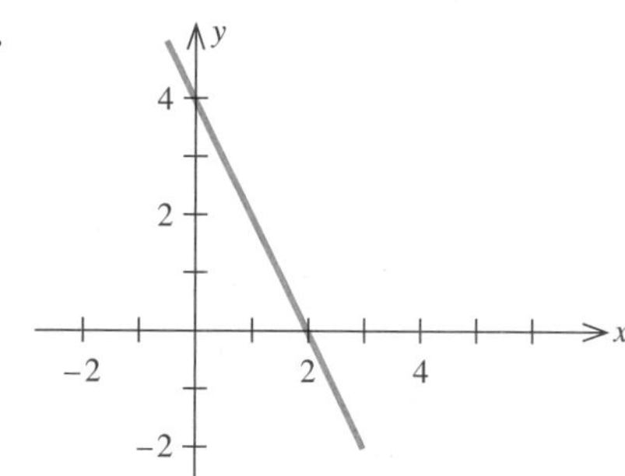

3.

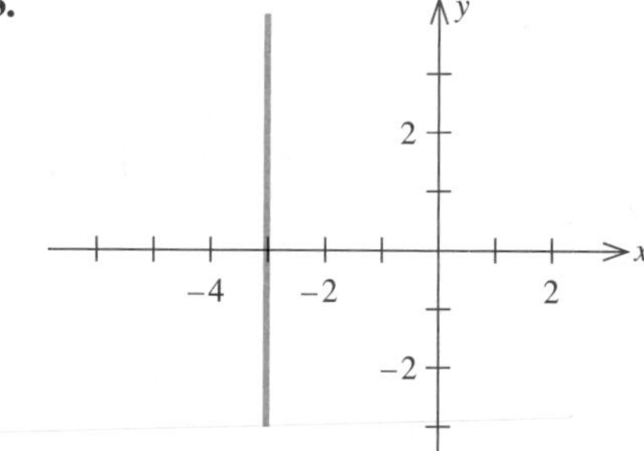

4.

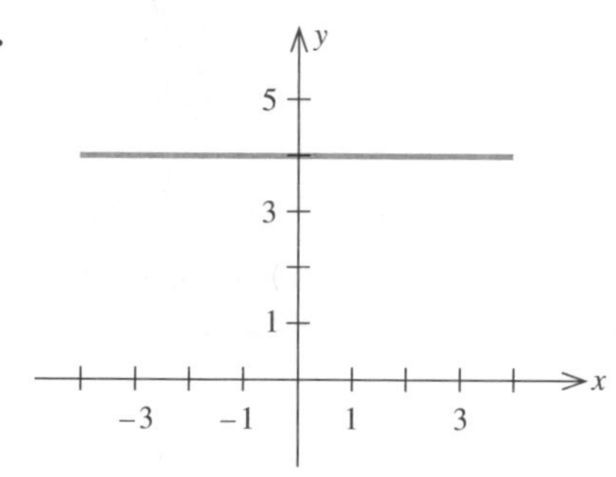

In Exercises 5–10, find the slope of the line that passes through the given pair of points.

5. $(4, 3)$ and $(5, 8)$

6. $(4, 5)$ and $(3, 8)$

7. $(-2, 3)$ and $(4, 8)$

8. $(-2, -2)$ and $(4, -4)$

9. (a, b) and (c, d)

10. $(-a + 1, b - 1)$ and $(a + 1, -b)$

11. Given the equation $y = 4x - 3$, answer the following questions.
    a. If x increases by 1 unit, what is the corresponding change in y?
    b. If x decreases by 2 units, what is the corresponding change in y?

12. Given the equation $2x + 3y = 4$, answer the following questions.
    a. Is the slope of the line described by this equation positive or negative?
    b. As x increases in value, does y increase or decrease?
    c. If x decreases by 2 units, what is the corresponding change in y?

In Exercises 13 and 14, determine whether the lines through the given pairs of points are parallel.

13. $A(1, -2)$, $B(-3, -10)$ and $C(1, 5)$, $D(-1, 1)$

14. $A(2, 3)$, $B(2, -2)$ and $C(-2, 4)$, $D(-2, 5)$

In Exercises 15 and 16, determine whether the lines through the given pairs of points are perpendicular.

15. $A(-2, 5)$, $B(4, 2)$ and $C(-1, -2)$, $D(3, 6)$

16. $A(2, 0)$, $B(1, -2)$ and $C(4, 2)$, $D(-8, 4)$

17. If the line passing through the points $(1, a)$ and $(4, -2)$ is parallel to the line passing through the points $(2, 8)$ and $(-7, a + 4)$, what is the value of a?

18. If the line passing through the points $(a, 1)$ and $(5, 8)$ is parallel to the line passing through the points $(4, 9)$ and $(a + 2, 1)$, what is the value of a?

19. Find an equation of the horizontal line that passes through $(-4, -3)$.

20. Find an equation of the vertical line that passes through $(0, 5)$.

In Exercises 21–26, match the statement with one of the graphs a–f.

21. The slope of the line is zero.

22. The slope of the line is undefined.

23. The slope of the line is positive, and its y-intercept is positive.

24. The slope of the line is positive, and its y-intercept is negative.

25. The slope of the line is negative, and its x-intercept is negative.

26. The slope of the line is negative, and its x-intercept is positive.

a.

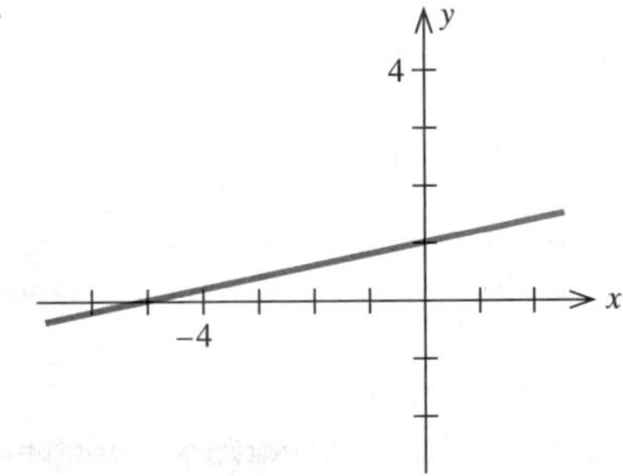

b.

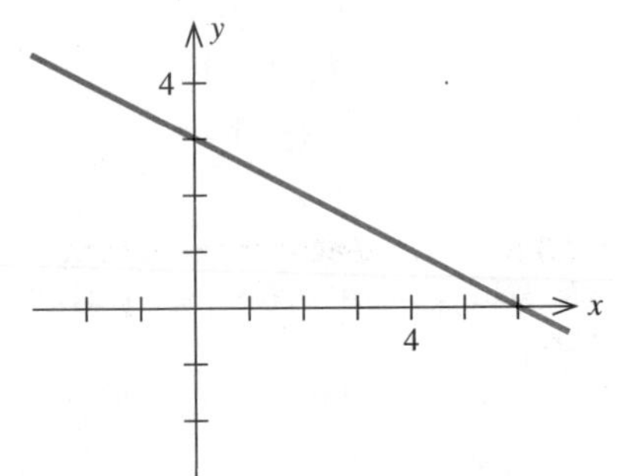

c.

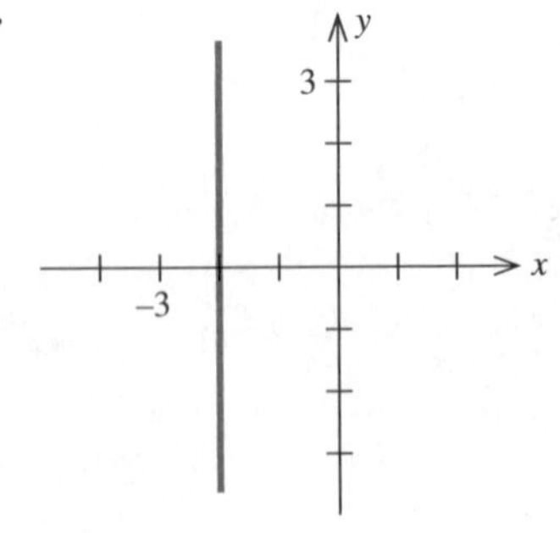

d.

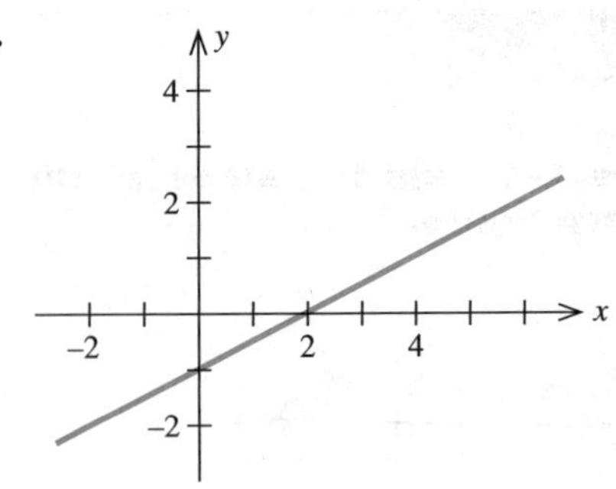

e.

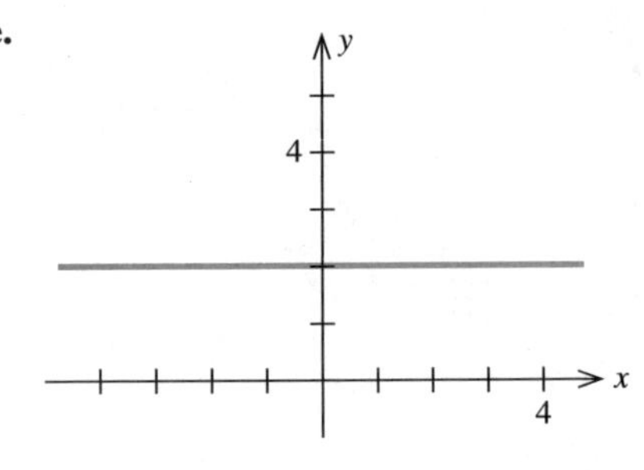

f.

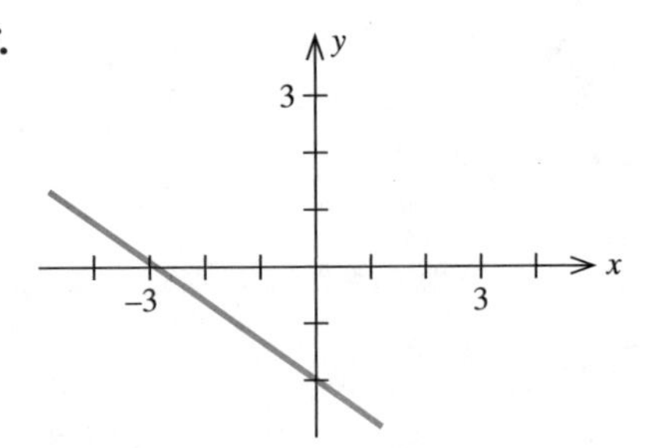

In Exercises 27–30, find an equation of the line that passes through the given point and has the indicated slope *m*.

27. $(3, -4)$; $m = 2$ **28.** $(2, 4)$; $m = -1$

29. $(-3, 2)$; $m = 0$ **30.** $(1, 2)$; $m = -1/2$

In Exercises 31–34, find an equation of the line that passes through the given points.

31. $(2, 4)$ and $(3, 7)$ **32.** $(2, 1)$ and $(2, 5)$

33. $(1, 2)$ and $(-3, -2)$ **34.** $(-1, -2)$ and $(3, -4)$

In Exercises 35–38, find an equation of the line that has slope *m* and *y*-intercept *b*.

35. $m = 3$; $b = 4$ **36.** $m = -2$; $b = -1$

37. $m = 0$; $b = 5$ **38.** $m = -1/2$; $b = 3/4$

In Exercises 39–44, write the given equation in the slope-intercept form and then find the slope and *y*-intercept of the corresponding line.

39. $x - 2y = 0$

40. $y - 2 = 0$

41. $2x - 3y - 9 = 0$

42. $3x - 4y + 8 = 0$

43. $2x + 4y = 14$

44. $5x + 8y - 24 = 0$

45. Find an equation of the line that passes through the point $(-2, 2)$ and is parallel to the line $2x - 4y - 8 = 0$.

46. Find an equation of the line that passes through the point $(2, 4)$ and is perpendicular to the line $3x + 4y - 22 = 0$.

In Exercises 47–54, find an equation of the line that satisfies the given condition.

47. The line parallel to the x-axis and 6 units below it

48. The line passing through the origin and parallel to the line passing through the points $(2, 4)$ and $(4, 7)$

49. The line passing through the point (a, b) with slope equal to zero

50. The line passing through $(-3, 4)$ and parallel to the x-axis

51. The line passing through $(-5, -4)$ and parallel to the line passing through $(-3, 2)$ and $(6, 8)$

52. The line passing through (a, b) with undefined slope

53. Given that the point $P(-3, 5)$ lies on the line $kx + 3y + 9 = 0$, find k.

54. Given that the point $P(2, -3)$ lies on the line $-2x + ky + 10 = 0$, find k.

In Exercises 55–60, sketch the straight line defined by the given linear equation by finding the *x*- and *y*-intercepts.

Hint: See Example 12.

55. $3x - 2y + 6 = 0$

56. $2x - 5y + 10 = 0$

57. $x + 2y - 4 = 0$

58. $2x + 3y - 15 = 0$

59. $y + 5 = 0$

60. $-2x - 8y + 24 = 0$

61. Show that an equation of a line through the points $(a, 0)$ and $(0, b)$ with $a \neq 0$ and $b \neq 0$ can be written in the form

$$\frac{x}{a} + \frac{y}{b} = 1$$

(Recall that the numbers a and b are the x- and y-intercepts, respectively, of the line. This form of an equation of a line is called the **intercept form.**)

In Exercises 62–65, use the results of Exercise 61 to find an equation of a line with the given *x*- and *y*-intercepts.

62. x-intercept 3; y-intercept 4

63. x-intercept -2; y-intercept -4

64. x-intercept $-1/2$; y-intercept $3/4$

65. x-intercept 4; y-intercept $-1/2$

In Exercises 66 and 67, determine whether the given points lie on a straight line.

66. $A(-1, 7)$, $B(2, -2)$, and $C(5, -9)$

67. $A(-2, 1)$, $B(1, 7)$, and $C(4, 13)$

68. **Social Security Contributions** For wages less than the maximum taxable wage base, Social Security contributions by employees are 7.65% of the employee's wages.
a. Find an equation that expresses the relationship between the wages earned (x) and the Social Security taxes paid (y) by an employee who earns less than the maximum taxable wage base.
b. For each additional dollar that an employee earns, by how much is his or her Social Security contribution increased? (Assume that the employee's wages are less than the maximum taxable wage base.)
c. What Social Security contributions will an employee who earns $35,000 (which is less than the maximum taxable wage base) be required to make?

69. **College Admissions** Using data compiled by the Admissions Office at Faber University, college admissions officers estimate that 55% of the students who are offered admission to the freshman class at the university will actually enroll.
a. Find an equation that expresses the relationship between the number of students who actually enroll (y) and the number of students who are offered admission to the university (x).
b. If the desired freshman class size for the upcoming academic year is 1100 students, how many students should be admitted?

70. **Weight of Whales** The equation $W = 3.51L - 192$, expressing the relationship between the length L (in feet) and the expected weight W (in British tons) of adult

blue whales, was adopted in the late 1960s by the International Whaling Commission.
a. What is the expected weight of an 80-ft blue whale?
b. Sketch the straight line that represents the equation.

71. **The Narrowing Gender Gap** Since the founding of the Equal Employment Opportunity Commission and the passage of equal-pay laws, the gulf between men's and women's earnings has continued to close gradually. At the beginning of 1990 ($t = 0$), women's wages were 68% of men's wages. However, women's wages are projected to be 80% of men's wages by the beginning of the year 2000 ($t = 10$). If this gap between women's and men's wages continues to narrow *linearly*, what percentage of men's wages can we expect women's wages to be at the beginning of 2002?
Source: Journal of Economic Perspectives

72. **Ideal Heights and Weights for Women** The Venus Health Club for Women provides its members with the following table, which gives the average desirable weight for women of a certain height:

Height, x (in in.)	60	63	66	69	72
Weight, y (in lb)	108	118	129	140	152

a. Plot the weight (y) versus the height (x).
b. Draw a straight line L through the points corresponding to heights of 5 ft and 6 ft.
c. Derive an equation of the line L.
d. Using the equation of part (c), estimate the average desirable weight for a woman who is 5 ft, 5 in. tall.

73. **Cost of a Commodity** A manufacturer obtained the following data relating the cost y (in dollars) to the number of units (x) of a commodity produced:

Number of Units Produced, x	0	20	40	60	80	100
Cost in Dollars, y	200	208	222	230	242	250

a. Plot the cost (y) versus the quantity produced (x).
b. Draw the straight line through the points (0, 200) and (100, 250).
c. Derive an equation of the straight line of part (b).
d. Taking this equation to be an approximation of the relationship between the cost and the level of production, estimate the cost of producing 54 units of the commodity.

74. **Digital TV Services** The percentage of homes with digital TV services stood at 5% at the beginning of 1999 ($t = 0$) and is projected to grow linearly so that at the beginning of 2003 ($t = 4$) the percentage of such homes is projected to be 25%.
a. Derive an equation of the line passing through the points $A(0, 5)$ and $B(4, 25)$.
b. Plot the line with the equation found in part (a).
c. Using the equation found in part (a) find the percentage of homes with digital TV services at the beginning of 2001.
Source: Paul Kagan Associates

75. **Sales Growth** Metro Department Store's annual sales (in millions of dollars) during the past 5 years were:

Annual Sales, y	5.8	6.2	7.2	8.4	9.0
Year, x	1	2	3	4	5

a. Plot the annual sales (y) versus the year (x).
b. Draw a straight line L through the points corresponding to the first and fifth years.
c. Derive an equation of the line L.
d. Using the equation found in part (c), estimate Metro's annual sales 4 years from now ($x = 9$).

76. Is there a difference between the statements "The slope of a straight line is zero" and "The slope of a straight line does not exist (is not defined)"? Explain your answer.

77. Consider the slope-intercept form of a straight line $y = mx + b$. Describe the family of straight lines obtained by keeping:
a. The value of m fixed and allowing the value of b to vary.
b. The value of b fixed and allowing the value of m to vary.

In Exercises 78–82, determine whether the statement is true or false. If it is true, explain why it is true. If it is false, give an example to show why it is false.

78. Suppose the slope of a line L is $-1/2$ and P is a given point on L. If Q is the point on L lying 4 units to the left of P, then Q is situated 2 units above P.

79. The line with equation $Ax + By + C = 0$ $(B \neq 0)$ and the line with equation $ax + by + c = 0$ $(b \neq 0)$ are parallel if $Ab - aB = 0$.

80. If the slope of the line L_1 is positive, then the slope of a line L_2 perpendicular to L_1 may be positive or negative.

81. The lines with equation $ax + by + c_1 = 0$ and $bx - ay + c_2 = 0$, where $a \neq 0$ and $b \neq 0$, are perpendicular to each other.

82. If L is the line with equation $Ax + By + C = 0$, where $A \neq 0$, then L crosses the x-axis at the point $(-C/A, 0)$.

83. Show that two distinct lines with equations $a_1x + b_1y + c_1 = 0$ and $a_2x + b_2y + c_2 = 0$, respectively, are parallel if and only if $a_1b_2 - b_1a_2 = 0$.

Hint: Write each equation in the slope-intercept form and compare.

84. Prove that if a line L_1 with slope m_1 is perpendicular to a line L_2 with slope m_2, then $m_1m_2 = -1$.

Hint: Refer to the accompanying figure. Show that $m_1 = b$ and $m_2 = c$. Next, apply the Pythagorean theorem and the distance formula to the triangles OAC, OCB, and OBA to show that $1 = -bc$.

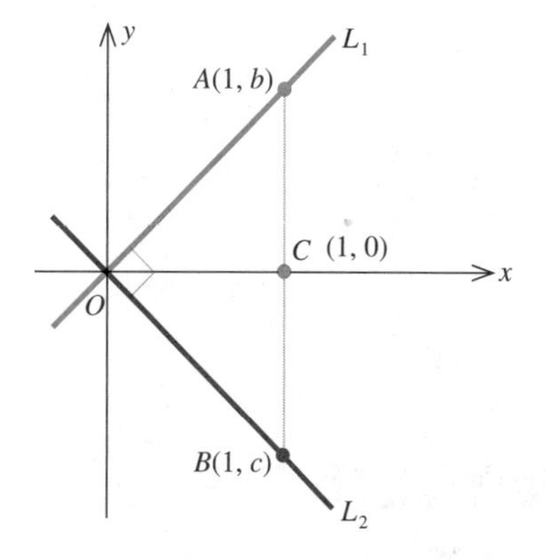

SOLUTIONS TO SELF-CHECK EXERCISES 1.2

1. The slope of the line that passes through the points $(a, 2)$ and $(3, 6)$ is

$$\begin{aligned} m &= \frac{6-2}{3-a} \\ &= \frac{4}{3-a} \end{aligned}$$

Since this line is parallel to a line with slope 4, m must be equal to 4; that is,

$$\frac{4}{3-a} = 4$$

or, upon multiplying both sides of the equation by $3 - a$,

$$\begin{aligned} 4 &= 4(3-a) \\ 4 &= 12 - 4a \\ 4a &= 8 \\ a &= 2 \end{aligned}$$

2. Since the required line L is perpendicular to a line with slope $-\frac{1}{2}$, the slope of L is

$$-\frac{1}{-\frac{1}{2}} = 2$$

Next, using the point-slope form of the equation of a line, we have

$$\begin{aligned} y - (-1) &= 2(x-3) \\ y + 1 &= 2x - 6 \\ y &= 2x - 7 \end{aligned}$$

Corresponding technology sections for CAS may be found at the Brooks/Cole Web site: http://www.brookscole.com/math/authors/tans/

GRAPHING A STRAIGHT LINE

The first step in plotting a straight line with a graphing utility is to select a suitable viewing rectangle. We usually do this by experimenting. For example, you might first plot the straight line using the **standard viewing rectangle** $[-10, 10] \times [-10, 10]$. If necessary, you then might adjust the viewing rectangle by enlarging it or reducing it to obtain a sufficiently complete view of the line or at least the portion of the line that is of interest.

EXAMPLE 1 Plot the straight line $2x + 3y - 6 = 0$ in the standard viewing rectangle.

SOLUTION ✓ The straight line in the standard viewing rectangle is shown in Figure T1.

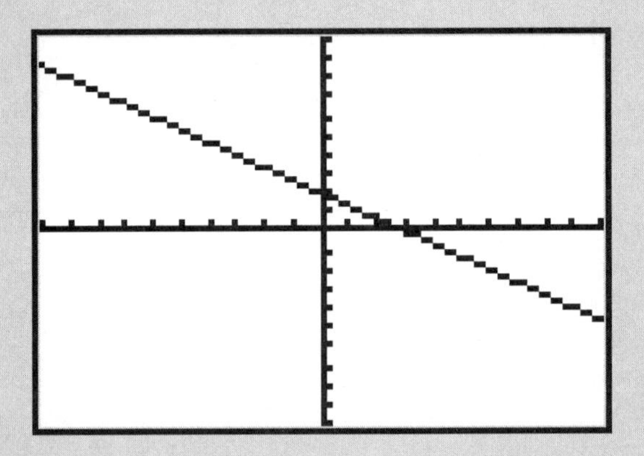

FIGURE T1
The straight line $2x + 3y - 6 = 0$ in the standard viewing rectangle

EXAMPLE 2 Plot the straight line $2x + 3y - 30 = 0$ in (a) the standard viewing rectangle and (b) the viewing rectangle $[-5, 20] \times [-5, 20]$.

SOLUTION ✓ **a.** The straight line in the standard viewing rectangle is shown in Figure T2a.

b. The straight line in the viewing rectangle $[-5, 20] \times [-5, 20]$ is shown in Figure T2b. This figure certainly gives a more complete view of the straight line.

FIGURE T2

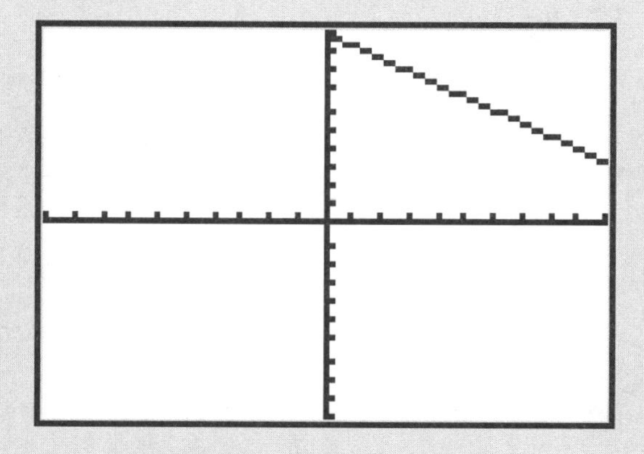

(a) The graph of $2x + 3y - 30 = 0$ in the standard viewing rectangle

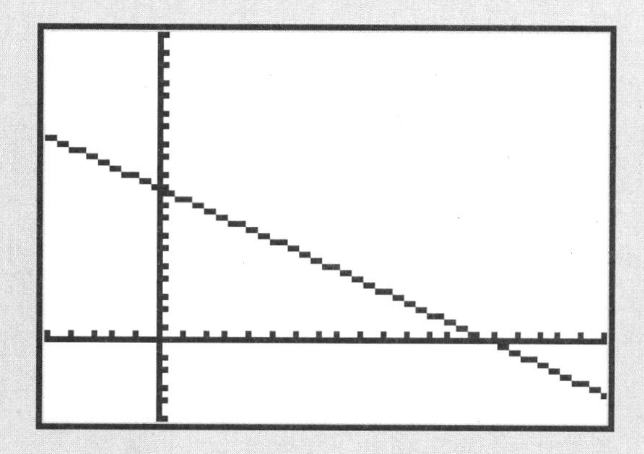

(b) The graph of $2x + 3y - 30 = 0$ in the viewing rectangle $[-5, 20] \times [-5, 20]$

Exercises

In Exercises 1–6, plot the straight line with the given equation in the standard viewing rectangle.

1. $3.2x + 2.1y - 6.72 = 0$
2. $2.3x - 4.1y - 9.43 = 0$
3. $1.6x + 5.1y = 8.16$
4. $-3.2x + 2.1y = 6.72$
5. $2.8x = -1.6y + 4.48$
6. $3.3y = 4.2x - 13.86$

In Exercises 7–10, plot the straight line with the given equation in (a) the standard viewing rectangle and (b) the indicated viewing rectangle.

7. $12.1x + 4.1y - 49.61 = 0$; $[-10, 10] \times [-10, 20]$
8. $4.1x - 15.2y - 62.32 = 0$; $[-10, 20] \times [-10, 10]$
9. $20x + 16y = 300$; $[-10, 20] \times [-10, 30]$
10. $32.2x + 21y = 676.2$; $[-10, 30] \times [-10, 40]$

In Exercises 11–18, plot the straight line with the given equation in an appropriate viewing window. (*Note:* The answer is not unique.)

11. $20x + 30y = 600$
12. $30x - 20y = 600$
13. $22.4x + 16.1y - 352 = 0$
14. $18.2x - 15.1y = 274.8$
15. $1.2x + 20y = 24$
16. $30x - 2.1y = 63$
17. $-4x + 12y = 50$
18. $20x - 12.2y = 240$

3. Substituting $x = 3$ and $y = -3$ into the left-hand side of the given equation, we find

$$2(3) - 3(-3) - 12 = 3$$

which is not equal to zero (the right-hand side). Therefore, $(3, -3)$ does not lie on the line with equation $2x - 3y - 12 = 0$.

Setting $x = 0$, we find $y = -4$, the y-intercept. Next, setting $y = 0$ gives $x = 6$, the x-intercept. We now draw the line passing through the points $(0, -4)$ and $(6, 0)$, as shown.

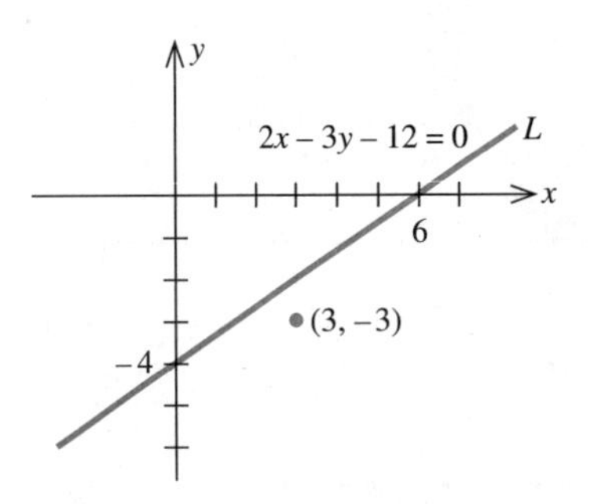

4. a. and **b.** See the accompanying figure.

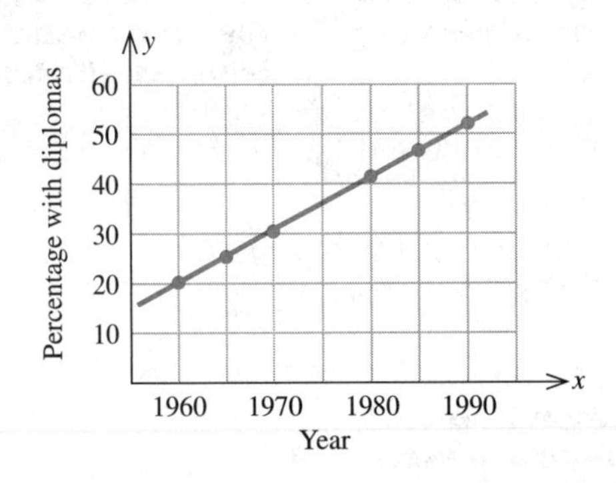

c. The slope of L is

$$m = \frac{52 - 20}{1990 - 1960} = \frac{32}{30} = \frac{16}{15}$$

Using the point-slope form of the equation of a line with the point (1960, 20), we find

$$y - 20 = \frac{16}{15}(x - 1960) = \frac{16}{15}x - \frac{(16)(1960)}{15}$$

$$y = \frac{16}{15}x - \frac{6272}{3} + 20$$

$$= \frac{16}{15}x - \frac{6212}{3}$$

d. To estimate the percentage of people over age 65 who will have high school diplomas by the year 2005, let $x = 2005$ in the equation obtained in part (c). Thus, the required estimate is

$$y = \frac{16}{15}(2005) - \frac{6212}{3} \approx 68.00$$

or approximately 68%.

1.3 Linear Functions and Mathematical Models

Mathematical Models

Before we can apply mathematics to solving real-world problems, we need to formulate these problems in the language of mathematics. This process is referred to as **mathematical modeling.** A mathematical model may give a precise description of the problem under consideration, or, more likely than not, it may provide only an acceptable approximation of the problem. For example, the accumulated amount A at the end of t years when a sum of P dollars is deposited in a fixed bank account and earns interest at the rate of r percent per year compounded m times a year is given *exactly* by the formula (model)

$$A = P\left(1 + \frac{r}{m}\right)^{mt}$$

On the other hand, the size of a cancer tumor may be approximated by the volume of a sphere,

$$V = \frac{4}{3}\pi r^3$$

where r is the radius of the tumor in centimeters.

The many techniques used in constructing mathematical models of practical problems range from theoretical considerations of a problem on the one extreme to an interpretation of data associated with the problem on the other. The model giving the accumulated amount for a fixed bank account mentioned earlier may be derived theoretically (see Chapter 5). In Section 1.5 we will see how linear equations (models) can be constructed from a given set of data points. Also, in the ensuing chapters we will see how other mathematical models, including statistical and probability models, are used to describe and analyze real-world situations.

We will now look at an important way of describing the relationship between two quantities using the notion of a *function*.

FUNCTIONS

A manufacturer would like to know how his company's profit is related to its production level; a biologist would like to know how the population of a certain culture of bacteria will change with time; a psychologist would like to know the relationship between the learning time of an individual and the length of a vocabulary list; and a chemist would like to know how the initial speed of a chemical reaction is related to the amount of substrate used. In each instance, we are concerned with the same question: How does one quantity depend on another? The relationship between two quantities is conveniently described in mathematics by using the concept of a function.

Function

A **function** f is a rule that assigns to each value of x one and only one value of y.

The number y is normally denoted by $f(x)$, read "f of x," emphasizing the dependency of y on x.

An example of a function may be drawn from the familiar relationship between the area of a circle and its radius. Letting x and y denote the radius and area of a circle, respectively, we have, from elementary geometry,

$$y = \pi x^2$$

This equation defines y as a function of x, since for each admissible value of x (a nonnegative number representing the radius of a certain circle) there corresponds precisely one number $y = \pi x^2$ giving the area of the circle. This *area function* may be written as

$$f(x) = \pi x^2 \tag{7}$$

For example, to compute the area of a circle with a radius of 5 inches, we simply replace x in Equation (7) by the number 5. Thus, the area of the circle is

$$f(5) = \pi 5^2 = 25\pi$$

or 25π square inches.

Suppose we are given the function $y = f(x)$.* The variable x is referred to as the **independent variable,** and the variable y is called the **dependent variable.** The set of all values that may be assumed by x is called the **domain** of the function f, and the set comprising all the values assumed by $y = f(x)$ as x takes on all possible values in its domain is called the **range** of the function f. For the area in Function (7), the domain of f is the set of all nonnegative numbers x, and the range of f is the set of all nonnegative numbers y.

We now focus our attention on an important class of functions known as linear functions. Recall that a linear equation in x and y has the form $Ax + By + C = 0$, where A, B, and C are constants and A and B are not both zero. If $B \neq 0$, the equation can always be solved for y in terms of x; in fact, as we saw in Section 1.2, the equation may be cast in the slope-intercept form:

$$y = mx + b \qquad (m, b \text{ constants}) \tag{8}$$

---

* It is customary to refer to a function f as $f(x)$.

Equation (8) defines y as a function of x. The domain and range of this function is the set of all real numbers. Furthermore, the graph of this function, as we saw in Section 1.2, is a straight line in the plane. For this reason, the function $f(x) = mx + b$ is called a linear function.

Linear Function

The function f defined by

$$f(x) = mx + b$$

where m and b are constants, is called a **linear function.**

Linear functions play an important role in the quantitative analysis of business and economic problems. First, many problems arising in these and other fields are *linear* in nature or are *linear* in the intervals of interest and thus can be formulated in terms of linear functions. Second, because linear functions are relatively easy to work with, assumptions involving linearity are often made in the formulation of problems. In many cases these assumptions are justified, and acceptable mathematical models are obtained that approximate real-life situations.

In the rest of this section, we will look at several applications that can be modeled using linear functions.

Simple Depreciation

We first discussed linear depreciation in Section 1.2 as a real-world application of straight lines. The following example illustrates how to derive an equation describing the book value of an asset being depreciated linearly.

EXAMPLE 1

A printing machine has an original value of \$100,000 and is to be depreciated linearly over 5 years with a \$30,000 scrap value. Find an expression giving the book value at the end of year t. What will be the book value of the machine at the end of the second year? What is the rate of depreciation of the printing machine?

SOLUTION ✓

Let V denote the printing machine's book value at the end of the tth year. Since the depreciation is linear, V is a linear function of t. Equivalently, the graph of the function is a straight line. Now, to find an equation of the straight line, observe that $V = 100{,}000$ when $t = 0$; this tells us that the line passes through the point (0, 100,000). Similarly, the condition that $V = 30{,}000$ when $t = 5$ says that the line also passes through the point (5, 30,000). The slope of the line is given by

$$m = \frac{100{,}000 - 30{,}000}{0 - 5} = -\frac{70{,}000}{5} = -14{,}000$$

Using the point-slope form of the equation of a line with the point (0, 100,000) and the slope $m = -14{,}000$, we have

$$V - 100{,}000 = -14{,}000(t - 0)$$
$$V = -14{,}000t + 100{,}000$$

the required expression. The book value at the end of the second year is given by

$$V = -14{,}000(2) + 100{,}000 = 72{,}000$$

or \$72,000. The rate of depreciation of the machine is given by the negative of the slope of the depreciation line. Since the slope of the line is $a = -14{,}000$, the rate of depreciation is \$14,000 per year. The graph of $V = -14{,}000t + 100{,}000$ is sketched in Figure 1.28.

FIGURE 1.28
Linear depreciation of an asset

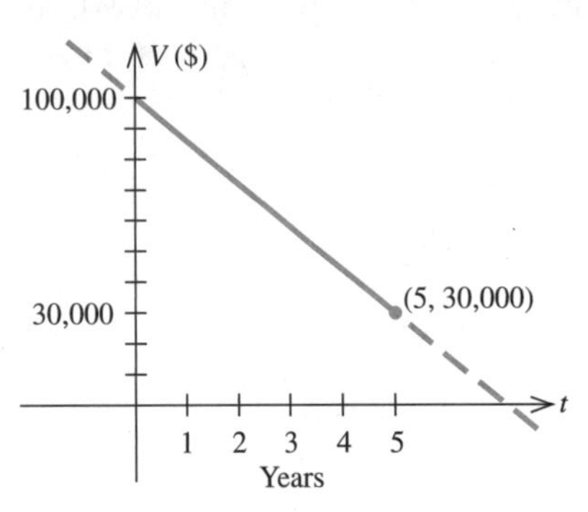

LINEAR COST, REVENUE, AND PROFIT FUNCTIONS

Whether a business is a sole proprietorship or a large corporation, the owner or chief executive must constantly keep track of operating costs, revenue resulting from the sale of products or services, and perhaps most important, the profits realized. Three functions provide management with a measure of these quantities: the total cost function, the revenue function, and the profit function.

Cost, Revenue, and Profit Functions

Let x denote the number of units of a product manufactured or sold. Then, the **total cost function** is

$C(x)$ = Total cost of manufacturing x units of the product

The **revenue function** is

$R(x)$ = Total revenue realized from the sale of x units of the product

The **profit function** is

$P(x)$ = Total profit realized from manufacturing and selling x units of the product

Portfolio

CAROL BUSA

TITLE: Assistant Vice President Financial Center Manager
INSTITUTION: A major Boston bank

"All banks provide more or less the same services. What distinguishes one bank from another is convenience," says Carol Busa. An experienced banker in Boston's highly competitive financial district, Busa has learned that banks must deliver more than basic checking and savings accounts. Customers want a broad range of financial products and the knowledge and service to back them up. Busa exhorts her staff to work closely with customers to make banking as trouble-free as possible.

Busa deals with a wide assortment of customers, such as a junior executive looking for a car loan and a middle-aged mother seeking to tap her home equity for a child's college education.

Working with each customer, Busa examines several loan scenarios, entering different variables into the computer to determine the best financial mix for the customer's circumstance.

Although the computer frees Busa from doing manual calculations, she has to be able to explain how each variable affects the equation. For example, given the junior executive's modest salary, she might recommend a smaller car loan based on his ability to handle monthly payments. A $10,000 loan at 8% interest would require payments of $244.13 per month over a 4-year period, whereas an $8000 loan would reduce the monthly obligation to $195.30—a more manageable amount.

To finance tuition bills, Busa might suggest an equity line of credit versus a home equity loan. The line of credit would allow the mother to draw on the money as needed rather than borrowing a large sum at one time. The one disadvantage in borrowing from a line of credit is its variable interest rate. Customers are usually charged the current prime rate (the interest banks charge their best customers) plus 1.5%. As the prime fluctuates, it becomes impossible to forecast how much interest will accrue over a 20-year repayment cycle. The best Busa can do is calculate the accrued interest during the first one or two years, hoping the prime rate will remain constant for that long.

Forecasting interest payments is only part of Busa's job. She is often called on to help customers with more mundane tasks, such as balancing their checking accounts. "You'd be surprised how many people can't balance a checkbook," she says. Busa usually asks them to try one more time with an up-to-date computer statement. If they still can't balance, then she or her staff help them find the error.

If customers sometimes feel overwhelmed reconciling their accounts each month, they can take some comfort in the fact that Busa has to balance her branch bank accounts each day. Even more daunting is the bank's responsibility for balancing accounts in all 300-plus branches daily, exponentially increasing the chance of arithmetic error.

As a branch manager for a major Boston bank, Busa wears many hats in a typical day—supervisor, financial counselor, diplomat—juggling them all with equal dexterity. She is responsible for the successful operation of the branch, overseeing a staff of eight, and she must be familiar with the full range of the bank's financial products. She also must deal with irate customers as well as those who just want a little extra attention.

Generally speaking, the total cost, revenue, and profit functions associated with a company are more likely than not to be nonlinear (these functions are best studied using the tools of calculus). But *linear* cost, revenue, and profit functions do arise in practice, and we will consider such functions in this section. Before deriving explicit forms of these functions, we need to recall some common terminology.

The costs incurred in operating a business are usually classified into two categories. Costs that remain more or less constant regardless of the firm's activity level are called **fixed costs.** Examples of fixed costs are rental fees and executive salaries. Costs that vary with production or sales are called **variable costs.** Examples of variable costs are wages and costs for raw materials.

Suppose a firm has a fixed cost of F dollars, a production cost of c dollars per unit, and a selling price of s dollars per unit. Then the *cost function* $C(x)$, the *revenue function* $R(x)$, and the *profit function* $P(x)$ for the firm are given by

$$\begin{aligned} C(x) &= cx + F \\ R(x) &= sx \\ P(x) &= R(x) - C(x) \qquad \text{(Revenue − cost)} \\ &= (s - c)x - F \end{aligned}$$

where x denotes the number of units of the commodity produced and sold. The functions C, R, and P are linear functions of x.

EXAMPLE 2 Puritron, a manufacturer of water filters, has a monthly fixed cost of \$20,000, a production cost of \$20 per unit, and a selling price of \$30 per unit. Find the cost function, the revenue function, and the profit function for Puritron.

SOLUTION ✔ Let x denote the number of units produced and sold. Then,

$$\begin{aligned} C(x) &= 20x + 20{,}000 \\ R(x) &= 30x \\ P(x) &= R(x) - C(x) \\ &= 30x - (20x + 20{,}000) \\ &= 10x - 20{,}000 \end{aligned}$$

■■■■

Linear Demand and Supply Curves

In a free-market economy, consumer demand for a particular commodity depends on the commodity's unit price. A **demand equation** expresses this relationship between the unit price and the quantity demanded. The corresponding graph of the demand equation is called a **demand curve.** In general, the quantity demanded of a commodity decreases as its unit price increases, and vice versa. Accordingly, a **demand function,** defined by $p = f(x)$, where p measures the unit price and x measures the number of units of the commodity, is generally characterized as a decreasing function of x; that is, $p = f(x)$ decreases as x increases.

The simplest demand function is defined by a linear equation in x and p, where both x and p assume only nonnegative values. Its graph is a straight line having a negative slope. Thus, the demand curve in this case is that part of the graph of a straight line that lies in the first quadrant (Figure 1.29).

FIGURE 1.29
A graph of a linear demand function

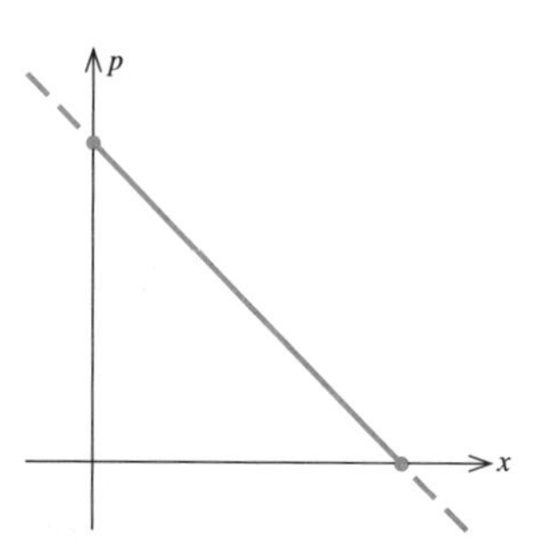

EXAMPLE 3 The quantity demanded of the Sentinel alarm clock is 48,000 units when the unit price is \$8. At \$12 per unit, the quantity demanded drops to 32,000 units. Find the demand equation, assuming that it is linear. What is the unit price corresponding to a quantity demanded of 40,000 units? What is the quantity demanded if the unit price is \$14?

SOLUTION ✔ Let p denote the unit price of an alarm clock (in dollars) and let x (in units of 1000) denote the quantity demanded when the unit price of the clocks is \$$p$. When $p = 8$, $x = 48$ and the point (48, 8) lies on the demand curve. Similarly, when $p = 12$, $x = 32$ and the point (32, 12) also lies on the demand curve. Since the demand equation is linear, its graph is a straight line. The slope of the required line is given by

$$m = \frac{12 - 8}{32 - 48} = \frac{4}{-16} = -\frac{1}{4}$$

So, using the point-slope form of an equation of a line with the point (48, 8), we find that

$$p - 8 = -\frac{1}{4}(x - 48)$$

$$p = -\frac{1}{4}x + 20$$

is the required equation. The demand curve is shown in Figure 1.30. If the

FIGURE 1.30
The graph of the demand equation $p = -\frac{1}{4}x + 20$

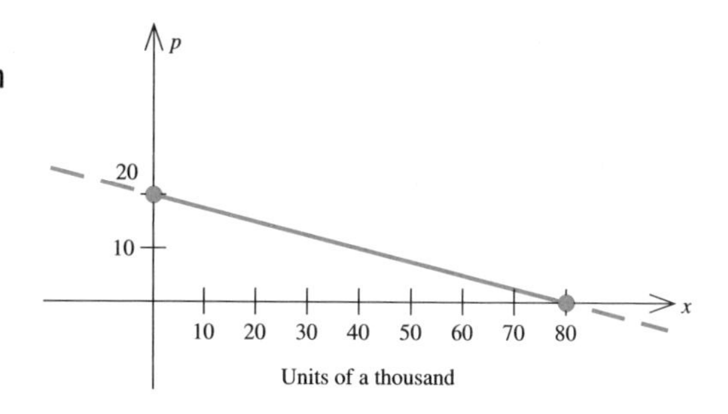

quantity demanded is 40,000 units ($x = 40$), the demand equation yields

$$y = -\frac{1}{4}(40) + 20 = 10$$

and we see that the corresponding unit price is \$10. Next, if the unit price is \$14 ($p = 14$), the demand equation yields

$$14 = -\frac{1}{4}x + 20$$

$$\frac{1}{4}x = 6$$

$$x = 24$$

so the quantity demanded will be 24,000 units. ■■■■

In a competitive market, a relationship also exists between the unit price of a commodity and its availability in the market. In general, an increase in a commodity's unit price will induce the manufacturer to increase the supply of that commodity. Conversely, a decrease in the unit price generally leads to a drop in the supply. An equation that expresses the relationship between the unit price and the quantity supplied is called a **supply equation,** and the corresponding graph is called a **supply curve.** A **supply function,** defined by $p = f(x)$, is generally characterized by an increasing function of x; that is, $p = f(x)$ increases as x increases.

As in the case of a demand equation, the simplest supply equation is a linear equation in p and x, where p and x have the same meaning as before but the line has a positive slope. The supply curve corresponding to a linear supply function is that part of the straight line that lies in the first quadrant (Figure 1.31).

FIGURE 1.31
A graph of a linear supply function

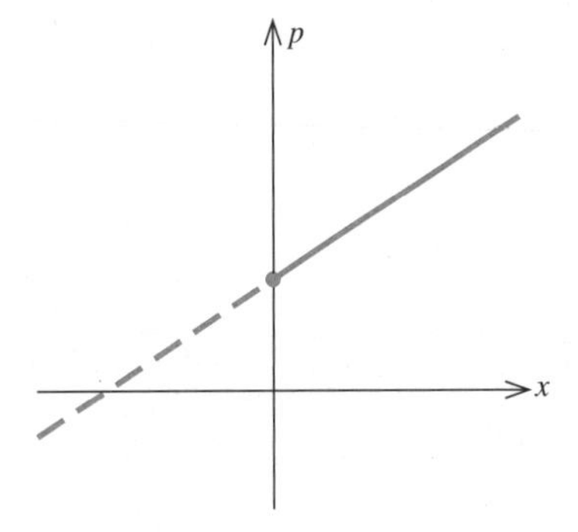

EXAMPLE 4 The supply equation for a commodity is given by $4p - 5x = 120$, where p is measured in dollars and x is measured in units of 100.

a. Sketch the corresponding curve.

b. How many units will be marketed when the unit price is \$55?

FIGURE 1.32
The graph of the supply equation $4p - 5x = 120$

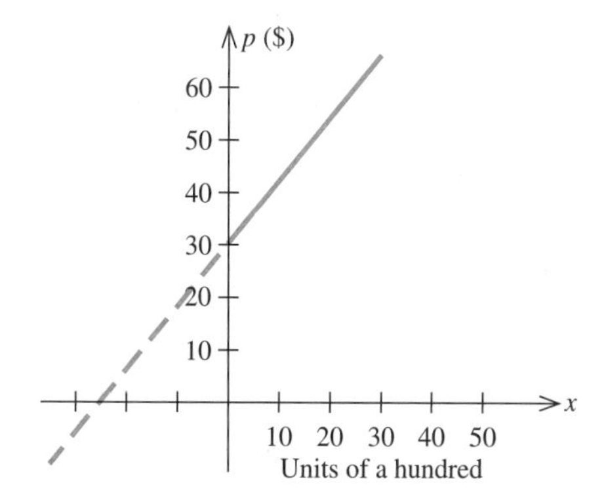

SOLUTION ✓

a. Setting $x = 0$ and $p = 0$, we find that the p- and x-intercepts are 30 and -24, respectively. The supply curve is sketched in Figure 1.32.

b. Substituting $p = 55$ in the supply equation, we have $4(55) - 5x = 120$, or $x = 20$, so the amount marketed will be 2000 units. ■■■■

Self-Check Exercises 1.3

1. A manufacturer has a monthly fixed cost of \$60,000 and a production cost of \$10 for each unit produced. The product sells for \$15/unit.
   a. What is the cost function?
   b. What is the revenue function?
   c. What is the profit function?
   d. Compute the profit (loss) corresponding to production levels of 10,000 and 14,000 units.

2. The quantity demanded for a certain make of 30-in. × 52-in. area rug is 500 when the unit price is \$100. For each \$20 decrease in the unit price, the quantity demanded increases by 500 units. Find the demand equation and sketch its graph.

Solutions to Self-Check Exercises 1.3 can be found on page 46.

1.3 Exercises

In Exercises 1–10, determine whether the equation defines y as a linear function of x. If so, write it in the form $y = mx + b$.

1. $2x + 3y = 6$

2. $-2x + 4y = 7$

3. $x = 2y - 4$

4. $2x = 3y + 8$

5. $2x - 4y + 9 = 0$

6. $3x - 6y + 7 = 0$

7. $2x^2 - 8y + 4 = 0$

8. $3\sqrt{x} + 4y = 0$

9. $2x - 3y^2 + 8 = 0$

10. $2x + \sqrt{y} - 4 = 0$

11. A manufacturer has a monthly fixed cost of \$40,000 and a production cost of \$8 for each unit produced. The product sells for \$12/unit.
   a. What is the cost function?
   b. What is the revenue function?
   c. What is the profit function?
   d. Compute the profit (loss) corresponding to production levels of 8000 and 12,000 units.

12. A manufacturer has a monthly fixed cost of \$100,000 and a production cost of \$14 for each unit produced. The product sells for \$20/unit.
a. What is the cost function?
b. What is the revenue function?
c. What is the profit function?
d. Compute the profit (loss) corresponding to production levels of 12,000 and 20,000 units.

13. Find the constants m and b in the linear function $f(x) = mx + b$ so that $f(0) = 2$ and $f(3) = -1$.

14. Find the constants m and b in the linear function $f(x) = mx + b$ so that $f(2) = 4$ and the straight line represented by f has slope -1.

15. **Linear Depreciation** An office building worth \$1 million when completed in 2000 is being depreciated linearly over 50 years. What will be the book value of the building in 2005? In 2010? (Assume the scrap value is \$0.)

16. **Linear Depreciation** An automobile purchased for use by the manager of a firm at a price of \$14,000 is to be depreciated using the straight-line method over 5 years. What will be the book value of the automobile at the end of 3 years? (Assume the scrap value is \$0.)

17. **Consumption Functions** A certain economy's consumption function is given by the equation

$$C(x) = 0.75x + 6$$

where $C(x)$ is the personal consumption expenditure in billions of dollars and x is the disposable personal income in billions of dollars. Find $C(0)$, $C(50)$, and $C(100)$.

18. **Sales Tax** In a certain state, the sales tax T on the amount of taxable goods is 6% of the value of the goods purchased (x), where both T and x are measured in dollars.
a. Express T as a function of x.
b. Find $T(200)$ and $T(5.60)$.

19. **Social Security Benefits** Social Security recipients receive an automatic cost-of-living adjustment (COLA) once each year. Their monthly benefit is increased by the same percentage that consumer prices have increased during the preceding year. Suppose consumer prices have increased by 5.3% during the preceding year.
a. Express the adjusted monthly benefit of a Social Security recipient as a function of his or her current monthly benefit.
b. If Carlos Garcia's monthly Social Security benefit is now \$620, what will be his adjusted monthly benefit?

20. **Profit Functions** Auto-Time, a manufacturer of 24-hour variable timers, has a monthly fixed cost of \$48,000 and a production cost of \$8 for each timer manufactured. The timers sell for \$14 each.
a. What is the cost function?
b. What is the revenue function?
c. What is the profit function?
d. Compute the profit (loss) corresponding to production levels of 4000, 6000, and 10,000 timers, respectively.

21. **Profit Functions** The management of TMI, Inc., finds that the monthly fixed costs attributable to the company's blank-tape division amount to \$12,100. If the cost for producing each reel of tape is \$.60 and each reel of tape sells for \$1.15, find the company's cost function, revenue function, and profit function.

22. **Linear Depreciation** In 2000, the National Textile Company installed a new machine in one of its factories at a cost of \$250,000. The machine is depreciated linearly over 10 years with a scrap value of \$10,000.
a. Find an expression for the machine's book value in the tth year of use $(0 \le t \le 10)$.
b. Sketch the graph of the function of part (a).
c. Find the machine's book value in 2004.
d. Find the rate at which the machine is being depreciated.

23. **Linear Depreciation** A minicomputer purchased at a cost of \$60,000 in 1999 has a scrap value of \$12,000 at the end of 4 years. If the straight-line method of depreciation is used,
a. Find the rate of depreciation.
b. Find the linear equation expressing the minicomputer's book value at the end of t years.
c. Sketch the graph of the function of part (b).
d. Find the minicomputer's book value at the end of the third year.

24. **Linear Depreciation** Suppose an asset has an original value of \$$C$ and is depreciated linearly over N years with a scrap value of \$$S$. Show that the asset's book value at the end of the tth year is described by the function

$$V(t) = C - \left(\frac{C - S}{N}\right)t$$

Hint: Find an equation of the straight line passing through the points $(0, C)$ and (N, S). (Why?)

25. Rework Exercise 15 using the formula derived in Exercise 24.

26. Rework Exercise 16 using the formula derived in Exercise 24.

27. **Drug Dosages** A method sometimes used by pediatricians to calculate the dosage of medicine for children is based on the child's surface area. If a denotes the adult dosage (in milligrams), and if S is the child's surface area (in square meters), then the child's dosage is given by

$$D(S) = \frac{Sa}{1.7}$$

a. Show that D is a linear function of S.

Hint: Think of D as having the form $D(S) = mS + b$. What is the slope m and the y-intercept b?

b. If the adult dose of a drug is 500 mg, how much should a child whose surface area is 0.4 m² receive?

28. **DRUG DOSAGES** Cowling's rule is a method for calculating pediatric drug dosages. If a denotes the adult dosage (in milligrams) and if t is the child's age (in years), then the child's dosage is given by

$$D(t) = \left(\frac{t+1}{24}\right)a$$

a. Show that D is a linear function of t.

Hint: Think of $D(t)$ as having the form $D(t) = mt + b$. What is the slope m and the y-intercept b?

b. If the adult dose of a drug is 500 mg, how much should a 4-year-old child receive?

29. **CELSIUS AND FAHRENHEIT TEMPERATURES** The relationship between temperature measured in the Celsius scale and the Fahrenheit scale is linear. The freezing point is 0°C and 32°F and the boiling point is 100°C and 212°F, respectively.
a. Find an equation giving the relationship between the temperature F measured in the Fahrenheit scale and the temperature C measured in the Celsius scale.
b. Find F as a function of C and use this formula to determine the temperature in Fahrenheit corresponding to a temperature of 20°C.
c. Find C as a function of F and use this formula to determine the temperature in Celsius corresponding to a temperature of 70°F.

30. **CRICKET CHIRPING AND TEMPERATURE** Entomologists have discovered that there is a linear relationship between the number of chirps of crickets of a certain species and the air temperature. When the temperature is 70°F, the crickets chirp at the rate of 120 times/minute, and when the temperature is 80°F, they chirp at the rate of 160 times/minute.
a. Find an equation giving the relationship between the air temperature T and the number of chirps per minute N of the crickets.
b. Find N as a function of T and use this formula to determine the rate at which the crickets chirp when the temperature is 102°F.

31. **DEMAND FOR CLOCK RADIOS** In the accompanying figure, L_1 is the demand curve for the model A clock radio manufactured by Ace Radio Inc., and L_2 is the demand curve for the model B clock radio. Which line has the greater slope? Interpret your results.

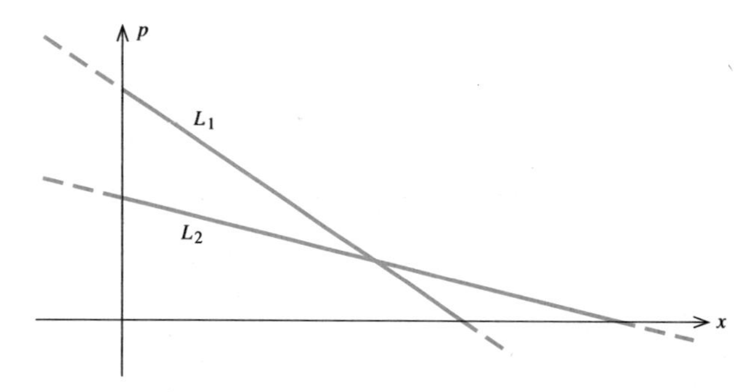

32. **SUPPLY OF CLOCK RADIOS** In the accompanying figure, L_1 is the supply curve for the model A clock radio manufactured by Ace Radio Inc., and L_2 is the supply curve for the model B clock radio. Which line has the greater slope? Interpret your results.

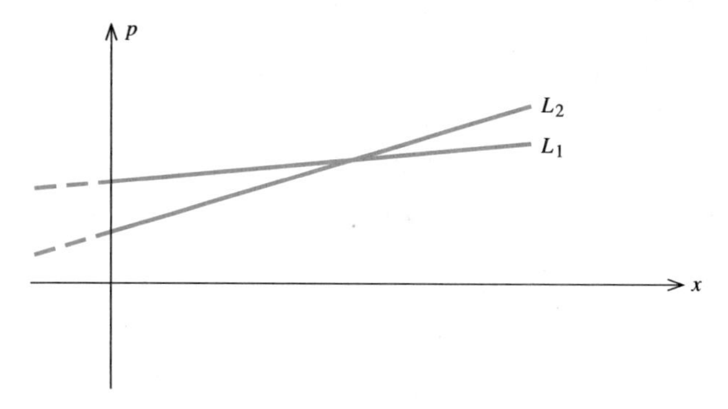

For each of the demand equations in Exercises 33–36, where x represents the quantity demanded in units of 1000 and p is the unit price in dollars, (a) sketch the demand curve; (b) determine the quantity demanded corresponding to the given unit price p.

33. $2x + 3p - 18 = 0$; $p = 4$

34. $5p + 4x - 80 = 0$; $p = 10$

35. $p = -3x + 60$; $p = 30$

36. $p = -0.4x + 120$; $p = 80$

37. **DEMAND FUNCTIONS** At a unit price of \$55, the quantity demanded of a certain commodity is 1000 units. At a unit price of \$85, the demand drops to 600 units. Given

(continued on p. 46)

Using Technology

EVALUATING A FUNCTION

A graphing utility can be used to find the value of a function f at a given point with minimal effort. However, to find the value of y for a given value of x in a linear equation such as $Ax + By + C = 0$, the equation must first be cast in the slope-intercept form $y = mx + b$, thus revealing the desired rule $f(x) = mx + b$ for y as a function of x.

EXAMPLE 1 Consider the equation $2x + 5y = 7$.

a. Plot the straight line with the given equation in the standard viewing rectangle.

b. Find the value of y when $x = 2$ and verify your result by direct computation.

c. Find the value of y when $x = 1.732$.

SOLUTION ✓ **a.** The straight line with equation $2x + 5y = 7$ or, equivalently, $y = -\frac{2}{5}x + \frac{7}{5}$ in the standard viewing rectangle is shown in Figure T1.

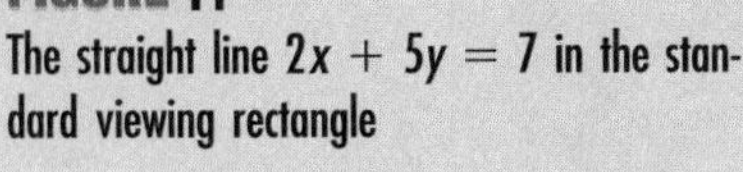

FIGURE T1
The straight line $2x + 5y = 7$ in the standard viewing rectangle

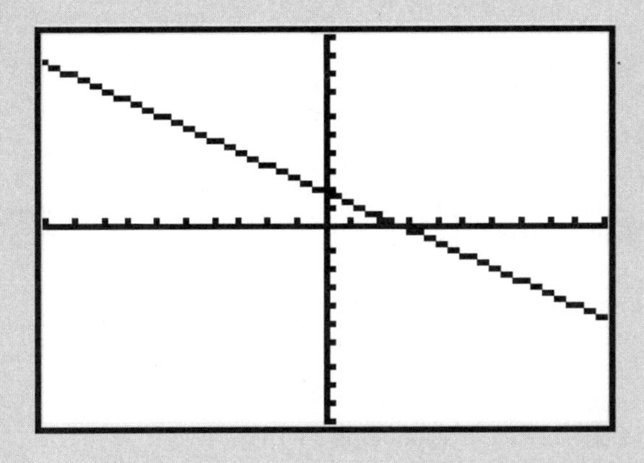

b. Using the evaluation function of the graphing utility and the value of 2 for x, we find $y = 0.6$. This result is verified by computing

$$y = -\frac{2}{5}(2) + \frac{7}{5} = -\frac{4}{5} + \frac{7}{5} = \frac{3}{5} = 0.6$$

when $x = 2$.

c. Once again using the evaluation function of the graphing utility, this time with the value 1.732 for x, we find $y = 0.7072$. The efficacy of the graphing utility is clearly demonstrated here!

When evaluating $f(x)$ at $x = a$, remember that the number a must lie between xMin and xMax.

EXAMPLE 2 According to Pacific Gas and Electric, the nation's largest utility company, the demand for electricity (in megawatts) in year t is approximately

$$D(t) = 295t + 328 \qquad (0 \leq t \leq 10)$$

with $t = 0$ corresponding to 1990.

a. Plot the graph of D in the viewing window $[0, 10] \times [0, 3500]$.

b. What was the demand for electricity in 1999?

Source: Pacific Gas and Electric

SOLUTION ✓ **a.** The graph of D is shown in Figure T2.

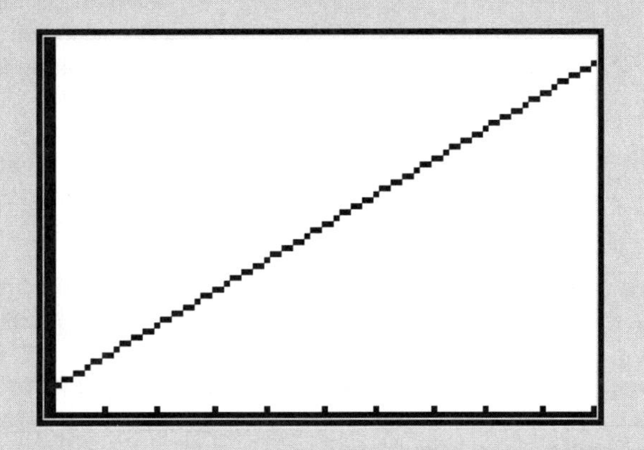

FIGURE T2
The graph of D in the viewing window $[0, 10] \times [0, 3500]$

b. Evaluating the function at $t = 9$, we find $y = 2983$. Therefore, the demand for electricity in 1999 was approximately 2983 megawatts.

Exercises

In Exercises 1–8, use the evaluation function of your graphing calculator to find the value of *y* corresponding to the given value of *x*.

1. $3.1x + 2.4y - 12 = 0; x = 2.1$

2. $1.2x - 3.2y + 8.2 = 0; x = 1.2$

3. $2.8x + 4.2y = 16.3; x = 1.5$

4. $-1.8x + 3.2y - 6.3 = 0; x = -2.1$

5. $22.1x + 18.2y - 400 = 0; x = 12.1$

6. $17.1x - 24.31y - 512 = 0; x = -8.2$

7. $2.8x = 1.41y - 2.64; x = 0.3$

8. $0.8x = 3.2y - 4.3; x = -0.4$

that it is linear, find the demand equation. Above what price will there be no demand? What quantity would be demanded if the commodity were free?

38. **DEMAND FUNCTIONS** The quantity demanded for a certain commodity is 200 units when the unit price is set at \$90. The quantity demanded is 1200 units when the unit price is \$40. Find the demand equation and sketch its graph.

39. **DEMAND FUNCTIONS** Assume that a certain commodity's demand equation has the form $p = ax + b$, where x is the quantity demanded and p is the unit price in dollars. Suppose the quantity demanded is 1000 units when the unit price is \$9 and 6000 when the unit price is \$4. What is the quantity demanded when the unit price is \$7.50?

40. **DEMAND FUNCTIONS** The demand equation for the Sicard wristwatch is

$$p = -0.025x + 50$$

where x is the quantity demanded per week and p is the unit price in dollars. Sketch the graph of the demand equation. What is the highest price (theoretically) anyone would pay for the watch?

For each supply equation in Exercises 41–44, where x is the quantity supplied in units of 1000 and p is the unit price in dollars, (a) sketch the supply curve; (b) determine the number of units of the commodity the supplier will make available in the market at the given unit price.

41. $3x - 4p + 24 = 0;\ p = 8$

42. $\frac{1}{2}x - \frac{2}{3}p + 12 = 0;\ p = 24$

43. $p = 2x + 10;\ p = 14$

44. $p = \frac{1}{2}x + 20;\ p = 28$

45. **SUPPLY FUNCTIONS** Suppliers of 15-in. black-and-white television sets will make 10,000 available in the market if the unit price is \$45. At a unit price of \$50, 20,000 units will be made available. Assuming that the relationship between the unit price and the quantity supplied is linear, derive the supply equation. Sketch the supply curve and determine the quantity suppliers will make available when the unit price is \$70.

46. **SUPPLY FUNCTIONS** Producers will make 2000 refrigerators available when the unit price is \$330. At a unit price of \$390, 6000 refrigerators will be marketed. Find the equation relating the unit price of a refrigerator to the quantity supplied if the equation is known to be linear. How many refrigerators will be marketed when the unit price is \$450? What is the lowest price at which a refrigerator will be marketed?

In Exercises 47 and 48, determine whether the statement is true or false. If it is true, explain why it is true. If it is false, give an example to show why it is false.

47. Suppose $C(x) = cx + F$ and $R(x) = sx$ are the cost and revenue functions of a certain firm. Then, the firm is making a profit if its level of production is less than $F/(s - c)$.

48. If $p = mx + b$ is a linear demand curve, then it is generally true that $m < 0$.

SOLUTIONS TO SELF-CHECK EXERCISES 1.3

1. Let x denote the number of units produced and sold. Then
   a. $C(x) = 10x + 60{,}000$
   b. $R(x) = 15x$
   c. $P(x) = R(x) - C(x) = 15x - (10x + 60{,}000)$
   $= 5x - 60{,}000$
   d. $P(10{,}000) = 5(10{,}000) - 60{,}000$
   $= -10{,}000$
   or a loss of \$10,000.
   $P(14{,}000) = 5(14{,}000) - 60{,}000$
   $= 10{,}000$
   or a profit of \$10,000.

2. Let p denote the price of a rug (in dollars) and let x denote the quantity of rugs demanded when the unit price is $\$p$. The condition that the quantity demanded is 500 when the unit price is \$100 tells us that the demand curve passes through the point (500, 100). Next, the condition that for each \$20 decrease in the unit price the quantity demanded increases by 500 tells us that the demand curve is linear and that its slope is given by $-20/500$, or $-1/25$. Therefore, letting $m = -1/25$ in the demand equation

$$p = mx + b$$

we find

$$p = -\frac{1}{25}x + b$$

To determine b, use the fact that the straight line passes through (500, 100) to obtain

$$100 = -\frac{1}{25}(500) + b$$

or $b = 120$. Therefore, the required equation is

$$p = -\frac{1}{25}x + 120$$

The demand curve is sketched in the accompanying figure.

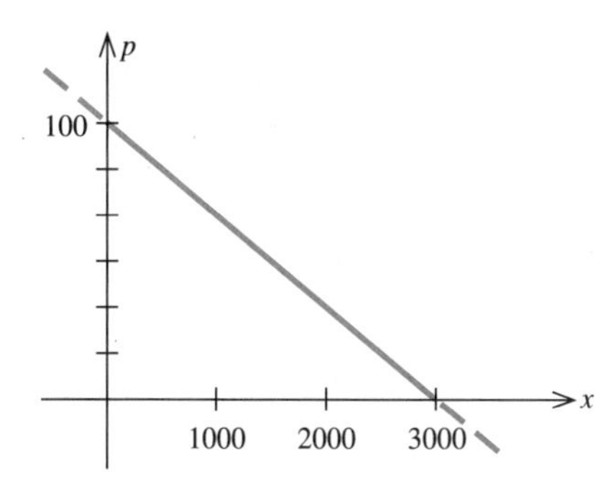

The graph of the demand curve $p = -\frac{1}{25}x + 120$

1.4 Intersection of Straight Lines

FINDING THE POINT OF INTERSECTION

The solution of certain practical problems involves finding the point of intersection of two straight lines. To see how such a problem may be solved algebraically, suppose we are given two straight lines L_1 and L_2 with equations

$$y = m_1x + b_1 \qquad \text{and} \qquad y = m_2x + b_2$$

(where m_1, b_1, m_2, and b_2 are constants) that intersect at the point $P(x, y)$ (Figure 1.33).

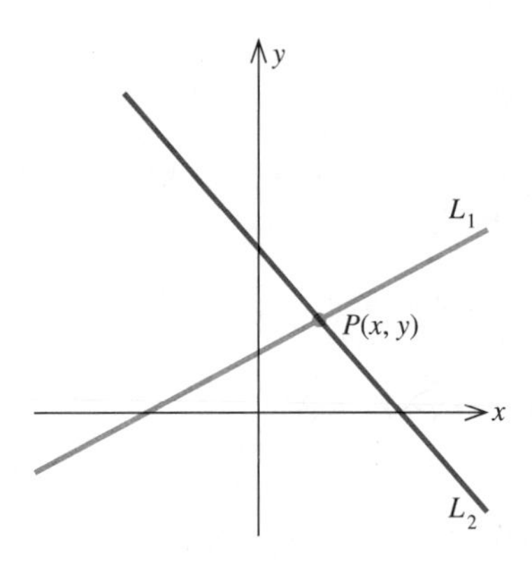

FIGURE 1.33
L_1 and L_2 intersect at the point $P(x, y)$.

The point $P(x, y)$ lies on the line L_1 and so satisfies the equation $y = m_1x + b_1$. It also lies on the line L_2 and so satisfies the equation $y = m_2x + b_2$. Therefore, to find the point of intersection $P(x, y)$ of the lines L_1 and L_2, we solve the system composed of the two equations

$$y = m_1x + b_1 \qquad \text{and} \qquad y = m_2x + b_2$$

for x and y.

EXAMPLE 1 Find the point of intersection of the straight lines that have equations $y = x + 1$ and $y = -2x + 4$.

SOLUTION ✔ We solve the given simultaneous equations. Substituting the value y as given in the first equation into the second, we obtain

$$\begin{aligned} x + 1 &= -2x + 4 \\ 3x &= 3 \\ x &= 1 \end{aligned}$$

Substituting this value of x into either one of the given equations yields $y = 2$. Therefore, the required point of intersection is (1, 2) (Figure 1.34).

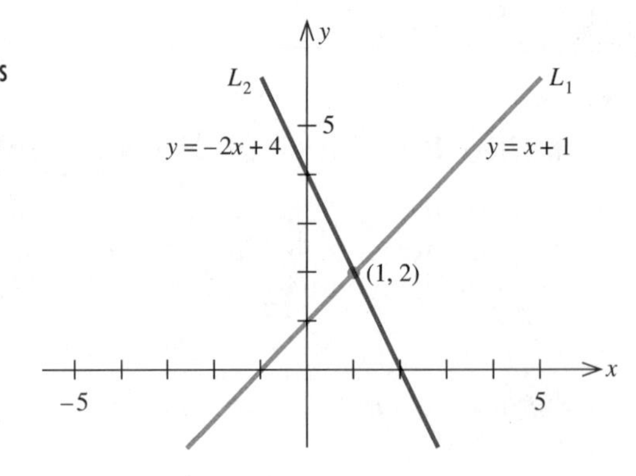

FIGURE 1.34
The point of intersection of L_1 and L_2 is (1, 2).

Exploring with Technology

1. Use a graphing utility to plot the straight lines L_1 and L_2 with equations $y = 3x - 2$ and $y = -2x + 3$, respectively, on the same set of axes using the standard viewing rectangle. Then use TRACE and ZOOM to find the point of intersection of L_1 and L_2. Repeat using the "intersection" function of your graphing utility.
2. Find the point of intersection of L_1 and L_2 algebraically.
3. Comment on the effectiveness of each method.

We now turn to some applications involving the intersections of pairs of straight lines.

Break-Even Analysis

Consider a firm with (linear) cost function $C(x)$, revenue function $R(x)$, and profit function $P(x)$ given by

$$C(x) = cx + F$$
$$R(x) = sx$$
$$P(x) = R(x) - C(x) = (s - c)x - F$$

where c denotes the unit cost of production, s denotes the selling price per unit, F denotes the fixed cost incurred by the firm, and x denotes the level of production and sales. The level of production at which the firm neither makes a profit nor sustains a loss is called the **break-even level of operation** and may be determined by solving the equations $p = C(x)$ and $p = R(x)$ simultaneously. For at this level of production, x_0, the profit is zero, so

$$P(x_0) = R(x_0) - C(x_0) = 0$$
$$R(x_0) = C(x_0)$$

FIGURE 1.35
P_0 is the break-even point.

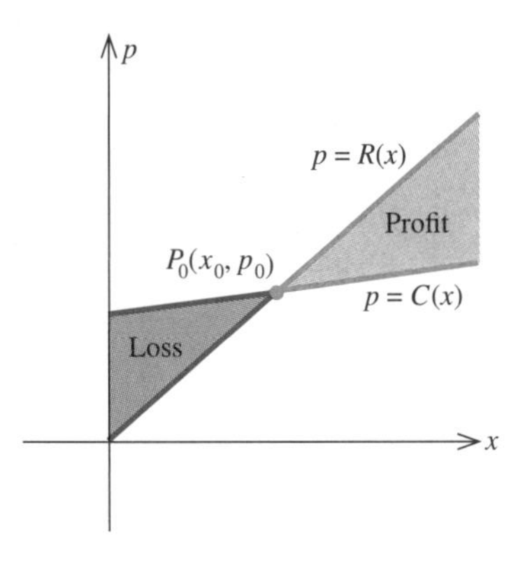

The point $P_0(x_0, p_0)$, the solution of the simultaneous equations $p = R(x)$ and $p = C(x)$, is referred to as the **break-even point;** the number x_0 and the number p_0 are called the **break-even quantity** and the **break-even revenue,** respectively.

Geometrically, the break-even point $P_0(x_0, p_0)$ is just the point of intersection of the straight lines representing the cost and revenue functions, respectively. This follows because $P_0(x_0, p_0)$, being the solution of the simultaneous equations $p = R(x)$ and $p = C(x)$, must lie on both these lines simultaneously (Figure 1.35).

Note that if $x < x_0$, then $R(x) < C(x)$ so that $P(x) = R(x) - C(x) < 0$ and thus the firm sustains a loss at this level of production. On the other hand, if $x > x_0$, then $P(x) > 0$ and the firm operates at a profitable level.

EXAMPLE 2 Prescott, Inc., manufactures its products at a cost of \$4 per unit and sells them for \$10 per unit. If the firm's fixed cost is \$12,000 per month, determine the firm's break-even point.

SOLUTION ✓ The cost function C and the revenue function R are given by $C(x) = 4x + 12{,}000$ and $R(x) = 10x$, respectively (Figure 1.36).

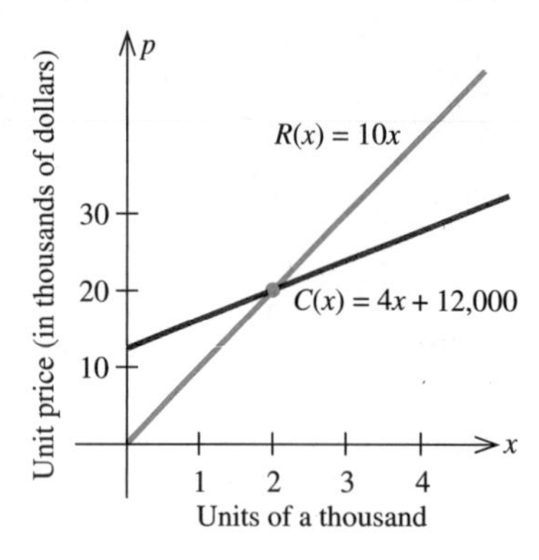

FIGURE 1.36
The point at which $R(x) = C(x)$ is the break-even point.

Setting $R(x) = C(x)$, we obtain

$$\begin{aligned} 10x &= 4x + 12{,}000 \\ 6x &= 12{,}000 \\ x &= 2000 \end{aligned}$$

Substituting this value of x into $R(x) = 10x$ gives

$$R(2000) = (10)(2000) = 20{,}000$$

So, for a break-even operation, the firm should manufacture 2000 units of its product, resulting in a break-even revenue of \$20,000. ■■■■

EXAMPLE 3 Using the data given in Example 2, answer the following questions:

a. What is the loss sustained by the firm if only 1500 units are produced and sold per month?

b. What is the profit if 3000 units are produced and sold per month?

c. How many units should the firm produce in order to realize a minimum monthly profit of \$9000?

SOLUTION ✓ The profit function P is given by the rule

$$\begin{aligned} P(x) &= R(x) - C(x) \\ &= 10x - (4x + 12{,}000) \\ &= 6x - 12{,}000 \end{aligned}$$

a. If 1500 units are produced and sold per month, we have

$$P(1500) = 6(1500) - 12{,}000 = -3000$$

so the firm will sustain a loss of \$3000 per month.

b. If 3000 units are produced and sold per month, we have

$$P(3000) = 6(3000) - 12{,}000 = 6000$$

or a monthly profit of \$6000.

c. Substituting 9000 for $P(x)$ in the equation $P(x) = 6x - 12{,}000$, we obtain

$$\begin{aligned} 9000 &= 6x - 12{,}000 \\ 6x &= 21{,}000 \\ x &= 3500 \end{aligned}$$

Thus, the firm should produce at least 3500 units in order to realize a \$9000 minimum monthly profit. ■■■■

EXAMPLE 4 The management of the Robertson Controls Company must decide between two manufacturing processes for its model C electronic thermostat. The monthly cost of the first process is given by $C_1(x) = 20x + 10{,}000$ dollars, where x is the number of thermostats produced; the monthly cost of the second process is given by $C_2(x) = 10x + 30{,}000$ dollars. If the projected sales are 800 thermostats at a unit price of \$40, which process should management choose in order to maximize the company's profit?

SOLUTION ✓ The break-even level of operation using the first process is obtained by solving the equation

$$\begin{aligned} 40x &= 20x + 10{,}000 \\ 20x &= 10{,}000 \\ x &= 500 \end{aligned}$$

giving an output of 500 units. Next, we solve the equation

$$\begin{aligned} 40x &= 10x + 30{,}000 \\ 30x &= 30{,}000 \\ x &= 1000 \end{aligned}$$

giving an output of 1000 units for a break-even operation using the second process. Since the projected sales are 800 units, we conclude that management should choose the first process, which will give the firm a profit. ■■■■

EXAMPLE 5 Referring to Example 4, decide which process Robertson's management should choose if the projected sales are (a) 1500 units and (b) 3000 units.

SOLUTION ✓ In both cases the production is past the break-even level. Since the revenue is the same regardless of which process is employed, the decision will be based on how much each process costs.

a. If $x = 1500$, then

$$\begin{aligned} C_1(x) &= (20)(1500) + 10{,}000 = 40{,}000 \\ C_2(x) &= (10)(1500) + 30{,}000 = 45{,}000 \end{aligned}$$

Hence, management should choose the first process.

b. If $x = 3000$, then

$$C_1(x) = (20)(3000) + 10{,}000 = 70{,}000$$
$$C_2(x) = (10)(3000) + 30{,}000 = 60{,}000$$

In this case, management should choose the second process. ■■■■

Exploring with Technology

1. Use a graphing utility to plot the straight lines L_1 and L_2 with equations $y = 2x - 1$ and $y = 2.1x + 3$, respectively, on the same set of axes using the standard viewing rectangle. Do the lines appear to intersect?
2. Plot the straight lines L_1 and L_2 using the viewing rectangle $[-100, 100] \times [-100, 100]$. Do the lines appear to intersect? Can you find the point of intersection using TRACE and ZOOM? Using the "intersection" function of your graphing utility?
3. Find the point of intersection of L_1 and L_2 algebraically.
4. Comment on the effectiveness of the solution methods in parts (2) and (3).

MARKET EQUILIBRIUM

Under pure competition, the price of a commodity eventually settles at a level dictated by the condition that the supply of the commodity be equal to the demand for it. If the price is too high, consumers will be more reluctant to buy, and if the price is too low, the supplier will be more reluctant to make the product available in the marketplace. **Market equilibrium** is said to prevail when the quantity produced is equal to the quantity demanded. The quantity produced at market equilibrium is called the **equilibrium quantity,** and the corresponding price is called the **equilibrium price.**

FIGURE 1.37
Market equilibrium is represented by the point (x_0, p_0).

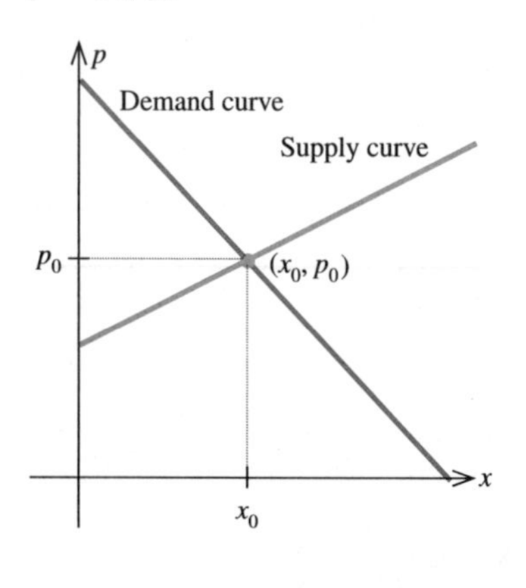

From a geometric point of view, market equilibrium corresponds to the point at which the demand curve and the supply curve intersect. In Figure 1.37, x_0 represents the equilibrium quantity and p_0 the equilibrium price. The point (x_0, p_0) lies on the supply curve and therefore satisfies the supply equation. At the same time, it also lies on the demand curve and therefore satisfies the demand equation. Thus, to find the point (x_0, p_0), and hence the equilibrium quantity and price, we solve the demand and supply equations simultaneously for x and p. For meaningful solutions, x and p must both be positive.

The management of the Thermo-Master Company, which manufactures an indoor–outdoor thermometer in its Mexico subsidiary, has determined that the demand equation for its product is

$$5x + 3p - 30 = 0$$

where p is the price of a thermometer in dollars and x is the quantity demanded in units of a thousand. The supply equation for these thermometers is

$$52x - 30p + 45 = 0$$

where x (measured in thousands) is the quantity Thermo-Master will make available in the market at p dollars each. Find the equilibrium quantity and price.

SOLUTION ✓ We need to solve the system of equations

$$5x + 3p - 30 = 0$$
$$52x - 30p + 45 = 0$$

for x and p. Let us use the *method of substitution* to solve it. As the name suggests, this method calls for choosing one of the equations in the system, solving for one variable in terms of the other, and then substituting the resulting expression into the other equation. This gives an equation in one variable that can then be solved in the usual manner.

Let's solve the first equation for p in terms of x. Thus,

$$3p = -5x + 30$$
$$p = -\frac{5}{3}x + 10$$

Next, we substitute this value of p into the second equation, obtaining

$$52x - 30\left(-\frac{5}{3}x + 10\right) + 45 = 0$$
$$52x + 50x - 300 + 45 = 0$$
$$102x - 255 = 0$$
$$x = \frac{255}{102} = \frac{5}{2}$$

The corresponding value of p is found by substituting this value of x into the equation for p obtained earlier. Thus,

$$p = -\frac{5}{3}\left(\frac{5}{2}\right) + 10 = -\frac{25}{6} + 10$$
$$= \frac{35}{6} \approx 5.83$$

We conclude that the equilibrium quantity is 2500 units (remember that x is measured in units of a thousand) and the equilibrium price is $5.83 per thermometer. ■■■■

EXAMPLE 7 The quantity demanded of a certain model of videocassette recorder (VCR) is 8000 units when the unit price is $260. At a unit price of $200, the quantity demanded increases to 10,000 units. The manufacturer will not market any VCRs if the price is $100 or lower. However, for each $50 increase in the unit price above $100, the manufacturer will market an additional 1000 units. Both the demand and the supply equations are known to be linear.

a. Find the demand equation.

b. Find the supply equation.

c. Find the equilibrium quantity and price.

SOLUTION ✓ Let p denote the unit price in hundreds of dollars and let x denote the number of units of VCRs in thousands.

a. Since the demand function is linear, the demand curve is a straight line passing through the points (8, 2.6) and (10, 2). Its slope is

$$m = \frac{2 - 2.6}{10 - 8} = -0.3$$

Using the point (10, 2) and the slope $m = -0.3$ in the point-slope form of the equation of a line, we see that the required demand equation is

$$p - 2 = -0.3(x - 10)$$
$$p = -0.3x + 5 \qquad \text{(Figure 1.38)}$$

b. The supply curve is the straight line passing through the points (0, 1) and (1, 1.5). Its slope is

$$m = \frac{1.5 - 1}{1 - 0} = 0.5$$

Using the point (0, 1) and the slope $m = 0.5$ in the point-slope form of the equation of a line, we see that the required supply equation is

$$p - 1 = 0.5(x - 0)$$
$$p = 0.5x + 1 \qquad \text{(Figure 1.38)}$$

c. To find the market equilibrium, we solve simultaneously the system comprising the demand and supply equations obtained in parts (a) and (b)—that is, the system

$$p = -0.3x + 5$$
$$p = 0.5x + 1$$

Subtracting the first equation from the second gives

$$0.8x - 4 = 0$$

and $x = 5$. Substituting this value of x in the second equation gives $p = 3.5$. Thus, the equilibrium quantity is 5000 units and the equilibrium price is \$350 (Figure 1.38).

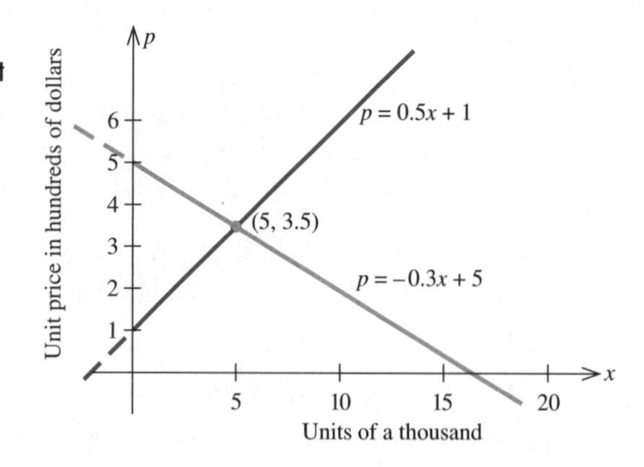

FIGURE 1.38
Market equilibrium occurs at the point (5, 3.5).

■■■■

SELF-CHECK EXERCISES 1.4

1. Find the point of intersection of the straight lines with equations $2x + 3y = 6$ and $x - 3y = 4$.
2. There is no demand for a certain make of videocassette tape when the unit price is \$12. However, when the unit price is \$8, the quantity demanded is 8000/week. The suppliers will not market any tapes if the unit price is \$2 or lower. At \$4/tape, however, the manufacturer will make available 5000 tapes/week. Both the demand and supply functions are known to be linear.
   a. Find the demand equation.
   b. Find the supply equation.
   c. Find the equilibrium quantity and price.

Solutions to Self-Check Exercises 1.4 can be found on page 57.

1.4 Exercises

In Exercises 1–6, find the point of intersection of each of the given pairs of straight lines.

1. $y = 3x + 4$, $y = -2x + 14$
2. $y = -4x - 7$, $-y = 5x + 10$
3. $2x - 3y = 6$, $3x + 6y = 16$
4. $2x + 4y = 11$, $-5x + 3y = 5$
5. $y = \frac{1}{4}x - 5$, $2x - \frac{3}{2}y = 1$
6. $y = \frac{2}{3}x - 4$, $x + 3y + 3 = 0$

In Exercises 7–10, find the break-even point for the firm whose cost function *C* and revenue function *R* are given.

7. $C(x) = 5x + 10{,}000$; $R(x) = 15x$
8. $C(x) = 15x + 12{,}000$; $R(x) = 21x$
9. $C(x) = 0.2x + 120$; $R(x) = 0.4x$
10. $C(x) = 150x + 20{,}000$; $R(x) = 270x$

11. BREAK-EVEN ANALYSIS Auto-Time, a manufacturer of 24-hour variable timers, has a monthly fixed cost of \$48,000 and a production cost of \$8 for each timer manufactured. The units sell for \$14 each.
    a. Sketch the graphs of the cost function and the revenue function and hence find the break-even point graphically.
    b. Find the break-even point algebraically.
    c. Sketch the graph of the profit function.
    d. At what point does the graph of the profit function cross the x-axis? Interpret your result.

12. BREAK-EVEN ANALYSIS A division of Carter Enterprises produces "Personal Income Tax" diaries. Each diary sells for \$8. The monthly fixed costs incurred by the division are \$25,000, and the variable cost of producing each diary is \$3.
    a. Find the break-even point for the division.
    b. What should be the level of sales in order for the division to realize a 15% profit over the cost of making the diaries?

13. BREAK-EVEN ANALYSIS A division of the Gibson Corporation manufactures bicycle pumps. Each pump sells for \$9, and the variable cost of producing each unit is 40% of the selling price. The monthly fixed costs incurred by the division are \$50,000. What is the break-even point for the division?

14. LEASING The Ace Truck Leasing Company leases a certain size truck for \$30/day and 15 cents/mile, whereas the Acme Truck Leasing Company leases the same size truck for \$25/day and 20 cents/mile.
    a. Find the functions describing the daily cost of leasing from each company.
    b. Sketch the graphs of the two functions on the same set of axes.
    c. If a customer plans to drive at most 70 mi, which company should he rent a truck from for 1 day?

15. DECISION ANALYSIS A product may be made using machine I or machine II. The manufacturer estimates that the monthly fixed costs of using machine I are \$18,000, whereas the monthly fixed costs of using machine II are \$15,000. The variable costs of manufacturing one unit

of the product using machine I and machine II are \$15 and \$20, respectively. The product sells for \$50 each.
a. Find the cost functions associated with using each machine.
b. Sketch the graphs of the cost functions of part (a) and the revenue functions on the same set of axes.
c. Which machine should management choose to maximize their profit if the projected sales are 450 units? 550 units? 650 units?
d. What is the profit for each case in part (c)?

16. The annual sales of the Crimson Drug Store are expected to be given by $S = 2.3 + 0.4t$ million dollars t years from now, whereas the annual sales of the Cambridge Drug Store are expected to be given by $S = 1.2 + 0.6t$ million dollars t years from now. When will the annual sales of the Cambridge Drug Store first surpass the annual sales of the Crimson Drug Store?

For each pair of supply and demand equations in Exercises 17–20, where x represents the quantity demanded in units of 1000 and p is the unit price in dollars, find the equilibrium quantity and the equilibrium price.

17. $4x + 3p - 59 = 0$ and $5x - 6p + 14 = 0$

18. $2x + 7p - 56 = 0$ and $3x - 11p + 45 = 0$

19. $p = -2x + 22$ and $p = 3x + 12$

20. $p = -0.3x + 6$ and $p = 0.15x + 1.5$

21. Equilibrium Quantity and Price The quantity demanded of a certain brand of videocassette recorder (VCR) is 3000/week when the unit price is \$485. For each decrease in unit price of \$20 below \$485, the quantity demanded increases by 250 units. The suppliers will not market any VCRs if the unit price is \$300 or lower. But at a unit price of \$525, they are willing to make available 2500 units in the market. The supply equation is also known to be linear.
a. Find the demand equation.
b. Find the supply equation.
c. Find the equilibrium quantity and price.

22. Equilibrium Quantity and Price The demand equation for the Miramar Heat Machine, a ceramic heater, is $x - 4p - 800 = 0$, where x is the quantity demanded per week and p is the wholesale unit price in dollars. The supply equation is $x - 20p + 1000 = 0$, where x is the quantity the supplier will make available in the market when the wholesale price is p dollars each. Find the equilibrium quantity and the equilibrium price for the Miramar heaters.

23. Equilibrium Quantity and Price The demand equation for the Schmidt-3000 fax machine is $3x + p - 1500 = 0$, where x is the quantity demanded per week and p is the unit price in dollars. The supply equation is $2x - 3p + 1200 = 0$, where x is the quantity the supplier will make available in the market when the unit price is p dollars. Find the equilibrium quantity and the equilibrium price for the fax machines.

24. Equilibrium Quantity and Price The quantity demanded per month of Russo Espresso Makers is 250 when the unit price is \$140. The quantity demanded per month is 1000 when the unit price is \$110. The suppliers will market 750 espresso makers if the unit price is \$60 or lower. At a unit price of \$80 they are willing to make available 2250 units in the market. Both the demand and supply equations are known to be linear.
a. Find the demand equation.
b. Find the supply equation.
c. Find the equilibrium quantity and the equilibrium price.

25. Suppose the demand and supply equations for a certain commodity are given by $p = ax + b$ and $p = cx + d$, respectively, where $a < 0$, $c > 0$, and $b > d > 0$ (see the accompanying figure).

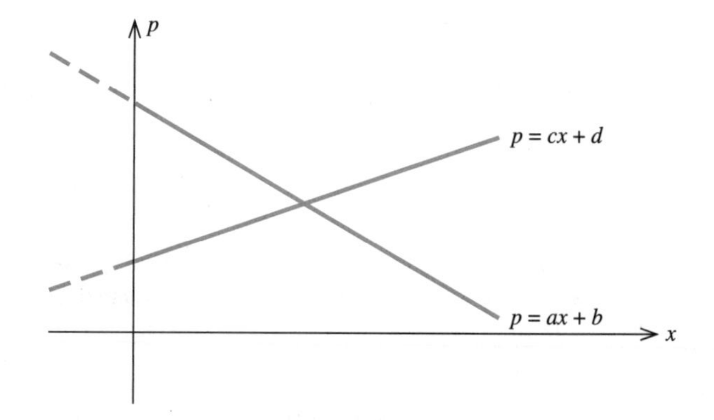

a. Find the equilibrium quantity and equilibrium price in terms of a, b, c, and d.
b. Use part (a) to determine what happens to the market equilibrium if c is increased while a, b, and d remain fixed. Interpret your answer in economic terms.
c. Use part (a) to determine what happens to the market equilibrium if b is decreased while a, c, and d remain fixed. Interpret your answer in economic terms.

26. Suppose the cost function associated with a product is $C(x) = cx + F$ dollars and the revenue function is $R(x) = sx$, where c denotes the unit cost of production, s denotes the unit selling price, F denotes the fixed cost incurred by the firm, and x denotes the level of production and sales. Find the break-even quantity and the break-even revenue in terms of the constants c, s, and F and interpret your results in economic terms.

In Exercises 27 and 28, determine whether the statement is true or false. If it is true, explain why it is true. If it is false, give an example to show why it is false.

27. Suppose $C(x) = cx + F$ and $R(x) = sx$ are the cost and revenue functions of a certain firm. Then, the firm is operating at a break-even level of production if its level of production is $F/(s - c)$.

28. If both the demand equation and the supply equation for a certain commodity are linear, then there must be at least one equilibrium point.

29. Let L_1 and L_2 be two nonvertical straight lines in the plane with equations $y = m_1x + b_1$ and $y = m_2x + b_2$, respectively. Find conditions on m_1, m_2, b_1, and b_2 so that (a) L_1 and L_2 do not intersect, (b) L_1 and L_2 intersect at one and only one point, and (c) L_1 and L_2 intersect at infinitely many points.

30. Find conditions on a_1, a_2, b_1, b_2, c_1, and c_2 so that the system of linear equations

$$\begin{aligned} a_1x + b_1y &= c_1 \\ a_2x + b_2y &= c_2 \end{aligned}$$

has (a) no solution, (b) a unique solution, and (c) infinitely many solutions.

Hint: Use the results of Exercise 29.

Solutions to Self-Check Exercises 1.4

1. The point of intersection of the two straight lines is found by solving the system of linear equations

$$\begin{aligned} 2x + 3y &= 6 \\ x - 3y &= 4 \end{aligned}$$

Solving the first equation for y in terms of x, we obtain

$$y = -\frac{2}{3}x + 2$$

Substituting this expression for y into the second equation, we obtain

$$\begin{aligned} x - 3\left(-\frac{2}{3}x + 2\right) &= 4 \\ x + 2x - 6 &= 4 \\ 3x &= 10 \end{aligned}$$

or $x = \frac{10}{3}$. Substituting this value of x into the expression for y obtained earlier, we find

$$y = -\frac{2}{3}\left(\frac{10}{3}\right) + 2 = -\frac{2}{9}$$

Therefore, the point of intersection is $\left(\frac{10}{3}, -\frac{2}{9}\right)$.

Finding the Points of Intersection of Two Graphs

A graphing utility can be used to find the point(s) of intersection of the graphs of two functions. Once again, it is important to remember that if the graphs are straight lines, the linear equations defining these lines must first be recast in the slope-intercept form.

EXAMPLE 1 Find the points of intersection of the straight lines with equations $2x + 3y = 6$ and $3x - 4y - 5 = 0$.

SOLUTION ✓ Solving each equation for y in terms of x, we obtain

$$y = -\frac{2}{3}x + 2 \qquad \text{and} \qquad y = \frac{3}{4}x - \frac{5}{4}$$

as the respective equations in the slope-intercept form. The graphs of the two straight lines in the standard viewing rectangle are shown in Figure T1.

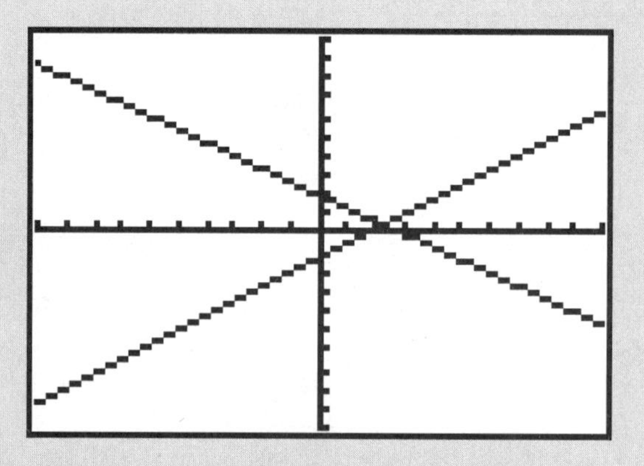

FIGURE T1
The straight lines $2x + 3y = 6$ and $3x - 4y - 5 = 0$

Then, using **TRACE** and **ZOOM** or the function for finding the point of intersection of two graphs, we find that the point of intersection, accurate to four decimal places, is (2.2941, 0.4706). ■■■■

Exercises

In Exercises 1–6, find the point of intersection of the pair of straight lines with the given equations. Express your answers accurate to four decimal places.

1. $y = 2x + 5$ and $y = -3x + 8$

2. $y = 1.2x + 6.2$ and $y = -4.3x + 9.1$

3. $2x - 5y = 7$ and $3x + 4y = 12$

4. $1.4x - 6.2y = 8.4$ and $4.1x + 7.3y = 14.4$

5. $2.1x = 5.1y + 71$ and $3.2x = 8.4y + 16.8$

6. $8.3x = 6.2y + 9.3$ and $12.4x = 12.3y + 24.6$

2. **a.** Let p denote the price per cassette and x the quantity demanded. The given conditions imply that $x = 0$ when $p = 12$ and $x = 8000$ when $p = 8$. Since the demand equation is linear, it has the form

$$p = mx + b$$

Now, the first condition implies that

$$12 = m(0) + b \qquad \text{or} \qquad b = 12$$

Therefore,

$$p = mx + 12$$

Using the second condition, we find

$$8 = 8000m + 12$$

$$m = -\frac{4}{8000} = -0.0005$$

Therefore, the required demand equation is

$$p = -0.0005x + 12$$

b. Let p denote the price per cassette and x the quantity made available at that price. Then, since the supply equation is linear, it also has the form

$$p = mx + b$$

The first condition implies that $x = 0$ when $p = 2$, so we have

$$2 = m(0) + b \qquad \text{or} \qquad b = 2$$

Therefore,

$$p = mx + 2$$

Next, using the second condition, $x = 5000$ when $p = 4$, we find

$$4 = 5000m + 2$$

giving $m = 0.0004$. So the required supply equation is

$$p = 0.0004x + 2$$

c. The equilibrium quantity and price are found by solving the system of linear equations

$$p = -0.0005x + 12$$
$$p = 0.0004x + 2$$

Equating the two equations yields

$$-0.0005x + 12 = 0.0004x + 2$$
$$0.0009x = 10$$

or $x \approx 11{,}111$. Substituting this value of x into either equation in the system yields

$$p = 6.44$$

Therefore, the equilibrium quantity is 11,111 and the equilibrium price is \$6.44.

1.5 The Method of Least Squares (Optional)

THE METHOD OF LEAST SQUARES

In Example 10, Section 1.2, we saw how a linear equation may be used to approximate the sales trend for a local sporting goods store. The *trend line*, as we saw, may be used to predict the store's future sales. Recall that we obtained the trend line in Example 10 by requiring that the line pass through two data points, the rationale being that such a line seems to *fit* the data reasonably well.

In this section we describe a general method known as the **method of least squares** for determining a straight line that, in some sense, best fits a set of data points when the points are scattered about a straight line. To illustrate the principle behind the method of least squares, suppose, for simplicity, that we are given five data points,

$$P_1(x_1, y_1), \quad P_2(x_2, y_2), \quad P_3(x_3, y_3), \quad P_4(x_4, y_4), \quad P_5(x_5, y_5)$$

describing the relationship between the two variables x and y. By plotting these data points, we obtain a graph called a **scatter diagram** (Figure 1.39).

If we try to *fit* a straight line to these data points, the line will miss the first, second, third, fourth, and fifth data points by the amounts d_1, d_2, d_3, d_4, and d_5, respectively (Figure 1.40). We can think of the amounts $d_1, d_2, \ldots, d_5$ as the errors made when the values $y_1, y_2, \ldots, y_5$ are approximated by the corresponding values of y lying on the straight line L.

The **principle of least squares** states that the straight line L that fits the data points *best* is the one chosen by requiring that the sum of the squares of $d_1, d_2, \ldots, d_5$—that is,

$$d_1^2 + d_2^2 + d_3^2 + d_4^2 + d_5^2$$

FIGURE 1.39
A scatter diagram

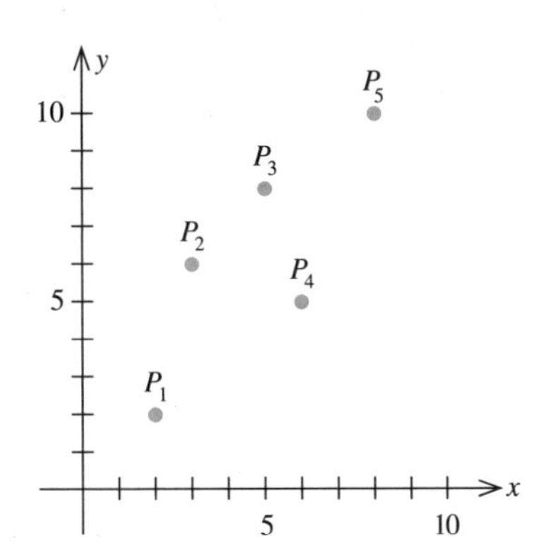

FIGURE 1.40
d_i is the distance between the straight line and a given data point.

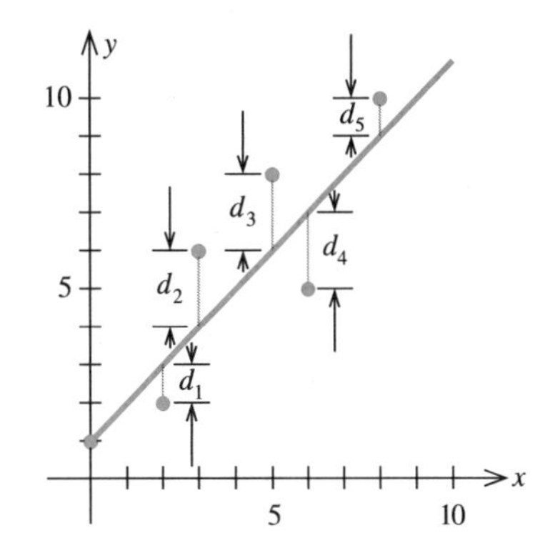

be made as small as possible. In other words, the least-squares criterion calls for minimizing the sum of the squares of the errors. The line L obtained in this manner is called the **least-squares line,** or *regression line.*

The method for computing the least-squares lines that best fits a set of data points is contained in the following result, which we state without proof.

The Method of Least Squares

Suppose we are given n data points

$$P_1(x_1, y_1), \quad P_2(x_2, y_2), \quad P_3(x_3, y_3), \ldots, P_n(x_n, y_n)$$

Then, the least-squares (regression) line for the data is given by the linear equation (function)

$$y = f(x) = mx + b$$

where the constants m and b satisfy the **normal equations**

$$nb + (x_1 + x_2 + \cdots + x_n)m = y_1 + y_2 + \cdots + y_n \qquad \textbf{(9)}$$

$$(x_1 + x_2 + \cdots + x_n)b + (x_1^2 + x_2^2 + \cdots + x_n^2)m$$
$$= x_1y_1 + x_2y_2 + \cdots + x_ny_n \qquad \textbf{(10)}$$

simultaneously.

EXAMPLE 1 Find the least-squares line for the data

$$P_1(1, 1), \quad P_2(2, 3), \quad P_3(3, 4), \quad P_4(4, 3), \quad P_5(5, 6)$$

SOLUTION ✓ Here, we have $n = 5$ and

$$x_1 = 1, \quad x_2 = 2, \quad x_3 = 3, \quad x_4 = 4, \quad x_5 = 5$$
$$y_1 = 1, \quad y_2 = 3, \quad y_3 = 4, \quad y_4 = 3, \quad y_5 = 6$$

Before using Equations (9) and (10), it is convenient to summarize this data in the form of a table:

	x	y	x^2	xy
	1	1	1	1
	2	3	4	6
	3	4	9	12
	4	3	16	12
	5	6	25	30
Sum	15	17	55	61

Using this table and (9) and (10), we obtain the normal equations

$$5b + 15m = 17 \tag{11}$$

$$15b + 55m = 61 \tag{12}$$

Solving Equation (11) for b gives

$$b = -3m + \frac{17}{5} \tag{13}$$

which upon substitution into Equation (12) gives

$$15\left(-3m + \frac{17}{5}\right) + 55m = 61$$

$$-45m + 51 + 55m = 61$$

$$10m = 10$$

$$m = 1$$

Substituting this value of m into Equation (13) gives

$$b = -3 + \frac{17}{5} = \frac{2}{5} = 0.4$$

Therefore, the required least-squares line is

$$y = x + 0.4$$

The scatter diagram and the least-squares line are shown in Figure 1.41.

FIGURE 1.41
The least-squares line $y = x + 0.4$ and the given data points

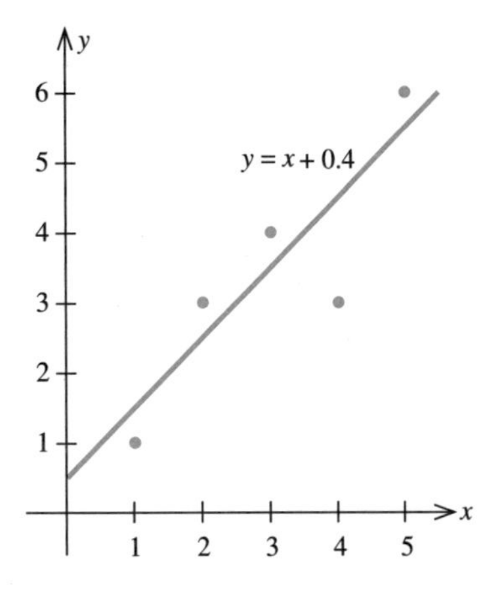

APPLICATION

EXAMPLE 2 The proprietor of the Leisure Travel Service compiled the following data relating the annual profit of the firm to its annual advertising expenditure (both measured in thousands of dollars):

Annual Advertising Expenditure, x	12	14	17	21	26	30
Annual Profit, y	60	70	90	100	100	120

a. Determine the equation of the least-squares line for these data.

b. Draw a scatter diagram and the least-squares line for these data.

c. Use the result obtained in part (a) to predict Leisure Travel's annual profit if the annual advertising budget is $20,000.

SOLUTION ✓

a. The calculations required for obtaining the normal equations are summarized in the accompanying table.

	x	y	x^2	xy
	12	60	144	720
	14	70	196	980
	17	90	289	1,530
	21	100	441	2,100
	26	100	676	2,600
	30	120	900	3,600
Sum	120	540	2,646	11,530

The normal equations are

$$6b + 120m = 540 \tag{14}$$

$$120b + 2646m = 11{,}530 \tag{15}$$

Solving Equation (14) for b gives

$$b = -20m + 90 \tag{16}$$

which upon substitution into Equation (15) gives

$$\begin{aligned} 120(-20m + 90) + 2646m &= 11{,}530 \\ -2400m + 10{,}800 + 2646m &= 11{,}530 \\ 246m &= 730 \\ m &= 2.97 \end{aligned}$$

Substituting this value of m into Equation (16) gives

$$\begin{aligned} b &= -20(2.97) + 90 \\ &= 30.6 \end{aligned}$$

Therefore, the required least-squares line is given by

$$y = f(x) = 2.97x + 30.6$$

b. The scatter diagram and the least-squares line are shown in Figure 1.42.

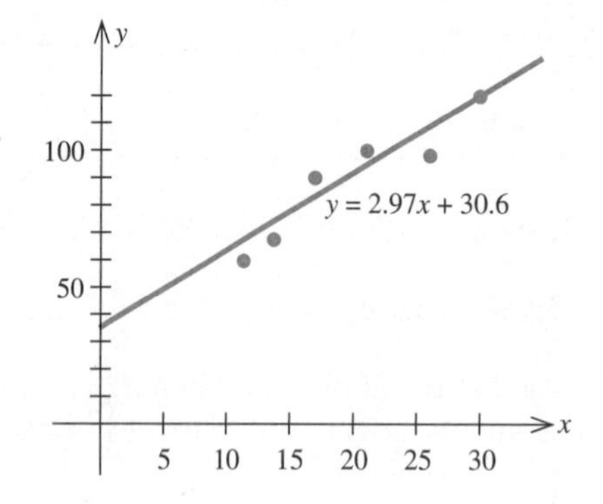

FIGURE 1.42
Profit versus advertising expenditure

c. Leisure Travel's predicted annual profit corresponding to an annual budget of \$20,000 is given by

$$f(20) = 2.97(20) + 30.6 = 90$$

or \$90,000. ■■■■

SELF-CHECK EXERCISES 1.5

1. Find an equation of the least-squares line for the data

x	1	3	4	5	7
y	4	10	11	12	16

2. In a market research study for the Century Communications Company, the data were provided based on the projected monthly sales x (in thousands) of a videocassette version of a box-office-hit adventure movie with a proposed wholesale unit price of p dollars.

x	2.2	5.4	7.0	11.5	14.6
p	38	36	34.5	30	28.5

Find the demand equation if the demand curve is the least-squares line for these data.

Solutions to Self-Check Exercises 1.5 can be found on page 71.

1.5 Exercises

A calculator is recommended for this exercise set.

In Exercises 1–6, (a) find the equation of the least-squares line for the given data and (b) draw a scatter diagram for the given data and graph the least-squares line.

1.

x	1	2	3	4
y	4	6	8	11

2.

x	1	3	5	7	9
y	9	8	6	3	2

3.

x	1	2	3	4	4	6
y	4.5	5	3	2	3.5	1

4.

x	1	1	2	3	4	4	5
y	2	3	3	3.5	3.5	4	5

5. $P_1(1, 3)$, $P_2(2, 5)$, $P_3(3, 5)$, $P_4(4, 7)$, $P_5(5, 8)$

6. $P_1(1, 8)$, $P_2(2, 6)$, $P_3(5, 6)$, $P_4(7, 4)$, $P_5(10, 1)$

7. **COLLEGE ADMISSIONS** The accompanying data were compiled by the admissions office at Faber College during the past 5 years. The data relate the number of college brochures and follow-up letters (x) sent to a preselected list of high school juniors who had taken the PSAT and the number of completed applications (y) received from these students (both measured in units of 1000).

x	4	4.5	5	5.5	6
y	0.5	0.6	0.8	0.9	1.2

a. Determine the equation of the least-squares line for these data.
b. Draw a scatter diagram and the least-squares line for these data.
c. Use the result obtained in part (a) to predict the number of completed applications expected if 6400 brochures and follow-up letters are sent out during the next year.

8. **NET SALES** The management of Kaldor, Inc., a manufacturer of electric motors, submitted the accompanying data in its annual stockholders report. The table shows the net sales (in millions of dollars) during the 5 years that have elapsed since the new management team took over. (The first year the firm operated under the new management corresponds to the time period $x = 1$, and the four subsequent years correspond to $x = 2, 3, 4, 5$.)

Year, x	1	2	3	4	5
Net Sales, y	426	437	460	473	477

a. Determine the equation of the least-squares line for these data.
b. Draw a scatter diagram and the least-squares line for these data.
c. Use the result obtained in part (a) to predict the net sales for the upcoming year.

9. **SAT VERBAL SCORES** The accompanying data were compiled by the superintendent of schools in a large metropolitan area. The table shows the average SAT verbal scores of high school seniors during the 5 years since the district implemented its "back-to-basics" program.

Year, x	1	2	3	4	5
Average Score, y	436	438	428	430	426

a. Determine the equation of the least-squares line for these data.
b. Draw a scatter diagram and the least-squares line for these data.
c. Use the result obtained in part (a) to predict the average SAT verbal score of high school seniors 2 years from now ($x = 7$).

10. **AUTO OPERATING COSTS** The accompanying figures were compiled by Clarke, Kingsley, and Company, a consulting firm that specializes in auto operating costs, relating the annual mileage (in thousands of miles) that an average new compact car is driven to the cost per mile (in cents) of operating the car.

Annual Mileage, x	5	10	15
Cost per Mile, y	50.3	34.8	30.1
Annual Mileage, x	20	25	30
Cost per Mile, y	27.4	25.6	23.5

a. Determine an equation of the least-squares line for these data.
b. Draw a scatter diagram and the least-squares line for these data.
c. Use the result obtained in part (a) to estimate the cost per mile of operating a new company car if it is driven 8000 miles during the first year of ownership.

11. **SIZE OF AVERAGE FARM** The size of the average farm in the United States has been growing steadily over the years. The accompanying data, obtained from the U.S. Department of Agriculture, gives the size of the average farm y (in acres) from 1940 through 1997. (Here, $x = 0$ corresponds to the beginning of the year 1940.)

Year, x	0	10	20
Number of Acres, y	168	213	297
Year, x	30	40	57
Number of Acres, y	374	427	471

a. Find the equation of the least-squares line for these data.
b. Use the result of part (a) to estimate the size of the average farm in the year 2003.
Source: The World Almanac

12. **Welfare Costs** According to the Massachusetts Department of Welfare, the spending (in billions of dollars) by Medicaid, the national health-care plan for the poor, over the 5-year period from 1988 to 1992 is summarized in the accompanying table. (Here, $x = 0$ represents the beginning of the year 1988.)

Year, x	0	1	2
Expenditure, y	1.550	1.662	1.786

Year, x	3	4
Expenditure, y	1.888	2.009

a. Find an equation of the least-squares line for these data.
b. Use the result of part (a) to estimate Medicaid spending for the year 1996, assuming the trend continued.
Source: Massachusetts Department of Welfare

13. **Mass-Transit Subsidies** The accompanying table gives the projected state subsidies (in millions of dollars) to the Massachusetts Bay Transit Authority (MBTA) over a 5-year period.

Year, x	1	2	3	4	5
Subsidy, y	20	24	26	28	32

a. Find an equation of the least-squares line for these data.
b. Use the result of part (a) to estimate the state subsidy to the MBTA for the eighth year ($x = 8$).
Source: Massachusetts Bay Transit Authority

14. **Social Security Wage Base** The Social Security (FICA) wage base (in thousands of dollars) from 1993 to 1999 is given in the accompanying table.

Year	1994	1995	1996
Wage Base, y	60.6	61.2	62.7

Year	1997	1998	1999
Wage Base, y	65.4	68.4	72.6

a. Find an equation of the least-squares line for these data. (Let $x = 1$ represent the year 1994.)
b. Use the result of part (a) to estimate the FICA wage base in the year 2003.
Source: The World Almanac

15. **Production of All-Aluminum Cans** Steel has been playing a decreasing role in the manufacture of beverage cans in the United States. According to the Can Manufacture Institute, the use of bimetallic cans has been dwindling while the use of all-aluminum cans has been growing steadily. The accompanying table gives the production (in billions) of all-aluminum cans over the period from 1975 through 1989.

Year	1975	1977	1979	1981
Number of Cans, y	16.7	26	33.3	48.3

Year	1983	1985	1987	1989
Number of Cans, y	57	65.8	74.2	83.3

a. Find an equation of the least-squares line for these data. (Let $x = 1$ represent 1975.)
b. Use the result of part (a) to estimate the number of cans produced in 1993, assuming the trend continued.
Source: Can Manufacturing Institute

16. **On-Line Banking** According to industry sources, on-line banking is expected to take off in the near future. The projected number of households (in millions) using this service is given in the following table. (Here, $x = 0$ corresponds to the beginning of 1997.)

Year, x	0	1	2	3	4	5
Number of Households, y	4.5	7.5	10.0	13.0	15.6	18.0

a. Find an equation of the least-squares line for these data.
b. Use the result of part (a) to estimate the number of households using on-line banking at the beginning of 2003, assuming the projection is accurate.
Source: Jupiter Communications, Forrester Research Inc.

17. **Net-Connected Computers in Europe** The projected number of computers (in millions) connected to the Internet in Europe from 1998 through 2002 is summarized in the accompanying table. (Here, $x = 0$ corresponds to the beginning of 1998.)

(continued on p. 70)

Using Technology

FINDING AN EQUATION OF A LEAST-SQUARES LINE

A graphing utility is especially useful in calculating an equation of the least-squares line for a set of data. We simply enter the given data in the form of lists into the calculator and then use the linear regression function to obtain the coefficients of the required equation.

EXAMPLE 1 Find an equation of the least-squares line for the data

x	1.1	2.3	3.2	4.6	5.8	6.7	8
y	−5.8	−5.1	−4.8	−4.4	−3.7	−3.2	−2.5

SOLUTION ✔ First we enter the data as

$$x_1 = 1.1, \quad y_1 = -5.8, \quad x_2 = 2.3, \quad y_2 = -5.1, \quad x_3 = 3.2,$$
$$y_3 = -4.8, \quad x_4 = 4.6, \quad y_4 = -4.4, \quad x_5 = 5.8, \quad y_5 = -3.7,$$
$$x_6 = 6.7, \quad y_6 = -3.2, \quad x_7 = 8, \quad y_7 = -2.5$$

Then, using the linear regression function from the statistics menu, we find

$$a = -6.29996900666, \qquad b = 0.460560979389,$$
$$\text{corr} = 0.994488871079, \qquad n = 7$$

Therefore, an equation of the least-squares line ($y = a + bx$) is

$$y = -6.3 + 0.46x$$

The correlation coefficient of 0.99449 attests to the excellent fit of the regression line.

EXAMPLE 2 According to Pacific Gas and Electric, the nation's largest utility company, the demand for electricity from 1990 through the year 2000 is summarized in the following table:

t	0	2	4	6	8	10
y	333	917	1500	2117	2667	3292

Here $t = 0$ corresponds to 1990 and y gives the amount of electricity demanded in year t measured in megawatts. Find an equation of the least-squares line for these data.

Source: Pacific Gas and Electric

SOLUTION ✓ First, we enter the data as

$$x_1 = 0, \quad y_1 = 333, \quad x_2 = 2, \quad y_2 = 917, \quad x_3 = 4, \quad y_3 = 1500,$$
$$x_4 = 6, \quad y_4 = 2117, \quad x_5 = 8, \quad y_5 = 2667, \quad x_6 = 10, \quad y_6 = 3292$$

Then, using the linear regression function from the statistics menu, we find

$$a = 328.476190476 \quad \text{and} \quad b = 295.171428571$$

Therefore, an equation of the least-squares line is

$$y = 328 + 295t$$

Exercises

In Exercises 1–4, find an equation of the least-squares line for the data.

1.

x	2.1	3.4	4.7	5.6	6.8	7.2
y	8.8	12.1	14.8	16.9	19.8	21.1

2.

x	1.1	2.4	3.2	4.7	5.6	7.2
y	−0.5	1.2	2.4	4.4	5.7	8.1

3.

x	−2.1	−1.1	0.1	1.4	2.5	4.2	5.1
y	6.2	4.7	3.5	1.9	0.4	−1.4	−2.5

4.

x	−1.12	0.1	1.24	2.76	4.21	6.82
y	7.61	4.9	2.74	−0.47	−3.51	−8.94

5. **WASTE GENERATION** According to data from the Council on Environmental Quality, the amount of waste (in millions of tons per year) generated in the United States from 1960 to 1990 was:

Year	1960	1965	1970	1975
Amount, y	81	100	120	124

Year	1980	1985	1990
Amount, y	140	152	164

a. Find an equation of the least-squares line for these data. (Let x be in units of 5 and let $x = 1$ represent 1960.)
b. Use the result of part (a) to estimate the amount of waste generated in the year 2000, assuming the trend continued.
Source: Council on Environmental Quality

6. **MEDIAN PRICE OF HOMES** According to data from the Association of Realtors, the median price (in thousands of dollars) of existing homes in a certain metropolitan area from 1982 to 1992 was:

Year	1982	1983	1984	1985	1986	1987
Price, y	66.4	69.8	72.8	76.0	79.6	83.1

Year	1988	1989	1990	1991	1992
Price, y	86.3	89.5	92.3	96.0	99.5

a. Find an equation of the least-squares line for these data. (Let $x = 1$ represent 1982.)
b. Use the result of part (a) to estimate the median price of a house in the year 1997, assuming the trend continued.
Source: Association of Realtors

Year, x	0	1	2	3	4
Net-Connected Computers, y	21.7	32.1	45.0	58.3	69.6

a. Find an equation of the least-squares line for these data.
b. Use the result of part (a) to estimate the projected number of computers connected to the Internet in Europe at the beginning of 2003, assuming the trend continues.
Source: Dataquest Inc.

18. **Portable Phone Services** The projected number of wireless subscribers y (in millions) from the year 2000 through 2006 is summarized in the accompanying table. (Here, $x = 0$ corresponds to the beginning of the year 2000.)

Year, x	0	1	2	3	4	5	6
Number of Subscribers, y	90.4	100.0	110.4	120.4	130.8	140.4	150.0

a. Find an equation of the least-squares line for these data.
b. Use the result of (a) to estimate the projected number of wireless subscribers at the beginning of 2006. How does this result compare with the given data for that year?
Source: BancAmerica Robertson Stephens

19. **Health-Care Spending** The following data, compiled by the Organization for Economic Cooperation and Development (OECD) in 1990, give the per capita Gross Domestic Product (GDP) (in dollars) and the corresponding per capita spending (in thousands of dollars) on health care for selected countries.

Country	Turkey	Spain	Netherlands	Sweden	Switzerland	Canada
GDP, x	4.25	10	14	15.5	17.8	19.5
Health-Care Spending, y	178	667	1194	1500	1388	1640

a. Letting x denote a country's GDP (in thousands of dollars per capita) and y denote the per capita health-care spending (in dollars), find an equation of the least-squares line for these data giving the typical relationship between GDP and health-care spending for the selected countries.
b. The per capita GDP of the United States in 1990 was $20,000. If the health-care spending of the United States were in line with that of these sample OECD countries, what would it have been? (*Note:* The actual per capita health-care spending of the United States in 1990 was $2444.)
Source: Organization for Economic Cooperation and Development

In Exercises 20 and 21, determine whether the statement is true or false. If it is true, explain why it is true. If it is false, give an example to show why it is false.

20. The least-squares line must pass through at least one data point.

21. The error incurred in approximating n data points using the least-squares linear function is zero if and only if the n data points lie along a straight line.

SOLUTIONS TO SELF-CHECK EXERCISES 1.5

1. The calculations required for obtaining the normal equations may be summarized as follows:

	x	y	x^2	xy
	1	4	1	4
	3	10	9	30
	4	11	16	44
	5	12	25	60
	7	16	49	112
Sum	20	53	100	250

The normal equations are

$$5b + 20m = 53$$
$$20b + 100m = 250$$

Solving the first equation for b gives

$$b = -4m + \frac{53}{5}$$

which, upon substitution into the second equation, yields

$$20\left(-4m + \frac{53}{5}\right) + 100m = 250$$
$$-80m + 212 + 100m = 250$$
$$20m = 38$$
$$m = 1.9$$

Substituting this value of m into the expression for b found earlier, we find

$$b = -4(1.9) + \frac{53}{5} = 3$$

Therefore, the required least-squares line is

$$y = 1.9x + 3$$

2. The calculations required for obtaining the normal equations may be summarized as follows:

	x	p	x^2	xy
	2.2	38.0	4.84	83.6
	5.4	36.0	29.16	194.4
	7.0	34.5	49.00	241.5
	11.5	30.0	132.25	345.0
	14.6	28.5	213.16	416.1
Sum	40.7	167.0	428.41	1280.6

The normal equations are

$$5b + 40.7m = 167$$
$$40.7b + 428.41m = 1280.6$$

Solving this system of linear equations simultaneously, we find that

$$m = -0.81 \quad \text{and} \quad b = 39.99$$

Therefore, the required least-squares line is given by

$$p = f(x) = -0.81x + 39.99$$

which is the required demand equation provided $0 \leq x \leq 49.37$.

Group projects for each chapter can be found at the Brooks/Cole Web site: http://www.brookscole.com/math/authors/tans/

CHAPTER 1 Summary of Principal Formulas and Terms

Formulas

1. Distance between two points	$d = \sqrt{(x_2 - x_1)^2 + (y_2 - y_1)^2}$
2. Equation of a circle	$(x - h)^2 + (y - k)^2 = r^2$
3. Slope of a nonvertical line	$m = \dfrac{y_2 - y_1}{x_2 - x_1}$
4. Equation of a vertical line	$x = a$
5. Equation of a horizontal line	$y = b$
6. Point-slope form of the equation of a line	$y - y_1 = m(x - x_1)$
7. Slope-intercept form of the equation of a line	$y = mx + b$
8. General equation of a line	$Ax + By + C = 0$

Terms

Cartesian coordinate system
ordered pair
coordinates
parallel lines
perpendicular lines
function
independent variable
dependent variable
domain
range
linear function
total cost function
revenue function
profit function
demand function
supply function
break-even point
market equilibrium
equilibrium quantity
equilibrium price

CHAPTER 1 REVIEW EXERCISES

In Exercises 1–4, find the distance between the two given points.

1. (2, 1) and (6, 4)
2. (9, 6) and (6, 2)
3. (−2, −3) and (1, −7)
4. $(1/2, \sqrt{3})$ and $(-1/2, 2\sqrt{3})$

In Exercises 5–10, find an equation of the line *L* that passes through the point (−2, 4) and satisfies the given condition.

5. L is a vertical line.
6. L is a horizontal line.
7. L passes through the point (3, 7/2).
8. The x-intercept of L is 3.
9. L is parallel to the line $5x - 2y = 6$.
10. L is perpendicular to the line $4x + 3y = 6$.
11. Find an equation of the line with slope $-1/2$ and y-intercept -3.
12. Find the slope and y-intercept of the line with equation $3x - 5y = 6$.
13. Find an equation of the line passing through the point (2, 3) and parallel to the line with equation $3x + 4y - 8 = 0$.
14. Find an equation of the line passing through the point (−1, 3) and parallel to the line joining the points (−3, 4) and (2, 1).
15. Find an equation of the line passing through the point (−2, −4) that is perpendicular to the line with equation $2x - 3y - 24 = 0$.

In Exercises 16 and 17, sketch the graph of the given equation.

16. $3x - 4y = 24$
17. $-2x + 5y = 15$
18. Sales of a certain clock radio are approximated by the relationship

$$S(x) = 6000x + 30{,}000 \qquad (0 \le x \le 5)$$

where $S(x)$ denotes the number of clock radios sold in year x ($x = 0$ corresponds to the year 1995). Find the number of clock radios expected to be sold in 2000.

19. A company's total sales (in millions of dollars) are approximately linear as a function of time (in years). Sales in 1994 were \$2.4 million, whereas sales in 1999 amounted to \$7.4 million.
a. Find an equation giving the company's sales as a function of time.
b. What were the sales in 1997?
20. Show that the triangle with vertices $A(1, 1)$, $B(5, 3)$, and $C(4, 5)$ is a right triangle.
21. Clark's rule is a method for calculating pediatric drug dosages based on a child's weight. If a denotes the adult dosage (in milligrams) and if w is the child's weight (in pounds), then the child's dosage is given by

$$D(w) = \frac{aw}{150}$$

a. Show that D is a linear function of w.
b. If the adult dose of a substance is 500 mg, how much should a 35-lb child receive?
22. An office building worth \$6 million when it was completed in 1995 is being depreciated linearly over 30 years.
a. What is the rate of depreciation?
b. What will the book value of the building be in 2005?
23. In 1996 a manufacturer installed a new machine in her factory at a cost of \$300,000. The machine is depreciated linearly over 12 years with a scrap value of \$30,000.
a. What is the rate of depreciation of the machine per year?
b. Find an expression for the book value of the machine in year t $(0 \le t \le 12)$.
24. A company has a fixed cost of \$30,000 and a production cost of \$6 for each unit it manufactures. A unit sells for \$10.
a. What is the cost function?
b. What is the revenue function?
c. What is the profit function?
d. Compute the profit (loss) corresponding to production levels of 6000, 8000, and 12,000 units, respectively.
25. There is no demand for a certain commodity when the unit price is \$200 or more, but for each \$10 decrease in price below \$200, the quantity demanded increases by 200 units. Find the demand equation and sketch its graph.
26. Bicycle suppliers will make 200 bicycles available in the market per month when the unit price is \$50 and 2000 bicycles available per month when the unit price is \$100. Find the supply equation if it is known to be linear.

In Exercises 27 and 28, find the point of intersection of the lines with the given equations.

27. $3x + 4y = -6$ and $2x + 5y = -11$

28. $y = \frac{3}{4}x + 6$ and $3x - 2y + 3 = 0$

29. The cost function and the revenue function for a certain firm are given by $C(x) = 12x + 20{,}000$ and $R(x) = 20x$, respectively. Find the break-even point for the company.

30. Given the demand equation $3x + p - 40 = 0$ and the supply equation $2x - p + 10 = 0$, where p is the unit price in dollars and x represents the quantity in units of a thousand, determine the equilibrium quantity and the equilibrium price.

31. The accompanying data were compiled by the Admissions Office of Carter College during the past 5 years. The data relate the number of college brochures and follow-up letters (x) sent to a preselected list of high school juniors who took the PSAT and the number of completed applications (y) received from these students (both measured in thousands).

Number of Brochures Sent, x	1.8	2	3.2	4	4.8
Number of Completed Applications, y	0.4	0.5	0.7	1	1.3

a. Derive an equation of the straight line L that passes through the points (2, 0.5) and (4, 1).
b. Use this equation to predict the number of completed applications that might be expected if 6400 brochures and follow-up letters are sent out during the next year.

Additional study hints and sample chapter tests for each chapter can be found at the Brooks/Cole Web site: http://www.brookscole.com/math/authors/tans/

2 SYSTEMS OF LINEAR EQUATIONS AND MATRICES

The linear equations in two variables studied in Chapter 1 are readily extended to the case involving more than two variables. For example, a linear equation in three variables represents a plane in three-dimensional space. In this chapter we see how some real-world problems can be formulated in terms of systems of linear equations, and we also develop two methods for solving these equations.

In addition, we see how *matrices* (ordered rectangular arrays of numbers) can be used to write systems of linear equations in compact form. We then go on to consider some real-life applications of matrices. Finally, we show how matrices can be used to describe the Leontief input–output model, an important tool used by economists. For his work in formulating this model, Wassily Leontief was awarded the Nobel Prize in 1973.

How fast is the traffic moving? The flow of downtown traffic is controlled by traffic lights installed at the intersections. One of the roads is to be resurfaced. In Example 5, page 111, you will see how the flow patterns must be altered in order to ensure a smooth flow of traffic even during rush hour.

2.1 Systems of Linear Equations—Introduction

SYSTEMS OF EQUATIONS

Recall that in Section 1.4 we had to solve two simultaneous linear equations in order to find the *break-even point* and the *equilibrium point.* These are two examples of situations in which solving a real-world problem calls for solving a **system of linear equations** in two or more variables. In this chapter we take up a more systematic study of such systems.

We begin by considering a system of two linear equations in two variables. Recall that such a system may be written in the general form

$$\begin{aligned} ax + by &= h \\ cx + dy &= k \end{aligned} \tag{1}$$

where a, b, c, d, h, and k are real constants and neither a and b nor c and d are both zero.

Now let's study the nature of the **solution of a system of linear equations** in more detail. Recall that the graph of each equation in System (1) is a straight line in the plane, so that geometrically the solution to the system is the point(s) of intersection of the two straight lines L_1 and L_2, represented by the first and second equations of the system.

Figure 2.1 depicts each of the three cases that may occur. The two lines L_1 and L_2 may

a. intersect at exactly one point,
b. be parallel and coincident, or
c. be parallel and distinct.

Note that one and only one of these cases must occur. In the first case the system has a unique solution corresponding to the single point of intersec-

FIGURE 2.1

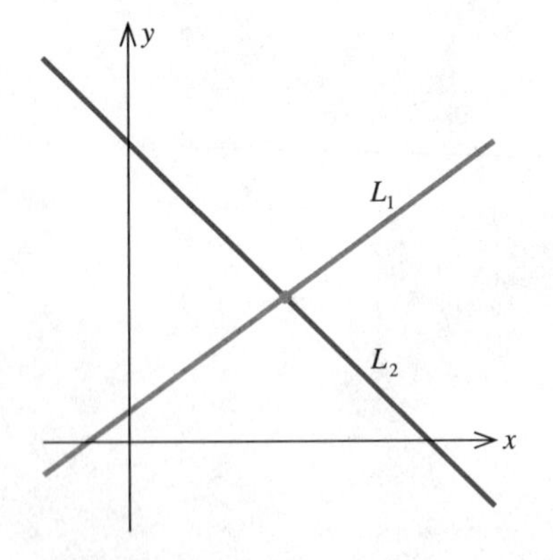

(a) Unique solution

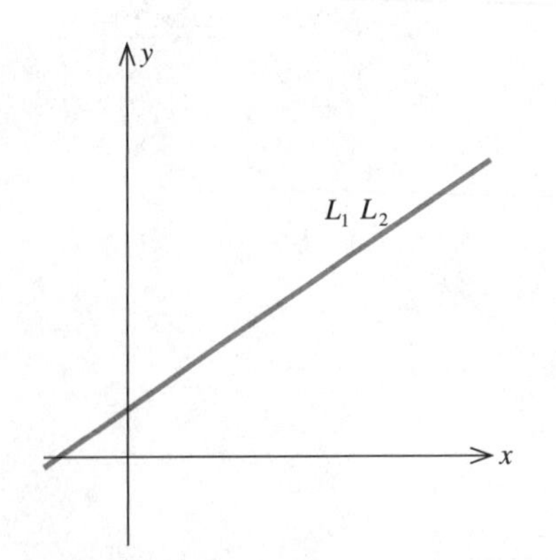

(b) Infinitely many solutions

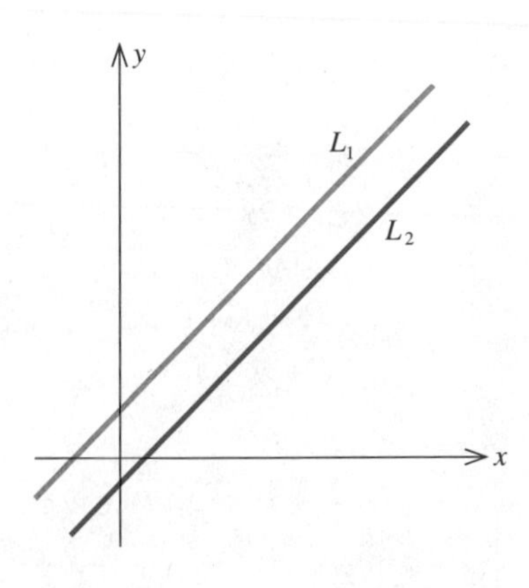

(c) No solution

Group Discussion

Generalize the discussion on page 78 to the case where there are three straight lines in the plane defined by three linear equations. What if there are n lines defined by n equations?

tion of the two lines. In the second case the system has infinitely many solutions corresponding to the points lying on the same line. Finally, in the third case, the system has no solution, since the two lines do not intersect.

Let's illustrate each of these possibilities by considering some specific examples.

1. A system of equations with exactly one solution Consider the system

$$\begin{aligned} 2x - y &= 1 \\ 3x + 2y &= 12 \end{aligned}$$

Solving the first equation for y in terms of x, we obtain the equation

$$y = 2x - 1$$

Substituting this expression for y into the second equation yields

$$\begin{aligned} 3x + 2(2x - 1) &= 12 \\ 3x + 4x - 2 &= 12 \\ 7x &= 14 \\ x &= 2 \end{aligned}$$

FIGURE 2.2
A system of equations with one solution

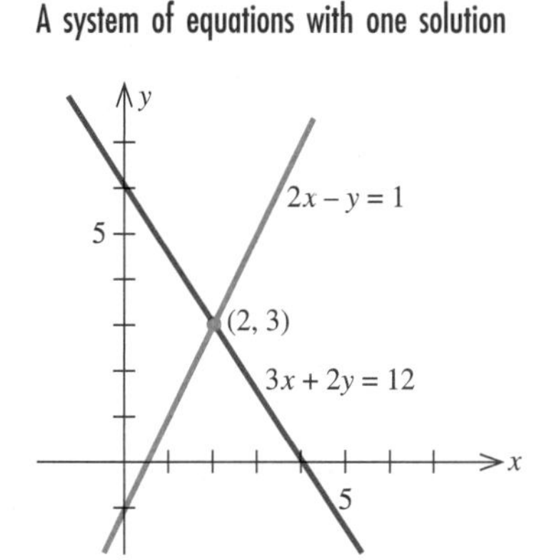

Finally, substituting this value of x into the expression for y obtained earlier gives

$$y = 2(2) - 1 = 3$$

Therefore, the unique solution of the system is given by $x = 2$ and $y = 3$. Geometrically, the two lines represented by the two linear equations that make up the system intersect at the point (2, 3) (Figure 2.2).

REMARK We can check our result by substituting the values $x = 2$ and $y = 3$ into the equations. Thus,

$$\begin{aligned} 2(2) - (3) &= 1 \qquad (\checkmark) \\ 3(2) + 2(3) &= 12 \qquad (\checkmark) \end{aligned}$$

From the geometrical point of view, we have just verified that the point (2, 3) lies on both lines.

2. **A system of equations with infinitely many solutions** Consider the system

$$\begin{aligned} 2x - y &= 1 \\ 6x - 3y &= 3 \end{aligned}$$

Solving the first equation for y in terms of x, we obtain the equation

$$y = 2x - 1$$

Substituting this expression for y into the second equation gives

$$\begin{aligned} 6x - 3(2x - 1) &= 3 \\ 6x - 6x + 3 &= 3 \\ 0 &= 0 \end{aligned}$$

This result simply tells us that the second equation is equivalent to the first. (To see this, just multiply both sides of the first equation by 3.) Our computations have revealed that the system of two equations is equivalent to the single equation $2x - y = 1$. Thus, any ordered pair of numbers (x, y) satisfying the equation $2x - y = 1$ (or $y = 2x - 1$) constitutes a solution to the system.

In particular, by assigning the value t to x, where t is any real number, we find that $y = 2t - 1$, so the ordered pair $(t, 2t - 1)$ is a solution of the system. The variable t is called a **parameter.** For example, setting $t = 0$ gives the point $(0, -1)$ as a solution, and setting $t = 1$ gives the point $(1, 1)$ as another solution of the system. Since t represents any real number, there are infinitely many solutions to the system. Geometrically, the two equations in the system represent the same line, and all solutions of the system are points lying on the line (Figure 2.3). Such a system is said to be **dependent.**

FIGURE 2.3
A system of equations with infinitely many solutions; any point on the line is a solution.

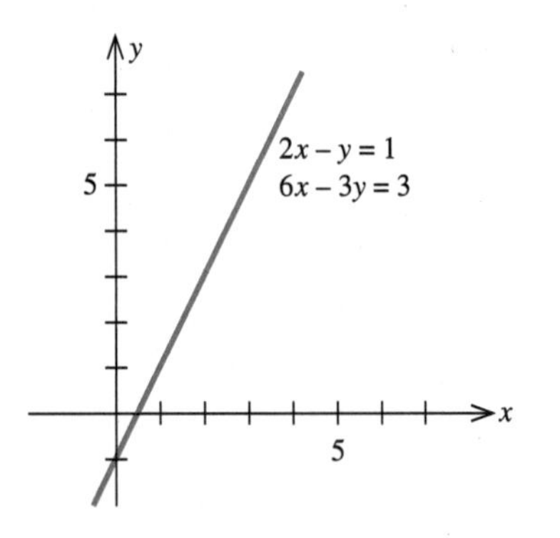

3. **A system of equations that has no solution** Consider the system

$$\begin{aligned} 2x - y &= 1 \\ 6x - 3y &= 12 \end{aligned}$$

The first equation is equivalent to $y = 2x - 1$. Substituting this expression for y into the second equation gives

$$\begin{aligned} 6x - 3(2x - 1) &= 12 \\ 6x - 6x + 3 &= 12 \\ 0 &= 9 \end{aligned}$$

which is clearly impossible. Thus, there is no solution to the system of equations. To interpret this situation geometrically, cast both equations in the slope-intercept form, obtaining

$$\begin{aligned} y &= 2x - 1 \\ y &= 2x - 4 \end{aligned}$$

Group Discussion

1. Consider a system composed of two linear equations in two variables. Can the system have exactly two solutions? Exactly three solutions? Exactly a finite number of solutions?

2. Suppose at least one of the equations in a system composed of two equations in two variables is nonlinear. Can the system have no solution? Exactly one solution? Exactly two solutions? Exactly a finite number of solutions? Infinitely many solutions? Illustrate each answer with a sketch.

FIGURE 2.4
A system of equations with no solution

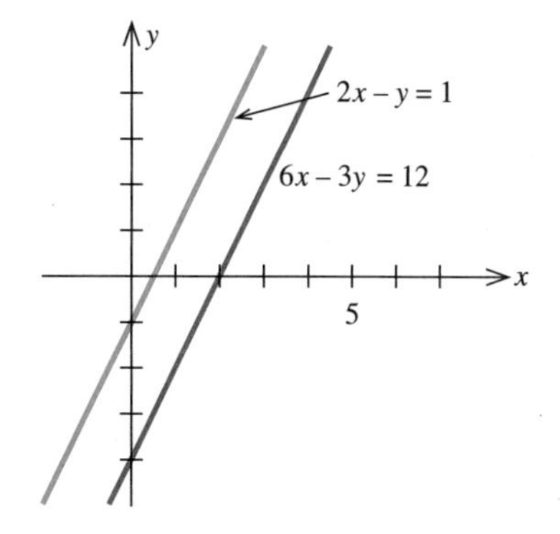

We see at once that the lines represented by these equations are parallel (each has slope 2) and distinct since the first has intercept -1 and the second has intercept -4 (Figure 2.4). Systems with no solutions, such as this one, are said to be **inconsistent.**

REMARK We have used the method of substitution in solving each of these systems. If you are familiar with the method of elimination, you might want to re-solve each of these systems using this method. We will study the method of elimination in detail in Section 2.2.

APPLICATION

In Section 1.4 we presented some real-world applications of systems involving two linear equations in two variables. Here is an example involving a system of three linear equations in three variables.

EXAMPLE 1

The Ace Novelty Company wishes to produce three types of souvenirs: types A, B, and C. To manufacture a type-A souvenir requires 2 minutes on machine I, 1 minute on machine II, and 2 minutes on machine III. A type-B souvenir requires 1 minute on machine I, 3 minutes on machine II, and 1 minute on machine III. A type-C souvenir requires 1 minute on machine I and 2 minutes each on machines II and III. There are 3 hours available on machine I, 5 hours available on machine II, and 4 hours available on machine III for processing the order. How many souvenirs of each type should Ace Novelty make in order to use all of the available time? Formulate but do not solve the problem. (We will solve this problem in Example 7, Section 2.2.)

SOLUTION ✓

The given information may be tabulated as follows:

	Type A	Type B	Type C	Time Available (min)
Machine I	2	1	1	180
Machine II	1	3	2	300
Machine III	2	1	2	240

We have to determine the number of each of *three* types of souvenirs to be made. So, let x, y, and z denote the respective numbers of type-A, type-B, and type-C souvenirs to be made. The total amount of time that machine I is used is given by $2x + y + z$ minutes and must equal 180 minutes. This leads to the equation

$$2x + y + z = 180 \qquad \text{(Time spent on machine I)}$$

Similar considerations on the use of machines II and III lead to the following equations:

$$\begin{aligned} x + 3y + 2z &= 300 \qquad \text{(Time spent on machine II)} \\ 2x + y + 2z &= 240 \qquad \text{(Time spent on machine III)} \end{aligned}$$

Since the variables x, y, and z must satisfy simultaneously the three conditions represented by the three equations, the solution to the problem is found by solving the following system of linear equations:

$$\begin{aligned} 2x + y + z &= 180 \\ x + 3y + 2z &= 300 \\ 2x + y + 2z &= 240 \end{aligned}$$

■■■■

Solutions of Systems of Equations

We will complete the solution of the problem posed in Example 1 later on (page 97). For the moment, let's look at the geometrical interpretation of a system of linear equations, such as the system in Example 1, in order to gain some insight into the nature of the solution.

A linear system composed of three linear equations in three variables x, y, and z has the general form

$$\begin{aligned} a_1x + b_1y + c_1z &= d_1 \\ a_2x + b_2y + c_2z &= d_2 \\ a_3x + b_3y + c_3z &= d_3 \end{aligned} \tag{2}$$

Just as a linear equation in two variables represents a straight line in the plane, it can be shown that a linear equation $ax + by + cz = d$ (a, b, and c not simultaneously equal to zero) in three variables represents a plane in three-dimensional space. Thus, each equation in System (2) represents a *plane* in three-dimensional space, and the *solution(s) of the system* is precisely the point(s) of intersection of the three planes defined by the three linear equations that make up the system. As before, the system has one and only one solution, infinitely many solutions, or no solution, depending on whether and how the planes intersect one another. Figure 2.5 illustrates each of these possibilities.

In Figure 2.5a, the three planes intersect at a point corresponding to the situation in which System (2) has a unique solution. Figure 2.5b depicts the situation in which there are infinitely many solutions to the system. Here, the three planes intersect along a line, and the solutions are represented

FIGURE 2.5

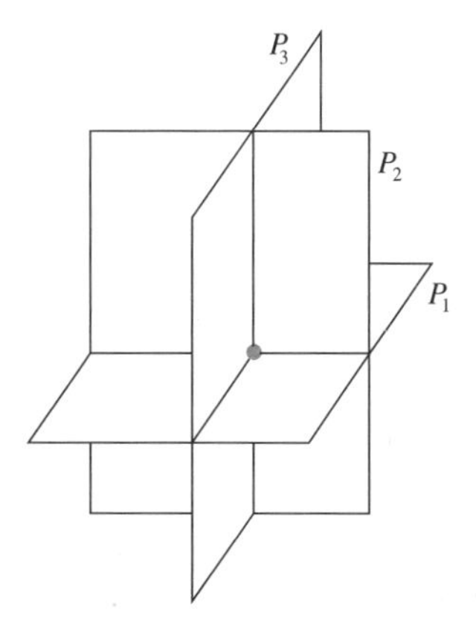

(a) A unique solution

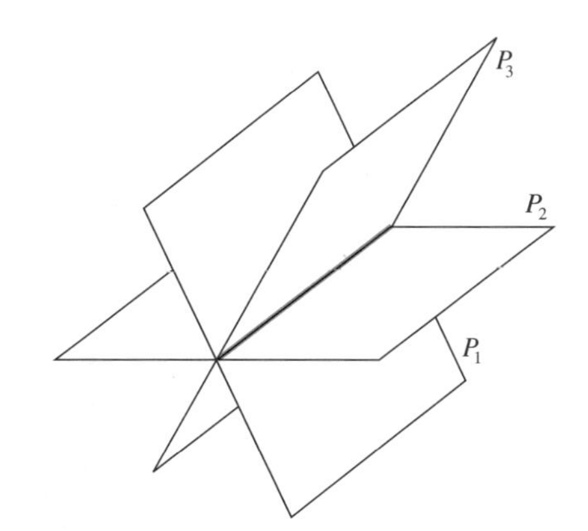

(b) Infinitely many solutions

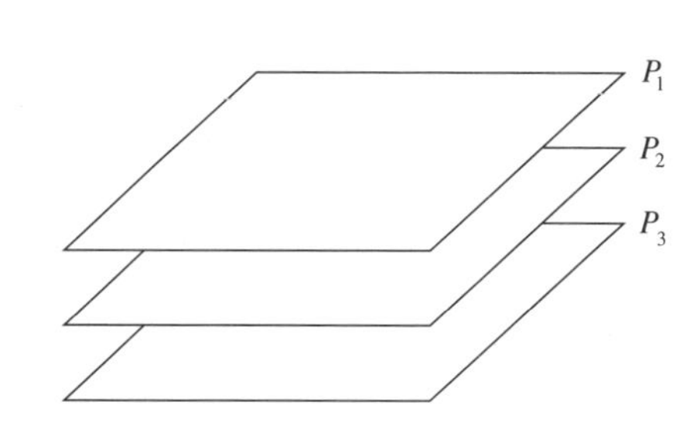

(c) No solution

by the infinitely many points lying on this line. In Figure 2.5c, the three planes are parallel and distinct, so there is no point in common to all three planes, and System (2) has no solution in this case.

REMARK The situations depicted in Figure 2.5 are by no means exhaustive. You may consider various other orientations of the three planes that would illustrate the three possible outcomes in solving a system of linear equations involving three variables.

Linear Equations in n Variables

A linear equation in n variables $x_1, x_2, \ldots, x_n$ is one of the form

$$a_1x_1 + a_2x_2 + \cdots + a_nx_n = c$$

where $a_1, a_2, \ldots, a_n$ (not all zero) and c are constants.

For example, the equation

$$3x_1 + 2x_2 - 4x_3 + 6x_4 = 8$$

is a linear equation in the four variables x_1, x_2, x_3, and x_4.

When the number of variables involved in a linear equation exceeds three, we no longer have the geometrical interpretation we had for the lower-dimensional spaces. Nevertheless, the algebraic concepts of the lower-dimensional spaces generalize to higher dimensions. For this reason, a linear equation in n variables $a_1x_1 + a_2x_2 + \cdots + a_nx_n = c$, where $a_1, a_2, \ldots, a_n$ are not all zero, is referred to as an *n-dimensional hyperplane.* We may interpret the solution(s) to a system comprising a finite number of such linear equations to be the *point(s) of intersection* of the hyperplanes defined by the equations that make up the system. As in the case of systems involving two or three

Group Discussion

Refer to the REMARK on this page.

Using the orientations of three planes, illustrate the outcomes in solving a system of three linear equations in three variables that result in no solution or infinitely many solutions.

variables, it can be shown that only three possibilities exist regarding the nature of the solution of such a system: (1) a unique solution, (2) infinitely many solutions, and (3) no solution.

SELF-CHECK EXERCISES 2.1

1. Determine whether the system of linear equations

$$\begin{aligned} 2x - 3y &= 12 \\ x + 2y &= 6 \end{aligned}$$

has (a) a unique solution, (b) infinitely many solutions, or (c) no solution. Find all solutions whenever they exist. Make a sketch of the set of lines described by the system.

2. A farmer has 200 acres of land suitable for cultivating crops A, B, and C. The cost per acre of cultivating crops A, B, and C is $40, $60, and $80, respectively. The farmer has $12,600 available for cultivation. Each acre of crop A requires 20 hr of labor, each acre of crop B requires 25 hr of labor, and each acre of crop C requires 40 hr of labor. The farmer has a maximum of 5950 hr of labor available. If he wishes to use all of his cultivatable land, the entire budget, and all the labor available, how many acres of each crop should he plant? Formulate but do not solve the problem.

Solutions to Self-Check Exercises 2.1 can be found on page 86.

2.1 Exercises

In Exercises 1–12, determine whether each system of linear equations has (a) one and only one solution, (b) infinitely many solutions, or (c) no solution. Find all solutions whenever they exist.

1. $\begin{aligned} x - 3y &= -1 \\ 4x + 3y &= 11 \end{aligned}$

2. $\begin{aligned} 2x - 4y &= 5 \\ 3x + 2y &= 6 \end{aligned}$

3. $\begin{aligned} x + 4y &= 7 \\ \frac{1}{2}x + 2y &= 5 \end{aligned}$

4. $\begin{aligned} 3x - 4y &= 7 \\ 9x - 12y &= 14 \end{aligned}$

5. $\begin{aligned} x + 2y &= 7 \\ 2x - y &= 4 \end{aligned}$

6. $\begin{aligned} \frac{3}{2}x - 2y &= 4 \\ x + \frac{1}{3}y &= 2 \end{aligned}$

7. $\begin{aligned} 2x - 5y &= 10 \\ 6x - 15y &= 30 \end{aligned}$

8. $\begin{aligned} 5x - 6y &= 8 \\ 10x - 12y &= 16 \end{aligned}$

9. $\begin{aligned} 4x - 5y &= 14 \\ 2x + 3y &= -4 \end{aligned}$

10. $\begin{aligned} \frac{5}{4}x - \frac{2}{3}y &= 3 \\ \frac{1}{4}x + \frac{5}{3}y &= 6 \end{aligned}$

11. $\begin{aligned} 2x - 3y &= 6 \\ 6x - 9y &= 12 \end{aligned}$

12. $\begin{aligned} \frac{2}{3}x + y &= 5 \\ \frac{1}{2}x + \frac{3}{4}y &= \frac{15}{4} \end{aligned}$

13. Determine the value of k for which the system of linear equations

$$\begin{aligned} 2x - y &= 3 \\ 4x + ky &= 4 \end{aligned}$$

has no solution.

14. Determine the value of k for which the system of linear equations

$$\begin{aligned} 3x + 4y &= 12 \\ x + ky &= 4 \end{aligned}$$

has infinitely many solutions. Then find all solutions corresponding to this value of k.

In Exercises 15–25, formulate, but do not solve, the problem. You will be asked to solve these problems in the next section.

15. AGRICULTURE The Johnson Farm has 500 acres of land allotted for cultivating corn and wheat. The cost of cultivating corn and wheat (including seeds and labor) is \$42 and \$30/acre, respectively. Jacob Johnson has \$18,600 available for cultivating these crops. If he wishes to use all the allotted land and his entire budget for cultivating these two crops, how many acres of each crop should he plant?

16. INVESTMENTS Michael Perez has a total of \$2000 on deposit with two savings institutions. One pays interest at the rate of 6%/year, whereas the other pays interest at the rate of 8%/year. If Michael earned a total of \$144 in interest during a single year, how much does he have on deposit in each institution?

17. MIXTURES The Coffee Shoppe sells a coffee blend made from two coffees, one costing \$2.50/lb and the other costing \$3.00/lb. If the blended coffee sells for \$2.80/lb, find how much of each coffee is used to obtain the desired blend. (Assume the weight of the blended coffee is 100 lb.)

18. INVESTMENTS Kelly Fisher has a total of \$30,000 invested in two municipal bonds that have yields of 8% and 10% interest per year, respectively. If the interest Kelly receives from the bonds in a year is \$2640, how much does she have invested in each bond?

19. RIDERSHIP The total number of passengers riding a certain city bus during the morning shift is 1000. If the child's fare is 25 cents, the adult fare is 75 cents, and the total revenue from the fares in the morning shift is \$650, how many children and how many adults rode the bus during the morning shift?

20. REAL ESTATE Cantwell Associates, a real estate developer, is planning to build a new apartment complex consisting of one-bedroom units and two- and three-bedroom townhouses. A total of 192 units is planned, and the number of family units (two- and three-bedroom townhouses) will equal the number of one-bedroom units. If the number of one-bedroom units will be three times the number of three-bedroom units, find how many units of each type will be in the complex.

21. INVESTMENT PLANNING The annual interest on Sid Carrington's three investments amounted to \$21,600: 6% on a savings account, 8% on mutual funds, and 12% on bonds. If the amount of Sid's investment in bonds was twice the amount of his investment in the savings account, and the interest earned from his investment in bonds was equal to the dividends he received from his investment in mutual funds, find how much money he placed in each type of investment.

22. BOX-OFFICE RECEIPTS A theater has a seating capacity of 900 and charges \$2 for children, \$3 for students, and \$4 for adults. At a certain screening with full attendance, there were half as many adults as children and students combined. The receipts totaled \$2800. How many children attended the show?

23. MANAGEMENT DECISIONS The management of Hartman Rent-A-Car has allocated \$1.25 million to buy a fleet of new automobiles consisting of compact, intermediate, and full-size cars. Compacts cost \$10,000 each, intermediate-size cars cost \$15,000 each, and full-size cars cost \$20,000 each. If Hartman purchases twice as many compacts as intermediate-size cars and the total number of cars to be purchased is 100, determine how many cars of each type will be purchased. (Assume that the entire budget will be used.)

24. INVESTMENT CLUBS The management of a private investment club has a fund of \$200,000 earmarked for investment in stocks. To arrive at an acceptable overall level of risk, the stocks that management is considering have been classified into three categories: high-risk, medium-risk, and low-risk. Management estimates that high-risk stocks will have a rate of return of 15%/year; medium-risk stocks, 10%/year; and low-risk stocks, 6%/year. The investment in low-risk stocks is to be twice the sum of the investments in stocks of the other two categories. If the investment goal is to have an average rate of return of 9%/year on the total investment, determine how much the club should invest in each type of stock.

25. DIET PLANNING A dietitian wishes to plan a meal around three foods. The percentage of the daily requirements of proteins, carbohydrates, and iron contained in each ounce of the three foods is summarized in the accompanying table. Determine how many ounces of each food the dietitian should include in the meal to meet exactly the daily requirement of proteins, carbohydrates, and iron (100% of each).

	Food I	Food II	Food III
Percentage of Proteins	10	6	8
Percentage of Carbohydrates	10	12	6
Percentage of Iron	5	4	12

In Exercises 26–28, determine whether the statement is true or false. If it is true, explain why it is true. If it is false, give an example to show why it is false.

26. A system composed of two linear equations must have at least one solution if the straight lines represented by these equations are nonparallel.

27. Suppose the straight lines represented by a system of three linear equations in two variables are parallel to each other. Then, the system has no solution, or it has infinitely many solutions.

28. If at least two of the three lines represented by a system composed of three linear equations in two variables are parallel, then the system has no solution.

SOLUTIONS TO SELF-CHECK EXERCISES 2.1

1. Solving the first equation for y in terms of x, we obtain

$$y = \frac{2}{3}x - 4$$

Next, substituting this result into the second equation of the system, we find

$$\begin{aligned} x + 2\left(\frac{2}{3}x - 4\right) &= 6 \\ x + \frac{4}{3}x - 8 &= 6 \\ \frac{7}{3}x &= 14 \\ x &= 6 \end{aligned}$$

Substituting this value of x into the expression for y obtained earlier, we have

$$y = \frac{2}{3}(6) - 4 = 0$$

Therefore, the system has the unique solution $x = 6$ and $y = 0$. Both lines are shown in the accompanying figure.

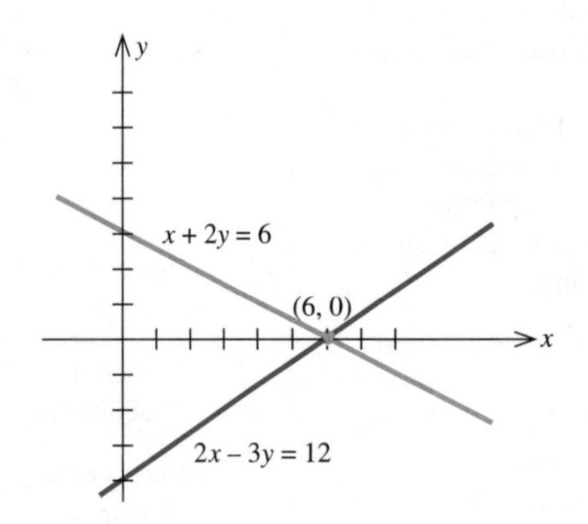

2. Let x, y, and z denote the number of acres of crop A, crop B, and crop C, respectively, to be cultivated. Then, the condition that all the cultivatable land be used translates into the equation

$$x + y + z = 200$$

Next, the total cost incurred in cultivating all three crops is $40x + 60y + 80z$ dollars, and since the entire budget is to be expended, we have

$$40x + 60y + 80z = 12{,}600$$

Finally, the amount of labor required to cultivate all three crops is $20x + 25y + 40z$ hr, and since all the available labor is to be used, we have

$$20x + 25y + 40z = 5950$$

Thus, the solution is found by solving the following system of linear equations:

$$\begin{aligned} x + \quad y + \quad z &= \quad 200 \\ 40x + 60y + 80z &= 12{,}600 \\ 20x + 25y + 40z &= \ \ 5{,}950 \end{aligned}$$

2.2 Solving Systems of Linear Equations I

THE GAUSS–JORDAN METHOD

The method of substitution used in Section 2.1 is well suited to solving a system of linear equations when the number of linear equations and variables is small. But for large systems, the steps involved in the procedure become difficult to manage.

A suitable technique for solving systems of linear equations of any size is the **Gauss–Jordan elimination method.** One advantage of this technique is its adaptability to the computer. This method involves a sequence of operations on a system of linear equations to obtain at each stage an **equivalent system**—that is, a system having the same solution as the original system. The reduction is complete when the original system has been transformed so that it is in a certain standard form from which the solution can be easily read.

The operations of the Gauss–Jordan elimination method are:

1. Interchange any two equations.
2. Replace an equation by a nonzero constant multiple of itself.
3. Replace an equation by the sum of that equation and a constant multiple of any other equation.

To illustrate the Gauss–Jordan elimination method for solving systems of linear equations, let's apply it to the solution of the following system:

$$\begin{aligned} 2x + 4y &= 8 \\ 3x - 2y &= 4 \end{aligned}$$

We begin by working with the first, or x, column. First, we transform the system into an equivalent system in which the coefficient of x in the first equation is 1:

$$\begin{aligned} 2x + 4y &= 8 \\ 3x - 2y &= 4 \end{aligned} \tag{3a}$$

$$\begin{aligned} x + 2y &= 4 \\ 3x - 2y &= 4 \end{aligned} \tag{3b}$$

[Multiply the first equation in (3a) by $\frac{1}{2}$ (operation 2).]

Next, we eliminate x from the second equation.

$$\begin{aligned} x + 2y &= 4 \\ -8y &= -8 \end{aligned} \tag{3c}$$

[Replace second equation in (3b) by the sum of -3 times first equation + second equation (operation 3).]

$$\begin{array}{r} -3x - 6y = -12 \\ 3x - 2y = 4 \\ \hline -8y = -8 \end{array}$$

Then, we obtain the following equivalent system in which the coefficient of y in the second equation is 1.

$$\begin{aligned} x + 2y &= 4 \\ y &= 1 \end{aligned} \tag{3d}$$

[Multiply second equation in (3c) by $-\frac{1}{8}$ (operation 2).]

Next, we eliminate y in the first equation.

$$\begin{aligned} x \qquad &= 2 \\ y &= 1 \end{aligned}$$

[Replace first equation in (3d) by the sum of -2 times the second equation + the first equation (operation 3).]

$$\begin{array}{r} x + 2y = 4 \\ -2y = -2 \\ \hline x \qquad = 2 \end{array}$$

This system is now in standard form, and we can read off the solution to (3a) as $x = 2$ and $y = 1$. We can also express this solution as (2, 1) and interpret it geometrically as the point of intersection of the two lines represented by the two linear equations that make up the given system of equations.

Let's consider another example involving a system of three linear equations and three variables.

EXAMPLE 1 Solve the following system of equations:

$$\begin{aligned} 2x + 4y + 6z &= 22 \\ 3x + 8y + 5z &= 27 \\ -x + y + 2z &= 2 \end{aligned}$$

SOLUTION ✓ First, we transform this system into an equivalent system in which the coefficient of x in the first equation is 1:

$$\begin{aligned} 2x + 4y + 6z &= 22 \\ 3x + 8y + 5z &= 27 \\ -x + y + 2z &= 2 \end{aligned} \tag{4a}$$

$$\begin{aligned} x + 2y + 3z &= 11 \\ 3x + 8y + 5z &= 27 \\ -x + y + 2z &= 2 \end{aligned} \tag{4b}$$

[Multiply first equation in (4a) by $\frac{1}{2}$.]

Next, we eliminate the variable x from all equations except the first:

$$\begin{aligned} x + 2y + 3z &= 11 \\ 2y - 4z &= -6 \\ -x + y + 2z &= 2 \end{aligned} \tag{4c}$$

[Replace second equation in (4b) by the sum of -3 times first equation + second equation.]

$$\begin{array}{r} -3x - 6y - 9z = -33 \\ 3x + 8y + 5z = 27 \\ \hline 2y - 4z = -6 \end{array}$$

$$\begin{aligned} x + 2y + 3z &= 11 \\ 2y - 4z &= -6 \\ 3y + 5z &= 13 \end{aligned} \tag{4d}$$

[Replace the third equation in (4c) by the sum of first equation + third equation.]

$$\begin{array}{r} x + 2y + 3z = 11 \\ -x + y + 2z = 2 \\ \hline 3y + 5z = 13 \end{array}$$

Then we transform System (4d) into yet another equivalent system, in which the coefficient of y in the second equation is 1:

$$\begin{aligned} x + 2y + 3z &= 11 \\ y - 2z &= -3 \\ 3y + 5z &= 13 \end{aligned} \tag{4e}$$

[Multiply second equation in (4d) by $\frac{1}{2}$.]

We now eliminate y from all equations except the second using operation 3 of the elimination method.

$$\begin{aligned} x \qquad + 7z &= 17 \\ y - 2z &= -3 \\ 3y + 5z &= 13 \end{aligned} \tag{4f}$$

[Replace first equation in (4e) by the sum of first equation + (-2) times second equation.]

$$\begin{array}{r} x + 2y + 3z = 11 \\ -2y + 4z = 6 \\ \hline x \qquad + 7z = 17 \end{array}$$

$$\begin{aligned} x \qquad + 7z &= 17 \\ y - 2z &= -3 \\ 11z &= 22 \end{aligned} \tag{4g}$$

[Replace third equation in (4f) by the sum of (-3) times second equation + third equation.]

$$\begin{array}{r} -3y + 6z = 9 \\ 3y + 5z = 13 \\ \hline 11z = 22 \end{array}$$

Finally, multiplying the third equation by 1/11 in (4g) leads to the system

$$\begin{aligned} x \qquad + 7z &= 17 \\ y - 2z &= -3 \\ z &= 2 \end{aligned}$$

Eliminating z from all equations except the third (try it!) then leads to the system

$$\begin{aligned} x \qquad\quad &= 3 \\ y \quad &= 1 \\ z &= 2 \end{aligned} \tag{4h}$$

In its final form, the solution to the given system of equations can be easily read off! We have $x = 3$, $y = 1$, and $z = 2$. Geometrically, the point (3, 1, 2) lies in the intersection of the three planes described by the three equations comprising the given system. ■■■■

Augmented Matrices

Observe from the preceding example that in each step of the reduction process the variables x, y, and z play no significant role except as a reminder of the position of each coefficient in the system. With the aid of **matrices,** which are rectangular arrays of numbers, we can eliminate writing the variables at each step of the reduction and thus save ourselves a great deal of work. For example, the system

$$\begin{aligned} 2x + 4y + 6z &= 22 \\ 3x + 8y + 5z &= 27 \\ -x + y + 2z &= 2 \end{aligned} \tag{5}$$

may be represented by the matrix

$$\left[\begin{array}{rrr|r} 2 & 4 & 6 & 22 \\ 3 & 8 & 5 & 27 \\ -1 & 1 & 2 & 2 \end{array}\right] \tag{6}$$

The augmented matrix representing System (5)

The submatrix, consisting of the first three columns of Matrix (6), is called the **coefficient matrix** of System (5). The matrix itself, (6), is referred to as the **augmented matrix** of System (5) since it is obtained by joining the matrix of coefficients to the column (matrix) of constants. The vertical line separates the column of constants from the matrix of coefficients.

The next example shows how much work you can save by using matrices instead of the standard representation of the systems of linear equations.

EXAMPLE 2 Write the augmented matrix corresponding to each equivalent system given in (4a) through (4h).

SOLUTION ✓ The required sequence of augmented matrices follows.

	Equivalent system	Augmented matrix	
a.	$\begin{aligned} 2x + 4y + 6z &= 22 \\ 3x + 8y + 5z &= 27 \\ -x + y + 2z &= 2 \end{aligned}$	$\left[\begin{array}{rrr\|r} 2 & 4 & 6 & 22 \\ 3 & 8 & 5 & 27 \\ -1 & 1 & 2 & 2 \end{array}\right]$	**(7a)**
b.	$\begin{aligned} x + 2y + 3z &= 11 \\ 3x + 8y + 5z &= 27 \\ -x + y + 2z &= 2 \end{aligned}$	$\left[\begin{array}{rrr\|r} 1 & 2 & 3 & 11 \\ 3 & 8 & 5 & 27 \\ -1 & 1 & 2 & 2 \end{array}\right]$	**(7b)**
c.	$\begin{aligned} x + 2y + 3z &= 11 \\ 2y - 4z &= -6 \\ -x + y + 2z &= 2 \end{aligned}$	$\left[\begin{array}{rrr\|r} 1 & 2 & 3 & 11 \\ 0 & 2 & -4 & -6 \\ -1 & 1 & 2 & 2 \end{array}\right]$	**(7c)**
d.	$\begin{aligned} x + 2y + 3z &= 11 \\ 2y - 4z &= -6 \\ 3y + 5z &= 13 \end{aligned}$	$\left[\begin{array}{rrr\|r} 1 & 2 & 3 & 11 \\ 0 & 2 & -4 & -6 \\ 0 & 3 & 5 & 13 \end{array}\right]$	**(7d)**
e.	$\begin{aligned} x + 2y + 3z &= 11 \\ y - 2z &= -3 \\ 3y + 5z &= 13 \end{aligned}$	$\left[\begin{array}{rrr\|r} 1 & 2 & 3 & 11 \\ 0 & 1 & -2 & -3 \\ 0 & 3 & 5 & 13 \end{array}\right]$	**(7e)**
f.	$\begin{aligned} x + \quad 7z &= 17 \\ y - 2z &= -3 \\ 3y + 5z &= 13 \end{aligned}$	$\left[\begin{array}{rrr\|r} 1 & 0 & 7 & 17 \\ 0 & 1 & -2 & -3 \\ 0 & 3 & 5 & 13 \end{array}\right]$	**(7f)**
g.	$\begin{aligned} x \quad + 7z &= 17 \\ y - 2z &= -3 \\ 11z &= 22 \end{aligned}$	$\left[\begin{array}{rrr\|r} 1 & 0 & 7 & 17 \\ 0 & 1 & -2 & -3 \\ 0 & 0 & 11 & 22 \end{array}\right]$	**(7g)**
h.	$\begin{aligned} x \quad &= 3 \\ y \quad &= 1 \\ z &= 2 \end{aligned}$	$\left[\begin{array}{rrr\|r} 1 & 0 & 0 & 3 \\ 0 & 1 & 0 & 1 \\ 0 & 0 & 1 & 2 \end{array}\right]$	**(7h)**

The augmented matrix in (7h) is an example of a matrix in row-reduced form. In general, an augmented matrix with m rows and n columns (called an $m \times n$ matrix) is in row-reduced form if it satisfies the following conditions.

Row-Reduced Form

1. Each row consisting entirely of zeros lies below any other row having nonzero entries.
2. The first nonzero entry in each row is 1 (called a **leading 1**).
3. In any two successive (nonzero) rows, the leading 1 in the lower row lies to the right of the leading 1 in the upper row.
4. If a column contains a leading 1, then the other entries in that column are zeros.

EXAMPLE 3 Determine which of the following matrices are in row-reduced form. If a matrix is not in row-reduced form, state which condition is violated.

a. $\left[\begin{array}{ccc|c} 1 & 0 & 0 & 0 \\ 0 & 1 & 0 & 0 \\ 0 & 0 & 1 & 3 \end{array}\right]$ **b.** $\left[\begin{array}{ccc|c} 1 & 0 & 0 & 4 \\ 0 & 1 & 0 & 3 \\ 0 & 0 & 0 & 0 \end{array}\right]$ **c.** $\left[\begin{array}{ccc|c} 1 & 2 & 0 & 0 \\ 0 & 0 & 1 & 0 \\ 0 & 0 & 0 & 1 \end{array}\right]$

d. $\left[\begin{array}{ccc|c} 0 & 1 & 2 & -2 \\ 1 & 0 & 0 & 3 \\ 0 & 0 & 1 & 2 \end{array}\right]$ **e.** $\left[\begin{array}{ccc|c} 1 & 2 & 0 & 0 \\ 0 & 0 & 1 & 3 \\ 0 & 0 & 2 & 1 \end{array}\right]$ **f.** $\left[\begin{array}{cc|c} 1 & 0 & 4 \\ 0 & 3 & 0 \\ 0 & 0 & 0 \end{array}\right]$

g. $\left[\begin{array}{ccc|c} 0 & 0 & 0 & 0 \\ 1 & 0 & 0 & 3 \\ 0 & 1 & 0 & 2 \end{array}\right]$

SOLUTION ✓ The matrices in parts **(a)–(c)** are in row-reduced form.

d. This matrix is not in row-reduced form. Conditions 3 and 4 are violated: The leading 1 in row 2 lies to the left of the leading 1 in row 1. Also, column 3 contains a leading 1 in row 3 and a nonzero element above it.

e. This matrix is not in row-reduced form. Conditions 2 and 4 are violated: The first nonzero entry in row 3 is a 2, not a 1. Also, column 3 contains a leading 1 and has a nonzero entry below it.

f. This matrix is not in row-reduced form. Condition 2 is violated: The first nonzero entry in row 2 is not a leading 1.

g. This matrix is not in row-reduced form. Condition 1 is violated: Row 1 consists of all zeros and does not lie below the other nonzero rows. ■■■■

The foregoing discussion suggests the following adaptation of the Gauss–Jordan elimination method in solving systems of linear equations using matrices. First, the three operations on the equations of a system (see page 87) translate into the following row operations on the corresponding augmented matrices.

Row Operations

1. Interchange any two rows.
2. Replace any row by a nonzero constant multiple of itself.
3. Replace any row by the sum of that row and a constant multiple of any other row.

We obtained the augmented matrices in Example 2 by using the same operations that we used on the equivalent system of equations in Example 1.

In order to help us describe the Gauss–Jordan elimination method using matrices, let us introduce some terminology. We begin by defining what is meant by a **unit column.**

Unit Column

A column in a coefficient matrix is in unit form if one of the entries in the column is a 1 and the other entries are zeros.

For example, in the coefficient matrix of (7d), only the first column is in unit form; in the coefficient matrix of (7h), all three columns are in unit form. Now, the sequence of row operations that transforms the augmented matrix (7a) into the equivalent matrix (7d) in which the first column

$$\begin{matrix} 2 \\ 3 \\ -1 \end{matrix}$$

of (7a) is transformed into the unit column

$$\begin{matrix} 1 \\ 0 \\ 0 \end{matrix}$$

is called *pivoting* the matrix about the element (number) 2. Similarly, we have pivoted about the element 2 in the second column of (7d), shown circled,

$$\begin{matrix} 2 \\ \textcircled{2} \\ 3 \end{matrix}$$

in order to obtain the augmented matrix (7g). Finally, pivoting about the element 11 in column 3 of (7g)

$$\begin{matrix} 7 \\ -2 \\ \textcircled{11} \end{matrix}$$

leads to the augmented matrix (7h), in which all columns are in unit form. The element about which a matrix is pivoted is called the *pivot element.*

Before looking at the next example, let's introduce the following notation for the three types of row operations.

Notation for Row Operations

Letting R_i denote the ith row of a matrix, we write:

Operation 1 $R_i \leftrightarrow R_j$ to mean: Interchange row i with row j.

Operation 2 cR_i to mean: Replace row i with c times row i.

Operation 3 $R_i + aR_j$ to mean: Replace row i with the sum of row i and a times row j.

EXAMPLE 4

Pivot the matrix about the circled element.

$$\left[\begin{array}{cc|c} \textcircled{3} & 5 & 9 \\ 2 & 3 & 5 \end{array}\right]$$

SOLUTION ✓

Using the notation just introduced, we obtain

$$\left[\begin{array}{cc|c} 3 & 5 & 9 \\ 2 & 3 & 5 \end{array}\right] \xrightarrow{\frac{1}{3}R_1} \left[\begin{array}{cc|c} 1 & \frac{5}{3} & 3 \\ 2 & 3 & 5 \end{array}\right] \xrightarrow{R_2 - 2R_1} \left[\begin{array}{cc|c} 1 & \frac{5}{3} & 3 \\ 0 & -\frac{1}{3} & -1 \end{array}\right]$$

The first column, which originally contained the entry 3, is now in unit form, with a 1 where the pivot element used to be, and we are done.

ALTERNATE SOLUTION ✓

In the first solution, we used operation 2 to obtain a 1 where the pivot element was originally. Alternatively, we can use operation 3 as follows:

$$\left[\begin{array}{cc|c} 3 & 5 & 9 \\ 2 & 3 & 5 \end{array}\right] \xrightarrow{R_1 - R_2} \left[\begin{array}{cc|c} 1 & 2 & 4 \\ 2 & 3 & 5 \end{array}\right] \xrightarrow{R_2 - 2R_1} \left[\begin{array}{cc|c} 1 & 2 & 4 \\ 0 & -1 & -3 \end{array}\right]$$

■■■■

REMARK In Example 4, the two matrices

$$\left[\begin{array}{cc|c} 1 & \frac{5}{3} & 3 \\ 0 & -\frac{1}{3} & -1 \end{array}\right] \quad \text{and} \quad \left[\begin{array}{cc|c} 1 & 2 & 4 \\ 0 & -1 & -3 \end{array}\right]$$

look quite different, but they are in fact equivalent. You can verify this by observing that they represent the systems of equations

$$\begin{aligned} x + \frac{5}{3}y &= 3 \\ -\frac{1}{3}y &= -1 \end{aligned} \quad \text{and} \quad \begin{aligned} x + 2y &= 4 \\ -y &= -3 \end{aligned}$$

respectively, and both have the same solution: $x = -2$ and $y = 3$. Example 4 also shows that we can sometimes avoid working with fractions by using the appropriate row operation. ■■■

A summary of the Gauss–Jordan method follows.

The Gauss–Jordan Elimination Method

1. Write the augmented matrix corresponding to the linear system.
2. Interchange rows (operation 1), if necessary, to obtain an augmented matrix in which the first entry in the first row is nonzero. Then pivot the matrix about this entry.
3. Interchange the second row with any row below it, if necessary, to obtain an augmented matrix in which the second entry in the second row is nonzero. Pivot the matrix about this entry.
4. Continue until the final matrix is in row-reduced form.

Before writing the augmented matrix, be sure to write all equations with the variables on the left and constant terms on the right of the equals sign. Also, make sure that the variables are in the same order in all equations.

EXAMPLE 5 Solve the system of linear equations given by

$$\begin{aligned} 3x - 2y + 8z &= 9 \\ -2x + 2y + z &= 3 \\ x + 2y - 3z &= 8 \end{aligned} \tag{8}$$

SOLUTION ✔ Using the Gauss–Jordan elimination method, we obtain the following sequence of equivalent augmented matrices:

$$\left[\begin{array}{rrr|r} \textcircled{3} & -2 & 8 & 9 \\ -2 & 2 & 1 & 3 \\ 1 & 2 & -3 & 8 \end{array}\right] \xrightarrow{R_1 + R_2} \left[\begin{array}{rrr|r} 1 & 0 & 9 & 12 \\ -2 & 2 & 1 & 3 \\ 1 & 2 & -3 & 8 \end{array}\right]$$

$$\xrightarrow[R_3 - R_1]{R_2 + 2R_1} \left[\begin{array}{rrr|r} 1 & 0 & 9 & 12 \\ 0 & 2 & 19 & 27 \\ 0 & 2 & -12 & -4 \end{array}\right]$$

$$\xrightarrow{R_2 \leftrightarrow R_3} \left[\begin{array}{rrr|r} 1 & 0 & 9 & 12 \\ 0 & \textcircled{2} & -12 & -4 \\ 0 & 2 & 19 & 27 \end{array}\right]$$

$$\xrightarrow{\frac{1}{2}R_2} \left[\begin{array}{rrr|r} 1 & 0 & 9 & 12 \\ 0 & 1 & -6 & -2 \\ 0 & 2 & 19 & 27 \end{array}\right]$$

$$\xrightarrow{R_3 - 2R_2} \left[\begin{array}{rrr|r} 1 & 0 & 9 & 12 \\ 0 & 1 & -6 & -2 \\ 0 & 0 & \textcircled{31} & 31 \end{array}\right]$$

$$\xrightarrow{\frac{1}{31}R_3} \left[\begin{array}{rrr|r} 1 & 0 & 9 & 12 \\ 0 & 1 & -6 & -2 \\ 0 & 0 & 1 & 1 \end{array}\right]$$

$$\xrightarrow[R_2 + 6R_3]{R_1 - 9R_3} \left[\begin{array}{rrr|r} 1 & 0 & 0 & 3 \\ 0 & 1 & 0 & 4 \\ 0 & 0 & 1 & 1 \end{array}\right]$$

The solution to System (8) is given by $x = 3$, $y = 4$, and $z = 1$ and may be verified by substitution into System (8) as follows:

$$\begin{aligned} 3(3) - 2(4) + 8(1) &= 9 \quad (\surd) \\ -2(3) + 2(4) + 1 \quad &= 3 \quad (\surd) \\ 3 + 2(4) - 3(1) &= 8 \quad (\surd) \end{aligned}$$

■■■■

When searching for an element to serve as a pivot, it is important to keep in mind that you may work only with the row containing the potential pivot or any row *below* it. To see what can go wrong if this caution is not heeded, consider the following augmented matrix for some linear system:

$$\left[\begin{array}{ccc|c} 1 & 1 & 2 & 3 \\ 0 & 0 & 3 & 1 \\ 0 & 2 & 1 & -2 \end{array}\right]$$

Observe that column 1 is in unit form. The next step in the Gauss–Jordan elimination procedure calls for obtaining a nonzero element in the second position of row 2. If you use row 1 (which is *above* the row under consideration) to help you obtain the pivot, you might proceed as follows:

$$\left[\begin{array}{ccc|c} 1 & 1 & 2 & 3 \\ 0 & 0 & 3 & 1 \\ 0 & 2 & 1 & -2 \end{array}\right] \xrightarrow{R_2 \leftrightarrow R_1} \left[\begin{array}{ccc|c} 0 & 0 & 3 & 1 \\ 1 & 1 & 2 & 3 \\ 0 & 2 & 1 & -2 \end{array}\right]$$

As you can see, not only have we obtained a nonzero element to serve as the next pivot, but it is already a 1, thus obviating the next step. This seems like a good move. But beware, we have undone some of our earlier work: Column 1 is no longer in the unit form where a 1 appears first. The correct move in this case is to interchange row 2 with row 3.

Group Discussion

1. Can the phrase "a nonzero constant multiple of itself" in a type-2 row operation be replaced by "any constant multiple of itself"? Explain.

2. Can a row of an augmented matrix be replaced by one obtained by adding a constant to every element in that row without changing the solution of the system of linear equations? Explain.

The next example illustrates how to handle a situation in which the entry in row 1 of the augmented matrix is zero.

EXAMPLE 6 Solve the system of linear equations given by

$$\begin{aligned} 2y + 3z &= 7 \\ 3x + 6y - 12z &= -3 \\ 5x - 2y + 2z &= -7 \end{aligned}$$

SOLUTION ✔ Using the Gauss–Jordan elimination method, we obtain the following sequence of equivalent augmented matrices:

$$\left[\begin{array}{ccc|c} 0 & 2 & 3 & 7 \\ 3 & 6 & -12 & -3 \\ 5 & -2 & 2 & -7 \end{array}\right] \xrightarrow{R_1 \leftrightarrow R_2} \left[\begin{array}{ccc|c} \textcircled{3} & 6 & -12 & -3 \\ 0 & 2 & 3 & 7 \\ 5 & -2 & 2 & -7 \end{array}\right] \xrightarrow{\frac{1}{3}R_1} \left[\begin{array}{ccc|c} 1 & 2 & -4 & -1 \\ 0 & 2 & 3 & 7 \\ 5 & -2 & 2 & -7 \end{array}\right] \xrightarrow{R_3 - 5R_1}$$

$$\left[\begin{array}{ccc|c} 1 & 2 & -4 & -1 \\ 0 & \textcircled{2} & 3 & 7 \\ 0 & -12 & 22 & -2 \end{array}\right] \xrightarrow{\frac{1}{2}R_2} \left[\begin{array}{ccc|c} 1 & 2 & -4 & -1 \\ 0 & 1 & \frac{3}{2} & \frac{7}{2} \\ 0 & -12 & 22 & -2 \end{array}\right] \xrightarrow[R_3 + 12R_2]{R_1 - 2R_2} \left[\begin{array}{ccc|c} 1 & 0 & -7 & -8 \\ 0 & 1 & \frac{3}{2} & \frac{7}{2} \\ 0 & 0 & \textcircled{40} & 40 \end{array}\right] \xrightarrow{\frac{1}{40}R_3}$$

$$\left[\begin{array}{ccc|c} 1 & 0 & -7 & -8 \\ 0 & 1 & \frac{3}{2} & \frac{7}{2} \\ 0 & 0 & 1 & 1 \end{array}\right] \xrightarrow[R_2 - \frac{3}{2}R_3]{R_1 + 7R_3} \left[\begin{array}{ccc|c} 1 & 0 & 0 & -1 \\ 0 & 1 & 0 & 2 \\ 0 & 0 & 1 & 1 \end{array}\right]$$

The solution to the system is given by $x = -1$, $y = 2$, and $z = 1$ and may be verified by substitution into the system. ■■■■

APPLICATION

EXAMPLE 7 Complete the solution to Example 1 in Section 2.1, page 81.

SOLUTION ✓ To complete the solution of the problem posed in Example 1, recall that the mathematical formulation of the problem led to the following system of linear equations:

$$\begin{aligned} 2x + y + z &= 180 \\ x + 3y + 2z &= 300 \\ 2x + y + 2z &= 240 \end{aligned}$$

where x, y, and z denote the respective numbers of type-A, type-B, and type-C souvenirs to be made.

Solving the foregoing system of linear equations by the Gauss–Jordan elimination method, we obtain the following sequence of equivalent augmented matrices:

$$\left[\begin{array}{ccc|c} \textcircled{2} & 1 & 1 & 180 \\ 1 & 3 & 2 & 300 \\ 2 & 1 & 2 & 240 \end{array}\right] \xrightarrow{R_1 \leftrightarrow R_2} \left[\begin{array}{ccc|c} 1 & 3 & 2 & 300 \\ 2 & 1 & 1 & 180 \\ 2 & 1 & 2 & 240 \end{array}\right]$$

$$\xrightarrow[R_3 - 2R_1]{R_2 - 2R_1} \left[\begin{array}{ccc|c} 1 & 3 & 2 & 300 \\ 0 & \textcircled{-5} & -3 & -420 \\ 0 & -5 & -2 & -360 \end{array}\right]$$

$$\xrightarrow{-\frac{1}{5}R_2} \left[\begin{array}{ccc|c} 1 & 3 & 2 & 300 \\ 0 & 1 & \frac{3}{5} & 84 \\ 0 & -5 & -2 & -360 \end{array}\right]$$

$$\xrightarrow[R_3 + 5R_2]{R_1 - 3R_2} \left[\begin{array}{ccc|c} 1 & 0 & \frac{1}{5} & 48 \\ 0 & 1 & \frac{3}{5} & 84 \\ 0 & 0 & \textcircled{1} & 60 \end{array}\right]$$

$$\xrightarrow[R_2 - \frac{3}{5}R_3]{R_1 - \frac{1}{5}R_3} \left[\begin{array}{ccc|c} 1 & 0 & 0 & 36 \\ 0 & 1 & 0 & 48 \\ 0 & 0 & 1 & 60 \end{array}\right]$$

Thus, $x = 36$, $y = 48$, and $z = 60$; that is, Ace Novelty should make 36 type-A souvenirs, 48 type-B souvenirs, and 60 type-C souvenirs in order to use all available machine time. ■■■■

Self-Check Exercises 2.2

1. Solve the system of linear equations

$$\begin{aligned} 2x + 3y + z &= 6 \\ x - 2y + 3z &= -3 \\ 3x + 2y - 4z &= 12 \end{aligned}$$

using the Gauss–Jordan elimination method.

2. A farmer has 200 acres of land suitable for cultivating crops A, B, and C. The cost per acre of cultivating crop A, crop B, and crop C is \$40, \$60, and \$80, respectively. The farmer has \$12,600 available for land cultivation. Each acre of crop A requires 20 hr of labor, each acre of crop B requires 25 hr of labor, and each acre of crop C requires 40 hr of labor. The farmer has a maximum of 5950 hr of labor available. If he wishes to use all of his cultivatable land, the entire budget, and all the labor available, how many acres of each crop should he plant?

Solutions to Self-Check Exercises 2.2 can be found on page 105.

2.2 Exercises

In Exercises 1–4, write the augmented matrix corresponding to the given system of equations.

1. $\begin{aligned} 2x - 3y &= 7 \\ 3x + y &= 4 \end{aligned}$

2. $\begin{aligned} 3x + 7y - 8z &= 5 \\ x \qquad + 3z &= -2 \\ 4x - 3y \qquad &= 7 \end{aligned}$

3. $\begin{aligned} -y + 2z &= 6 \\ 2x + 2y - 8z &= 7 \\ 3y + 4z &= 0 \end{aligned}$

4. $\begin{aligned} 3x_1 + 2x_2 \qquad &= 0 \\ x_1 - x_2 + 2x_3 &= 4 \\ 2x_2 - 3x_3 &= 5 \end{aligned}$

In Exercises 5–8, write the system of equations corresponding to the given augmented matrix.

5. $\left[\begin{array}{cc|c} 3 & 2 & -4 \\ 1 & -1 & 5 \end{array}\right]$

6. $\left[\begin{array}{ccc|c} 0 & 3 & 2 & 4 \\ 1 & -1 & -2 & -3 \\ 4 & 0 & 3 & 2 \end{array}\right]$

7. $\left[\begin{array}{ccc|c} 1 & 3 & 2 & 4 \\ 2 & 0 & 0 & 5 \\ 3 & -3 & 2 & 6 \end{array}\right]$

8. $\left[\begin{array}{ccc|c} 2 & 3 & 1 & 6 \\ 4 & 3 & 2 & 5 \\ 0 & 0 & 0 & 0 \end{array}\right]$

In Exercises 9–18, indicate whether the matrix is in row-reduced form.

9. $\left[\begin{array}{cc|c} 1 & 0 & 3 \\ 0 & 1 & -2 \end{array}\right]$

10. $\left[\begin{array}{cc|c} 1 & 1 & 3 \\ 0 & 0 & 0 \end{array}\right]$

11. $\left[\begin{array}{cc|c} 0 & 1 & 3 \\ 1 & 0 & 5 \end{array}\right]$

12. $\left[\begin{array}{cc|c} 0 & 1 & 3 \\ 0 & 0 & 5 \end{array}\right]$

13. $\left[\begin{array}{ccc|c} 1 & 0 & 0 & 3 \\ 0 & 1 & 0 & 4 \\ 0 & 0 & 1 & 5 \end{array}\right]$

14. $\left[\begin{array}{ccc|c} 1 & 0 & 0 & -1 \\ 0 & 1 & 0 & -2 \\ 0 & 0 & 2 & -3 \end{array}\right]$

15. $\left[\begin{array}{ccc|c} 1 & 0 & 1 & 3 \\ 0 & 1 & 0 & 4 \\ 0 & 0 & -1 & 6 \end{array}\right]$

16. $\left[\begin{array}{cc|c} 1 & 0 & -10 \\ 0 & 1 & 2 \\ 0 & 0 & 0 \end{array}\right]$

17. $\left[\begin{array}{ccc|c} 0 & 0 & 0 & 0 \\ 0 & 1 & 2 & 4 \\ 0 & 0 & 0 & 0 \end{array}\right]$

18. $\left[\begin{array}{ccc|c} 1 & 0 & 0 & 3 \\ 0 & 1 & 0 & 6 \\ 0 & 0 & 0 & 4 \\ 0 & 0 & 1 & 5 \end{array}\right]$

In Exercises 19–26, pivot the given system about the circled element.

19. $\left[\begin{array}{cc|c} \textcircled{2} & 4 & 8 \\ 3 & 1 & 2 \end{array}\right]$

20. $\left[\begin{array}{cc|c} 3 & 2 & 6 \\ \textcircled{4} & 2 & 5 \end{array}\right]$

21. $\left[\begin{array}{cc|c} \textcircled{-1} & 2 & 3 \\ 6 & 4 & 2 \end{array}\right]$

22. $\left[\begin{array}{cc|c} \textcircled{1} & 3 & 4 \\ 2 & 4 & 6 \end{array}\right]$

23. $\left[\begin{array}{ccc|c} \textcircled{2} & 4 & 6 & 12 \\ 2 & 3 & 1 & 5 \\ 3 & -1 & 2 & 4 \end{array}\right]$

24. $\left[\begin{array}{ccc|c} 1 & 3 & 2 & 4 \\ \textcircled{2} & 4 & 8 & 6 \\ -1 & 2 & 3 & 4 \end{array}\right]$

25. $\left[\begin{array}{ccc|c} 0 & 1 & 3 & 4 \\ 2 & 4 & \textcircled{1} & 3 \\ 5 & 6 & 2 & -4 \end{array}\right]$

26. $\left[\begin{array}{ccc|c} 1 & 2 & 3 & 5 \\ 0 & \textcircled{-3} & 3 & 2 \\ 0 & 4 & -1 & 3 \end{array}\right]$

In Exercises 27–30, fill in the missing entries by performing the indicated row operations to obtain the row-reduced matrices.

27. $\left[\begin{array}{cc|c} 3 & 9 & 6 \\ 2 & 1 & 4 \end{array}\right] \xrightarrow{\frac{1}{3}R_1} \left[\begin{array}{cc|c} \cdot & \cdot & \cdot \\ 2 & 1 & 4 \end{array}\right] \xrightarrow{R_2 - 2R_1}$

$\left[\begin{array}{cc|c} 1 & 3 & 2 \\ \cdot & \cdot & \cdot \end{array}\right] \xrightarrow{-\frac{1}{5}R_2} \left[\begin{array}{cc|c} 1 & 3 & 2 \\ \cdot & \cdot & \cdot \end{array}\right] \xrightarrow{R_1 - 3R_2}$

$\left[\begin{array}{cc|c} 1 & 0 & 2 \\ 0 & 1 & 0 \end{array}\right]$

28. $\left[\begin{array}{cc|c} 1 & 2 & 1 \\ 2 & 3 & -1 \end{array}\right] \xrightarrow{R_2 - 2R_1} \left[\begin{array}{cc|c} 1 & 2 & 1 \\ \cdot & \cdot & \cdot \end{array}\right] \xrightarrow{-R_2}$

$\left[\begin{array}{cc|c} 1 & 2 & 1 \\ \cdot & \cdot & \cdot \end{array}\right] \xrightarrow{R_1 - 2R_2} \left[\begin{array}{cc|c} 1 & 0 & -5 \\ 0 & 1 & 3 \end{array}\right]$

29. $\left[\begin{array}{ccc|c} 1 & 3 & 1 & 3 \\ 3 & 8 & 3 & 7 \\ 2 & -3 & 1 & -10 \end{array}\right] \xrightarrow[R_3 - 2R_1]{R_2 - 3R_1}$

$\left[\begin{array}{ccc|c} 1 & 3 & 1 & 3 \\ \cdot & \cdot & \cdot & \cdot \\ \cdot & \cdot & \cdot & \cdot \end{array}\right] \xrightarrow{-R_2}$

$\left[\begin{array}{ccc|c} 1 & 3 & 1 & 3 \\ \cdot & \cdot & \cdot & \cdot \\ 0 & -9 & -1 & -16 \end{array}\right] \xrightarrow[R_3 + 9R_2]{R_1 - 3R_2}$

$\left[\begin{array}{ccc|c} \cdot & \cdot & \cdot & \cdot \\ 0 & 1 & 0 & 2 \\ \cdot & \cdot & \cdot & \cdot \end{array}\right] \xrightarrow[-R_3]{R_1 + R_3} \left[\begin{array}{ccc|c} 1 & 0 & 0 & -1 \\ 0 & 1 & 0 & 2 \\ 0 & 0 & 1 & -2 \end{array}\right]$

30. $\left[\begin{array}{ccc|c} 0 & 1 & 3 & -4 \\ 1 & 2 & 1 & 7 \\ 1 & -2 & 0 & 1 \end{array}\right] \xrightarrow{R_1 \leftrightarrow R_2}$

$\left[\begin{array}{ccc|c} \cdot & \cdot & \cdot & \cdot \\ \cdot & \cdot & \cdot & \cdot \\ 1 & -2 & 0 & 1 \end{array}\right] \xrightarrow{R_3 - R_1} \left[\begin{array}{ccc|c} 1 & 2 & 1 & 7 \\ 0 & 1 & 3 & -4 \\ \cdot & \cdot & \cdot & \cdot \end{array}\right]$

$\xrightarrow[R_3 + 4R_2]{R_1 + \frac{1}{2}R_2} \left[\begin{array}{ccc|c} \cdot & \cdot & \cdot & \cdot \\ 0 & 1 & 3 & -4 \\ \cdot & \cdot & \cdot & \cdot \end{array}\right] \xrightarrow{\frac{1}{11}R_3}$

$\left[\begin{array}{ccc|c} 1 & 0 & \frac{1}{2} & 4 \\ 0 & 1 & 3 & -4 \\ \cdot & \cdot & \cdot & \cdot \end{array}\right] \xrightarrow[R_2 - 3R_3]{R_1 - \frac{1}{2}R_3} \left[\begin{array}{ccc|c} 1 & 0 & 0 & 5 \\ 0 & 1 & 0 & 2 \\ 0 & 0 & 1 & -2 \end{array}\right]$

31. Write a system of linear equations for the augmented matrix of Exercise 27. Using the results of Exercise 27, determine the solution of the system.

32. Repeat Exercise 31 for the augmented matrix of Exercise 28.

33. Repeat Exercise 31 for the augmented matrix of Exercise 29.

34. Repeat Exercise 31 for the augmented matrix of Exercise 30.

In Exercises 35–50, solve the system of linear equations using the Gauss–Jordan elimination method.

35. $\begin{aligned} x - 2y &= 8 \\ 3x + 4y &= 4 \end{aligned}$

36. $\begin{aligned} 3x + y &= 1 \\ -7x - 2y &= -1 \end{aligned}$

37. $\begin{aligned} 2x - 3y &= -8 \\ 4x + y &= -2 \end{aligned}$

38. $\begin{aligned} 5x + 3y &= 9 \\ -2x + y &= -8 \end{aligned}$

39. $\begin{aligned} x + y + z &= 0 \\ 2x - y + z &= 1 \\ x + y - 2z &= 2 \end{aligned}$

40. $\begin{aligned} 2x + y - 2z &= 4 \\ x + 3y - z &= -3 \\ 3x + 4y - z &= 7 \end{aligned}$

41. $\begin{aligned} 2x + 2y + z &= 9 \\ x + z &= 4 \\ 4y - 3z &= 17 \end{aligned}$

42. $\begin{aligned} 2x + 3y - 2z &= 10 \\ 3x - 2y + 2z &= 0 \\ 4x - y + 3z &= -1 \end{aligned}$

43. $\begin{aligned} -x_2 + x_3 &= 2 \\ 4x_1 - 3x_2 + 2x_3 &= 16 \\ 3x_1 + 2x_2 + x_3 &= 11 \end{aligned}$

44. $\begin{aligned} 2x + 4y - 6z &= 38 \\ x + 2y + 3z &= 7 \\ 3x - 4y + 4z &= -19 \end{aligned}$

45. $\begin{aligned} x_1 - 2x_2 + x_3 &= 6 \\ 2x_1 + x_2 - 3x_3 &= -3 \\ x_1 - 3x_2 + 3x_3 &= 10 \end{aligned}$

46. $\begin{aligned} 2x + 3y - 6z &= -11 \\ x - 2y + 3z &= 9 \\ 3x + y &= 7 \end{aligned}$

47. $\begin{aligned} 2x + 3z &= -1 \\ 3x - 2y + z &= 9 \\ x + y + 4z &= 4 \end{aligned}$

48. $\begin{aligned} 2x_1 - x_2 + 3x_3 &= -4 \\ x_1 - 2x_2 + x_3 &= -1 \\ x_1 - 5x_2 + 2x_3 &= -3 \end{aligned}$

49. $\begin{aligned} x_1 - x_2 + 3x_3 &= 14 \\ x_1 + x_2 + x_3 &= 6 \\ -2x_1 - x_2 + x_3 &= -4 \end{aligned}$

50. $\begin{aligned} 2x_1 - x_2 - x_3 &= 0 \\ 3x_1 + 2x_2 + x_3 &= 7 \\ x_1 + 2x_2 + 2x_3 &= 5 \end{aligned}$

The problems in Exercises 51–61 correspond to those in Exercises 15–25, Section 2.1. Use the results of your previous work to help you solve these problems.

51. AGRICULTURE The Johnson Farm has 500 acres of land allotted for cultivating corn and wheat. The cost of cultivating corn and wheat (including seeds and labor) is $42 and $30/acre, respectively. Jacob Johnson has $18,600 available for cultivating these crops. If he wishes to use all the allotted land and his entire budget for cultivating these two crops, how many acres of each crop should he plant?

52. INVESTMENTS Michael Perez has a total of $2000 on deposit with two savings institutions. One pays interest at the rate of 6%/year, whereas the other pays interest at the rate of 8%/year. If Michael earned a total of $144 in interest during a single year, how much does he have on deposit in each institution?

53. MIXTURES The Coffee Shoppe sells a coffee blend made from two coffees, one costing $2.50/lb and the other costing $3.00/lb. If the blended coffee sells for $2.80/lb, find how much of each coffee is used to obtain the desired blend. (Assume the weight of the blended coffee is 100 lb.)

54. INVESTMENTS Kelly Fisher has a total of $30,000 invested in two municipal bonds that have yields of 8% and 10% interest per year, respectively. If the interest Kelly receives from the bonds in a year is $2640, how much does she have invested in each bond?

55. RIDERSHIP The total number of passengers riding a certain city bus during the morning shift is 1000. If the child's fare is 25 cents, the adult fare is 75 cents, and the total revenue from the fares in the morning shift is $650, how many children and how many adults rode the bus during the morning shift?

56. REAL ESTATE Cantwell Associates, a real estate developer, is planning to build a new apartment complex consisting of one-bedroom units and two- and three-bedroom townhouses. A total of 192 units is planned, and the number of family units (two- and three-bedroom townhouses) will equal the number of one-bedroom units. If the number of one-bedroom units will be three times the number of three-bedroom units, find how many units of each type will be in the complex.

57. INVESTMENT PLANNING The annual interest on Sid Carrington's three investments amounted to $21,600: 6% on a savings account, 8% on mutual funds, and 12% on bonds. If the amount of Sid's investment in bonds was twice the amount of his investment in the savings account, and the interest earned from his investment in bonds was equal to the dividends he received from his investment in mutual funds, find how much money he placed in each type of investment.

58. BOX-OFFICE RECEIPTS A theater has a seating capacity of 900 and charges $2 for children, $3 for students, and $4 for adults. At a certain screening with full attendance there were half as many adults as children and students combined. The receipts totaled $2800. How many children attended the show?

59. MANAGEMENT DECISIONS The management of Hartman Rent-A-Car has allocated $1.25 million to buy a fleet of new automobiles consisting of compact, intermediate, and full-size cars. Compacts cost $10,000 each, intermediate-size cars cost $15,000 each, and full-size cars cost $20,000 each. If Hartman purchases twice as many compacts as intermediate-size cars and the total number of cars to be purchased is 100, determine how many cars of each type will be purchased. (Assume that the entire budget will be used.)

60. **INVESTMENT CLUBS** The management of a private investment club has a fund of $200,000 earmarked for investment in stocks. To arrive at an acceptable overall level of risk, the stocks that management is considering have been classified into three categories: high-risk, medium-risk, and low-risk. Management estimates that high-risk stocks will have a rate of return of 15%/year; medium-risk stocks, 10%/year; and low-risk stocks, 6%/year. The investment in low-risk stocks is to be twice the sum of the investments in stocks of the other two categories. If the investment goal is to have an average rate of return of 9%/year on the total investment, determine how much the club should invest in each type of stock.

61. **DIET PLANNING** A dietitian wishes to plan a meal around three foods. The percentage of the daily requirements of proteins, carbohydrates, and iron contained in each ounce of the three foods is summarized in the accompanying table.

	Food I	Food II	Food III
Percentage of Proteins	10	6	8
Percentage of Carbohydrates	10	12	6
Percentage of Iron	5	4	12

Determine how many ounces of each food the dietitian should include in the meal to meet exactly the daily requirement of proteins, carbohydrates, and iron (100% of each).

62. **INVESTMENTS** Mr. and Mrs. Garcia have a total of $100,000 to be invested in stocks, bonds, and a money-market account. The stocks have a rate of return of 12%/year, while the bonds and the money-market account pay 8% and 4%/year, respectively. They have stipulated that the amount invested in the money-market account should be equal to the sum of 20% of the amount invested in stocks and 10% of the amount invested in bonds. How should the Garcias allocate their resources if they require an annual income of $10,000 from their investments?

63. **BOX-OFFICE RECEIPTS** For the opening night at the Opera House, a total of 1000 tickets were sold. Front orchestra seats cost $80 apiece, rear orchestra seats cost $60 apiece, and front balcony seats cost $50 apiece. The combined number of tickets sold for the front orchestra and rear orchestra exceeded twice the number of front balcony tickets sold by 400. The total receipts for the performance were $62,800. Determine how many tickets of each type were sold.

64. **PRODUCTION SCHEDULING** A manufacturer of women's blouses makes three types of blouses: sleeveless, short-sleeve, and long-sleeve. The time (in minutes) required by each department to produce a dozen blouses of each type is shown in the accompanying table.

Department	Sleeveless	Short-Sleeve	Long-Sleeve
Cutting	9	12	15
Sewing	22	24	28
Packaging	6	8	8

The cutting, sewing, and packaging departments have available a maximum of 80, 160, and 48 labor-hours, respectively, per day. How many dozens of each type of blouse can be produced each day if the plant is operated at full capacity?

In Exercises 65 and 66, determine whether the statement is true or false. If it is true, explain why it is true. If it is false, give an example to show why it is false.

65. An equivalent system of linear equations can be obtained from a system of equations by replacing one of its equations by any constant multiple of itself.

66. If the augmented matrix corresponding to a system of three linear equations in three variables has a row of the form $[0 \quad 0 \quad 0 \mid a]$, where a is a nonzero number, then the system has no solution.

SOLVING SYSTEMS OF LINEAR EQUATIONS I

SOLVING A SYSTEM OF LINEAR EQUATIONS USING THE GAUSS–JORDAN METHOD

The three matrix operations can be performed on a matrix using a graphing calculator. The commands are summarized below:

Operation	Calculator Function (TI-83)	(TI-86)	
$R_i \leftrightarrow R_j$	**rowSwap**([A], i, j)	**rSwap**(A, i, j)	or equivalent
cR_i	***row**(c, [A], i)	**multR**(c, A, i)	or equivalent
$R_i + aR_j$	***row+**(a, [A], j, i)	**mRAdd**(a, A, j, i)	or equivalent

When a row operation is performed on a matrix, the result is stored as an answer in the calculator. If another operation is performed on this matrix, then the matrix is erased. Should a mistake be made in the operation, then the previous matrix is lost. For this reason, you should store the results of each operation. We do this by pressing STO, followed by the name of a matrix, and then ENTER. We use this process in the following example.

EXAMPLE 1 Use a graphing calculator to solve the following system of linear equations by the Gauss–Jordan method (see Example 5 in Section 2.2):

$$\begin{aligned} 3x - 2y + 8z &= 9 \\ -2x + 2y + z &= 3 \\ x + 2y - 3z &= 8 \end{aligned}$$

SOLUTION ✓ Using the Gauss–Jordan method, we obtain the following sequence of equivalent matrices.

$$\left[\begin{array}{rrr|r} 3 & -2 & 8 & 9 \\ -2 & 2 & 1 & 3 \\ 1 & 2 & -3 & 8 \end{array}\right] \xrightarrow{\text{*row+}(1,\ [A],\ 2,\ 1)\ \blacktriangleright\ B}$$

$$\left[\begin{array}{rrr|r} 1 & 0 & 9 & 12 \\ -2 & 2 & 1 & 3 \\ 1 & 2 & -3 & 8 \end{array}\right] \xrightarrow{\text{*row+}(2,\ [B],\ 1,\ 2)\ \blacktriangleright\ C}$$

$$\left[\begin{array}{rrr|r} 1 & 0 & 9 & 12 \\ 0 & 2 & 19 & 27 \\ 1 & 2 & -3 & 8 \end{array}\right] \xrightarrow{\text{*row+}(-1,\ [C],\ 1,\ 3)\ \blacktriangleright\ B}$$

$$\left[\begin{array}{ccc|c} 1 & 0 & 9 & 12 \\ 0 & 2 & 19 & 27 \\ 0 & 2 & -12 & -4 \end{array}\right] \xrightarrow{\text{*row}(\frac{1}{2},\ [B],\ 2) \blacktriangleright C}$$

$$\left[\begin{array}{ccc|c} 1 & 0 & 9 & 12 \\ 0 & 1 & 9.5 & 13.5 \\ 0 & 2 & -12 & -4 \end{array}\right] \xrightarrow{\text{*row+}(-2,\ [C],\ 2,\ 3) \blacktriangleright B}$$

$$\left[\begin{array}{ccc|c} 1 & 0 & 9 & 12 \\ 0 & 1 & 9.5 & 13.5 \\ 0 & 0 & -31 & -31 \end{array}\right] \xrightarrow{\text{*row}(-\frac{1}{31},\ [B],\ 3) \blacktriangleright C}$$

$$\left[\begin{array}{ccc|c} 1 & 0 & 9 & 12 \\ 0 & 1 & 9.5 & 13.5 \\ 0 & 0 & 1 & 1 \end{array}\right] \xrightarrow{\text{*row+}(-9,\ [C],\ 3,\ 1) \blacktriangleright B}$$

$$\left[\begin{array}{ccc|c} 1 & 0 & 0 & 3 \\ 0 & 1 & 9.5 & 13.5 \\ 0 & 0 & 1 & 1 \end{array}\right] \xrightarrow{\text{*row+}(-9.5,\ [B],\ 3,\ 2) \blacktriangleright C} \left[\begin{array}{ccc|c} 1 & 0 & 0 & 3 \\ 0 & 1 & 0 & 4 \\ 0 & 0 & 1 & 1 \end{array}\right]$$

The last matrix is in row-reduced form, and we see that the solution of the system is $x = 3$, $y = 4$, and $z = 1$. ■■■■

USING rref (TI-83 and TI-86) TO SOLVE A SYSTEM OF LINEAR EQUATIONS

The operation **rref** (or equivalent function in your calculator, if there is one) will transform an augmented matrix into one that is in row-reduced form. For example, using **rref,** we find

$$\left[\begin{array}{ccc|c} 3 & -2 & 8 & 9 \\ -2 & 2 & 1 & 3 \\ 1 & 2 & -3 & 8 \end{array}\right] \xrightarrow{\text{rref}} \left[\begin{array}{ccc|c} 1 & 0 & 0 & 3 \\ 0 & 1 & 0 & 4 \\ 0 & 0 & 1 & 1 \end{array}\right]$$

as obtained earlier!

USING SIMULT (TI-86) TO SOLVE A SYSTEM OF EQUATIONS

The operation SIMULT (or equivalent operation on your calculator, if there is one) of a graphing utility can be used to solve a system of n linear equations in n variables, where n is an integer between 2 and 30.

EXAMPLE 2 Use the SIMULT operation to solve the system of Example 1.

SOLUTION ✔ Call for the SIMULT operation. Since the system under consideration has three equations in three variables, enter $n = 3$. Next, enter a1, 1 = 3, a1, 2 = −2, a1, 3 = 8, . . . , b1 = 9, a2, 1 = −2, . . . , b3 = 8. Select ⟨SOLVE⟩ and the display

x1 = 3

x2 = 4

x3 = 1

appears on the screen giving $x = 3$, $y = 4$, and $z = 1$ as the required solution.

Exercises

In Exercises 1–6, use a graphing utility to solve the system of equations (a) by the Gauss-Jordan method, (b) using the rref operation, and (c) using SIMULT.

1.
$$\begin{aligned} x_1 - 2x_2 + 2x_3 - 3x_4 &= -7 \\ 3x_1 + 2x_2 - x_3 + 5x_4 &= 22 \\ 2x_1 - 3x_2 + 4x_3 - x_4 &= -3 \\ 3x_1 - 2x_2 - x_3 + 2x_4 &= 12 \end{aligned}$$

2.
$$\begin{aligned} 2x_1 - x_2 + 3x_3 - 2x_4 &= -2 \\ x_1 - 2x_2 + x_3 - 3x_4 &= 2 \\ x_1 - 5x_2 + 2x_3 + 3x_4 &= -6 \\ -3x_1 + 3x_2 - 4x_3 - 4x_4 &= 9 \end{aligned}$$

3.
$$\begin{aligned} 2x_1 + x_2 + 3x_3 - x_4 &= 9 \\ -x_1 - 2x_2 \qquad\quad - 3x_4 &= -1 \\ x_1 \qquad\quad - 3x_3 + x_4 &= 10 \\ x_1 - x_2 - x_3 - x_4 &= 8 \end{aligned}$$

4.
$$\begin{aligned} x_1 - 2x_2 - 2x_3 + x_4 &= 1 \\ 2x_1 - x_2 + 2x_3 + 3x_4 &= -2 \\ -x_1 - 5x_2 + 7x_3 - 2x_4 &= 3 \\ 3x_1 - 4x_2 + 3x_3 + 4x_4 &= -4 \end{aligned}$$

5.
$$\begin{aligned} 2x_1 - 2x_2 + 3x_3 - x_4 + 2x_5 &= 16 \\ 3x_1 + x_2 - 2x_3 + x_4 - 3x_5 &= -11 \\ x_1 + 3x_2 - 4x_3 + 3x_4 - x_5 &= -13 \\ 2x_1 - x_2 + 3x_3 - 2x_4 + 2x_5 &= 15 \\ 3x_1 + 4x_2 - 3x_3 + 5x_4 - x_5 &= -10 \end{aligned}$$

6.
$$\begin{aligned} 2.1x_1 - 3.2x_2 + 6.4x_3 + 7x_4 - 3.2x_5 &= 54.3 \\ 4.1x_1 + 2.2x_2 - 3.1x_3 - 4.2x_4 + 3.3x_5 &= -20.81 \\ 3.4x_1 - 6.2x_2 + 4.7x_3 + 2.1x_4 - 5.3x_5 &= 24.7 \\ 4.1x_1 + 7.3x_2 + 5.2x_3 + 6.1x_4 - 8.2x_5 &= 29.25 \\ 2.8x_1 + 5.2x_2 + 3.1x_3 + 5.4x_4 + 3.8x_5 &= 43.72 \end{aligned}$$

SOLUTIONS TO SELF-CHECK EXERCISES 2.2

1. We obtain the following sequence of equivalent augmented matrices:

$$\left[\begin{array}{rrr|r} 2 & 3 & 1 & 6 \\ 1 & -2 & 3 & -3 \\ 3 & 2 & -4 & 12 \end{array}\right] \xrightarrow{R_1 \leftrightarrow R_2} \left[\begin{array}{rrr|r} \textcircled{1} & -2 & 3 & -3 \\ 2 & 3 & 1 & 6 \\ 3 & 2 & -4 & 12 \end{array}\right] \xrightarrow[R_3 - 3R_1]{R_2 - 2R_1}$$

$$\left[\begin{array}{rrr|r} 1 & -2 & 3 & -3 \\ 0 & 7 & -5 & 12 \\ 0 & 8 & -13 & 21 \end{array}\right] \xrightarrow{R_2 \leftrightarrow R_3} \left[\begin{array}{rrr|r} 1 & -2 & 3 & -3 \\ 0 & \textcircled{8} & -13 & 21 \\ 0 & 7 & -5 & 12 \end{array}\right] \xrightarrow{R_2 - R_3}$$

$$\left[\begin{array}{rrr|r} 1 & -2 & 3 & -3 \\ 0 & 1 & -8 & 9 \\ 0 & 7 & -5 & 12 \end{array}\right] \xrightarrow[R_3 - 7R_2]{R_1 + 2R_2} \left[\begin{array}{rrr|r} 1 & 0 & -13 & 15 \\ 0 & 1 & -8 & 9 \\ 0 & 0 & 51 & -51 \end{array}\right] \xrightarrow{\frac{1}{51}R_3}$$

$$\left[\begin{array}{rrr|r} 1 & 0 & -13 & 15 \\ 0 & 1 & -8 & 9 \\ 0 & 0 & \textcircled{1} & -1 \end{array}\right] \xrightarrow[R_2 + 8R_3]{R_1 + 13R_3} \left[\begin{array}{rrr|r} 1 & 0 & 0 & 2 \\ 0 & 1 & 0 & 1 \\ 0 & 0 & 1 & -1 \end{array}\right]$$

The solution to the system is $x = 2$, $y = 1$, and $z = -1$.

2. Referring to the solution of Exercise 2, Self-Check Exercises 2.1, we see that the problem reduces to solving the following system of linear equations:

$$\begin{aligned} x + \quad y + \quad z &= \quad 200 \\ 40x + 60y + 80z &= 12{,}600 \\ 20x + 25y + 40z &= \ \ 5{,}950 \end{aligned}$$

Using the Gauss–Jordan elimination method, we have

$$\left[\begin{array}{rrr|r} 1 & 1 & 1 & 200 \\ 40 & 60 & 80 & 12{,}600 \\ 20 & 25 & 40 & 5{,}950 \end{array}\right] \xrightarrow[R_3 - 20R_1]{R_2 - 40R_1} \left[\begin{array}{rrr|r} 1 & 1 & 1 & 200 \\ 0 & 20 & 40 & 4600 \\ 0 & 5 & 20 & 1950 \end{array}\right] \xrightarrow{\frac{1}{20}R_2}$$

$$\left[\begin{array}{rrr|r} 1 & 1 & 1 & 200 \\ 0 & 1 & 2 & 230 \\ 0 & 5 & 20 & 1950 \end{array}\right] \xrightarrow[R_3 - 5R_2]{R_1 - R_2} \left[\begin{array}{rrr|r} 1 & 0 & -1 & -30 \\ 0 & 1 & 2 & 230 \\ 0 & 0 & 10 & 800 \end{array}\right] \xrightarrow{\frac{1}{10}R_3}$$

$$\left[\begin{array}{rrr|r} 1 & 0 & -1 & -30 \\ 0 & 1 & 2 & 230 \\ 0 & 0 & 1 & 80 \end{array}\right] \xrightarrow[R_2 - 2R_3]{R_1 + R_3} \left[\begin{array}{rrr|r} 1 & 0 & 0 & 50 \\ 0 & 1 & 0 & 70 \\ 0 & 0 & 1 & 80 \end{array}\right]$$

From the last augmented matrix in reduced form, we see that $x = 50$, $y = 70$, and $z = 80$. Therefore, the farmer should plant 50 acres of crop A, 70 acres of crop B, and 80 acres of crop C.

2.3 Solving Systems of Linear Equations II

In this section we continue our study of systems of linear equations. More specifically, we look at systems that have infinitely many solutions and those that have no solution. We also study systems of linear equations in which the number of variables is not equal to the number of equations in the system.

Solution(s) of Linear Equations

Our first example illustrates the situation in which a system of linear equations has infinitely many solutions.

EXAMPLE 1

A System of Equations with an Infinite Number of Solutions

Solve the system of linear equations given by

$$\begin{aligned} x + 2y - 3z &= -2 \\ 3x - y - 2z &= 1 \\ 2x + 3y - 5z &= -3 \end{aligned} \qquad \textbf{(9)}$$

SOLUTION ✓ Using the Gauss–Jordan elimination method, we obtain the following sequence of equivalent augmented matrices:

$$\left[\begin{array}{ccc|c} \textcircled{1} & 2 & -3 & -2 \\ 3 & -1 & -2 & 1 \\ 2 & 3 & -5 & -3 \end{array}\right] \xrightarrow[R_3 - 2R_1]{R_2 - 3R_1} \left[\begin{array}{ccc|c} 1 & 2 & -3 & -2 \\ 0 & \textcircled{-7} & 7 & 7 \\ 0 & -1 & 1 & 1 \end{array}\right] \xrightarrow{-\frac{1}{7}R_2}$$

$$\left[\begin{array}{ccc|c} 1 & 2 & -3 & -2 \\ 0 & 1 & -1 & -1 \\ 0 & -1 & 1 & 1 \end{array}\right] \xrightarrow[R_3 + R_2]{R_1 - 2R_2} \left[\begin{array}{ccc|c} 1 & 0 & -1 & 0 \\ 0 & 1 & -1 & -1 \\ 0 & 0 & 0 & 0 \end{array}\right]$$

The last augmented matrix is in row-reduced form. Interpreting it as a system of linear equations gives

$$\begin{aligned} x - z &= 0 \\ y - z &= -1 \end{aligned}$$

a system of two equations in the three variables x, y, and z.

Let's now single out one variable—say, z—and solve for x and y in terms of it. We obtain

$$\begin{aligned} x &= z \\ y &= z - 1 \end{aligned}$$

If we assign a particular value to z—say, $z = 0$—we obtain $x = 0$ and $y = -1$, giving the solution $(0, -1, 0)$ to System (9). By setting $z = 1$, we obtain the solution (1, 0, 1). In general, if we set $z = t$, where t represents some real number, we obtain a solution given by $(t, t - 1, t)$. Since the parameter t may be any real number, we see that System (9) has infinitely many solutions. Geometrically, the solutions of System (9) lie on the straight line in three-dimensional space given by the intersection of the three planes determined by the three equations in the system. ■■■■

REMARK In Example 1 we chose the parameter to be z because it is more convenient to solve for x and y (both the x- and y-columns are in unit form) in terms of z. ■■■

The next example shows what happens in the elimination procedure when the system does not have a solution.

EXAMPLE 2

A System of Equations That Has No Solution

Solve the system of linear equations given by

$$\begin{aligned} x + y + z &= 1 \\ 3x - y - z &= 4 \\ x + 5y + 5z &= -1 \end{aligned} \qquad (10)$$

SOLUTION ✓

Using the Gauss–Jordan elimination method, we obtain the following sequence of equivalent augmented matrices:

$$\left[\begin{array}{ccc|c} ① & 1 & 1 & 1 \\ 3 & -1 & -1 & 4 \\ 1 & 5 & 5 & -1 \end{array}\right] \xrightarrow[R_3 - R_1]{R_2 - 3R_1} \left[\begin{array}{ccc|c} 1 & 1 & 1 & 1 \\ 0 & -4 & -4 & 1 \\ 0 & 4 & 4 & -2 \end{array}\right]$$

$$\xrightarrow{R_3 + R_2} \left[\begin{array}{ccc|c} 1 & 1 & 1 & 1 \\ 0 & -4 & -4 & 1 \\ 0 & 0 & 0 & -1 \end{array}\right]$$

Observe that row 3 in the last matrix reads $0x + 0y + 0z = -1$—that is, $0 = -1$! We conclude therefore that System (10) is inconsistent and has no solution. Geometrically, we have a situation in which two of the planes intersect in a straight line but the third plane is parallel to this line of intersection of the two planes and does not intersect it. Consequently, there is no point of intersection of the three planes. ■■■■

Example 2 illustrates the following more general result of using the Gauss–Jordan elimination procedure.

Systems with No Solution

If there is a row in the augmented matrix containing all zeros to the left of the vertical line and a nonzero entry to the right of the line, then the system of equations has no solution.

It may have dawned on you that in all the previous examples we have dealt only with systems involving exactly the same number of linear equations as there are variables. However, systems in which the number of equations is different from the number of variables also occur in practice. Indeed, we will consider such systems in Examples 3 and 4.

The following theorem provides us with some preliminary information on a system of linear equations.

THEOREM 1

a. If the number of equations is greater than or equal to the number of variables in a linear system, then one of the following is true:
  i. The system has no solution
  ii. The system has exactly one solution
  iii. The system has infinitely many solutions.

b. If there are fewer equations than variables in a linear system, then the system either has no solution or it has infinitely many solutions.

REMARK Theorem 1 may be used to tell us, before we even begin to solve a problem, what the nature of the solution may be.

Although we will not prove this theorem, you should recall that we have illustrated geometrically part (a) for the case in which there are exactly as many equations (three) as there are variables. To show the validity of part (b), let us once again consider the case in which a system has three variables. Now, if there is only one equation in the system, then it is clear that there are infinitely many solutions corresponding geometrically to all the points lying on the plane represented by the equation.

Next, if there are two equations in the system, then *only* the following possibilities exist:

1. The two planes are parallel and distinct.
2. The two planes intersect in a straight line.
3. The two planes are coincident (the two equations define the same plane) (Figure 2.6).

FIGURE 2.6

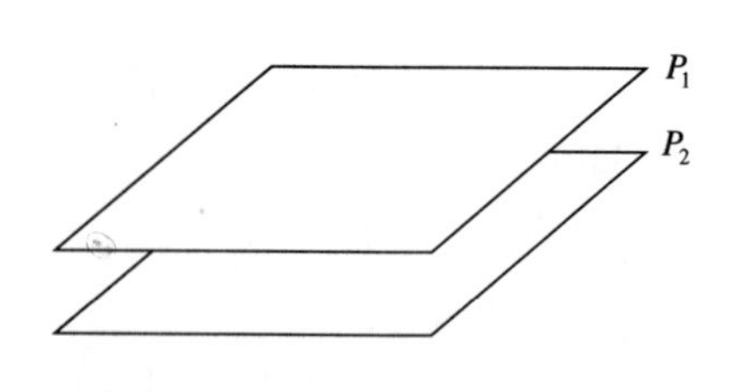

(a) No solution

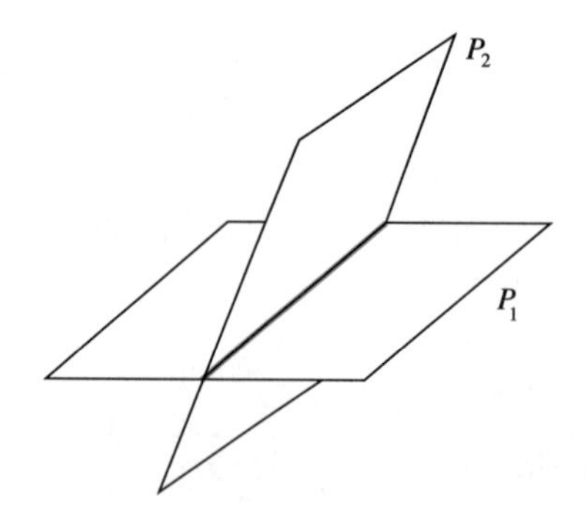

(b) Infinitely many solutions

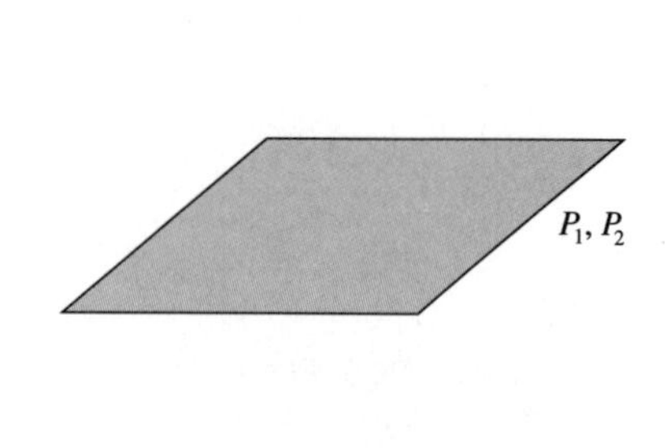

(c) Infinitely many solutions

Group Discussion

Give a geometric interpretation of Theorem 1 for a linear system composed of equations involving two variables. Specifically, illustrate what can happen if there are three linear equations in the system (the case involving two linear equations has already been discussed in Section 2.1). What if there are four linear equations? What if there is only one linear equation in the system?

Thus, either there is no solution or there are infinitely many solutions corresponding to the points lying on a line of intersection of the two planes or on a single plane determined by the two equations. In the case where two planes intersect in a straight line, the solutions will involve one parameter, and in the case where the two planes are coincident, the solutions will involve two parameters.

EXAMPLE 3

A System with More Equations Than Variables

Solve the following system of linear equations:

$$\begin{aligned} x + 2y &= 4 \\ x - 2y &= 0 \\ 4x + 3y &= 12 \end{aligned}$$

SOLUTION ✓

We obtain the following sequence of equivalent augmented matrices:

$$\left[\begin{array}{cc|c} \textcircled{1} & 2 & 4 \\ 1 & -2 & 0 \\ 4 & 3 & 12 \end{array}\right] \xrightarrow[R_3 - 4R_1]{R_2 - R_1} \left[\begin{array}{cc|c} 1 & 2 & 4 \\ 0 & \textcircled{-4} & -4 \\ 0 & -5 & -4 \end{array}\right] \xrightarrow{-\frac{1}{4}R_2}$$

$$\left[\begin{array}{cc|c} 1 & 2 & 4 \\ 0 & 1 & 1 \\ 0 & -5 & -4 \end{array}\right] \xrightarrow[R_3 + 5R_2]{R_1 - 2R_2} \left[\begin{array}{cc|c} 1 & 0 & 2 \\ 0 & 1 & 1 \\ 0 & 0 & 1 \end{array}\right]$$

The last row of the row-reduced augmented matrix implies that $0 = 1$, which is impossible, so we conclude that the given system has no solution. Geometrically, the three lines defined by the three equations in the system do not intersect at a point. (To see this for yourself, draw the graphs of these equations.) ■■■■

EXAMPLE 4

A System with More Variables Than Equations

Solve the following system of linear equations:

$$\begin{aligned} x + 2y - 3z + w &= -2 \\ 3x - y - 2z - 4w &= 1 \\ 2x + 3y - 5z + w &= -3 \end{aligned}$$

SOLUTION ✓ First, observe that the given system consists of three equations in four variables, and so, by Theorem 1(b), either the system has no solution or it has infinitely many solutions. To solve it we use the Gauss–Jordan method and obtain the following sequence of equivalent augmented matrices:

$$\left[\begin{array}{cccc|c} \textcircled{1} & 2 & -3 & 1 & -2 \\ 3 & -1 & -2 & -4 & 1 \\ 2 & 3 & -5 & 1 & -3 \end{array}\right] \xrightarrow[R_3 - 2R_1]{R_2 - 3R_1} \left[\begin{array}{cccc|c} 1 & 2 & -3 & 1 & -2 \\ 0 & \textcircled{-7} & 7 & -7 & 7 \\ 0 & -1 & 1 & -1 & 1 \end{array}\right] \xrightarrow{-\frac{1}{7}R_2}$$

$$\left[\begin{array}{cccc|c} 1 & 2 & -3 & 1 & -2 \\ 0 & 1 & -1 & 1 & -1 \\ 0 & -1 & 1 & -1 & 1 \end{array}\right] \xrightarrow[R_3 + R_2]{R_1 - 2R_2} \left[\begin{array}{cccc|c} 1 & 0 & -1 & -1 & 0 \\ 0 & 1 & -1 & 1 & -1 \\ 0 & 0 & 0 & 0 & 0 \end{array}\right]$$

The last augmented matrix is in row-reduced form. Observe that the given system is equivalent to the system

$$\begin{aligned} x - z - w &= 0 \\ y - z + w &= -1 \end{aligned}$$

of two equations in four variables. Thus, we may solve for two of the variables in terms of the other two. Letting $z = s$ and $w = t$ (s, t, parameters), we find that

$$\begin{aligned} x &= s + t \\ y &= s - t - 1 \\ z &= s \\ w &= t \end{aligned}$$

The solutions may be written in the form $(s + t, s - t - 1, s, t)$, where s and t are any real numbers. Geometrically, the three equations in the system represent three hyperplanes in four-dimensional space (since there are four variables) and their "points" of intersection lie in a two-dimensional subspace of four-space (since there are two parameters). ■■■■

REMARK In Example 4 we assigned parameters to z and w rather than to x and y because x and y are readily solved in terms of z and w. ■■■

APPLICATION

The following example illustrates a situation in which a system of linear equations has infinitely many solutions.

EXAMPLE 5

Figure 2.7 shows the flow of downtown traffic in a certain city during the rush hours on a typical weekday. The arrows indicate the direction of traffic flow on each one-way road, and the average number of vehicles entering and leaving each intersection per hour appears beside each road. Fifth and Sixth Avenues can each handle up to 2000 vehicles per hour without causing congestion, whereas the maximum capacity of each of the two streets is 1000 vehicles per hour. The flow of traffic is controlled by traffic lights installed at each of the four intersections.

FIGURE 2.7

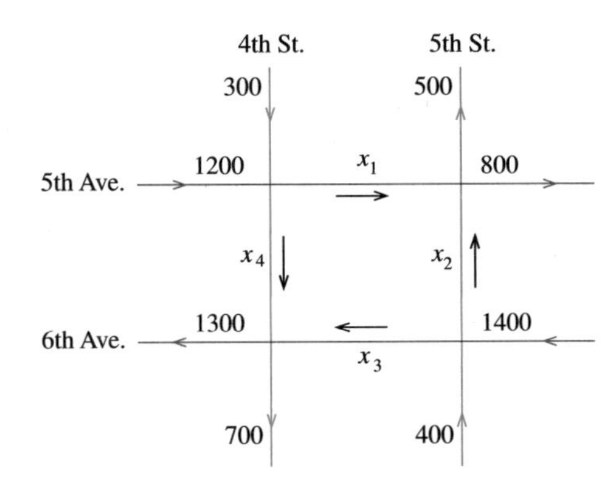

a. Write a general expression involving the rates of flow—x_1, x_2, x_3, x_4—and suggest two possible flow patterns that will ensure that there is no traffic congestion.

b. Suppose that the part of 4th Street between 5th Avenue and 6th Avenue is to be resurfaced and that traffic flow between the two junctions has to be slowed to at most 300 vehicles per hour. Find two possible flow patterns that will result in a smooth flow of traffic.

SOLUTION ✓

a. To avoid congestion, all traffic entering an intersection must also leave that intersection. Applying this condition to each of the four intersections in a clockwise direction beginning with the 5th Avenue and 4th Street intersection, we obtain the following equations:

$$1500 = x_1 + x_4$$
$$1300 = x_1 + x_2$$
$$1800 = x_2 + x_3$$
$$2000 = x_3 + x_4$$

This system of four linear equations in the four variables x_1, x_2, x_3, x_4 may be rewritten in the more standard form

$$\begin{aligned} x_1 \qquad\qquad\quad + x_4 &= 1500 \\ x_1 + x_2 \qquad\qquad\quad &= 1300 \\ x_2 + x_3 \qquad &= 1800 \\ x_3 + x_4 &= 2000 \end{aligned}$$

Using the Gauss–Jordan elimination method to solve the system, we obtain

$$\left[\begin{array}{cccc|c} 1 & 0 & 0 & 1 & 1500 \\ 1 & 1 & 0 & 0 & 1300 \\ 0 & 1 & 1 & 0 & 1800 \\ 0 & 0 & 1 & 1 & 2000 \end{array}\right] \xrightarrow{R_2 - R_1} \left[\begin{array}{cccc|c} 1 & 0 & 0 & 1 & 1500 \\ 0 & 1 & 0 & -1 & -200 \\ 0 & 1 & 1 & 0 & 1800 \\ 0 & 0 & 1 & 1 & 2000 \end{array}\right]$$

$$\xrightarrow{R_3 - R_2} \left[\begin{array}{cccc|c} 1 & 0 & 0 & 1 & 1500 \\ 0 & 1 & 0 & -1 & -200 \\ 0 & 0 & 1 & 1 & 2000 \\ 0 & 0 & 1 & 1 & 2000 \end{array}\right]$$

$$\xrightarrow{R_4 - R_3} \left[\begin{array}{cccc|c} 1 & 0 & 0 & 1 & 1500 \\ 0 & 1 & 0 & -1 & -200 \\ 0 & 0 & 1 & 1 & 2000 \\ 0 & 0 & 0 & 0 & 0 \end{array}\right]$$

The last augmented matrix is in row-reduced form and is equivalent to a system of three linear equations in the four variables x_1, x_2, x_3, x_4. Thus, we may express three of the variables—say, x_1, x_2, x_3—in terms of x_4. Setting $x_4 = t$ (t, a parameter), we may write the infinitely many solutions of the system as

$$\begin{aligned} x_1 &= 1500 - t \\ x_2 &= -200 + t \\ x_3 &= 2000 - t \\ x_4 &= t \end{aligned}$$

Observe that for a meaningful solution, $200 \leq t \leq 1000$, since x_1, x_2, x_3, and x_4 must all be nonnegative. For example, picking $t = 300$ gives the flow pattern

$$x_1 = 1200, \quad x_2 = 100, \quad x_3 = 1700, \quad x_4 = 300$$

Selecting $t = 500$ gives the flow pattern

$$x_1 = 1000, \quad x_2 = 300, \quad x_3 = 1500, \quad x_4 = 500$$

b. In this case, x_4 must not exceed 300. Again, using the results of part (a), we find, upon setting $x_4 = t = 300$, the flow pattern

$$x_1 = 1200, \quad x_2 = 100, \quad x_3 = 1700, \quad x_4 = 300$$

obtained earlier. Picking $t = 250$ gives the flow pattern

$$x_1 = 1250, \quad x_2 = 50, \quad x_3 = 1750, \quad x_4 = 250$$

■■■■

SELF-CHECK EXERCISES 2.3

1. The following augmented matrix in row-reduced form is equivalent to the augmented matrix of a certain system of linear equations. Use this result to solve the system of equations.

$$\left[\begin{array}{ccc|c} 1 & 0 & -1 & 3 \\ 0 & 1 & 5 & -2 \\ 0 & 0 & 0 & 0 \end{array}\right]$$

2. Solve the system of linear equations

$$\begin{aligned} 2x - 3y + z &= 6 \\ x + 2y + 4z &= -4 \\ x - 5y - 3z &= 10 \end{aligned}$$

using the Gauss–Jordan elimination method.

3. Solve the system of linear equations

$$\begin{aligned} x - 2y + 3z &= 9 \\ 2x + 3y - z &= 4 \\ x + 5y - 4z &= 2 \end{aligned}$$

using the Gauss–Jordan elimination method.

Solutions to Self-Check Exercises 2.3 can be found on page 118.

2.3 Exercises

In Exercises 1–12, given that the augmented matrix in row-reduced form is equivalent to the augmented matrix of a system of linear equations, (a) determine whether the system has a solution and (b) find the solution or solutions to the system, if they exist.

1. $\left[\begin{array}{ccc|c} 1 & 0 & 0 & 3 \\ 0 & 1 & 0 & -1 \\ 0 & 0 & 1 & 2 \end{array}\right]$

2. $\left[\begin{array}{ccc|c} 1 & 0 & 0 & 3 \\ 0 & 1 & 0 & -2 \\ 0 & 0 & 1 & 1 \end{array}\right]$

3. $\left[\begin{array}{cc|c} 1 & 0 & 2 \\ 0 & 1 & 4 \\ 0 & 0 & 0 \end{array}\right]$

4. $\left[\begin{array}{ccc|c} 1 & 0 & 0 & 3 \\ 0 & 1 & 0 & 1 \\ 0 & 0 & 0 & 0 \end{array}\right]$

5. $\left[\begin{array}{ccc|c} 1 & 0 & 1 & 4 \\ 0 & 1 & 0 & -2 \end{array}\right]$

6. $\left[\begin{array}{cccc|c} 1 & 0 & 0 & 0 & 3 \\ 0 & 1 & 1 & 0 & -1 \\ 0 & 0 & 0 & 1 & 2 \end{array}\right]$

7. $\left[\begin{array}{cccc|c} 1 & 0 & 0 & 0 & 2 \\ 0 & 1 & 0 & 0 & 1 \\ 0 & 0 & 1 & 0 & 3 \\ 0 & 0 & 0 & 0 & 1 \end{array}\right]$

8. $\left[\begin{array}{ccc|c} 1 & 0 & 0 & 4 \\ 0 & 1 & 0 & -1 \\ 0 & 0 & 1 & 3 \\ 0 & 0 & 0 & 1 \end{array}\right]$

9. $\left[\begin{array}{cccc|c} 1 & 0 & 0 & 0 & 2 \\ 0 & 1 & 0 & 0 & -1 \\ 0 & 0 & 1 & 1 & 2 \\ 0 & 0 & 0 & 0 & 0 \end{array}\right]$

10. $\left[\begin{array}{cccc|c} 0 & 1 & 0 & 1 & 3 \\ 0 & 0 & 1 & -2 & 4 \\ 0 & 0 & 0 & 0 & 0 \\ 0 & 0 & 0 & 0 & 0 \end{array}\right]$

11. $\left[\begin{array}{cccc|c} 1 & 0 & 3 & 0 & 2 \\ 0 & 1 & -1 & 0 & 1 \\ 0 & 0 & 0 & 0 & 0 \\ 0 & 0 & 0 & 0 & 0 \end{array}\right]$

12. $\left[\begin{array}{cccc|c} 1 & 0 & 3 & -1 & 4 \\ 0 & 1 & -2 & 3 & 2 \\ 0 & 0 & 0 & 0 & 0 \\ 0 & 0 & 0 & 0 & 0 \end{array}\right]$

In Exercises 13–32, solve the system of linear equations using the Gauss–Jordan elimination method.

13. $\begin{aligned} 2x - y &= 3 \\ x + 2y &= 4 \\ 2x + 3y &= 7 \end{aligned}$

14. $\begin{aligned} x + 2y &= 3 \\ 2x - 3y &= -8 \\ x - 4y &= -9 \end{aligned}$

15. $\begin{aligned} 3x - 2y &= -3 \\ 2x + y &= 3 \\ x - 2y &= -5 \end{aligned}$

16. $\begin{aligned} 2x + 3y &= 2 \\ x + 3y &= -2 \\ x - y &= 3 \end{aligned}$

17. $\begin{aligned} 3x - 2y &= 5 \\ -x + 3y &= -4 \\ 2x - 4y &= 6 \end{aligned}$

18. $\begin{aligned} 4x + 6y &= 8 \\ 3x - 2y &= -7 \\ x + 3y &= 5 \end{aligned}$

19. $\begin{aligned} x - 2y &= 2 \\ 7x - 14y &= 14 \\ 3x - 6y &= 6 \end{aligned}$

20. $\begin{aligned} x + 2y + z &= -2 \\ -2x - 3y - z &= 1 \\ 2x + 4y + 2z &= -4 \end{aligned}$

21. $\begin{aligned} 3x + 2y &= 4 \\ -\tfrac{3}{2}x - y &= -2 \\ 6x + 4y &= 8 \end{aligned}$

22. $\begin{aligned} 3y + 2z &= 4 \\ 2x - y - 3z &= 3 \\ 2x + 2y - z &= 7 \end{aligned}$

23. $\begin{aligned} 2x_1 - x_2 + x_3 &= -4 \\ 3x_1 - \tfrac{3}{2}x_2 + \tfrac{3}{2}x_3 &= -6 \\ -6x_1 + 3x_2 - 3x_3 &= 12 \end{aligned}$

24. $\begin{aligned} x + y - 2z &= -3 \\ 2x - y + 3z &= 7 \\ x - 2y + 5z &= 0 \end{aligned}$

25. $\begin{aligned} x - 2y + 3z &= 4 \\ 2x + 3y - z &= 2 \\ x + 2y - 3z &= -6 \end{aligned}$

26. $\begin{aligned} x_1 - 2x_2 + x_3 &= -3 \\ 2x_1 + x_2 - 2x_3 &= 2 \\ x_1 + 3x_2 - 3x_3 &= 5 \end{aligned}$

27. $\begin{aligned} 4x + y - z &= 4 \\ 8x + 2y - 2z &= 8 \end{aligned}$

28. $\begin{aligned} x_1 + 2x_2 + 4x_3 &= 2 \\ x_1 + x_2 + 2x_3 &= 1 \end{aligned}$

29. $\begin{aligned} 2x + y - 3z &= 1 \\ x - y + 2z &= 1 \\ 5x - 2y + 3z &= 6 \end{aligned}$

30. $\begin{aligned} 3x - 9y + 6z &= -12 \\ x - 3y + 2z &= -4 \\ 2x - 6y + 4z &= 8 \end{aligned}$

31. $\begin{aligned} x + 2y - z &= -4 \\ 2x + y + z &= 7 \\ x + 3y + 2z &= 7 \\ x - 3y + z &= 9 \end{aligned}$

32. $\begin{aligned} 3x - 2y + z &= 4 \\ x + 3y - 4z &= -3 \\ 2x - 3y + 5z &= 7 \\ x - 8y + 9z &= 10 \end{aligned}$

33. MANAGEMENT DECISIONS The management of Hartman Rent-A-Car has allocated \$840,000 to purchase 60 new automobiles to add to their existing fleet of rental cars. The company will choose from compact, mid-sized, and full-sized cars costing \$10,000, \$16,000, and \$22,000 each, respectively. Find formulas giving the options available to the company. Give two specific options. *Note:* Your answers will not be unique.

34. NUTRITION A dietitian wishes to plan a meal around three foods. The meal is to include 8800 units of vitamin A, 3380 units of vitamin C, and 1020 units of calcium. The number of units of the vitamins and calcium in each ounce of the foods is summarized in the accompanying table.

	Food I	Food II	Food III
Vitamin A	400	1200	800
Vitamin C	110	570	340
Calcium	90	30	60

Determine the amount of each food the dietitian should include in the meal in order to meet the vitamin and calcium requirements.

35. NUTRITION Refer to Exercise 34. In planning for another meal, the dietitian changes the requirement of vitamin C to 2160 units instead of 3380 units. All other requirements remain the same. Show that such a meal cannot be planned around the same foods.

36. **Investments** Mr. and Mrs. Garcia have a total of \$100,000 to be invested in stocks, bonds, and a money-market account. The stocks have a rate of return of 12%/year, while the bonds and the money-market account pay 8% and 4%/year, respectively. They have stipulated that the amount invested in stocks should be equal to the sum of the amount invested in bonds and three times the amount invested in the money-market account. How should the Garcias allocate their resources if they require an annual income of \$10,000 from their investments?

37. **Traffic Control** The accompanying figure shows the flow of traffic near a city's Civic Center during the rush hours on a typical weekday. Each road can handle a maximum of 1000 cars/hour without causing congestion. The flow of traffic is controlled by traffic lights at each of the five intersections.

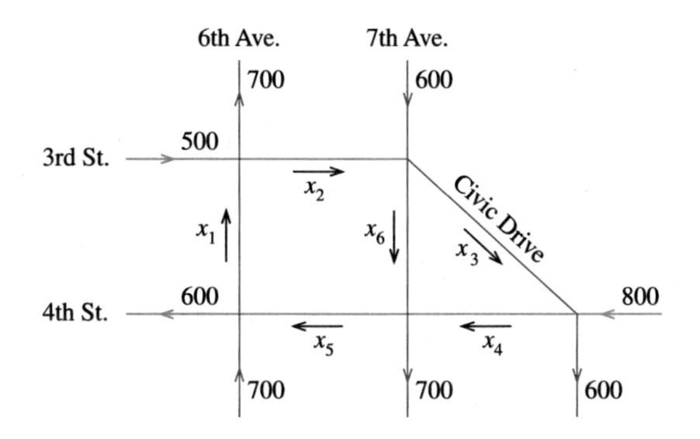

a. Set up a system of linear equations describing the traffic flow.
b. Solve the system devised in part (a) and suggest two possible traffic-flow patterns that will ensure that there is no traffic congestion.
c. Suppose that 7th Avenue between 3rd and 4th Streets is soon to be closed for road repairs. Find one possible flow pattern that will result in a smooth flow of traffic.

38. **Traffic Control** The accompanying figure shows the flow of downtown traffic during the rush hours on a typical weekday. Each avenue can handle up to 1500 vehicles/hour without causing congestion, whereas the maximum capacity of each street is 1000 vehicles/hour. The flow of traffic is controlled by traffic lights at each of the six intersections.
a. Set up a system of linear equations describing the traffic flow.

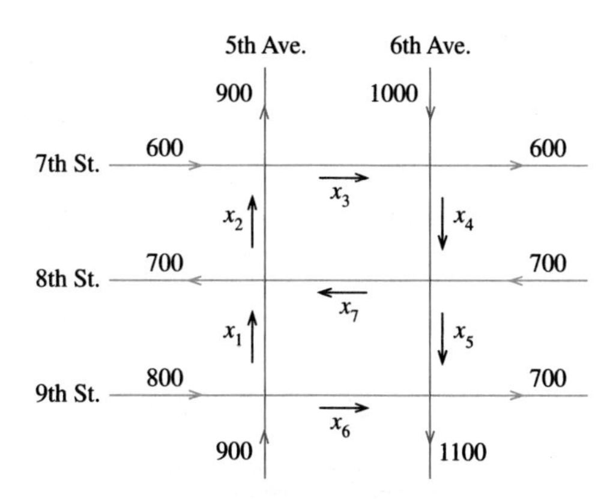

b. Solve the system devised in part (a) and suggest two possible traffic-flow patterns that will ensure there is no traffic congestion.
c. Suppose that the traffic flow along 9th Street between 5th and 6th Avenues, x_6, is restricted due to sewer construction. What is the minimum permissible traffic flow along this road that will not result in traffic congestion?

39. Determine the value of k so that the following system of linear equations has a solution and then find the solution:

$$\begin{aligned} 2x + 3y &= 2 \\ x + 4y &= 6 \\ 5x + ky &= 2 \end{aligned}$$

40. Determine the value of k so that the following system of linear equations has infinitely many solutions and then find the solutions:

$$\begin{aligned} 3x - 2y + 4z &= 12 \\ -9x + 6y - 12z &= k \end{aligned}$$

In Exercises 41 and 42, determine whether the statement is true or false. If it is true, explain why it is true. If it is false, give an example to show why it is false.

41. A system of linear equations having fewer equations than variables has no solution, a unique solution, or infinitely many solutions.

42. A system of linear equations having more equations than variables has no solution, a unique solution, or infinitely many solutions.

Solving Systems of Linear Equations II

We can use the row operations of a graphing utility to solve a system of m linear equations in n unknowns by the Gauss–Jordan method, as we did in the previous technology section. We can also use the **rref** or equivalent operation to obtain the row-reduced form without going through all the steps of the Gauss–Jordan method. The SIMULT function, however, cannot be used to solve a system where the number of equations and the number of variables are not the same.

EXAMPLE 1 Solve the system

$$\begin{aligned} x_1 - 2x_2 + 4x_3 &= 2 \\ 2x_1 + x_2 - 2x_3 &= -1 \\ 3x_1 - x_2 + 2x_3 &= 1 \\ 2x_1 + 6x_2 - 12x_3 &= -6 \end{aligned}$$

SOLUTION ✓ First, we enter the augmented matrix A into the calculator as

$$A = \left[\begin{array}{rrr|r} 1 & -2 & 4 & 2 \\ 2 & 1 & -2 & -1 \\ 3 & -1 & 2 & 1 \\ 2 & 6 & -12 & -6 \end{array}\right]$$

Then using the **rref** or equivalent operation, we obtain the equivalent matrix

$$\left[\begin{array}{rrr|r} 1 & 0 & 0 & 0 \\ 0 & 1 & -2 & -1 \\ 0 & 0 & 0 & 0 \\ 0 & 0 & 0 & 0 \end{array}\right]$$

in reduced form. Thus, the given system is equivalent to

$$\begin{aligned} x_1 \qquad\quad &= 0 \\ x_2 - 2x_3 &= -1 \end{aligned}$$

Letting $x_3 = t$, where t is a parameter, we see that the solutions are $(0, 2t - 1, t)$. ■■■■

Exercises

In Exercises 1–6, use a graphing utility to solve the system of equations using the rref or equivalent operation.

1.
$$\begin{aligned} 2x_1 - x_2 - x_3 &= 0 \\ 3x_1 - 2x_2 - x_3 &= -1 \\ -x_1 + 2x_2 - x_3 &= 3 \\ 2x_2 - 2x_3 &= 4 \end{aligned}$$

2.
$$\begin{aligned} 3x_1 + x_2 - 4x_3 &= 5 \\ 2x_1 - 3x_2 + 2x_3 &= -4 \\ -x_1 - 2x_2 + 4x_3 &= 6 \\ 4x_1 + 3x_2 - 5x_3 &= 9 \end{aligned}$$

3.
$$\begin{aligned} 2x_1 + 3x_2 + 2x_3 + x_4 &= -1 \\ x_1 - x_2 + x_3 - 2x_4 &= -8 \\ 5x_1 + 6x_2 - 2x_3 + 2x_4 &= 11 \\ x_1 + 3x_2 + 8x_3 + x_4 &= -14 \end{aligned}$$

4.
$$\begin{aligned} x_1 - x_2 + 3x_3 - 6x_4 &= 2 \\ x_1 + x_2 + x_3 - 2x_4 &= 2 \\ -2x_1 - x_2 + x_3 + 2x_4 &= 0 \end{aligned}$$

5.
$$\begin{aligned} x_1 + x_2 - x_3 - x_4 &= -1 \\ x_1 - x_2 + x_3 + 4x_4 &= -6 \\ 3x_1 + x_2 - x_3 + 2x_4 &= -4 \\ 5x_1 + x_2 - 3x_3 + x_4 &= -9 \end{aligned}$$

6.
$$\begin{aligned} 1.2x_1 - 2.3x_2 + 4.2x_3 + 5.4x_4 - 1.6x_5 &= 4.2 \\ 2.3x_1 + 1.4x_2 - 3.1x_3 + 3.3x_4 - 2.4x_5 &= 6.3 \\ 1.7x_1 + 2.6x_2 - 4.3x_3 + 7.2x_4 - 1.8x_5 &= 7.8 \\ 2.6x_1 - 4.2x_2 + 8.3x_3 - 1.6x_4 + 2.5x_5 &= 6.4 \end{aligned}$$

SOLUTIONS TO SELF-CHECK EXERCISES 2.3

1. Let x, y, and z denote the variables. Then, the given row-reduced augmented matrix tells us that the system of linear equations is equivalent to the two equations

$$\begin{aligned} x \qquad - z &= 3 \\ y + 5z &= -2 \end{aligned}$$

Letting $z = t$, where t is a parameter, we find the infinitely many solutions given by

$$\begin{aligned} x &= t + 3 \\ y &= -5t - 2 \\ z &= t \end{aligned}$$

2. We obtain the following sequence of equivalent augmented matrices:

$$\left[\begin{array}{rrr|r} 2 & -3 & 1 & 6 \\ 1 & 2 & 4 & -4 \\ 1 & -5 & -3 & 10 \end{array}\right] \xrightarrow{R_1 \leftrightarrow R_2}$$

$$\left[\begin{array}{rrr|r} 1 & 2 & 4 & -4 \\ 2 & -3 & 1 & 6 \\ 1 & -5 & -3 & 10 \end{array}\right] \xrightarrow[R_3 - R_1]{R_2 - 2R_1}$$

$$\left[\begin{array}{rrr|r} 1 & 2 & 4 & -4 \\ 0 & -7 & -7 & 14 \\ 0 & -7 & -7 & 14 \end{array}\right] \xrightarrow{-\frac{1}{7}R_2}$$

$$\left[\begin{array}{rrr|r} 1 & 2 & 4 & -4 \\ 0 & 1 & 1 & -2 \\ 0 & -7 & -7 & 14 \end{array}\right] \xrightarrow[R_3 + 7R_2]{R_1 - 2R_2} \left[\begin{array}{rrr|r} 1 & 0 & 2 & 0 \\ 0 & 1 & 1 & -2 \\ 0 & 0 & 0 & 0 \end{array}\right]$$

The last augmented matrix, which is in row-reduced form, tells us that the given system of linear equations is equivalent to the following system of two equations:

$$\begin{aligned} x \qquad + 2z &= 0 \\ y + z &= -2 \end{aligned}$$

Letting $z = t$, where t is a parameter, we see that the infinitely many solutions are given by

$$\begin{aligned} x &= -2t \\ y &= -t - 2 \\ z &= t \end{aligned}$$

3. We obtain the following sequence of equivalent augmented matrices:

$$\left[\begin{array}{ccc|c} ① & -2 & 3 & 9 \\ 2 & 3 & -1 & 4 \\ 1 & 5 & -4 & 2 \end{array}\right] \xrightarrow[R_3 - R_1]{R_2 - 2R_1}$$

$$\left[\begin{array}{ccc|c} 1 & -2 & 3 & 9 \\ 0 & 7 & -7 & -14 \\ 0 & 7 & -7 & -7 \end{array}\right] \xrightarrow{R_3 - R_2} \left[\begin{array}{ccc|c} 1 & -2 & 3 & 9 \\ 0 & 7 & -7 & -14 \\ 0 & 0 & 0 & 7 \end{array}\right]$$

Since the last row of the final augmented matrix is equivalent to the equation $0 = 7$, a contradiction, we conclude that the given system has no solution.

2.4 Matrices

Using Matrices to Represent Data

Many practical problems are solved by using arithmetic operations on the data associated with the problems. By properly organizing the data into *blocks* of numbers, we can then carry out these arithmetic operations in an orderly and efficient manner. In particular, this systematic approach enables us to use the computer to full advantage.

Let us begin by considering how the monthly output data of a manufacturer may be organized. The Acrosonic Company manufactures four different loudspeaker systems in three separate locations. The company's May output is summarized in Table 2.1.

Table 2.1

	Model A	Model B	Model C	Model D
Location I	320	280	460	280
Location II	480	360	580	0
Location III	540	420	200	880

Now, if we agree to preserve the relative location of each entry in Table 2.1, we can summarize the set of data further, as follows:

$$\begin{bmatrix} 320 & 280 & 460 & 280 \\ 480 & 360 & 580 & 0 \\ 540 & 420 & 200 & 880 \end{bmatrix}$$

A matrix summarizing the data in Table 2.1

The array of numbers displayed here is an example of a *matrix.* Observe that the numbers in row 1 give the output of models A, B, C, and D of Acrosonic loudspeaker systems manufactured in location I; similarly, the numbers in rows 2 and 3 give the respective outputs of these loudspeaker systems in locations II and III. The numbers in each column of the matrix give the outputs of a particular model of loudspeaker system manufactured in each of the company's three manufacturing locations.

More generally, a matrix is an ordered rectangular array of real numbers. For example, each of the following arrays is a matrix:

$$A = \begin{bmatrix} 3 & 0 & -1 \\ 2 & 1 & 4 \end{bmatrix}, \quad B = \begin{bmatrix} 3 & 2 \\ 0 & 1 \\ -1 & 4 \end{bmatrix}, \quad C = \begin{bmatrix} 1 \\ 2 \\ 4 \\ 0 \end{bmatrix}, \quad D = [1 \quad 3 \quad 0 \quad 1]$$

The real numbers that make up the array are called the **entries,** or *elements,* of the matrix. The entries in a row in the array are referred to as a **row** of the matrix, whereas the entries in a column in the array are referred to as a **column** of the matrix. Matrix A, for example, has two rows and three columns, which may be identified as follows:

$$\begin{array}{cccc} & \text{Column 1} & \text{Column 2} & \text{Column 3} \\ \text{Row 1} & 3 & 0 & -1 \\ \text{Row 2} & 2 & 1 & 4 \end{array}$$

A 2 × 3 matrix

The **size,** or *dimension,* of a matrix is described in terms of the number of rows and columns of the matrix. For example, matrix A has two rows and three columns and is said to have size 2 by 3, denoted 2×3. In general, a matrix having m rows and n columns is said to have size $m \times n$.

Matrix

A **matrix** is an ordered rectangular array of numbers. A matrix with m rows and n columns has size $m \times n$. The entry in the ith row and jth column is denoted by a_{ij}.

A matrix of size $1 \times n$—a matrix having one row and n columns—is referred to as a **row matrix,** or *row vector,* of dimension n. For example, the matrix D is a row vector of dimension 4. Similarly, a matrix having m rows and one column is referred to as a **column matrix,** or *column vector,* of dimension m. The matrix C is a column vector of dimension 4. Finally, an $n \times n$ matrix—that is, a matrix having the same number of rows as columns—is called a **square matrix.** For example, the matrix

$$\begin{bmatrix} -3 & 8 & 6 \\ 2 & \frac{1}{4} & 4 \\ 1 & 3 & 2 \end{bmatrix}$$

A 3 × 3 square matrix

is a square matrix of size 3×3, or simply of size 3.

EXAMPLE 1 Consider the matrix

$$P = \begin{bmatrix} 320 & 280 & 460 & 280 \\ 480 & 360 & 580 & 0 \\ 540 & 420 & 200 & 880 \end{bmatrix}$$

representing the output of loudspeaker systems of the Acrosonic Company discussed earlier (see Table 2.1).

a. What is the size of the matrix P?

b. Find a_{24} (the entry in row 2 and column 4 of the matrix P) and give an interpretation of this number.

c. Find the sum of the entries that make up row 1 of P and interpret the result.

d. Find the sum of the entries that make up column 4 of P and interpret the result.

SOLUTION ✓ **a.** The matrix P has three rows and four columns and hence has size 3×4.

b. The required entry lies in row 2 and column 4 and is the number 0. This means that no model-D loudspeaker system was manufactured in location II in May.

c. The required sum is given by

$$320 + 280 + 460 + 280 = 1340$$

which gives the total number of loudspeaker systems manufactured in location I in May as 1340 units.

d. The required sum is given by

$$280 + 0 + 880 = 1160$$

giving the output of model-D loudspeaker systems in all locations of the company in May as 1160 units. ■■■■

EQUALITY OF MATRICES

Two matrices are said to be *equal* if they have the same size and their corresponding entries are equal. For example,

$$\begin{bmatrix} 2 & 3 & 1 \\ 4 & 6 & 2 \end{bmatrix} = \begin{bmatrix} (3-1) & 3 & 1 \\ 4 & (4+2) & 2 \end{bmatrix}$$

Also,

$$\begin{bmatrix} 1 & 3 & 5 \\ 2 & 4 & 3 \end{bmatrix} \neq \begin{bmatrix} 1 & 2 \\ 3 & 4 \\ 5 & 3 \end{bmatrix}$$

since the matrix on the left has size 2 × 3, whereas the matrix on the right has size 3 × 2, and

$$\begin{bmatrix} 2 & 3 \\ 4 & 6 \end{bmatrix} \neq \begin{bmatrix} 2 & 3 \\ 4 & 7 \end{bmatrix}$$

since the corresponding elements in row 2 and column 2 of the two matrices are not equal.

Equality of Matrices

Two matrices are equal if they have the same size and their corresponding entries are equal.

EXAMPLE 2 Solve the following matrix equation for x, y, and z:

$$\begin{bmatrix} 1 & x & 3 \\ 2 & y-1 & 2 \end{bmatrix} = \begin{bmatrix} 1 & 4 & z \\ 2 & 1 & 2 \end{bmatrix}$$

SOLUTION ✔ Since the corresponding elements of the two matrices must be equal, we find that $x = 4$, $z = 3$, and $y - 1 = 1$, or $y = 2$.

ADDITION AND SUBTRACTION

Two matrices A and B of the *same size* can be added or subtracted to produce a matrix of the same size. This is done by adding or subtracting the corresponding entries in the two matrices. For example,

$$\begin{bmatrix} 1 & 3 & 4 \\ -1 & 2 & 0 \end{bmatrix} + \begin{bmatrix} 1 & 4 & 3 \\ 6 & 1 & -2 \end{bmatrix} = \begin{bmatrix} 1+1 & 3+4 & 4+3 \\ -1+6 & 2+1 & 0+(-2) \end{bmatrix} = \begin{bmatrix} 2 & 7 & 7 \\ 5 & 3 & -2 \end{bmatrix}$$

Adding two matrices of the same size

and

$$\begin{bmatrix} 1 & 2 \\ -1 & 3 \\ 4 & 0 \end{bmatrix} - \begin{bmatrix} 2 & -1 \\ 3 & 2 \\ -1 & 0 \end{bmatrix} = \begin{bmatrix} 1-2 & 2-(-1) \\ -1-3 & 3-2 \\ 4-(-1) & 0-0 \end{bmatrix} = \begin{bmatrix} -1 & 3 \\ -4 & 1 \\ 5 & 0 \end{bmatrix}$$

Subtracting two matrices of the same size

Addition and Subtraction of Matrices

If A and B are two matrices of the same size,

1. The *sum* $A + B$ is the matrix obtained by adding the corresponding entries in the two matrices.
2. The *difference* $A - B$ is the matrix obtained by subtracting the corresponding entries in B from A.

EXAMPLE 3 The total output of the Acrosonic Company for June is summarized in Table 2.2.

Table 2.2

	Model A	Model B	Model C	Model D
Location I	210	180	330	180
Location II	400	300	450	40
Location III	420	280	180	740

The output for May was given earlier in Table 2.1. Find the total output of the company for May and June.

SOLUTION ✓ As we saw earlier, the production matrix for the Acrosonic Company in May is given by

$$A = \begin{bmatrix} 320 & 280 & 460 & 280 \\ 480 & 360 & 580 & 0 \\ 540 & 420 & 200 & 880 \end{bmatrix}$$

Next, from Table 2.2, we see that the production matrix for June is given by

$$B = \begin{bmatrix} 210 & 180 & 330 & 180 \\ 400 & 300 & 450 & 40 \\ 420 & 280 & 180 & 740 \end{bmatrix}$$

Finally, the total output of the Acrosonic Company for May and June is given by the matrix

$$A + B = \begin{bmatrix} 320 & 280 & 460 & 280 \\ 480 & 360 & 580 & 0 \\ 540 & 420 & 200 & 880 \end{bmatrix} + \begin{bmatrix} 210 & 180 & 330 & 180 \\ 400 & 300 & 450 & 40 \\ 420 & 280 & 180 & 740 \end{bmatrix}$$

$$= \begin{bmatrix} 530 & 460 & 790 & 460 \\ 880 & 660 & 1030 & 40 \\ 960 & 700 & 380 & 1620 \end{bmatrix}$$

■■■■

The following laws hold for matrix addition.

Laws for Matrix Addition

If A, B, and C are matrices of the same size, then

1. $A + B = B + A$ (Commutative law)

2. $(A + B) + C = A + (B + C)$ (Associative law)

The *commutative law* for matrix addition states that the order in which matrix addition is performed is immaterial. The *associative law* states that when adding three matrices together, we may first add A and B and then add the resulting sum to C. Equivalently, we can add A to the sum of B and C.

A *zero matrix* is one in which all entries are zero. The zero matrix O has the property that

$$A + O = O + A = A$$

for any matrix A having the same size as that of O. For example, the zero matrix of size 3×2 is

$$O = \begin{bmatrix} 0 & 0 \\ 0 & 0 \\ 0 & 0 \end{bmatrix}$$

If A is any 3×2 matrix, then

$$A + O = \begin{bmatrix} a_{11} & a_{12} \\ a_{21} & a_{22} \\ a_{31} & a_{32} \end{bmatrix} + \begin{bmatrix} 0 & 0 \\ 0 & 0 \\ 0 & 0 \end{bmatrix} = \begin{bmatrix} a_{11} & a_{12} \\ a_{21} & a_{22} \\ a_{31} & a_{32} \end{bmatrix} = A$$

where a_{ij} denotes the entry in the ith row and jth column of the matrix A.

The matrix obtained by interchanging the rows and columns of a given matrix A is called the *transpose* of A and is denoted A^T. For example, if

$$A = \begin{bmatrix} 1 & 2 & 3 \\ 4 & 5 & 6 \\ 7 & 8 & 9 \end{bmatrix}$$

then

$$A^T = \begin{bmatrix} 1 & 4 & 7 \\ 2 & 5 & 8 \\ 3 & 6 & 9 \end{bmatrix}$$

Transpose of a Matrix

If A is an $m \times n$ matrix with elements a_{ij}, then the **transpose** of A is the $n \times m$ matrix A^T with elements a_{ji}.

Scalar Multiplication

A matrix A may be multiplied by a real number, called a **scalar** in the context of matrix algebra. The scalar product, denoted by cA, is a matrix obtained by multiplying each entry of A by c. For example, the scalar product of the matrix

$$A = \begin{bmatrix} 3 & -1 & 2 \\ 0 & 1 & 4 \end{bmatrix}$$

and the scalar 3 is the matrix

$$3A = 3\begin{bmatrix} 3 & -1 & 2 \\ 0 & 1 & 4 \end{bmatrix} = \begin{bmatrix} 9 & -3 & 6 \\ 0 & 3 & 12 \end{bmatrix}$$

Scalar Product

If A is a matrix and c is a real number, then the **scalar product** cA is the matrix obtained by multiplying each entry of A by c.

EXAMPLE 4 Given

$$A = \begin{bmatrix} 3 & 4 \\ -1 & 2 \end{bmatrix} \quad \text{and} \quad B = \begin{bmatrix} 3 & 2 \\ -1 & 2 \end{bmatrix}$$

find the matrix X satisfying the *matrix equation* $2X + B = 3A$.

SOLUTION ✓ From the given equation $2X + B = 3A$, we find that

$$\begin{aligned} 2X &= 3A - B \\ &= 3\begin{bmatrix} 3 & 4 \\ -1 & 2 \end{bmatrix} - \begin{bmatrix} 3 & 2 \\ -1 & 2 \end{bmatrix} \\ &= \begin{bmatrix} 9 & 12 \\ -3 & 6 \end{bmatrix} - \begin{bmatrix} 3 & 2 \\ -1 & 2 \end{bmatrix} = \begin{bmatrix} 6 & 10 \\ -2 & 4 \end{bmatrix} \end{aligned}$$

Thus,

$$X = \frac{1}{2}\begin{bmatrix} 6 & 10 \\ -2 & 4 \end{bmatrix} = \begin{bmatrix} 3 & 5 \\ -1 & 2 \end{bmatrix}$$

■■■■

EXAMPLE 5 The management of the Acrosonic Company has decided to increase its July production of loudspeaker systems by 10% (over its June output). Find a matrix giving the targeted production for July.

SOLUTION ✓ From the results of Example 3, we see that Acrosonic's total output for June may be represented by the matrix

$$B = \begin{bmatrix} 210 & 180 & 330 & 180 \\ 400 & 300 & 450 & 40 \\ 420 & 280 & 180 & 740 \end{bmatrix}$$

The required matrix is given by

$$(1.1)B = 1.1 \begin{bmatrix} 210 & 180 & 330 & 180 \\ 400 & 300 & 450 & 40 \\ 420 & 280 & 180 & 740 \end{bmatrix}$$

$$= \begin{bmatrix} 231 & 198 & 363 & 198 \\ 440 & 330 & 495 & 44 \\ 462 & 308 & 198 & 814 \end{bmatrix}$$

and is interpreted in the usual manner. ■■■■

SELF-CHECK EXERCISES 2.4

1. Perform the indicated operations:

$$\begin{bmatrix} 1 & 3 & 2 \\ -1 & 4 & 7 \end{bmatrix} - 3\begin{bmatrix} 2 & 1 & 0 \\ 1 & 3 & 4 \end{bmatrix}$$

2. Solve the following matrix equation for x, y, and z:

$$\begin{bmatrix} x & 3 \\ z & 2 \end{bmatrix} + \begin{bmatrix} 2-y & z \\ 2-z & -x \end{bmatrix} = \begin{bmatrix} 3 & 7 \\ 2 & 0 \end{bmatrix}$$

3. Jack owns two gas stations, one downtown and the other in the Wilshire district. Over 2 consecutive days his gas stations recorded gasoline sales represented by the following matrices:

$A =$

	Regular-unleaded	Unleaded-plus	Super-unleaded
Downtown	1200	750	650
Wilshire	1100	850	600

and

$B =$

	Regular-unleaded	Unleaded-plus	Super-unleaded
Downtown	1250	825	550
Wilshire	1150	750	750

Find a matrix representing the total sales of the two gas stations over the 2-day period.

Solutions to Self-Check Exercises 2.4 can be found on page 129.

Spreadsheet examples and exercises for this section that may be solved using the Microsoft® Excel program are given at the Brooks/Cole Web site: http://www.brookscole.com/math/authors/tans/

2.4 Exercises

In Exercises 1–6, refer to the following matrices:

$$A = \begin{bmatrix} 2 & -3 & 9 & -4 \\ -11 & 2 & 6 & 7 \\ 6 & 0 & 2 & 9 \\ 5 & 1 & 5 & -8 \end{bmatrix}$$

$$B = \begin{bmatrix} 3 & -1 & 2 \\ 0 & 1 & 4 \\ 3 & 2 & 1 \\ -1 & 0 & 8 \end{bmatrix}$$

$$C = [1 \quad 0 \quad 3 \quad 4 \quad 5]$$

$$D = \begin{bmatrix} 1 \\ 3 \\ -2 \\ 0 \end{bmatrix}$$

1. What is the size of A? Of B? Of C? Of D?
2. Find a_{14}, a_{21}, a_{31}, and a_{43}.
3. Find b_{13}, b_{31}, and b_{43}.
4. Identify the row matrix. What is its transpose?
5. Identify the column matrix. What is its transpose?
6. Identify the square matrix. What is its transpose?

In Exercises 7–12, refer to the following matrices:

$$A = \begin{bmatrix} -1 & 2 \\ 3 & -2 \\ 4 & 0 \end{bmatrix}, \qquad B = \begin{bmatrix} 2 & 4 \\ 3 & 1 \\ -2 & 2 \end{bmatrix},$$

$$C = \begin{bmatrix} 3 & -1 & 0 \\ 2 & -2 & 3 \\ 4 & 6 & 2 \end{bmatrix}, \qquad D = \begin{bmatrix} 2 & -2 & 4 \\ 3 & 6 & 2 \\ -2 & 3 & 1 \end{bmatrix}$$

7. What is the size of A? Of B? Of C? Of D?
8. Explain why the matrix $A + C$ does not exist.
9. Compute $A + B$.
10. Compute $2A - 3B$.
11. Compute $C - D$.
12. Compute $4D - 2C$.

In Exercises 13–20, perform the indicated operations.

13. $\begin{bmatrix} 6 & 3 & 8 \\ 4 & 5 & 6 \end{bmatrix} - \begin{bmatrix} 3 & -2 & -1 \\ 0 & -5 & -7 \end{bmatrix}$

14. $\begin{bmatrix} 2 & -3 & 4 & -1 \\ 3 & 1 & 0 & 0 \end{bmatrix} + \begin{bmatrix} 4 & 3 & -2 & -4 \\ 6 & 2 & 0 & -3 \end{bmatrix}$

15. $\begin{bmatrix} 1 & 4 & -5 \\ 3 & -8 & 6 \end{bmatrix} + \begin{bmatrix} 4 & 0 & -2 \\ 3 & 6 & 5 \end{bmatrix} - \begin{bmatrix} 2 & 8 & 9 \\ -11 & 2 & -5 \end{bmatrix}$

16. $3\begin{bmatrix} 1 & 1 & -3 \\ 3 & 2 & 3 \\ 7 & -1 & 6 \end{bmatrix} + 4\begin{bmatrix} -2 & -1 & 8 \\ 4 & 2 & 2 \\ 3 & 6 & 3 \end{bmatrix}$

17. $\begin{bmatrix} 1.2 & 4.5 & -4.2 \\ 8.2 & 6.3 & -3.2 \end{bmatrix} - \begin{bmatrix} 3.1 & 1.5 & -3.6 \\ 2.2 & -3.3 & -4.4 \end{bmatrix}$

18. $\begin{bmatrix} 0.06 & 0.12 \\ 0.43 & 1.11 \\ 1.55 & -0.43 \end{bmatrix} - \begin{bmatrix} 0.77 & -0.75 \\ 0.22 & -0.65 \\ 1.09 & -0.57 \end{bmatrix}$

19. $\frac{1}{2}\begin{bmatrix} 1 & 0 & 0 & -4 \\ 3 & 0 & -1 & 6 \\ -2 & 1 & -4 & 2 \end{bmatrix} + \frac{4}{3}\begin{bmatrix} 3 & 0 & -1 & 4 \\ -2 & 1 & -6 & 2 \\ 8 & 2 & 0 & -2 \end{bmatrix} - \frac{1}{3}\begin{bmatrix} 3 & -9 & -1 & 0 \\ 6 & 2 & 0 & -6 \\ 0 & 1 & -3 & 1 \end{bmatrix}$

20. $0.5\begin{bmatrix} 1 & 3 & 5 \\ 5 & 2 & -1 \\ -2 & 0 & 1 \end{bmatrix} - 0.2\begin{bmatrix} 2 & 3 & 4 \\ -1 & 1 & -4 \\ 3 & 5 & -5 \end{bmatrix} + 0.6\begin{bmatrix} 3 & 4 & -1 \\ 4 & 5 & 1 \\ 1 & 0 & 0 \end{bmatrix}$

In Exercises 21–24, solve for *u*, *x*, *y*, and *z* in the matrix equation.

21. $\begin{bmatrix} 2x-2 & 3 & 2 \\ 2 & 4 & y-2 \\ 2z & -3 & 2 \end{bmatrix} = \begin{bmatrix} 3 & u & 2 \\ 2 & 4 & 5 \\ 4 & -3 & 2 \end{bmatrix}$

22. $\begin{bmatrix} x & -2 \\ 3 & y \end{bmatrix} + \begin{bmatrix} -2 & z \\ -1 & 2 \end{bmatrix} = \begin{bmatrix} 4 & -2 \\ 2u & 4 \end{bmatrix}$

23. $\begin{bmatrix} 1 & x \\ 2y & -3 \end{bmatrix} - 4\begin{bmatrix} 2 & -2 \\ 0 & 3 \end{bmatrix} = \begin{bmatrix} 3z & 10 \\ 4 & -u \end{bmatrix}$

24. $\begin{bmatrix} 1 & 2 \\ 3 & 4 \\ x & -1 \end{bmatrix} - 3\begin{bmatrix} y-1 & 2 \\ 1 & 2 \\ 4 & 2z+1 \end{bmatrix} = 2\begin{bmatrix} -4 & -u \\ 0 & -1 \\ 4 & 4 \end{bmatrix}$

In Exercises 25 and 26, let

$$A = \begin{bmatrix} 2 & -4 & 3 \\ 4 & 2 & 1 \end{bmatrix}, \quad B = \begin{bmatrix} 4 & -3 & 2 \\ 1 & 0 & 4 \end{bmatrix}, \quad C = \begin{bmatrix} 1 & 0 & 2 \\ 3 & -2 & 1 \end{bmatrix}$$

25. Verify by direct computation the validity of the commutative law for matrix addition.

26. Verify by direct computation the validity of the associative law for matrix addition.

In Exercises 27–30, let

$$A = \begin{bmatrix} 3 & 1 \\ 2 & 4 \\ -4 & 0 \end{bmatrix} \quad \text{and} \quad B = \begin{bmatrix} 1 & 2 \\ -1 & 0 \\ 3 & 2 \end{bmatrix}$$

Verify each equation by direct computation.

27. $(3 + 5)A = 3A + 5A$

28. $2(4A) = (2 \cdot 4)A = 8A$

29. $4(A + B) = 4A + 4B$

30. $2(A - 3B) = 2A - 6B$

In Exercises 31–34, find the transpose of the given matrix.

31. $\begin{bmatrix} 3 & 2 & -1 & 5 \end{bmatrix}$

32. $\begin{bmatrix} 4 & 2 & 0 & -1 \\ 3 & 4 & -1 & 5 \end{bmatrix}$

33. $\begin{bmatrix} 1 & -1 & 2 \\ 3 & 4 & 2 \\ 0 & 1 & 0 \end{bmatrix}$

34. $\begin{bmatrix} 1 & 2 & 6 & 4 \\ 2 & 3 & 2 & 5 \\ 6 & 2 & 3 & 0 \\ 4 & 5 & 0 & 2 \end{bmatrix}$

35. CHOLESTEROL LEVELS Mr. Cross, Mr. Jones, and Mr. Smith each suffer from coronary heart disease. As part of their treatment, they were put on special low-cholesterol diets: Cross on diet I, Jones on diet II, and Smith on diet III. Progressive records of each patient's cholesterol level were kept. At the beginning of the first, second, third, and fourth months, the cholesterol levels of the three patients were:

Cross: 220, 215, 210, and 205

Jones: 220, 210, 200, and 195

Smith: 215, 205, 195, and 190

Represent this information in a 3×4 matrix.

36. BOOKSTORE INVENTORIES The Campus Bookstore's inventory of books is:

Hardcover: Textbooks, 5280; Fiction, 1680; Nonfiction, 2320; Reference, 1890

Paperback: Fiction, 2810; Nonfiction, 1490; Reference, 2070; Textbooks, 1940

The College Bookstore's inventory of books is:

Hardcover: Textbooks, 6340; Fiction, 2220; Nonfiction, 1790; Reference, 1980

Paperback: Fiction, 3100; Nonfiction, 1720; Reference, 2710; Textbooks, 2050

a. Represent Campus's inventory as a matrix *A*.
b. Represent College's inventory as a matrix *B*.
c. The two companies decide to merge, so now write a matrix *C* that represents the total inventory of the newly amalgamated company.

37. BANKING The numbers of three types of bank accounts on January 1 in the Central Bank and its branches are represented by matrix *A*.

$A =$	Checking accounts	Savings accounts	Fixed deposit accounts
Main office	2820	1470	1120
West side branch	1030	520	480
East side branch	1170	540	460

The number and types of accounts opened during the first quarter are represented by matrix *B*, and the number and types of accounts closed during the same period are represented by matrix C. Thus,

$$B = \begin{bmatrix} 260 & 120 & 110 \\ 140 & 60 & 50 \\ 120 & 70 & 50 \end{bmatrix} \quad \text{and} \quad C = \begin{bmatrix} 120 & 80 & 80 \\ 70 & 30 & 40 \\ 60 & 20 & 40 \end{bmatrix}$$

a. Find matrix D, which represents the number of each type of account at the end of the first quarter at each location.
b. Because a new manufacturing plant is opening in the immediate area, it is anticipated that there will be a 10% increase in the number of accounts at each location during the second quarter. Write a matrix E to reflect this anticipated increase.

In Exercises 38–40, determine whether the statement is true or false. If it is true, explain why it is true. If it is false, give an example to show why it is false.

38. If A and B are matrices of the same order and c is a scalar, then $c(A + B) = cA + cB$.

39. If A is a matrix and c is a nonzero scalar, then $(cA)^T = (1/c)\,A^T$.

40. If A is a matrix, then $(A^T)^T = A$.

SOLUTIONS TO SELF-CHECK EXERCISES 2.4

1. $$\begin{bmatrix} 1 & 3 & 2 \\ -1 & 4 & 7 \end{bmatrix} - 3\begin{bmatrix} 2 & 1 & 0 \\ 1 & 3 & 4 \end{bmatrix} = \begin{bmatrix} 1 & 3 & 2 \\ -1 & 4 & 7 \end{bmatrix} - \begin{bmatrix} 6 & 3 & 0 \\ 3 & 9 & 12 \end{bmatrix}$$
$$= \begin{bmatrix} -5 & 0 & 2 \\ -4 & -5 & -5 \end{bmatrix}$$

2. We are given

$$\begin{bmatrix} x & 3 \\ z & 2 \end{bmatrix} + \begin{bmatrix} 2-y & z \\ 2-z & -x \end{bmatrix} = \begin{bmatrix} 3 & 7 \\ 2 & 0 \end{bmatrix}$$

Performing the indicated operation on the left-hand side, we obtain

$$\begin{bmatrix} 2+x-y & 3+z \\ 2 & 2-x \end{bmatrix} = \begin{bmatrix} 3 & 7 \\ 2 & 0 \end{bmatrix}$$

By the equality of matrices, we have

$$\begin{aligned} 2 + x - y &= 3 \\ 3 + z &= 7 \\ 2 - x &= 0 \end{aligned}$$

from which we deduce that $x = 2$, $y = 1$, and $z = 4$.

3. The required matrix is

$$A + B = \begin{bmatrix} 1200 & 750 & 650 \\ 1100 & 850 & 600 \end{bmatrix} + \begin{bmatrix} 1250 & 825 & 550 \\ 1150 & 750 & 750 \end{bmatrix}$$
$$= \begin{bmatrix} 2450 & 1575 & 1200 \\ 2250 & 1600 & 1350 \end{bmatrix}$$

Using Technology

MATRIX ADDITION AND SUBTRACTION, SCALAR MULTIPLICATION, AND THE TRANSPOSE OF A MATRIX

The graphing utility can be used to perform matrix addition, matrix subtraction, and scalar multiplication. It can also be used to find the transpose of a matrix.

EXAMPLE 1 Let

$$A = \begin{bmatrix} 1.2 & 3.1 \\ -2.1 & 4.2 \\ 3.1 & 4.8 \end{bmatrix} \quad \text{and} \quad B = \begin{bmatrix} 4.1 & 3.2 \\ 1.3 & 6.4 \\ 1.7 & 0.8 \end{bmatrix}$$

Find $(2.1A + 3.2B)^T$.

SOLUTION ✓ We first enter the matrices A and B into the calculator. Then, using the matrix operations, we enter the expression $(2.1A + 3.2B)^T$ and obtain the desired matrix:

$$(2.1A + 3.2B)^T = \begin{bmatrix} 15.64 & -.25 & 11.95 \\ 16.75 & 29.3 & 12.64 \end{bmatrix}$$

■■■■

Exercises

In Exercises 1–8, refer to the following matrices and use a graphing utility to perform the indicated operations.

$$A = \begin{bmatrix} 1.2 & 3.1 & -5.4 & 2.7 \\ 4.1 & 3.2 & 4.2 & -3.1 \\ 1.7 & 2.8 & -5.2 & 8.4 \end{bmatrix}$$

$$B = \begin{bmatrix} 6.2 & -3.2 & 1.4 & -1.2 \\ 3.1 & 2.7 & -1.2 & 1.7 \\ 1.2 & -1.4 & -1.7 & 2.8 \end{bmatrix}$$

1. $12.5A$ **2.** $-8.4B$ **3.** $A - B$

4. $B - A$ **5.** $1.3A + 2.4B$ **6.** $2.1A - 1.7B$

7. $(A + B)^T$ **8.** $3A^T + 4B^T$

2.5 Multiplication of Matrices

MATRIX PRODUCT

In Section 2.4 we saw how matrices of the same size may be added or subtracted and how a matrix may be multiplied by a scalar (real number), an operation referred to as scalar multiplication. In this section we see how, with certain restrictions, one matrix may be multiplied by another matrix.

To define matrix multiplication, let's consider the following problem: On a certain day, Al's Service Station sold 1600 gallons of regular-unleaded, 1000 gallons of unleaded-plus, and 800 gallons of super-unleaded gasoline. If the price of gasoline on this day was $1.14 for regular-unleaded, $1.29 for unleaded-plus, and $1.39 for super-unleaded gasoline, find the total revenue realized by Al's for that day.

The day's sale of gasoline may be represented by the matrix

$$A = \begin{bmatrix} 1600 & 1000 & 800 \end{bmatrix} \qquad \text{[Row matrix } (1 \times 3)\text{]}$$

Next, we let the unit selling price of regular-unleaded, unleaded-plus, and super-unleaded gasoline be the entries in the matrix

$$B = \begin{bmatrix} 1.14 \\ 1.29 \\ 1.39 \end{bmatrix} \qquad \text{[Column matrix } (3 \times 1)\text{]}$$

The first entry in matrix A gives the number of gallons of regular-unleaded gasoline sold, and the first entry in matrix B gives the selling price for each gallon of regular-unleaded gasoline, so their product (1600)(1.14) gives the revenue realized from the sale of regular-unleaded gasoline for the day. A similar interpretation of the second and third entries in the two matrices suggests that we multiply the corresponding entries to obtain the respective revenues realized from the sale of regular-unleaded, unleaded-plus, and super-unleaded gasoline. Finally, the total revenue realized by Al's from the sale of gasoline is given by adding these products to obtain

$$(1600)(1.14) + (1000)(1.29) + (800)(1.39) = 4226$$

or $4226.

This example suggests that if we have a row matrix of size $1 \times n$,

$$A = [a_1 \quad a_2 \quad a_3 \cdots a_n]$$

and a column matrix of size $n \times 1$,

$$B = \begin{bmatrix} b_1 \\ b_2 \\ b_3 \\ \vdots \\ b_n \end{bmatrix}$$

we may define the **matrix product** of A and B, written AB, by

$$AB = [a_1 \quad a_2 \quad a_3 \quad \cdots \quad a_n] \begin{bmatrix} b_1 \\ b_2 \\ b_3 \\ \vdots \\ b_n \end{bmatrix} = a_1b_1 + a_2b_2 + a_3b_3 + \cdots + a_nb_n \qquad \textbf{(11)}$$

EXAMPLE 1 Let

$$A = [1, \quad -2, \quad 3, \quad 5] \qquad \text{and} \qquad B = \begin{bmatrix} 2 \\ 3 \\ 0 \\ -1 \end{bmatrix}$$

Then,

$$AB = [1, \quad -2, \quad 3, \quad 5] \begin{bmatrix} 2 \\ 3 \\ 0 \\ -1 \end{bmatrix} = (1)(2) + (-2)(3) + (3)(0) + (5)(-1) = -9$$

■■■■

EXAMPLE 2 Judy's stock holdings are given by the matrix

$$\begin{matrix} & \text{GM} & \text{IBM} & \text{BAC} \\ A = & [700 & 400 & 200] \end{matrix}$$

At the close of trading on a certain day, the prices (in dollars per share) of these stocks are

$$B = \begin{bmatrix} 50 \\ 120 \\ 42 \end{bmatrix} \begin{matrix} \text{GM} \\ \text{IBM} \\ \text{BAC} \end{matrix}$$

What is the total value of Judy's holdings as of that day?

SOLUTION ✔ Judy's holdings are worth

$$AB = [700 \quad 400 \quad 200] \begin{bmatrix} 50 \\ 120 \\ 42 \end{bmatrix} = (700)(50) + (400)(120) + (200)(42)$$

or \$91,400.

■■■■

Returning once again to the matrix product AB in Equation (11), observe that the number of columns of the row matrix A is *equal* to the number of rows of the column matrix B. Observe further that the product matrix AB has size 1×1 (a real number may be thought of as a 1×1 matrix). Schematically,

Size of A Size of B

$$(1 \times n) \qquad (n \times 1)$$

$$(1 \times 1)$$

Size of AB

More generally, if A is a matrix of size $m \times n$ and B is a matrix of size $n \times p$ (the number of columns of A equals the numbers of rows of B), then the *matrix product* of A and B, AB, is defined and is a matrix of size $m \times p$. Schematically,

Size of A Size of B

$$(m \times n) \qquad (n \times p)$$

$$(m \times p)$$

Size of AB

Next, let us illustrate the mechanics of matrix multiplication by computing the product of a 2×3 matrix A and a 3×4 matrix B. Suppose

$$A = \begin{bmatrix} a_{11} & a_{12} & a_{13} \\ a_{21} & a_{22} & a_{23} \end{bmatrix}$$

$$B = \begin{bmatrix} b_{11} & b_{12} & b_{13} & b_{14} \\ b_{21} & b_{22} & b_{23} & b_{24} \\ b_{31} & b_{32} & b_{33} & b_{34} \end{bmatrix}$$

From the schematic

Same

Size of A (2×3) (3×4) Size of B

$$(2 \times 4)$$

Size of AB

we see that the matrix product $C = AB$ is defined (the number of columns of A equals the number of rows of B) and has size 2×4. Thus,

$$C = \begin{bmatrix} c_{11} & c_{12} & c_{13} & c_{14} \\ c_{21} & c_{22} & c_{23} & c_{24} \end{bmatrix}$$

The entries of C are computed as follows: The entry c_{11} (the entry in the *first* row, *first* column of C) is the product of the row matrix composed of the entries from the *first* row of A and the column matrix composed of the *first* column of B. Thus,

$$c_{11} = [a_{11} \quad a_{12} \quad a_{13}] \begin{bmatrix} b_{11} \\ b_{21} \\ b_{31} \end{bmatrix} = a_{11}b_{11} + a_{12}b_{21} + a_{13}b_{31}$$

The entry c_{12} (the entry in the *first* row, *second* column of C) is the product of the row matrix composed of the *first* row of A and the column matrix composed of the *second* column of B. Thus,

$$c_{12} = [a_{11} \quad a_{12} \quad a_{13}] \begin{bmatrix} b_{12} \\ b_{22} \\ b_{32} \end{bmatrix} = a_{11}b_{12} + a_{12}b_{22} + a_{13}b_{32}$$

The other entries in C are computed in a similar manner.

EXAMPLE 3 Let

$$A = \begin{bmatrix} 3 & 1 & 4 \\ -1 & 2 & 3 \end{bmatrix} \quad \text{and} \quad B = \begin{bmatrix} 1 & 3 & -3 \\ 4 & -1 & 2 \\ 2 & 4 & 1 \end{bmatrix}$$

Compute AB.

SOLUTION ✓ The size of matrix A is 2×3, and the size of matrix B is 3×3. Since the number of columns of matrix A is equal to the number of rows of matrix B, the matrix product $C = AB$ is defined. Furthermore, the size of matrix C is 2×3. Thus,

$$\begin{bmatrix} 3 & 1 & 4 \\ -1 & 2 & 3 \end{bmatrix} \begin{bmatrix} 1 & 3 & -3 \\ 4 & -1 & 2 \\ 2 & 4 & 1 \end{bmatrix} = \begin{bmatrix} c_{11} & c_{12} & c_{13} \\ c_{21} & c_{22} & c_{23} \end{bmatrix}$$

It remains now to determine the entries c_{11}, c_{12}, c_{13}, c_{21}, c_{22}, and c_{23}. We have

$$c_{11} = [3 \quad 1 \quad 4] \begin{bmatrix} 1 \\ 4 \\ 2 \end{bmatrix} = (3)(1) + (1)(4) + (4)(2) = 15$$

$$c_{12} = [3 \quad 1 \quad 4] \begin{bmatrix} 3 \\ -1 \\ 4 \end{bmatrix} = (3)(3) + (1)(-1) + (4)(4) = 24$$

$$c_{13} = [3 \quad 1 \quad 4] \begin{bmatrix} -3 \\ 2 \\ 1 \end{bmatrix} = (3)(-3) + (1)(2) + (4)(1) = -3$$

$$c_{21} = [-1 \quad 2 \quad 3] \begin{bmatrix} 1 \\ 4 \\ 2 \end{bmatrix} = (-1)(1) + (2)(4) + (3)(2) = 13$$

$$c_{22} = [-1 \quad 2 \quad 3] \begin{bmatrix} 3 \\ -1 \\ 4 \end{bmatrix} = (-1)(3) + (2)(-1) + (3)(4) = 7$$

$$c_{23} = [-1 \quad 2 \quad 3]\begin{bmatrix} -3 \\ 2 \\ 1 \end{bmatrix} = (-1)(-3) + (2)(2) + (3)(1) = 10$$

so the required product AB is given by

$$AB = \begin{bmatrix} 15 & 24 & -3 \\ 13 & 7 & 10 \end{bmatrix}$$

EXAMPLE 4 Let

$$A = \begin{bmatrix} 3 & 2 & 1 \\ -1 & 2 & 3 \\ 3 & 1 & 4 \end{bmatrix} \quad \text{and} \quad B = \begin{bmatrix} 1 & 3 & 4 \\ 2 & 4 & 1 \\ -1 & 2 & 3 \end{bmatrix}$$

Then,

$$AB = \begin{bmatrix} 3\cdot 1 + 2\cdot 2 + 1\cdot(-1) & 3\cdot 3 + 2\cdot 4 + 1\cdot 2 & 3\cdot 4 + 2\cdot 1 + 1\cdot 3 \\ (-1)\cdot 1 + 2\cdot 2 + 3\cdot(-1) & (-1)\cdot 3 + 2\cdot 4 + 3\cdot 2 & (-1)\cdot 4 + 2\cdot 1 + 3\cdot 3 \\ 3\cdot 1 + 1\cdot 2 + 4\cdot(-1) & 3\cdot 3 + 1\cdot 4 + 4\cdot 2 & 3\cdot 4 + 1\cdot 1 + 4\cdot 3 \end{bmatrix}$$

$$= \begin{bmatrix} 6 & 19 & 17 \\ 0 & 11 & 7 \\ 1 & 21 & 25 \end{bmatrix}$$

$$BA = \begin{bmatrix} 1\cdot 3 + 3\cdot(-1) + 4\cdot 3 & 1\cdot 2 + 3\cdot 2 + 4\cdot 1 & 1\cdot 1 + 3\cdot 3 + 4\cdot 4 \\ 2\cdot 3 + 4\cdot(-1) + 1\cdot 3 & 2\cdot 2 + 4\cdot 2 + 1\cdot 1 & 2\cdot 1 + 4\cdot 3 + 1\cdot 4 \\ (-1)\cdot 3 + 2\cdot(-1) + 3\cdot 3 & (-1)\cdot 2 + 2\cdot 2 + 3\cdot 1 & (-1)\cdot 1 + 2\cdot 3 + 3\cdot 4 \end{bmatrix}$$

$$= \begin{bmatrix} 12 & 12 & 26 \\ 5 & 13 & 18 \\ 4 & 5 & 17 \end{bmatrix}$$

As the last example shows, in general, $AB \neq BA$ for any two square matrices A and B. However, the following laws are valid for matrix multiplication.

Laws for Matrix Multiplication

If the products and sums are defined for the matrices A, B, and C, then

1. $(AB)C = A(BC)$ (Associative law)

2. $A(B + C) = AB + AC$ (Distributive law)

The square matrix of size n having 1s along the main diagonal and zeros elsewhere is called the identity matrix of size n.

Identity Matrix

The **identity matrix** of size n is given by

$$I_n = \begin{bmatrix} 1 & 0 & . & . & . & 0 \\ 0 & 1 & . & . & . & 0 \\ . & . & & & & . \\ . & . & & & & . \\ . & . & & & & . \\ 0 & 0 & . & . & . & 1 \end{bmatrix} \quad n \text{ rows}$$

n columns

The identity matrix has the property that $I_nA = A$ for any $n \times r$ matrix A, and $BI_n = B$ for any $s \times n$ matrix B. In particular, if A is a square matrix of size n, then

$$I_nA = AI_n = A$$

EXAMPLE 5 Let

$$A = \begin{bmatrix} 1 & 3 & 1 \\ -4 & 3 & 2 \\ 1 & 0 & 1 \end{bmatrix}$$

Then,

$$I_3A = \begin{bmatrix} 1 & 0 & 0 \\ 0 & 1 & 0 \\ 0 & 0 & 1 \end{bmatrix}\begin{bmatrix} 1 & 3 & 1 \\ -4 & 3 & 2 \\ 1 & 0 & 1 \end{bmatrix} = \begin{bmatrix} 1 & 3 & 1 \\ -4 & 3 & 2 \\ 1 & 0 & 1 \end{bmatrix} = A$$

$$AI_3 = \begin{bmatrix} 1 & 3 & 1 \\ -4 & 3 & 2 \\ 1 & 0 & 1 \end{bmatrix}\begin{bmatrix} 1 & 0 & 0 \\ 0 & 1 & 0 \\ 0 & 0 & 1 \end{bmatrix} = \begin{bmatrix} 1 & 3 & 1 \\ -4 & 3 & 2 \\ 1 & 0 & 1 \end{bmatrix} = A$$

so $I_3A = AI_3$, confirming our result for this special case. ■■■■

APPLICATION

EXAMPLE 6 The Ace Novelty Company received an order from Magic World Amusement Park for 900 "Giant Pandas," 1200 "Saint Bernards," and 2000 "Big Birds." Ace's management decided that 500 Giant Pandas, 800 Saint Bernards, and 1300 Big Birds could be manufactured in their Los Angeles plant, and the balance of the order could be filled by their Seattle plant. Each Panda requires 1.5 square yards of plush, 30 cubic feet of stuffing, and 5 pieces of trim; each Saint Bernard requires 2 square yards of plush, 35 cubic feet of stuffing, and 8 pieces of trim; and each Big Bird requires 2.5 square yards of plush, 25 cubic feet of stuffing, and 15 pieces of trim. The plush costs \$4.50 per square yard, the stuffing costs 10 cents per cubic foot, and the trim costs 25 cents per unit.

a. Find how much of each type of material must be purchased for each plant.

b. What is the total cost of materials incurred by each plant and the total cost of materials incurred by Ace Novelty in filling the order?

SOLUTION ✓ The quantities of each type of stuffed animal to be produced at each plant location may be expressed as a 2×3 *production matrix P*. Thus,

$$P = \begin{array}{l} \text{L.A.} \\ \text{Seattle} \end{array} \overset{\begin{array}{ccc} \text{Pandas} & \text{St. Bernards} & \text{Birds} \end{array}}{\begin{bmatrix} 500 & 800 & 1300 \\ 400 & 400 & 700 \end{bmatrix}}$$

Similarly, we may represent the amount and type of material required to manufacture each type of animal by a 3×3 *activity matrix A*. Thus,

$$A = \begin{array}{l} \text{Pandas} \\ \text{St. Bernards} \\ \text{Birds} \end{array} \overset{\begin{array}{ccc} \text{Plush} & \text{Stuffing} & \text{Trim} \end{array}}{\begin{bmatrix} 1.5 & 30 & 5 \\ 2 & 35 & 8 \\ 2.5 & 25 & 15 \end{bmatrix}}$$

Finally, the unit cost for each type of material may be represented by the 3×1 *cost matrix C*:

$$C = \begin{array}{l} \text{Plush} \\ \text{Stuffing} \\ \text{Trim} \end{array} \begin{bmatrix} 4.50 \\ 0.10 \\ 0.25 \end{bmatrix}$$

a. The amount of each type of material required for each plant is given by the matrix PA. Thus,

$$PA = \begin{bmatrix} 500 & 800 & 1300 \\ 400 & 400 & 700 \end{bmatrix} \begin{bmatrix} 1.5 & 30 & 5 \\ 2 & 35 & 8 \\ 2.5 & 25 & 15 \end{bmatrix}$$

$$= \begin{array}{l} \text{L.A.} \\ \text{Seattle} \end{array} \overset{\begin{array}{ccc} \text{Plush} & \text{Stuffing} & \text{Trim} \end{array}}{\begin{bmatrix} 5600 & 75{,}500 & 28{,}400 \\ 3150 & 43{,}500 & 15{,}700 \end{bmatrix}}$$

b. The total cost of materials for each plant is given by the matrix PAC:

$$PAC = \begin{bmatrix} 5600 & 75{,}500 & 28{,}400 \\ 3150 & 43{,}500 & 15{,}700 \end{bmatrix} \begin{bmatrix} 4.50 \\ 0.10 \\ 0.25 \end{bmatrix}$$

$$= \begin{array}{l} \text{L.A.} \\ \text{Seattle} \end{array} \begin{bmatrix} 39{,}850 \\ 22{,}450 \end{bmatrix}$$

or \$39,850 for the L.A. plant and \$22,450 for the Seattle plant. Thus, the total cost of materials incurred by Ace Novelty is \$62,300. ■■■■

MATRIX REPRESENTATION

Example 7 shows how a system of linear equations may be written in a compact form with the help of matrices. (We will use this matrix equation representation in Section 2.6.)

EXAMPLE 7 Write the following system of linear equations in matrix form.

$$\begin{aligned} 2x - 4y + z &= 6 \\ -3x + 6y - 5z &= -1 \\ x - 3y + 7z &= 0 \end{aligned}$$

SOLUTION ✓ Let's write

$$A = \begin{bmatrix} 2 & -4 & 1 \\ -3 & 6 & -5 \\ 1 & -3 & 7 \end{bmatrix}, \quad X = \begin{bmatrix} x \\ y \\ z \end{bmatrix}, \quad B = \begin{bmatrix} 6 \\ -1 \\ 0 \end{bmatrix}$$

Note that A is just the 3×3 matrix of coefficients of the system, X is the 3×1 column matrix of unknowns (variables), and B is the 3×1 column matrix of constants. We now show that the required matrix representation of the system of linear equations is

$$AX = B$$

To see this, observe that

$$AX = \begin{bmatrix} 2 & -4 & 1 \\ -3 & 6 & -5 \\ 1 & -3 & 7 \end{bmatrix} \begin{bmatrix} x \\ y \\ z \end{bmatrix} = \begin{bmatrix} 2x - 4y + z \\ -3x + 6y - 5z \\ x - 3y + 7z \end{bmatrix}$$

Equating this 3×1 matrix with matrix B now gives

$$\begin{bmatrix} 2x - 4y + z \\ -3x + 6y - 5z \\ x - 3y + 7z \end{bmatrix} = \begin{bmatrix} 6 \\ -1 \\ 0 \end{bmatrix}$$

which, by matrix equality, is easily seen to be equivalent to the given system of linear equations. ■■■■

Self-Check Exercises 2.5

1. Compute

$$\begin{bmatrix} 1 & 3 & 0 \\ 2 & 4 & -1 \end{bmatrix}\begin{bmatrix} 3 & 1 & 4 \\ 2 & 0 & 3 \\ 1 & 2 & -1 \end{bmatrix}$$

2. Write the following system of linear equations in matrix form:

$$\begin{aligned} y - 2z &= 1 \\ 2x - y + 3z &= 0 \\ x \qquad + 4z &= 7 \end{aligned}$$

3. On June 1, the stock holdings of Ash and Joan Robinson were given by the matrix

$$A = \begin{array}{c} \text{Ash} \\ \text{Joan} \end{array} \overset{\begin{array}{cccc} \text{AT\&T} & \text{AOL} & \text{IBM} & \text{GM} \end{array}}{\begin{bmatrix} 2000 & 1000 & 500 & 5000 \\ 1000 & 2500 & 2000 & 0 \end{bmatrix}}$$

and the closing prices of AT&T, AOL, IBM, and GM were \$54, \$113, \$112, and \$70/share, respectively. Use matrix multiplication to determine the separate values of Ash's and Joan's stock holdings as of that date.

Solutions to Self-Check Exercises 2.5 can be found on page 147.

Spreadsheet examples and exercises for this section that may be solved using the Microsoft EXCEL program are given at the Brooks/Cole website: http://www.brookscole.com/math/authors/tans/

2.5 Exercises

In Exercises 1–4, the sizes of matrices *A* and *B* are given. Find the size of *AB* and *BA* whenever they are defined.

1. A is of size 2×3, and B is of size 3×5.

2. A is of size 3×4, and B is of size 4×3.

3. A is of size 1×7, and B is of size 7×1.

4. A is of size 4×4, and B is of size 4×4.

5. Let A be a matrix of size $m \times n$ and B be a matrix of size $s \times t$. Find conditions on m, n, s, and t so that both matrix products AB and BA are defined.

6. Find condition(s) on the size of a matrix A so that A^2 (that is, AA) is defined.

In Exercises 7–24, compute the indicated products.

7. $\begin{bmatrix} 1 & 2 \\ 3 & 0 \end{bmatrix}\begin{bmatrix} 1 \\ -1 \end{bmatrix}$

8. $\begin{bmatrix} -1 & 3 \\ 5 & 0 \end{bmatrix}\begin{bmatrix} 7 \\ 2 \end{bmatrix}$

9. $\begin{bmatrix} 3 & 1 & 2 \\ -1 & 2 & 4 \end{bmatrix}\begin{bmatrix} 4 \\ 1 \\ -2 \end{bmatrix}$

10. $\begin{bmatrix} 3 & 2 & -1 \\ 4 & -1 & 0 \\ -5 & 2 & 1 \end{bmatrix}\begin{bmatrix} 3 \\ -2 \\ 0 \end{bmatrix}$

11. $\begin{bmatrix} -1 & 2 \\ 3 & 1 \end{bmatrix}\begin{bmatrix} 2 & 4 \\ 3 & 1 \end{bmatrix}$

12. $\begin{bmatrix} 1 & 3 \\ -1 & 2 \end{bmatrix}\begin{bmatrix} 1 & 3 & 0 \\ 3 & 0 & 2 \end{bmatrix}$

13. $\begin{bmatrix} 2 & 1 & 2 \\ 3 & 2 & 4 \end{bmatrix}\begin{bmatrix} -1 & 2 \\ 4 & 3 \\ 0 & 1 \end{bmatrix}$

14. $\begin{bmatrix} -1 & 2 \\ 4 & 3 \\ 0 & 1 \end{bmatrix}\begin{bmatrix} 2 & 1 & 2 \\ 3 & 2 & 4 \end{bmatrix}$

15. $\begin{bmatrix} 0.1 & 0.9 \\ 0.2 & 0.8 \end{bmatrix}\begin{bmatrix} 1.2 & 0.4 \\ 0.5 & 2.1 \end{bmatrix}$

16. $\begin{bmatrix} 1.2 & 0.3 \\ 0.4 & 0.5 \end{bmatrix}\begin{bmatrix} 0.2 & 0.6 \\ 0.4 & -0.5 \end{bmatrix}$

17. $\begin{bmatrix} 6 & -3 & 0 \\ -2 & 1 & -8 \\ 4 & -4 & 9 \end{bmatrix}\begin{bmatrix} 1 & 0 & 0 \\ 0 & 1 & 0 \\ 0 & 0 & 1 \end{bmatrix}$

18. $\begin{bmatrix} 2 & 4 \\ -1 & -5 \\ 3 & -1 \end{bmatrix}\begin{bmatrix} 2 & -2 & 4 \\ 1 & 3 & -1 \end{bmatrix}$

19. $\begin{bmatrix} 3 & 0 & -2 & 1 \\ 1 & 2 & 0 & -1 \end{bmatrix}\begin{bmatrix} 2 & 1 & -1 \\ -1 & 2 & 0 \\ 0 & 0 & 1 \\ -1 & -2 & 2 \end{bmatrix}$

20. $\begin{bmatrix} 2 & 1 & -3 & 0 \\ 4 & -2 & -1 & 1 \\ -1 & 2 & 0 & 1 \end{bmatrix}\begin{bmatrix} 2 & -1 \\ 1 & 4 \\ 3 & -3 \\ 0 & -5 \end{bmatrix}$

21. $4\begin{bmatrix} 1 & -2 & 0 \\ 2 & -1 & 1 \\ 3 & 0 & -1 \end{bmatrix}\begin{bmatrix} 1 & 3 & 1 \\ 1 & 4 & 0 \\ 0 & 1 & -2 \end{bmatrix}$

22. $3\begin{bmatrix} 2 & -1 & 0 \\ 2 & 1 & 2 \\ 1 & 0 & -1 \end{bmatrix}\begin{bmatrix} 2 & 3 & 1 \\ 3 & -3 & 0 \\ 0 & 1 & -1 \end{bmatrix}$

23. $\begin{bmatrix} 1 & 0 \\ 0 & 1 \end{bmatrix}\begin{bmatrix} 4 & -3 & 2 \\ 7 & 1 & -5 \end{bmatrix}\begin{bmatrix} 1 & 0 & 0 \\ 0 & 1 & 0 \\ 0 & 0 & 1 \end{bmatrix}$

24. $2\begin{bmatrix} 3 & 2 & -1 \\ 0 & 1 & 3 \\ 2 & 0 & 3 \end{bmatrix}\begin{bmatrix} 1 & 0 & 0 \\ 0 & 1 & 0 \\ 0 & 0 & 1 \end{bmatrix}\begin{bmatrix} 1 & 2 & 0 \\ 0 & -1 & -2 \\ 1 & 3 & 1 \end{bmatrix}$

In Exercises 25 and 26, let

$$A = \begin{bmatrix} 1 & 0 & -2 \\ 1 & -3 & 2 \\ -2 & 1 & 1 \end{bmatrix}, \quad B = \begin{bmatrix} 3 & 1 & 0 \\ 2 & 2 & 0 \\ 1 & -3 & -1 \end{bmatrix},$$

$$C = \begin{bmatrix} 2 & -1 & 0 \\ 1 & -1 & 2 \\ 3 & -2 & 1 \end{bmatrix}$$

25. Verify the validity of the associative law for matrix multiplication.

26. Verify the validity of the distributive law for matrix multiplication.

27. Let

$$A = \begin{bmatrix} 1 & 2 \\ 3 & 4 \end{bmatrix} \quad \text{and} \quad B = \begin{bmatrix} 2 & 1 \\ 4 & 3 \end{bmatrix}$$

Compute AB and BA and hence deduce that matrix multiplication is, in general, not commutative.

28. Let

$$A = \begin{bmatrix} 0 & 3 & 0 \\ 1 & 0 & 1 \\ 0 & 2 & 0 \end{bmatrix}, \quad B = \begin{bmatrix} 2 & 4 & 5 \\ 3 & -1 & -6 \\ 4 & 3 & 4 \end{bmatrix},$$

$$C = \begin{bmatrix} 4 & 5 & 6 \\ 3 & -1 & -6 \\ 2 & 2 & 3 \end{bmatrix}$$

a. Compute AB.
b. Compute AC.
c. Using the results of parts (a) and (b), conclude that $AB = AC$ does *not* imply that $B = C$.

29. Let

$$A = \begin{bmatrix} 3 & 0 \\ 8 & 0 \end{bmatrix} \quad \text{and} \quad B = \begin{bmatrix} 0 & 0 \\ 4 & 5 \end{bmatrix}$$

Show that $AB = O$, thereby demonstrating that for matrix multiplication the equation $AB = O$ does not imply that one or both of the matrices A and B must be the zero matrix.

30. Let

$$A = \begin{bmatrix} 2 & 2 \\ -2 & -2 \end{bmatrix}$$

Show that $A^2 = O$. Compare this with the equation $a^2 = 0$, where a is a real number.

31. Find the matrix A such that

$$A\begin{bmatrix} 1 & 0 \\ -1 & 3 \end{bmatrix} = \begin{bmatrix} -1 & -3 \\ 3 & 6 \end{bmatrix}$$

Hint: Let $A = \begin{bmatrix} a & b \\ c & d \end{bmatrix}$.

32. Let

$$A = \begin{bmatrix} 3 & 1 \\ 0 & 2 \end{bmatrix} \quad \text{and} \quad B = \begin{bmatrix} 4 & -2 \\ 2 & 1 \end{bmatrix}$$

a. Compute $(A + B)^2$.
b. Compute $A^2 + 2AB + B^2$.
c. From the results of parts (a) and (b), show that in general $(A + B)^2 \neq A^2 + 2AB + B^2$.

33. Let

$$A = \begin{bmatrix} 2 & 4 \\ 5 & -6 \end{bmatrix} \quad \text{and} \quad B = \begin{bmatrix} 4 & 8 \\ -7 & 3 \end{bmatrix}$$

a. Find A^T and show that $(A^T)^T = A$.
b. Show that $(A + B)^T = A^T + B^T$.
c. Show that $(AB)^T = B^T A^T$.

34. Let

$$A = \begin{bmatrix} 1 & 3 \\ -2 & -1 \end{bmatrix} \quad \text{and} \quad B = \begin{bmatrix} 3 & -4 \\ 2 & -2 \end{bmatrix}$$

a. Find A^T and show that $(A^T)^T = A$.
b. Show that $(A + B)^T = A^T + B^T$.
c. Show that $(AB)^T = B^T A^T$.

In Exercises 35–40, write the given system of linear equations in matrix form.

35. $\begin{aligned} 2x - 3y &= 7 \\ 3x - 4y &= 8 \end{aligned}$

36. $\begin{aligned} 2x &= 7 \\ 3x - 2y &= 12 \end{aligned}$

37. $\begin{aligned} 2x - 3y + 4z &= 6 \\ 2y - 3z &= 7 \\ x - y + 2z &= 4 \end{aligned}$

38. $\begin{aligned} x - 2y + 3z &= -1 \\ 3x + 4y - 2z &= 1 \\ 2x - 3y + 7z &= 6 \end{aligned}$

39. $\begin{aligned} -x_1 + x_2 + x_3 &= 0 \\ 2x_1 - x_2 - x_3 &= 2 \\ -3x_1 + 2x_2 + 4x_3 &= 4 \end{aligned}$

40. $\begin{aligned} 3x_1 - 5x_2 + 4x_3 &= 10 \\ 4x_1 + 2x_2 - 3x_3 &= -12 \\ -x_1 + x_3 &= -2 \end{aligned}$

41. INVESTMENTS William and Michael's stock holdings are given by the matrix

$$A = \begin{array}{c} \\ \text{William} \\ \text{Michael} \end{array} \begin{array}{c} \begin{array}{cccc} \text{BAC} & \text{GM} & \text{IBM} & \text{TRW} \end{array} \\ \begin{bmatrix} 200 & 300 & 100 & 200 \\ 100 & 200 & 400 & 0 \end{bmatrix} \end{array}$$

At the close of trading on a certain day, the prices (in dollars per share) of the stocks are given by the matrix

$$B = \begin{array}{c} \text{BAC} \\ \text{GM} \\ \text{IBM} \\ \text{TRW} \end{array} \begin{bmatrix} 54 \\ 48 \\ 98 \\ 82 \end{bmatrix}$$

a. Find AB.
b. Explain the meaning of the entries in the matrix AB.

42. BOX-OFFICE RECEIPTS Four theaters comprise the Cinema Center: cinemas I, II, III, and IV. The admission price for one feature at the Center is \$2 for children, \$3 for students, and \$4 for adults. The attendance for the Sunday matinee is given by the matrix

$$A = \begin{array}{c} \\ \text{Cinema I} \\ \text{Cinema II} \\ \text{Cinema III} \\ \text{Cinema IV} \end{array} \begin{array}{c} \begin{array}{ccc} \text{Children} & \text{Students} & \text{Adults} \end{array} \\ \begin{bmatrix} 225 & 110 & 50 \\ 75 & 180 & 225 \\ 280 & 85 & 110 \\ 0 & 250 & 225 \end{bmatrix} \end{array}$$

Write a column vector B representing the admission prices. Then compute AB, the column vector showing the gross receipts for each theater. Finally, find the total revenue collected at the Cinema Center for admission that Sunday afternoon.

43. REAL ESTATE Bond Brothers, Inc., a real estate developer, builds houses in three states. The projected number of units of each model to be built in each state is given by the matrix

$$A = \begin{array}{c} \\ \\ \text{N.Y.} \\ \text{Conn.} \\ \text{Mass.} \end{array} \begin{array}{c} \text{Model} \\ \begin{array}{cccc} \text{I} & \text{II} & \text{III} & \text{IV} \end{array} \\ \begin{bmatrix} 60 & 80 & 120 & 40 \\ 20 & 30 & 60 & 10 \\ 10 & 15 & 30 & 5 \end{bmatrix} \end{array}$$

The profits to be realized are \$20,000, \$22,000, \$25,000, and \$30,000, respectively, for each model I, II, III, and IV house sold.
a. Write a column matrix B representing the profit for each type of house.
b. Find the total profit Bond Brothers expects to earn in each state if all the houses are sold.

44. 401K RETIREMENT PLANS Three network consultants, Alan, Maria, and Steven, each received a year-end bonus of \$10,000, which they decided to invest in a 401K retirement plan sponsored by their employer. Under this plan, each employee is allowed to place their investments in three funds—an equity index fund (I), a growth fund (II), and a global equity fund (III). The allocations of the investments (in dollars) of the three employees at the beginning of the year are summarized in the matrix

$$A = \begin{array}{l} \text{Alan} \\ \text{Maria} \\ \text{Steven} \end{array} \overset{\begin{array}{ccc} \text{I} & \text{II} & \text{III} \end{array}}{\begin{bmatrix} 4000 & 3000 & 3000 \\ 2000 & 5000 & 3000 \\ 2000 & 3000 & 5000 \end{bmatrix}}$$

The returns of the three funds after 1 yr are given in the matrix

$$B = \begin{array}{c} \text{I} \\ \text{II} \\ \text{III} \end{array} \begin{bmatrix} 0.18 \\ 0.24 \\ 0.12 \end{bmatrix}$$

Which employee realized the best returns on his or her investment for the year in question? The worst return?

45. **POLITICS: VOTER AFFILIATION** Matrix A gives the percentage of eligible voters in the city of Newton, classified according to party affiliation and age group.

$$A = \begin{array}{l} \text{Under 30} \\ \text{30 to 50} \\ \text{Over 50} \end{array} \overset{\begin{array}{ccc} \text{Dem.} & \text{Rep.} & \text{Ind.} \end{array}}{\begin{bmatrix} 0.50 & 0.30 & 0.20 \\ 0.45 & 0.40 & 0.15 \\ 0.40 & 0.50 & 0.10 \end{bmatrix}}$$

The population of eligible voters in the city by age group is given by the matrix B:

$$B = \overset{\begin{array}{ccc} \text{Under 30} & \text{30 to 50} & \text{Over 50} \end{array}}{[30{,}000 \quad 40{,}000 \quad 20{,}000]}$$

Find a matrix giving the total number of eligible voters in the city who will vote Democratic, Republican, and Independent.

46. **PRODUCTION PLANNING** Refer to Example 6 in this section. Suppose Ace Novelty received an order from another amusement park for 1200 Pink Panthers, 1800 Giant Pandas, and 1400 Big Birds. The quantity of each type of stuffed animal to be produced at each plant is shown in the following production matrix:

$$P = \begin{array}{l} \text{L.A.} \\ \text{Seattle} \end{array} \overset{\begin{array}{ccc} \text{Panthers} & \text{Pandas} & \text{Birds} \end{array}}{\begin{bmatrix} 700 & 1000 & 800 \\ 500 & 800 & 600 \end{bmatrix}}$$

Each Panther requires 1.3 yd^2 of plush, 20 ft^3 of stuffing, and 12 pieces of trim. Assume the materials required to produce the other two stuffed animals and the unit cost for each type of material are the same as those given in Example 6.

a. How much of each type of material must be purchased for each plant?

b. What is the total cost of materials that will be incurred at each plant?

c. What is the total cost of materials incurred by Ace Novelty in filling the order?

47. **COLLEGE ADMISSIONS** A university admissions committee anticipates an enrollment of 8000 students in its freshman class next year. To satisfy admission quotas, incoming students have been categorized according to their sex and place of residence. The number of students in each category is given by the matrix

$$A = \begin{array}{l} \text{In-state} \\ \text{Out-of-state} \\ \text{Foreign} \end{array} \overset{\begin{array}{cc} \text{Male} & \text{Female} \end{array}}{\begin{bmatrix} 2700 & 3000 \\ 800 & 700 \\ 500 & 300 \end{bmatrix}}$$

By using data accumulated in previous years, the admissions committee has determined that these students will elect to enter the College of Letters and Science, the College of Fine Arts, the School of Business Administration, and the School of Engineering according to the percentages that appear in the following matrix:

$$B = \begin{array}{l} \text{Male} \\ \text{Female} \end{array} \overset{\begin{array}{cccc} \text{L. \& S.} & \text{Fine Arts} & \text{Bus. Ad.} & \text{Eng.} \end{array}}{\begin{bmatrix} 0.25 & 0.20 & 0.30 & 0.25 \\ 0.30 & 0.35 & 0.25 & 0.10 \end{bmatrix}}$$

Find the matrix AB that shows the number of in-state, out-of-state, and foreign students expected to enter each discipline.

48. **COMPUTING PHONE BILLS** Cindy regularly makes long distance phone calls to three foreign cities—London, Tokyo, and Hong Kong. The matrices A and B give the lengths (in minutes) of her calls during peak and nonpeak hours, respectively, to each of these three cities during the month of June.

$$A = \overset{\begin{array}{ccc} \text{London} & \text{Tokyo} & \text{Hong Kong} \end{array}}{[\,80 \quad 60 \quad 40\,]}$$

and

$$B = \overset{\begin{array}{ccc} \text{London} & \text{Tokyo} & \text{Hong Kong} \end{array}}{[\,300 \quad 150 \quad 250\,]}$$

(continued on p. 146)

Using Technology

MATRIX MULTIPLICATION

The graphing utility can be used to perform matrix multiplication.

EXAMPLE 1 Let

$$A = \begin{bmatrix} 1.2 & 3.1 & -1.4 \\ 2.7 & 4.2 & 3.4 \end{bmatrix}$$

$$B = \begin{bmatrix} 0.8 & 1.2 & 3.7 \\ 6.2 & -0.4 & 3.3 \end{bmatrix}$$

$$C = \begin{bmatrix} 1.2 & 2.1 & 1.3 \\ 4.2 & -1.2 & 0.6 \\ 1.4 & 3.2 & 0.7 \end{bmatrix}$$

Find $(1.1A + 2.3B)C^T$.

SOLUTION ✓ First, we enter the matrices A, B, and C into the calculator. Then, using matrix operations, we enter the expression $(1.1A + 2.3B)C^T$. We obtain the desired matrix:

$$(1.1A + 2.3B)C^T = \begin{bmatrix} 25.81 & 10.05 & 29.047 \\ 43.175 & 74.724 & 43.893 \end{bmatrix}$$

■■■■

Exercises

In Exercises 1–8, refer to the following matrices and use a graphing utility to perform the indicated operations. Express your answers accurate to two decimal places.

$$A = \begin{bmatrix} 1.2 & 3.1 & -1.2 & 4.3 \\ 7.2 & 6.3 & 1.8 & -2.1 \\ 0.8 & 3.2 & -1.3 & 2.8 \end{bmatrix}$$

$$B = \begin{bmatrix} 0.7 & 0.3 & 1.2 & -0.8 \\ 1.2 & 1.7 & 3.5 & 4.2 \\ -3.3 & -1.2 & 4.2 & 3.2 \end{bmatrix}$$

$$C = \begin{bmatrix} 0.8 & 7.1 & 6.2 \\ 3.3 & -1.2 & 4.8 \\ 1.3 & 2.8 & -1.5 \\ 2.1 & 3.2 & -8.4 \end{bmatrix}$$

1. AC

2. CB

3. $(A + B)C$

4. $(2A + 3B)C$

5. $(2A - 3.1B)C$

6. $C(2.1A + 3.2B)$

7. $(AC)^T$

8. $(CA)^T$

In Exercises 9–12, refer to the following matrices and use a graphing utility to perform the indicated operations. Express your answers accurate to two decimal places.

$$A = \begin{bmatrix} 2 & 5 & -4 & 2 & 8 \\ 6 & 7 & 2 & 9 & 6 \\ 4 & 5 & 4 & 4 & 4 \\ 9 & 6 & 8 & 3 & 2 \end{bmatrix} \qquad B = \begin{bmatrix} 2 & 6 & 7 & 5 \\ 3 & 4 & 6 & 2 \\ -5 & 8 & 4 & 3 \\ 8 & 6 & 9 & 5 \\ 4 & 7 & 8 & 8 \end{bmatrix}$$

$$C = \begin{bmatrix} 6.2 & 7.3 & -4.0 & 7.1 & 9.3 \\ 4.8 & 6.5 & 8.4 & -6.3 & 8.4 \\ 5.4 & 3.2 & 6.3 & 9.1 & -2.8 \\ 8.2 & 7.3 & 6.5 & 4.1 & 9.8 \\ 10.3 & 6.8 & 4.8 & -9.1 & 20.4 \end{bmatrix}$$

$$D = \begin{bmatrix} 4.6 & 3.9 & 8.4 & 6.1 & 9.8 \\ 2.4 & -6.8 & 7.9 & 11.4 & 2.9 \\ 7.1 & 9.4 & 6.3 & 5.7 & 4.2 \\ 3.4 & 6.1 & 5.3 & 8.4 & 6.3 \\ 7.1 & -4.2 & 3.9 & -6.4 & 7.1 \end{bmatrix}$$

9. Find AB and BA.

10. Find CD and DC. Is $CD = DC$?

11. Find $AC + AD$.

12. Find
 a. AC
 b. AD
 c. $A(C + D)$
 d. Is $A(C + D) = AC + AD$?

The costs for the calls (in dollars per minute) for the peak and nonpeak periods in the month in question are given, respectively, by the matrices

$$C = \begin{matrix} \text{London} \\ \text{Tokyo} \\ \text{Hong Kong} \end{matrix} \begin{bmatrix} .34 \\ .42 \\ .48 \end{bmatrix} \quad \text{and} \quad D = \begin{matrix} \text{London} \\ \text{Tokyo} \\ \text{Hong Kong} \end{matrix} \begin{bmatrix} .24 \\ .31 \\ .35 \end{bmatrix}$$

Compute the matrix $AC + BD$ and explain what it represents.

49. **Production Planning** The total output of loudspeaker systems of the Acrosonic Company in their three production facilities for May and June is given by the matrices A and B, respectively, where

$$A = \begin{matrix} \\ \text{Location I} \\ \text{Location II} \\ \text{Location II} \end{matrix} \begin{matrix} \text{Model A} & \text{Model B} & \text{Model C} & \text{Model D} \\ \left[\begin{matrix} 320 \\ 480 \\ 540 \end{matrix}\right. & \begin{matrix} 280 \\ 360 \\ 420 \end{matrix} & \begin{matrix} 460 \\ 580 \\ 200 \end{matrix} & \left.\begin{matrix} 280 \\ 0 \\ 880 \end{matrix}\right] \end{matrix}$$

$$B = \begin{matrix} \\ \text{Location I} \\ \text{Location II} \\ \text{Location III} \end{matrix} \begin{matrix} \text{Model A} & \text{Model B} & \text{Model C} & \text{Model D} \\ \left[\begin{matrix} 210 \\ 400 \\ 420 \end{matrix}\right. & \begin{matrix} 180 \\ 300 \\ 280 \end{matrix} & \begin{matrix} 330 \\ 450 \\ 180 \end{matrix} & \left.\begin{matrix} 180 \\ 40 \\ 740 \end{matrix}\right] \end{matrix}$$

The unit production costs and selling prices for these loudspeakers are given by matrices C and D, respectively, where

$$C = \begin{matrix} \text{Model A} \\ \text{Model B} \\ \text{Model C} \\ \text{Model D} \end{matrix} \begin{bmatrix} 120 \\ 180 \\ 260 \\ 500 \end{bmatrix} \quad \text{and} \quad D = \begin{matrix} \text{Model A} \\ \text{Model B} \\ \text{Model C} \\ \text{Model D} \end{matrix} \begin{bmatrix} 160 \\ 250 \\ 350 \\ 700 \end{bmatrix}$$

Compute the following matrices and explain the meaning of the entries in each matrix.

a. AC **b.** AD **c.** BC **d.** BD
e. $(A + B)C$ **f.** $(A + B)D$
g. $A(D - C)$ **h.** $B(D - C)$
i. $(A + B)(D - C)$

50. **Diet Planning** A dietitian plans a meal around three foods. The number of units of vitamin A, vitamin C, and calcium in each ounce of these foods is represented by the matrix M, where

$$M = \begin{matrix} \\ \text{Vitamin A} \\ \text{Vitamin C} \\ \text{Calcium} \end{matrix} \begin{matrix} \text{Food I} & \text{Food II} & \text{Food III} \\ \left[\begin{matrix} 400 \\ 110 \\ 90 \end{matrix}\right. & \begin{matrix} 1200 \\ 570 \\ 30 \end{matrix} & \left.\begin{matrix} 800 \\ 340 \\ 60 \end{matrix}\right] \end{matrix}$$

The matrices A and B represent the amount of each food (in ounces) consumed by a girl at two different meals, where

$$A = \begin{matrix} \text{Food I} & \text{Food II} & \text{Food III} \\ [7 & 1 & 6] \end{matrix}$$

$$B = \begin{matrix} \text{Food I} & \text{Food II} & \text{Food III} \\ [9 & 3 & 2] \end{matrix}$$

Calculate the following matrices and explain the meaning of the entries in each matrix:

a. MA^T **b.** MB^T **c.** $M(A + B)^T$

In Exercises 51 and 52, determine whether the statement is true or false. If it is true, explain why it is true. If it is false, give an example to show why it is false.

51. If A and B are matrices such that AB and BA are both defined, then A and B must be square matrices of the same order.

52. If A and B are matrices such that AB is defined and if c is a scalar, then $(cA)B = A(cB) = cAB$.

SOLUTIONS TO SELF-CHECK EXERCISES 2.5

1. We compute

$$\begin{bmatrix} 1 & 3 & 0 \\ 2 & 4 & -1 \end{bmatrix} \begin{bmatrix} 3 & 1 & 4 \\ 2 & 0 & 3 \\ 1 & 2 & -1 \end{bmatrix}$$

$$= \begin{bmatrix} 1(3)+3(2)+0(1) & 1(1)+3(0)+0(2) & 1(4)+3(3)+0(-1) \\ 2(3)+4(2)-1(1) & 2(1)+4(0)-1(2) & 2(4)+4(3)-1(-1) \end{bmatrix}$$

$$= \begin{bmatrix} 9 & 1 & 13 \\ 13 & 0 & 21 \end{bmatrix}$$

2. Let

$$A = \begin{bmatrix} 0 & 1 & -2 \\ 2 & -1 & 3 \\ 1 & 0 & 4 \end{bmatrix}, \quad X = \begin{bmatrix} x \\ y \\ z \end{bmatrix}, \quad \text{and} \quad B = \begin{bmatrix} 1 \\ 0 \\ 7 \end{bmatrix}$$

Then, the given system may be written as the matrix equation

$$AX = B$$

3. Write

$$B = \begin{bmatrix} 54 \\ 113 \\ 112 \\ 70 \end{bmatrix} \begin{matrix} \text{AT\&T} \\ \text{AOL} \\ \text{IBM} \\ \text{GM} \end{matrix}$$

and compute

$$AB = \begin{matrix} \text{Ash} \\ \text{Joan} \end{matrix} \begin{bmatrix} 2000 & 1000 & 500 & 5000 \\ 1000 & 2500 & 2000 & 0 \end{bmatrix} \begin{bmatrix} 54 \\ 113 \\ 112 \\ 70 \end{bmatrix}$$

$$= \begin{bmatrix} 627{,}000 \\ 560{,}500 \end{bmatrix} \begin{matrix} \text{Ash} \\ \text{Joan} \end{matrix}$$

We conclude that Ash's stock holdings were worth \$627,000 and Joan's stock holdings were worth \$560,500 on June 1.

2.6 The Inverse of a Square Matrix

THE INVERSE OF A SQUARE MATRIX

In this section we discuss a procedure for finding the inverse of a matrix and show how the inverse can be used to help us solve a system of linear equations. The inverse of a matrix also plays a central role in the Leontief input–output model, which we will discuss in Section 2.7.

Recall that if a is a nonzero real number, then there exists a unique real number a^{-1} (that is, $1/a$) such that

$$a^{-1}a = \left(\frac{1}{a}\right)(a) = 1$$

The use of the (multiplicative) inverse of a real number enables us to solve algebraic equations of the form

$$ax = b \tag{12}$$

For if $a \neq 0$, then $a^{-1} = 1/a$. Upon multiplying both sides of (12) by a^{-1}, we have

$$a^{-1}(ax) = a^{-1}b$$

$$\left(\frac{1}{a}\right)(ax) = \frac{1}{a}(b)$$

$$x = \frac{b}{a}$$

For example, since the inverse of 2 is $2^{-1} = 1/2$, we can solve the equation

$$2x = 5$$

by multiplying both sides of the equation by $2^{-1} = 1/2$, giving

$$2^{-1}(2x) = 2^{-1} \cdot 5$$

$$x = \frac{5}{2}$$

We can use a similar procedure to solve the matrix equation

$$AX = B$$

where A, X, and B are matrices of the proper sizes. To do this we need the matrix equivalent of the inverse of a real number. Such a matrix, whenever it exists, is called the *inverse of a matrix*. More specifically, we have the following definition.

Inverse of a Matrix

Let A be a square matrix of size n. A square matrix A^{-1} of size n such that

$$A^{-1}A = AA^{-1} = I_n$$

is called the **inverse** of A.

Group Discussion
In defining the inverse of a matrix A, why is it necessary to require that A be a square matrix?

Let's show that the matrix

$$A = \begin{bmatrix} 1 & 2 \\ 3 & 4 \end{bmatrix}$$

has as its inverse

$$A^{-1} = \begin{bmatrix} -2 & 1 \\ \frac{3}{2} & -\frac{1}{2} \end{bmatrix}$$

Since

$$AA^{-1} = \begin{bmatrix} 1 & 2 \\ 3 & 4 \end{bmatrix}\begin{bmatrix} -2 & 1 \\ \frac{3}{2} & -\frac{1}{2} \end{bmatrix} = \begin{bmatrix} 1 & 0 \\ 0 & 1 \end{bmatrix}$$

$$A^{-1}A = \begin{bmatrix} -2 & 1 \\ \frac{3}{2} & -\frac{1}{2} \end{bmatrix}\begin{bmatrix} 1 & 2 \\ 3 & 4 \end{bmatrix} = \begin{bmatrix} 1 & 0 \\ 0 & 1 \end{bmatrix} = I$$

we see that A^{-1} is the inverse of A, as asserted.

Not every square matrix has an inverse. A matrix that does not have an inverse is called **singular.** An example of a singular matrix is given by

$$B = \begin{bmatrix} 0 & 1 \\ 0 & 0 \end{bmatrix}$$

If B had an inverse given by

$$B^{-1} = \begin{bmatrix} a & b \\ c & d \end{bmatrix}$$

where a, b, c, and d are some appropriate numbers, then, by the definition of an inverse, we would have $BB^{-1} = I$; that is,

$$\begin{bmatrix} 0 & 1 \\ 0 & 0 \end{bmatrix}\begin{bmatrix} a & b \\ c & d \end{bmatrix} = \begin{bmatrix} 1 & 0 \\ 0 & 1 \end{bmatrix}$$

$$\begin{bmatrix} c & d \\ 0 & 0 \end{bmatrix} = \begin{bmatrix} 1 & 0 \\ 0 & 1 \end{bmatrix}$$

which implies that $0 = 1$—an impossibility! This contradiction shows that B does not have an inverse.

A Method for Finding the Inverse of a Square Matrix

The methods of Section 2.5 can be used to find the inverse of a nonsingular matrix. To discover such an algorithm, let's find the inverse of the matrix A, given by

$$A = \begin{bmatrix} 1 & 2 \\ -1 & 3 \end{bmatrix}$$

Suppose A^{-1} exists and is given by

$$A^{-1} = \begin{bmatrix} a & b \\ c & d \end{bmatrix}$$

where a, b, c, and d are to be determined. By the definition of an inverse, we have $AA^{-1} = I$; that is,

$$\begin{bmatrix} 1 & 2 \\ -1 & 3 \end{bmatrix}\begin{bmatrix} a & b \\ c & d \end{bmatrix} = \begin{bmatrix} 1 & 0 \\ 0 & 1 \end{bmatrix}$$

which simplifies to

$$\begin{bmatrix} a + 2c & b + 2d \\ -a + 3c & -b + 3d \end{bmatrix} = \begin{bmatrix} 1 & 0 \\ 0 & 1 \end{bmatrix}$$

But this matrix equation is equivalent to the two systems of linear equations

$$\left.\begin{aligned} a + 2c &= 1 \\ -a + 3c &= 0 \end{aligned}\right\} \quad \text{and} \quad \left.\begin{aligned} b + 2d &= 0 \\ -b + 3d &= 1 \end{aligned}\right\}$$

with augmented matrices given by

$$\left[\begin{array}{rr|r} 1 & 2 & 1 \\ -1 & 3 & 0 \end{array}\right] \quad \text{and} \quad \left[\begin{array}{rr|r} 1 & 2 & 0 \\ -1 & 3 & 1 \end{array}\right]$$

Note that the matrices of coefficients of the two systems are identical. This suggests that we solve the two systems of simultaneous linear equations by writing the following augmented matrix, which we obtain by joining the coefficient matrix and the two columns of constants:

$$\left[\begin{array}{rr|rr} 1 & 2 & 1 & 0 \\ -1 & 3 & 0 & 1 \end{array}\right]$$

Using the Gauss–Jordan elimination method, we obtain the following sequence of equivalent matrices:

$$\left[\begin{array}{rr|rr} 1 & 2 & 1 & 0 \\ -1 & 3 & 0 & 1 \end{array}\right] \xrightarrow{R_2 + R_1} \left[\begin{array}{rr|rr} 1 & 2 & 1 & 0 \\ 0 & 5 & 1 & 1 \end{array}\right] \xrightarrow{\frac{1}{5}R_2}$$

$$\left[\begin{array}{rr|rr} 1 & 2 & 1 & 0 \\ 0 & 1 & \frac{1}{5} & \frac{1}{5} \end{array}\right] \xrightarrow{R_1 - 2R_2} \left[\begin{array}{rr|rr} 1 & 0 & \frac{3}{5} & -\frac{2}{5} \\ 0 & 1 & \frac{1}{5} & \frac{1}{5} \end{array}\right]$$

Thus, $a = 3/5$, $c = 1/5$, $b = -2/5$, and $d = 1/5$, giving

$$A^{-1} = \begin{bmatrix} \frac{3}{5} & -\frac{2}{5} \\ \frac{1}{5} & \frac{1}{5} \end{bmatrix}$$

The following computations verify that A^{-1} is indeed the inverse of A:

$$\begin{bmatrix} 1 & 2 \\ -1 & 3 \end{bmatrix} \begin{bmatrix} \frac{3}{5} & -\frac{2}{5} \\ \frac{1}{5} & \frac{1}{5} \end{bmatrix} = \begin{bmatrix} 1 & 0 \\ 0 & 1 \end{bmatrix} = \begin{bmatrix} \frac{3}{5} & -\frac{2}{5} \\ \frac{1}{5} & \frac{1}{5} \end{bmatrix} \begin{bmatrix} 1 & 2 \\ -1 & 3 \end{bmatrix}$$

The preceding example suggests the general algorithm for computing the inverse of a square matrix of size n when it exists.

Finding the Inverse of a Matrix

Given the $n \times n$ matrix A:

1. Adjoin the $n \times n$ identity matrix I to obtain the augmented matrix

$$[A \mid I]$$

2. Use a sequence of row operations to reduce $[A \mid I]$ to the form

$$[I \mid B]$$

if possible.

The matrix B is the inverse of A.

REMARK Although matrix multiplication is not generally commutative, it is possible to prove that if $AB = I$, then $BA = I$ also. Hence, to verify that B is the inverse of A, it suffices to show that $AB = I$. ■■■

EXAMPLE 1 Find the inverse of the matrix

$$A = \begin{bmatrix} 2 & 1 & 1 \\ 3 & 2 & 1 \\ 2 & 1 & 2 \end{bmatrix}$$

SOLUTION ✓ We form the augmented matrix

$$\left[\begin{array}{ccc|ccc} 2 & 1 & 1 & 1 & 0 & 0 \\ 3 & 2 & 1 & 0 & 1 & 0 \\ 2 & 1 & 2 & 0 & 0 & 1 \end{array}\right]$$

and use the Gauss–Jordan elimination method to reduce it to the form $[I \mid B]$:

$$\left[\begin{array}{ccc|ccc} 2 & 1 & 1 & 1 & 0 & 0 \\ 3 & 2 & 1 & 0 & 1 & 0 \\ 2 & 1 & 2 & 0 & 0 & 1 \end{array}\right] \xrightarrow{R_1 - R_2} \left[\begin{array}{ccc|ccc} -1 & -1 & 0 & 1 & -1 & 0 \\ 3 & 2 & 1 & 0 & 1 & 0 \\ 2 & 1 & 2 & 0 & 0 & 1 \end{array}\right]$$

$$\xrightarrow[\substack{R_2 + 3R_1 \\ R_3 + 2R_1}]{-R_1} \left[\begin{array}{ccc|ccc} 1 & 1 & 0 & -1 & 1 & 0 \\ 0 & -1 & 1 & 3 & -2 & 0 \\ 0 & -1 & 2 & 2 & -2 & 1 \end{array}\right]$$

$$\xrightarrow[\substack{-R_2 \\ R_3 - R_2}]{R_1 + R_2} \left[\begin{array}{ccc|ccc} 1 & 0 & 1 & 2 & -1 & 0 \\ 0 & 1 & -1 & -3 & 2 & 0 \\ 0 & 0 & 1 & -1 & 0 & 1 \end{array}\right]$$

$$\xrightarrow[R_2 + R_3]{R_1 - R_3} \left[\begin{array}{ccc|ccc} 1 & 0 & 0 & 3 & -1 & -1 \\ 0 & 1 & 0 & -4 & 2 & 1 \\ 0 & 0 & 1 & -1 & 0 & 1 \end{array}\right]$$

The inverse of A is the matrix

$$A^{-1} = \begin{bmatrix} 3 & -1 & -1 \\ -4 & 2 & 1 \\ -1 & 0 & 1 \end{bmatrix}$$

We leave it to you to verify these results. ■■■■

Example 2 illustrates what happens to the reduction process when a matrix A does not have an inverse.

EXAMPLE 2 Find the inverse of the matrix

$$A = \begin{bmatrix} 1 & 2 & 3 \\ 2 & 1 & 2 \\ 3 & 3 & 5 \end{bmatrix}$$

SOLUTION ✔ We form the augmented matrix

$$\left[\begin{array}{ccc|ccc} 1 & 2 & 3 & 1 & 0 & 0 \\ 2 & 1 & 2 & 0 & 1 & 0 \\ 3 & 3 & 5 & 0 & 0 & 1 \end{array}\right]$$

and use the Gauss–Jordan elimination method:

$$\left[\begin{array}{ccc|ccc} 1 & 2 & 3 & 1 & 0 & 0 \\ 2 & 1 & 2 & 0 & 1 & 0 \\ 3 & 3 & 5 & 0 & 0 & 1 \end{array}\right] \xrightarrow[R_3 - 3R_1]{R_2 - 2R_1} \left[\begin{array}{ccc|ccc} 1 & 2 & 3 & 1 & 0 & 0 \\ 0 & -3 & -4 & -2 & 1 & 0 \\ 0 & -3 & -4 & -3 & 0 & 1 \end{array}\right]$$

$$\xrightarrow[R_3 - R_2]{-R_2} \left[\begin{array}{ccc|ccc} 1 & 2 & 3 & 1 & 0 & 0 \\ 0 & 3 & 4 & 2 & -1 & 0 \\ 0 & 0 & 0 & -1 & -1 & 1 \end{array}\right]$$

Since all entries in the last row of the 3×3 submatrix that comprises the left-hand side of the augmented matrix just obtained are all equal to zero, the latter cannot be reduced to the form $[I \mid B]$. Accordingly, we draw the conclusion that A is singular—that is, does not have an inverse. ■■■■

More generally, we have the following criterion for determining when the inverse of a matrix does not exist.

Matrices That Have No Inverses

If there is a row to the left of the vertical line in the augmented matrix containing all zeros, then the matrix does not have an inverse.

A Formula for the Inverse of a 2 × 2 Matrix

Before turning to some applications, we show an alternative method that employs a formula for finding the inverse of a 2×2 matrix. This method will prove useful in many situations—we will see an application in Example 5. The derivation of this formula is left as an exercise (Exercise 43).

Formula for the Inverse of a 2 × 2 Matrix

Let

$$A = \begin{bmatrix} a & b \\ c & d \end{bmatrix}$$

Suppose $D = ad - bc$ is not equal to zero. Then, A^{-1} exists and is given by

$$A^{-1} = \frac{1}{D}\begin{bmatrix} d & -b \\ -c & a \end{bmatrix} \qquad \textbf{(13)}$$

REMARK As an aid to memorizing the formula, note that D is the product of the elements along the main diagonal minus the product of the elements along the other diagonal:

$$\begin{bmatrix} a & b \\ c & d \end{bmatrix} \qquad (D = ad - bc)$$

Main diagonal

Next, the matrix

$$\begin{bmatrix} d & -b \\ -c & a \end{bmatrix}$$

is obtained by interchanging a and d and reversing the signs of b and c. Finally, A^{-1} is obtained by dividing this matrix by D. ■■■

Group Discussion

Suppose A is a square matrix with the property that one of its rows is a nonzero constant multiple of another row. What can you say about the existence or nonexistence of A^{-1}? Explain your answer.

EXAMPLE 3 Find the inverse of

$$A = \begin{bmatrix} 1 & 2 \\ 3 & 4 \end{bmatrix}$$

SOLUTION ✓ We first compute $D = (1)(4) - (3)(2) = 4 - 6 = -2$. Next, we write the matrix

$$\begin{bmatrix} 4 & -2 \\ -3 & 1 \end{bmatrix}$$

Finally, dividing this matrix by D, we obtain

$$A^{-1} = \frac{1}{-2}\begin{bmatrix} 4 & -2 \\ -3 & 1 \end{bmatrix} = \begin{bmatrix} -2 & 1 \\ \frac{3}{2} & -\frac{1}{2} \end{bmatrix}$$

■■■■

Solving Systems of Equations with Inverses

We now show how the inverse of a matrix may be used to solve certain systems of linear equations in which the number of equations in the system is equal to the number of variables. For simplicity, let's illustrate the process for a system of three linear equations in three variables:

$$\begin{aligned} a_{11}x_1 + a_{12}x_2 + a_{13}x_3 &= c_1 \\ a_{21}x_1 + a_{22}x_2 + a_{23}x_3 &= c_2 \\ a_{31}x_1 + a_{32}x_2 + a_{33}x_3 &= c_3 \end{aligned} \qquad \textbf{(14)}$$

Let's write

$$A = \begin{bmatrix} a_{11} & a_{12} & a_{13} \\ a_{21} & a_{22} & a_{23} \\ a_{31} & a_{32} & a_{33} \end{bmatrix}, \quad X = \begin{bmatrix} x_1 \\ x_2 \\ x_3 \end{bmatrix}, \quad \text{and} \quad B = \begin{bmatrix} b_1 \\ b_2 \\ b_3 \end{bmatrix}$$

You should verify that System (14) of linear equations may be written in the form of the matrix equation

$$AX = B \qquad \textbf{(15)}$$

If A is nonsingular, then the method of this section may be used to compute A^{-1}. Next, multiplying both sides of Equation (15) by A^{-1} (on the left), we obtain

$$A^{-1}AX = A^{-1}B \quad \text{or} \quad IX = A^{-1}B \quad \text{or} \quad X = A^{-1}B$$

the desired solution to the problem.

In the case of a system of n equations with n unknowns, we have the following, more general result.

Using Inverses to Solve Systems of Equations

If $AX = B$ is a linear system of n equations in n unknowns and if A^{-1} exists, then

$$X = A^{-1}B$$

is the unique solution of the system.

The use of inverses to solve systems of equations is particularly advantageous when we are required to solve more than one system of equations, $AX = B$, involving the same coefficient matrix, A, and different matrices of constants, B. As you will see in Examples 4 and 5, we need to compute A^{-1} just once in each case.

APPLICATIONS

EXAMPLE 4

Solve the following systems of linear equations:

a.
$$\begin{aligned} 2x + y + z &= 1 \\ 3x + 2y + z &= 2 \\ 2x + y + 2z &= -1 \end{aligned}$$

b.
$$\begin{aligned} 2x + y + z &= 2 \\ 3x + 2y + z &= -3 \\ 2x + y + 2z &= 1 \end{aligned}$$

SOLUTION ✔ We may write the given systems of equations in the form

$$AX = B \quad \text{and} \quad AX = C$$

respectively, where

$$A = \begin{bmatrix} 2 & 1 & 1 \\ 3 & 2 & 1 \\ 2 & 1 & 2 \end{bmatrix}, \quad X = \begin{bmatrix} x \\ y \\ z \end{bmatrix}, \quad B = \begin{bmatrix} 1 \\ 2 \\ -1 \end{bmatrix}, \quad C = \begin{bmatrix} 2 \\ -3 \\ 1 \end{bmatrix}$$

The inverse of the matrix A,

$$A^{-1} = \begin{bmatrix} 3 & -1 & -1 \\ -4 & 2 & 1 \\ -1 & 0 & 1 \end{bmatrix}$$

was found in Example 1. Using this result, we find that the solution of the first system (a) is

$$X = A^{-1}B = \begin{bmatrix} 3 & -1 & -1 \\ -4 & 2 & 1 \\ -1 & 0 & 1 \end{bmatrix} \begin{bmatrix} 1 \\ 2 \\ -1 \end{bmatrix}$$

$$= \begin{bmatrix} (3)(1) + (-1)(2) + (-1)(-1) \\ (-4)(1) + (2)(2) + (1)(-1) \\ (-1)(1) + (0)(2) + (1)(-1) \end{bmatrix} = \begin{bmatrix} 2 \\ -1 \\ -2 \end{bmatrix}$$

or $x = 2$, $y = -1$, and $z = -2$.

The solution of the second system (b) is

$$X = A^{-1}C = \begin{bmatrix} 3 & -1 & -1 \\ -4 & 2 & 1 \\ -1 & 0 & 1 \end{bmatrix} \begin{bmatrix} 2 \\ -3 \\ 1 \end{bmatrix} = \begin{bmatrix} 8 \\ -13 \\ -1 \end{bmatrix}$$

or $x = 8$, $y = -13$, and $z = -1$.

EXAMPLE 5 The management of Checkers Rent-A-Car plans to expand its fleet of rental cars for the next quarter by purchasing compact and full-size cars. The average cost of a compact is $10,000, and the average cost of a full-size car is $24,000.

a. If a total of 800 cars is to be purchased with a budget of $12 million, how many cars of each size will be acquired?

b. If the predicted demand calls for a total purchase of 1000 cars with a budget of $14 million, how many cars of each type will be acquired?

SOLUTION ✓ Let x and y denote the number of compact and full-size cars to be purchased. Furthermore, let n denote the total number of cars to be acquired and b the amount of money budgeted for the purchase of these cars. Then,

$$\begin{aligned} x + y &= n \\ 10{,}000x + 24{,}000y &= b \end{aligned}$$

This system of two equations in two variables may be written in the matrix form

$$AX = B$$

where

$$A = \begin{bmatrix} 1 & 1 \\ 10{,}000 & 24{,}000 \end{bmatrix}, \qquad X = \begin{bmatrix} x \\ y \end{bmatrix}, \qquad \text{and} \qquad B = \begin{bmatrix} n \\ b \end{bmatrix}$$

Therefore,

$$X = A^{-1}B$$

Since A is a 2×2 matrix, its inverse may be found by using Formula (13). We find $D = (1)(24{,}000) - (10{,}000)(1) = 14{,}000$, so

$$A^{-1} = \frac{1}{14{,}000}\begin{bmatrix} 24{,}000 & -1 \\ -10{,}000 & 1 \end{bmatrix} = \begin{bmatrix} \frac{24{,}000}{14{,}000} & -\frac{1}{14{,}000} \\ -\frac{10{,}000}{14{,}000} & \frac{1}{14{,}000} \end{bmatrix}$$

Thus,

$$X = \begin{bmatrix} \frac{12}{7} & -\frac{1}{14{,}000} \\ -\frac{5}{7} & \frac{1}{14{,}000} \end{bmatrix} \begin{bmatrix} n \\ b \end{bmatrix}$$

a. Here, $n = 800$ and $b = 12{,}000{,}000$, so

$$X = A^{-1}B = \begin{bmatrix} \frac{12}{7} & -\frac{1}{14{,}000} \\ -\frac{5}{7} & \frac{1}{14{,}000} \end{bmatrix} \begin{bmatrix} 800 \\ 12{,}000{,}000 \end{bmatrix} = \begin{bmatrix} 514.3 \\ 285.7 \end{bmatrix}$$

Therefore, 514 compact cars and 286 full-size cars will be acquired in this case.

b. Here, $n = 1000$ and $b = 14{,}000{,}000$, so

$$X = A^{-1}B = \begin{bmatrix} \frac{12}{7} & -\frac{1}{14{,}000} \\ -\frac{5}{7} & \frac{1}{14{,}000} \end{bmatrix} \begin{bmatrix} 1000 \\ 14{,}000{,}000 \end{bmatrix} = \begin{bmatrix} 714.3 \\ 285.7 \end{bmatrix}$$

Therefore, 714 compact cars and 286 full-size cars will be purchased in this case.

SELF-CHECK EXERCISES 2.6

1. Find the inverse of the matrix

$$A = \begin{bmatrix} 2 & 1 & -1 \\ 1 & 1 & -1 \\ -1 & -2 & 3 \end{bmatrix}$$

if it exists.

2. Solve the system of linear equations

$$\begin{aligned} 2x + y - z &= b_1 \\ x + y - z &= b_2 \\ -x - 2y + 3z &= b_3 \end{aligned}$$

where (a) $b_1 = 5$, $b_2 = 4$, $b_3 = -8$ and (b) $b_1 = 2$, $b_2 = 0$, $b_3 = 5$, by finding the inverse of the coefficient matrix.

3. Grand Canyon Tours, Inc., offers air and ground scenic tours of the Grand Canyon. Tickets for the $7\frac{1}{2}$-hr tour cost \$169 for an adult and \$129 for a child, and each tour group is limited to 19 people. On three recent fully booked tours, total receipts were \$2931 for the first tour, \$3011 for the second tour, and \$2771 for the third tour. Determine how many adults and how many children were in each tour.

Solutions to Self-Check Exercises 2.6 can be found on page 162.

Spreadsheet examples and exercises for this section that may be solved using the Microsoft® Excel program are given at the Brooks/Cole Web site:
http://www.brookscole.com/math/authors/tans/

2.6 Exercises

In Exercises 1–4, show that the given matrices are inverses of each other by showing that their product is the identity matrix I.

1. $\begin{bmatrix} 1 & -3 \\ 1 & -2 \end{bmatrix}$ and $\begin{bmatrix} -2 & 3 \\ -1 & 1 \end{bmatrix}$

2. $\begin{bmatrix} 4 & 5 \\ 2 & 3 \end{bmatrix}$ and $\begin{bmatrix} \frac{3}{2} & -\frac{5}{2} \\ -1 & 2 \end{bmatrix}$

3. $\begin{bmatrix} 3 & 2 & 3 \\ 2 & 2 & 1 \\ 2 & 1 & 1 \end{bmatrix}$ and $\begin{bmatrix} -\frac{1}{3} & -\frac{1}{3} & \frac{4}{3} \\ 0 & 1 & -1 \\ \frac{2}{3} & -\frac{1}{3} & -\frac{2}{3} \end{bmatrix}$

4. $\begin{bmatrix} 2 & 4 & -2 \\ -4 & -6 & 1 \\ 3 & 5 & -1 \end{bmatrix}$ and $\begin{bmatrix} \frac{1}{2} & -3 & -4 \\ -\frac{1}{2} & 2 & 3 \\ -1 & 1 & 2 \end{bmatrix}$

In Exercises 5–16, find the inverse of the given matrix, if it exists. Verify your answer.

5. $\begin{bmatrix} 2 & 5 \\ 1 & 3 \end{bmatrix}$

6. $\begin{bmatrix} 2 & 3 \\ 3 & 5 \end{bmatrix}$

7. $\begin{bmatrix} 3 & -3 \\ -2 & 2 \end{bmatrix}$

8. $\begin{bmatrix} 4 & 2 \\ 6 & 3 \end{bmatrix}$

9. $\begin{bmatrix} 2 & -3 & -4 \\ 0 & 0 & -1 \\ 1 & -2 & 1 \end{bmatrix}$

10. $\begin{bmatrix} 1 & -1 & 3 \\ 2 & 1 & 2 \\ -2 & -2 & 1 \end{bmatrix}$

11. $\begin{bmatrix} 4 & 2 & 2 \\ -1 & -3 & 4 \\ 3 & -1 & 6 \end{bmatrix}$

12. $\begin{bmatrix} 1 & 2 & 0 \\ -3 & 4 & -2 \\ -5 & 0 & -2 \end{bmatrix}$

13. $\begin{bmatrix} 1 & 4 & -1 \\ 2 & 3 & -2 \\ -1 & 2 & 3 \end{bmatrix}$

14. $\begin{bmatrix} 3 & -2 & 7 \\ -2 & 1 & 4 \\ 6 & -5 & 8 \end{bmatrix}$

15. $\begin{bmatrix} 1 & 1 & -1 & 1 \\ 2 & 1 & 1 & 0 \\ 2 & 1 & 0 & 1 \\ 2 & -1 & -1 & 3 \end{bmatrix}$

16. $\begin{bmatrix} 1 & 1 & 2 & 3 \\ 2 & 3 & 0 & -1 \\ 0 & 2 & -1 & 1 \\ 1 & 2 & 1 & 1 \end{bmatrix}$

In Exercises 17–24, (a) write a matrix equation that is equivalent to the given system of linear equations and (b) solve the system using the inverses found in Exercises 5–16.

17. $$\begin{aligned} 2x + 5y &= 3 \\ x + 3y &= 2 \end{aligned}$$
(See Exercise 5.)

18. $$\begin{aligned} 2x + 3y &= 5 \\ 3x + 5y &= 8 \end{aligned}$$
(See Exercise 6.)

19. $$\begin{aligned} 2x - 3y - 4z &= 4 \\ -z &= 3 \\ x - 2y + z &= -8 \end{aligned}$$
(See Exercise 9.)

20. $$\begin{aligned} x_1 - x_2 + 3x_3 &= 2 \\ 2x_1 + x_2 + 2x_3 &= 2 \\ -2x_1 - 2x_2 + x_3 &= 3 \end{aligned}$$
(See Exercise 10.)

21. $$\begin{aligned} x + 4y - z &= 3 \\ 2x + 3y - 2z &= 1 \\ -x + 2y + 3z &= 7 \end{aligned}$$
(See Exercise 13.)

22. $$\begin{aligned} 3x_1 - 2x_2 + 7x_3 &= 6 \\ -2x_1 + x_2 + 4x_3 &= 4 \\ 6x_1 - 5x_2 + 8x_3 &= 4 \end{aligned}$$
(See Exercise 14.)

23. $$\begin{aligned} x_1 + x_2 - x_3 + x_4 &= 6 \\ 2x_1 + x_2 + x_3 &= 4 \\ 2x_1 + x_2 + x_4 &= 7 \\ 2x_1 - x_2 - x_3 + 3x_4 &= 9 \end{aligned}$$
(See Exercise 15.)

24. $$\begin{aligned} x_1 + x_2 + 2x_3 + 3x_4 &= 4 \\ 2x_1 + 3x_2 - x_4 &= 11 \\ 2x_2 - x_3 + x_4 &= 7 \\ x_1 + 2x_2 + x_3 + x_4 &= 6 \end{aligned}$$
(See Exercise 16.)

In Exercises 25–32, (a) write each system of equations as a matrix equation and (b) solve the system of equations by using the inverse of the coefficient matrix.

25. $$\begin{aligned} x + 2y &= b_1 \\ 2x - y &= b_2 \end{aligned}$$
where (i) $b_1 = 14, b_2 = 5$
and (ii) $b_1 = 4, b_2 = -1$

26. $$\begin{aligned} 3x - 2y &= b_1 \\ 4x + 3y &= b_2 \end{aligned}$$
where (i) $b_1 = -6, b_2 = 10$
and (ii) $b_1 = 3, b_2 = -2$

27.
$$\begin{aligned} x + 2y + z &= b_1 \\ x + y + z &= b_2 \\ 3x + y + z &= b_3 \end{aligned}$$
where (i) $b_1 = 7, b_2 = 4, b_3 = 2$
and (ii) $b_1 = 5, b_2 = -3, b_3 = -1$

28.
$$\begin{aligned} x_1 + x_2 + x_3 &= b_1 \\ x_1 - x_2 + x_3 &= b_2 \\ x_1 - 2x_2 - x_3 &= b_3 \end{aligned}$$
where (i) $b_1 = 5, b_2 = -3, b_3 = -1$
and (ii) $b_1 = 1, b_2 = 4, b_3 = -2$

29.
$$\begin{aligned} 3x + 2y - z &= b_1 \\ 2x - 3y + z &= b_2 \\ x - y - z &= b_3 \end{aligned}$$
where (i) $b_1 = 2, b_2 = -2, b_3 = 4$
and (ii) $b_1 = 8, b_2 = -3, b_3 = 6$

30.
$$\begin{aligned} 2x_1 + x_2 + x_3 &= b_1 \\ x_1 - 3x_2 + 4x_3 &= b_2 \\ -x_1 \qquad + x_3 &= b_3 \end{aligned}$$
where (i) $b_1 = 1, b_2 = 4, b_3 = -3$
and (ii) $b_1 = 2, b_2 = -5, b_3 = 0$

31.
$$\begin{aligned} x_1 + x_2 + x_3 + x_4 &= b_1 \\ x_1 - x_2 - x_3 + x_4 &= b_2 \\ x_2 + 2x_3 + 2x_4 &= b_3 \\ x_1 + 2x_2 + x_3 - 2x_4 &= b_4 \end{aligned}$$
where (i) $b_1 = 1, b_2 = -1, b_3 = 4, b_4 = 0$
and (ii) $b_1 = 2, b_2 = 8, b_3 = 4, b_4 = -1$

32.
$$\begin{aligned} x_1 + x_2 + 2x_3 + x_4 &= b_1 \\ 4x_1 + 5x_2 + 9x_3 + x_4 &= b_2 \\ 3x_1 + 4x_2 + 7x_3 + x_4 &= b_3 \\ 2x_1 + 3x_2 + 4x_3 + 2x_4 &= b_4 \end{aligned}$$
where (i) $b_1 = 3, b_2 = 6, b_3 = 5, b_4 = 7$
and (ii) $b_1 = 1, b_2 = -1, b_3 = 0, b_4 = -4$

33. Let
$$A = \begin{bmatrix} 2 & 3 \\ -4 & -5 \end{bmatrix}$$

a. Find A^{-1}. **b.** Show that $(A^{-1})^{-1} = A$.

34. Let
$$A = \begin{bmatrix} 6 & -4 \\ -4 & 3 \end{bmatrix} \quad \text{and} \quad B = \begin{bmatrix} 3 & -5 \\ 4 & -7 \end{bmatrix}$$

a. Find AB, A^{-1}, and B^{-1}.
b. Show that $(AB)^{-1} = B^{-1}A^{-1}$.

35. Let
$$A = \begin{bmatrix} 2 & -5 \\ 1 & -3 \end{bmatrix}$$
$$B = \begin{bmatrix} 4 & 3 \\ 1 & 1 \end{bmatrix}$$
$$C = \begin{bmatrix} 2 & 3 \\ -2 & 1 \end{bmatrix}$$

a. Find ABC, A^{-1}, B^{-1}, and C^{-1}.
b. Show that $(ABC)^{-1} = C^{-1}B^{-1}A^{-1}$.

36. TICKET REVENUES Rainbow Harbor Cruises charges \$8/adult and \$4/child for a round-trip ticket. The records show that, on a certain weekend, 1000 people took the cruise on Saturday and 800 people took the cruise on Sunday. The total receipts for Saturday were \$6400, and the total receipts for Sunday were \$4800. Determine how many adults and children took the cruise on Saturday and on Sunday.

37. PRICING Bel Air Publishing Company publishes a deluxe leather edition and a standard edition of its Daily Organizer. The company's marketing department estimates that x copies of the deluxe edition and y copies of the standard edition will be demanded per month when the unit prices are p dollars and q dollars, respectively, where x, y, p, and q are related by the following system of linear equations:
$$\begin{aligned} 5x + y &= 1000(70 - p) \\ x + 3y &= 1000(40 - q) \end{aligned}$$
Find the monthly demand for the deluxe edition and the standard edition when the unit prices are set according to the following schedules:
a. $p = 50$ and $q = 25$ **b.** $p = 45$ and $q = 25$
c. $p = 45$ and $q = 20$

38. NUTRITION/DIET PLANNING Bob, a nutritionist attached to the University Medical Center, has been asked to prepare special diets for two patients, Susan and Tom. Bob has decided that Susan's meals should contain at least 400 mg of calcium, 20 mg of iron, and 50 mg of vitamin C, whereas Tom's meals should contain at least 350 mg of calcium, 15 mg of iron, and 40 mg of vitamin C. Bob has also decided that the meals are to be prepared from three basic foods: food A, food B, and food C. The special nutritional contents of these foods are summarized in the accompanying table. Find how many ounces of each type of food should be used in a meal so that the

(continued on p. 162)

Using Technology

FINDING THE INVERSE OF A SQUARE MATRIX

The graphing utility can be used to find the inverse of a square matrix.

EXAMPLE 1 Use a graphing utility to find the inverse of

$$\begin{bmatrix} 1 & 3 & 5 \\ -2 & 2 & 4 \\ 5 & 1 & 3 \end{bmatrix}$$

SOLUTION ✓ We first enter the given matrix as

$$A = \begin{bmatrix} 1 & 3 & 5 \\ -2 & 2 & 4 \\ 5 & 1 & 3 \end{bmatrix}$$

Then, recalling the matrix A and using the $\boxed{x^{-1}}$ key, we find

$$A^{-1} = \begin{bmatrix} 0.1 & -0.2 & 0.1 \\ 1.3 & -1.1 & -0.7 \\ -0.6 & 0.7 & 0.4 \end{bmatrix}$$

■■■■

Exercises

In Exercises 1–6, use a graphing utility to find the inverse of the matrix. Express your answers accurate to two decimal places.

1. $\begin{bmatrix} 1.2 & 3.1 & -2.1 \\ 3.4 & 2.6 & 7.3 \\ -1.2 & 3.4 & -1.3 \end{bmatrix}$

2. $\begin{bmatrix} 4.2 & 3.7 & 4.6 \\ 2.1 & -1.3 & -2.3 \\ 1.8 & 7.6 & -2.3 \end{bmatrix}$

3. $\begin{bmatrix} 1.1 & 2.3 & 3.1 & 4.2 \\ 1.6 & 3.2 & 1.8 & 2.9 \\ 4.2 & 1.6 & 1.4 & 3.2 \\ 1.6 & 2.1 & 2.8 & 7.2 \end{bmatrix}$

4. $\begin{bmatrix} 2.1 & 3.2 & -1.4 & -3.2 \\ 6.2 & 7.3 & 8.4 & 1.6 \\ 2.3 & 7.1 & 2.4 & -1.3 \\ -2.1 & 3.1 & 4.6 & 3.7 \end{bmatrix}$

5. $\begin{bmatrix} 2 & -1 & 3 & 2 & 4 \\ 3 & 2 & -1 & 4 & 1 \\ 3 & 2 & 6 & 4 & -1 \\ 2 & 1 & -1 & 4 & 2 \\ 3 & 4 & 2 & 5 & 6 \end{bmatrix}$

6. $\begin{bmatrix} 1 & 4 & 2 & 3 & 1.4 \\ 6 & 2.4 & 5 & 1.2 & 3 \\ 4 & 1 & 2 & 3 & 1.2 \\ -1 & 2 & -3 & 4 & 2 \\ 1.1 & 2.2 & 3 & 5.1 & 4 \end{bmatrix}$

minimum requirements of calcium, iron, and vitamin C are met for each patient's meals.

	Contents in mg/oz		
	Calcium	**Iron**	**Vitamin C**
Food A	30	1	2
Food B	25	1	5
Food C	20	2	4

39. **RESEARCH FUNDING** The Carver Foundation funds three nonprofit organizations engaged in alternate-energy research activities. From past data, the proportion of funds spent by each organization in research on solar energy, energy from harnessing the wind, and energy from the motion of ocean tides is given in the accompanying table.

	Proportion of Money Spent		
	Solar	**Wind**	**Tides**
Organization I	0.6	0.3	0.1
Organization II	0.4	0.3	0.3
Organization III	0.2	0.6	0.2

Find the amount awarded to each organization if the total amount spent by all three organizations on solar, wind, and tidal research is
a. \$9.2 million, \$9.6 million, and \$5.2 million, respectively.
b. \$8.2 million, \$7.2 million, and \$3.6 million, respectively.

In Exercises 40–42, determine whether the statement is true or false. If it is true, explain why it is true. If it is false, give an example to show why it is false.

40. If A is a square matrix with inverse A^{-1} and c is a nonzero real number, then

$$(cA)^{-1} = (1/c)A^{-1}.$$

41. The matrix

$$A = \begin{bmatrix} a & b \\ c & d \end{bmatrix}$$

has an inverse if and only if $ad - bc = 0$.

42. If A^{-1} does not exist, then the system $AX = B$ of n linear equations in n unknowns does not have a unique solution.

43. Let

$$A = \begin{bmatrix} a & b \\ c & d \end{bmatrix}$$

a. Find A^{-1}.
b. Find the necessary condition for A to be nonsingular.
c. Verify that $AA^{-1} = A^{-1}A = I$.

SOLUTIONS TO SELF-CHECK EXERCISES 2.6

1. We form the augmented matrix

$$\left[\begin{array}{rrr|rrr} 2 & 1 & -1 & 1 & 0 & 0 \\ 1 & 1 & -1 & 0 & 1 & 0 \\ -1 & -2 & 3 & 0 & 0 & 1 \end{array}\right]$$

and row-reduce as follows:

$$\left[\begin{array}{rrr|rrr} 2 & 1 & -1 & 1 & 0 & 0 \\ 1 & 1 & -1 & 0 & 1 & 0 \\ -1 & -2 & 3 & 0 & 0 & 1 \end{array}\right] \xrightarrow{R_1 \leftrightarrow R_2}$$

$$\left[\begin{array}{rrr|rrr} 1 & 1 & -1 & 0 & 1 & 0 \\ 2 & 1 & -1 & 1 & 0 & 0 \\ -1 & -2 & 3 & 0 & 0 & 1 \end{array}\right] \xrightarrow[R_3 + R_1]{R_2 - 2R_1}$$

$$\left[\begin{array}{rrr|rrr} 1 & 1 & -1 & 0 & 1 & 0 \\ 0 & -1 & 1 & 1 & -2 & 0 \\ 0 & -1 & 2 & 0 & 1 & 1 \end{array}\right] \xrightarrow[\substack{-R_2 \\ R_3 - R_2}]{R_1 + R_2}$$

$$\left[\begin{array}{rrr|rrr} 1 & 0 & 0 & 1 & -1 & 0 \\ 0 & 1 & -1 & -1 & 2 & 0 \\ 0 & 0 & 1 & -1 & 3 & 1 \end{array}\right] \xrightarrow{R_2 + R_3}$$

$$\left[\begin{array}{rrr|rrr} 1 & 0 & 0 & 1 & -1 & 0 \\ 0 & 1 & 0 & -2 & 5 & 1 \\ 0 & 0 & 1 & -1 & 3 & 1 \end{array}\right]$$

From the preceding results, we see that

$$A^{-1} = \begin{bmatrix} 1 & -1 & 0 \\ -2 & 5 & 1 \\ -1 & 3 & 1 \end{bmatrix}$$

2. a. We write the systems of linear equations in the matrix form

$$AX = B_1$$

where

$$A = \begin{bmatrix} 2 & 1 & -1 \\ 1 & 1 & -1 \\ -1 & -2 & 3 \end{bmatrix}, \quad X = \begin{bmatrix} x \\ y \\ z \end{bmatrix}, \quad B_1 = \begin{bmatrix} 5 \\ 4 \\ -8 \end{bmatrix}$$

Now, using the results of Exercise 1, we have

$$X = \begin{bmatrix} x \\ y \\ z \end{bmatrix} = A^{-1}B_1 = \begin{bmatrix} 1 & -1 & 0 \\ -2 & 5 & 1 \\ -1 & 3 & 1 \end{bmatrix} \begin{bmatrix} 5 \\ 4 \\ -8 \end{bmatrix} = \begin{bmatrix} 1 \\ 2 \\ -1 \end{bmatrix}$$

Therefore, $x = 1$, $y = 2$, and $z = -1$.

b. Here A and X are as in part (a), but

$$B_2 = \begin{bmatrix} 2 \\ 0 \\ 5 \end{bmatrix}$$

Therefore,

$$X = \begin{bmatrix} x \\ y \\ z \end{bmatrix} = A^{-1}B_2 = \begin{bmatrix} 1 & -1 & 0 \\ -2 & 5 & 1 \\ -1 & 3 & 1 \end{bmatrix} \begin{bmatrix} 2 \\ 0 \\ 5 \end{bmatrix} = \begin{bmatrix} 2 \\ 1 \\ 3 \end{bmatrix}$$

or $x = 2$, $y = 1$, and $z = 3$.

3. Let x denote the number of adults and y the number of children in a tour. Since the tours are filled to capacity, we have

$$x + y = 19$$

Next, using the fact that the total receipts for the first tour were \$2931 leads to the equation

$$169x + 129y = 2931$$

Therefore, the number of adults and the number of children in the first tour is found by solving the system of linear equations

$$\begin{aligned} x + \quad y &= \quad 19 \\ 169x + 129y &= 2931 \end{aligned} \qquad \textbf{(a)}$$

Similarly, we see that the number of adults and the number of children in the second and third tours are found by solving the systems

$$\begin{aligned} x + \quad y &= \quad 19 \\ 169x + 129y &= 3011 \end{aligned} \qquad \textbf{(b)}$$

$$\begin{aligned} x + \quad y &= \quad 19 \\ 169x + 129y &= 2771 \end{aligned} \qquad \textbf{(c)}$$

These systems may be written in the form

$$AX = B_1, \qquad AX = B_2, \qquad AX = B_3$$

where

$$A = \begin{bmatrix} 1 & 1 \\ 169 & 129 \end{bmatrix} \qquad X = \begin{bmatrix} x \\ y \end{bmatrix}$$

$$B_1 = \begin{bmatrix} 19 \\ 2931 \end{bmatrix} \qquad B_2 = \begin{bmatrix} 19 \\ 3011 \end{bmatrix}$$

$$B_3 = \begin{bmatrix} 19 \\ 2771 \end{bmatrix}$$

To solve these systems, we first find A^{-1}. Using Formula (13), we obtain

$$A^{-1} = \begin{bmatrix} -\frac{129}{40} & \frac{1}{40} \\ \frac{169}{40} & -\frac{1}{40} \end{bmatrix}$$

Then, solving each system, we find

$$\begin{aligned} X = \begin{bmatrix} x \\ y \end{bmatrix} &= A^{-1}B_1 \\ &= \begin{bmatrix} -\frac{129}{40} & \frac{1}{40} \\ \frac{169}{40} & -\frac{1}{40} \end{bmatrix} \begin{bmatrix} 19 \\ 2931 \end{bmatrix} = \begin{bmatrix} 12 \\ 7 \end{bmatrix} \end{aligned} \qquad \textbf{(a)}$$

$$X = \begin{bmatrix} x \\ y \end{bmatrix} = A^{-1}B_2$$
$$= \begin{bmatrix} -\frac{129}{40} & \frac{1}{40} \\ \frac{169}{40} & -\frac{1}{40} \end{bmatrix} \begin{bmatrix} 19 \\ 3011 \end{bmatrix}$$
$$= \begin{bmatrix} 14 \\ 5 \end{bmatrix} \qquad \textbf{(b)}$$

$$X = \begin{bmatrix} x \\ y \end{bmatrix} = A^{-1}B_3$$
$$= \begin{bmatrix} -\frac{129}{40} & \frac{1}{40} \\ \frac{169}{40} & -\frac{1}{40} \end{bmatrix} \begin{bmatrix} 19 \\ 2771 \end{bmatrix} = \begin{bmatrix} 8 \\ 11 \end{bmatrix} \qquad \textbf{(c)}$$

We conclude that there were

a. 12 adults and 7 children on the first tour.
b. 14 adults and 5 children on the second tour.
c. 8 adults and 11 children on the third tour.

2.7 Leontief Input–Output Model (Optional)

Input–Output Analysis

One of the many important applications of matrix theory to the field of economics is the study of the relationship between industrial production and consumer demand. At the heart of this analysis is the Leontief input–output model, pioneered by Wassily Leontief, who was awarded a Nobel Prize in economics in 1973 for his contributions to the field.

To illustrate this concept, let's consider an oversimplified economy consisting of three sectors: agriculture (A), manufacturing (M), and service (S). In general, part of the output of one sector is absorbed by another sector through interindustry purchases, with the excess available to fulfill consumer demands. The relationship governing both intraindustrial and interindustrial sales and purchases is conveniently represented by means of an **input–output matrix:**

$$\begin{array}{ccc} & & \text{Output (amount produced)} \\ & & \begin{array}{ccc} A & M & S \end{array} \\ \begin{array}{c} \\ \text{Input} \\ \text{(amount used in production)} \end{array} & \begin{array}{c} A \\ M \\ S \end{array} & \begin{bmatrix} 0.2 & 0.2 & 0.1 \\ 0.2 & 0.4 & 0.1 \\ 0.1 & 0.2 & 0.3 \end{bmatrix} \end{array} \qquad \textbf{(16)}$$

The first column (read from top to bottom) tells us that the production of 1 unit of agricultural products requires the consumption of 0.2 unit of agricultural products, 0.2 unit of manufactured goods, and 0.1 unit of services. The second column tells us that the production of 1 unit of manufactured products requires the consumption of 0.2 unit of agricultural products, 0.4 unit of manufactured products, and 0.2 unit of services. Finally, the third column tells us that the production of 1 unit of services requires the consumption of 0.1 unit each of agricultural goods and manufactured products, and 0.3 unit of services.

EXAMPLE 1 Refer to the input–output matrix (16).

a. If the units are measured in millions of dollars, determine the amount of agricultural products consumed in the production of \$100 million worth of manufactured goods.

b. Determine the dollar amount of manufactured products required to produce \$200 million worth of all goods and services in the economy.

SOLUTION ✓ **a.** The production of 1 unit—that is, \$1 million worth of manufactured goods—requires the consumption of 0.2 unit of agricultural products. Thus, the amount of agricultural products consumed in the production of \$100 million worth of manufactured goods is given by (100)(0.2), or \$20 million.

b. The amount of manufactured goods required to produce 1 unit of all goods and services in the economy is given by adding the numbers of the second row of the input–output matrix—that is, $0.2 + 0.4 + 0.1$, or 0.7 unit. Therefore, the production of \$200 million worth of all goods and services in the economy requires 200(0.7), or \$140 million worth, of manufactured products. ■■■■

Next, suppose the total output of goods of the agriculture and manufacturing sectors and the total output from the service sector of the economy are given by x, y, and z units, respectively. What is the value of agricultural products consumed in the internal process of producing this total output of various goods and services?

To answer this question, we first note, by examining the input–output matrix

$$\begin{array}{cc} & \text{Output} \\ & \begin{array}{ccc} A & M & S \end{array} \\ \text{Input} \begin{array}{c} A \\ M \\ S \end{array} & \begin{bmatrix} 0.2 & 0.2 & 0.1 \\ 0.2 & 0.4 & 0.1 \\ 0.1 & 0.2 & 0.3 \end{bmatrix} \end{array}$$

that 0.2 unit of agricultural products is required to produce 1 unit of agricultural products, so the amount of agricultural goods required to produce x units of agricultural products is given by $0.2x$ unit. Next, again referring to the input–output matrix, we see that 0.2 unit of agricultural products is required to produce 1 unit of manufactured products, so the requirement for producing y units of the latter is $0.2y$ unit of agricultural products. Finally, we see that 0.1 unit of agricultural goods is required to produce 1 unit of services, so the value of agricultural products required to produce z units of services is $0.1z$ unit. Thus, the total amount of agricultural products required to produce the total output of goods and services in the economy is

$$0.2x + 0.2y + 0.1z$$

units. In a similar manner, we see that the total amount of manufactured products and the total value of services to produce the total output of goods and services in the economy are given by

$$0.2x + 0.4y + 0.1z$$
$$0.1x + 0.2y + 0.3z$$

respectively.

These results could also be obtained using matrix multiplication. To see this, write the total output of goods and services x, y, and z as a 3×1 matrix

$$X = \begin{bmatrix} x \\ y \\ z \end{bmatrix} \qquad \text{(Gross production matrix)}$$

The matrix X is called the **gross production matrix.** Letting A denote the input–output matrix, we have

$$A = \begin{bmatrix} 0.2 & 0.2 & 0.1 \\ 0.2 & 0.4 & 0.1 \\ 0.1 & 0.2 & 0.3 \end{bmatrix} \qquad \text{(Input–output matrix)}$$

Then, the product

$$AX = \begin{bmatrix} 0.2 & 0.2 & 0.1 \\ 0.2 & 0.4 & 0.1 \\ 0.1 & 0.2 & 0.3 \end{bmatrix} \begin{bmatrix} x \\ y \\ z \end{bmatrix}$$
$$= \begin{bmatrix} 0.2x + 0.2y + 0.1z \\ 0.2x + 0.4y + 0.1z \\ 0.1x + 0.2y + 0.3z \end{bmatrix} \qquad \text{(Internal consumption matrix)}$$

is a 3×1 matrix whose entries represent the respective values of the agricultural products, manufactured products, and services consumed in the internal process of production. The matrix AX is referred to as the **internal consumption matrix.**

Now, since X gives the total production of goods and services in the economy, and AX, as we have just seen, gives the amount of products and services consumed in the production of these goods and services, the 3×1 matrix $X - AX$ gives the net output of goods and services that is exactly enough to satisfy consumer demands. Letting matrix D represent these consumer demands, we are led to the following matrix equation:

$$X - AX = D$$
$$(I - A)X = D$$

where I is the 3×3 identity matrix.

Assuming that the inverse of $(I - A)$ exists, multiplying both sides of the last equation by $(I - A)^{-1}$ yields

$$X = (I - A)^{-1}D$$

Leontief Input–Output Model

In a **Leontief input–output model,** the matrix equation giving the net output of goods and services needed to satisfy consumer demand is

$$\underset{\text{Total output}}{X} - \underset{\text{Internal consumption}}{AX} = \underset{\text{Consumer demand}}{D}$$

where X is the total output matrix, A is the input–output matrix, and D is the matrix representing consumer demand.

The solution to this equation is

$$X = (I - A)^{-1}D \qquad \text{[Assuming that } (I - A)^{-1} \text{ exists]} \qquad \textbf{(17)}$$

which gives the amount of goods and services that must be produced to satisfy consumer demand.

APPLICATIONS

Equation (17) gives us a means of finding the amount of goods and services to be produced in order to satisfy a given level of consumer demand, as illustrated by the following example.

EXAMPLE 2 For the three-sector economy with input–output matrix given by (16), which is reproduced here,

$$\begin{bmatrix} 0.2 & 0.2 & 0.1 \\ 0.2 & 0.4 & 0.1 \\ 0.1 & 0.2 & 0.3 \end{bmatrix} \qquad \text{(Each unit equals \$1 million.)}$$

a. Find the gross output of goods and services needed to satisfy a consumer demand of \$100 million worth of agricultural products, \$80 million worth of manufactured products, and \$50 million worth of services.

b. Find the value of the goods and services consumed in the internal process of production in order to meet this gross output.

SOLUTION ✔ **a.** We are required to determine the gross production matrix

$$X = \begin{bmatrix} x \\ y \\ z \end{bmatrix}$$

where x, y, and z denote the value of the agricultural products, the manufactured products, and services. The matrix representing the consumer demand is given by

$$D = \begin{bmatrix} 100 \\ 80 \\ 50 \end{bmatrix}$$

Next, we compute

$$I - A = \begin{bmatrix} 1 & 0 & 0 \\ 0 & 1 & 0 \\ 0 & 0 & 1 \end{bmatrix} - \begin{bmatrix} 0.2 & 0.2 & 0.1 \\ 0.2 & 0.4 & 0.1 \\ 0.1 & 0.2 & 0.3 \end{bmatrix} = \begin{bmatrix} 0.8 & -0.2 & -0.1 \\ -0.2 & 0.6 & -0.1 \\ -0.1 & -0.2 & 0.7 \end{bmatrix}$$

Using the method of Section 2.6, we find (to two decimal places)

$$(I - A)^{-1} = \begin{bmatrix} 1.43 & 0.57 & 0.29 \\ 0.54 & 1.96 & 0.36 \\ 0.36 & 0.64 & 1.57 \end{bmatrix}$$

Finally, using Equation (17), we find

$$X = (I - A)^{-1}D = \begin{bmatrix} 1.43 & 0.57 & 0.29 \\ 0.54 & 1.96 & 0.36 \\ 0.36 & 0.64 & 1.57 \end{bmatrix} \begin{bmatrix} 100 \\ 80 \\ 50 \end{bmatrix} = \begin{bmatrix} 203.1 \\ 228.8 \\ 165.7 \end{bmatrix}$$

To fulfill consumer demand, \$203.1 million worth of agricultural products, \$228.8 million worth of manufactured products, and \$165.7 million worth of services should be produced.

b. The amount of goods and services consumed in the internal process of production is given by AX, or equivalently by $X - D$. In this case it is more convenient to use the latter, which gives the required result of

$$\begin{bmatrix} 203.1 \\ 228.8 \\ 165.7 \end{bmatrix} - \begin{bmatrix} 100 \\ 80 \\ 50 \end{bmatrix} = \begin{bmatrix} 103.1 \\ 148.8 \\ 115.7 \end{bmatrix}$$

or \$103.1 million worth of agricultural products, \$148.8 million worth of manufactured products, and \$115.7 million worth of services. ■■■■

EXAMPLE 3 The TKK Corporation, a large conglomerate, has three subsidiaries engaged in producing raw rubber, manufacturing tires, and manufacturing other rubber-based goods. The production of 1 unit of raw rubber requires the consumption of 0.08 unit of rubber, 0.04 unit of tires, and 0.02 unit of other rubber-based goods. To produce 1 unit of tires requires 0.6 unit of raw rubber, 0.02 unit of tires, and 0 units of other rubber-based goods. To produce 1 unit of other rubber-based goods requires 0.3 unit of raw rubber, 0.01 unit of tires, and 0.06 unit of other rubber-based goods. Market research indicates that the demand for the following year will be \$200 million for raw rubber, \$800 million for tires, and \$120 million for other rubber-based products. Find the level of production for each subsidiary in order to satisfy this demand.

SOLUTION ✔ View the corporation as an economy having three sectors, with an input–output matrix given by

$$A = \begin{array}{l} \text{Raw rubber} \\ \text{Tires} \\ \text{Goods} \end{array} \overset{\begin{array}{ccc} \text{Raw rubber} & \text{Tires} & \text{Goods} \end{array}}{\begin{bmatrix} 0.08 & 0.60 & 0.30 \\ 0.04 & 0.02 & 0.01 \\ 0.02 & 0 & 0.06 \end{bmatrix}}$$

Using Equation (17), we find that the required level of production is given by

$$X = \begin{bmatrix} x \\ y \\ z \end{bmatrix} = (I - A)^{-1}D$$

where x, y, and z denote the outputs of raw rubber, tires, and other rubber-based goods, and

$$D = \begin{bmatrix} 200 \\ 800 \\ 120 \end{bmatrix}$$

Now,

$$I - A = \begin{bmatrix} 0.92 & -0.60 & -0.30 \\ -0.04 & 0.98 & -0.01 \\ -0.02 & 0 & 0.94 \end{bmatrix}$$

You are asked to verify that

$$(I - A)^{-1} = \begin{bmatrix} 1.13 & 0.69 & 0.37 \\ 0.05 & 1.05 & 0.03 \\ 0.02 & 0.02 & 1.07 \end{bmatrix} \qquad \text{(See Exercise 7.)}$$

Therefore,

$$X = (I - A)^{-1}D = \begin{bmatrix} 1.13 & 0.69 & 0.37 \\ 0.05 & 1.05 & 0.03 \\ 0.02 & 0.02 & 1.07 \end{bmatrix} \begin{bmatrix} 200 \\ 800 \\ 120 \end{bmatrix} = \begin{bmatrix} 822.4 \\ 853.6 \\ 148.4 \end{bmatrix}$$

To fulfill the predicted demand, \$822.4 million worth of raw rubber, \$853.6 million worth of tires, and \$148.4 million worth of other rubber-based goods should be produced. ■■■■

SELF-CHECK EXERCISES 2.7

1. Solve the matrix equation $(I - A)X = D$ for x and y given that

$$A = \begin{bmatrix} 0.4 & 0.1 \\ 0.2 & 0.2 \end{bmatrix}, \quad X = \begin{bmatrix} x \\ y \end{bmatrix}, \quad D = \begin{bmatrix} 50 \\ 10 \end{bmatrix}$$

2. A simple economy consists of two sectors: agriculture (A) and transportation (T). The input–output matrix for this economy is given by

$$A = \begin{array}{c} \\ A \\ T \end{array} \begin{array}{c} \begin{array}{cc} A & T \end{array} \\ \begin{bmatrix} 0.4 & 0.1 \\ 0.2 & 0.2 \end{bmatrix} \end{array}$$

a. Find the gross output of agricultural products needed to satisfy a consumer demand for \$50 million worth of agricultural products and \$10 million worth of transportation.
b. Find the value of agricultural products and transportation consumed in the internal process of production in order to meet the gross output.

Solutions to Self-Check Exercises 2.7 can be found on page 175.

Spreadsheet examples and exercises for this section that may be solved using the Microsoft® Excel program are given at the Brooks/Cole Web site:
http://www.brookscole.com/math/authors/tans/

2.7 Exercises

1. **AN INPUT–OUTPUT MATRIX FOR A THREE-SECTOR ECONOMY** A simple economy consists of three sectors: agriculture (A), manufacturing (M), and transportation (T). The input–output matrix for this economy is given by

$$\begin{array}{c} \\ A \\ M \\ T \end{array} \begin{array}{c} \begin{array}{ccc} A & M & T \end{array} \\ \begin{bmatrix} 0.4 & 0.1 & 0.1 \\ 0.1 & 0.4 & 0.3 \\ 0.2 & 0.2 & 0.2 \end{bmatrix} \end{array}$$

a. If the units are measured in millions of dollars, determine the amount of agricultural products consumed in the production of \$100 million worth of manufactured goods.
b. Determine the dollar amount of manufactured products required to produce \$200 million worth of all goods in the economy.
c. Which sector consumes the greatest amount of agricultural products in the production of a unit of goods in that sector? The least?

2. **AN INPUT–OUTPUT MATRIX FOR A FOUR-SECTOR ECONOMY** The relationship governing the intraindustrial and interindustrial sales and purchases of four basic industries—agriculture (A), manufacturing (M), transportation (T), and energy (E)—of a certain economy is given by the following input–output matrix.

$$\begin{array}{c} \\ A \\ M \\ T \\ E \end{array} \begin{array}{c} \begin{array}{cccc} A & M & T & E \end{array} \\ \begin{bmatrix} 0.3 & 0.2 & 0 & 0.1 \\ 0.2 & 0.3 & 0.2 & 0.1 \\ 0.2 & 0.2 & 0.1 & 0.3 \\ 0.1 & 0.2 & 0.3 & 0.2 \end{bmatrix} \end{array}$$

a. How many units of energy are required to produce 1 unit of manufactured goods?
b. How many units of energy are required to produce 3 units of all goods in the economy?
c. Which sector of the economy is least dependent on the cost of energy?

(continued on p. 174)

The Leontief Input–Output Model

Since the solution to a problem involving a Leontief input–output model often involves several matrix operations, the graphing utility can be used to facilitate the necessary computations.

EXAMPLE 1 Suppose the input–output matrix associated with an economy is given by A and the matrix D is a demand vector, where

$$A = \begin{bmatrix} 0.2 & 0.4 & 0.15 \\ 0.3 & 0.1 & 0.4 \\ 0.25 & 0.4 & 0.2 \end{bmatrix} \quad \text{and} \quad D = \begin{bmatrix} 20 \\ 15 \\ 40 \end{bmatrix}$$

Find the final outputs of each industry so that the demands of both industry and the open sector are met.

SOLUTION ✓ First, we enter the matrices I (the identity matrix), A, and D. We are required to compute the output matrix $X = (I - A)^{-1}D$. Using the matrix operations of the graphing utility, we find

$$X = (I - A)^{-1} * D = \begin{bmatrix} 110.28 \\ 116.95 \\ 142.94 \end{bmatrix}$$

So, the final outputs of the first, second, and third industries are 110.28, 116.95, and 142.94 units, respectively. ■■■■

Exercises

In Exercises 1–4, *A* is an input–output matrix associated with an economy, and *D* (in units of dollars) is a demand vector. Using a graphing calculator, find the final outputs of each industry so that the demands of both industry and the open sector are met.

1.
$$A = \begin{bmatrix} 0.3 & 0.2 & 0.4 & 0.1 \\ 0.2 & 0.1 & 0.2 & 0.3 \\ 0.3 & 0.1 & 0.2 & 0.3 \\ 0.4 & 0.2 & 0.1 & 0.2 \end{bmatrix} \quad \text{and} \quad D = \begin{bmatrix} 40 \\ 60 \\ 70 \\ 20 \end{bmatrix}$$

2.
$$A = \begin{bmatrix} 0.12 & 0.31 & 0.40 & 0.05 \\ 0.31 & 0.22 & 0.12 & 0.20 \\ 0.18 & 0.32 & 0.05 & 0.15 \\ 0.32 & 0.14 & 0.22 & 0.05 \end{bmatrix} \quad \text{and} \quad D = \begin{bmatrix} 50 \\ 20 \\ 40 \\ 60 \end{bmatrix}$$

3.
$$A = \begin{bmatrix} 0.2 & 0.2 & 0.3 & 0.05 \\ 0.1 & 0.1 & 0.2 & 0.3 \\ 0.3 & 0.2 & 0.1 & 0.4 \\ 0.2 & 0.05 & 0.2 & 0.1 \end{bmatrix} \quad \text{and} \quad D = \begin{bmatrix} 25 \\ 30 \\ 50 \\ 40 \end{bmatrix}$$

4.
$$A = \begin{bmatrix} 0.2 & 0.4 & 0.3 & 0.1 \\ 0.1 & 0.2 & 0.1 & 0.3 \\ 0.2 & 0.1 & 0.4 & 0.05 \\ 0.3 & 0.1 & 0.2 & 0.05 \end{bmatrix} \quad \text{and} \quad D = \begin{bmatrix} 40 \\ 20 \\ 30 \\ 60 \end{bmatrix}$$

d. Which sector of the economy has the smallest intra-industry purchases (sales)?

A calculator is recommended for the remainder of this exercise set.

In Exercises 3–6, solve the matrix equation $(I - A)X = D$ for the given matrices A and D.

3. $A = \begin{bmatrix} 0.4 & 0.2 \\ 0.3 & 0.1 \end{bmatrix}$ and $D = \begin{bmatrix} 10 \\ 12 \end{bmatrix}$

4. $A = \begin{bmatrix} 0.2 & 0.3 \\ 0.5 & 0.2 \end{bmatrix}$ and $D = \begin{bmatrix} 4 \\ 8 \end{bmatrix}$

5. $A = \begin{bmatrix} 0.5 & 0.2 \\ 0.2 & 0.5 \end{bmatrix}$ and $D = \begin{bmatrix} 10 \\ 20 \end{bmatrix}$

6. $A = \begin{bmatrix} 0.6 & 0.2 \\ 0.1 & 0.4 \end{bmatrix}$ and $D = \begin{bmatrix} 8 \\ 12 \end{bmatrix}$

7. Let

$$A = \begin{bmatrix} 0.08 & 0.60 & 0.30 \\ 0.04 & 0.02 & 0.01 \\ 0.02 & 0 & 0.06 \end{bmatrix}$$

Show that

$$(I - A)^{-1} = \begin{bmatrix} 1.13 & 0.69 & 0.37 \\ 0.05 & 1.05 & 0.03 \\ 0.02 & 0.02 & 1.07 \end{bmatrix}$$

8. An Input–Output Model for a Two-Sector Economy A simple economy consists of two industries: agriculture and manufacturing. The production of 1 unit of agricultural products requires the consumption of 0.2 unit of agricultural products and 0.3 unit of manufactured goods. The production of 1 unit of manufactured products requires the consumption of 0.4 unit of agricultural products and 0.3 unit of manufactured goods.
a. Find the gross output of goods needed to satisfy a consumer demand for \$100 million worth of agricultural products and \$150 million worth of manufactured products.
b. Find the value of the goods consumed in the internal process of production in order to meet the gross output.

9. Rework Exercise 8 if the consumer demand for the output of agricultural goods and the consumer demand for manufactured products are \$120 million and \$140 million, respectively.

10. Refer to Example 3. Suppose the demand for raw rubber increases by 10%, the demand for tires increases by 20%, and the demand for rubber-based products decreases by 10%. Find the level of production for each subsidiary in order to meet this demand.

11. An Input–Output Model for a Three-Sector Economy Consider the economy of Exercise 1, consisting of three sectors: agriculture (A), manufacturing (M), and transportation (T), with an input–output matrix given by

$$\begin{array}{c} \\ A \\ M \\ T \end{array} \begin{array}{c} \begin{array}{ccc} A & M & T \end{array} \\ \begin{bmatrix} 0.4 & 0.1 & 0.1 \\ 0.1 & 0.4 & 0.3 \\ 0.2 & 0.2 & 0.2 \end{bmatrix} \end{array}$$

a. Find the gross output of goods needed to satisfy a consumer demand for \$200 million worth of agricultural products, \$100 million worth of manufactured products, and \$60 million worth of transportation.
b. Find the value of goods and transportation consumed in the internal process of production in order to meet this gross output.

12. An Input–Output Model for a Three-Sector Economy Consider a simple economy consisting of three sectors: food, clothing, and shelter. The production of 1 unit of food requires the consumption of 0.4 unit of food, 0.2 unit of clothing, and 0.2 unit of shelter. The production of 1 unit of clothing requires the consumption of 0.1 unit of food, 0.2 unit of clothing, and 0.3 unit of shelter. The production of 1 unit of shelter requires the consumption of 0.3 unit of food, 0.1 unit of clothing, and 0.1 unit of shelter. Find the level of production for each sector in order to satisfy the demand for \$100 million worth of food, \$30 million worth of clothing, and \$250 million worth of shelter.

In Exercises 13–16, matrix A is an input–output matrix associated with an economy, and matrix D (units in millions of dollars) is a demand vector. In each problem, find the final outputs of each industry so that the demands of both industry and the open sector are met.

13. $A = \begin{bmatrix} 0.4 & 0.2 \\ 0.3 & 0.5 \end{bmatrix}$ and $D = \begin{bmatrix} 12 \\ 24 \end{bmatrix}$

14. $A = \begin{bmatrix} 0.1 & 0.4 \\ 0.3 & 0.2 \end{bmatrix}$ and $D = \begin{bmatrix} 5 \\ 10 \end{bmatrix}$

15. $A = \begin{bmatrix} \frac{1}{5} & \frac{2}{5} & \frac{1}{5} \\ \frac{1}{2} & 0 & \frac{1}{2} \\ 0 & \frac{1}{5} & 0 \end{bmatrix}$ and $D = \begin{bmatrix} 10 \\ 5 \\ 15 \end{bmatrix}$

16. $A = \begin{bmatrix} 0.2 & 0.4 & 0.1 \\ 0.3 & 0.2 & 0.1 \\ 0.1 & 0.2 & 0.2 \end{bmatrix}$ and $D = \begin{bmatrix} 6 \\ 8 \\ 10 \end{bmatrix}$

SOLUTIONS TO SELF-CHECK EXERCISES 2.7

1. Multiplying both sides of the given equation on the left by $(I - A)^{-1}$, we see that

$$X = (I - A)^{-1}D$$

Now,

$$I - A = \begin{bmatrix} 1 & 0 \\ 0 & 1 \end{bmatrix} - \begin{bmatrix} 0.4 & 0.1 \\ 0.2 & 0.2 \end{bmatrix} = \begin{bmatrix} 0.6 & -0.1 \\ -0.2 & 0.8 \end{bmatrix}$$

Next, we use the Gauss–Jordan procedure to compute $(I - A)^{-1}$ (to two decimal places):

$$\left[\begin{array}{cc|cc} 0.6 & -0.1 & 1 & 0 \\ -0.2 & 0.8 & 0 & 1 \end{array}\right] \xrightarrow{\frac{1}{0.6}R_1}$$

$$\left[\begin{array}{cc|cc} 1 & -0.17 & 1.67 & 0 \\ -0.2 & 0.8 & 0 & 1 \end{array}\right] \xrightarrow{R_2 + 0.2R_1}$$

$$\left[\begin{array}{cc|cc} 1 & -0.17 & 1.67 & 0 \\ 0 & 0.77 & 0.33 & 1 \end{array}\right] \xrightarrow{\frac{1}{0.77}R_2}$$

$$\left[\begin{array}{cc|cc} 1 & -0.17 & 1.67 & 0 \\ 0 & 1 & 0.43 & 1.30 \end{array}\right] \xrightarrow{R_1 + 0.17R_2}$$

$$\left[\begin{array}{cc|cc} 1 & 0 & 1.74 & 0.22 \\ 0 & 1 & 0.43 & 1.30 \end{array}\right]$$

giving

$$(I - A)^{-1} = \begin{bmatrix} 1.74 & 0.22 \\ 0.43 & 1.30 \end{bmatrix}$$

Therefore,

$$X = \begin{bmatrix} x \\ y \end{bmatrix} = (I - A)^{-1}D = \begin{bmatrix} 1.74 & 0.22 \\ 0.43 & 1.30 \end{bmatrix}\begin{bmatrix} 50 \\ 10 \end{bmatrix} = \begin{bmatrix} 89.2 \\ 34.5 \end{bmatrix}$$

or $x = 89.2$ and $y = 34.5$.

2. a. Let

$$X = \begin{bmatrix} x \\ y \end{bmatrix}$$

denote the gross production matrix, where x denotes the value of the agricultural products and y the value of transportation. Also, let

$$D = \begin{bmatrix} 50 \\ 10 \end{bmatrix}$$

denote the consumer demand. Then,

$$(I - A)X = D$$

or equivalently,

$$X = (I - A)^{-1}D$$

Using the results of Exercise 1, we find that $x = 89.2$ and $y = 34.5$. That is, to fulfill consumer demands, \$89.2 million worth of agricultural products must be produced and \$34.5 million worth of transportation services must be used.
b. The amount of agricultural products consumed and transportation services used is given by

$$X - D = \begin{bmatrix} 89.2 \\ 34.5 \end{bmatrix} - \begin{bmatrix} 50 \\ 10 \end{bmatrix} = \begin{bmatrix} 39.2 \\ 24.5 \end{bmatrix}$$

or \$39.2 million worth of agricultural products and \$24.5 million worth of transportation services.

Group projects for each chapter can be found at the Brooks/Cole Web site: http://www.brookscole.com/math/authors/tans/

CHAPTER 2 Summary of Principal Formulas and Terms

Formulas

1. Laws for matrix addition

a. Commutative law $A + B = B + A$

b. Associative law $(A + B) + C = A + (B + C)$

2. Laws for matrix multiplication

a. Associative law $(AB)C = A(BC)$

b. Distributive law $A(B + C) = AB + AC$

3. Inverse of a 2×2 matrix

If $A = \begin{bmatrix} a & b \\ c & d \end{bmatrix}$

and $D = ad - bc \neq 0$

then $A^{-1} = \dfrac{1}{D}\begin{bmatrix} d & -b \\ -c & a \end{bmatrix}$

4. Solution of system $AX = B$ (A, nonsingular) $X = A^{-1}B$

Terms

system of linear equations
solution of a system of linear equations
parameter
dependent system
inconsistent system
Gauss–Jordan elimination method
equivalent system
matrix
coefficient matrix
augmented matrix
row-reduced form of a matrix
row operations
unit column
column matrix
square matrix
transpose of a matrix
scalar
scalar product
matrix product
identity matrix
inverse of a matrix
singular matrix
input–output matrix
gross production matrix
internal consumption matrix
Leontief input–output model

CHAPTER 2 REVIEW EXERCISES

In Exercises 1–4, perform the given operations, if possible.

1. $\begin{bmatrix} 1 & 2 \\ -1 & 3 \\ 2 & 1 \end{bmatrix} + \begin{bmatrix} 1 & 0 \\ 0 & 1 \\ 1 & 2 \end{bmatrix}$

2. $\begin{bmatrix} -1 & 2 \\ 3 & 4 \end{bmatrix} - \begin{bmatrix} 1 & 2 \\ 5 & -2 \end{bmatrix}$

3. $\begin{bmatrix} -3 & 2 & 1 \end{bmatrix} \begin{bmatrix} 2 & 1 \\ -1 & 0 \\ 2 & 1 \end{bmatrix}$

4. $\begin{bmatrix} 1 & 3 & 2 \\ -1 & 2 & 3 \end{bmatrix} \begin{bmatrix} 1 \\ 4 \\ 2 \end{bmatrix}$

In Exercises 5–8, find the values of the variables.

5. $\begin{bmatrix} 1 & x \\ y & 3 \end{bmatrix} = \begin{bmatrix} z & 2 \\ 3 & w \end{bmatrix}$

6. $\begin{bmatrix} 3 & x \\ y & 3 \end{bmatrix} \begin{bmatrix} 1 \\ 2 \end{bmatrix} = \begin{bmatrix} 7 \\ 4 \end{bmatrix}$

7. $\begin{bmatrix} 3 & a+3 \\ -1 & b \\ c+1 & d \end{bmatrix} = \begin{bmatrix} 3 & 6 \\ e+2 & 4 \\ -1 & 2 \end{bmatrix}$

8. $\begin{bmatrix} x & 3 & 1 \\ 0 & y & 2 \end{bmatrix} \begin{bmatrix} 1 & 1 \\ 3 & z \\ 4 & 2 \end{bmatrix} = \begin{bmatrix} 12 & 4 \\ 2 & 2 \end{bmatrix}$

In Exercises 9–16, compute the given expressions, if possible, given that

$$A = \begin{bmatrix} 1 & 3 & 1 \\ -2 & 1 & 3 \\ 4 & 0 & 2 \end{bmatrix}$$

$$B = \begin{bmatrix} 2 & 1 & 3 \\ -2 & -1 & -1 \\ 1 & 4 & 2 \end{bmatrix}$$

$$C = \begin{bmatrix} 3 & -1 & 2 \\ 1 & 6 & 4 \\ 2 & 1 & 3 \end{bmatrix}$$

9. $2A + 3B$

10. $3A - 2B$

11. $2(3A)$

12. $2(3A - 4B)$

13. $A(B - C)$

14. $AB + AC$

15. $A(BC)$

16. $(\frac{1}{2})(CA - CB)$

In Exercises 17–23, solve the system of linear equations using the Gauss–Jordan elimination method.

17. $\begin{aligned} 2x - 3y &= 5 \\ 3x + 4y &= -1 \end{aligned}$

18. $\begin{aligned} 3x + 2y &= 3 \\ 2x - 4y &= -14 \end{aligned}$

19. $\begin{aligned} x - y + 2z &= 5 \\ 3x + 2y + z &= 10 \\ 2x - 3y - 2z &= -10 \end{aligned}$

20. $\begin{aligned} 3x - 2y + 4z &= 16 \\ 2x + y - 2z &= -1 \\ x + 4y - 8z &= -18 \end{aligned}$

21. $\begin{aligned} 3x - 2y + 4z &= 11 \\ 2x - 4y + 5z &= 4 \\ x + 2y - z &= 10 \end{aligned}$

22. $\begin{aligned} x - 2y + 3z + 4w &= 17 \\ 2x + y - 2z - 3w &= -9 \\ 3x - y + 2z - 4w &= 0 \\ 4x + 2y - 3z + w &= -2 \end{aligned}$

23. $\begin{aligned} 3x - 2y + z &= 4 \\ x + 3y - 4z &= -3 \\ 2x - 3y + 5z &= 7 \\ x - 8y + 9z &= 10 \end{aligned}$

In Exercises 24–31, find the inverse of the given matrix (if it exists).

24. $A = \begin{bmatrix} 2 & 4 \\ 1 & 6 \end{bmatrix}$

25. $A = \begin{bmatrix} 3 & 1 \\ 1 & 2 \end{bmatrix}$

26. $A = \begin{bmatrix} 2 & 4 \\ 1 & -2 \end{bmatrix}$

27. $A = \begin{bmatrix} 3 & 4 \\ 2 & 2 \end{bmatrix}$

28. $A = \begin{bmatrix} 1 & 2 & 4 \\ 2 & 1 & 3 \\ -1 & 0 & 2 \end{bmatrix}$

29. $A = \begin{bmatrix} 2 & 3 & 1 \\ 1 & -1 & 2 \\ 1 & 2 & 1 \end{bmatrix}$

30. $A = \begin{bmatrix} 2 & 1 & -3 \\ 1 & 2 & -4 \\ 3 & 1 & -2 \end{bmatrix}$

31. $A = \begin{bmatrix} 1 & 2 & 4 \\ 3 & 1 & 2 \\ 1 & 0 & -6 \end{bmatrix}$

In Exercises 32–35, compute the given expressions, if possible, given that

$$A=\begin{bmatrix}1 & 2\\ -1 & 2\end{bmatrix},\quad B=\begin{bmatrix}3 & 1\\ 4 & 2\end{bmatrix},\quad C=\begin{bmatrix}1 & 1\\ -1 & 2\end{bmatrix}$$

32. $(ABC)^{-1}$

33. $(A^{-1}B)^{-1}$

34. $(A + B)^{-1}$

35. $(2A - C)^{-1}$

In Exercises 36–39, write each system of linear equations in the form $AX = C$. Find A^{-1} and use the result to solve the system.

36. $\begin{aligned} x - 3y &= -1\\ 2x + 4y &= 8\end{aligned}$

37. $\begin{aligned} 2x + 3y &= -8\\ x - 2y &= 3\end{aligned}$

38. $\begin{aligned} 2x - 3y + 4z &= 17\\ x + 2y - 4z &= -7\\ 3x - y + 2z &= 14\end{aligned}$

39. $\begin{aligned} x - 2y + 4z &= 13\\ 2x + 3y - 2z &= 0\\ x + 4y - 6z &= -15\end{aligned}$

40. Jack Spaulding bought 10,000 shares of stock X, 20,000 shares of stock Y, and 30,000 shares of stock Z at a unit price of $20, $30, and $50/share, respectively. Six months later, the closing prices of stocks X, Y, and Z were $22, $35, and $51/share, respectively. Jack made no other stock transactions during the period in question. Compare the value of Jack's stock holdings at the time of purchase and six months later.

41. Gloria Newburg operates three self-service gasoline stations in different parts of town. On a certain day, station A sold 600 gal of premium, 1000 gal of regular, 800 gal of super-unleaded, and 1400 gal of regular-unleaded gasoline; station B sold 700 gal of premium, 800 gal of regular, 600 gal of super-unleaded, and 1200 gal of regular-unleaded gasoline; station C sold 1200 gal of premium, 800 gal of regular, 1000 gal of super-unleaded, and 900 gal of regular-unleaded gasoline. Assume that the price of gasoline on that day was $1.60/gal for premium, $1.20/gal for regular, $1.50/gal for super-unleaded, and $1.30/gal for regular-unleaded gasoline, and use matrix algebra to find the total revenue at each station.

42. The Wildcat Oil Company has two refineries—one located in Houston and the other in Tulsa. The Houston refinery ships 60% of its petroleum to a Chicago distributor and 40% of its petroleum to a Los Angeles distributor. The Tulsa refinery ships 30% of its petroleum to the Chicago distributor and 70% of its petroleum to the Los Angeles distributor. Assume that, over the year, the Chicago distributor received 240,000 gal of petroleum and the Los Angeles distributor received 460,000 gal of petroleum. Find the amount of petroleum produced at each of Wildcat's refineries.

43. Desmond Jewelry, Inc., wishes to produce three types of pendants: type A, type B, and type C. To manufacture a type-A pendant requires 2 min on machines I and II and 3 min on machine III. A type-B pendant requires 2 min on machine I, 3 min on machine II, and 4 min on machine III. A type-C pendant requires 3 min on machine I, 4 min on machine II, and 3 min on machine III. There are $3\frac{1}{2}$ hr available on machine I, $4\frac{1}{2}$ hr available on machine II, and 5 hr available on machine III. How many pendants of each type should Desmond make in order to use all the available time?

Additional study hints and sample chapter tests for each chapter can be found at the Brooks/Cole Web site: http://www.brookscole.com/math/authors/tans/

3

LINEAR PROGRAMMING: A GEOMETRIC APPROACH

Many practical problems involve maximizing or minimizing a function subject to certain constraints. For example, we may wish to maximize a profit function subject to certain limitations on the amount of material and labor available. Maximization or minimization problems that can be formulated in terms of a *linear* objective function and constraints in the form of linear inequalities are called *linear programming problems.* In this chapter we look at linear programming problems involving two variables. These problems are amenable to geometric analysis, and the method of solution introduced here will shed much light on the basic nature of a linear programming problem.

How should the aircraft engines be shipped? Curtis-Roe Aviation Industries manufactures jet engines in two different locations. These engines are to be shipped to the company's two main assembly plants. In Example 3, page 194, we will show how many engines should be produced and shipped from each manufacturing plant to each assembly plant in order to minimize shipping costs.

3.1 Graphing Systems of Linear Inequalities in Two Variables

GRAPHING LINEAR INEQUALITIES

In Chapter 1 we saw that a linear equation in two variables x and y

$$ax + by + c = 0 \qquad (a, b \text{ not both equal to zero})$$

has a *solution set* that may be exhibited graphically as points on a straight line in the xy-plane. We now show that there is also a simple graphical representation for **linear inequalities** in two variables:

$$ax + by + c < 0 \qquad ax + by + c \le 0$$
$$ax + by + c > 0 \qquad ax + by + c \ge 0$$

Before turning to a general procedure for graphing such inequalities, let's consider a specific example. Suppose we wish to graph

$$2x + 3y < 6 \tag{1}$$

We first graph the equation $2x + 3y = 6$, which is obtained by replacing the given inequality "<" with an equality "=" (Figure 3.1).

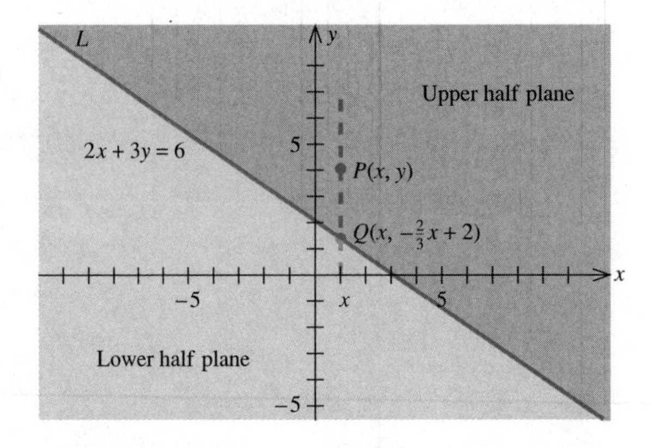

FIGURE 3.1
A straight line divides the xy-plane into two half planes.

Observe that this line divides the xy-plane into two half planes: an upper half plane and a lower half plane. Let's show that the upper half plane is the graph of the linear inequality

$$2x + 3y > 6 \tag{2}$$

whereas the lower half plane is the graph of the linear inequality

$$2x + 3y < 6 \tag{3}$$

To see this, let's write Equations (2) and (3) in the equivalent forms

$$y > -\frac{2}{3}x + 2 \tag{4}$$

and

$$y < -\frac{2}{3}x + 2 \tag{5}$$

The equation of the line itself is

$$y = -\frac{2}{3}x + 2 \tag{6}$$

Now pick any point $P(x, y)$ lying above the line L. Let Q be the point lying on L and directly below P (see Figure 3.1). Since Q lies on L, its coordinates must satisfy Equation (6). In other words, Q has representation $Q(x, -\frac{2}{3}x + 2)$. Comparing the y-coordinates of P and Q and recalling that P lies above Q so that its y-coordinate must be larger than that of Q, we have

$$y > -\frac{2}{3}x + 2$$

But this inequality is just Equation (4) or equivalently, Equation (2). Similarly, we can show that any point lying below L must satisfy Equation (5) and therefore (3).

This analysis shows that the lower half plane provides a solution to our problem (Figure 3.2). (The dashed line shows that the points on L do not belong to the solution set.) Observe that the two half planes in question are mutually exclusive; that is, they do not have any points in common. Because of this, there is an alternative and easier method of determining the solution to the problem.

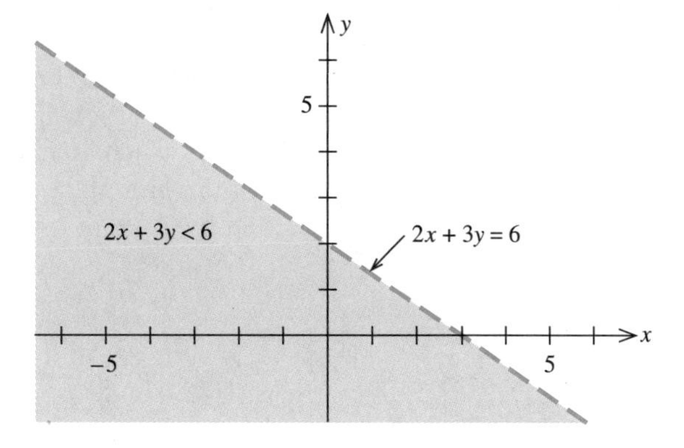

FIGURE 3.2
The set of points lying below the dashed line satisfies the given inequality.

To determine the required half plane, let's pick *any* point lying in one of the half planes. For simplicity, pick the origin (0, 0), which lies in the lower half plane. Substituting $x = 0$ and $y = 0$ (the coordinates of this point) into the given Inequality (1), we find

$$2(0) + 3(0) < 6$$

or $0 < 6$, which is certainly true. This tells us that the required half plane is the half plane containing the test point—namely, the lower half plane.

Next, let's see what happens if we choose the point (2, 3), which lies in the upper half plane. Substituting $x = 2$ and $y = 3$ into the given inequality, we find

$$2(2) + 3(3) < 6$$

or $13 < 6$, which is false. This tells us that the upper half plane is *not* the required half plane, as expected. Note, too, that no point (x, y) lying on the line constitutes a solution to our problem, because of the *strict* inequality "$<$".

This discussion suggests the following procedure for graphing a linear inequality in two variables.

Procedure for Graphing Linear Inequalities

1. Draw the graph of the equation obtained for the given inequality by replacing the inequality sign with an equals sign. Use a dashed or dotted line if the problem involves a strict inequality, "$<$" or "$>$". Otherwise, use a solid line to indicate that the line itself constitutes part of the solution.
2. Pick a test point lying in one of the half planes determined by the line sketched in step 1 and substitute the values of x and y into the given inequality. Use the origin whenever possible.
3. If the inequality is satisfied, the graph of the inequality includes the half plane containing the test point. Otherwise, the solution includes the half plane not containing the test point.

EXAMPLE 1 Determine the solution set for the inequality $2x + 3y \geq 6$.

SOLUTION ✓ Replacing the inequality "$\geq$" with an equality "$=$", we obtain the equation $2x + 3y = 6$, whose graph is the straight line shown in Figure 3.3. Instead of a dashed line as before, we use a solid line to show that all points on the line are also solutions to the problem. Picking the origin as our test point, we find $2(0) + 3(0) \geq 6$, or $0 \geq 6$, which is impossible. So we conclude that the solution set is made up of the half plane not containing the origin, including, in this case, the line given by $2x + 3y = 6$.

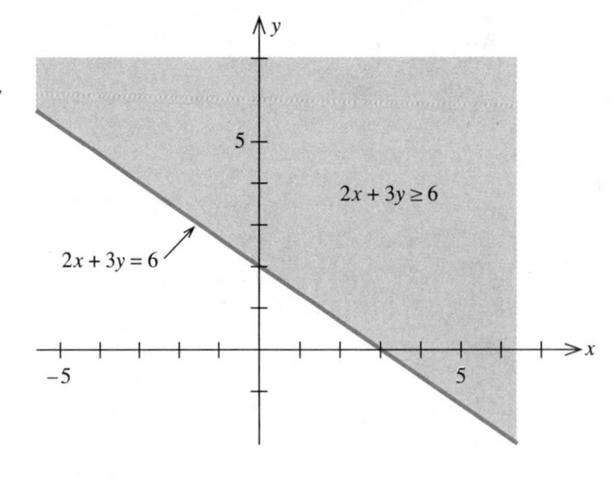

FIGURE 3.3
The set of points lying on the line and in the upper half plane satisfies the given inequality.

■■■■

EXAMPLE 2 Graph $x \leq -1$.

SOLUTION ✓ The graph of $x = -1$ is the vertical line shown in Figure 3.4. Picking the origin (0, 0) as a test point, we find $0 \leq -1$, which is false. Therefore, the required solution is the *left* half plane, which does not contain the origin.

FIGURE 3.4
The set of points lying on the line $x = -1$ and in the left half plane satisfies the given inequality.

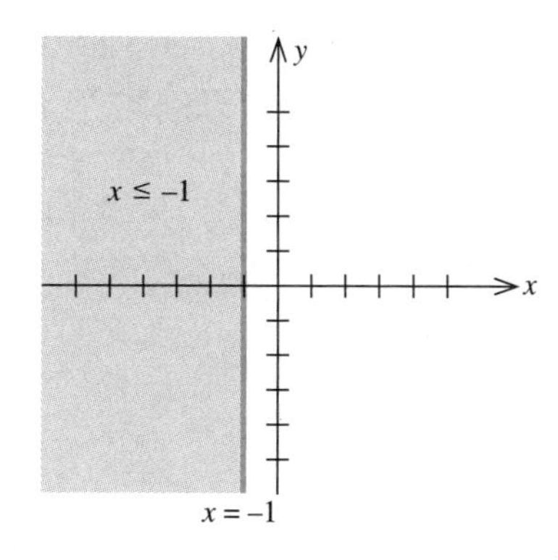

■■■■

EXAMPLE 3 Graph $x - 2y > 0$.

SOLUTION ✓ We first graph the equation $x - 2y = 0$, or $y = (\frac{1}{2})x$ (Figure 3.5). Since the origin lies on the line, we may not use it as a test point. (Why?) Let's pick (1, 2) as a test point. Substituting $x = 1$ and $y = 2$ into the given inequality, we find $1 - 2(2) > 0$, or $-3 > 0$, which is false. Therefore, the required solution is the half plane that does not contain the test point—namely, the lower half plane.

FIGURE 3.5
The set of points in the lower half plane satisfies $x - 2y > 0$.

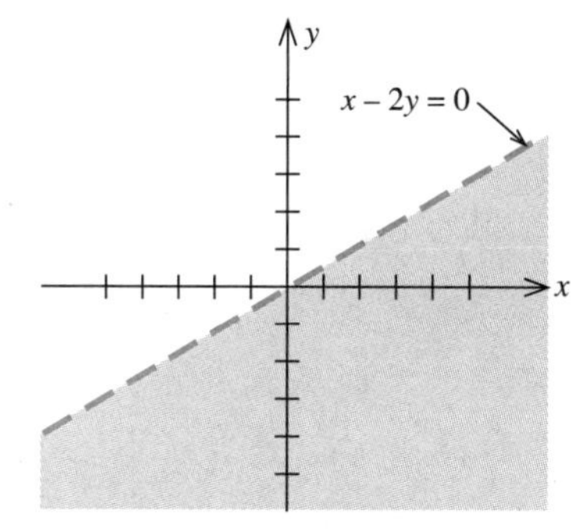

■■■■

GRAPHING SYSTEMS OF LINEAR INEQUALITIES

By the **solution set of a system of linear inequalities** in the two variables x and y, we mean the set of all points (x, y) satisfying each inequality of the system. The graphical solution of such a system may be obtained by graphing the solution set for each inequality independently and then determining the region in common with each solution set.

EXAMPLE 4 Determine the solution set for the system

$$\begin{aligned} 4x + 3y &\geq 12 \\ x - y &\leq 0 \end{aligned}$$

SOLUTION ✔ Proceeding as in the previous examples, you should have no difficulty locating the half planes determined by each of the linear inequalities that make up the system. These half planes are shown in Figure 3.6. The intersection of the

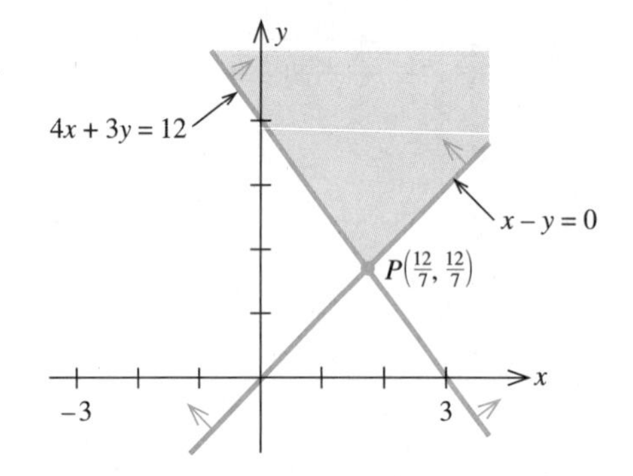

FIGURE 3.6
The set of points in the shaded area satisfies the system

$$\begin{aligned} 4x + 3y &\geq 12 \\ x - y &\leq 0 \end{aligned}$$

two half planes is the shaded region. A point in this region is an element of the solution set for the given system. The point P, the intersection of the two straight lines determined by the equations, is found by solving the simultaneous equations

$$\begin{aligned} 4x + 3y &= 12 \\ x - y &= 0 \end{aligned}$$

■■■■

EXAMPLE 5 Sketch the solution set for the system

$$\begin{aligned} x &\geq 0 \\ y &\geq 0 \\ x + y - 6 &\leq 0 \\ 2x + y - 8 &\leq 0 \end{aligned}$$

SOLUTION ✔ The first inequality in the system defines the right half plane—all points to the right of the y-axis plus all points lying on the y-axis itself. The second inequality in the system defines the upper half plane, including the x-axis. The half planes defined by the third and fourth inequalities are indicated by arrows in Figure 3.7. Thus, the required region, the intersection of the four half planes defined by the four inequalities in the given system of linear inequalities, is the shaded region. The point P is found by solving the simultaneous equations $x + y - 6 = 0$ and $2x + y - 8 = 0$.

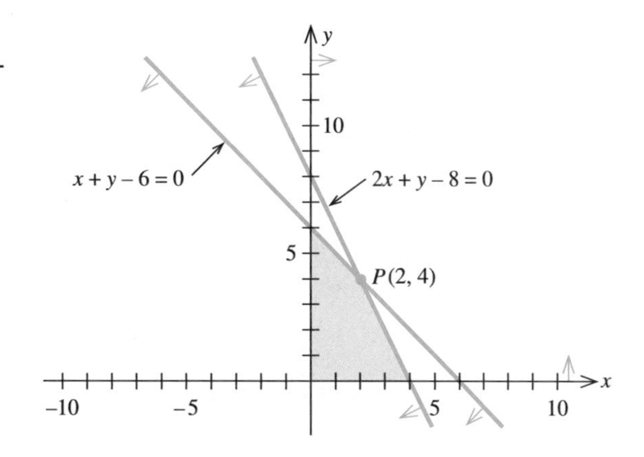

FIGURE 3.7
The set of points in the shaded region, including the x- and y-axes, satisfies the given inequalities.

The solution set found in Example 5 is an example of a *bounded set.* Observe that the set can be enclosed by a circle. For example, if you draw a circle of radius 10 with center at the origin, you will see that the set lies entirely inside the circle. On the other hand, the solution set of Example 4 cannot be enclosed by a circle and is said to be *unbounded.*

Bounded and Unbounded Solution Sets

A solution set of a system of linear inequalities is **bounded** if it can be enclosed by a circle. Otherwise, it is **unbounded.**

EXAMPLE 6 Determine the graphical solution set for the following system of linear inequalities:

$$
\begin{aligned}
2x + y &\geq 50 \\
x + 2y &\geq 40 \\
x &\geq 0 \\
y &\geq 0
\end{aligned}
$$

SOLUTION ✓ The required solution set is the unbounded region shown in Figure 3.8.

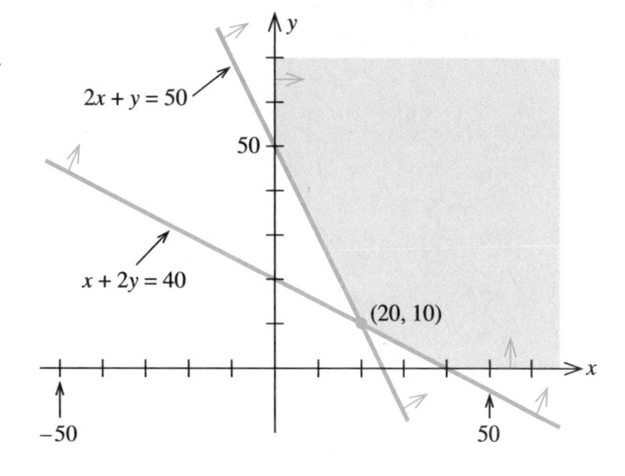

FIGURE 3.8
The solution set is an unbounded region.

SELF-CHECK EXERCISES 3.1

1. Determine graphically the solution set for the following system of inequalities:

$$x + 2y \leq 10$$
$$5x + 3y \leq 30$$
$$x \geq 0, y \geq 0$$

2. Determine graphically the solution set for the following system of inequalities:

$$5x + 3y \geq 30$$
$$x - 3y \leq 0$$
$$x \geq 2$$

Solutions to Self-Check Exercises 3.1 can be found on page 190.

3.1 Exercises

In Exercises 1–10, find the graphical solution of each inequality.

1. $4x - 8 < 0$
2. $3y + 2 > 0$
3. $x - y \leq 0$
4. $3x + 4y \leq -2$
5. $x \leq -3$
6. $y \geq -1$
7. $2x + y \leq 4$
8. $-3x + 6y \geq 12$
9. $4x - 3y \leq -24$
10. $5x - 3y \geq 15$

In Exercises 11–18, write a system of linear inequalities that describes the shaded region.

11.

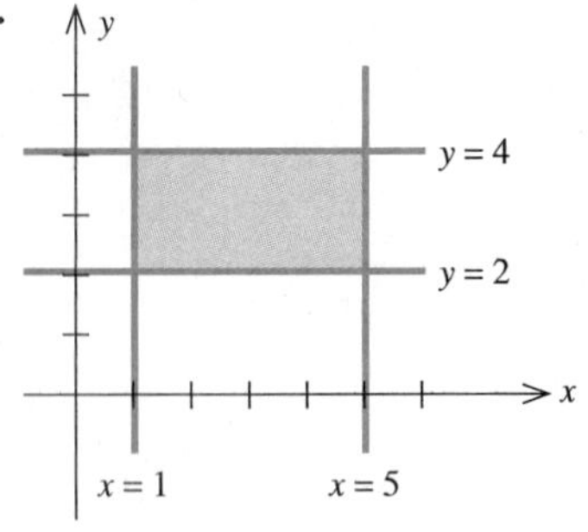

12.

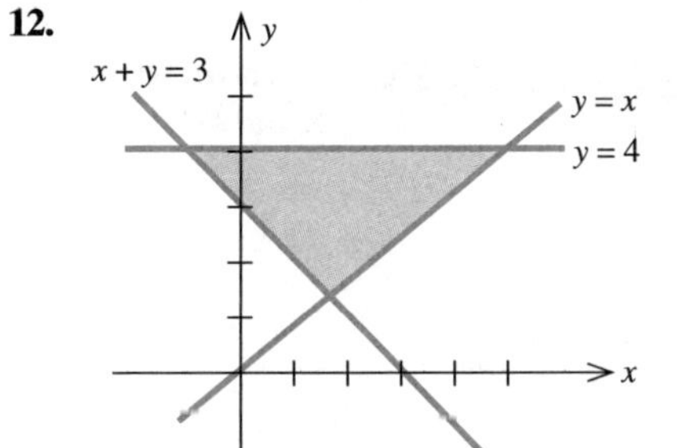

13.

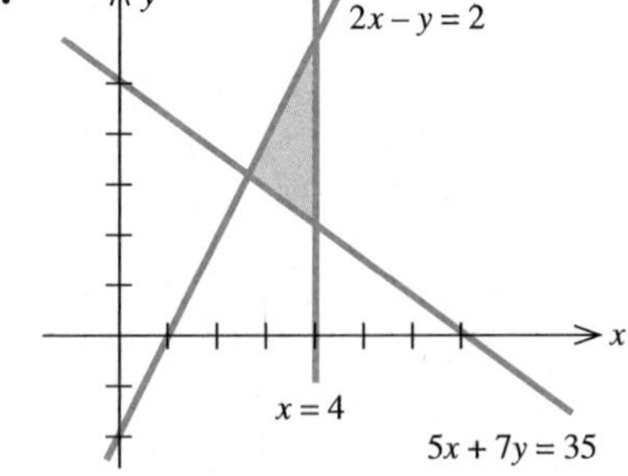

14.

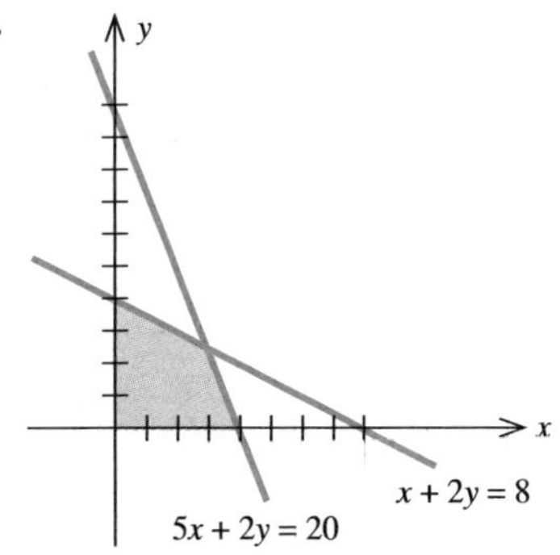

15.

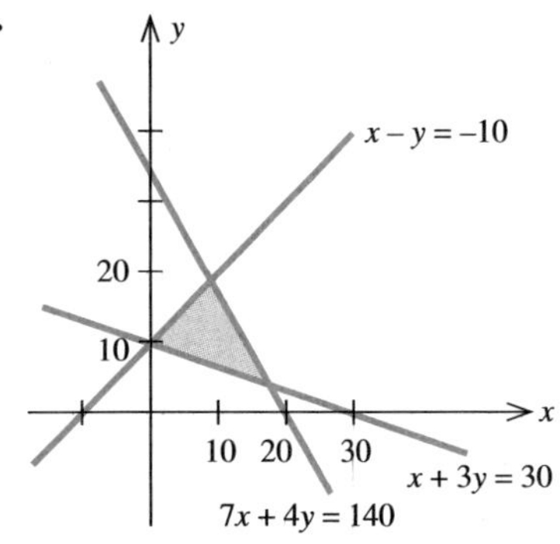

16.

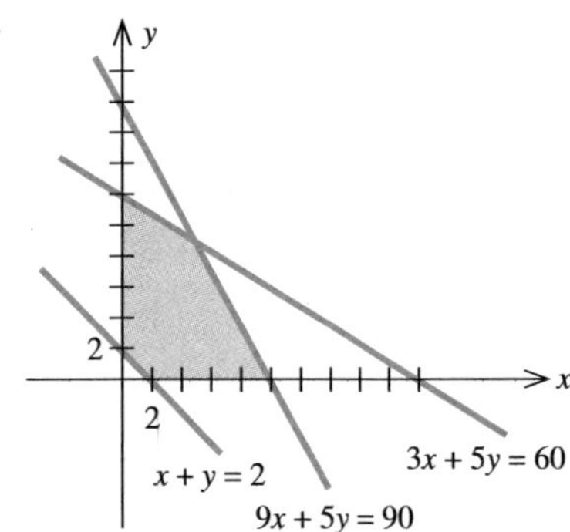

17.

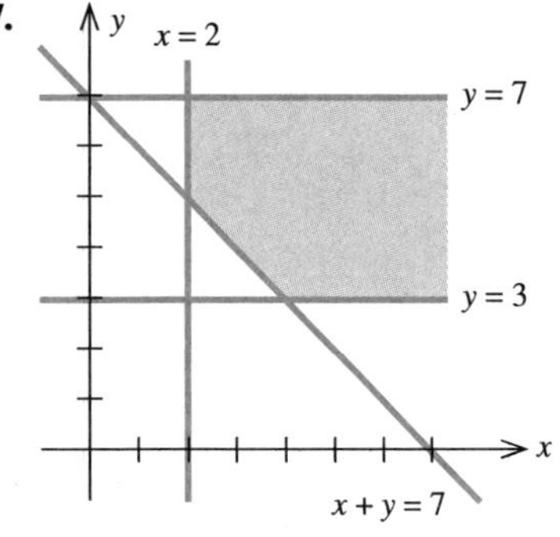

18.

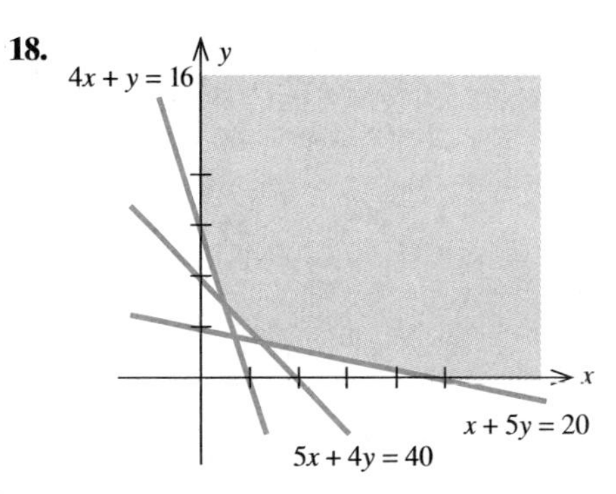

In Exercises 19–36, determine graphically the solution set for each system of inequalities and indicate whether the solution set is bounded or unbounded.

19. $2x + 4y > 16$, $-x + 3y \geq 7$

20. $3x - 2y > -13$, $-x + 2y > 5$

21. $x - y \leq 0$, $2x + 3y \geq 10$

22. $x + y \geq -2$, $3x - y \leq 6$

23. $x + 2y \geq 3$, $2x + 4y \leq -2$

24. $2x - y \geq 4$, $4x - 2y < -2$

25. $x + y \leq 6$, $0 \leq x \leq 3$, $y \geq 0$

26. $4x - 3y \leq 12$, $5x + 2y \leq 10$, $x \geq 0, y \geq 0$

27. $3x - 6y \leq 12$, $-x + 2y \leq 4$, $x \geq 0, y \geq 0$

28. $x + y \geq 20$, $x + 2y \geq 40$, $x \geq 0, y \geq 0$

29. $3x - 7y \geq -24$, $x + 3y \geq 8$, $x \geq 0, y \geq 0$

30. $3x + 4y \geq 12$, $2x - y \geq -2$, $0 \leq y \leq 3$, $x \geq 0$

31. $x + 2y \geq 3$, $5x - 4y \leq 16$, $0 \leq y \leq 2$, $x \geq 0$

32. $x + y \leq 4$, $2x + y \leq 6$, $2x - y \geq -1$, $x \geq 0, y \geq 0$

33. $6x + 5y \leq 30$, $3x + y \geq 6$, $x + y \geq 4$, $x \geq 0, y \geq 0$

34. $6x + 7y \leq 84$, $12x - 11y \leq 18$, $6x - 7y \leq 28$, $x \geq 0, y \geq 0$

35. $x - y \geq -6$, $x - 2y \leq -2$, $x + 2y \geq 6$, $x - 2y \geq -14$, $x \geq 0, y \geq 0$

36. $x - 3y \geq -18$, $3x - 2y \geq 2$, $x - 3y \leq -4$, $3x - 2y \leq 16$, $x \geq 0, y \geq 0$

In Exercises 37–39, determine whether the statement is true or false. If it is true, explain why it is true. If it is false, give an example to show why it is false.

37. The solution set of a linear inequality involving two variables is either a half plane or a straight line.

38. The solution set of the inequality $ax + by + c \leq 0$ is either a left half plane or a lower half plane.

39. The solution set of a system of linear inequalities in two variables is bounded if it can be enclosed by a rectangle.

Solutions to Self-Check Exercises 3.1

1. The required solution set is shown in the following figure:

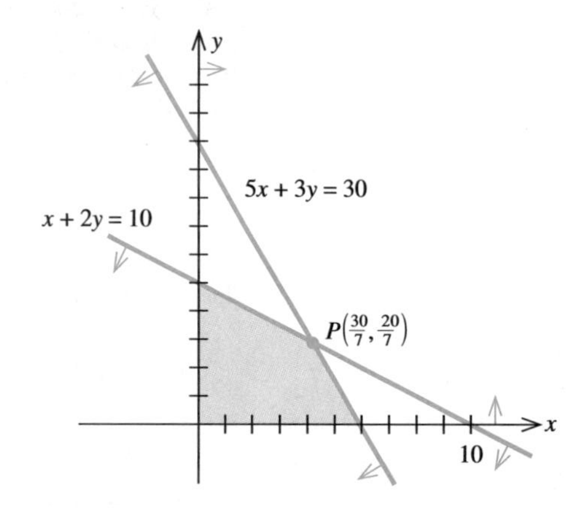

The point P is found by solving the system of equations

$$\begin{aligned} x + 2y &= 10 \\ 5x + 3y &= 30 \end{aligned}$$

Solving the first equation for x in terms of y gives

$$x = 10 - 2y$$

Substituting this value of x into the second equation of the system gives

$$\begin{aligned} 5(10 - 2y) + 3y &= 30 \\ 50 - 10y + 3y &= 30 \\ -7y &= -20 \end{aligned}$$

so $y = 20/7$. Substituting this value of y into the expression for x found earlier, we obtain

$$x = 10 - 2\left(\frac{20}{7}\right) = \frac{30}{7}$$

giving the point of intersection as (30/7, 20/7).

2. The required solution set is shown in the following figure:

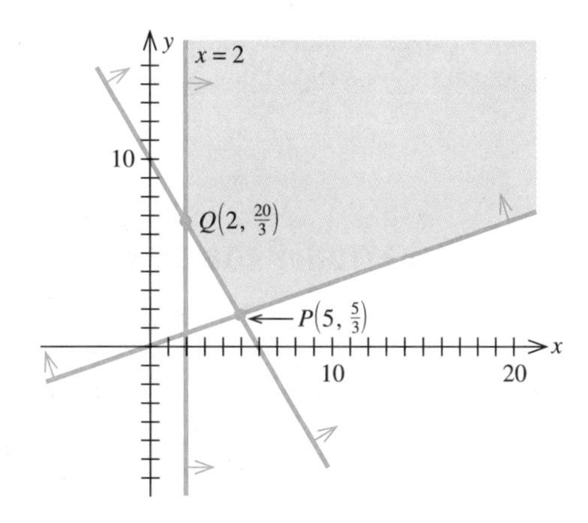

To find the coordinates of P, we solve the system

$$
\begin{aligned}
5x + 3y &= 30 \\
x - 3y &= 0
\end{aligned}
$$

Solving the second equation for x in terms of y and substituting this value of x in the first equation gives

$$5(3y) + 3y = 30$$

or $y = 5/3$. Substituting this value of y into the second equation gives $x = 5$. Next, the coordinates of Q are found by solving the system

$$
\begin{aligned}
5x + 3y &= 30 \\
x &= 2
\end{aligned}
$$

obtaining $x = 2$ and $y = 20/3$.

3.2 Linear Programming Problems

Many business and economic problems are concerned with optimizing (maximizing or minimizing) a function subject to a system of equalities or inequalities. The function to be optimized is called the **objective function.** Profit functions and cost functions are examples of objective functions. The system of equalities or inequalities to which the objective function is subjected reflects the constraints (for example, limitations on resources such as materials and labor) imposed on the solution(s) to the problem. Problems of this nature are called **mathematical programming problems.** In particular, problems in which both the objective function and the constraints are expressed as linear equations or inequalities are called linear programming problems.

A Linear Programming Problem

A **linear programming problem** consists of a linear objective function to be maximized or minimized subject to certain constraints in the form of linear equalities or inequalities.

MAXIMIZATION PROBLEMS

As an example of a linear programming problem in which the objective function is to be maximized, let's consider the following simplified version of a production problem involving two variables.

EXAMPLE 1 The Ace Novelty Company wishes to produce two types of souvenirs: type A and type B. Each type-A souvenir will result in a profit of \$1, and each type-B souvenir will result in a profit of \$1.20. To manufacture a type-A souvenir requires 2 minutes on machine I and 1 minute on machine II. A type-B souvenir requires 1 minute on machine I and 3 minutes on machine II. There are 3 hours available on machine I and 5 hours available on machine II for processing the order. How many souvenirs of each type should Ace make in order to maximize profit?

SOLUTION ✔ As a first step toward the mathematical formulation of this problem, we tabulate the given information, as shown in Table 3.1.

Table 3.1

	Type A	Type B	Time Available
Machine I	2 min	1 min	180 min
Machine II	1 min	3 min	300 min
Profit per Unit	\$1	\$1.20	

Let x be the number of type-A souvenirs and y be the number of type-B souvenirs to be made. Then, the total profit P (in dollars) is given by

$$P = x + 1.2y$$

which is the objective function to be maximized.

The total amount of time that machine I is used is given by $2x + y$ minutes and must not exceed 180 minutes. Thus, we have the inequality

$$2x + y \leq 180$$

Similarly, the total amount of time that machine II is used is $x + 3y$ minutes, which cannot exceed 300 minutes, so we are led to the inequality

$$x + 3y \leq 300$$

Finally, neither x nor y can be negative, so

$$x \geq 0$$
$$y \geq 0$$

To summarize, the problem at hand is one of maximizing the objective function $P = x + 1.2y$ subject to the system of inequalities

$$\begin{aligned} 2x + y &\leq 180 \\ x + 3y &\leq 300 \\ x &\geq 0 \\ y &\geq 0 \end{aligned}$$

The solution to this problem will be completed in Example 1, Section 3.3.

MINIMIZATION PROBLEMS

In the following example of a linear programming problem, the objective function is to be minimized.

EXAMPLE 2 A nutritionist advises an individual who is suffering from iron- and vitamin-B deficiency to take at least 2400 milligrams (mg) of iron, 2100 mg of vitamin B_1 (thiamine), and 1500 mg of vitamin B_2 (riboflavin) over a period of time. Two vitamin pills are suitable, brand A and brand B. Each brand A pill contains 40 mg of iron, 10 mg of vitamin B_1, and 5 mg of vitamin B_2, and costs 6 cents. Each brand B pill contains 10 mg of iron and 15 mg each of vitamins B_1 and B_2, and costs 8 cents (Table 3.2). What combination of pills should the individual purchase in order to meet the minimum iron and vitamin requirements at the lowest cost?

Table 3.2

	Brand A	Brand B	Minimum Requirement
Iron	40 mg	10 mg	2400 mg
Vitamin B_1	10 mg	15 mg	2100 mg
Vitamin B_2	5 mg	15 mg	1500 mg
Cost per Pill	6¢	8¢	

SOLUTION ✓ Let x be the number of brand A pills and y be the number of brand B pills to be purchased. The cost C (in cents) is given by

$$C = 6x + 8y$$

and is the objective function to be minimized.

The amount of iron contained in x brand A pills and y brand B pills is given by $40x + 10y$ mg, and this must be greater than or equal to 2400 mg. This translates into the inequality

$$40x + 10y \geq 2400$$

Similar considerations involving the minimum requirements of vitamins B_1 and B_2 lead to the inequalities

$$\begin{aligned} 10x + 15y &\geq 2100 \\ 5x + 15y &\geq 1500 \end{aligned}$$

respectively. Thus, the problem here is to minimize $C = 6x + 8y$ subject to

$$\begin{aligned} 40x + 10y &\geq 2400 \\ 10x + 15y &\geq 2100 \\ 5x + 15y &\geq 1500 \\ x \geq 0, y &\geq 0 \end{aligned}$$

The solution to this problem will be completed in Example 2, Section 3.3.

■■■■

A Transportation Problem

EXAMPLE 3

Curtis-Roe Aviation Industries has two plants, I and II, that produce the Zephyr jet engines used in their light commercial airplanes. The maximum production capacities of these two plants are 100 units and 110 units per month, respectively. The engines are shipped to two of Curtis-Roe's main assembly plants, A and B. The shipping costs (in dollars) per engine from plants I and II to the main assembly plants A and B are as follows:

From	To Assembly Plant A	B
Plant I	100	60
Plant II	120	70

In a certain month, assembly plant A needs 80 engines, whereas assembly plant B needs 70 engines. Find how many engines should be shipped from each plant to each main assembly plant if shipping costs are to be kept to a minimum.

SOLUTION ✓

Let x denote the number of engines shipped from plant I to assembly plant A and let y denote the number of engines shipped from plant I to assembly plant B. Since the requirements of assembly plants A and B are 80 and 70 engines, respectively, the number of engines shipped from plant II to assembly plants A and B are $(80 - x)$ and $(70 - y)$, respectively. These numbers may be displayed in a schematic. With the aid of the following schematic and the shipping cost schedule, we find that the total shipping costs incurred by Curtis-Roe are given by

$$\begin{aligned} C &= 100x + 60y + 120(80 - x) + 70(70 - y) \\ &= 14{,}500 - 20x - 10y \end{aligned}$$

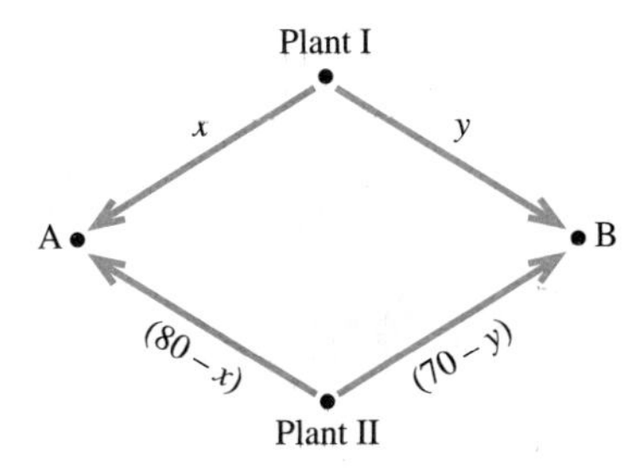

Next, the production constraints on plants I and II lead to the inequalities

$$\begin{aligned} x + y &\leq 100 \\ (80 - x) + (70 - y) &\leq 110 \end{aligned}$$

The last inequality simplifies to

$$x + y \geq 40$$

Also, the requirements of the two main assembly plants lead to the inequalities

$$x \geq 0, \qquad y \geq 0, \qquad 80 - x \geq 0, \qquad 70 - y \geq 0$$

The last two may be written as $x \leq 80$ and $y \leq 70$.

Summarizing, we have the following linear programming problem: Minimize the objective (cost) function $C = 14{,}500 - 20x - 10y$ subject to the constraints

$$\begin{aligned} x + y &\geq 40 \\ x + y &\leq 100 \\ x &\leq 80 \\ y &\leq 70 \end{aligned}$$

where $x \geq 0$ and $y \geq 0$.

You will be asked to complete the solution to this problem in Exercise 31, Section 3.3.

A WAREHOUSE PROBLEM

EXAMPLE 4

The Acrosonic Company manufactures its model F loudspeaker systems in two separate locations, plant I and plant II. The output at plant I is at most 400 per month, whereas the output at plant II is at most 600 per month. These loudspeaker systems are shipped to three warehouses that serve as distribution centers for the company. For the warehouses to meet their orders, the minimum monthly requirements of warehouses A, B, and C are 200, 300, and 400, respectively. Shipping costs from plant I to warehouses A, B, and C are \$20, \$8, and \$10 per loudspeaker system, respectively, and shipping costs from plant II to each of these warehouses are \$12, \$22, and \$18, respectively. What should the shipping schedule be if Acrosonic wishes to meet the requirements of the distribution centers and at the same time keep its shipping costs to a minimum?

SOLUTION ✓ The respective shipping costs (in dollars) per loudspeaker system may be tabulated as in Table 3.3.

Table 3.3

	Warehouse		
Plant	A	B	C
I	20	8	10
II	12	22	18

Letting x_1 denote the number of loudspeaker systems shipped from plant I to warehouse A, x_2 the number shipped from plant I to warehouse B, and so on leads to Table 3.4.

Table 3.4

	Warehouse			
Plant	A	B	C	Max. Prod.
I	x_1	x_2	x_3	400
II	x_4	x_5	x_6	600
Min. req.	200	300	400	

From Tables 3.3 and 3.4 we see that the cost of shipping x_1 loudspeaker systems from plant I to warehouse A is $\$20x_1$, the cost of shipping x_2 loudspeaker systems from plant I to warehouse B is $\$8x_2$, and so on. Thus, the total monthly shipping cost incurred by Acrosonic is given by

$$C = 20x_1 + 8x_2 + 10x_3 + 12x_4 + 22x_5 + 18x_6$$

Next, the production constraints on plants I and II lead to the inequalities

$$x_1 + x_2 + x_3 \leq 400$$
$$x_4 + x_5 + x_6 \leq 600$$

(see Table 3.4). Also, the minimum requirements of each of the three warehouses lead to the three inequalities

$$x_1 + x_4 \geq 200$$
$$x_2 + x_5 \geq 300$$
$$x_3 + x_6 \geq 400$$

Summarizing, we have the following linear programming problem:

$$\begin{aligned} \text{Minimize} \quad & C = 20x_1 + 8x_2 + 10x_3 + 12x_4 + 22x_5 + 18x_6 \\ \text{subject to} \quad & x_1 + x_2 + x_3 \le 400 \\ & x_4 + x_5 + x_6 \le 600 \\ & x_1 + x_4 \ge 200 \\ & x_2 + x_5 \ge 300 \\ & x_3 + x_6 \ge 400 \\ & x_1 \ge 0, x_2 \ge 0, \ldots, x_6 \ge 0 \end{aligned}$$

The solution to this problem will be completed in Section 4.2, Example 5.

SELF-CHECK EXERCISE 3.2

Gino Balduzzi, proprietor of Luigi's Pizza Palace, allocates \$9000 a month for advertising in two newspapers, the *City Tribune* and the *Daily News*. The *City Tribune* charges \$300 for a certain advertisement, whereas the *Daily News* charges \$100 for the same ad. Gino has stipulated that the ad is to appear in at least 15 but no more than 30 editions of the *Daily News* per month. The *City Tribune* has a daily circulation of 50,000, and the *Daily News* has a circulation of 20,000. Under these conditions, determine how many ads Gino should place in each newspaper in order to reach the largest number of readers. Formulate but do not solve the problem. (The solution to this problem can be found in Exercise 3 of Solutions to Self-Check Exercises 3.3.)

Solution to Self-Check Exercise 3.2 can be found on page 201.

3.2 Exercises

Formulate but do not solve each of the following exercises as a linear programming problem.

1. MANUFACTURING—PRODUCTION SCHEDULING A company manufactures two products, A and B, on two machines I and II. It has been determined that the company will realize a profit of \$3 on each unit of product A and a profit of \$4 on each unit of product B. To manufacture a unit of product A requires 6 min on machine I and 5 min on machine II. To manufacture a unit of product B requires 9 min on machine I and 4 min on machine II. There are 5 hr of machine time available on machine I and 3 hr of machine time available on machine II in each work shift. How many units of each product should be produced in each shift to maximize the company's profit?

2. MANUFACTURING—PRODUCTION SCHEDULING Kane Manufacturing has a division that produces two models of fireplace grates, model A and model B. To produce each model A grate requires 3 lb of cast iron and 6 min of labor. To produce each model B grate requires 4 lb of cast iron and 3 min of labor. The profit for each model A grate is \$2, and the profit for each model B grate is \$1.50. If 1000 lb of cast iron and 20 hr of labor are available for the production of grates per day, how many grates of each model should the division produce per day in order to help maximize Kane's profits?

3. MANUFACTURING—PRODUCTION SCHEDULING Refer to Exercise 2. Because of a backlog of orders on model A grates, the manager of Kane Manufacturing has decided to produce at least 150 of these models a day. Operating under this additional constraint, how many grates of each model should Kane produce to maximize profit?

4. FINANCE—ALLOCATION OF FUNDS Madison Finance has a total of \$20 million earmarked for homeowner and auto loans. On the average, homeowner loans have a 10% annual rate of return, whereas auto loans yield a 12% annual rate of return. Management has also stipulated

Portfolio

LEANNE JENKINS

TITLE: Associate Media Director/National Broadcast
INSTITUTION: Hill, Holliday, Connors, Cosmopulos, Inc.

Leanne Jenkins contrasts the freedom she enjoys when making broadcasting media buys with the approach taken in print advertising. Whereas print buyers have to deal with a publication's rate card—which prices ads by size and color—broadcast buyers tell the networks "This is what I'll pay you" for a typical 30-second spot. Like a tourist in a Middle East bazaar, Jenkins enjoys haggling over the price of a prime-time commercial.

Spalding, a sporting goods manufacturer, is a Hill Holliday client. To advertise golf balls, it budgets approximately $10 million for commercial time as well as print ads to reach its primary target audience: men age 35 and older. To get Spalding's message across, Jenkins first determines what percentage of the budget must be allocated to television versus print in order to compete with other golf ball manufacturers. To blanket the Professional Golfers' Association of America (PGA) broadcasts as well as other sports programs, Jenkins figures she will need approximately $6 million for commercials.

As she studies the PGA ratings from previous years, Jenkins estimates that next year's broadcasts will achieve an average of 3 rating points each. On a given weekend afternoon, 3%, or 1,650,000 men (out of 55,000,000 men in the target audience), will watch golf and have the opportunity to see Spalding's commercials.

In a typical scenario, Jenkins contacts CBS to purchase a block of 50 commercials based on her cost-per-point estimate of $14,000 for each rating point. Spread out over an entire year, Spalding's commercials will run three or four times during each broadcast. Jenkins multiplies the cost per point by 50 commercials at 3.5 points per commercial to determine a total cost of $2.625 million. Over 40% of Spalding's television advertising budget will go for commercials on CBS alone. The balance will be divided among golf programs on other networks as well as other sports broadcasts.

Knowing how popular golf is, the network will argue that its coverage will achieve at least 3.5 rating points and is worth at least $15,000 per point. Because the demand for sports events is high, advertisers usually must accept slightly higher costs—although Jenkins works to keep expenses down.

When negotiating for commercials on regularly scheduled programs that are not special events, Jenkins frequently gets lower prices. Popular programs (such as *Felicity*), which are shown weekly, mean higher supply and lower demand. With special programming such as the U.S. Open Golf Championship, however, Spalding may have to pay the network's price. The Open happens only once a year, which means very limited supply with much higher demand.

For Jenkins and the networks, ratings, as well as their associated costs, are the key ingredients for reaching millions of viewers each week.

Hill Holliday, New England's largest advertising agency, handles a diverse group of clients from beer makers to insurance companies. As associate media director for national broadcasting, Jenkins purchases blocks of commercial time from ABC, CBS, NBC, FOX, and over ten cable networks. Based on demographic studies, she determines whether a particular program will attract a specific audience interested in a client's goods or services. If a match occurs, Jenkins negotiates the lowest possible price to reach the highest number of viewers.

that the total amount of homeowner loans should be greater than or equal to four times the total amount of automobile loans. Determine the total amount of loans of each type Madison should extend to each category in order to maximize its returns.

5. **MANUFACTURING—PRODUCTION SCHEDULING** The Acoustical Company manufactures a CD storage cabinet that can be bought fully assembled or as a kit. Each cabinet is processed in the fabrications department and the assembly department. If the fabrication department only manufactures fully assembled cabinets, then it can produce 200 units/day; but if it only manufactures kits, it can produce 200 units/day. If the assembly department only produces fully assembled cabinets, then it can produce 100 units/day; but if it only produces kits, then it can produce 300 units/day. Each fully assembled cabinet contributes $50 to the profits of the company, whereas each kit contributes $40 to its profits. How many fully assembled units and how many kits should the company produce per day in order to maximize its profits?

6. **AGRICULTURE—CROP PLANNING** A farmer has 150 acres of land suitable for cultivating crops A and B. The cost of cultivating crop A is $40/acre, whereas that of crop B is $60/acre. The farmer has a maximum of $7400 available for land cultivation. Each acre of crop A requires 20 hr of labor, and each acre of crop B requires 25 hr of labor. The farmer has a maximum of 3300 hr of labor available. If he expects to make a profit of $150/acre on crop A and $200/acre on crop B, how many acres of each crop should he plant in order to maximize his profit?

7. **NUTRITION—DIET PLANNING** A nutritionist at the Medical Center has been asked to prepare a special diet for certain patients. She has decided that the meals should contain a minimum of 400 mg of calcium, 10 mg of iron, and 40 mg of vitamin C. She has further decided that the meals are to be prepared from foods A and B. Each ounce of food A contains 30 mg of calcium, 1 mg of iron, 2 mg of vitamin C, and 2 mg of cholesterol. Each ounce of food B contains 25 mg of calcium, 0.5 mg of iron, 5 mg of vitamin C, and 5 mg of cholesterol. Find how many ounces of each type of food should be used in a meal so that the cholesterol content is minimized and the minimum requirements of calcium, iron, and vitamin C are met.

8. **SOCIAL PROGRAMS PLANNING** AntiFam, a hunger-relief organization, has earmarked between $2 and $2.5 million, inclusive, for aid to two African countries, country A and country B. Country A is to receive between $1 and $1.5 million, inclusive, in aid, and country B is to receive at least $0.75 million in aid. It has been estimated that each dollar spent in country A will yield an effective return of $.60, whereas a dollar spent in country B will yield an effective return of $.80. How should the aid be allocated if the money is to be utilized most effectively according to these criteria?

Hint: If x and y denote the amount of money to be given to country A and country B, respectively, then the objective function to be maximized is $P = 0.6x + 0.8y$.

9. **MANUFACTURING—SHIPPING COSTS** The TMA Company manufactures 19-in. color television picture tubes in two separate locations, location I and location II. The output at location I is at most 6000 tubes/month, whereas the output at location II is at most 5000/month. TMA is the main supplier of picture tubes to the Pulsar Corporation, its holding company, which has priority in having all its requirements met. In a certain month, Pulsar placed orders for 3000 and 4000 picture tubes to be shipped to two of its factories located in city A and city B, respectively. The shipping costs (in dollars) per picture tube from the two TMA plants to the two Pulsar factories are as follows:

Shipping Costs per Picture Tube

	To Pulsar Factories	
From TMA	**City A**	**City B**
Location I	$3	$2
Location II	$4	$5

Find a shipping schedule that meets the requirements of both companies while keeping costs to a minimum.

10. **ADVERTISING** As part of a campaign to promote its annual clearance sale, the Excelsior Company decided to buy television advertising time on Station KAOS. Excelsior's advertising budget is $102,000. Morning time costs $3000/minute, afternoon time costs $1000/minute, and evening (prime) time costs $12,000/minute. Because of previous commitments, KAOS cannot offer Excelsior more than 6 min of prime time or more than a total of 25 min of advertising time over the 2 weeks in which the commercials are to be run. KAOS estimates that morning commercials are seen by 200,000 people, afternoon commercials are seen by 100,000 people, and evening commercials are seen by 600,000 people. How much morning, afternoon, and evening advertising time should Excelsior buy to maximize exposure of its commercials?

11. **MANUFACTURING—PRODUCTION SCHEDULING** A company manufactures products A, B, and C. Each product is processed in three departments: I, II, and III. The total available labor-hours per week for departments I, II, and III are 900, 1080, and 840, respectively. The time

requirements (in hours per unit) and profit per unit for each product are as follows:

	Product A	Product B	Product C
Dept. I	2	1	2
Dept. II	3	1	2
Dept. III	2	2	1
Profit	$18	$12	$15

How many units of each product should the company produce in order to maximize its profit?

12. MANUFACTURING—SHIPPING COSTS The Acrosonic Company of Example 4 also manufactures a model G loudspeaker system in plants I and II. The output at plant I is at most 800 systems/month, whereas the output at plant II is at most 600/month. These loudspeaker systems are also shipped to the three warehouses—A, B, and C—whose minimum monthly requirements are 500, 400, and 400, respectively. Shipping costs from plant I to warehouse A, warehouse B, and warehouse C are $16, $20, and $22/loudspeaker system, respectively, and shipping costs from plant II to each of these warehouses are $18, $16, and $14, respectively. What shipping schedule will enable Acrosonic to meet the warehouses' requirements and at the same time keep its shipping costs to a minimum?

13. MANUFACTURING—PRODUCTION SCHEDULING The Custom Office Furniture Company is introducing a new line of executive desks made from a specially selected grade of walnut. Initially, three different models—A, B, and C—are to be marketed. Each model A desk requires $1\frac{1}{4}$ hr for fabrication, 1 hr for assembly, and 1 hr for finishing; each model B desk requires $1\frac{1}{2}$ hr for fabrication, 1 hr for assembly, and 1 hr for finishing; each model C desk requires $1\frac{1}{2}$ hr, $\frac{3}{4}$ hr, and $\frac{1}{2}$ hr for fabrication, assembly, and finishing, respectively. The profit on each model A desk is $26, the profit on each model B desk is $28, and the profit on each model C desk is $24. The total time available in the fabrication department, the assembly department, and the finishing department in the first month of production is 310 hr, 205 hr, and 190 hr, respectively. To maximize Custom's profit, how many desks of each model should be made in the month?

14. MANUFACTURING—PREFABRICATED HOUSING PRODUCTION Boise Lumber has decided to enter the lucrative prefabricated housing business. Initially, it plans to offer three models: standard, deluxe, and luxury. Each house is prefabricated and partially assembled in the factory, and the final assembly is completed on site. The dollar amount of building material required, the amount of labor required in the factory for prefabrication and partial assembly, the amount of on-site labor required, and the profit per unit are as follows:

	Standard Model	Deluxe Model	Luxury Model
Material ($)	6,000	8,000	10,000
Factory Labor (hr)	240	220	200
On-site Labor (hr)	180	210	300
Profit ($)	3,400	4,000	5,000

For the first year's production, a sum of $8,200,000 is budgeted for the building material; the number of labor-hours available for work in the factory (for prefabrication and partial assembly) is not to exceed 218,000 hr; and the amount of labor for on-site work is to be less than or equal to 237,000 labor-hours. Determine how many houses of each type Boise should produce (market research has confirmed that there should be no problems with sales) to maximize its profit from this new venture.

15. MANUFACTURING—SHIPPING COSTS Steinwelt Piano manufactures uprights and consoles in two plants, plant I and plant II. The output of plant I is at most 300/month, whereas the output of plant II is at most 250/month. These pianos are shipped to three warehouses that serve as distribution centers for the company. To fill current and projected future orders, warehouse A requires a minimum of 200 pianos/month, warehouse B requires at least 150 pianos/month, and warehouse C requires at least 200 pianos/month. The shipping cost of each piano from plant I to warehouse A, warehouse B, and warehouse C is $60, $60, and $80, respectively, and the shipping cost of each piano from plant II to warehouse A, warehouse B, and warehouse C is $80, $70, and $50, respectively. What shipping schedule will enable Steinwelt to meet the warehouses' requirements while keeping shipping costs to a minimum?

16. MANUFACTURING—COLD FORMULA PRODUCTION Bayer Pharmaceutical produces three kinds of cold formulas: formula I, formula II, and formula III. It takes 2.5 hr to produce 1000 bottles of formula I, 3 hr to produce 1000 bottles of formula II, and 4 hr to produce 1000 bottles of formula III. The profits for each 1000 bottles of formula I, formula II, and formula III are $180, $200, and $300, respectively. Suppose that for a certain production run there are enough ingredients on hand to make at most 9000 bottles of formula I, 12,000 bottles of formula II, and 6000 bottles of formula III. Furthermore, the time for the production run is limited to a maximum of 70 hr. How many bottles of each formula should be produced in this production run so that the profit is maximized?

In Exercises 17 and 18, determine whether the statement is true or false. If it is true, explain why it is true. If it is false, give an example to show why it is false.

17. The problem

$$\begin{aligned} \text{Maximize} \quad & P = xy \\ \text{subject to} \quad & 2x + 3y \leq 12 \\ & 2x + y \leq 8 \\ & x \geq 0, y \geq 0 \end{aligned}$$

is a linear programming problem.

18. The problem

$$\begin{aligned} \text{Minimize} \quad & C = 2x + 3y \\ \text{subject to} \quad & 2x + 3y \leq 6 \\ & x - y = 0 \\ & x \geq 0, y \geq 0 \end{aligned}$$

is a linear programming problem.

Solution to Self-Check Exercise 3.2

Let x denote the number of ads to be placed in the *City Tribune* and y the number to be placed in the *Daily News*. The total cost for placing x ads in the *City Tribune* and y ads in the *Daily News* is $300x + 100y$ dollars, and since the monthly budget is \$9000, we must have

$$300x + 100y \leq 9000$$

Next, the condition that the ad must appear in at least 15 but no more than 30 editions of the *Daily News* translates into the inequalities

$$\begin{aligned} y &\geq 15 \\ y &\leq 30 \end{aligned}$$

Finally, the objective function to be maximized is

$$P = 50{,}000x + 20{,}000y$$

To summarize, we have the following linear programming problem:

$$\begin{aligned} \text{Maximize} \quad & P = 50{,}000x + 20{,}000y \\ \text{subject to} \quad & 300x + 100y \leq 9000 \\ & y \geq 15 \\ & y \leq 30 \\ & x \geq 0, y \geq 0 \end{aligned}$$

3.3 Graphical Solution of Linear Programming Problems

The Graphical Method

Linear programming problems in two variables have relatively simple geometrical interpretations. For example, the system of linear constraints associated with a two-dimensional linear programming problem, unless it is inconsistent, defines a planar region whose boundary is composed of straight-line segments and/or half lines. Such problems are therefore amenable to graphical analysis.

Consider the following two-dimensional linear programming problem:

$$\begin{aligned} \text{Maximize} \quad & P = 3x + 2y \\ \text{subject to} \quad & 2x + 3y \leq 12 \\ & 2x + y \leq 8 \\ & x \geq 0, y \geq 0 \end{aligned} \tag{7}$$

The system of linear inequalities (7) defines the planar region S shown in Figure 3.9. Each point in S is a candidate for the solution of the problem at hand and is referred to as a **feasible solution.** The set S itself is referred to as a **feasible set.** Our goal is to find, from among all the points in the set S, the point(s) that optimizes the objective function P. Such a feasible solution is called an **optimal solution** and constitutes the solution to the linear programming problem under consideration.

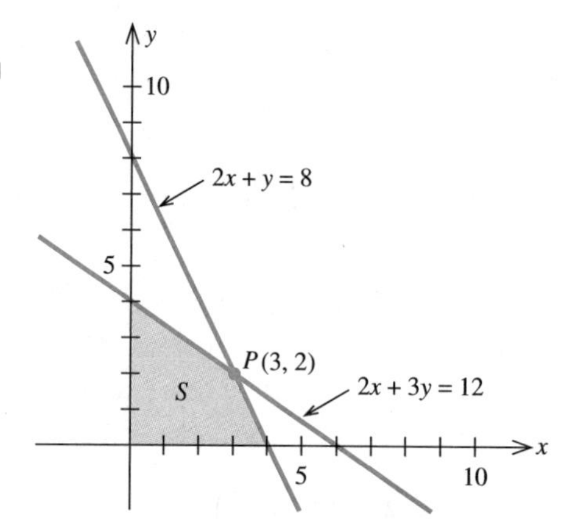

FIGURE 3.9
Each point in the feasible set S is a candidate for the optimal solution.

As noted earlier, each point $P(x, y)$ in S is a candidate for the optimal solution to the problem at hand. For example, the point (1, 3) is easily seen to lie in S and is therefore in the running. The value of the objective function P at the point (1, 3) is given by $P = 3(1) + 2(3) = 9$. Now, if we could compute the value of P corresponding to each point in S, then the point(s) in S that gave the largest value to P would constitute the solution set sought. Unfortunately, in most problems, the number of candidates is either too large or, as in this problem, the number is infinite. Thus, this method is at best unwieldy and at worst impractical.

Let's turn the question around. Instead of asking for the value of the objective function P at a feasible point, let's assign a value to the objective function P and ask whether there are feasible points that would correspond to the given value of P. To this end, suppose we assign a value of 6 to P. Then, the objective function P becomes $3x + 2y = 6$, a linear equation in x and y, and therefore has a graph that is a straight line L_1 in the plane. In

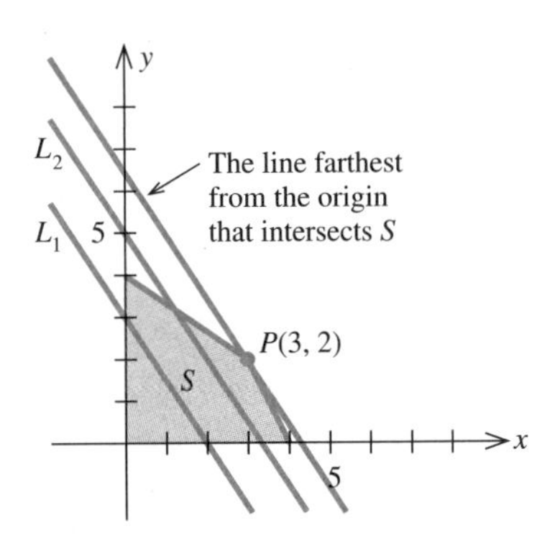

FIGURE 3.10
A family of parallel lines that intersect the feasible set S

Figure 3.10 we have drawn the graph of this straight line superimposed on the feasible set S.

It is clear that each point on the straight-line segment given by the intersection of the straight line L_1 and the feasible set S corresponds to the given value, 6, of P. For this reason the line L_1 is called an **isoprofit line.** Let's repeat the process, this time assigning a value of 10 to P. We obtain the equation $3x + 2y = 10$ and the line L_2 (see Figure 3.10), which suggests that there are feasible points that correspond to a larger value of P. Observe that the line L_2 is parallel to the line L_1 since both lines have slope equal to $-3/2$, which is easily seen by casting the corresponding equations in the slope-intercept form.

In general, by assigning different values to the objective function, we obtain a family of parallel lines, each with slope equal to $-3/2$. Furthermore, a line corresponding to a larger value of P lies farther away from the origin than one with a smaller value of P. The implication is clear. To obtain the optimal solution(s) to the problem at hand, find the straight line, from this family of straight lines, that is farthest from the origin and still intersects the feasible set S. The required line is the one that passes through the point $P(3, 2)$ (see Figure 3.10), so the solution to the problem is given by $x = 3$, $y = 2$, resulting in a maximum value of $P = 3(3) + 2(2) = 13$.

That the optimal solution to this problem was found to occur at a vertex of the feasible set S is no accident. In fact, the result is a consequence of the following basic theorem on linear programming, which we state without proof.

THEOREM 1

Linear Programming

If a linear programming problem has a solution, then it must occur at a vertex, or corner point, of the feasible set S associated with the problem. Furthermore, if the objective function P is optimized at two adjacent vertices of S, then it is optimized at every point on the line segment joining these vertices, in which case there are infinitely many solutions to the problem.

Theorem 1 tells us that our search for the solution(s) to a linear programming problem may be restricted to the examination of the set of vertices of the feasible set S associated with the problem. Since a feasible set S has finitely many vertices, the theorem suggests that the solution(s) to the linear programming problem may be found by inspecting the values of the objective function P at these vertices.

Although Theorem 1 sheds some light on the nature of the solution of a linear programming problem, it does not tell us when a linear programming problem has a solution. The following theorem states some conditions that will guarantee when a linear programming problem has a solution.

THEOREM 2

Existence of a Solution

Suppose we are given a linear programming problem with a feasible set S and an objective function $P = ax + by$.

1. If S is bounded, then P has both a maximum and a minimum value on S.
2. If S is unbounded and both a and b are nonnegative, then P has a minimum value on S provided that the constraints defining S include the inequalities $x \geq 0$ and $y \geq 0$.
3. If S is the empty set, then the linear programming problem has no solution; that is, P has neither a maximum nor a minimum value.

The **method of corners,** a simple procedure for solving linear programming problems based on Theorem 1, follows.

The Method of Corners

1. Graph the feasible set.
2. Find the coordinates of all corner points (vertices) of the feasible set.
3. Evaluate the objective function at each corner point.
4. Find the vertex that renders the objective function a maximum (minimum). If there is only one such vertex, then this vertex constitutes a unique solution to the problem. If the objective function is maximized (minimized) at two adjacent corner points of S, there are infinitely many optimal solutions given by the points on the line segment determined by these two vertices.

APPLICATIONS

EXAMPLE 1

We are now in a position to complete the solution to the production problem posed in Example 1, Section 3.2. Recall that the mathematical formulation led to the following linear programming problem.

$$\begin{aligned} \text{Maximize} \quad & P = x + 1.2y \\ \text{subject to} \quad & 2x + y \leq 180 \\ & x + 3y \leq 300 \\ & x \geq 0, y \geq 0 \end{aligned}$$

SOLUTION ✓

The feasible set S for the problem is shown in Figure 3.11.

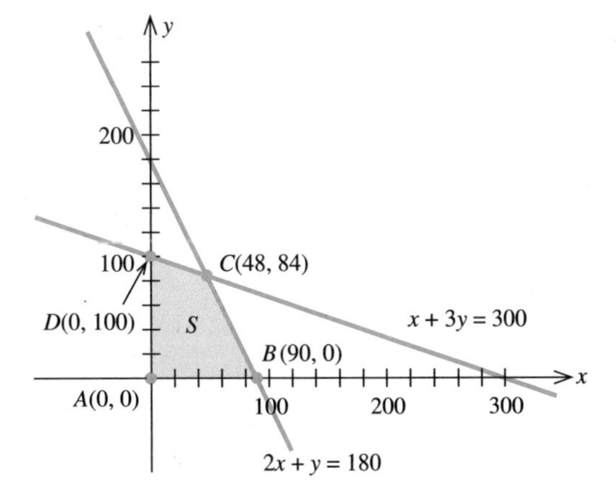

FIGURE 3.11
The corner point that yields the maximum profit is $C(48, 84)$.

The vertices of the feasible set are $A(0, 0)$, $B(90, 0)$, $C(48, 84)$, and $D(0, 100)$. The values of P at these vertices may be tabulated as follows:

Vertex	$P = x + 1.2y$
$A(0, 0)$	0
$B(90, 0)$	90
$C(48, 84)$	148.8
$D(0, 100)$	120

From the table, we see that the maximum of $P = x + 1.2y$ occurs at the vertex (48, 84) and has a value of 148.8. Recalling what the symbols x, y, and P represent, we conclude that Ace Novelty would maximize its profit (a figure of \$148.80) by producing 48 type A souvenirs and 84 type B souvenirs.

■■■■

Group Discussion

Consider the linear programming problem:

$$\begin{aligned} \text{Maximize} \quad & P = 4x + 3y \\ \text{subject to} \quad & 2x + y \le 10 \\ & 2x + 3y \le 18 \\ & x \ge 0, y \ge 0 \end{aligned}$$

1. Sketch the feasible set S for the linear programming problem.
2. Draw the isoprofit lines superimposed on S corresponding to $P = 12$, 16, 20, and 24 and show that these lines are parallel to each other.
3. Show that the solution to the linear programming problem is $x = 3$ and $y = 4$. Is this result the same as that found using the method of corners?

EXAMPLE 2 Complete the solution of the nutrition problem posed in Example 2, Section 3.2.

SOLUTION ✓ Recall that the mathematical formulation of the problem led to the following linear programming problem in two variables:

$$\begin{aligned} \text{Minimize} \quad & C = 6x + 8y \\ \text{subject to} \quad & 40x + 10y \geq 2400 \\ & 10x + 15y \geq 2100 \\ & 5x + 15y \geq 1500 \\ & x \geq 0, y \geq 0 \end{aligned}$$

The feasible set S defined by the system of constraints is shown in Figure 3.12. The vertices of the feasible set S are $A(0, 240)$, $B(30, 120)$, $C(120, 60)$, and $D(300, 0)$. The values of the objective function C at these vertices are given in the following table:

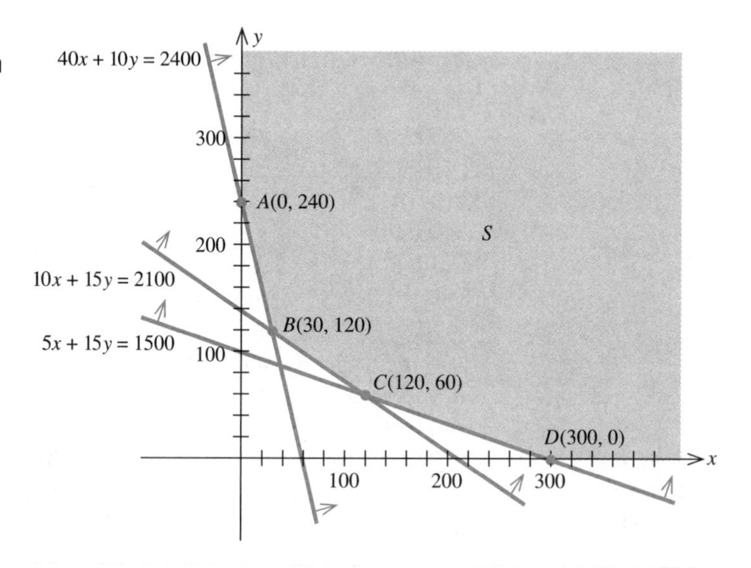

FIGURE 3.12
The corner point that yields the minimum cost is $B(30, 120)$.

Vertex	$C = 6x + 8y$
$A(0, 240)$	1920
$B(30, 120)$	1140
$C(120, 60)$	1200
$D(300, 0)$	1800

From the table, we can see that the minimum for the objective function $C = 6x + 8y$ occurs at the vertex $B(30, 120)$ and has a value of 1140. Thus, the individual should purchase 30 brand A pills and 120 brand B pills at a minimum cost of \$11.40.

■■■■

EXAMPLE 3 Find the maximum and minimum of $P = 2x + 3y$ subject to the following system of linear inequalities:

$$\begin{aligned} 2x + 3y &\leq 30 \\ y - x &\leq 5 \\ x + y &\geq 5 \\ x &\leq 10 \\ x \geq 0, y &\geq 0 \end{aligned}$$

SOLUTION ✓ The feasible set S is shown in Figure 3.13. The vertices of the feasible set S are $A(5, 0)$, $B(10, 0)$, $C(10, 10/3)$, $D(3, 8)$, and $E(0, 5)$. The values of the objective function P at these vertices are given in the following table.

Vertex	$P = 2x + 3y$
$A(5, 0)$	10
$B(10, 0)$	20
$C(10, 10/3)$	30
$D(3, 8)$	30
$E(0, 5)$	15

From the table, we see that the maximum for the objective function $P = 2x + 3y$ occurs at the vertices $C(10, 10/3)$ and $D(3, 8)$. This tells us that every point on the line segment joining the points $C(10, 10/3)$ and $D(3, 8)$ maximizes P, giving it a value of 30 at each of these points. From the table, it is also clear that P is minimized at the point $(5, 0)$, where it attains a value of 10.

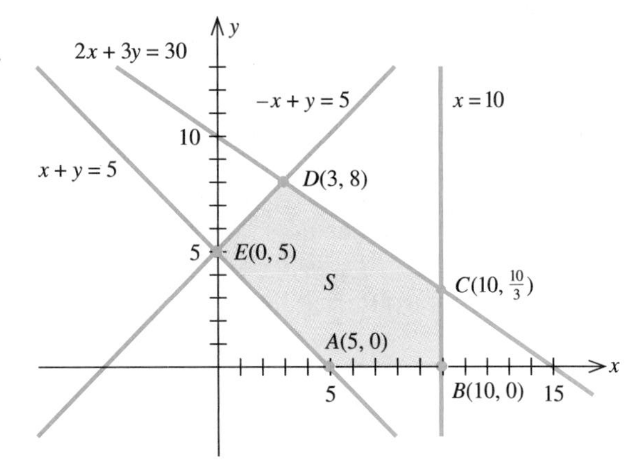

FIGURE 3.13
Every point lying on the line segment joining C and D maximizes P.

Group Discussion

Consider the linear programming problem:

$$\begin{aligned} \text{Maximize} \quad & P = 2x + 3y \\ \text{subject to} \quad & 2x + y \leq 10 \\ & 2x + 3y \leq 18 \\ & x \geq 0, y \geq 0 \end{aligned}$$

1. Sketch the feasible set S for the linear programming problem.
2. Draw the isoprofit lines superimposed on S corresponding to $P = 6$, 8, 12, and 18 and show that these lines are parallel to each other.
3. Show that there are infinitely many solutions to the problem. Is this result as predicted by the method of corners?

We close this section by examining two situations in which a linear programming problem may have no solution.

EXAMPLE 4 Solve the following linear programming problem:

$$\begin{aligned} \text{Maximize} \quad & P = x + 2y \\ \text{subject to} \quad & -2x + y \leq 4 \\ & x - 3y \leq 3 \\ & x \geq 0, y \geq 0 \end{aligned}$$

SOLUTION ✓ The feasible set S for this problem is shown in Figure 3.14. Since the set S is unbounded (both x and y can take on arbitrarily large positive values), we see that we can make P as large as we please by making x and y large enough.

FIGURE 3.14
The maximization problem has no solution, because the feasible set is unbounded.

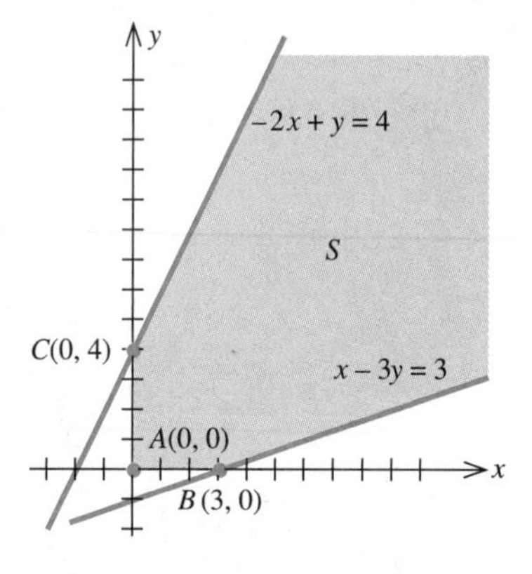

The problem has no solution. and we say that the solution is unbounded. ■

EXAMPLE 5 Solve the following linear programming problem:

$$\begin{aligned} \text{Maximize} \quad & P = x + 2y \\ \text{subject to} \quad & x + 2y \leq 4 \\ & 2x + 3y \geq 12 \\ & x \geq 0, y \geq 0 \end{aligned}$$

SOLUTION ✓

The half planes described by the constraints (inequalities) have no points in common (Figure 3.15). Therefore, there are no feasible points, and the problem has no solution. In this situation, we say that this problem is **infeasible,** or **inconsistent.** (These situations are unlikely to occur in well-posed problems arising from practical applications of linear programming.) ■■■■

FIGURE 3.15
The problem is inconsistent because there is no point that satisfies all given inequalities.

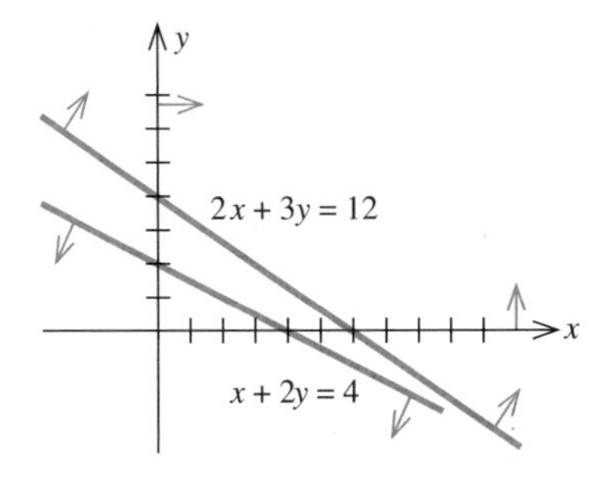

The method of corners is particularly effective in solving two-variable linear programming problems with a small number of constraints, as the preceding examples have amply demonstrated. Its effectiveness, however, decreases rapidly as the number of variables and/or constraints increases. For example, it may be shown that a linear programming problem in three variables and five constraints may have up to 10 feasible corner points. The determination of the feasible corner points calls for the solution of ten 3×3 systems of linear equations and then the verification, by the substitution of each of these solutions into the system of constraints, that it is in fact a feasible point. When the number of variables and constraints goes up to five and ten, respectively (still a very small system from the standpoint of applications in economics), the number of vertices to be found and checked for feasible corner points increases dramatically to 252, and each of these vertices is found by solving a 5×5 linear system! For this reason, the method of corners is seldom used to solve linear programming problems; its redeeming value lies in the fact that much insight is gained into the nature of the solutions of linear programming problems through its use in solving two-variable problems.

SELF-CHECK EXERCISES 3.3

1. Use the method of corners to solve the following linear programming problem:

$$\begin{aligned} \text{Maximize} \quad & P = 4x + 5y \\ \text{subject to} \quad & x + 2y \leq 10 \\ & 5x + 3y \leq 30 \\ & x \geq 0, y \geq 0 \end{aligned}$$

2. Use the method of corners to solve the following linear programming problem:

$$\begin{aligned} \text{Minimize} \quad & C = 5x + 3y \\ \text{subject to} \quad & 5x + 3y \geq 30 \\ & x - 3y \leq 0 \\ & x \geq 2 \end{aligned}$$

3. Gino Balduzzi, proprietor of Luigi's Pizza Palace, allocates \$9000 a month for advertising in two newspapers, the *City Tribune* and the *Daily News.* The *City Tribune* charges \$300 for a certain advertisement, whereas the *Daily News* charges \$100 for the same ad. Gino has stipulated that the ad is to appear in at least 15 but no more than 30 editions of the *Daily News* per month. The *City Tribune* has a daily circulation of 50,000, and the *Daily News* has a circulation of 20,000. Under these conditions, determine how many ads Gino should place in each newspaper in order to reach the largest number of readers.

Solutions to Self-Check Exercises 3.3 can be found on page 214.

3.3 Exercises

In Exercises 1–6, find the optimal value(s) of the given objective function on the feasible set S.

1. $Z = 2x + 3y$

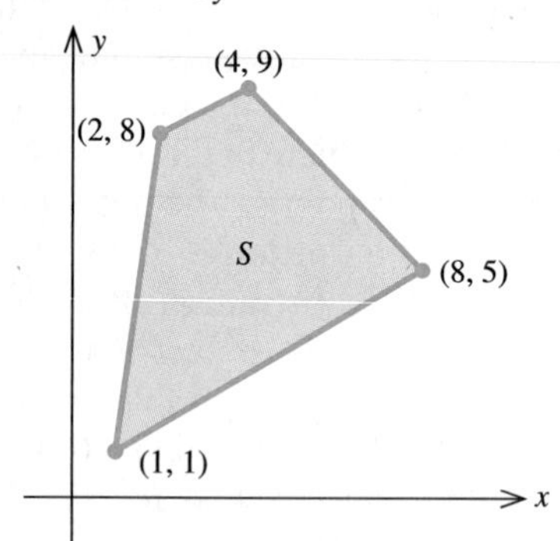

2. $Z = 3x - y$

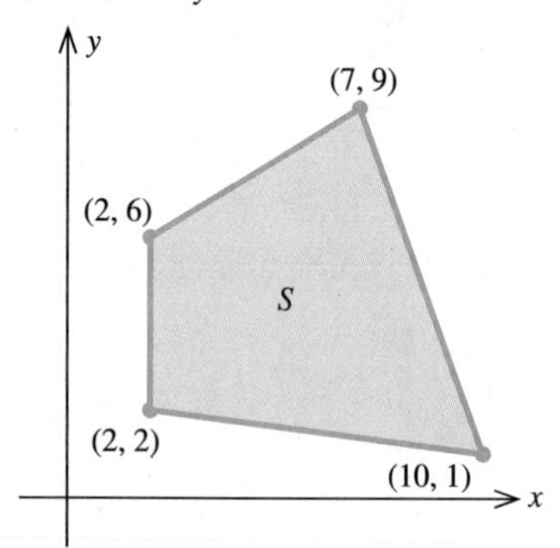

3. $Z = 3x + 4y$

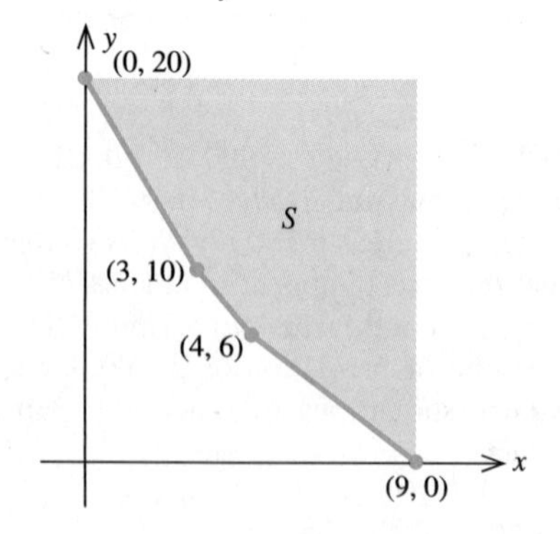

4. $Z = 7x + 9y$

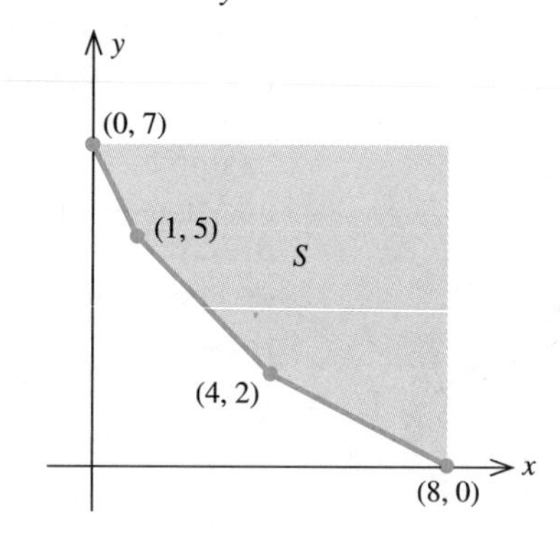

5. $Z = x + 4y$

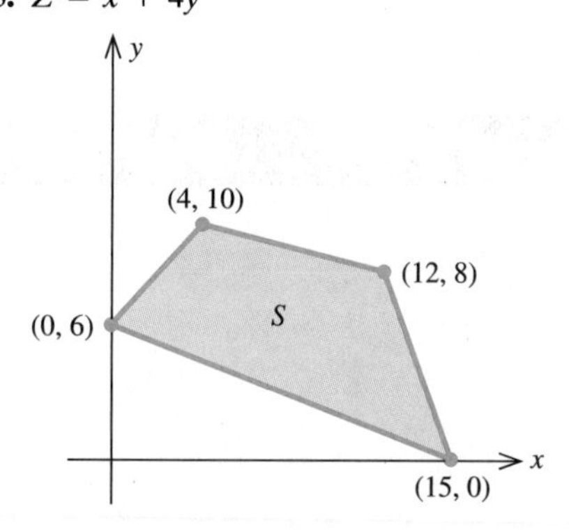

6. $Z = 3x + 2y$

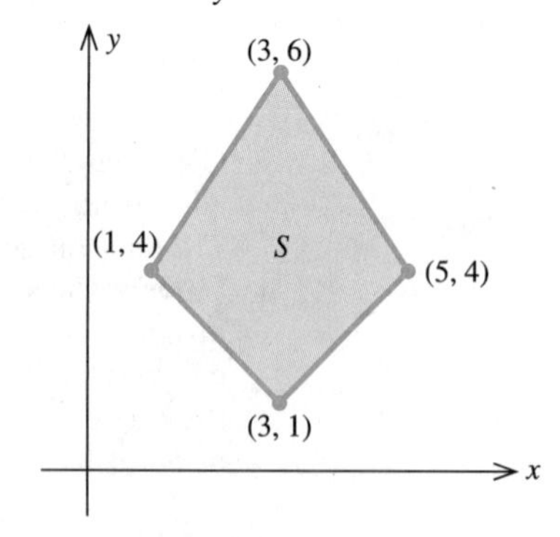

In Exercises 7–21, solve each linear programming problem by the method of corners.

7. Maximize $P = 2x + 3y$
subject to
$$\begin{aligned} x + y &\le 6 \\ x &\le 3 \\ x \ge 0, y &\ge 0 \end{aligned}$$

8. Maximize $P = 3x - 4y$
subject to
$$\begin{aligned} x + 3y &\le 15 \\ 4x + y &\le 16 \\ x \ge 0, y &\ge 0 \end{aligned}$$

9. Minimize $C = 2x + 10y$
subject to
$$\begin{aligned} 5x + 2y &\ge 40 \\ x + 2y &\ge 20 \\ y \ge 3, x &\ge 0 \end{aligned}$$

10. Minimize $C = 2x + 5y$
subject to
$$\begin{aligned} 4x + y &\ge 40 \\ 2x + y &\ge 30 \\ x + 3y &\ge 30 \\ x \ge 0, y &\ge 0 \end{aligned}$$

11. Minimize $C = 6x + 3y$ subject to the constraints of Exercise 10.

12. Maximize $P = 2x + 5y$
subject to
$$\begin{aligned} 2x + y &\le 16 \\ 2x + 3y &\le 24 \\ y &\le 6 \\ x \ge 0, y &\ge 0 \end{aligned}$$

13. Minimize $C = 10x + 15y$
subject to
$$\begin{aligned} x + y &\le 10 \\ 3x + y &\ge 12 \\ -2x + 3y &\ge 3 \\ x \ge 0, y &\ge 0 \end{aligned}$$

14. Maximize $P = 2x + 5y$ subject to the constraints of Exercise 13.

15. Maximize $P = 3x + 4y$
subject to
$$\begin{aligned} x + 2y &\le 50 \\ 5x + 4y &\le 145 \\ 2x + y &\ge 25 \\ y \ge 5, x &\ge 0 \end{aligned}$$

16. Maximize $P = 4x - 3y$ subject to the constraints of Exercise 15.

17. Maximize $P = 2x + 3y$
subject to
$$\begin{aligned} x + y &\le 48 \\ x + 3y &\ge 60 \\ 9x + 5y &\le 320 \\ x \ge 10, y &\ge 0 \end{aligned}$$

18. Minimize $C = 5x + 3y$ subject to the constraints of Exercise 17.

19. Find the maximum and minimum of $P = 10x + 12y$ subject to
$$\begin{aligned} 5x + 2y &\ge 63 \\ x + y &\ge 18 \\ 3x + 2y &\le 51 \\ x \ge 0, y &\ge 0 \end{aligned}$$

20. Find the maximum and minimum of $P = 4x + 3y$ subject to
$$\begin{aligned} 3x + 5y &\ge 20 \\ 3x + y &\le 16 \\ -2x + y &\le 1 \\ x \ge 0, y &\ge 0 \end{aligned}$$

21. Find the maximum and minimum of $P = 2x + 4y$ subject to
$$\begin{aligned} x + y &\le 20 \\ -x + y &\le 10 \\ x &\le 10 \\ x + y &\ge 5 \\ y \ge 5, x &\ge 0 \end{aligned}$$

22. MANUFACTURING—PRODUCTION SCHEDULING A company manufactures two products, A and B, on two machines I and II. It has been determined that the company will realize a profit of \$3 on each unit of product A and a profit of \$4 on each unit of product B. To manufacture a unit of product A requires 6 min on machine I and 5 min on machine II. To manufacture a unit of product B requires 9 min on machine I and 4 min on machine II. There are 5 hr of machine time available on machine I and 3 hr of machine time available on machine II in each work shift. How many units of each product should be produced in each shift to maximize the company's profit?

23. MANUFACTURING—PRODUCTION SCHEDULING The National Business Machines Corporation manufactures two models of fax machines: A and B. Each model A costs \$200 to make, and each model B costs \$300. The profits are

$25 for each model A and $40 for each model B fax machine. If the total number of fax machines demanded per month does not exceed 2500 and the company has earmarked no more than $600,000/month for manufacturing costs, how many units of each model should National make each month in order to maximize its monthly profits?

24. MANUFACTURING—PRODUCTION SCHEDULING The Bata Aerobics Company manufactures two models of steppers used for aerobic exercises. To manufacture each luxury model requires 10 lb of plastic and 10 min of labor. To manufacture each standard model requires 16 lb of plastic and 8 min of labor. The profit for each luxury model is $40, and the profit for each standard model is $30. If 6000 lb of plastic and 60 hr of labor are available for the production of the steppers per day, how many steppers of each model should Bata produce in order to maximize its profits?

25. MANUFACTURING—PRODUCTION SCHEDULING Kane Manufacturing has a division that produces two models of fireplace grates, model A and model B. To produce each model A grate requires 3 lb of cast iron and 6 min of labor. To produce each model B grate requires 4 lb of cast iron and 3 min of labor. The profit for each model A grate is $2, and the profit for each model B grate is $1.50. If 1000 lb of cast iron and 20 hr of labor are available for the production of fireplace grates per day, how many grates of each model should the division produce in order to help maximize Kane's profits?

26. MANUFACTURING—PRODUCTION SCHEDULING Refer to Exercise 25. Because of a backlog of orders for model A grates, Kane's manager had decided to produce at least 150 of these models a day. Operating under this additional constraint, how many grates of each model should Kane produce to maximize profit?

27. FINANCE—ALLOCATION OF FUNDS Madison Finance has a total of $20 million earmarked for homeowner and auto loans. On the average, homeowner loans have a 10% annual rate of return, whereas auto loans yield a 12% annual rate of return. Management has also stipulated that the total amount of homeowner loans should be greater than or equal to four times the total amount of automobile loans. Determine the total amount of loans of each type Madison should extend to each category in order to maximize its returns.

28. MANUFACTURING—PRODUCTION SCHEDULING The Acoustical Company manufactures a CD storage cabinet that can be bought fully assembled or as a kit. Each cabinet is processed in the fabrications department and the assembly department. If the fabrication department only manufactures fully assembled cabinets, then it can produce 200 units/day; but if it only manufactures kits, it can produce 200 units/day. If the assembly department only produces fully assembled cabinets, then it can produce 100 units/day; but if it only produces kits, then it can produce 300 units/day. Each fully assembled cabinet contributes $50 to the profits of the company, whereas each kit contributes $40 to its profits. How many fully assembled units and how many kits should the company produce per day in order to maximize its profits?

29. AGRICULTURE—CROP PLANNING A farmer has 150 acres of land suitable for cultivating crops A and B. The cost of cultivating crop A is $40/acre, whereas that of crop B is $60/acre. The farmer has a maximum of $7400 available for land cultivation. Each acre of crop A requires 20 hr of labor, and each acre of crop B requires 25 hr of labor. The farmer has a maximum of 3300 hr of labor available. If he expects to make a profit of $150/acre on crop A and $200/acre on crop B, how many acres of each crop should he plant in order to maximize his profit?

30. NUTRITION—DIET PLANNING A nutritionist at the Medical Center has been asked to prepare a special diet for certain patients. She has decided that the meals should contain a minimum of 400 mg of calcium, 10 mg of iron, and 40 mg of vitamin C. She has further decided that the meals are to be prepared from foods A and B. Each ounce of food A contains 30 mg of calcium, 1 mg of iron, 2 mg of vitamin C, and 2 mg of cholesterol. Each ounce of food B contains 25 mg of calcium, 0.5 mg of iron, 5 mg of vitamin C, and 5 mg of cholesterol. Find how many ounces of each type of food should be used in a meal so that the cholesterol content is minimized and the minimum requirements of calcium, iron, and vitamin C are met.

31. Complete the solution to Example 3, Section 3.2.

32. SOCIAL PROGRAMS PLANNING AntiFam, a hunger-relief organization, has earmarked between $2 and $2.5 million, inclusive, for aid to two African countries, country A and country B. Country A is to receive between $1 and $1.5 million, inclusive, in aid, and country B is to receive at least $ 0.75 million in aid. It has been estimated that each dollar spent in country A will yield an effective return of $.60, whereas a dollar spent in country B will yield an effective return of $.80. How should the aid be allocated if the money is to be utilized most effectively according to these criteria?

Hint: If x and y denote the amount of money to be given to country A and country B, respectively, then the objective function to be maximized is $P = 0.6x + 0.8y$.

33. MANUFACTURING—SHIPPING COSTS The TMA Company manufactures 19-in. color television picture tubes in two separate locations, locations I and II. The output at location I is at most 6000 tubes/month, whereas the output at location II is at most 5000/month. TMA is the main supplier of picture tubes to the Pulsar Corporation, its

holding company, which has priority in having all its requirements met. In a certain month, Pulsar placed orders for 3000 and 4000 picture tubes to be shipped to two of its factories located in city A and city B, respectively. The shipping costs (in dollars) per picture tube from the two TMA plants to the two Pulsar factories are as follows:

Shipping Costs per Picture Tube

From	To Pulsar Factories City A	City B
TMA (Loc. I)	\$3	\$2
TMA (Loc. II)	\$4	\$5

Find a shipping schedule that meets the requirements of both companies while keeping costs to a minimum.

34. **Veterinary Science** A veterinarian has been asked to prepare a diet for a group of dogs to be used in a nutrition study at the School of Animal Science. It has been stipulated that each serving should be no larger than 8 oz and must contain at least 29 units of nutrient I and 20 units of nutrient II. The vet has decided that the diet may be prepared from two brands of dog food: brand A and brand B. Each ounce of brand A contains 3 units of nutrient I and 4 units of nutrient II. Each ounce of brand B contains 5 units of nutrient I and 2 units of nutrient II. Brand A costs 3 cents/ounce and brand B costs 4 cents/ounce. Determine how many ounces of each brand of dog food should be used per serving to meet the given requirements at a minimum cost.

35. **Investment Planning** Patricia has at most \$30,000 to invest in securities in the form of corporate stocks. She has narrowed her choices to two groups of stocks: growth stocks that she assumes will yield a 15% return (dividends and capital appreciation) within a year and speculative stocks that she assumes will yield a 25% return (mainly in capital appreciation) within a year. Determine how much she should invest in each group of stocks in order to maximize the return on her investments within a year if she has decided to invest at least three times as much in growth stocks as in speculative stocks.

36. **Market Research** The Trendex Corporation, a telephone survey company, has been hired to conduct a television-viewing poll among urban and suburban families in the Los Angeles area. The client has stipulated that a maximum of 1500 families is to be interviewed. At least 500 urban families must be interviewed, and at least half of the total number of families interviewed must be from the suburban area. For this service, Trendex will be paid \$3000 plus \$4 for each completed interview. From previous experience, Trendex has determined that it will incur an expense of \$2.20 for each successful interview with an urban family and \$2.50 for each successful interview with a suburban family. How many urban and suburban families Trendex should interview in order to maximize its profit?

In Exercises 37–40, determine whether the statement is true or false. If it is true, explain why it is true. If it is false, give an example to show why it is false.

37. An optimal solution of a linear programming problem is a feasible solution, but a feasible solution of a linear programming problem need not be an optimal solution.

38. A linear programming problem can have exactly three (optimal) solutions.

39. If a maximization problem has no solution, then the feasible set associated with the linear programming problem must be unbounded.

40. Suppose you are given the following linear programming problem: Maximize $P = ax + by$ on the unbounded feasible set S shown in the accompanying figure.

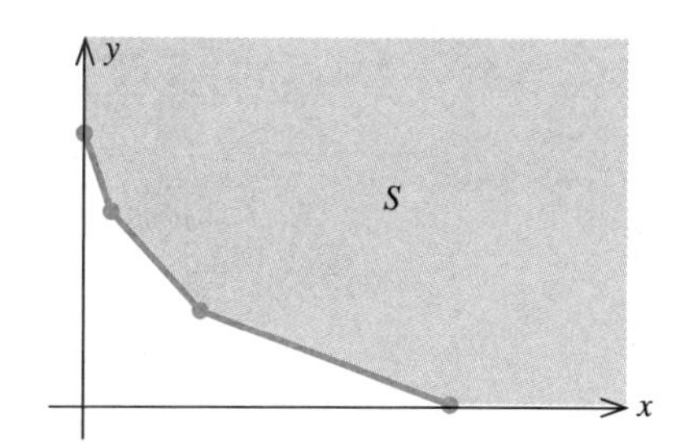

a. If $a > 0$ or $b > 0$, then the linear programming problem has no optimal solution.
b. If $a \leq 0$ and $b \leq 0$, then the linear programming problem has at least one optimal solution.

41. Suppose you are given the following linear programming problem: Maximize $P = ax + by$, where $a > 0$ and $b > 0$, on the feasible set S shown in the accompanying figure.

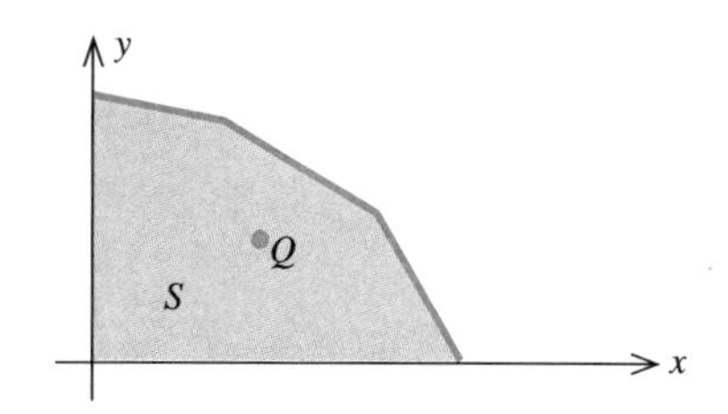

Explain, without using Theorem 1, why the optimal solution of the linear programming problem cannot occur at the point Q.

42. Suppose you are given the following linear programming problem: Maximize $P = ax + by$, where $a > 0$ and $b > 0$, on the feasible set S shown in the accompanying figure.

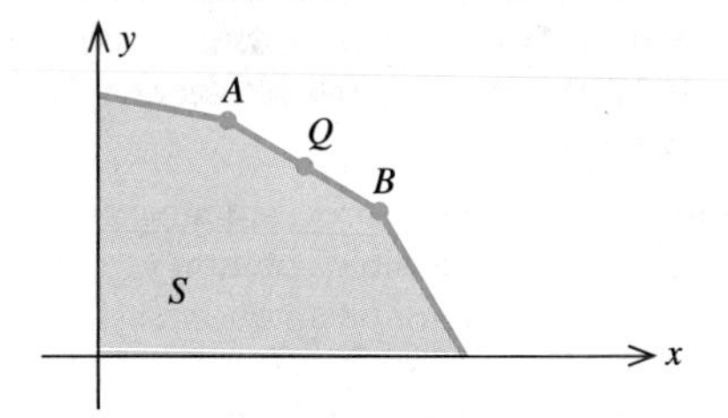

Explain, without using Theorem 1, why the optimal solution of the linear programming problem cannot occur at the point Q unless the problem has infinitely many solutions lying along the line segment joining the vertices A and B.
Hint: Let $A(x_1, y_1)$ and $B(x_2, y_2)$. Then $Q(\bar{x}, \bar{y})$, where $\bar{x} = x_1 + (x_2 - x_1)t$ and $\bar{y} = y_2 + (y_2 - y_1)t$ with $0 < t < 1$. Study the value of P at and near Q.

43. Consider the linear programming problem:

$$\begin{aligned} \text{Minimize} \quad & C = -2x + 5y \\ \text{subject to} \quad & x + y \le 3 \\ & 2x + y \le 4 \\ & 5x + 8y \ge 40 \end{aligned}$$

a. Sketch the feasible set.
b. Find the solution(s) of the linear programming problem, if it exists.

44. Consider the linear programming problem:

$$\begin{aligned} \text{Maximize} \quad & P = 2x + 7y \\ \text{subject to} \quad & 2x + y \ge 8 \\ & x + y \ge 6 \\ & x \ge 0, y \ge 0 \end{aligned}$$

a. Sketch the feasible set S.
b. Find the corner points of S.
c. Find the values of P at the corner points of S found in part (b).
d. Show that the linear programming problem has no optimal) solution. Does this contradict Theorem 1?

SOLUTIONS TO SELF-CHECK EXERCISES 3.3

1. The feasible set S for the problem was graphed in the solution to Exercise 1, Self-Check Exercises 3.1. It is reproduced in the accompanying figure.

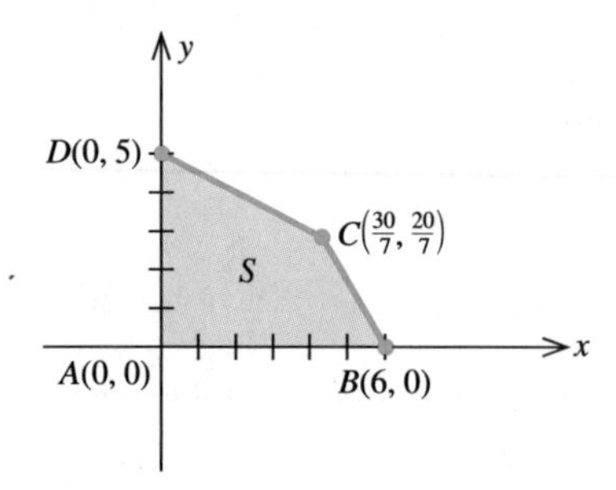

The values of the objective function P at the vertices of S are summarized in the accompanying table.

Vertex	$P = 4x + 5y$
$A(0, 0)$	0
$B(6, 0)$	24
$C(\frac{30}{7}, \frac{20}{7})$	$\frac{220}{7} = 31\frac{3}{7}$
$D(0, 5)$	25

From the table, we see that the maximum for the objective function P is attained at the vertex $C(30/7, 20/7)$. Therefore, the solution to the problem is $x = 30/7$, $y = 20/7$, and $P = 31\frac{3}{7}$.

2. The feasible set S for the problem was graphed in the solution to Exercise 2, Self-Check Exercises 3.1. It is reproduced in the accompanying figure.

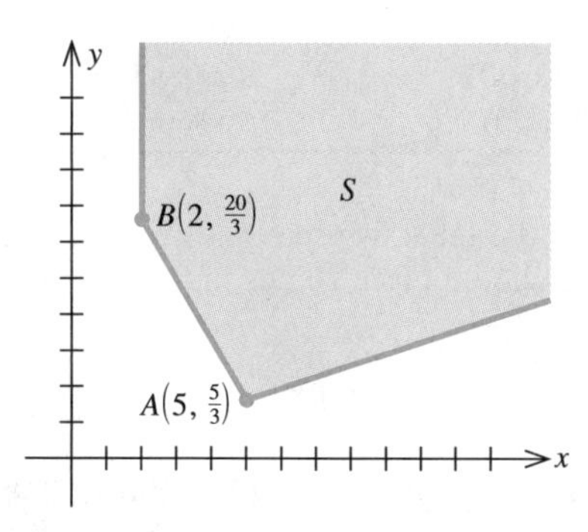

Evaluating the objective function $C = 5x + 3y$ at each corner point, we obtain the table

Vertex	$C = 5x + 3y$
$A(5, \frac{5}{3})$	30
$B(2, \frac{20}{3})$	30

We conclude that the objective function is minimized at every point on the line segment joining the points $(5, 5/3)$ and $(2, 20/3)$ and the minimum value of C is 30.

3. Refer to Self-Check Exercise 3.2. The problem is to maximize $P = 50{,}000x + 20{,}000y$ subject to

$$\begin{aligned} 300x + 100y &\leq 9000 \\ y &\geq 15 \\ y &\leq 30 \\ x \geq 0, y &\geq 0 \end{aligned}$$

The feasible set S for the problem is shown in the accompanying figure.

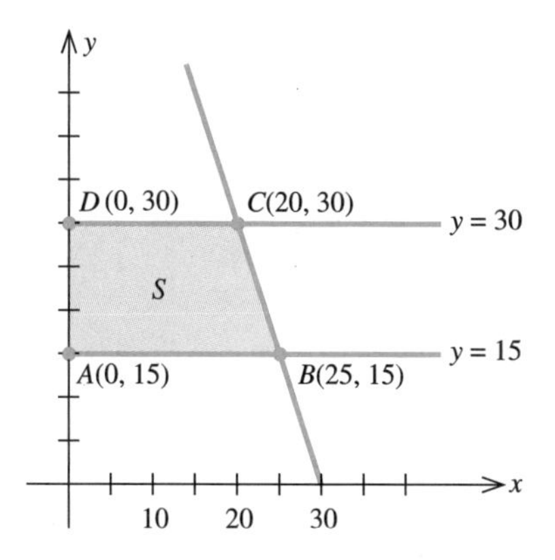

Evaluating the objective function $P = 50{,}000x + 20{,}000y$ at each vertex of S, we obtain

Vertex	$P = 50{,}000x + 20{,}000y$
$A(0, 15)$	300,000
$B(25, 15)$	1,550,000
$C(20, 30)$	1,600,000
$D(0, 30)$	600,000

From the table, we see that P is maximized when $x = 20$ and $y = 30$. Therefore, Gino should place 20 ads in the *City Tribune* and 30 with the *Daily News.*

CHAPTER 3 Summary of Principal Terms

Terms

- solution set of a system of linear inequalities
- bounded solution set
- unbounded solution set
- objective function
- linear programming problem
- feasible solution
- feasible set
- optimal solution
- method of corners

CHAPTER 3 REVIEW EXERCISES

In Exercises 1 and 2, find the optimal value(s) of the given objective function on the feasible set *S*.

1. $Z = 2x + 3y$

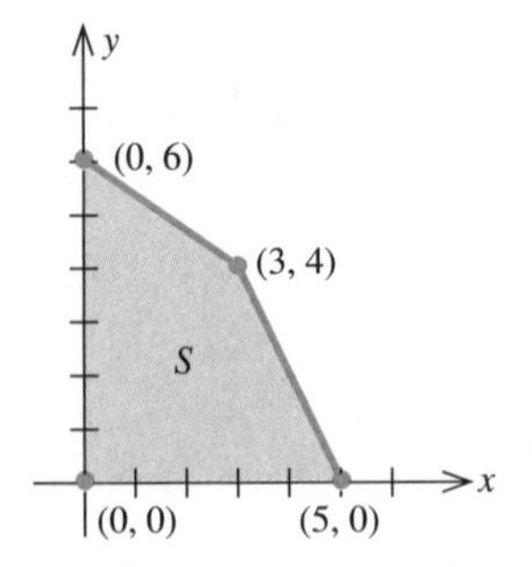

2. $Z = 4x + 3y$

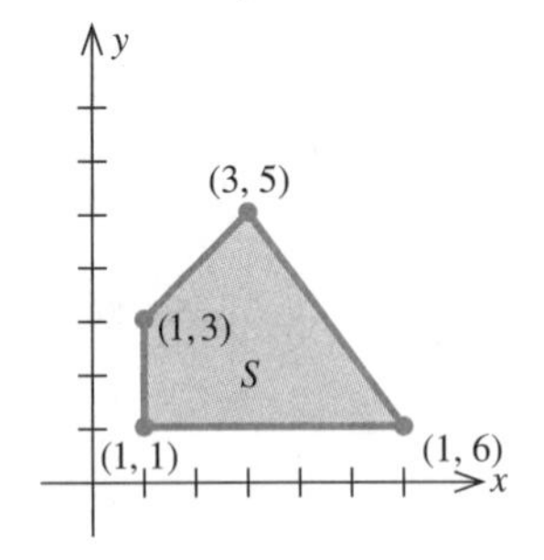

In Exercises 3–12, use the method of corners to solve the given linear programming problem.

3. Maximize $P = 3x + 5y$
subject to
$$2x + 3y \le 12$$
$$x + y \le 5$$
$$x \ge 0, y \ge 0$$

4. Maximize $P = 2x + 3y$
subject to
$$2x + y \le 12$$
$$x - 2y \le 1$$
$$x \ge 0, y \ge 0$$

5. Minimize $C = 2x + 5y$
subject to
$$x + 3y \le 15$$
$$4x + y \le 16$$
$$x \ge 0, y \ge 0$$

6. Minimize $C = 3x + 4y$
subject to
$$2x + y \ge 4$$
$$2x + 5y \ge 10$$
$$x \ge 0, y \ge 0$$

7. Maximize $P = 3x + 2y$
subject to
$$2x + y \le 16$$
$$2x + 3y \le 36$$
$$4x + 5y \ge 28$$
$$x \ge 0, y \ge 0$$

8. Maximize $P = 6x + 2y$
subject to
$$x + 2y \le 12$$
$$x + y \le 8$$
$$2x - 3y \ge 6$$
$$x \ge 0, y \ge 0$$

9. Minimize $C = 2x + 7y$
subject to
$$3x + 5y \ge 45$$
$$3x + 10y \ge 60$$
$$x \ge 0, y \ge 0$$

10. Minimize $C = 4x + y$
subject to
$$6x + y \ge 18$$
$$2x + y \ge 10$$
$$x + 4y \ge 12$$
$$x \ge 0, y \ge 0$$

11. Find the maximum and minimum of $Q = x + y$ subject to
$$5x + 2y \ge 20$$
$$x + 2y \ge 8$$
$$x + 4y \le 22$$
$$x \ge 0, y \ge 0$$

12. Find the maximum and minimum of $Q = 2x + 5y$ subject to
$$x + y \ge 4$$
$$-x + y \le 6$$
$$x + 3y \le 30$$
$$x \le 12$$
$$x \ge 0, y \ge 0$$

13. FINANCIAL ANALYSIS An investor has decided to commit no more than $80,000 to the purchase of the common stocks of two companies, company A and company B. He has also estimated that there is a chance of a 1% capital loss on his investment in company A and a chance of a 4% loss on his investment in company B, and he has decided that these losses should not exceed $2000. On the other hand, he expects to make a 14% profit from his investment in company A and a 20% profit from his investment in company B. Determine how much he should invest in the stock of each company in order to maximize his investment returns.

14. MANUFACTURING—PRODUCTION SCHEDULING The Soundex Company produces two models of clock radios. Model A requires 15 min of work on assembly line I and 10 min of work on assembly line II. Model B requires 10 min of work on assembly line I and 12 min of work on assembly line II. At most, 25 hr of assembly time on line I and 22 hr of assembly time on line II are available per day. It is anticipated that Soundex will realize a profit of $12 on model A and $10 on model B. How many clock radios of each model should be produced per day in order to maximize Soundex's profit?

15. MANUFACTURING—PRODUCTION SCHEDULING Kane Manufacturing has a division that produces two models of grates, model A and model B. To produce each model A grate requires 3 lb of cast iron and 6 min of labor. To produce each model B grate requires 4 lb of cast iron and 3 min of labor. The profit for each model A grate is $2, and the profit for each model B grate is $1.50. Available for grate production each day are 1000 lb of cast iron and 20 hr of labor. Because of a backlog of orders for model B grates, Kane's manager has decided to produce at least 180 model B grates per day. How many grates of each model should Kane produce to maximize its profits?

4 LINEAR PROGRAMMING: AN ALGEBRAIC APPROACH

The geometric approach introduced in the previous chapter may be used to solve linear programming problems involving two or even three variables. But for linear programming problems involving more than two variables, an algebraic approach is preferred. One such technique, the *simplex method,* was developed by George Dantzig in the late 1940s and remains in wide use to this day.

We begin Chapter 4 by developing the simplex method for solving *standard maximization problems.* We then see how, thanks to the principle of duality discovered by the great mathematician John von Neumann, this method can be used to solve a restricted class of *standard minimization problems.* Finally, we see how the simplex method can be adapted to solve *nonstandard problems*—problems that do not belong to the other aforementioned categories.

How much profit? The Ace Novelty Company produces three types of souvenirs. Each type requires a certain amount of time on each of three different machines. Each machine may be operated for a certain amount of time per day. In Example 5, page 234, we will determine how many souvenirs of each type Ace Novelty should make per day in order to maximize its daily profits.

4.1 The Simplex Method: Standard Maximization Problems

THE SIMPLEX METHOD

As mentioned in Chapter 3, the method of corners is not suitable for solving linear programming problems when the number of variables or constraints is large. Its major shortcoming is that a knowledge of all the corner points of the feasible set S associated with the problem is required. What we need is a method of solution that is based on a judicious selection of the corner points of the feasible set S, thereby reducing the number of points to be inspected. One such technique, called the *simplex method,* was developed in the late 1940s by George Dantzig and is based on the Gauss–Jordan elimination method. The simplex method is readily adaptable to the computer, which makes it ideally suitable for solving linear programming problems involving large numbers of variables and constraints.

Basically, the simplex method is an iterative procedure; that is, it is repeated over and over again. Beginning at some initial feasible solution (a corner point of the feasible set S, usually the origin) each iteration brings us to another corner point of S with an improved (but certainly no worse) value of the objective function. The iteration is terminated when the optimal solution is reached (if it exists).

In this section we describe the simplex method for solving a large class of problems that are referred to as **standard maximization problems.**

Before stating a formal procedure for solving standard linear programming problems based on the simplex method, let's consider the following analysis of a two-variable problem. The ensuing discussion will clarify the general procedure and at the same time enhance our understanding of the simplex method by examining the motivation that led to the steps of the procedure.

A Standard Linear Programming Problem

A **standard maximization problem** is one in which

1. The objective function is to be maximized.
2. All the variables involved in the problem are nonnegative.
3. Each linear constraint may be written so that the expression involving the variables is less than or equal to a nonnegative constant.

Consider the linear programming problem presented at the beginning of Section 3.3.

$$\text{Maximize} \quad P = 3x + 2y \qquad (1)$$

$$\begin{aligned} \text{subject to} \quad 2x + 3y &\le 12 \\ 2x + y &\le 8 \qquad (2) \\ x \ge 0, y &\ge 0 \end{aligned}$$

FIGURE 4.1
The optimal solution occurs at $C(3, 2)$.

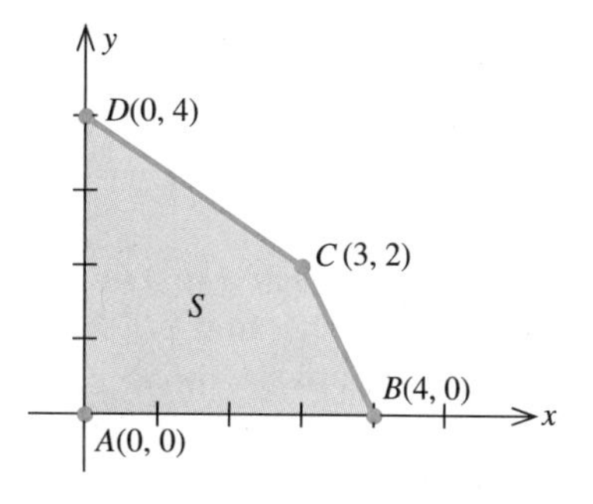

You can easily verify that this is a standard maximization problem. The feasible set S associated with this problem is reproduced in Figure 4.1, where we have labeled the four feasible corner points $A(0, 0)$, $B(4, 0)$, $C(3, 2)$, and $D(0, 4)$. Recall that the optimal solution to the problem occurs at the corner point $C(3, 2)$.

As a first step in the solution using the simplex method, we replace the system of inequality constraints (2) with a system of equality constraints. This may be accomplished by using nonnegative variables called **slack variables.** Let's begin by considering the inequality

$$2x + 3y \leq 12$$

Observe that the left-hand side of this equation is always less than or equal to the right-hand side. Therefore, by adding a nonnegative variable u to the left-hand side to compensate for this difference, we obtain the equality

$$2x + 3y + u = 12$$

For example, if $x = 1$ and $y = 1$ [you can see by referring to Figure 4.1 that the point (1, 1) is a feasible point of S], then $u = 7$. Thus,

$$2(1) + 3(1) + 7 = 12$$

If $x = 2$ and $y = 1$ [the point (2, 1) is also a feasible point of S], then $u = 5$. Thus,

$$2(2) + 3(1) + 5 = 12$$

The variable u is a *slack variable.*

Similarly, the inequality $2x + y \leq 8$ is converted into the equation $2x + y + v = 8$ through the introduction of the slack variable v. System (2) of linear inequalities may now be viewed as the system of linear equations

$$\begin{aligned} 2x + 3y + u \phantom{{}+v} &= 12 \\ 2x + y \phantom{{}+u} + v &= 8 \end{aligned}$$

where x, y, u, and v are all nonnegative.

Finally, rewriting the objective function (1) in the form $-3x - 2y + P = 0$, where the coefficient of P is $+1$, we are led to the following system of linear equations:

$$\begin{aligned} 2x + 3y + u \phantom{{}+v+P} &= 12 \\ 2x + y \phantom{{}+u} + v \phantom{{}+P} &= 8 \\ -3x - 2y \phantom{{}+u+v} + P &= 0 \end{aligned} \qquad (3)$$

Since System (3) consists of three linear equations in the five variables x, y, u, v, and P, we may solve for three of the variables in terms of the other two. Thus, there are infinitely many solutions to this system expressible in terms of two parameters. Our linear programming problem is now seen to be equivalent to the following: From among all the solutions of System (3) for which x, y, u, and v are nonnegative (such solutions are called **feasible solutions**), determine the solution(s) that renders P a maximum.

The augmented matrix associated with System (3) is

Nonbasic variables: x, y — Basic variables: u, v, P — Column of constants

$$\begin{array}{ccccc|c} x & y & u & v & P & \\ 2 & 3 & 1 & 0 & 0 & 12 \\ 2 & 1 & 0 & 1 & 0 & 8 \\ -3 & -2 & 0 & 0 & 1 & 0 \end{array} \quad (4)$$

Observe that each of the u-, v-, and P-columns of the augmented matrix (4) is a unit column (see page 93). The variables associated with unit columns are called **basic variables;** all other variables are called **nonbasic variables.**

Now, the configuration of the augmented matrix (4) suggests that we solve for the basic variables u, v, and P in terms of the nonbasic variables x and y, obtaining

$$\begin{aligned} u &= 12 - 2x - 3y \\ v &= 8 - 2x - y \\ P &= 3x + 2y \end{aligned} \quad (5)$$

Of the infinitely many feasible solutions obtainable by assigning arbitrary nonnegative values to the parameters x and y, a particular solution is obtained by letting $x = 0$ and $y = 0$. In fact, this solution is given by

$$x = 0, \quad y = 0, \quad u = 12, \quad v = 8, \quad P = 0$$

Such a solution, obtained by setting the nonbasic variables equal to zero, is called a **basic solution** of the system. This particular solution corresponds to the corner point $A(0, 0)$ of the feasible set associated with the linear programming problem (see Figure 4.1). Observe that $P = 0$ at this point.

Now, if the value of P cannot be increased, we have found the optimal solution to the problem at hand. To determine whether the value of P can in fact be improved, let's turn our attention to the objective function in (1). Since both the coefficients of x and y are positive, the value of P can be improved by increasing x and/or y—that is, by moving away from the origin. Note that we arrive at the same conclusion by observing that the last row of the augmented matrix (4) contains entries that are *negative.* (Compare the original objective function, $P = 3x + 2y$, with the rewritten objective function, $-3x - 2y + P = 0$.)

Continuing our quest for an optimal solution, our next task is to determine whether it is more profitable to increase the value of x or that of y (increasing x and y simultaneously is more difficult). Since the coefficient of x is greater than that of y, a unit increase in the x-direction will result in a greater increase in the value of the objective function P than a unit increase in the y-direction. Thus, we should increase the value of x while holding y constant. How much can x be increased while holding $y = 0$? Upon setting $y = 0$ in the first two equations of (5), we see that

$$\begin{aligned} u &= 12 - 2x \\ v &= 8 - 2x \end{aligned} \quad (6)$$

Since u must be nonnegative, the first equation of (6) implies that x cannot exceed 12/2, or 6. The second equation of (6) and the nonnegativity of v implies that x cannot exceed 8/2, or 4. Thus, we conclude that x can be increased by at most 4.

Now, if we set $y = 0$ and $x = 4$ in System (5), we obtain the solution

$$x = 4, \quad y = 0, \quad u = 4, \quad v = 0, \quad P = 12$$

which is a basic solution to System (3), this time with y and v as nonbasic variables. (Recall that the nonbasic variables are precisely the variables that are set equal to zero.)

Let's see how this basic solution may be found by working with the augmented matrix of the system. Since x is to replace v as a basic variable, our aim is to find an augmented matrix that is equivalent to the matrix (4) and has a configuration in which the x-column is in the unit form

$$\begin{bmatrix} 0 \\ 1 \\ 0 \end{bmatrix}$$

replacing what is presently the form of the v-column in (4). Now, this may be accomplished by pivoting about the circled number 2.

$$\begin{array}{c} \begin{array}{cccccc} x & y & u & v & P & \text{Const.} \end{array} \\ \left[\begin{array}{ccccc|c} 2 & 3 & 1 & 0 & 0 & 12 \\ \textcircled{2} & 1 & 0 & 1 & 0 & 8 \\ -3 & -2 & 0 & 0 & 1 & 0 \end{array}\right] \end{array} \xrightarrow{\frac{1}{2}R_2} \begin{array}{c} \begin{array}{cccccc} x & y & u & v & P & \text{Const.} \end{array} \\ \left[\begin{array}{ccccc|c} 2 & 3 & 1 & 0 & 0 & 12 \\ \textcircled{1} & \frac{1}{2} & 0 & \frac{1}{2} & 0 & 4 \\ -3 & -2 & 0 & 0 & 1 & 0 \end{array}\right] \end{array} \qquad \textbf{(7)}$$

$$\xrightarrow[R_3 + 3R_2]{R_1 - 2R_2} \begin{array}{c} \begin{array}{cccccc} x & y & u & v & P & \text{Const.} \end{array} \\ \left[\begin{array}{ccccc|c} 0 & 2 & 1 & -1 & 0 & 4 \\ 1 & \frac{1}{2} & 0 & \frac{1}{2} & 0 & 4 \\ 0 & -\frac{1}{2} & 0 & \frac{3}{2} & 1 & 12 \end{array}\right] \end{array} \qquad \textbf{(8)}$$

Using (8), we now solve for the basic variables x, u, and P in terms of the nonbasic variables y and v, obtaining

$$\begin{aligned} x &= 4 - \left(\frac{1}{2}\right)y - \left(\frac{1}{2}\right)v \\ u &= 4 - 2y + v \\ P &= 12 + \left(\frac{1}{2}\right)y - \left(\frac{3}{2}\right)v \end{aligned}$$

Setting the nonbasic variables y and v equal to zero gives

$$x = 4, \quad y = 0, \quad u = 4, \quad v = 0, \quad \text{and} \quad P = 12$$

as before.

FIGURE 4.2
One iteration has taken us from $A(0, 0)$, where $P = 0$, to $B(4, 0)$, where $P = 12$.

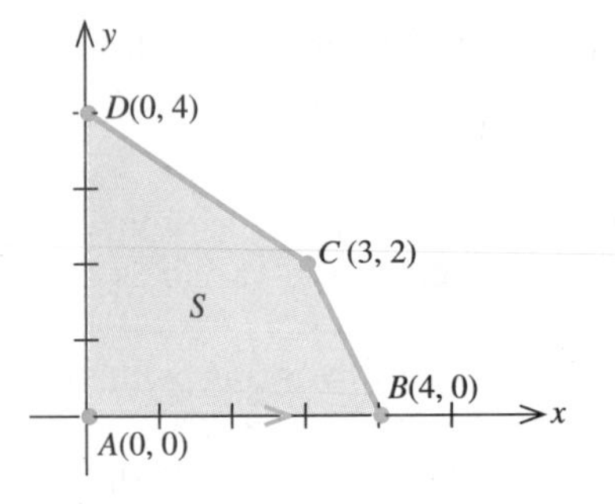

We have now completed one iteration of the simplex procedure, and our search has brought us from the feasible corner point $A(0, 0)$, where $P = 0$, to the feasible corner point $B(4, 0)$, where P attained a value of 12, which is certainly an improvement! (See Figure 4.2.)

Before going on, let's introduce the following terminology. The circled element 2 in the first augmented matrix of (7), which was to be converted into a 1, is called a *pivot element.* The column containing the pivot element is called the *pivot column.* The pivot column is associated with a nonbasic variable that is to be converted to a basic variable. Note that *the last entry in the pivot column is the negative number with the largest absolute value to the left of the vertical line in the last row*—precisely the criterion for choosing the direction of maximum increase in P.

The row containing the pivot element is called the *pivot row.* The pivot row can also be found by dividing each positive number in the pivot column into the corresponding number in the last column (the column of constants). *The pivot row is the one with the smallest ratio.* In the augmented matrix (7), the pivot row is the second row since the ratio 8/2, or 4, is less than the ratio 12/2, or 6. (Compare this with the earlier analysis pertaining to the determination of the largest permissible increase in the value of x.)

The following is a summary of the procedure for selecting the pivot element.

Selecting the Pivot Element

1. Select the pivot column: Locate the most negative entry to the left of the vertical line in the last row. The column containing this entry is the **pivot column.** (If there is more than one such column, choose any one.)
2. Select the pivot row: Divide each positive entry in the pivot column into its corresponding entry in the column of constants. The **pivot row** is the row corresponding to the smallest ratio thus obtained. (If there is more than one such entry, choose any one.)
3. The **pivot element** is the element common to both the pivot column and the pivot row.

Continuing with the solution to our problem, we observe that the last row of the augmented matrix (8) contains a negative number—namely, $-1/2$. This indicates that P is not maximized at the feasible corner point $B(4, 0)$, and another iteration is required. Without once again going into a detailed analysis, we proceed immediately to the selection of a pivot element. In accordance with the rules, we perform the necessary row operations as follows:

$$\begin{array}{r} \\ \text{Pivot row} \rightarrow \\ \\ \\ \end{array}
\begin{array}{c}
\begin{array}{ccccc} x & y & u & v & P \end{array} \\
\left[\begin{array}{ccccc|c}
0 & \textcircled{2} & 1 & -1 & 0 & 4 \\
1 & \frac{1}{2} & 0 & \frac{1}{2} & 0 & 4 \\
0 & -\frac{1}{2} & 0 & \frac{3}{2} & 1 & 12
\end{array}\right]
\end{array}
\quad
\begin{array}{c}
\text{ratio} \\
\frac{4}{2} = 2 \\
\frac{4}{1/2} = 8 \\
\\
\end{array}$$

↑ Pivot column

FIGURE 4.3
The next iteration has taken us from $B(4, 0)$, where $P = 12$, to $C(3, 2)$, where $P = 13$.

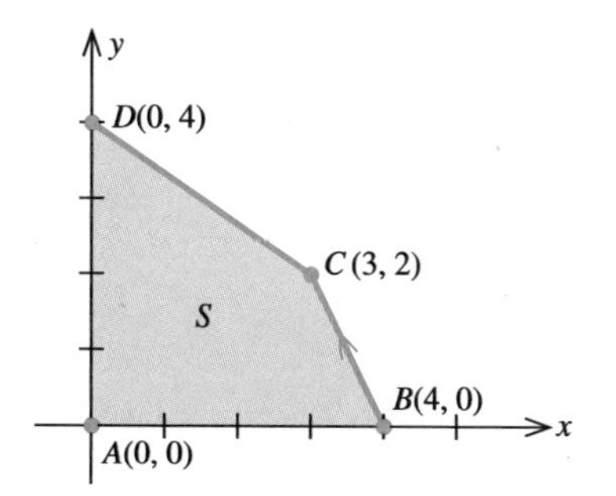

$$\xrightarrow{\frac{1}{2}R_1} \begin{array}{c} \begin{array}{cccccc} x & y & u & v & P & \end{array} \\ \left[\begin{array}{ccccc|c} 0 & \textcircled{1} & \frac{1}{2} & -\frac{1}{2} & 0 & 2 \\ 1 & \frac{1}{2} & 0 & \frac{1}{2} & 0 & 4 \\ 0 & -\frac{1}{2} & 0 & \frac{3}{2} & 1 & 12 \end{array}\right] \end{array}$$

$$\xrightarrow[R_3 + \frac{1}{2}R_1]{R_2 - \frac{1}{2}R_1} \begin{array}{c} \begin{array}{cccccc} x & y & u & v & P & \end{array} \\ \left[\begin{array}{ccccc|c} 0 & 1 & \frac{1}{2} & -\frac{1}{2} & 0 & 2 \\ 1 & 0 & -\frac{1}{4} & \frac{3}{4} & 0 & 3 \\ 0 & 0 & \frac{1}{4} & \frac{5}{4} & 1 & 13 \end{array}\right] \end{array}$$

Interpreting the last augmented matrix in the usual fashion, we find the basic solution $x = 3$, $y = 2$, and $P = 13$. Since there are no negative entries in the last row, the solution is optimal and P cannot be increased further. The optimal solution is the feasible corner point $C(3, 2)$ (Figure 4.3). Observe that this agrees with the solution we found using the method of corners in Section 3.3.

Having seen how the simplex method works, let us list the steps involved in the procedure. The first step is to set up the initial **simplex tableau.**

Setting Up the Initial Simplex Tableau

1. Transform the system of linear inequalities into a system of linear equations by introducing slack variables.
2. Rewrite the objective function

$$P = c_1x_1 + c_2x_2 + \cdots + c_nx_n$$

in the form

$$-c_1x_1 - c_2x_2 - \cdots - c_nx_n + P = 0$$

where all the variables are on the left and the coefficient of P is $+1$. Write this equation below the equations of step 1.
3. Write the augmented matrix associated with this system of linear equations.

EXAMPLE 1 Set up the initial simplex tableau for the linear programming problem posed in Example 1, Section 3.2.

SOLUTION ✓ The problem at hand is to maximize

$$P = x + 1.2y$$

or, equivalently,

$$P = x + \frac{6}{5}y$$

subject to

$$\begin{aligned} 2x + y &\le 180 \\ x + 3y &\le 300 \\ x \ge 0, y &\ge 0 \end{aligned} \tag{9}$$

This is a standard maximization problem and may be solved by the simplex method. Since System (9) has two linear inequalities (other than $x \geq 0, y \geq 0$), we introduce the two slack variables u and v to convert it to a system of linear equations:

$$\begin{aligned} 2x + y + u \quad &= 180 \\ x + 3y \quad + v &= 300 \end{aligned}$$

Next, by rewriting the objective function in the form

$$-x - \frac{6}{5}y + P = 0$$

where the coefficient of P is $+1$, and placing it below the system of equations, we obtain the system of linear equations

$$\begin{aligned} 2x + y + u \qquad &= 180 \\ x + 3y \quad + v \qquad &= 300 \\ -x - \frac{6}{5}y \qquad + P &= 0 \end{aligned}$$

The initial simplex tableau associated with this system is

x	y	u	v	P	Constant
2	1	1	0	0	180
1	3	0	1	0	300
−1	$-\frac{6}{5}$	0	0	1	0

■■■■

Before completing the solution to the problem posed in Example 1, let's summarize the main steps of the **simplex method.**

The Simplex Method

1. *Set up the initial simplex tableau.*
2. *Determine whether the optimal solution has been reached by examining all entries in the last row to the left of the vertical line:*
   a. If all the entries are nonnegative, the optimal solution has been reached. Proceed to step 4.
   b. If there are one or more negative entries, the optimal solution has not been reached. Proceed to step 3.
3. *Perform the pivot operation:* Locate the pivot element and convert it to a 1 by dividing all the elements in the pivot row by the pivot element. Using row operations, convert the pivot column into a unit column by adding suitable multiples of the pivot row to each of the other rows as required. Return to step 2.
4. *Determine the optimal solution(s):* The value of the variable heading each unit column is given by the entry lying in the column of constants in the row containing the 1. The variables heading columns not in unit form are assigned the value zero.

EXAMPLE 2 Complete the solution to the problem discussed in Example 1.

SOLUTION ✓ The first step in our procedure, setting up the initial simplex tableau, was completed in Example 1. We continue with step 2.

Step 2 *Determine whether the optimal solution has been reached.* First, refer to the initial simplex tableau:

x	y	u	v	P	Constant
2	1	1	0	0	180
1	3	0	1	0	300
−1	$-\frac{6}{5}$	0	0	1	0

(10)

Since there are negative entries in the last row of the initial simplex tableau, the initial solution is not optimal. We proceed to step 3.

Step 3 *Perform the following iterations.* First, locate the pivot element:
a. Since the entry −6/5 is the most negative entry to the left of the vertical line in the last row of the initial simplex tableau, the second column in the tableau is the pivot column.
b. Divide each positive number of the pivot column into the corresponding entry in the column of constants and compare the ratios thus obtained. We see that the ratio 300/3 is less than the ratio 180/1, so row 2 in the tableau is the pivot row.
c. The entry 3 lying in the pivot column and the pivot row is the pivot element.

Next, we convert this pivot element into a 1 by multiplying all the entries in the pivot row by 1/3. Then, using elementary row operations, we complete the conversion of the pivot column into a unit column.

The details of the iteration are recorded as follows:

	x	y	u	v	P	Constant	ratio
	2	1	1	0	0	180	$\frac{180}{1} = 180$
Pivot row →	1	③	0	1	0	300	$\frac{300}{3} = 100$
	−1	$-\frac{6}{5}$	0	0	1	0	
		↑ Pivot column					

$\xrightarrow{\frac{1}{3}R_2}$

x	y	u	v	P	Constant
2	1	1	0	0	180
$\frac{1}{3}$	①	0	$\frac{1}{3}$	0	100
−1	$-\frac{6}{5}$	0	0	1	0

$$\xrightarrow[R_3 + \frac{6}{5}R_2]{R_1 - R_2} \quad \begin{array}{ccccc|c} x & y & u & v & P & \text{Constant} \\ \hline \frac{5}{3} & 0 & 1 & -\frac{1}{3} & 0 & 80 \\ \frac{1}{3} & 1 & 0 & \frac{1}{3} & 0 & 100 \\ \hline -\frac{3}{5} & 0 & 0 & \frac{2}{5} & 1 & 120 \end{array} \qquad \textbf{(11)}$$

This completes one iteration. The last row of the simplex tableau contains a negative number, so an optimal solution has not been reached. Therefore, we repeat the iterative step once again, as follows:

$$\begin{array}{l} \text{Pivot} \\ \text{row} \to \end{array} \begin{array}{ccccc|c} x & y & u & v & P & \text{Constant} \\ \hline \textcircled{\tfrac{5}{3}} & 0 & 1 & -\frac{1}{3} & 0 & 80 \\ \frac{1}{3} & 1 & 0 & \frac{1}{3} & 0 & 100 \\ \hline -\frac{3}{5} & 0 & 0 & \frac{2}{5} & 1 & 120 \\ \uparrow & & & & & \end{array} \quad \begin{array}{c} \text{ratio} \\ \frac{80}{5/3} = 48 \\ \frac{100}{1/3} = 300 \end{array}$$

Pivot column

$$\xrightarrow{\frac{3}{5}R_1} \quad \begin{array}{ccccc|c} x & y & u & v & P & \text{Constant} \\ \hline \textcircled{1} & 0 & \frac{3}{5} & -\frac{1}{5} & 0 & 48 \\ \frac{1}{3} & 1 & 0 & \frac{1}{3} & 0 & 100 \\ \hline -\frac{3}{5} & 0 & 0 & \frac{2}{5} & 1 & 120 \end{array}$$

$$\xrightarrow[R_3 + \frac{3}{5}R_1]{R_2 - \frac{1}{3}R_1} \quad \begin{array}{ccccc|c} x & y & u & v & P & \text{Constant} \\ \hline 1 & 0 & \frac{3}{5} & -\frac{1}{5} & 0 & 48 \\ 0 & 1 & -\frac{1}{5} & \frac{2}{5} & 0 & 84 \\ \hline 0 & 0 & \frac{9}{25} & \frac{7}{25} & 1 & 148\frac{4}{5} \end{array} \qquad \textbf{(12)}$$

The last row of the simplex tableau (12) contains no negative numbers, and we therefore conclude that the optimal solution has been reached.

Step 4 *Determine the optimal solution.* Locate the basic variables in the final tableau. In this case the basic variables (those heading unit columns) are x, y, and P. The value assigned to the basic variable x is the number 48, which is the entry lying in the column of constants and in row 1 (the row that contains the 1).

x	y	u	v	P	Constant
①	0	$\frac{3}{5}$	$-\frac{1}{5}$	0	48 ←
0	①	$-\frac{1}{5}$	$\frac{2}{5}$	0	84 ←
0	0	$\frac{9}{25}$	$\frac{7}{25}$	①	$148\frac{4}{5}$ ←

Similarly, we conclude that $y = 84$ and $P = 148.8$. Next, we note that the variables u and v are nonbasic and are accordingly assigned the values $u = 0$ and $v = 0$. These results agree with those obtained in Example 1, Section 3.3.

EXAMPLE 3

$$\begin{aligned} \text{Maximize} \quad & P = 2x + 2y + z \\ \text{subject to} \quad & 2x + y + 2z \leq 14 \\ & 2x + 4y + z \leq 26 \\ & x + 2y + 3z \leq 28 \\ & x \geq 0, y \geq 0, z \geq 0 \end{aligned}$$

SOLUTION ✔ Introducing the slack variables u, v, and w and rewriting the objective function in the standard form gives the system of linear equations

$$\begin{aligned} 2x + y + 2z + u \quad\quad\quad\quad &= 14 \\ 2x + 4y + z \quad + v \quad\quad &= 26 \\ x + 2y + 3z \quad\quad + w \quad &= 28 \\ -2x - 2y - z \quad\quad\quad + P &= 0 \end{aligned}$$

The initial simplex tableau is given by

x	y	z	u	v	w	P	Constant
2	1	2	1	0	0	0	14
2	4	1	0	1	0	0	26
1	2	3	0	0	1	0	28
−2	−2	−1	0	0	0	1	0

Since the most negative entry in the last row (−2) occurs twice, we may choose either the x- or the y-column as the pivot column. Choosing the x-column as the pivot column and proceeding with the first iteration, we obtain the following sequence of tableaus:

	x	y	z	u	v	w	P	Constant	ratio
Pivot row →	(2)	1	2	1	0	0	0	14	$\frac{14}{2} = 7$
	2	4	1	0	1	0	0	26	$\frac{26}{2} = 13$
	1	2	3	0	0	1	0	28	$\frac{28}{1} = 28$
	−2	−2	−1	0	0	0	1	0	
	↑ Pivot column								

	x	y	z	u	v	w	P	Constant
	(1)	$\frac{1}{2}$	1	$\frac{1}{2}$	0	0	0	7
$\xrightarrow{\frac{1}{2}R_1}$	2	4	1	0	1	0	0	26
	1	2	3	0	0	1	0	28
	−2	−2	−1	0	0	0	1	0

	x	y	z	u	v	w	P	Constant
	1	$\frac{1}{2}$	1	$\frac{1}{2}$	0	0	0	7
$\xrightarrow{R_2 - 2R_1}$	0	3	−1	−1	1	0	0	12
$R_3 - R_1$	0	$\frac{3}{2}$	2	$-\frac{1}{2}$	0	1	0	21
$R_4 + 2R_1$	0	−1	1	1	0	0	1	14

Since there is a negative number in the last row of the simplex tableau, we perform another iteration, as follows:

	x	y	z	u	v	w	P	Constant	ratio
	1	$\frac{1}{2}$	1	$\frac{1}{2}$	0	0	0	7	$\frac{7}{1/2} = 14$
Pivot row →	0	(3)	−1	−1	1	0	0	12	$\frac{12}{3} = 4$
	0	$\frac{3}{2}$	2	$-\frac{1}{2}$	0	1	0	21	$\frac{21}{3/2} = 14$
	0	−1	1	1	0	0	1	14	
		↑ Pivot column							

	x	y	z	u	v	w	P	Constant
	1	$\frac{1}{2}$	1	$\frac{1}{2}$	0	0	0	7
$\xrightarrow{\frac{1}{3}R_2}$	0	(1)	$-\frac{1}{3}$	$-\frac{1}{3}$	$\frac{1}{3}$	0	0	4
	0	$\frac{3}{2}$	2	$-\frac{1}{2}$	0	1	0	21
	0	−1	1	1	0	0	1	14

	x	y	z	u	v	w	P	Constant
	1	0	$\frac{7}{6}$	$\frac{2}{3}$	$-\frac{1}{6}$	0	0	5
$\xrightarrow{R_1 - \frac{1}{2}R_2}$	0	1	$-\frac{1}{3}$	$-\frac{1}{3}$	$\frac{1}{3}$	0	0	4
$R_3 - \frac{3}{2}R_2$	0	0	$\frac{5}{2}$	0	$-\frac{1}{2}$	1	0	15
$R_4 + R_2$	0	0	$\frac{2}{3}$	$\frac{2}{3}$	$\frac{1}{3}$	0	1	18

All entries in the last row are nonnegative, so we have reached the optimal solution. We conclude that $x = 5$, $y = 4$, $z = 0$, $u = 0$, $v = 0$, $w = 15$, and $P = 18$. ■■■■

The following example is constructed to illustrate the geometry associated with the simplex method when used to solve a problem in 3-dimensional space. We sketch the feasible set for the problem and show the path dictated by the simplex method in arriving at the optimal solution for the problem.

Group Discussion

Consider the linear programming problem

$$\begin{aligned} \text{Maximize} \quad & P = x + 2y \\ \text{subject to} \quad & -2x + y \leq 4 \\ & x - 3y \leq 3 \\ & x \geq 0, y \geq 0 \end{aligned}$$

1. Sketch the feasible set S for the linear programming problem and explain why the problem has an unbounded solution.
2. Use the simplex method to solve the problem as follows:
   a. Perform one iteration on the initial simplex tableau. Interpret your result. Indicate the point on S corresponding to this (nonoptimal) solution.
   b. Show that the simplex procedure breaks down when you attempt to perform another iteration by demonstrating that there is no pivot element.
   c. Describe what happens if you violate the rule for finding the pivot element by allowing the ratios to be negative and proceeding with the iteration.

The use of a calculator will help in the arithmetic operations if you wish to verify the steps.

EXAMPLE 4

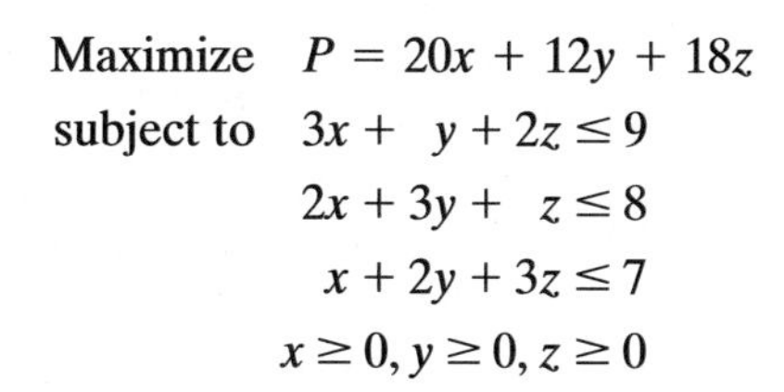

$$\begin{aligned} \text{Maximize} \quad & P = 20x + 12y + 18z \\ \text{subject to} \quad & 3x + y + 2z \leq 9 \\ & 2x + 3y + z \leq 8 \\ & x + 2y + 3z \leq 7 \\ & x \geq 0, y \geq 0, z \geq 0 \end{aligned}$$

SOLUTION ✓ Introducing the slack variables u, v, and w and rewriting the objective function in standard form gives the system of linear equations:

$$\begin{aligned} 3x + y + 2z + u \qquad\qquad &= 9 \\ 2x + 3y + z \qquad + v \qquad &= 8 \\ x + 2y + 3z \qquad\qquad + w &= 7 \\ -20x - 12y - 18z \qquad\qquad + P &= 0 \end{aligned}$$

The initial simplex tableau is given by

x	y	z	u	v	w	P	Constant
3	1	2	1	0	0	0	9
2	3	1	0	1	0	0	8
1	2	3	0	0	1	0	7
−20	−12	−18	0	0	0	1	0

Since the most negative entry in the last row (−20) occurs in the x-column, we choose the x-column as the pivot column. Proceeding with the first iteration, we obtain the following sequence of tableaus:

	x	y	z	u	v	w	P	Constant	ratio
Pivot row →	(3)	1	2	1	0	0	0	9	$\frac{9}{3} = 3$
	2	3	1	0	1	0	0	8	$\frac{8}{2} = 4$
	1	2	3	0	0	1	0	7	$\frac{7}{1} = 7$
	−20	−12	−18	0	0	0	1	0	
	↑ Pivot column								

$\xrightarrow{\frac{1}{3}R_1}$

x	y	z	u	v	w	P	Constant
(1)	$\frac{1}{3}$	$\frac{2}{3}$	$\frac{1}{3}$	0	0	0	3
2	3	1	0	1	0	0	8
1	2	3	0	0	1	0	7
−20	−12	−18	0	0	0	1	0

$\xrightarrow[R_4 + 20R_1]{R_2 - 2R_1,\ R_3 - R_1}$

	x	y	z	u	v	w	P	Constant	ratio
	1	$\frac{1}{3}$	$\frac{2}{3}$	$\frac{1}{3}$	0	0	0	3	9
Pivot row →	0	($\frac{7}{3}$)	$-\frac{1}{3}$	$-\frac{2}{3}$	1	0	0	2	$\frac{6}{7}$
	0	$\frac{5}{3}$	$\frac{7}{3}$	$-\frac{1}{3}$	0	1	0	4	$\frac{12}{5}$
	0	$-\frac{16}{3}$	$-\frac{14}{3}$	$\frac{20}{3}$	0	0	1	60	
		↑ Pivot column							

After one iteration we are at the point (3, 0, 0) with $P = 60$. (See Figure 4.5 on page 234.) Since the most negative entry in the last row is $-16/3$, we choose the y-column as the pivot column. Proceeding with this iteration, we obtain

$\xrightarrow{\frac{3}{7}R_2}$

x	y	z	u	v	w	P	Constant
1	$\frac{1}{3}$	$\frac{2}{3}$	$\frac{1}{3}$	0	0	0	3
0	(1)	$-\frac{1}{7}$	$-\frac{2}{7}$	$\frac{3}{7}$	0	0	$\frac{6}{7}$
0	$\frac{5}{3}$	$\frac{7}{3}$	$-\frac{1}{3}$	0	1	0	4
0	$-\frac{16}{3}$	$-\frac{14}{3}$	$\frac{20}{3}$	0	0	1	60

$\xrightarrow[R_4 + \frac{16}{3}R_2]{R_1 - \frac{1}{3}R_2,\ R_3 - \frac{5}{3}R_2}$

x	y	z	u	v	w	P	Constant	ratio
1	0	$\frac{5}{7}$	$\frac{3}{7}$	$-\frac{1}{7}$	0	0	$\frac{19}{7}$	$\frac{19}{5}$
0	1	$-\frac{1}{7}$	$-\frac{2}{7}$	$\frac{3}{7}$	0	0	$\frac{6}{7}$	—
0	0	($\frac{18}{7}$)	$\frac{1}{7}$	$-\frac{5}{7}$	1	0	$\frac{18}{7}$	1
0	0	$-\frac{38}{7}$	$\frac{36}{7}$	$\frac{16}{7}$	0	1	$64\frac{4}{7}$	
		↑ Pivot column						

The second iteration brings us to the point $(\frac{19}{7}, \frac{6}{7}, 0)$ with $P = 64\frac{4}{7}$. (See Figure 4.5 on page 234.) Since there is a negative number in the last row of the simplex tableau, we perform another iteration, as follows:

	x	y	z	u	v	w	P	Constant
	1	0	$\frac{5}{7}$	$\frac{3}{7}$	$-\frac{1}{7}$	0	0	$\frac{19}{7}$
$\xrightarrow{\frac{7}{18}R_3}$	0	1	$-\frac{1}{7}$	$-\frac{2}{7}$	$\frac{3}{7}$	0	0	$\frac{6}{7}$
	0	0	⓵	$\frac{1}{18}$	$-\frac{5}{18}$	$\frac{7}{18}$	0	1
	0	0	$-\frac{38}{7}$	$\frac{36}{7}$	$\frac{16}{7}$	0	1	$64\frac{4}{7}$

	x	y	z	u	v	w	P	Constant
$R_1 - \frac{5}{7}R_3$	1	0	0	$\frac{7}{18}$	$\frac{1}{18}$	$-\frac{5}{18}$	0	2
$\xrightarrow{R_2 + \frac{1}{7}R_3}$	0	1	0	$-\frac{5}{18}$	$\frac{7}{18}$	$\frac{1}{18}$	0	1
$R_4 + \frac{38}{7}R_3$	0	0	1	$\frac{1}{18}$	$-\frac{5}{18}$	$\frac{7}{18}$	0	1
	0	0	0	$\frac{49}{9}$	$\frac{7}{9}$	$\frac{19}{9}$	1	70

All entries in the last row are nonnegative, so we have reached the optimal solution. We conclude that $x = 2, y = 1, z = 1, u = 0, v = 0, w = 0$, and $P = 70$.

The feasible set S for the problem is the hexahedron shown in Figure 4.4. It is the intersection of the half spaces determined by the planes P_1, P_2, and P_3 with equations $3x + y + 2z = 9$, $2x + 3y + z = 8$, $x + 2y + 3z = 7$, respectively, and the coordinate planes $x = 0$, $y = 0$, and $z = 0$. That portion of the figure showing the feasible set S is shown in Figure 4.5. Observe that the first iteration of the simplex method brings us from $A(0, 0, 0)$ with $P = 0$

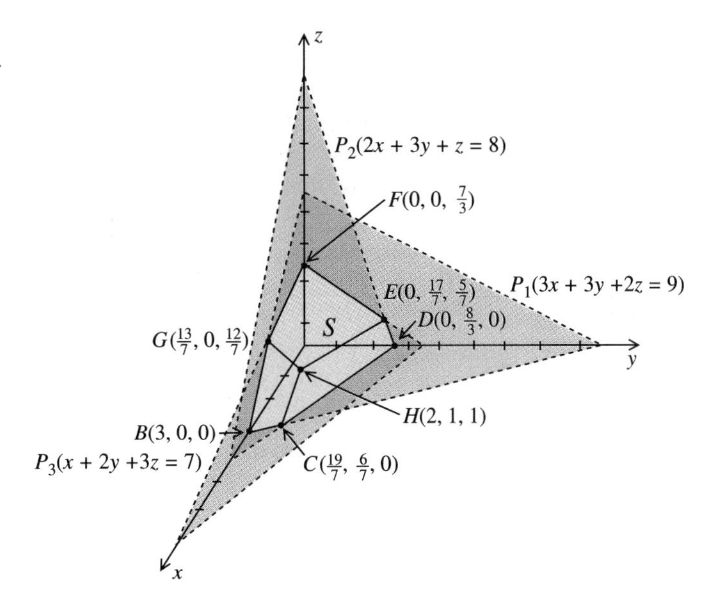

FIGURE 4.4
The feasible set S is obtained from the intersection of the half spaces determined by P_1, P_2, and P_3 with the coordinate planes $x = 0$, $y = 0$, and $z = 0$.

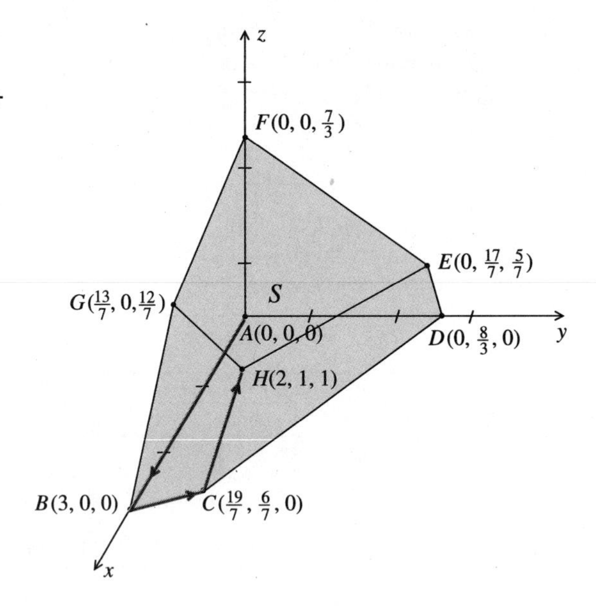

FIGURE 4.5
The simplex method brings us from the point A to the point H, at which the objective function is maximized.

to $B(3, 0, 0)$ with $P - 60$. The second iteration brings us from $B(3, 0, 0)$ to $C(\frac{19}{7}, \frac{6}{7}, 0)$ with $P = 64\frac{4}{7}$, and the third iteration brings us from $C(\frac{19}{7}, \frac{6}{7}, 0)$ to the point $H(2, 1, 1)$ with an optimal value of 70 for P.

APPLICATION

EXAMPLE 5

The Ace Novelty Company has determined that the profits are \$6, \$5, and \$4 for each type-A, type-B, and type-C souvenir that it plans to produce. To manufacture a type-A souvenir requires 2 minutes on machine I, 1 minute on machine II, and 2 minutes on machine III. A type-B souvenir requires 1 minute on machine I, 3 minutes on machine II, and 1 minute on machine III. A type-C souvenir requires 1 minute on machine I and 2 minutes on each of machines II and III. Each day there are 3 hours available on machine I, 5 hours available on machine II, and 4 hours available on machine III for manufacturing these souvenirs. How many souvenirs of each type should Ace Novelty make per day in order to maximize its profit? (Compare with Example 1, Section 2.1.)

SOLUTION ✓

The given information may be tabulated as follows:

	Type A	Type B	Type C	Time Available (min)
Machine I	2	1	1	180
Machine II	1	3	2	300
Machine III	2	1	2	240
Profit per Unit (\$)	6	5	4	

Let x, y, and z denote the respective numbers of type-A, type-B, and type-C souvenirs to be made. The total amount of time that machine I is used is given by $2x + y + z$ minutes and must not exceed 180 minutes. Thus, we have the inequality

$$2x + y + z \leq 180$$

Similar considerations on the use of machines II and III lead to the inequalities

$$\begin{aligned} x + 3y + 2z &\leq 300 \\ 2x + y + 2z &\leq 240 \end{aligned}$$

The profit resulting from the sale of the souvenirs produced is given by

$$P = 6x + 5y + 4z$$

The mathematical formulation of this problem has led to the following standard linear programming problem: Maximize the objective (profit) function $P = 6x + 5y + 4z$ subject to

$$\begin{aligned} 2x + y + z &\leq 180 \\ x + 3y + 2z &\leq 300 \\ 2x + y + 2z &\leq 240 \\ x \geq 0, y \geq 0, z &\geq 0 \end{aligned}$$

Introducing the slack variables u, v, and w gives the system of linear equations

$$\begin{aligned} 2x + y + z + u \qquad\qquad &= 180 \\ x + 3y + 2z \quad + v \qquad &= 300 \\ 2x + y + 2z \qquad + w &= 240 \\ -6x - 5y - 4z \qquad\quad + P &= 0 \end{aligned}$$

The tableaus resulting from the use of the simplex algorithm are

	x	y	z	u	v	w	P	Constant	ratio
Pivot row →	(2)	1	1	1	0	0	0	180	$\frac{180}{2} = 90$
	1	3	2	0	1	0	0	300	$\frac{300}{1} = 300$
	2	1	2	0	0	1	0	240	$\frac{240}{2} = 120$
	−6	−5	−4	0	0	0	1	0	
	↑ Pivot column								

$\xrightarrow{\frac{1}{2}R_1}$

x	y	z	u	v	w	P	Constant
(1)	$\frac{1}{2}$	$\frac{1}{2}$	$\frac{1}{2}$	0	0	0	90
1	3	2	0	1	0	0	300
2	1	2	0	0	1	0	240
−6	−5	−4	0	0	0	1	0

$\xrightarrow[R_4+6R_1]{R_2-R_1 \quad R_3-2R_1}$

	x	y	z	u	v	w	P	Constant	ratio
	1	$\frac{1}{2}$	$\frac{1}{2}$	$\frac{1}{2}$	0	0	0	90	$\frac{90}{1/2}=180$
Pivot row →	0	$\textcircled{\frac{5}{2}}$	$\frac{3}{2}$	$-\frac{1}{2}$	1	0	0	210	$\frac{210}{5/2}=84$
	0	0	1	-1	0	1	0	60	
	0	-2	-1	3	0	0	1	540	
		↑ Pivot column							

$\xrightarrow{\frac{2}{5}R_2}$

x	y	z	u	v	w	P	Constant
1	$\frac{1}{2}$	$\frac{1}{2}$	$\frac{1}{2}$	0	0	0	90
0	①	$\frac{3}{5}$	$-\frac{1}{5}$	$\frac{2}{5}$	0	0	84
0	0	1	-1	0	1	0	60
0	-2	-1	3	0	0	1	540

$\xrightarrow[R_4+2R_2]{R_1-\frac{1}{2}R_2}$

x	y	z	u	v	w	P	Constant
1	0	$\frac{1}{5}$	$\frac{3}{5}$	$-\frac{1}{5}$	0	0	48
0	1	$\frac{3}{5}$	$-\frac{1}{5}$	$\frac{2}{5}$	0	0	84
0	0	1	-1	0	1	0	60
0	0	$\frac{1}{5}$	$\frac{13}{5}$	$\frac{4}{5}$	0	1	708

From the final simplex tableau, we read off the solution

$$x = 48, \quad y = 84, \quad z = 0, \quad u = 0, \quad v = 0, \quad w = 60, \quad P = 708$$

Thus, in order to maximize its profit, Ace Novelty should produce 48 type-A souvenirs, 84 type-B souvenirs, and no type-C souvenirs. The resulting profit is $708 per day. The value of the slack variable $w = 60$ tells us that 1 hour of the available time on machine III is left unused. ■■■■

Interpreting Our Results It is instructive to compare the results obtained here with those obtained in Example 7, Section 2.2. Recall that in order to use all of the machine time available on each of the three machines, Ace Novelty had to produce 36 type-A, 48 type-B, and 60 type-C souvenirs. This would have resulted in a profit of $696. Example 5 shows how, through the optimal use of equipment, a company may boost its profit and at the same time reduce machine wear!

Multiple Solutions and Unbounded Solutions

As we saw in Section 3.3, a linear programming problem may have infinitely many solutions. We also saw that a linear programming problem may have no solution or an unbounded solution. How do we spot each of these phenomena when using the simplex method to solve a problem?

A linear programming problem may have infinitely many solutions if and only if the last row to the left of the vertical line of the final simplex tableau has a zero in a column that is not a unit column. Next, a linear programming problem will have no solution if the simplex method breaks down at some stage. For example, if at some stage there are no nonnegative ratios in our computation, then the linear programming problem has no solution.

Group Discussion

Consider the linear programming problem

$$\begin{aligned} \text{Maximize} \quad & P = 4x + 6y \\ \text{subject to} \quad & 2x + y \le 10 \\ & 2x + 3y \le 18 \\ & x \ge 0, y \ge 0 \end{aligned}$$

1. Sketch the feasible set for the linear programming problem.
2. Use the method of corners to show that there are infinitely many optimal solutions. What are they?
3. Use the simplex method to solve the problem as follows:
 a. Perform one iteration on the initial simplex tableau and conclude that you have arrived at an optimal solution. What is the value of P and where is it attained? Compare this result with that obtained in step 2.
 b. Observe that the tableau obtained in part (a) indicates that there are infinitely many solutions (see the comment on multiple solutions on this page). Now perform another iteration on the simplex tableau using the x-column as the pivot column. Interpret the final tableau.

SELF-CHECK EXERCISES 4.1

1. Solve the following linear programming problem by the simplex method:

$$\begin{aligned} \text{Maximize} \quad & P = 2x + 3y + 6z \\ \text{subject to} \quad & 2x + 3y + z \le 10 \\ & x + y + 2z \le 8 \\ & 2y + 3z \le 6 \\ & x \ge 0, y \ge 0, z \ge 0 \end{aligned}$$

2. The LaCrosse Iron Works makes two models of cast-iron fireplace grates, model A and model B. Producing one model A grate requires 20 lb of cast iron and 20 min of labor, whereas producing one model B grate requires 30 lb of cast iron and 15 min of labor. The profit for a model A grate is \$6, and the profit for a model B grate is \$8. There are 7200 lb of cast iron and 100 hr of labor available per week. Because of a surplus from the previous week, the proprietor has decided that he should make no more than 150 units of model A grates this week. Determine how many of each model he should make in order to maximize his profits.

Solutions to Self-Check Exercises 4.1 can be found on page 244.

4.1 Exercises

In Exercises 1–9, determine whether the given simplex tableau is in final form. If so, find the solution to the associated regular linear programming problem. If not, find the pivot element to be used in the next iteration of the simplex method.

1.

x	y	u	v	P	Constant
0	1	$\frac{5}{7}$	$-\frac{1}{7}$	0	$\frac{20}{7}$
1	0	$-\frac{3}{7}$	$\frac{2}{7}$	0	$\frac{30}{7}$
0	0	$\frac{13}{7}$	$\frac{3}{7}$	1	$\frac{220}{7}$

2.

x	y	u	v	P	Constant
1	1	1	0	0	6
1	0	−1	1	0	2
3	0	5	0	1	30

3.

x	y	u	v	P	Constant
0	$\frac{1}{2}$	1	$-\frac{1}{2}$	0	2
1	$\frac{1}{2}$	0	$\frac{1}{2}$	0	4
0	$-\frac{1}{2}$	0	$\frac{3}{2}$	1	12

4.

x	y	z	u	v	w	P	Constant
3	0	5	1	1	0	0	28
2	1	3	0	1	0	0	16
2	0	8	0	3	0	1	48

5.

x	y	z	u	v	w	P	Constant
1	$-\frac{1}{3}$	0	$\frac{1}{3}$	0	$-\frac{2}{3}$	0	$\frac{1}{3}$
0	2	0	0	1	1	0	6
0	$\frac{2}{3}$	1	$\frac{1}{3}$	0	$\frac{1}{3}$	0	$\frac{13}{3}$
0	4	0	1	0	2	1	17

6.

x	y	z	u	v	w	P	Constant
$\frac{1}{2}$	0	$\frac{1}{4}$	1	$-\frac{1}{4}$	0	0	$\frac{19}{2}$
$\frac{1}{2}$	1	$\frac{3}{4}$	0	$\frac{1}{4}$	0	0	$\frac{21}{2}$
2	0	3	0	0	1	0	30
−1	0	$-\frac{1}{2}$	6	$\frac{3}{2}$	0	1	63

7.

x	y	z	s	t	u	v	P	Constant
$\frac{5}{2}$	3	0	1	0	0	−4	0	46
1	0	0	0	1	0	0	0	9
0	1	0	0	0	1	0	0	12
0	0	1	0	0	0	1	0	6
−180	−200	0	0	0	0	300	1	1800

8.

x	y	z	s	t	u	v	P	Constant
1	0	0	$\frac{2}{5}$	0	$-\frac{6}{5}$	$-\frac{8}{5}$	0	4
0	0	0	$-\frac{2}{5}$	1	$\frac{6}{5}$	$\frac{8}{5}$	0	5
0	1	0	0	0	1	0	0	12
0	0	1	0	0	0	1	0	6
0	0	0	72	0	−16	12	1	4920

9.

x	y	z	u	v	P	Constant
1	1	$\frac{3}{5}$	0	$\frac{1}{5}$	0	30
0	0	$-\frac{19}{5}$	1	$-\frac{3}{5}$	0	10
0	0	$\frac{26}{5}$	0	$\frac{2}{5}$	1	60

In Exercises 10–23, solve each linear programming problem by the simplex method.

10. Maximize $P = 5x + 3y$

subject to
$$\begin{aligned} x + y &\leq 80 \\ 3x &\leq 90 \\ x \geq 0, y &\geq 0 \end{aligned}$$

11. Maximize $P = 10x + 12y$

subject to
$$\begin{aligned} x + 2y &\leq 12 \\ 3x + 2y &\leq 24 \\ x \geq 0, y &\geq 0 \end{aligned}$$

12. Maximize $P = 5x + 4y$

subject to
$$\begin{aligned} 3x + 5y &\leq 78 \\ 4x + y &\leq 36 \\ x \geq 0, y &\geq 0 \end{aligned}$$

13. Maximize $P = 4x + 6y$

subject to
$$\begin{aligned} 3x + y &\leq 24 \\ 2x + y &\leq 18 \\ x + 3y &\leq 24 \\ x \geq 0, y &\geq 0 \end{aligned}$$

14. Maximize $P = 15x + 12y$
subject to
$$\begin{aligned} x + y &\le 12 \\ 3x + y &\le 30 \\ 10x + 7y &\le 70 \\ x \ge 0, y &\ge 0 \end{aligned}$$

15. Maximize $P = 3x + 4y + 5z$
subject to
$$\begin{aligned} x + y + z &\le 8 \\ 3x + 2y + 4z &\le 24 \\ x \ge 0, y \ge 0, z &\ge 0 \end{aligned}$$

16. Maximize $P = 3x + 3y + 4z$
subject to
$$\begin{aligned} x + y + 3z &\le 15 \\ 4x + 4y + 3z &\le 65 \\ x \ge 0, y \ge 0, z &\ge 0 \end{aligned}$$

17. Maximize $P = 3x + 4y + z$
subject to
$$\begin{aligned} 3x + 10y + 5z &\le 120 \\ 5x + 2y + 8z &\le 6 \\ 8x + 10y + 3z &\le 105 \\ x \ge 0, y \ge 0, z &\ge 0 \end{aligned}$$

18. Maximize $P = x + 2y - z$
subject to
$$\begin{aligned} 2x + y + z &\le 14 \\ 4x + 2y + 3z &\le 28 \\ 2x + 5y + 5z &\le 30 \\ x \ge 0, y \ge 0, z &\ge 0 \end{aligned}$$

19. Maximize $P = 4x + 6y + 5z$
subject to
$$\begin{aligned} x + y + z &\le 20 \\ 2x + 4y + 3z &\le 42 \\ 2x \quad\quad + 3z &\le 30 \\ x \ge 0, y \ge 0, z &\ge 0 \end{aligned}$$

20. Maximize $P = x + 4y - 2z$
subject to
$$\begin{aligned} 3x + y - z &\le 80 \\ 2x + y - z &\le 40 \\ -x + y + z &\le 80 \\ x \ge 0, y \ge 0, z &\ge 0 \end{aligned}$$

21. Maximize $P = 12x + 10y + 5z$
subject to
$$\begin{aligned} 2x + y + z &\le 10 \\ 3x + 5y + z &\le 45 \\ 2x + 5y + z &\le 40 \\ x \ge 0, y \ge 0, z &\ge 0 \end{aligned}$$

22. Maximize $P = 2x + 6y + 6z$
subject to
$$\begin{aligned} 2x + y + 3z &\le 10 \\ 4x + y + 2z &\le 56 \\ 6x + 4y + 3z &\le 126 \\ 2x + y + z &\le 32 \\ x \ge 0, y \ge 0, z &\ge 0 \end{aligned}$$

23. Maximize $P = 24x + 16y + 23z$
subject to
$$\begin{aligned} 2x + y + 2z &\le 7 \\ 2x + 3y + z &\le 8 \\ x + 2y + 3z &\le 7 \\ x \ge 0, y \ge 0, z &\ge 0 \end{aligned}$$

24. Rework Example 3 using the y-column as the pivot column in the first iteration of the simplex method.

25. Show that the following linear programming problem

Maximize $P = 2x + 2y - 4z$
subject to
$$\begin{aligned} 3x + 3y - 2z &\le 100 \\ 5x + 5y + 3z &\le 150 \\ x \ge 0, y \ge 0, z &\ge 0 \end{aligned}$$

has optimal solutions $x = 30$, $y = 0$, $z = 0$, $P = 60$ and $x = 0$, $y = 30$, $z = 0$, and $P = 60$.

26. Manufacturing—Production Scheduling A company manufactures two products, A and B, on two machines, I and II. It has been determined that the company will realize a profit of $3/unit of product A and a profit of $4/unit of product B. To manufacture a unit of product A requires 6 min on machine I and 5 min on machine II. To manufacture a unit of product B requires 9 min on machine I and 4 min on machine II. There are 5 hr of machine time available on machine I and 3 hr of machine time available on machine II in each work shift. How many units of each product should be produced in each shift to maximize the company's profit?

27. Manufacturing—Production Scheduling The National Business Machines Corporation manufactures two models of fax machines: A and B. Each model A costs $200 to make, and each model B costs $300. The profits are $25 for each model A and $40 for each model B fax machine. If the total number of fax machines demanded per month does not exceed 2500 and the company has earmarked no more than $600,000/month for manufacturing costs, find how many units of each model National should make each month in order to maximize its monthly profits.

28. Manufacturing—Production Scheduling Kane Manufacturing has a division that produces two models of hibachis, model A and model B. To produce each model A hibachi requires 3 lb of cast iron and 6 min of labor. To produce each model B hibachi requires 4 lb of cast iron and 3 min of labor. The profit for each model A hibachi is $2, and the profit for each model B hibachi is $1.50. If 1000 lb of cast iron and 20 hr of labor are available for the production of hibachis per day, how many hibachis of each model should the division produce in order to help maximize Kane's profits?

29. **AGRICULTURE—CROP PLANNING** A farmer has 150 acres of land suitable for cultivating crops A and B. The cost of cultivating crop A is $40/acre, whereas that of crop B is $60/acre. The farmer has a maximum of $7400 available for land cultivation. Each acre of crop A requires 20 hr of labor, and each acre of crop B requires 25 hr of labor. The farmer has a maximum of 3300 hr of labor available. If he expects to make a profit of $150/acre on crop A and $200/acre on crop B, how many acres of each crop should he plant in order to maximize his profit?

30. **MANUFACTURING—PRODUCTION SCHEDULING** A company manufactures products A, B, and C. Each product is processed in three departments: I, II, and III. The total available labor-hours per week for departments I, II, and III are 900, 1080, and 840, respectively. The time requirements (in hours per unit) and profit per unit for each product are as follows:

	Product A	Product B	Product C
Dept. I	2	1	2
Dept. II	3	1	2
Dept. III	2	2	1
Profit	$18	$12	$15

How many units of each product should the company produce in order to maximize its profit?

31. **ADVERTISING—TELEVISION COMMERCIALS** As part of a campaign to promote its annual clearance sale, the Excelsior Company decided to buy television advertising time on Station KAOS. Excelsior's television advertising budget is $102,000. Morning time costs $3000/minute, afternoon time costs $1000/minute, and evening (prime) time costs $12,000/minute. Because of previous commitments, KAOS cannot offer Excelsior more than 6 min of prime time or more than a total of 25 min of advertising time over the 2 weeks in which the commercials are to be run. KAOS estimates that morning commercials are seen by 200,000 people, afternoon commercials are seen by 100,000 people, and evening commercials are seen by 600,000 people. How much morning, afternoon, and evening advertising time should Excelsior buy to maximize exposure of its commercials?

32. **INVESTMENTS—ASSET ALLOCATION** Sharon has a total of $200,000 to invest in three types of mutual funds: growth, balanced, and income funds. Growth funds have a rate of return of 12%/year, balanced funds have a rate of return of 10%/year, and income funds have a return of 6%/year. The growth, balanced, and income mutual funds are assigned risk factors of 0.1, 0.06, and 0.02, respectively. Sharon has decided that at least half of her total portfolio is to be in income funds and at least a quarter of it in balanced funds. She has also decided that the average risk factor for her investment should not exceed 0.05. Determine how much Sharon should invest in each type of fund in order to realize a maximum return on her investment.
Hint: The average risk factor for the investment is given by $0.1x + 0.06y + 0.02z \leq 0.05(x + y + z)$.

33. **MANUFACTURING—PRODUCTION CONTROL** The Custom Office Furniture Company is introducing a new line of executive desks made from a specially selected grade of walnut. Initially, three models—A, B, and C—are to be marketed. Each model A desk requires $1\frac{1}{4}$ hr for fabrication, 1 hr for assembly, and 1 hr for finishing; each model B desk requires $1\frac{1}{2}$ hr for fabrication, 1 hr for assembly, and 1 hr for finishing; each model C desk requires $1\frac{1}{2}$ hr, $\frac{3}{4}$ hr, and $\frac{1}{2}$ hr for fabrication, assembly, and finishing, respectively. The profit on each model A desk is $26, the profit on each model B desk is $28, and the profit on each model C desk is $24. The total time available in the fabrication department, the assembly department, and the finishing department in the first month of production is 310 hr, 205 hr, and 190 hr, respectively. To maximize Custom's profit, how many desks of each model should be made in the month?

34. **MANUFACTURING—PREFABRICATED HOUSING PRODUCTION** Boise Lumber has decided to enter the lucrative prefabricated housing business. Initially, it plans to offer three models: standard, deluxe, and luxury. Each house is prefabricated and partially assembled in the factory, and the final assembly is completed on site. The dollar amount of building material required, the amount of labor required in the factory for prefabrication and partial assembly, the amount of on-site labor required, and the profit per unit are as follows:

	Standard Model	Deluxe Model	Luxury Model
Material ($)	6,000	8,000	10,000
Factory Labor (hr)	240	220	200
On-Site Labor (hr)	180	210	300
Profit ($)	3,400	4,000	5,000

(continued on p. 244)

Portfolio

HARLEY LANCE KAPLAN

TITLE: Registered Investment Advisor, Certified Financial Planner
INSTITUTION: Beta Industries

Beta Industries' clients are people who "want to invest their money more efficiently," notes Harley Kaplan. He helps them "identify their available assets and define their financial objectives." He then assembles balanced portfolios to meet their needs.

Kaplan must frequently educate clients on the potential risks and rewards investors face. He patiently explains that, based on historical and statistical data and current risk factors, he can *project,* not predict, "possible future consequences." Even then, any investment strategy has to be tempered by an individual client's tolerance for risk.

Age often plays a pivotal role in a client's tolerance level. A 25-year-old investor with only $10,000 and quite a few working years remaining might be more willing to take greater risks, relying on either a growth or an aggressive growth vehicle. (For the client with over $250,000 to invest, Kaplan recommends a personal money manager rather than a vehicle such as a mutual fund, which has many investors contributing relatively small amounts.)

Data show that with a growth investment, a client risks gaining or losing as much as 10% to 25% of the principal. An aggressive growth investment can gain or lose up to 50%. Both types realize a solid rate of return over inflation, from a minimum of 10% to as much as 35% in the best scenarios. The key to success is keeping the funds in place for at least 5 to 7 years to ride out the cycles that typically affect the economy.

Kaplan's strategy for a 62-year-old, who has only a few more years to continue earning, might rely heavily on the safety and liquidity offered by NOW accounts (combined checking and savings) and income vehicles such as bonds with moderate risk and a 3% and 5% rate of return over inflation.

With a $120,000 principal, Kaplan might recommend that 70% be placed into safety and income, whereas the balance be invested as a hedge against inflation (20% into growth and 10% into aggressive growth). Kaplan stresses that inflation robs everyone of purchasing power, especially those on fixed incomes. For example, if inflation remains constant at 3% per year, $10,000 in interest earned by a retiree would buy only $7374 in goods and services after 10 years.

Investment vehicles vary widely. When choosing among investments, Kaplan uses a number of comparisons, including the net rate of return, how each investment affects a client's tax situation, and potential risk. For the client in the 40% tax bracket, the net return on a corporate bond paying 10% and a tax-free municipal paying 6% is the same. The only substantial difference may be the level of risk each bond carries. Everything else being equal, the safer bond is the logical choice.

Through the years, Kaplan has learned to weigh not just tangible assets but intangible emotions—to tailor an investment strategy suited to each individual's needs.

Beta Industries is an investment firm, not a broad-based manufacturing conglomerate as its name implies. Its clients range from individuals with significant net worth to Fortune 500 corporations and nonprofit organizations with growing endowments. Kaplan and his partners help clients meet their long-term financial goals by recommending sound investment strategies.

Using Technology

THE SIMPLEX METHOD: SOLVING MAXIMIZATION PROBLEMS

The graphing calculator can be used to solve a linear programming problem by the simplex method as illustrated in Example 1.

EXAMPLE 1 (Refer to Example 5, Section 4.1.) The problem reduces to the following linear programming problem:

$$\begin{aligned} \text{Maximize} \quad & P = 6x + 5y + 4z \\ \text{subject to} \quad & 2x + y + z \leq 180 \\ & x + 3y + 2z \leq 300 \\ & 2x + y + 2z \leq 240 \\ & x \geq 0, y \geq 0, z \geq 0 \end{aligned}$$

With u, v, and w as slack variables, we are led to the following sequence of simplex tableaus, where the first tableau is entered as the matrix A:

	x	y	z	u	v	w	P	Constant	ratio
Pivot row →	(2)	1	1	1	0	0	0	180	$\frac{180}{2} = 90$
	1	3	2	0	1	0	0	300	$\frac{300}{1} = 300$
	2	1	2	0	0	1	0	240	$\frac{240}{2} = 120$
	−6	−5	−4	0	0	0	1	0	
	↑ Pivot column								

$\xrightarrow{\text{*row}(\frac{1}{2}, A, 1) \blacktriangleright B}$

x	y	z	u	v	w	P	Constant
(1)	0.5	0.5	0.5	0	0	0	90
1	3	2	0	1	0	0	300
2	1	2	0	0	1	0	240
−6	−5	−4	0	0	0	1	0

$\xrightarrow[\substack{\text{*row+}(-2, C, 1, 3) \blacktriangleright B \\ \text{*row+}(6, B, 1, 4) \blacktriangleright C}]{\text{*row+}(-1, B, 1, 2) \blacktriangleright C}$

	x	y	z	u	v	w	P	Constant	ratio
	1	0.5	0.5	0.5	0	0	0	90	$\frac{90}{0.5} = 180$
Pivot row →	0	(2.5)	1.5	−0.5	1	0	0	210	$\frac{210}{2.5} = 84$
	0	0	1	−1	0	1	0	60	
	0	−2	−1	3	0	0	1	540	
		↑ Pivot column							

$\xrightarrow{\text{*row}(\frac{1}{2.5}, C, 2) \blacktriangleright B}$

x	y	z	u	v	w	P	Constant
1	0.5	0.5	0.5	0	0	0	90
0	①	0.6	−0.2	0.4	0	0	84
0	0	1	−1	0	1	0	60
0	−2	−1	3	0	0	1	540

***row+**(−0.5, *B*, 2, 1) ▶ *C*
***row+**(2, *C*, 2, 4) ▶ *B*

x	y	z	u	v	w	P	Constant
1	0	0.2	0.6	−0.2	0	0	48
0	1	0.6	−0.2	0.4	0	0	84
0	0	1	−1	0	1	0	60
0	0	0.2	2.6	0.8	0	1	708

The final simplex tableau is the same as the one obtained earlier. We see that $x = 48$, $y = 84$, $z = 0$, and $P = 708$. So, Ace Novelty should produce 48 type-A souvenirs, 84 type-B souvenirs, and no type-C souvenirs—resulting in a profit of $708 per day.

Exercises

In Exercises 1–4, use a graphing calculator to solve the linear programming problem by the simplex method.

1. Maximize $P = 2x + 3y + 4z + 2w$
   subject to
   $$\begin{aligned} x + 2y + 3z + 2w &\le 6 \\ 2x + 4y + z - w &\le 4 \\ 3x + 2y - 2z + 3w &\le 12 \\ x \ge 0, y \ge 0, z \ge 0, w &\ge 0 \end{aligned}$$

2. Maximize $P = 3x + 2y + 2z + w$
   subject to
   $$\begin{aligned} 2x + y - z + 2w &\le 8 \\ 2x - y + 2z + 3w &\le 20 \\ x + y + z + 2w &\le 8 \\ 4x - 2y + z + 3w &\le 24 \\ x \ge 0, y \ge 0, z \ge 0, w &\ge 0 \end{aligned}$$

3. Maximize $P = x + y + 2z + 3w$
   subject to
   $$\begin{aligned} 3x + 6y + 4z + 2w &\le 12 \\ x + 4y + 8z + 4w &\le 16 \\ 2x + y + 4z + w &\le 10 \\ x \ge 0, y \ge 0, z \ge 0, w &\ge 0 \end{aligned}$$

4. Maximize $P = 2x + 4y + 3z + 5w$
   subject to
   $$\begin{aligned} x - 2y + 3z + 4w &\le 8 \\ 2x + 2y + 4z + 6w &\le 12 \\ 3x + 2y + z + 5w &\le 10 \\ 2x + 8y - 2z + 6w &\le 24 \\ x \ge 0, y \ge 0, z \ge 0, w &\ge 0 \end{aligned}$$

For the first year's production, a sum of $8,200,000 is budgeted for the building material; the number of labor-hours available for work in the factory (for prefabrication and partial assembly) is not to exceed 218,000 hr; and the amount of labor for on-site work is to be less than or equal to 237,000 labor-hours. Determine how many houses of each type Boise should produce (market research has confirmed that there should be no problems with sales) to maximize its profit from this new venture.

35. **MANUFACTURING—COLD FORMULA PRODUCTION** Bayer Pharmaceutical produces three kinds of cold formulas: I, II, and III. It takes 2.5 hr to produce 1000 bottles of formula I, 3 hr to produce 1000 bottles of formula II, and 4 hr to produce 1000 bottles of formula III. The profits for each 1000 bottles of formula I, formula II, and formula III are $180, $200, and $300, respectively. Suppose, for a certain production run, there are enough ingredients on hand to make at most 9000 bottles of formula I, 12,000 bottles of formula II, and 6000 bottles of formula III. Furthermore, suppose the time for the production run is limited to a maximum of 70 hr. How many bottles of each formula should be produced in this production run so that the profit is maximized?

In Exercises 36–38, determine whether the statement is true or false. If it is true, explain why it is true. If it is false, give an example to show why it is false.

36. If at least one of the coefficients $a_1, a_2, \ldots, a_n$ of the objective function $P = a_1x_1 + a_2x_2 + \cdots + a_nx_n$ is positive, then $(0, 0, \ldots, 0)$ cannot be the optimal solution of the standard (maximization) linear programming problem.
37. Choosing the pivot row by requiring that the ratio associated with that row be the smallest ensures that the iteration will not take us from a feasible point to a nonfeasible point.
38. If, at any stage of an iteration of the simplex method, it is not possible to compute the ratios (division by zero) or the ratios are negative, then we can conclude that the standard linear programming problem may have no solution.

SOLUTIONS TO SELF-CHECK EXERCISES 4.1

1. Introducing the slack variables u, v, and w, we obtain the system of linear equations

$$\begin{aligned} 2x + 3y + z + u \qquad\qquad &= 10 \\ x + y + 2z \quad + v \qquad &= 8 \\ 2y + 3z \qquad + w \quad &= 6 \\ -2x - 3y - 6z \qquad\qquad + P &= 0 \end{aligned}$$

The initial simplex tableau and the successive tableaus resulting from the use of the simplex procedure follow:

	x	y	z	u	v	w	P	Constant	ratio
	2	3	1	1	0	0	0	10	$\frac{10}{1} = 10$
	1	1	2	0	1	0	0	8	$\frac{8}{2} = 4$
Pivot row →	0	2	(3)	0	0	1	0	6	$\frac{6}{3} = 2$
	−2	−3	−6	0	0	0	1	0	
			↑ Pivot column						

$\xrightarrow{\frac{1}{3}R_3}$

x	y	z	u	v	w	P	Constant
2	3	1	1	0	0	0	10
1	1	2	0	1	0	0	8
0	$\frac{2}{3}$	(1)	0	0	$\frac{1}{3}$	0	2
−2	−3	−6	0	0	0	1	0

$\xrightarrow[\substack{R_2 - 2R_3 \\ R_4 + 6R_3}]{R_1 - R_3}$

x	y	z	u	v	w	P	Constant	ratio
2	$\frac{7}{3}$	0	1	0	$-\frac{1}{3}$	0	8	$\frac{8}{2}=4$
(1)	$-\frac{1}{3}$	0	0	1	$-\frac{2}{3}$	0	4	$\frac{4}{1}=4$
0	$\frac{2}{3}$	1	0	0	$\frac{1}{3}$	0	2	—
−2	1	0	0	0	2	1	12	

Pivot row → second row; Pivot column ↑ x column; $\xrightarrow[R_4 + 2R_2]{R_1 - 2R_2}$

x	y	z	u	v	w	P	Constant
0	3	0	1	−2	1	0	0
1	$-\frac{1}{3}$	0	0	1	$-\frac{2}{3}$	0	4
0	$\frac{2}{3}$	1	0	0	$\frac{1}{3}$	0	2
0	$\frac{1}{3}$	0	0	2	$\frac{2}{3}$	1	20

All entries in the last row are nonnegative, and the tableau is final. We conclude that $x = 4$, $y = 0$, $z = 2$, and $P = 20$.

2. Let x denote the number of model A grates and y the number of model B grates to be made this week. Then the profit function to be maximized is given by

$$P = 6x + 8y$$

The limitations on the availability of material and labor may be expressed by the linear inequalities

$$20x + 30y \leq 7200 \quad \text{or} \quad 2x + 3y \leq 720$$
$$20x + 15y \leq 6000 \quad \text{or} \quad 4x + 3y \leq 1200$$

Finally, the condition that no more than 150 units of model A grates be made per week may be expressed by the linear inequality

$$x \leq 150$$

Thus, we are led to the following linear programming problem:

$$\begin{aligned} \text{Maximize} \quad & P = 6x + 8y \\ \text{subject to} \quad & 2x + 3y \leq 720 \\ & 4x + 3y \leq 1200 \\ & x \leq 150 \\ & x \geq 0, y \geq 0 \end{aligned}$$

To solve this problem, we introduce slack variables u, v, and w and use the simplex method, obtaining the following sequence of simplex tableaus:

x	y	u	v	w	P	Constant	ratio
2	(3)	1	0	0	0	720	$\frac{720}{3} = 240$
4	3	0	1	0	0	1200	$\frac{1200}{3} = 400$
1	0	0	0	1	0	150	—
−6	−8	0	0	0	1	0	

Pivot row → first row; Pivot column ↑ y column

$\xrightarrow{\frac{1}{3}R_1}$

x	y	u	v	w	P	Constant
$\frac{2}{3}$	(1)	$\frac{1}{3}$	0	0	0	240
4	3	0	1	0	0	1200
1	0	0	0	1	0	150
−6	−8	0	0	0	1	0

$\xrightarrow[R_4 + 8R_1]{R_2 - 3R_1}$

	x	y	u	v	w	P	Constant	ratio
	$\frac{2}{3}$	1	$\frac{1}{3}$	0	0	0	240	$\frac{240}{2/3} = 360$
	2	0	−1	1	0	0	480	$\frac{480}{2} = 240$
Pivot row →	(1)	0	0	0	1	0	150	$\frac{150}{1} = 150$
	$-\frac{2}{3}$	0	$\frac{8}{3}$	0	0	1	1920	
	↑ Pivot column							

$\xrightarrow[\substack{R_2 - 2R_3 \\ R_4 + \frac{2}{3}R_3}]{R_1 - \frac{2}{3}R_3}$

x	y	u	v	w	P	Constant
0	1	$\frac{1}{3}$	0	$-\frac{2}{3}$	0	140
0	0	−1	1	−2	0	180
1	0	0	0	1	0	150
0	0	$\frac{8}{3}$	0	$\frac{2}{3}$	1	2020

The last tableau is final, and we see that $x = 150$, $y = 140$, and $P = 2020$. Therefore, LaCrosse should make 150 model A grates and 140 model B grates this week. The profit will be \$2020.

4.2 The Simplex Method: Standard Minimization Problems

MINIMIZATION WITH ≤ CONSTRAINTS

In the last section we developed a procedure, called the simplex method, for solving standard linear programming problems. Recall that a standard maximization problem satisfies three conditions:

1. The objective function is to be maximized.
2. All the variables involved are nonnegative.
3. Each linear constraint may be written so that the expression involving the variables is less than or equal to a nonnegative constant.

In this section we see how the simplex method may be used to solve certain classes of problems that are not necessarily standard maximization problems. In particular, we see how a modified procedure may be used to solve problems involving the minimization of objective functions.

We begin by considering the class of linear programming problems that calls for the minimization of objective functions but otherwise satisfies Conditions 2 and 3 for standard maximization problems. The method used to solve these problems is illustrated in the following example.

EXAMPLE 1

$$\begin{aligned} \text{Minimize} \quad & C = -2x - 3y \\ \text{subject to} \quad & 5x + 4y \leq 32 \\ & x + 2y \leq 10 \\ & x \geq 0, y \geq 0 \end{aligned}$$

SOLUTION ✓ This problem involves the minimization of the objective function and is accordingly not a standard maximization problem. Note, however, that all other conditions for a standard maximization problem hold true. To solve a problem of this type, we observe that minimizing the objective function C is equivalent to maximizing the objective function $P = -C$. Thus, the solution to this problem may be found by solving the following associated standard maximization problem: Maximize $P = 2x + 3y$ subject to the given constraints. Using the simplex method with u and v as slack variables, we obtain the following sequence of simplex tableaus:

	x	y	u	v	P	Constant	ratio
	5	4	1	0	0	32	$\frac{32}{4} = 8$
Pivot row →	1	(2)	0	1	0	10	$\frac{10}{2} = 5$
	−2	−3	0	0	1	0	
		↑ Pivot column					

	x	y	u	v	P	Constant
$\xrightarrow{\frac{1}{2}R_2}$	5	4	1	0	0	32
	$\frac{1}{2}$	(1)	0	$\frac{1}{2}$	0	5
	−2	−3	0	0	1	0

	x	y	u	v	P	Constant	ratio
Pivot row →	(3)	0	1	−2	0	12	$\frac{12}{3} = 4$
$\xrightarrow[R_3 + 3R_2]{R_1 - 4R_2}$	$\frac{1}{2}$	1	0	$\frac{1}{2}$	0	5	$\frac{5}{1/2} = 10$
	$-\frac{1}{2}$	0	0	$\frac{3}{2}$	1	15	
	↑ Pivot column						

	x	y	u	v	P	Constant
$\xrightarrow{\frac{1}{3}R_1}$	(1)	0	$\frac{1}{3}$	$-\frac{2}{3}$	0	4
	$\frac{1}{2}$	1	0	$\frac{1}{2}$	0	5
	$-\frac{1}{2}$	0	0	$\frac{3}{2}$	1	15

Group Discussion

Refer to Example 1.

1. Sketch the feasible set S for the linear programming problem.
2. Solve the problem using the method of corners.
3. Indicate on S the corner points corresponding to each iteration of the simplex procedure and trace the path leading to the optimal solution.

$\xrightarrow[R_3 + \frac{1}{2}R_1]{R_2 - \frac{1}{2}R_1}$

x	y	u	v	P	Constant
1	0	$\frac{1}{3}$	$-\frac{2}{3}$	0	4
0	1	$-\frac{1}{6}$	$\frac{5}{6}$	0	3
0	0	$\frac{1}{6}$	$\frac{7}{6}$	1	17

The last tableau is in final form. The solution to the standard maximization problem associated with the given linear programming problem is $x = 4$, $y = 3$, and $P = 17$, so the required solution is given by $x = 4$, $y = 3$, and $C = -17$. You may verify that the solution is correct by using the method of corners. ■■■■

THE DUAL PROBLEM

Another special class of linear programming problems we encounter in practical applications is characterized by the following conditions:

1. The objective function is to be *minimized.*
2. All the variables involved are nonnegative.
3. Each linear constraint may be written so that the expression involving the variables is *greater than* or equal to a constant.

Such problems are called **standard minimization problems.**

A convenient method for solving this type of problem is based on the following observation. Each maximization linear programming problem is associated with a minimization problem, and vice versa. For the purpose of identification, the given problem is called the **primal problem;** the problem related to it is called the **dual problem.** The following example illustrates the technique for constructing the dual of a given linear programming problem.

EXAMPLE 2 Write the dual problem associated with the following problem:

Minimize the objective function $C = 6x + 8y$

subject to
$$\begin{aligned} 40x + 10y &\geq 2400 \\ 10x + 15y &\geq 2100 \\ 5x + 15y &\geq 1500 \\ x \geq 0, y &\geq 0 \end{aligned}$$

(Primal problem)

SOLUTION ✓ We first write down the following tableau for the given primal problem:

x	y	Constant
40	10	2400
10	15	2100
5	15	1500
6	8	

Next, we interchange the columns and rows of the foregoing tableau and head the three columns of the resulting array with the three variables u, v, and w, obtaining the tableau

u	v	w	Constant
40	10	5	6
10	15	15	8
2400	2100	1500	

Interpreting the last tableau as if it were part of the initial simplex tableau for a standard maximization problem, with the exception that the signs of the coefficients pertaining to the objective function are not reversed, we construct the required dual problem as follows:

Maximize the objective function $P = 2400u + 2100v + 1500w$

subject to
$$\begin{aligned} 40u + 10v + 5w &\leq 6 \\ 10u + 15v + 15w &\leq 8 \end{aligned}$$

where $u \geq 0$, $v \geq 0$, and $w \geq 0$.

Dual problem

■■■■

The connection between the solution of the primal problem and that of the dual problem is given by the following theorem. The theorem, attributed to John von Neumann (1903–1957), is stated without proof.

THEOREM

The Fundamental Theorem of Duality

A primal problem has a solution if and only if the corresponding dual problem has a solution. Furthermore, if a solution exists, then

1. The objective functions of both the primal and the dual problem attain the same optimal value.
2. The optimal solution to the primal problem appears under the slack variables in the last row of the final simplex tableau associated with the dual problem.

Armed with this theorem, we will solve the problem posed in Example 2.

EXAMPLE 3 Complete the solution to the problem posed in Example 2.

SOLUTION ✓ Observe that the dual problem associated with the given (primal) problem is a standard maximization problem. The solution may thus be found using the simplex algorithm. Introducing the slack variables x and y, we obtain the system of linear equations

$$\begin{aligned} 40u + 10v + 5w + x \quad &= 6 \\ 10u + 15v + 15w \quad + y &= 8 \\ -2400u - 2100v - 1500w \quad + P &= 0 \end{aligned}$$

Continuing with the simplex algorithm, we obtain the sequence of simplex tableaus

	u	v	w	x	y	P	Constant	ratio
Pivot row →	(40)	10	5	1	0	0	6	$\frac{6}{40} = \frac{3}{20}$
	10	15	15	0	1	0	8	$\frac{8}{10} = \frac{4}{5}$
	−2400	−2100	−1500	0	0	1	0	
	↑ Pivot column							

	u	v	w	x	y	P	Constant
	(1)	$\frac{1}{4}$	$\frac{1}{8}$	$\frac{1}{40}$	0	0	$\frac{3}{20}$
$\xrightarrow{\frac{1}{40}R_1}$	10	15	15	0	1	0	8
	−2400	−2100	−1500	0	0	1	0

	u	v	w	x	y	P	Constant	ratio
$R_2 - 10R_1$	1	$\frac{1}{4}$	$\frac{1}{8}$	$\frac{1}{40}$	0	0	$\frac{3}{20}$	$\frac{3/20}{1/4} = \frac{3}{5}$
$R_3 + 2400R_1$	0	$\left(\frac{25}{2}\right)$	$\frac{55}{4}$	$-\frac{1}{4}$	1	0	$\frac{13}{2}$	$\frac{13/2}{25/2} = \frac{13}{25}$
	0	−1500	−1200	60	0	1	360	

	u	v	w	x	y	P	Constant
$\xrightarrow{\frac{2}{25}R_2}$	1	$\frac{1}{4}$	$\frac{1}{8}$	$\frac{1}{40}$	0	0	$\frac{3}{20}$
	0	(1)	$\frac{11}{10}$	$-\frac{1}{50}$	$\frac{2}{25}$	0	$\frac{13}{25}$
	0	−1500	−1200	60	0	1	360

	u	v	w	x	y	P	Constant
$R_1 - \frac{1}{4}R_2$	1	0	$-\frac{3}{20}$	$\frac{3}{100}$	$-\frac{1}{50}$	0	$\frac{1}{50}$
$R_3 + 1500R_2$	0	1	$\frac{11}{10}$	$-\frac{1}{50}$	$\frac{2}{25}$	0	$\frac{13}{25}$
	0	0	450	30	120	1	1140

Solution for the primal problem (under x and y in the last row)

The last tableau is final. The fundamental theorem of duality tells us that the solution to the primal problem is $x = 30$ and $y = 120$ with a minimum value for C of 1140. Observe that the solution to the dual (maximization) problem may be read from the simplex tableau in the usual manner: $u = 1/50$, $v = 13/25$, $w = 0$, and $P = 1140$. Note that the maximum value of P is equal to the minimum value of C as guaranteed by the fundamental theorem of duality. The solution to the primal problem agrees with the solution of the same problem solved in Section 3.3, Example 2, using the method of corners.

■■■■

REMARKS

1. We leave it to you to demonstrate that the dual of a standard minimization problem is always a standard maximization problem provided that the coefficients of the objective function in the primal problem are all nonnegative. Such problems can always be solved by applying the simplex method to solve the dual problem.
2. Standard minimization problems in which the coefficients of the objective function are not all nonnegative do not necessarily have a dual problem that is a standard maximization problem. We will study such problems in Section 4.3.

EXAMPLE 4

Minimize $C = 3x + 2y$

subject to
$$\begin{aligned} 8x + y &\geq 80 \\ 8x + 5y &\geq 240 \\ x + 5y &\geq 100 \\ x \geq 0,\ y &\geq 0 \end{aligned}$$

SOLUTION ✓ We begin by writing the dual problem associated with the given primal problem. First, we write down the following tableau for the primal problem:

x	y	Constant
8	1	80
8	5	240
1	5	100
3	2	

Next, interchanging the columns and rows of this tableau and heading the three columns of the resulting array with the three variables u, v, and w, we obtain the tableau

u	v	w	Constant
8	8	1	3
1	5	5	2
80	240	100	

Interpreting the last tableau as if it were part of the initial simplex tableau for a standard maximization problem, with the exception that the signs of the coefficients pertaining to the objective function are not reversed, we construct the dual problem as follows: Maximize the objective function $P = 80u + 240v + 100w$ subject to the constraints

$$\begin{aligned} 8u + 8v + w &\leq 3 \\ u + 5v + 5w &\leq 2 \end{aligned}$$

where $u \geq 0$, $v \geq 0$, and $w \geq 0$. Having constructed the dual problem, which is a standard maximization problem, we now solve it using the simplex method.

Introducing the slack variables x and y, we obtain the system of linear equations

$$\begin{aligned} 8u + 8v + w + x &= 3 \\ u + 5v + 5w + y &= 2 \\ -80u - 240v - 100w + P &= 0 \end{aligned}$$

Continuing with the simplex algorithm, we obtain the sequence of simplex tableaus

	u	v	w	x	y	P	Constant	ratio
Pivot row →	8	(8)	1	1	0	0	3	$\frac{3}{8}$
	1	5	5	0	1	0	2	$\frac{2}{5}$
	−80	−240	−100	0	0	1	0	
		↑ Pivot column						

$\xrightarrow{\frac{1}{8}R_1}$	u	v	w	x	y	P	Constant
	1	(1)	$\frac{1}{8}$	$\frac{1}{8}$	0	0	$\frac{3}{8}$
	1	5	5	0	1	0	2
	−80	−240	−100	0	0	1	0

$\xrightarrow[R_3 + 240R_1]{R_2 - 5R_1}$	u	v	w	x	y	P	Constant	ratio
	1	1	$\frac{1}{8}$	$\frac{1}{8}$	0	0	$\frac{3}{8}$	3
Pivot row →	−4	0	$(\frac{35}{8})$	$-\frac{5}{8}$	1	0	$\frac{1}{8}$	$\frac{1}{35}$
	160	0	−70	30	0	1	90	
			↑ Pivot column					

$\xrightarrow{\frac{8}{35}R_2}$	u	v	w	x	y	P	Constant
	1	1	$\frac{1}{8}$	$\frac{1}{8}$	0	0	$\frac{3}{8}$
	$-\frac{32}{35}$	0	(1)	$-\frac{1}{7}$	$\frac{8}{35}$	0	$\frac{1}{35}$
	160	0	−70	30	0	1	90

$\xrightarrow[R_3 + 70R_2]{R_1 - \frac{1}{8}R_2}$	u	v	w	x	y	P	Constant
	$\frac{39}{35}$	1	0	$\frac{1}{7}$	$-\frac{1}{35}$	0	$\frac{13}{35}$
	$-\frac{32}{35}$	0	1	$-\frac{1}{7}$	$\frac{8}{35}$	0	$\frac{1}{35}$
	96	0	0	20	16	1	92

Solution for the primal problem (x, y columns: 20, 16)

The last tableau is final. The fundamental theorem of duality tells us that the solution to the primal problem is $x = 20$ and $y = 16$ with a minimum value for C of 92. ■■■■

APPLICATION

Our last example illustrates how the warehouse problem posed in Section 3.2 may be solved by duality.

EXAMPLE 5 Complete the solution to the warehouse problem given in Section 3.2, Example 4 (page 195). Minimize

$$C = 20x_1 + 8x_2 + 10x_3 + 12x_4 + 22x_5 + 18x_6 \qquad \textbf{(13)}$$

subject to

$$\begin{aligned} x_1 + x_2 + x_3 \qquad\qquad\qquad &\leq 400 \\ x_4 + x_5 + x_6 &\leq 600 \\ x_1 \qquad\qquad + x_4 \qquad\qquad &\geq 200 \\ x_2 \qquad\qquad + x_5 \qquad &\geq 300 \\ x_3 \qquad\qquad + x_6 &\geq 400 \\ x_1 \geq 0, x_2 \geq 0, \ldots, x_6 &\geq 0 \end{aligned} \qquad \textbf{(14)}$$

SOLUTION ✓ Upon multiplying each of the first two inequalities of (14) by -1, we obtain the following equivalent system of constraints in which each of the expressions involving the variables is greater than or equal to a constant:

$$\begin{aligned} -x_1 - x_2 - x_3 \qquad\qquad\qquad &\geq -400 \\ -x_4 - x_5 - x_6 &\geq -600 \\ x_1 \qquad\qquad + x_4 \qquad\qquad &\geq 200 \\ x_2 \qquad\qquad + x_5 \qquad &\geq 300 \\ x_3 \qquad\qquad + x_6 &\geq 400 \end{aligned}$$

The problem may now be solved by duality. First, we write the array of numbers:

x_1	x_2	x_3	x_4	x_5	x_6	Constant
−1	−1	−1	0	0	0	−400
0	0	0	−1	−1	−1	−600
1	0	0	1	0	0	200
0	1	0	0	1	0	300
0	0	1	0	0	1	400
20	8	10	12	22	18	

Interchanging the rows and columns of this array of numbers and heading the five columns of the resulting array of numbers by the variables u_1, u_2, u_3, u_4, and u_5 leads to

u_1	u_2	u_3	u_4	u_5	Constant
−1	0	1	0	0	20
−1	0	0	1	0	8
−1	0	0	0	1	10
0	−1	1	0	0	12
0	−1	0	1	0	22
0	−1	0	0	1	18
−400	−600	200	300	400	

from which we construct the associated dual problem: Maximize $P = -400u_1 - 600u_2 + 200u_3 + 300u_4 + 400u_5$ subject to

$$\begin{aligned}
-u_1 \quad + u_3 \quad\quad\quad &\leq 20 \\
-u_1 \quad\quad + u_4 \quad &\leq 8 \\
-u_1 \quad\quad\quad + u_5 &\leq 10 \\
- u_2 + u_3 \quad\quad\quad &\leq 12 \\
- u_2 \quad + u_4 \quad &\leq 22 \\
- u_2 \quad\quad + u_5 &\leq 18 \\
u_1 \geq 0, u_2 \geq 0, \ldots, u_5 &\geq 0
\end{aligned}$$

Solving the standard maximization problem by the simplex algorithm, we obtain the following sequence of tableaus ($x_1, x_2, \ldots, x_6$ are slack variables):

	u_1	u_2	u_3	u_4	u_5	x_1	x_2	x_3	x_4	x_5	x_6	P	Constant	ratio
	−1	0	1	0	0	1	0	0	0	0	0	0	20	—
	−1	0	0	1	0	0	1	0	0	0	0	0	8	—
Pivot row →	−1	0	0	0	(1)	0	0	1	0	0	0	0	10	10
	0	−1	1	0	0	0	0	0	1	0	0	0	12	—
	0	−1	0	1	0	0	0	0	0	1	0	0	22	—
	0	−1	0	0	1	0	0	0	0	0	1	0	18	18
	400	600	−200	−300	−400	0	0	0	0	0	0	1	0	
					↑ Pivot column									

Row operations: $R_6 - R_3$, $R_7 + 400R_3$

	u_1	u_2	u_3	u_4	u_5	x_1	x_2	x_3	x_4	x_5	x_6	P	Constant	ratio
	−1	0	1	0	0	1	0	0	0	0	0	0	20	—
Pivot row →	−1	0	0	①	0	0	1	0	0	0	0	0	8	8
	−1	0	0	0	1	0	0	1	0	0	0	0	10	—
	0	−1	1	0	0	0	0	0	1	0	0	0	12	—
	0	−1	0	1	0	0	0	0	0	1	0	0	22	22
	1	−1	0	0	0	0	0	−1	0	0	1	0	8	—
	0	600	−200	−300	0	0	0	400	0	0	0	1	4000	
				↑ Pivot column										

Row operations: $R_5 - R_2$, $R_7 + 300R_2$

	u_1	u_2	u_3	u_4	u_5	x_1	x_2	x_3	x_4	x_5	x_6	P	Constant	ratio
	−1	0	1	0	0	1	0	0	0	0	0	0	20	—
	−1	0	0	1	0	0	1	0	0	0	0	0	8	—
	−1	0	0	0	1	0	0	1	0	0	0	0	10	—
	0	−1	1	0	0	0	0	0	1	0	0	0	12	—
	1	−1	0	0	0	0	−1	0	0	1	0	0	14	14
Pivot row →	①	−1	0	0	0	0	0	−1	0	0	1	0	8	8
	−300	600	−200	0	0	0	300	400	0	0	0	1	6400	
	↑ Pivot column													

Row operations: $R_1 + R_6$, $R_2 + R_6$, $R_3 + R_6$, $R_5 - R_6$, $R_7 + 300R_6$

	u_1	u_2	u_3	u_4	u_5	x_1	x_2	x_3	x_4	x_5	x_6	P	Constant	ratio
	0	−1	1	0	0	1	0	−1	0	0	1	0	28	28
	0	−1	0	1	0	0	1	−1	0	0	1	0	16	—
	0	−1	0	0	1	0	0	0	0	0	1	0	18	—
Pivot row →	0	−1	①	0	0	0	0	0	1	0	0	0	12	12
	0	0	0	0	0	0	−1	1	0	1	−1	0	6	—
	1	−1	0	0	0	0	0	−1	0	0	1	0	8	—
	0	300	−200	0	0	0	300	100	0	0	300	1	8800	
			↑ Pivot column											

$\xrightarrow[R_7 + 200R_4]{R_1 - R_4}$

u_1	u_2	u_3	u_4	u_5	x_1	x_2	x_3	x_4	x_5	x_6	P	Constant
0	0	0	0	0	1	0	−1	−1	0	1	0	16
0	−1	0	1	0	0	1	−1	0	0	1	0	16
0	−1	0	0	1	0	0	0	0	0	1	0	18
0	−1	1	0	0	0	0	0	1	0	0	0	12
0	0	0	0	0	0	−1	1	0	1	−1	0	6
1	−1	0	0	0	0	0	−1	0	0	1	0	8
0	100	0	0	0	0	300	100	200	0	300	1	11,200

The last tableau is final, and we find that

$$x_1 = 0 \quad x_2 = 300 \quad x_3 = 100 \quad x_4 = 200$$
$$x_5 = 0 \quad x_6 = 300 \quad P = 11{,}200$$

Thus, to minimize shipping costs, Acrosonic should ship 300 loudspeaker systems from plant I to warehouse B, 100 systems from plant I to warehouse C, 200 systems from plant II to warehouse A, and 300 systems from plant II to warehouse C at a total cost of $11,200. ■■■■

SELF-CHECK EXERCISES 4.2

1. Write the dual problem associated with the following problem:

$$\begin{aligned} \text{Minimize} \quad & C = 2x + 5y \\ \text{subject to} \quad & 4x + y \geq 40 \\ & 2x + y \geq 30 \\ & x + 3y \geq 30 \\ & x \geq 0, y \geq 0 \end{aligned}$$

2. Solve the primal problem posed in Exercise 1.

Solutions to Self-Check Exercises 4.2 can be found on page 259.

4.2 Exercises

In Exercises 1–6, use the technique developed in this section to solve the given minimization problem.

1. Minimize $C = -2x + y$

$$\begin{aligned} \text{subject to} \quad & x + 2y \leq 6 \\ & 3x + 2y \leq 12 \\ & x \geq 0, y \geq 0 \end{aligned}$$

2. Minimize $C = -2x - 3y$

$$\begin{aligned} \text{subject to} \quad & 3x + 4y \leq 24 \\ & 7x - 4y \leq 16 \\ & x \geq 0, y \geq 0 \end{aligned}$$

3. Minimize $C = -3x - 2y$ subject to the constraints of Exercise 2.

4. Minimize $C = x - 2y + z$

subject to
$$\begin{aligned} x - 2y + 3z &\le 10 \\ 2x + y - 2z &\le 15 \\ 2x + y + 3z &\le 20 \\ x \ge 0, y \ge 0, z &\ge 0 \end{aligned}$$

5. Minimize $C = 2x - 3y - 4z$

subject to
$$\begin{aligned} -x + 2y - z &\le 8 \\ x - 2y + 2z &\le 10 \\ 2x + 4y - 3z &\le 12 \\ x \ge 0, y \ge 0, z &\ge 0 \end{aligned}$$

6. Minimize $C = -3x - 2y - z$ subject to the constraints of Exercise 5.

In Exercises 7–10, you are given the final simplex tableau for the dual problem. Give the solution to the primal problem and to the associated dual problem.

7. Problem: Minimize $C = 8x + 12y$

subject to
$$\begin{aligned} x + 3y &\ge 2 \\ 2x + 2y &\ge 3 \\ x \ge 0, y &\ge 0 \end{aligned}$$

Final tableau:

u	v	x	y	P	Constant
0	1	$\frac{3}{4}$	$-\frac{1}{4}$	0	3
1	0	$-\frac{1}{2}$	$\frac{1}{2}$	0	2
0	0	$\frac{5}{4}$	$\frac{1}{4}$	1	13

8. Problem: Minimize $C = 3x + 2y$

subject to
$$\begin{aligned} 5x + y &\ge 10 \\ 2x + 2y &\ge 12 \\ x + 4y &\ge 12 \\ x \ge 0, y &\ge 0 \end{aligned}$$

Final tableau:

u	v	w	x	y	P	Constant
1	0	$-\frac{3}{4}$	$\frac{1}{4}$	$-\frac{1}{4}$	0	$\frac{1}{4}$
0	1	$\frac{19}{8}$	$-\frac{1}{8}$	$\frac{5}{8}$	0	$\frac{7}{8}$
0	0	9	1	5	1	13

9. Problem: Minimize $C = 10x + 3y + 10z$

subject to
$$\begin{aligned} 2x + y + 5z &\ge 20 \\ 4x + y + z &\ge 30 \\ x \ge 0, y \ge 0, z &\ge 0 \end{aligned}$$

Final tableau:

u	v	x	y	z	P	Constant
0	1	$\frac{1}{2}$	-1	0	0	2
1	0	$-\frac{1}{2}$	2	0	0	1
0	0	2	-9	1	0	3
0	0	5	10	0	1	80

10. Problem: Minimize $C = 2x + 3y$

subject to
$$\begin{aligned} x + 4y &\ge 8 \\ x + y &\ge 5 \\ 2x + y &\ge 7 \\ x \ge 0, y &\ge 0 \end{aligned}$$

Final tableau:

u	v	w	x	y	P	Constant
0	1	$\frac{7}{3}$	$\frac{4}{3}$	$-\frac{1}{3}$	0	$\frac{5}{3}$
1	0	$-\frac{1}{3}$	$-\frac{1}{3}$	$\frac{1}{3}$	0	$\frac{1}{3}$
0	0	2	4	1	1	11

In Exercises 11–20, construct the dual problem associated with the given primal problem. Solve the primal problem.

11. Minimize $C = 2x + 5y$

subject to
$$\begin{aligned} x + 2y &\ge 4 \\ 3x + 2y &\ge 6 \\ x \ge 0, y &\ge 0 \end{aligned}$$

12. Minimize $C = 3x + 2y$

subject to
$$\begin{aligned} 2x + 3y &\ge 90 \\ 3x + 2y &\ge 120 \\ x \ge 0, y &\ge 0 \end{aligned}$$

13. Minimize $C = 6x + 4y$

subject to
$$\begin{aligned} 6x + y &\ge 60 \\ 2x + y &\ge 40 \\ x + y &\ge 30 \\ x \ge 0, y &\ge 0 \end{aligned}$$

14. Minimize $C = 10x + y$

subject to
$$\begin{aligned} 4x + y &\ge 16 \\ x + 2y &\ge 12 \\ x &\ge 2 \\ x \ge 0, y &\ge 0 \end{aligned}$$

15. Minimize $C = 200x + 150y + 120z$

subject to
$$\begin{aligned} 20x + 10y + z &\ge 10 \\ x + y + 2z &\ge 20 \\ x \ge 0, y \ge 0, z &\ge 0 \end{aligned}$$

16. Minimize $C = 40x + 30y + 11z$

subject to
$$\begin{aligned} 2x + y + z &\ge 8 \\ x + y - z &\ge 6 \\ x \ge 0, y \ge 0, z &\ge 0 \end{aligned}$$

17. Minimize $C = 6x + 8y + 4z$

subject to
$$\begin{aligned} x + 2y + 2z &\ge 10 \\ 2x + y + z &\ge 24 \\ x + y + z &\ge 16 \\ x \ge 0, y \ge 0, z &\ge 0 \end{aligned}$$

18. Minimize $C = 12x + 4y + 8z$

subject to
$$\begin{aligned} 2x + 4y + z &\ge 6 \\ 3x + 2y + 2z &\ge 2 \\ 4x + y + z &\ge 2 \\ x \ge 0, y \ge 0, z &\ge 0 \end{aligned}$$

19. Minimize $C = 30x + 12y + 20z$

subject to
$$\begin{aligned} 2x + 4y + 3z &\ge 6 \\ 6x \qquad + z &\ge 2 \\ 6y + 2z &\ge 4 \\ x \ge 0, y \ge 0, z &\ge 0 \end{aligned}$$

20. Minimize $C = 8x + 6y + 4z$

subject to
$$\begin{aligned} 2x + 3y + z &\ge 6 \\ x + 2y - 2z &\ge 4 \\ x + y + 2z &\ge 2 \\ x \ge 0, y \ge 0, z &\ge 0 \end{aligned}$$

21. SHIPPING COSTS The Acrosonic Company also manufactures a model G loudspeaker system in plants I and II. The output at plant I is at most 800/month, and the output at plant II is at most 600/month. Model G loudspeaker systems are also shipped to the three warehouses—A, B, and C—whose minimum monthly requirements are 500, 400, and 400, respectively. Shipping costs from plant I to warehouse A, warehouse B, and warehouse C are \$16, \$20, and \$22/loudspeaker system, respectively, and shipping costs from plant II to each of these warehouses are \$18, \$16, and \$14, respectively. What shipping schedule will enable Acrosonic to meet the requirements of the warehouses while keeping its shipping costs to a minimum?

22. SHIPPING COSTS The Steinwelt Piano Company manufactures uprights and consoles in two plants, plant I and plant II. The output of plant I is at most 300/month, and the output of plant II is at most 250/month. These pianos are shipped to three warehouses that serve as distribution centers for Steinwelt. To fill current and projected future orders, warehouse A requires a minimum of 200 pianos/month, warehouse B requires at least 150 pianos/month, and warehouse C requires at least 200 pianos/month. The shipping cost of each piano from plant I to warehouse A, warehouse B, and warehouse C is \$60, \$60, and \$80, respectively, and the shipping cost of each piano from plant II to warehouse A, warehouse B, and warehouse C is \$80, \$70, and \$50, respectively. What shipping schedule will enable Steinwelt to meet the requirements of the warehouses while keeping the shipping costs to a minimum?

23. NUTRITION—DIET PLANNING The owner of the Health Juice-Bar wishes to prepare a low-calorie fruit juice with a high vitamin A and C content by blending orange juice and pink grapefruit juice. Each glass of the blended juice is to contain at least 1200 International Units (IU) of vitamin A and 200 IU of vitamin C. One ounce of orange juice contains 60 IU of vitamin A, 16 IU of vitamin C, and 14 calories; each ounce of pink grapefruit juice contains 120 IU of vitamin A, 12 IU of vitamin C, and 11 calories. How many ounces of each juice should a glass of the blend contain if it is to meet the minimum vitamin requirements and at the same time contain a minimum number of calories?

24. PRODUCTION CONTROL An oil company operates two refineries in a certain city. Refinery I has an output of 200, 100, and 100 barrels of low-, medium-, and high-grade oil per day, respectively. Refinery II has an output of 100, 200, and 600 barrels of low-, medium-, and high-grade oil per day, respectively. The company wishes to produce at least 1000, 1400, and 3000 barrels of low-, medium-, and high-grade oil to fill an order. If it costs \$200/day to operate refinery I and \$300/day to operate refinery II, determine how many days each refinery should be operated to meet the requirements of the order at minimum cost to the company.

In Exercises 25 and 26, determine whether the statement is true or false. If it is true, explain why it is true. If it is false, give an example to show why it is false.

25. If a standard minimization linear programming problem has a unique solution, then so does the corresponding maximization problem with objective function $P = -C$, where $C = a_1x_1 + a_2x_2 + \cdots + a_nx_n$ is the objective function for the minimization problem.

26. The optimal value attained by the objective function of the primal problem may be different from that attained by the objective function of the dual problem.

SOLUTIONS TO SELF-CHECK EXERCISES 4.2

1. We first write down the following tableau for the given (primal) problem:

x	y	Constant
4	1	40
2	1	30
1	3	30
2	5	0

Next, we interchange the columns and rows of the tableau and head the three columns of the resulting array with the three variables u, v, and w, obtaining the tableau

u	v	w	Constant
4	2	1	2
1	1	3	5
40	30	30	0

Interpreting the last tableau as if it were the initial tableau for a standard linear programming problem, with the exception that the signs of the coefficients pertaining to the objective function are not reversed, we construct the required dual problem as follows:

$$\begin{aligned} \text{Maximize} \quad & P = 40u + 30v + 30w \\ \text{subject to} \quad & 4u + 2v + w \le 2 \\ & u + v + 3w \le 5 \\ & u \ge 0, v \ge 0, w \ge 0 \end{aligned}$$

2. We introduce slack variables x and y to obtain the system of linear equations

$$\begin{aligned} 4u + 2v + w + x \quad &= 2 \\ u + v + 3w \quad + y &= 5 \\ -40u - 30v - 30w \quad + P &= 0 \end{aligned}$$

(continued on p. 263)

THE SIMPLEX METHOD: SOLVING MINIMIZATION PROBLEMS

The graphing calculator can be used to solve minimization problems using the simplex method.

EXAMPLE 1

$$\begin{aligned} \text{Minimize} \quad & C = 2x + 3y \\ \text{subject to} \quad & 8x + y \geq 80 \\ & 3x + 2y \geq 100 \\ & x + 4y \geq 80 \\ & x \geq 0, y \geq 0 \end{aligned}$$

SOLUTION ✓ We begin by writing the dual problem associated with the given primal problem. From the tableau for the primal problem

x	y	Constant
8	1	80
3	2	100
1	4	80
2	3	

we find, upon changing the columns and rows of this tableau and heading the three columns of the resulting array with the variables u, v, and w, the tableau

u	v	w	Constant
8	3	1	2
1	2	4	3
80	100	80	

This tells us that the dual problem is:

$$\begin{aligned} \text{Maximize} \quad & P = 80u + 100v + 80w \\ \text{subject to} \quad & 8u + 3v + w \leq 2 \\ & u + 2u + 4w \leq 3 \\ & u \geq 0, v \geq 0, w \geq 0 \end{aligned}$$

To solve this standard maximization problem, we proceed as follows:

	u	v	w	x	y	P	Constant	ratio
Pivot row →	8	(3)	1	1	0	0	2	$\frac{2}{3}$
	1	2	4	0	1	0	3	$\frac{3}{2}$
	−80	−100	−80	0	0	1	0	
		↑ Pivot column						

$\xrightarrow{\text{*row}(\frac{1}{3}, A, 1) \blacktriangleright B}$

u	v	w	x	y	P	Constant
2.67	(1)	0.33	0.33	0	0	0.67
1	2	4	0	1	0	3
−80	−100	−80	0	0	1	0

$\xrightarrow[\text{*row+}(100, C, 1, 3) \blacktriangleright B]{\text{*row+}(-2, B, 1, 2) \blacktriangleright C}$

	u	v	w	x	y	P	Constant	ratio
	2.67	1	0.33	0.33	0	0	0.67	2
Pivot row →	−4.33	0	(3.33)	−0.67	1	0	1.67	0.5
	186.67	0	−46.67	33.33	0	1	66.67	
			↑ Pivot column					

$\xrightarrow{\text{*row}(\frac{1}{3.33}, B, 2) \blacktriangleright C}$

u	v	w	x	y	P	Constant
2.67	1	0.33	0.33	0	0	0.67
−1.30	0	1	−0.2	0.3	0	0.5
186.67	0	−46.67	33.33	0	1	66.67

$\xrightarrow[\text{*row+}(46.67, B, 2, 3) \blacktriangleright C]{\text{*row+}(-0.33, C, 2, 1) \blacktriangleright B}$

u	v	w	x	y	P	Constant
3.1	1	0	0.4	−0.1	0	0.50
−1.3	0	1	−0.2	0.3	0	0.50
125.93	0	0.05	23.99	14.02	1	90.03

Solution for the primal problem (23.99, 14.02)

From the last tableau, we see that $x = 23.99$, $y = 14.02$, and the minimum value of C is 90.03. ■■■■

Exercises

In Exercises 1–4, use a graphing calculator to solve the linear programming problem by the simplex method.

1. Minimize $C = x + y + 3z$
   subject to
   $$\begin{aligned} 2x + y + 3z &\geq 6 \\ x + 2y + 4z &\geq 8 \\ 3x + y - 2z &\geq 4 \\ x \geq 0, y \geq 0, z &\geq 0 \end{aligned}$$

2. Minimize $C = 2x + 4y + z$
   subject to
   $$\begin{aligned} x + 2y + 4z &\geq 7 \\ 3x + y - z &\geq 6 \\ x + 4y + 2z &\geq 24 \\ x \geq 0, y \geq 0, z &\geq 0 \end{aligned}$$

3. Minimize $C = x + 1.2y + 3.5z$
   subject to
   $$\begin{aligned} 2x + 3y + 5z &\geq 12 \\ 3x + 1.2y - 2.2z &\geq 8 \\ 1.2x + 3y + 1.8z &\geq 14 \\ x \geq 0, y \geq 0, z &\geq 0 \end{aligned}$$

4. Minimize $C = 2.1x + 1.2y + z$
   subject to
   $$\begin{aligned} x + y - z &\geq 5.2 \\ x - 2.1y + 4.2z &\geq 8.4 \\ x \geq 0, y \geq 0, z &\geq 0 \end{aligned}$$

Using the simplex algorithm, we obtain the sequence of simplex tableaus

	u	v	w	x	y	P	Constant	ratio
Pivot row →	(4)	2	1	1	0	0	2	$\frac{2}{4} = \frac{1}{2}$
	1	1	3	0	1	0	5	$\frac{5}{1} = 5$
	−40	−30	−30	0	0	1	0	

↑ Pivot column (u)

$\xrightarrow{\frac{1}{4}R_1}$

u	v	w	x	y	P	Constant
(1)	$\frac{1}{2}$	$\frac{1}{4}$	$\frac{1}{4}$	0	0	$\frac{1}{2}$
1	1	3	0	1	0	5
−40	−30	−30	0	0	1	0

$\xrightarrow[R_3 + 40R_1]{R_2 - R_1}$

	u	v	w	x	y	P	Constant	ratio
	1	$\frac{1}{2}$	$\frac{1}{4}$	$\frac{1}{4}$	0	0	$\frac{1}{2}$	$\frac{1/2}{1/4} = 2$
Pivot row →	0	$\frac{1}{2}$	($\frac{11}{4}$)	$-\frac{1}{4}$	1	0	$\frac{9}{2}$	$\frac{9/2}{11/4} = \frac{18}{11}$
	0	−10	−20	10	0	1	20	

↑ Pivot column (w)

$\xrightarrow{\frac{4}{11}R_2}$

u	v	w	x	y	P	Constant
1	$\frac{1}{2}$	$\frac{1}{4}$	$\frac{1}{4}$	0	0	$\frac{1}{2}$
0	$\frac{2}{11}$	(1)	$-\frac{1}{11}$	$\frac{4}{11}$	0	$\frac{18}{11}$
0	−10	−20	10	0	1	20

$\xrightarrow[R_3 + 20R_2]{R_1 - \frac{1}{4}R_2}$

	u	v	w	x	y	P	Constant	ratio
Pivot row →	1	($\frac{5}{11}$)	0	$\frac{3}{11}$	$-\frac{1}{11}$	0	$\frac{1}{11}$	$\frac{1/11}{5/11} = \frac{1}{5}$
	0	$\frac{2}{11}$	1	$-\frac{1}{11}$	$\frac{4}{11}$	0	$\frac{18}{11}$	$\frac{18/11}{2/11} = 9$
	0	$-\frac{70}{11}$	0	$\frac{90}{11}$	$\frac{80}{11}$	1	$\frac{580}{11}$	

↑ Pivot column (v)

$\xrightarrow{\frac{11}{5}R_1}$

u	v	w	x	y	P	Constant
$\frac{11}{5}$	(1)	0	$\frac{3}{5}$	$-\frac{1}{5}$	0	$\frac{1}{5}$
0	$\frac{2}{11}$	1	$-\frac{1}{11}$	$\frac{4}{11}$	0	$\frac{18}{11}$
0	$-\frac{70}{11}$	0	$\frac{90}{11}$	$\frac{80}{11}$	1	$\frac{580}{11}$

$\xrightarrow[R_3 + \frac{70}{11}R_1]{R_2 - \frac{2}{11}R_1}$

u	v	w	x	y	P	Constant
$\frac{11}{5}$	1	0	$\frac{3}{5}$	$-\frac{1}{5}$	0	$\frac{1}{5}$
$-\frac{2}{5}$	0	1	$-\frac{1}{5}$	$\frac{2}{5}$	0	$\frac{8}{5}$
14	0	0	12	6	1	54

Solution for the primal problem

The last tableau is final, and the solution to the primal problem is $x = 12$ and $y = 6$ with a minimum value for C of 54.

4.3 The Simplex Method: Nonstandard Problems (Optional)

Section 4.1 showed how the simplex method can be used to solve standard maximization problems, and Section 4.2 showed how, thanks to duality, it can be used to solve standard minimization problems provided that the coefficients in the objective function are all nonnegative.

In this section we see how the simplex method can be incorporated into a method for solving **nonstandard problems**—problems that do not fall into either of the two previous categories. We begin by recalling the characteristics of standard problems.

A. Standard Maximization Problem

1. The objective function is to be maximized.
2. All the variables involved in the problem are nonnegative.
3. Each linear constraint may be written so that the expression involving the variables is less than or equal to a nonnegative constant.

B. Standard Minimization Problem (restricted version—see Condition 4)

1. The objective function is to be minimized.
2. All the variables involved in the problem are nonnegative.
3. Each linear constraint may be written so that the expression involving the variables is greater than or equal to a constant.
4. *All the coefficients in the objective function are nonnegative.*

REMARK Recall that if all the coefficients in the objective function are nonnegative, then a standard minimization problem can be solved by using the simplex method to solve the associated dual problem. ■■■

We now give some examples of linear programming problems that do not fit into these two categories of problems.

EXAMPLE 1 Explain why the following linear programming problem is not a standard maximization problem.

$$\begin{aligned} \text{Maximize} \quad & P = x + 2y \\ \text{subject to} \quad & 4x + 3y \le 18 \\ & -x + 3y \ge 3 \\ & x \ge 0, y \ge 0 \end{aligned}$$

SOLUTION ✓ This is not a standard maximization problem because the second constraint inequality,

$$-x + 3y \ge 3$$

violates condition 3. Observe that by multiplying both sides of this inequality by -1, we obtain

$$x - 3y \le -3$$ (Recall that multiplying both sides of an inequality by a negative number reverses the inequality sign.)

Now, the last equation still violates condition 3 because the constant on the right is *negative*. ■■■■

Observe that the constraints in Example 1 involve both *less than or equal to constraints* ($\le$) and *greater than or equal to constraints* ($\ge$). Such constraints are called **mixed constraints.** We will solve the problem posed in Example 1 later.

EXAMPLE 2 Explain why the following linear programming problem is not a restricted standard minimization problem.

$$\begin{aligned} \text{Minimize} \quad & C = 2x - 3y \\ \text{subject to} \quad & x + y \le 5 \\ & x + 3y \ge 9 \\ & -2x + y \le 2 \\ & x \ge 0, y \ge 0 \end{aligned}$$

SOLUTION ✓ Observe that the coefficients in the objective function C are not all nonnegative. Therefore, the problem is not a restricted standard minimization problem. By constructing the dual problem, you can convince yourself that the latter is not a standard maximization problem and thus cannot be solved using the methods described in Sections 4.1 and 4.2. Again, we will solve this problem later. ■■■■

EXAMPLE 3 Explain why the following linear programming problem is not a standard maximization problem. Show that it cannot be rewritten as a restricted standard minimization problem.

$$\begin{aligned} \text{Maximize} \quad & P = x + 2y \\ \text{subject to} \quad & 2x + 3y \le 12 \\ & -x + 3y = 3 \\ & x \ge 0, y \ge 0 \end{aligned}$$

SOLUTION ✓ The constraint equation $-x + 3y = 3$ is equivalent to the two inequalities

$$-x + 3y \le 3 \qquad \text{and} \qquad -x + 3y \ge 3$$

By multiplying both sides of the second inequality by -1, it can be written in the form

$$x - 3y \le -3$$

Therefore, the two given constraints are equivalent to the three inequality constraints

$$\begin{aligned} 2x + 3y &\le 12 \\ -x + 3y &\le 3 \\ x - 3y &\le -3 \end{aligned}$$

The third inequality violates condition 3 for a standard maximization problem. Next, we see that the given problem is equivalent to the following:

$$\begin{aligned} \text{Minimize} \quad & C = -x - 2y \\ \text{subject to} \quad & -2x - 3y \ge -12 \\ & x - 3y \ge -3 \\ & -x + 3y \ge 3 \\ & x \ge 0, y \ge 0 \end{aligned}$$

Since the coefficients of the objective function are not all nonnegative, we conclude that the given problem cannot be rewritten as a restricted standard minimization problem. You will be asked to solve this problem in Exercise 11.

■■■■

THE SIMPLEX METHOD FOR SOLVING NONSTANDARD PROBLEMS

To describe a technique for solving nonstandard problems, let's consider the problem of Example 1:

$$\begin{aligned} \text{Maximize} \quad & P = x + 2y \\ \text{subject to} \quad & 4x + 3y \le 18 \\ & -x + 3y \ge 3 \\ & x \ge 0, y \ge 0 \end{aligned}$$

As a first step, we rewrite the inequality constraints so that the second constraint involves a $\le$ constraint. As in Example 1, we obtain

$$\begin{aligned} 4x + 3y &\le 18 \\ x - 3y &\le -3 \\ x \ge 0, y &\ge 0 \end{aligned}$$

Disregarding the fact that the constant on the right of the second inequality constraint is negative, let's attempt to solve the problem using the simplex method for problems in standard form. Introducing the slack variables u and v gives the system of linear equations

$$\begin{aligned} 4x + 3y + u \qquad\quad &= 18 \\ x - 3y \quad + v \quad &= -3 \\ -x - 2y \qquad\quad + P &= 0 \end{aligned}$$

The initial simplex tableau is

x	y	u	v	P	Constant
4	3	1	0	0	18
1	−3	0	1	0	−3
−1	−2	0	0	1	0

Interpreting the tableau in the usual fashion, we see that

$$x = 0, \quad y = 0, \quad u = 18, \quad v = -3$$

Since the value of the slack variable v is negative, we see that this cannot be a feasible solution (remember, all variables must be nonnegative). In fact, you can see from Figure 4.6 that the point (0, 0) does not lie in the feasible set associated with the given problem. Since we must start from a feasible point when using the simplex method for problems in standard form, we see that this method is not applicable at this juncture.

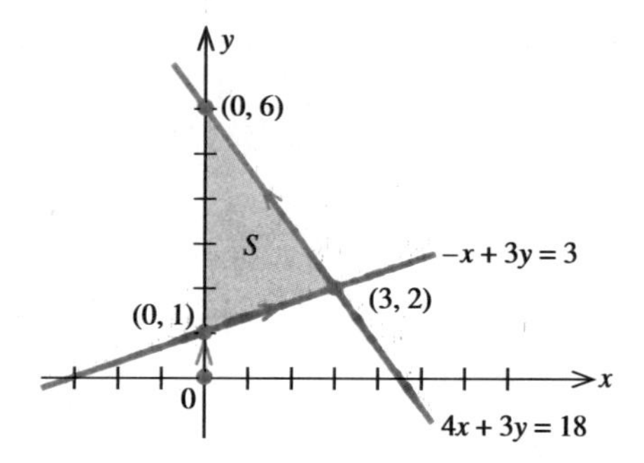

FIGURE 4.6
S is the feasible set for the problem.

Let's find a way to bring us from the nonfeasible point (0, 0) to *any* feasible point, after which we can switch to the simplex method for problems in standard form. This can be accomplished by pivoting as follows: Referring to the tableau,

	x	y	u	v	P	Constant	ratio
	4	3	1	0	0	18	$\frac{18}{3} = 6$
Pivot row →	1	(−3)	0	1	0	−3	$\frac{-3}{-3} = 1$
	−1	−2	0	0	1	0	
		↑ Pivot column					

notice the negative number -3 lying in the column of constants, above the lower horizontal line. Locate any negative number to the left of this number (there must always be at least one such number if the problem has a solution). For the problem under consideration there is only one such number, the number -3 in the y-column. This column is designated as the pivot column. To find the pivot element, we form the *positive* ratios of the numbers in the column of constants to the corresponding numbers in the pivot column (above the last row). The pivot row is the row corresponding to the smallest ratio, and the pivot element is the element common to both the pivot row and the pivot column (the number circled in the foregoing tableau). Pivoting about this element, we have

$\xrightarrow{-\frac{1}{3}R_2}$

x	y	u	v	P	Constant
4	3	1	0	0	18
$-\frac{1}{3}$	(1)	0	$-\frac{1}{3}$	0	1
-1	-2	0	0	1	0

$\xrightarrow[R_3 + 2R_2]{R_1 - 3R_2}$

x	y	u	v	P	Constant
5	0	1	1	0	15
$-\frac{1}{3}$	1	0	$-\frac{1}{3}$	0	1
$-\frac{5}{3}$	0	0	$-\frac{2}{3}$	1	2

Interpreting the last tableau in the usual fashion, we see that

$$x = 0, \quad y = 1, \quad u = 15, \quad v = 0, \quad P = 2$$

Observe that the point (0, 1) is a feasible point (see Figure 4.6). Our iteration has brought us from a nonfeasible point to a feasible point in one iteration. Observe, too, that all the constants in the column of constants are now nonnegative, reflecting the fact that (0, 1) is a feasible point, as we have just noted.

We can now use the simplex method for problems in standard form to complete the solution to our problem.

	x	y	u	v	P	Constant	ratio
Pivot row →	(5)	0	1	1	0	15	3
	$-\frac{1}{3}$	1	0	$-\frac{1}{3}$	0	1	—
	$-\frac{5}{3}$	0	0	$-\frac{2}{3}$	1	2	
	↑ Pivot column						

$\xrightarrow{\frac{1}{5}R_1}$

x	y	u	v	P	Constant
(1)	0	$\frac{1}{5}$	$\frac{1}{5}$	0	3
$-\frac{1}{3}$	1	0	$-\frac{1}{3}$	0	1
$-\frac{5}{3}$	0	0	$-\frac{2}{3}$	1	2

$\xrightarrow[R_3 + \frac{5}{3}R_1]{R_2 + \frac{1}{3}R_1}$

	x	y	u	v	P	Constant	ratio
Pivot row →	1	0	$\frac{1}{5}$	($\frac{1}{5}$)	0	3	15
	0	1	$\frac{1}{15}$	$-\frac{4}{15}$	0	2	—
	0	0	$\frac{1}{3}$	$-\frac{1}{3}$	1	7	
				↑ Pivot column			

$\xrightarrow{5R_1}$

x	y	u	v	P	Constant
5	0	1	①	0	15
0	1	$\frac{1}{15}$	$-\frac{4}{15}$	0	2
0	0	$\frac{1}{3}$	$-\frac{1}{3}$	1	7

$\xrightarrow[R_3 + \frac{1}{3}R_1]{R_2 + \frac{4}{15}R_1}$

x	y	u	v	P	Constant
5	0	1	1	0	15
$\frac{4}{3}$	1	$\frac{1}{3}$	0	0	6
$\frac{5}{3}$	0	$\frac{2}{3}$	0	1	12

All entries in the last row are nonnegative and the tableau is final. We see that the optimal solution is

$$x = 0, \quad y = 6, \quad u = 0, \quad v = 15, \quad P = 12$$

Observe that the maximum of P occurs at (0, 6) (see Figure 4.6). The arrows indicate the path that our search for the maximum of P has taken us on.

Before looking at further examples, let's summarize the method for solving nonstandard problems.

The Simplex Method for Solving Nonstandard Problems

1. If necessary, rewrite the problem as a maximization problem (recall that minimizing C is equivalent to maximizing $-C$).
2. If necessary, rewrite all inequality constraints (except $x \geq 0$, $y \geq 0$, $z \geq 0$, . . .) using less than or equal to ($\leq$) inequalities.
3. Introduce slack variables and set up the initial simplex tableau.
4. Scan the upper part of the column of constants of the tableau for negative entries.
   a. If there are no negative entries, complete the solution using the simplex method for problems in standard form.
   b. If there are negative entries, proceed to step 5.
5. **a.** Pick any negative entry in the row in which a negative entry in the column of constants occurs. The column containing this entry is the pivot column.
   b. Compute the positive ratios of the numbers in the column of constants to the corresponding numbers in the pivot column (above the last row). The pivot row corresponds to the smallest ratio. The intersection of the pivot column and the pivot row determines the pivot element.
   c. Pivot the tableau about the pivot element. Then return to step 4.

We now apply the method to solve the nonstandard problem posed in Example 2.

EXAMPLE 4 Solve the problem of Example 2:

$$\begin{aligned} \text{Minimize} \quad & C = 2x - 3y \\ \text{subject to} \quad & x + y \leq 5 \\ & x + 3y \geq 9 \\ & -2x + y \leq 2 \\ & x \geq 0, y \geq 0 \end{aligned}$$

SOLUTION ✔ We first rewrite the problem as a maximization problem with inequality constraints using ≤, obtaining the following equivalent problem:

$$\begin{aligned} \text{Maximize} \quad & P = -C = -2x + 3y \\ \text{subject to} \quad & x + y \leq 5 \\ & -x - 3y \leq -9 \\ & -2x + y \leq 2 \\ & x \geq 0, y \geq 0 \end{aligned}$$

Introducing slack variables u, v, and w and following the procedure for solving nonstandard problems outlined earlier, we obtain the following sequence of tableaus:

	x	y	u	v	w	P	Constant	ratio
	1	1	1	0	0	0	5	$\frac{5}{1} = 5$
	−1	−3	0	1	0	0	−9	$\frac{-9}{-3} = 3$
Pivot row →	−2	(1)	0	0	1	0	2	$\frac{2}{1} = 2$
	2	−3	0	0	0	1	0	
		↑ Pivot column						

(Column 1 could have been chosen as the pivot column as well.)

$\xrightarrow[R_4 + 3R_3]{R_1 - R_3 \quad R_2 + 3R_3}$

	x	y	u	v	w	P	Constant	ratio
	3	0	1	0	−1	0	3	$\frac{3}{3} = 1$
Pivot row →	(−7)	0	0	1	3	0	−3	$\frac{-3}{-7} = \frac{3}{7}$
	−2	1	0	0	1	0	2	
	−4	0	0	0	3	1	6	
	↑ Pivot column							

$\xrightarrow{-\frac{1}{7}R_2}$

x	y	u	v	w	P	Constant
3	0	1	0	−1	0	3
(1)	0	0	$-\frac{1}{7}$	$-\frac{3}{7}$	0	$\frac{3}{7}$
−2	1	0	0	1	0	2
−4	0	0	0	3	1	6

$\xrightarrow[R_4 + 4R_2]{R_1 - 3R_2 \quad R_3 + 2R_2}$

	x	y	u	v	w	P	Constant	ratio
Pivot row →	0	0	1	$(\frac{3}{7})$	$\frac{2}{7}$	0	$\frac{12}{7}$	4
	1	0	0	$-\frac{1}{7}$	$-\frac{3}{7}$	0	$\frac{3}{7}$	—
	0	1	0	$-\frac{2}{7}$	$\frac{1}{7}$	0	$\frac{20}{7}$	—
	0	0	0	$-\frac{4}{7}$	$\frac{9}{7}$	1	$\frac{54}{7}$	
				↑ Pivot column				

(We now use the simplex method for problems in standard form to complete the problem.)

$\xrightarrow{\frac{7}{3}R_1}$

x	y	u	v	w	P	Constant
0	0	$\frac{7}{3}$	①	$\frac{2}{3}$	0	4
1	0	0	$-\frac{1}{7}$	$-\frac{3}{7}$	0	$\frac{3}{7}$
0	1	0	$-\frac{2}{7}$	$\frac{1}{7}$	0	$\frac{20}{7}$
0	0	0	$-\frac{4}{7}$	$\frac{9}{7}$	1	$\frac{54}{7}$

$\xrightarrow[\substack{R_3 + \frac{2}{7}R_1 \\ R_4 + \frac{4}{7}R_1}]{R_2 + \frac{1}{7}R_1}$

x	y	u	v	w	P	Constant
0	0	$\frac{7}{3}$	1	$\frac{2}{3}$	0	4
1	0	$\frac{1}{3}$	0	$-\frac{1}{3}$	0	1
0	1	$\frac{2}{3}$	0	$\frac{1}{3}$	0	4
0	0	$\frac{4}{3}$	0	$\frac{5}{3}$	1	10

All the entries in the last row are nonnegative and the tableau is final. We see that the optimal solution is

$$x = 1, \quad y = 4, \quad u = 0, \quad v = 4, \quad w = 0, \quad C = -P = -10$$

The feasible set S for this problem is shown in Figure 4.7. The path leading from the nonfeasible initial point $(0, 0)$ to the optimal point $(1, 4)$ goes through the nonfeasible point $(0, 2)$ and the feasible point $(\frac{3}{7}, \frac{20}{7})$, in that order.

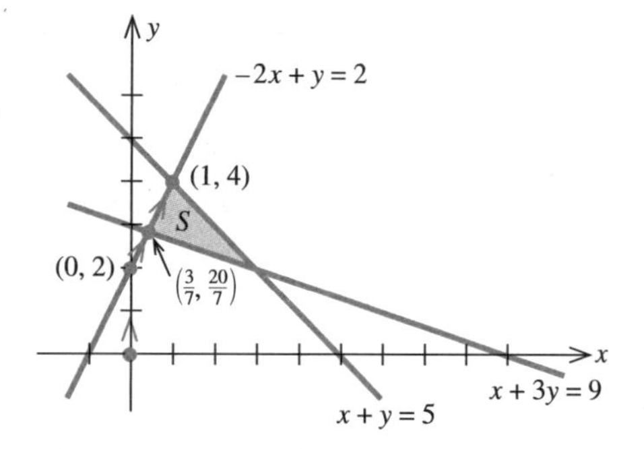

FIGURE 4.7
The feasible set S and the path leading from the initial nonfeasible point $(0, 0)$ to the optimal point $(1, 4)$

APPLICATION

EXAMPLE 5 The Rockford Company manufactures two models of exercise bicycles, a standard model and a deluxe model, in two separate plants—plant I and plant II. The maximum output at plant I is 1200 per month; the maximum output at plant II is 1000 per month. The profit per bike for standard and deluxe models manufactured at plant I is \$40 and \$60, respectively; the profit per bike for standard and deluxe models manufactured at plant II is \$45 and \$50, respectively.

For the month of May, Rockford received an order for 1000 standard models and 800 deluxe models. If prior commitments dictate that the number of deluxe models manufactured at plant I may not exceed the number of standard models manufactured there by more than 200, find how many of each model should be produced at each plant so as to satisfy the order and at the same time maximize Rockford's profit.

SOLUTION ✓

Let x and y denote the number of standard and deluxe models to be manufactured at plant I. Since the number of standard and deluxe models required are 1000 and 800, respectively, we see that the number of standard and deluxe models to be manufactured at plant II are $(1000 - x)$ and $(800 - y)$, respectively. Rockford's profit will then be

$$\begin{aligned} P &= 40x + 60y + 45(1000 - x) + 50(800 - y) \\ &= 85{,}000 - 5x + 10y \end{aligned}$$

Since the maximum output of plant I is 1200, we have the constraint

$$x + y \le 1200$$

Similarly, since the maximum output of plant II is 1000, we have

$$(1000 - x) + (800 - y) \le 1000$$

or, equivalently,

$$-x - y \le -800$$

Finally, the additional constraints placed on the production schedule at plants I and II translate into the inequalities

$$\begin{aligned} y - x &\le 200 \\ x &\le 1000 \\ y &\le 800 \end{aligned}$$

To summarize, the problem at hand is the following nonstandard problem:

$$\begin{aligned} \text{Maximize} \quad P &= 85{,}000 - 5x + 10y \\ \text{subject to} \quad x + y &\le 1200 \\ -x - y &\le -800 \\ -x + y &\le 200 \\ x &\le 1000 \\ y &\le 800 \\ x \ge 0, y &\ge 0 \end{aligned}$$

Let's introduce the slack variables, u, v, w, r, and s. Using the simplex method for nonstandard problems, we obtain the following sequence of tableaus:

	x	y	u	v	w	r	s	P	Constant	ratio
	1	1	1	0	0	0	0	0	1,200	$\frac{1200}{1} = 1200$
Pivot row →	(−1)	−1	0	1	0	0	0	0	−800	$\frac{-800}{-1} = 800$
	−1	1	0	0	1	0	0	0	200	—
	1	0	0	0	0	1	0	0	1,000	$\frac{1000}{1} = 1000$
	0	1	0	0	0	0	1	0	800	—
	5	−10	0	0	0	0	0	1	85,000	
	↑ Pivot column									

$\xrightarrow{-R_2}$

x	y	u	v	w	r	s	P	Constant
1	1	1	0	0	0	0	0	1,200
(1)	1	0	−1	0	0	0	0	800
−1	1	0	0	1	0	0	0	200
1	0	0	0	0	1	0	0	1,000
0	1	0	0	0	0	1	0	800
5	−10	0	0	0	0	0	1	85,000

$\xrightarrow[\substack{R_4 - R_2 \\ R_6 - 5R_2}]{\substack{R_1 - R_2 \\ R_3 + R_2}}$ (Pivot row → third row)

x	y	u	v	w	r	s	P	Constant	ratio
0	0	1	1	0	0	0	0	400	—
1	1	0	−1	0	0	0	0	800	$\frac{800}{1} = 800$
0	(2)	0	−1	1	0	0	0	1,000	$\frac{1000}{2} = 500$
0	−1	0	1	0	1	0	0	200	—
0	1	0	0	0	0	1	0	800	$\frac{800}{1} = 800$
0	−15	0	5	0	0	0	1	81,000	

Pivot row; ↑ Pivot column (y)

(We now use the simplex method for standard problems.)

$\xrightarrow{\frac{1}{2}R_3}$

x	y	u	v	w	r	s	P	Constant
0	0	1	1	0	0	0	0	400
1	1	0	−1	0	0	0	0	800
0	(1)	0	$-\frac{1}{2}$	$\frac{1}{2}$	0	0	0	500
0	−1	0	1	0	1	0	0	200
0	1	0	0	0	0	1	0	800
0	−15	0	5	0	0	0	1	81,000

$\xrightarrow[\substack{R_5 - R_3 \\ R_6 + 15R_3}]{\substack{R_2 - R_3 \\ R_4 + R_3}}$

x	y	u	v	w	r	s	P	Constant	ratio
0	0	1	(1)	0	0	0	0	400	$\frac{400}{1} = 400$
1	0	0	$-\frac{1}{2}$	$-\frac{1}{2}$	0	0	0	300	—
0	1	0	$-\frac{1}{2}$	$\frac{1}{2}$	0	0	0	500	—
0	0	0	$\frac{1}{2}$	$\frac{1}{2}$	1	0	0	700	$\frac{700}{1/2} = 1400$
0	0	0	$\frac{1}{2}$	$-\frac{1}{2}$	0	1	0	300	$\frac{300}{1/2} = 600$
0	0	0	$-\frac{5}{2}$	$\frac{15}{2}$	0	0	1	88,500	

↑ Pivot column (v)

$\xrightarrow[\substack{R_4 - \frac{1}{2}R_1 \\ R_5 - \frac{1}{2}R_1 \\ R_6 + \frac{5}{2}R_1}]{\substack{R_2 + \frac{1}{2}R_1 \\ R_3 + \frac{1}{2}R_1}}$

x	y	u	v	w	r	s	P	Constant
0	0	1	1	0	0	0	0	400
1	0	$\frac{1}{2}$	0	$-\frac{1}{2}$	0	0	0	500
0	1	$\frac{1}{2}$	0	$\frac{1}{2}$	0	0	0	700
0	0	$-\frac{1}{2}$	0	$\frac{1}{2}$	1	0	0	500
0	0	$-\frac{1}{2}$	0	$-\frac{1}{2}$	0	1	0	100
0	0	$\frac{5}{2}$	0	$\frac{15}{2}$	0	0	1	89,500

All the entries in the last row are nonnegative, and the tableau is final. We see that $x = 500$, $y = 700$, and $P = 89{,}500$. This tells us that plant I should manufacture 500 standard and 700 deluxe exercise bicycles and that plant II should manufacture $(1000 - 500)$, or 500, standard and $(800 - 700)$, or 100, deluxe models. Rockford's profit will then be \$89,500.

Group Discussion

Refer to Example 5.

1. Sketch the feasible set S for the linear programming problem.
2. Solve the problem using the method of corners.
3. Indicate on S the points (both nonfeasible and feasible) corresponding to each iteration of the simplex method and trace the path leading to the optimal solution.

SELF-CHECK EXERCISES 4.3

1. Solve the following nonstandard problem using the method of this section:

$$\begin{aligned} \text{Maximize} \quad & P = 2x + 3y \\ \text{subject to} \quad & x + y \leq 40 \\ & -x + 2y \leq -10 \\ & x \geq 0, y \geq 0 \end{aligned}$$

2. A farmer has 150 acres of land suitable for cultivating crops A and B. The cost of cultivating crop A is \$40/acre and that of crop B is \$60/acre. The farmer has a maximum of \$7400 available for land cultivation. Each acre of crop A requires 20 hr of labor, and each acre of crop B requires 25 hr of labor. The farmer has a maximum of 3300 hr of labor available. He has also decided that he will cultivate at least 70 acres of crop A. If he expects to make a profit of \$150/acre on crop A and \$200/acre on crop B, how many acres of each crop should he plant in order to maximize his profit?

Solutions to Self-Check Exercises 4.3 can be found on page 277.

4.3 Exercises

In Exercises 1–4, rewrite the given linear programming problem as a maximization problem with constraints involving inequalities of the form ≤ (with the exception of the inequalities $x \geq 0$, $y \geq 0$, and $z \geq 0$).

1. Minimize $P = 2x - 3y$
subject to
$$\begin{aligned} 3x + 5y &\geq 20 \\ 3x + y &\leq 16 \\ -2x + y &\leq 1 \\ x \geq 0, y &\geq 0 \end{aligned}$$

2. Minimize $C = 2x + 3y$
subject to
$$\begin{aligned} x + y &\leq 10 \\ x + 2y &\geq 12 \\ 2x + y &\geq 12 \\ x \geq 0, y &\geq 0 \end{aligned}$$

3. Minimize $C = 5x + 10y + z$
subject to
$$\begin{aligned} 2x + y + z &\geq 4 \\ x + 2y + 2z &\geq 2 \\ 2x + 4y + 3z &\leq 12 \\ x \geq 0, y \geq 0, z &\geq 0 \end{aligned}$$

4. Maximize $P = 2x + y - 2z$
subject to
$$\begin{aligned} x + 2y + z &\geq 10 \\ 3x + 4y + 2z &\geq 5 \\ 2x + 5y + 12z &\leq 20 \\ x \geq 0, y \geq 0, z &\geq 0 \end{aligned}$$

In Exercises 5–20, use the method of this section to solve the given linear programming problem.

5. Maximize $P = x + 2y$
subject to
$$\begin{aligned} 2x + 5y &\leq 20 \\ x - 5y &\leq -5 \\ x \geq 0, y &\geq 0 \end{aligned}$$

6. Maximize $P = 2x + 3y$
subject to
$$\begin{aligned} x + 2y &\leq 8 \\ x - y &\leq -2 \\ x \geq 0, y &\geq 0 \end{aligned}$$

7. Minimize $C = -2x + y$
subject to
$$\begin{aligned} x + 2y &\leq 6 \\ 3x + 2y &\leq 12 \\ x \geq 0, y &\geq 0 \end{aligned}$$

8. Minimize $C = -2x + 3y$
subject to
$$\begin{aligned} x + 3y &\leq 60 \\ 2x + y &\geq 45 \\ x &\leq 40 \\ x \geq 0, y &\geq 0 \end{aligned}$$

9. Maximize $P = x + 4y$
subject to
$$\begin{aligned} x + 3y &\leq 6 \\ -2x + 3y &\leq -6 \\ x \geq 0, y &\geq 0 \end{aligned}$$

10. Maximize $P = 5x + y$
subject to
$$\begin{aligned} 2x + y &\leq 8 \\ -x + y &\geq 2 \\ x \geq 0, y &\geq 0 \end{aligned}$$

11. Maximize $P = x + 2y$
subject to
$$\begin{aligned} 2x + 3y &\leq 12 \\ -x + 3y &= 3 \\ x \geq 0, y &\geq 0 \end{aligned}$$

12. Minimize $P = x + 2y$
subject to
$$\begin{aligned} 4x + 7y &\leq 70 \\ 2x + y &= 20 \\ x \geq 0, y &\geq 0 \end{aligned}$$

13. Maximize $P = 5x + 4y + 2z$
subject to
$$\begin{aligned} x + 2y + 3z &\leq 24 \\ x - y + z &\geq 6 \\ x \geq 0, y \geq 0, z &\geq 0 \end{aligned}$$

14. Maximize $P = x - 2y + z$
subject to
$$\begin{aligned} 2x + 3y + 2z &\leq 12 \\ x + 2y - 3z &\geq 6 \\ x \geq 0, y \geq 0, z &\geq 0 \end{aligned}$$

15. Minimize $C = x - 2y + z$

subject to
$$\begin{aligned} x - 2y + 3z &\le 10 \\ 2x + y - 2z &\le 15 \\ 2x + y + 3z &\le 20 \\ x \ge 0, y \ge 0, z &\ge 0 \end{aligned}$$

16. Minimize $C = 2x - 3y + 4z$

subject to
$$\begin{aligned} -x + 2y - z &\le 8 \\ x - 2y + 2z &\le 10 \\ 2x + 4y - 3z &\le 12 \\ x \ge 0, y \ge 0, z &\ge 0 \end{aligned}$$

17. Maximize $P = 2x + y + z$

subject to
$$\begin{aligned} x + 2y + 3z &\le 28 \\ 2x + 3y - z &\le 6 \\ x - 2y + z &\ge 4 \\ x \ge 0, y \ge 0, z &\ge 0 \end{aligned}$$

18. Minimize $C = 2x - y + 3z$

subject to
$$\begin{aligned} 2x + y + z &\ge 2 \\ x + 3y + z &\ge 6 \\ 2x + y + 2z &\le 12 \\ x \ge 0, y \ge 0, z &\ge 0 \end{aligned}$$

19. Maximize $P = x + 2y + 3z$

subject to
$$\begin{aligned} x + 2y + z &\le 20 \\ 3x + y \quad\; &\le 30 \\ 2x + y + z &= 10 \\ x \ge 0, y \ge 0, z &\ge 0 \end{aligned}$$

20. Minimize $C = 3x + 2y + z$

subject to
$$\begin{aligned} x + 2y + z &\le 20 \\ 3x + y \quad\; &\le 30 \\ 2x + y + z &= 10 \\ x \ge 0, y \ge 0, z &\ge 0 \end{aligned}$$

21. AGRICULTURE—CROP PLANNING A farmer has 150 acres of land suitable for cultivating crops A and B. The cost of cultivating crop A is \$40/acre and that of crop B is \$60/acre. The farmer has a maximum of \$7400 available for land cultivation. Each acre of crop A requires 20 hr of labor, and each acre of crop B requires 25 hr of labor. The farmer has a maximum of 3300 hr of labor available. He has also decided that he will cultivate at least 80 acres of crop A. If he expects to make a profit of \$150/acre on crop A and \$200/acre on crop B, how many acres of each crop should he plant in order to maximize his profit?

22. FINANCE—INVESTMENTS Natsano has at most \$50,000 to invest in the common stocks of two companies. He estimates that an investment in company A will yield a return of 10%, whereas an investment in company B, which he feels is a riskier investment, will yield a return of 20%. If he decides that his investment in the stocks of company A is to exceed his investment in the stocks of company B by at least \$20,000, determine how much he should invest in the stocks of each company in order to maximize the returns on his investment.

23. FINANCE—ALLOCATION OF FUNDS The First Street branch of the Capitol Bank has a sum of \$60 million earmarked for home and commercial-development loans. The bank expects to realize an 8% annual rate of return on the home loans and a 6% annual rate of return on the commercial-development loans. Management has decided that the total amount of home loans is to be greater than or equal to three times the total amount of commercial-development loans. Owing to prior commitments, at least \$10 million of the funds has been designated for commercial-development loans. Determine the amount of each type of loan the bank should extend in order to maximize its returns.

24. MANUFACTURING—PRODUCTION SCHEDULING The Wayland Company manufactures two models of its twin-size futons, standard and deluxe, in two locations, I and II. The maximum output at location I is 600/week, whereas the maximum output at location II is 400/week. The profit per futon for standard and deluxe models manufactured at location I is \$30 and \$20, respectively; the profit per futon for standard and deluxe models manufactured at location II is \$34 and \$18, respectively. For a certain week, the company has received an order for 600 standard models and 300 deluxe models. If prior commitments dictate that the number of deluxe models manufactured at location II may not exceed the number of standard models manufactured there by more than 50, find how many of each model should be manufactured at each location so as to satisfy the order and at the same time maximize Wayland's profit.

25. MANUFACTURING—PRODUCTION SCHEDULING A company manufactures products A, B, and C. Each product is processed in three departments: I, II, and III. The total available labor-hours per week for departments I, II, and III are 900, 1080, and 840, respectively. The time

requirements (in hours per unit) and the profit per unit for each product are as follows:

	Product A	Product B	Product C
Dept. I	2	1	2
Dept. II	3	1	2
Dept. III	2	2	1
Profit	\$18	\$12	\$15

If management decides that the number of units of product B manufactured must equal or exceed the number of units of products A and C manufactured, how many units of each product should the company produce in order to maximize its profit?

26. MANUFACTURING—SHIPPING COSTS Steinwelt Piano manufactures uprights and consoles in two plants, plant I and plant II. The output of plant I is at most 300/month, whereas the output of plant II is at most 250/month. These pianos are shipped to three warehouses that serve as distribution centers for the company. To fill current and projected future orders, warehouse A requires a minimum of 200 pianos/month, warehouse B requires at least 150 pianos/month, and warehouse C requires at least 200 pianos/month. The shipping cost of each piano from plant I to warehouse A, warehouse B, and warehouse C is \$60, \$60, and \$80, respectively, and the shipping cost of each piano from plant II to warehouse A, warehouse B, and warehouse C is \$80, \$70, and \$50, respectively. Use the method of this section to determine the shipping schedule that will enable Steinwelt to meet the warehouses' requirements while keeping the shipping costs to a minimum.

27. NUTRITION—DIET PLANNING A nutritionist at the Medical Center has been asked to prepare a special diet for certain patients. She has decided that the meals should contain a minimum of 400 mg of calcium, 10 mg of iron, and 40 mg of vitamin C. She has further decided that the meals are to be prepared from foods A and B. Each ounce of food A contains 30 mg of calcium, 1 mg of iron, 2 mg of vitamin C, and 2 mg of cholesterol. Each ounce of food B contains 25 mg of calcium, 0.5 mg of iron, 5 mg of vitamin C, and 5 mg of cholesterol. Use the method of this section to determine how many ounces of each type of food the nutritionist should use in a meal so the cholesterol content is minimized and the minimum requirements of calcium, iron, and vitamin C are met.

SOLUTIONS TO SELF-CHECK EXERCISES 4.3

1. We are given the problem

$$\begin{aligned} \text{Maximize} \quad & P = 2x + 3y \\ \text{subject to} \quad & x + y \le 40 \\ & -x + 2y \le -10 \\ & x \ge 0, y \ge 0 \end{aligned}$$

Using the method of this section and introducing the slack variables u and v, we obtain the following tableaus:

	x	y	u	v	P	Constant	ratio
	1	1	1	0	0	40	$\frac{40}{1} = 40$
Pivot row →	(−1)	2	0	1	0	−10	$\frac{-10}{-1} = 10$
	−2	−3	0	0	1	0	
	↑ Pivot column						

$\xrightarrow{-R_2}$

x	y	u	v	P	Constant
1	1	1	0	0	40
(1)	−2	0	−1	0	10
−2	−3	0	0	1	0

Pivot row →

$\xrightarrow[R_3 + 2R_2]{R_1 - R_2}$

x	y	u	v	P	Constant	ratio
0	(3)	1	1	0	30	$\frac{30}{3} = 10$
1	−2	0	−1	0	10	—
0	−7	0	−2	1	20	

↑ Pivot column

$\xrightarrow{\frac{1}{3}R_1}$

x	y	u	v	P	Constant
0	(1)	$\frac{1}{3}$	$\frac{1}{3}$	0	10
1	−2	0	−1	0	10
0	−7	0	−2	1	20

$\xrightarrow[R_3 + 7R_1]{R_2 + 2R_1}$

x	y	u	v	P	Constant
0	1	$\frac{1}{3}$	$\frac{1}{3}$	0	10
1	0	$\frac{2}{3}$	$-\frac{1}{3}$	0	30
0	0	$\frac{7}{3}$	$\frac{1}{3}$	1	90

The last tableau is in final form, and we obtain the solution

$$x = 30, \quad y = 10, \quad u = 0, \quad v = 0, \quad P = 90$$

2. Let x denote the number of acres of crop A to be cultivated and y the number of acres of crop B to be cultivated. Since there is a total of 150 acres of land available for cultivation, we have $x + y \leq 150$. Next, the restriction on the amount of money available for land cultivation implies that $40x + 60y \leq 7400$. Similarly, the restriction on the amount of time available for labor implies that $20x + 25y \leq 3300$. Also, since he will cultivate at least 70 acres of crop A, $x \geq 70$. Since the profit on each acre of crop A is \$150 and the profit on each acre of crop B is \$200, we see that the profit realizable by the farmer is $P = 150x + 200y$. Summarizing, we have the following linear programming problem:

$$\begin{aligned}
\text{Maximize} \quad & P = 150x + 200y \\
\text{subject to} \quad & x + y \leq 150 \\
& 40x + 60y \leq 7400 \\
& 20x + 25y \leq 3300 \\
& x \geq 70 \\
& x \geq 0, y \geq 0
\end{aligned}$$

To solve this nonstandard problem, we rewrite the fourth inequality in the form

$$-x \leq -70$$

Using the method of this section with u, v, w, and z as slack variables, we have the following tableaus:

	x	y	u	v	w	z	P	Constant	ratio
	1	1	1	0	0	0	0	150	$\frac{150}{1} = 150$
	40	60	0	1	0	0	0	7,400	$\frac{7400}{40} = 185$
	20	25	0	0	1	0	0	3,300	$\frac{3300}{20} = 165$
Pivot row →	(−1)	0	0	0	0	1	0	−70	$\frac{-70}{-1} = 70$
	−150	−200	0	0	0	0	1	0	
	↑ Pivot column								

$\xrightarrow{-R_4}$

x	y	u	v	w	z	P	Constant
1	1	1	0	0	0	0	150
40	60	0	1	0	0	0	7,400
20	25	0	0	1	0	0	3,300
(1)	0	0	0	0	−1	0	70
−150	−200	0	0	0	0	1	0

$\xrightarrow[R_3 - 20R_4 \\ R_5 + 150R_4]{R_1 - R_4 \\ R_2 - 40R_4}$

	x	y	u	v	w	z	P	Constant	ratio
	0	1	1	0	0	1	0	80	$\frac{80}{1} = 80$
	0	60	0	1	0	40	0	4,600	$\frac{4600}{60} = 76\frac{2}{3}$
Pivot row →	0	(25)	0	0	1	20	0	1,900	$\frac{1900}{25} = 76$
	1	0	0	0	0	−1	0	70	—
	0	−200	0	0	0	−150	1	10,500	
		↑ Pivot column							

$\xrightarrow{\frac{1}{25}R_3}$

x	y	u	v	w	z	P	Constant
0	1	1	0	0	1	0	80
0	60	0	1	0	40	0	4,600
0	(1)	0	0	$\frac{1}{25}$	$\frac{4}{5}$	0	76
1	0	0	0	0	−1	0	70
0	−200	0	0	0	−150	1	10,500

$$\xrightarrow[\substack{R_2 - 60R_3 \\ R_5 + 200R_3}]{R_1 - R_3}$$

x	y	u	v	w	z	P	Constant
0	0	1	0	$-\frac{1}{25}$	$\frac{1}{5}$	0	4
0	0	0	1	$-\frac{12}{5}$	-8	0	40
0	1	0	0	$\frac{1}{25}$	$\frac{4}{5}$	0	76
1	0	0	0	0	-1	0	70
0	0	0	0	8	10	1	25,700

This last tableau is in final form, and the solution is

$$x = 70, \quad y = 76, \quad u = 4, \quad v = 40, \quad w = 0, \quad z = 0, \quad P = 25{,}700$$

Thus, by cultivating 70 acres of crop A and 76 acres of crop B, the farmer will attain a maximum profit of \$25,700.

CHAPTER 4 Summary of Principal Terms

Terms

standard maximization problem
slack variable
basic variable
nonbasic variable
pivot column
pivot row
pivot element
simplex tableau
simplex method
standard minimization problem
primal problem
dual problem
nonstandard problem
mixed constraints

CHAPTER 4 REVIEW EXERCISES

In Exercises 1–12, use the simplex method to solve the given linear programming problem.

1. Maximize $P = 3x + 4y$
   subject to
   $x + 3y \le 15$
   $4x + y \le 16$
   $x \ge 0, y \ge 0$

2. Maximize $P = 2x + 5y$
   subject to
   $2x + y \le 16$
   $2x + 3y \le 24$
   $y \le 6$
   $x \ge 0, y \ge 0$

3. Maximize $P = 2x + 3y + 5z$
subject to
$$x + 2y + 3z \leq 12$$
$$x - 3y + 2z \leq 10$$
$$x \geq 0, y \geq 0, z \geq 0$$

4. Maximize $P = x + 2y + 3z$
subject to
$$2x + y + z \leq 14$$
$$3x + 2y + 4z \leq 24$$
$$2x + 5y - 2z \leq 10$$
$$x \geq 0, y \geq 0, z \geq 0$$

5. Minimize $C = 3x + 2y$
subject to
$$2x + 3y \geq 6$$
$$2x + y \geq 4$$
$$x \geq 0, y \geq 0$$

6. Minimize $C = x + 2y$
subject to
$$3x + y \geq 12$$
$$x + 4y \geq 16$$
$$x \geq 0, y \geq 0$$

7. Minimize $C = 24x + 18y + 24z$
subject to
$$3x + 2y + z \geq 4$$
$$x + y + 3z \geq 6$$
$$x \geq 0, y \geq 0, z \geq 0$$

8. Minimize $C = 4x + 2y + 6z$
subject to
$$x + 2y + z \geq 4$$
$$2x + y + 2z \geq 2$$
$$3x + 2y + z \geq 3$$
$$x \geq 0, y \geq 0, z \geq 0$$

9. Maximize $P = 3x - 4y$
subject to
$$x + y \leq 45$$
$$x - 2y \geq 10$$
$$x \geq 0, y \geq 0$$

10. Minimize $C = 2x + 3y$
subject to
$$x + y \leq 10$$
$$x + 2y \geq 12$$
$$2x + y \geq 12$$
$$x \geq 0, y \geq 0$$

11. Maximize $P = 2x + 3y$
subject to
$$2x + 5y \leq 20$$
$$-x + 5y \geq 5$$
$$x \geq 0, y \geq 0$$

12. Minimize $C = -3x - 4y$
subject to
$$x + y \leq 45$$
$$x + y \geq 15$$
$$x \leq 30$$
$$y \leq 25$$
$$x \geq 0, y \geq 0$$

13. A company manufactures three products, A, B, and C, on two machines, I and II. It has been determined that the company will realize a profit of \$4/unit of product A, \$6/unit of product B, and \$8/unit of product C. To manufacture a unit of product A requires 9 min on machine I and 6 min on machine II; to manufacture a unit of product B requires 12 min on machine I and 6 min on machine II; and to manufacture a unit of product C requires 18 min on machine I and 10 min on machine II. There are 6 hr of machine time available on machine I and 4 hr of machine time available on machine II in each work shift. How many units of each product should be produced in each shift in order to maximize the company's profit?

14. Jorge has decided to invest at most \$100,000 in securities in the form of corporate stocks. He has classified his options into three groups of stocks: blue-chip stocks that he assumes will yield a 10% return (dividends and capital appreciation) within a year, growth stocks that he assumes will yield a 15% return within a year, and speculative stocks that he assumes will yield a 20% return (mainly due to capital appreciation) within a year. Because of the relative risks involved in his investment, Jorge has further decided that no more than 30% of his investment should be in growth and speculative stocks and at least 50% of his investment should be in blue-chip and speculative stocks. Determine how much Jorge should invest in each group of stocks in the hope of maximizing the return on his investments.

15. Sandra has at most \$200,000 to invest in stocks, bonds, and money-market funds. She expects annual yields of 15%, 10%, and 8%, respectively, on these investments. If Sandra wants at least \$50,000 to be invested in money-market funds and requires that the amount invested in bonds be greater than or equal to the sum of her investments in stocks and money-market funds, determine how much she should invest in each vehicle in order to maximize the returns on her investments.

5 MATHEMATICS OF FINANCE

Interest that is periodically added to the principal and thereafter itself earns interest is called *compound interest.* Albert Einstein called compound interest the greatest invention of mankind. We begin this chapter by deriving the *compound interest formula,* which gives the amount of money accumulated when an initial amount of money is invested in an account for a fixed term and earns compound interest.

An *annuity* is a sequence of payments made at regular intervals. We derive formulas giving the *future value of an annuity* (what you end up with) and the *present value of an annuity* (the lump sum that, when invested now, will yield the same future value of the annuity). Then, using these formulas, we answer questions involving the amortization of certain types of installment loans and questions involving *sinking funds* (funds that are set up to be used for a specific purpose at a future date).

How much will the home mortgage payment be? The Blakelys received a bank loan to help finance the purchase of a house. They have agreed to repay the loan in equal monthly installments over a certain period of time. In Example 2, page 311, we will show how to determine the size of the monthly installment so that the loan is fully amortized at the end of the term.

5.1 Compound Interest

SIMPLE INTEREST

A natural application of linear functions to the business world is found in the computation of **simple interest**—interest that is computed on the original principal only. Thus, if I denotes the interest on a principal P (in dollars) at an interest rate of r per year for t years, then we have

$$I = Prt$$

The **accumulated amount** A, the sum of the principal and interest after t years, is given by

$$\begin{aligned} A &= P + I = P + Prt \\ &= P(1 + rt) \end{aligned}$$

and is a linear function of t (see Exercise 30 at the end of this section). In business applications we are normally interested only in the case where t is positive, so only that part of the line that lies in Quadrant I is of interest to us (Figure 5.1).

FIGURE 5.1
The accumulated amount is a linear function of t.

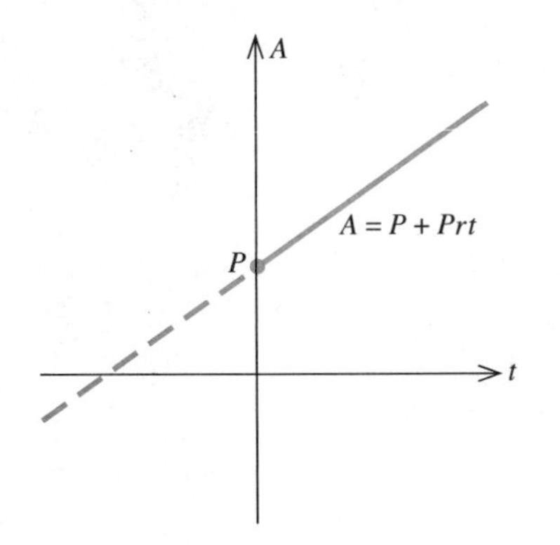

Simple Interest Formulas

Interest:	$I = Prt$	**(1a)**
Accumulated amount:	$A = P(1 + rt)$	**(1b)**

EXAMPLE 1 A bank pays simple interest at the rate of 8% per year for certain deposits. If a customer deposits \$1000 and makes no withdrawals for 3 years, what is the total amount on deposit at the end of 3 years? What is the interest earned in that period of time?

SOLUTION ✓ Using Equation (1b) with $P = 1000$, $r = 0.08$, and $t = 3$, we see that the total amount on deposit at the end of 3 years is given by

$$\begin{aligned} A &= P(1 + rt) \\ &= 1000[1 + (0.08)(3)] = 1240 \end{aligned}$$

or \$1240.

The interest earned over the 3-year period is given by

$$\begin{aligned} I &= Prt \qquad \text{(Using 1a)} \\ &= 1000(0.08)(3) = 240 \end{aligned}$$

or \$240. ■■■■

Exploring with Technology

Refer to Example 1. Use a graphing utility to plot the graph of the function $A = 1000\,(1 + 0.08t)$, using the viewing rectangle $[0, 10] \times [0, 2000]$.

1. What is the A-intercept of the straight line, and what does it represent?
2. What is the slope of the straight line, and what does it represent? (See Exercise 30.)

EXAMPLE 2 An amount of \$2000 is invested in a 10-year trust fund that pays 6% annual simple interest. What is the total amount of the trust fund at the end of 10 years?

SOLUTION ✓ The total amount of the trust fund at the end of 10 years is given by

$$\begin{aligned} A &= P(1 + rt) \\ &= 2000[1 + (0.06)(10)] = 3200 \end{aligned}$$

or \$3200. ■■■■

COMPOUND INTEREST

In contrast to simple interest, earned interest that is periodically added to the principal and thereafter itself earns interest at the same rate is called **compound interest.** To find a formula for the accumulated amount, let's consider a numerical example. Suppose \$1000 (the principal) is deposited in a bank for a term of 3 years, earning interest at the rate of 8% per year (called the **nominal,** or **stated, rate**) compounded annually. Then, using Equation (1b) with $P = 1000$, $r = 0.08$, and $t = 1$, we see that the accumulated amount at the end of the first year is

$$\begin{aligned} A_1 &= P(1 + rt) \\ &= 1000[1 + 0.08(1)] = 1000(1.08) = 1080 \end{aligned}$$

or \$1080.

To find the accumulated amount A_2 at the end of the second year, we use (1b) once again, this time with $P = A_1$. (Remember, the principal *and* interest now earn interest over the second year.) We obtain

$$\begin{aligned} A_2 &= P(1 + rt) = A_1(1 + rt) \\ &= 1000[1 + 0.08(1)][1 + 0.08(1)] \\ &= 1000[1 + 0.08]^2 = 1000(1.08)^2 = 1166.40 \end{aligned}$$

or \$1166.40.

Finally, the accumulated amount A_3 at the end of the third year is found using (1b) with $P = A_2$, giving

$$\begin{aligned} A_3 &= P(1 + rt) = A_2(1 + rt) \\ &= 1000[1 + 0.08(1)]^2[1 + 0.08(1)] \\ &= 1000[1 + 0.08]^3 = 1000(1.08)^3 \approx 1259.71 \end{aligned}$$

or approximately $1259.71.

If you reexamine our calculations, you will see that the accumulated amounts at the end of each year have the following form:

First year: $A_1 = 1000(1 + 0.08)$, or $A_1 = P(1 + r)$
Second year: $A_2 = 1000(1 + 0.08)^2$, or $A_2 = P(1 + r)^2$
Third year: $A_3 = 1000(1 + 0.08)^3$, or $A_3 = P(1 + r)^3$

These observations suggest the following general result: If P dollars is invested over a term of t years, earning interest at the rate of r per year compounded annually, then the accumulated amount is

$$A = P(1 + r)^t \qquad \textbf{(2)}$$

Formula (2) was derived under the assumption that interest was compounded *annually.* In practice, however, interest is usually compounded more than once a year. The interval of time between successive interest calculations is called the **conversion period.**

If interest at a nominal rate of r per year is compounded m times a year on a principal of P dollars, then the simple interest rate per conversion period is

$$i = \frac{r}{m} \qquad \frac{\text{(Annual interest rate)}}{\text{(Periods per year)}}$$

For example, if the nominal interest rate is 8% per year ($r = 0.08$) and interest is compounded quarterly ($m = 4$), then

$$i = \frac{r}{m} = \frac{0.08}{4} = 0.02$$

or 2% per period.

To find a general formula for the accumulated amount when a principal of P dollars is deposited in a bank for a term of t years and earns interest at the (nominal) rate of r per year compounded m times per year, we proceed as before, using (1b) repeatedly with the interest rate $i = r/m$. We see that the accumulated amount at the end of each period is

First period: $A_1 = P(1 + i)$
Second period: $A_2 = A_1(1 + i) = [P(1 + i)](1 + i) = P(1 + i)^2$
Third period: $A_3 = A_2(1 + i) = [P(1 + i)^2](1 + i) = P(1 + i)^3$
$\vdots$
nth period: $A_n = A_{n-1}(1 + i) = [P(1 + i)^{n-1}](1 + i) = P(1 + i)^n$

But there are $n = mt$ periods in t years (number of conversion periods times the term). Therefore, the accumulated amount at the end of t years is given by

$$A = P(1 + i)^n$$

Compound Interest Formula

$$A = P(1 + i)^n \tag{3}$$

where $i = r/m$, $n = mt$, and

A = Accumulated amount at the end of n conversion periods
P = Principal
r = Nominal interest rate per year
m = Number of conversion periods per year
t = Term (number of years)

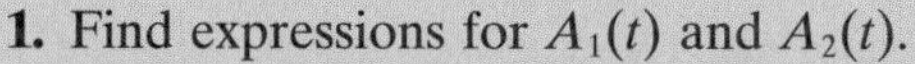

Exploring with Technology

Let $A_1(t)$ denote the accumulated amount of \$100 earning simple interest at the rate of 10% per year over t years, and let $A_2(t)$ denote the accumulated amount of \$100 earning interest at the rate of 10% per year compounded monthly over t years.

1. Find expressions for $A_1(t)$ and $A_2(t)$.
2. Use a graphing utility to plot the graphs of A_1 and A_2 on the same set of axes, using the viewing rectangle $[0, 20] \times [0, 800]$.
3. Comment on the growth of $A_1(t)$ and $A_2(t)$ by referring to the graphs of A_1 and A_2.

EXAMPLE 3 Find the accumulated amount after 3 years if \$1000 is invested at 8% per year compounded (a) annually, (b) semiannually, (c) quarterly, and (d) monthly.

SOLUTION ✓ **a.** Here, $P = 1000$, $r = 0.08$, and $m = 1$. Thus, $i = r = 0.08$ and $n = 3$, so Equation (3) gives

$$\begin{aligned} A &= 1000(1.08)^3 \\ &= 1259.71 \end{aligned}$$

or \$1259.71.

b. Here, $P = 1000$, $r = 0.08$, and $m = 2$. Thus, $i = 0.08/2 = 0.04$ and $n = (3)(2) = 6$, so Equation (3) gives

$$\begin{aligned} A &= 1000(1.04)^6 \\ &= 1265.32 \end{aligned}$$

or \$1265.32.

c. In this case, $P = 1000$, $r = 0.08$, and $m = 4$. Thus, $i = 0.08/4 = 0.02$ and $n = (3)(4) = 12$, so Equation (3) gives

$$\begin{aligned} A &= 1000(1.02)^{12} \\ &= 1268.24 \end{aligned}$$

or $1268.24.

d. Here, $P = 1000$, $r = 0.08$, and $m = 12$. Thus, $i = 0.08/12$ and $n = (3)(12) = 36$, so Equation (3) gives

$$\begin{aligned} A &= 1000\left(1 + \frac{0.08}{12}\right)^{36} \\ &= 1270.24 \end{aligned}$$

or $1270.24. These results are summarized in Table 5.1.

Table 5.1

Nominal Rate (r)	Conversion Period	Interest Rate/ Conversion Period	Initial Investment	Accumulated Amount
8%	Annual ($m = 1$)	8%	$1000	$1259.71
8	Semiannual ($m = 2$)	4	1000	1265.32
8	Quarterly ($m = 4$)	2	1000	1268.24
8	Monthly ($m = 12$)	2/3	1000	1270.24

■■■■

EFFECTIVE RATE OF INTEREST

Example 3 showed that the interest actually earned on an investment depends on the frequency with which the interest is compounded. Thus, the stated, or nominal, rate of 8% per year does not reflect the actual rate at which interest is earned. This suggests that we need to find a common basis for comparing interest rates. One such way of comparing interest rates is provided by the use of the *effective rate.* The **effective rate** is the *simple* interest rate that would produce the same accumulated amount in 1 year as the nominal rate compounded m times a year. The effective rate is also called the **effective annual yield.**

To derive a relationship between the nominal interest rate, r per year compounded m times, and its corresponding effective rate, R per year, let's assume an initial investment of P dollars. Then, the accumulated amount after 1 year at a simple interest rate of R per year is

$$A = P(1 + R)$$

Also, the accumulated amount after 1 year at an interest rate of r per year compounded m times a year is

$$A = P(1 + i)^n = P\left(1 + \frac{r}{m}\right)^m \qquad \text{(Since } i = r/m\text{)}$$

Exploring with Technology

Investments allowed to grow over time can increase in value surprisingly fast. Consider the potential growth of \$10,000 if earnings are reinvested. More specifically, suppose $A_1(t)$, $A_2(t)$, $A_3(t)$, $A_4(t)$, and $A_5(t)$ denote the accumulated values of an investment of \$10,000 over a term of t years and earning interest at the rate of 4%, 6%, 8%, 10%, and 12% per year compounded annually.

1. Find expressions for $A_1(t), A_2(t), \ldots, A_5(t)$.
2. Use a graphing utility to plot the graphs of $A_1, A_2, \ldots, A_5$ on the same set of axes, using the viewing rectangle $[0, 20] \times [0, 100{,}000]$.
3. Use **TRACE** to find $A_1(20), A_2(20), \ldots, A_5(20)$ and interpret your results.

Equating the two expressions gives

$$P(1 + R) = P\left(1 + \frac{r}{m}\right)^m$$

$$1 + R = \left(1 + \frac{r}{m}\right)^m \qquad \text{(Dividing both sides by } P\text{)}$$

or, upon solving for R, we obtain the following formula for computing the effective rate of interest.

Effective Rate of Interest Formula

$$r_{\text{eff}} = \left(1 + \frac{r}{m}\right)^m - 1 \qquad \textbf{(4)}$$

where

r_{eff} = Effective rate of interest
r = Nominal interest rate per year
m = Number of conversion periods per year

EXAMPLE 4 Find the effective rate of interest corresponding to a nominal rate of 8% per year compounded (a) annually, (b) semiannually, (c) quarterly, and (d) monthly.

SOLUTION ✓ **a.** The effective rate of interest corresponding to a nominal rate of 8% per year compounded annually is of course given by 8% per year. This result is also confirmed by using Equation (4) with $r = 0.08$ and $m = 1$. Thus,

$$r_{\text{eff}} = (1 + 0.08) - 1 = 0.08$$

b. Let $r = 0.08$ and $m = 2$. Then, Equation (4) yields

$$\begin{aligned} r_{\text{eff}} &= \left(1 + \frac{0.08}{2}\right)^2 - 1 \\ &= (1.04)^2 - 1 \\ &= 0.0816 \end{aligned}$$

so the required effective rate is 8.16% per year.

c. Let $r = 0.08$ and $m = 4$. Then, Equation (4) yields

$$\begin{aligned} r_{\text{eff}} &= \left(1 + \frac{0.08}{4}\right)^4 - 1 \\ &= (1.02)^4 - 1 \\ &= 0.08243 \end{aligned}$$

so the corresponding effective rate in this case is 8.243% per year.

d. Let $r = 0.08$ and $m = 12$. Then, Equation (4) yields

$$\begin{aligned} r_{\text{eff}} &= \left(1 + \frac{0.08}{12}\right)^{12} - 1 \\ &= 0.08300 \end{aligned}$$

so the corresponding effective rate in this case is 8.300% per year. ■■■■

Group Discussion

Recall the effective rate of interest formula:

$$r_{\text{eff}} = \left(1 + \frac{r}{m}\right)^m - 1$$

1. Show that

$$r = m[(1 + r_{\text{eff}})^{1/m} - 1]$$

2. A certificate of deposit (CD) is known to have an effective rate of 8.3%. If interest is compounded monthly, find the nominal rate of interest by using the result of part 1.

Now, if the effective rate of interest r_{eff} is known, the accumulated amount after t years on an investment of P dollars may be more readily computed by using the formula

$$A = P(1 + r_{\text{eff}})^t$$

The 1968 Truth in Lending Act passed by Congress requires that the effective rate of interest be disclosed in all contracts involving interest charges. The passage of this act has benefited consumers because they now have a common basis for comparing the various nominal rates quoted by different financial institutions. Furthermore, knowing the effective rate enables consumers to compute the actual charges involved in a transaction. Thus, if the effective rates of interest found in Example 4 were known, the accumulated values of Example 3, shown in Table 5.2, could have been readily found.

Table 5.2

Nominal Rate	Frequency of Interest Payment	Effective Rate	Initial Investment	Accumulated Amount After 3 Years
8%	Annually	8%	\$1000	$1000(1 + 0.08)^3 = \$1259.71$
8	Semiannually	8.16	1000	$1000(1 + 0.0816)^3 = 1265.32$
8	Quarterly	8.243	1000	$1000(1 + 0.08243)^3 = 1268.23$
8	Monthly	8.300	1000	$1000(1 + 0.08300)^3 = 1270.24$

PRESENT VALUE

Let's return to the compound interest Formula (3), which expresses the accumulated amount at the end of n periods when interest at the rate of r is compounded m times a year. The principal P in (3) is often referred to as the **present value,** and the accumulated value A is called the **future value,** since it is realized at a future date. In certain instances an investor may wish to determine how much money he should invest now, at a fixed rate of interest, so that he will realize a certain sum at some future date. This problem may be solved by expressing P in terms of A. Thus, from (3) we find

$$P = A(1 + i)^{-n}$$

Here, as before, $i = r/m$, where m is the number of conversion periods per year.

Present Value Formula for Compound Interest

$$P = A(1 + i)^{-n} \qquad (5)$$

How much money should be deposited in a bank paying interest at the rate of 6% per year compounded monthly so that at the end of 3 years the accumulated amount will be $20,000?

Here, $r = 0.06$ and $m = 12$, so $i = 0.06/12 = 0.005$ and $n = (3)(12) = 36$. Thus, the problem is to determine P given that $A = 20{,}000$. Using Equation (5) we obtain

$$\begin{aligned} P &= 20{,}000(1.005)^{-36} \\ &\approx 16{,}713 \end{aligned}$$

or $16,713. ■■■■

Find the present value of $49,158.60 due in 5 years at an interest rate of 10% per year compounded quarterly.

Using Formula (5) with $r = 0.1$ and $m = 4$, so that $i = 0.1/4 = 0.025$, $n = (4)(5) = 20$, and $A = 49{,}158.6$, we obtain

$$P = (49{,}158.6)(1.025)^{-20} = 30{,}000.07$$

or approximately $30,000. ■■■■

APPLICATIONS

The returns on certain investments such as zero coupon certificates of deposit (CDs) and zero coupon bonds are compared by quoting the time it takes for each investment to triple, or even quadruple. These calculations make use of the compound interest Formula (3).

EXAMPLE 7

Jane has narrowed her investment options down to two:

a. Purchase a CD that matures in 12 years and pays interest upon maturity at the rate of 10% per year compounded daily (assume 365 days in a year).
b. Purchase a zero coupon CD that will triple her investment in the same period.

Which option will optimize her investment?

SOLUTION ✓

Let's compute the accumulated amount under option (a). Here,

$$r = 0.10, \qquad m = 365, \qquad t = 12$$

so $n = 12(365) = 4380$ and $i = 0.10/365 \approx 0.0002740$. The accumulated amount at the end of 12 years (after 4380 conversion periods) is

$$A = P(1.0002740)^{4380} \approx 3.32P$$

or \$3.32P. If Jane chooses option (b), the accumulated amount of her investment after 12 years will be \$3P. Therefore, she should choose option (a). ■■■■

EXAMPLE 8

Moesha has an Individual Retirement Account (IRA) with a brokerage firm. Her money is invested in a money-market mutual fund that pays interest on a daily basis. Over a 2-year period in which no deposits or withdrawals were made, her account grew from \$4500 to \$5268.24. Find the effective rate at which Moesha's account was earning interest over that period (assume 365 days in a year).

SOLUTION ✓

Let r_{eff} denote the required effective rate of interest. We have

$$\begin{aligned} 5268.24 &= 4500(1 + r_{\text{eff}})^2 \\ (1 + r_{\text{eff}})^2 &= 1.17072 \\ 1 + r_{\text{eff}} &= 1.081998 \qquad \text{(Taking the square root on both sides)} \end{aligned}$$

or $r_{\text{eff}} = 0.081998$. Therefore, the required effective rate is 8.20% per year. ■■■■

SELF-CHECK EXERCISES 5.1

1. Find the present value of \$20,000 due in three years at an interest rate of 12%/year compounded monthly.
2. Paul is a retiree living on Social Security and the income from his investment. Currently, his \$100,000 investment in a 1-year CD is yielding 10.6% interest compounded daily. If he reinvests the principal (\$100,000) on the due date of the CD in another 1-year CD paying 9.2% interest compounded daily, find the net decrease in his yearly income from his investment.

Solutions to Self-Check Exercises 5.1 can be found on page 295.

5.1 Exercises

A calculator is recommended for this exercise set.

1. Find the simple interest on a \$500 investment made for 2 years at an interest rate of 8%/year. What is the accumulated amount?

2. Find the simple interest on a \$1000 investment made for 3 years at an interest rate of 5%/year. What is the accumulated amount?

3. Find the accumulated amount at the end of 9 months on an \$800 deposit in a bank paying simple interest at a rate of 6%/year.

4. Find the accumulated amount at the end of 8 months on a \$1200 bank deposit paying simple interest at a rate of 7%/year.

5. If the accumulated amount is \$1160 at the end of 2 years and the simple rate of interest is 8%/year, what is the principal?

6. A bank deposit paying simple interest at the rate of 5%/year grew to a sum of \$3100 in 10 months. Find the principal.

7. How many days will it take for a sum of \$1000 to earn \$20 interest if it is deposited in a bank paying ordinary simple interest at the rate of 5%/year? (Use a 365-day year.)

8. How many days will it take for a sum of \$1500 to earn \$25 interest if it is deposited in a bank paying 5%/year? (Use a 365-day year.)

9. A bank deposit paying simple interest grew from an initial sum of \$1000 to a sum of \$1075 in 9 months. Find the interest rate.

10. Determine the simple interest rate at which \$1200 will grow to \$1250 in 8 months.

In Exercises 11–20, find the accumulated amount *A* if the principal *P* is invested at the interest rate of *r* per year for *t* years.

11. $P = \$1000$, $r = 7\%$, $t = 8$, compounded annually

12. $P = \$1000$, $r = 8\frac{1}{2}\%$, $t = 6$, compounded annually

13. $P = \$2500$, $r = 7\%$, $t = 10$, compounded semiannually

14. $P = \$2500$, $r = 9\%$, $t = 10.5$, compounded semiannually

15. $P = \$12{,}000$, $r = 8\%$, $t = 10.5$, compounded quarterly

16. $P = \$42{,}000$, $r = 7\frac{3}{4}\%$, $t = 8$, compounded quarterly

17. $P = \$150{,}000$, $r = 14\%$, $t = 4$, compounded monthly

18. $P = \$180{,}000$, $r = 9\%$, $t = 6\frac{1}{4}$, compounded monthly

19. $P = \$150{,}000$, $r = 12\%$, $t = 3$, compounded daily

20. $P = \$200{,}000$, $r = 8\%$, $t = 4$, compounded daily

In Exercises 21–24, find the effective rate corresponding to the given nominal rate.

21. 10% compounded semiannually

22. 9% compounded quarterly

23. 8% compounded monthly

24. 8% compounded daily

In Exercises 25–28, find the present value of \$40,000 due in 4 years at the given rate of interest.

25. 8% compounded semiannually

26. 8% compounded quarterly

27. 7% compounded monthly

28. 9% compounded daily

29. CONSUMER DECISIONS Mitchell has been given the option of either paying his \$300 bill now or settling it for \$306 after 1 month. If he chooses to pay after 1 month, find the simple interest rate at which he would be charged.

30. Write Equation (1b) in the slope-intercept form and interpret the meaning of the slope and the A-intercept in terms of r and P.

    Hint: Refer to Figure 5.1.

31. HOSPITAL COSTS If the cost of a semiprivate room in a hospital was \$380/day 5 years ago and hospital costs have risen at the rate of 8%/year since that time, what rate would you expect to pay for a semiprivate room today?

32. FAMILY FOOD EXPENDITURE Today a typical family of four spends \$600 monthly for food. If inflation occurs at the

rate of 4%/year over the next 6 years, how much should the typical family of four expect to spend for food 6 years from now?

33. **HOUSING APPRECIATION** The Kwans are planning to buy a house 4 years from now. Housing experts in their area have estimated that the cost of a home will increase at a rate of 5%/year during that period. If this economic prediction holds true, how much can the Kwans expect to pay for a house that currently costs $150,000?

34. **ELECTRICITY CONSUMPTION** A utility company in a western city of the United States expects the consumption of electricity to increase by 8%/year during the next decade, due mainly to the expected increase in population. If consumption does increase at this rate, find the amount by which the utility company will have to increase its generating capacity to meet the needs of the area at the end of the decade.

35. **PENSION FUNDS** The managers of a pension fund have invested $1.5 million in U.S. government certificates of deposit that pay interest at the rate of 9.5%/year compounded semiannually over a period of 10 years. At the end of this period, how much will the investment be worth?

36. **TRUST FUNDS** A young man is the beneficiary of a trust fund established for him 21 years ago at his birth. If the original amount placed in trust was $10,000, how much will he receive if the money has earned interest at the rate of 8%/year compounded annually? Compounded quarterly? Compounded monthly?

37. **INVESTMENT PLANNING** Find how much money should be deposited in a bank paying interest at the rate of 8.5%/year compounded quarterly so that at the end of 5 years the accumulated amount will be $40,000.

38. **PROMISSORY NOTES** An individual purchased a 4-year, $10,000 promissory note with an interest rate of 8.5%/year compounded semiannually. How much did the note cost?

39. **FINANCING A COLLEGE EDUCATION** The parents of a child have just come into a large inheritance and wish to establish a trust fund for her college education. If they estimate that they will need $100,000 in 13 years, how much should they set aside in the trust now if they can invest the money at $8\frac{1}{2}$%/year compounded annually? Compounded semiannually? Compounded quarterly?

40. **INVESTMENTS** Anthony invested a sum of money 5 years ago in a savings account that has since paid interest at the rate of 8%/year compounded quarterly. His investment is now worth $22,289.22. How much did he originally invest?

41. **LOAN CONSOLIDATION** The proprietors of The Coachmen Inn secured two loans from the Union Bank: one for $8000 due in 3 years and one for $15,000 due in 6 years, both at an interest rate of 10%/year compounded semiannually. The bank has agreed to allow the two loans to be consolidated into one loan payable in five years at the same interest rate. What amount will the proprietors of the inn be required to pay the bank at the end of 5 years?

42. **EFFECTIVE RATE OF INTEREST** Find the effective rate of interest corresponding to a nominal rate of 9%/year compounded annually, semiannually, quarterly, and monthly.

43. **HOUSING APPRECIATION** Georgia purchased a house in 1994 for 100,000. In 2000 she sold the house and made a net profit of $28,000. Find the effective annual rate of return on her investment over the 6-year period.

44. **COMMON STOCK TRANSACTION** Steven purchased 1000 shares of a certain stock for $25,250 (including commissions). He sold the shares 2 years later and received $32,100 after deducting commissions. Find the effective annual rate of return on his investment over the 2-year period.

45. **ZERO COUPON BONDS** Nina purchased a zero coupon bond for $6595.37. The bond matures in 5 years and has a face value of $10,000. Find the effective rate of interest for the bond if interest is compounded semiannually.
Hint: Assume that the purchase price of the bond is the initial investment and that the face value of the bond is the accumulated amount.

46. **MONEY-MARKET MUTUAL FUNDS** Carlos invested $5000 in a money-market mutual fund that pays interest on a daily basis. The balance in his account at the end of 8 months (245 days) was $5347.09. Find the effective rate at which Carlos' account earned interest over this period (assume 365 days in a year).

47. The simple interest formula $A = P(1 + rt)$ [Formula (1b)] can be written in the form $A = Prt + P$, which is the slope-intercept form of a straight line with slope Pr and A-intercept P.
a. Describe the family of straight lines obtained by keeping the value of r fixed and allowing the value of P to vary. Interpret your results.
b. Describe the family of straight lines obtained by keeping the value of P fixed and allowing the value of r to vary. Interpret your results.

In Exercises 48–51, determine whether the statement is true or false. If it is true, explain why it is true. If it is false, give an example to show why it is false.

48. When simple interest is used, the accumulated amount is a linear function of t.

49. If compound interest is converted annually, then the accumulated amount after t years is the same as the accumulated amount under simple interest over t years.

50. If interest is compounded annually, then the effective rate is the same as the nominal rate.

51. Susan's salary increased from \$40,000/year to \$50,000/year over a 5-year period. Therefore, Susan got annual increases of 5% over that period.

Spreadsheet examples and exercises for this section that may be solved using the Microsoft® Excel program are given at the Brooks/Cole Web site: http://www.brookscole.com/math/authors/tans/

SOLUTIONS TO SELF-CHECK EXERCISES 5.1

1. Using Equation (5) with $r = 0.12$ and $m = 12$, so that

$$i = \frac{0.12}{12} = 0.01, \quad n = (12)(3) = 36, \quad \text{and} \quad A = 20{,}000$$

we find the required present value to be

$$P = 20{,}000(1.01)^{-36} = 13{,}978.50$$

or \$13,978.50.

2. The accumulated amount of Paul's current investment is found by using Equation (3) with $P = 100{,}000$, $r = 0.106$, and $m = 365$. Thus,

$$i = \frac{0.106}{365} \quad \text{and} \quad n = 365$$

so the required accumulated amount is

$$A = 100{,}000\left(1 + \frac{0.106}{365}\right)^{365} \approx 111{,}180.48$$

or \$111,180.48. Next, we compute the accumulated amount of Paul's reinvestment. Once again, using (3) with $P = 100{,}000$, $r = 0.092$, and $m = 365$ so that

$$i = \frac{0.092}{365} \quad \text{and} \quad n = 365$$

we find the required accumulated amount in this case to be

$$\bar{A} = 100{,}000\left(1 + \frac{0.092}{365}\right)^{365}$$

or \$109,635.21. Therefore, Paul can expect to experience a net decrease in yearly income of

$$111{,}180.48 - 109{,}635.21$$

or \$1545.27.

Using Technology

The graphing calculator can be programmed to facilitate the calculation of quantities encountered in the mathematics of finance such as the accumulated amount under compound interest and the effective rate of interest corresponding to a nominal rate.

Finding the Accumulated Amount of an Investment

The first program, which we will call **COMPINT**, enables us to compute the accumulated amount A when P dollars is invested for a term of t years and earns interest at the rate of r percent per year compounded m times a year [see Formula (3), page 287].

```
: PROGRAM: COMPINT
: Disp "P"
: Input P
: Disp "r"
: Input r
: Disp "t"
: Input t
: Disp "m"
: Input m
: P(1 + r/m)^(m*t)→A
: Disp "AMOUNT IS"
: Disp A
```

EXAMPLE 1 Find the accumulated amount after 10 years if $500 is invested at 10% per year compounded monthly.

SOLUTION ✓ First, we call the program **COMPINT**. Next, we enter 500 for the value of P, 0.1 for the value of r, 10 for the value of t, and 12 for the value of m. We find the accumulated amount to be $1353.52.

Finding the Effective Rate of Interest

The second program, which we will call **EFFRATE**, enables us to compute the effective rate of interest r_{eff} corresponding to a nominal interest rate r per year converted m times per year [see Formula (4), page 289].

```
: PROGRAM: EFFRATE
: Disp "r"
: Input r
: Disp "m"
: Input m
: (1 + r/m)^m - 1 → E
: Disp "EFFECTIVE RATE IS"
: Disp E
```

Find the effective rate of interest corresponding to a nominal rate of 10% per year compounded monthly.

SOLUTION ✓

First, we call the program **EFFRATE**. Then, we enter 0.1 for the nominal rate r and 12 for the number of conversion periods m. We find that the effective rate is 0.1047, or 10.47% per year.

Exercises

Use the COMPINT program to solve Exercises 1–4.

1. Find the accumulated amount A if \$5000 is invested at the interest rate of $5\frac{3}{8}$%/year compounded monthly for 3 years.

2. Find the accumulated amount A if \$2850 is invested at the interest rate of $6\frac{5}{8}$%/year compounded monthly for 4 years.

3. Find the accumulated amount A if \$327.35 is invested at the interest rate of $5\frac{1}{3}$%/year compounded daily for 7 years.

4. Find the accumulated amount A if \$327.35 is invested at the interest rate of $6\frac{7}{8}$%/year compounded daily for 8 years.

Use the EFFRATE program to solve Exercises 5–8.

5. Find the effective rate corresponding to $8\frac{2}{3}$%/year compounded quarterly.

6. Find the effective rate corresponding to $10\frac{5}{8}$%/year compounded monthly.

7. Find the effective rate corresponding to $9\frac{3}{4}$%/year compounded monthly.

8. Find the effective rate corresponding to $4\frac{3}{8}$%/year compounded quarterly.

9. Write a program to calculate the present value for compound interest [Formula (5), page 291].

Use the program you wrote for Exercise 9 to solve Exercises 10–13.

10. Find the present value of \$38,000 due in 3 years at $8\frac{1}{4}$%/year compounded quarterly.

11. Find the present value of \$150,000 due in 5 years at $9\frac{3}{8}$%/year compounded monthly.

12. Find the present value of \$67,456 due in 3 years at $7\frac{7}{8}$%/year compounded monthly.

13. Find the present value of \$111,000 due in 5 years at $11\frac{5}{8}$%/year compounded monthly.

5.2 Annuities

FUTURE VALUE OF AN ANNUITY

An **annuity** is a sequence of payments made at regular time intervals. The time period in which these payments are made is called the **term** of the annuity. Depending on whether the term is given by a *fixed time interval,* a time interval that begins at a definite date but extends indefinitely, or one that is not fixed in advance, an annuity is called an **annuity certain,** a *perpetuity,* or a *contingent annuity,* respectively. In general, the payments in an annuity need not be equal, but in many important applications they are equal. In this section we assume that annuity payments are equal. Examples of annuities are regular deposits to a savings account, monthly home mortgage payments, and monthly insurance payments.

Annuities are also classified by payment dates. An annuity in which the payments are made at the *end* of each payment period is called an **ordinary annuity,** whereas an annuity in which the payments are made at the beginning of each period is called an *annuity due.* Furthermore, an annuity in which the payment period coincides with the interest conversion period is called a **simple annuity,** whereas an annuity in which the payment period differs from the interest conversion period is called a *complex annuity.*

In this section we consider ordinary annuities that are certain and simple, with periodic payments that are equal in size. In other words, we study annuities that are subject to the following conditions:

1. The terms are given by fixed time intervals.
2. The periodic payments are equal in size.
3. The payments are made at the *end* of the payment periods.
4. The payment periods coincide with the interest conversion periods.

To find a formula for the accumulated amount S of an annuity, suppose a sum of \$100 is paid into an account at the end of each quarter over a period of 3 years. Furthermore, suppose the account earns interest on the deposit at the rate of 8% per year, compounded quarterly. Then, the first payment of \$100 made at the end of the first quarter earns interest at the rate of 8% per year compounded four times a year (or $8/4 = 2\%$ per quarter) over the remaining 11 quarters and therefore, by the compound interest formula, has an accumulated amount of

$$100\left(1 + \frac{0.08}{4}\right)^{11} \quad \text{or} \quad 100(1 + 0.02)^{11}$$

dollars at the end of the term of the annuity (Figure 5.2).

The second payment of \$100 made at the end of the second quarter earns interest at the same rate over the remaining 10 quarters and therefore has an accumulated amount of

$$100(1 + 0.02)^{10}$$

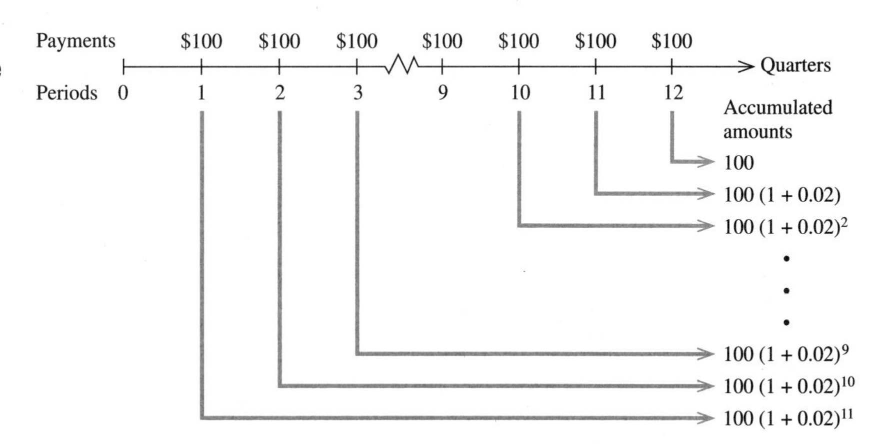

FIGURE 5.2
The sum of the accumulated amounts is the amount of the annuity.

dollars at the end of the term of the annuity, and so on. The last payment earns no interest since it is due at the end of the term. The amount of the annuity is obtained by adding all the terms in Figure 5.2. Thus,

$$S = 100 + 100(1 + 0.02) + 100(1 + 0.02)^2 + \cdots + 100(1 + 0.02)^{11}$$

The sum on the right is the sum of the first n terms of a *geometric progression* with first term R and common ratio $(1 + i)$. We will show in Section 5.4 that the sum S can be written in the more compact form

$$S = 100\left[\frac{(1 + 0.02)^{12} - 1}{0.02}\right]$$
$$\approx 1341.21$$

or approximately $1341.21.

To find a general formula for the accumulated amount S of an annuity, suppose a sum of $\$R$ is paid into an account at the end of each period for n periods and the account earns interest at the rate of i per period. Then, proceeding as we did with the numerical example, we see that

$$\begin{aligned} S &= R + R(1 + i) + R(1 + i)^2 + \cdots + R(1 + i)^{n-1} \\ &= R\left[\frac{(1 + i)^n - 1}{i}\right] \end{aligned} \tag{6}$$

The expression inside the brackets is commonly denoted by $s_{\overline{n}|i}$ (read "s angle n at i") and is called the **compound-amount factor.** Extensive tables have been constructed that give values of $s_{\overline{n}|i}$ for different values of i and n (such as Table 1, Appendix C). In terms of the compound-amount factor,

$$S = Rs_{\overline{n}|i} \tag{7}$$

The quantity S in Equations (6) and (7) is realizable at some future date and is accordingly called the future value of an annuity.

Future Value of an Annuity

The **future value S of an annuity** of n payments of R dollars each, paid at the end of each investment period into an account that earns interest at the rate of i per period, is

$$S = R\left[\frac{(1+i)^n - 1}{i}\right]$$

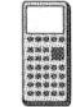

EXAMPLE 1

Find the amount of an ordinary annuity of 12 monthly payments of \$100 that earn interest at 12% per year compounded monthly.

SOLUTION ✓

Since i is the interest rate per *period* and interest is compounded monthly in this case, we have $i = 0.12/12 = 0.01$. Using Equation (6) with $R = 100$, $n = 12$, and $i = 0.01$, we have

$$\begin{aligned} S &= \frac{100[(1.01)^{12} - 1]}{0.01} \\ &= 1268.25 \qquad \text{(Using a calculator)} \end{aligned}$$

or \$1268.25. The same result is obtained by observing that

$$\begin{aligned} S &= 100 s_{\overline{12}|0.01} \\ &= 100(12.6825) \\ &= 1268.25 \qquad \text{(Using Table 1, Appendix C)} \end{aligned}$$

Group Discussion

Future value S of an annuity due:

1. Consider an annuity satisfying Conditions 1, 2, and 4 on page 298, but with Condition 3 replaced by the condition that payments are made at the beginning of the payment periods. By using an argument similar to that used to establish Formula (6), show that the future value S of an annuity due of n payments of R dollars each, paid at the *beginning* of each investment into an account that earns interest at the rate of i per period, is

$$S = R(1+i)\left[\frac{(1+i)^n - 1}{i}\right]$$

2. Use the result of part 1 to see how much your nest egg will be at age 65 if you start saving \$4000 annually at age 30, assuming a 10% average annual return; if you start saving at 35; if you start saving at 40. [Moral of the story: It is never too early to start saving!]

Exploring with Technology

Refer to the group discussion problem on page 300.

1. Show that if $R = 4000$ and $i = 0.1$, then $S = 44{,}000[(1.1)^n - 1]$. Using a graphing utility, plot the graph of $f(x) = 44{,}000[(1.1)^x - 1]]$, using the viewing rectangle $[0, 40] \times [0, 1{,}200{,}000]$.
2. Verify the results of part 1 of the group discussion problem by evaluating $f(35)$, $f(30)$, and $f(25)$ using the EVAL function.

PRESENT VALUE OF AN ANNUITY

In certain instances you may want to determine the current value P of a sequence of equal periodic payments that will be made over a certain period of time. After each payment is made, the new balance continues to earn interest at some nominal rate. The amount P is referred to as the *present value of an annuity.*

To derive a formula for determining the present value P of an annuity, we may argue as follows. The amount P invested now and earning interest at the rate of i per period will have an accumulated value of $P(1 + i)^n$ at the end of n periods. But this must be equal to the future value of the annuity S given by Formula (6). Therefore, equating the two expressions, we have

$$P(1 + i)^n = R\left[\frac{(1 + i)^n - 1}{i}\right]$$

Multiplying both sides of this equation by $(1 + i)^{-n}$ gives

$$\begin{aligned}
P &= R(1 + i)^{-n}\left[\frac{(1 + i)^n - 1}{i}\right] \\
&= R\left[\frac{(1 + i)^n(1 + i)^{-n} - (1 + i)^{-n}}{i}\right] \qquad [(1 + i)^n(1 + i)^{-n} = 1] \\
&= R\left[\frac{1 - (1 + i)^{-n}}{i}\right] \\
&= Ra_{\overline{n}|i}
\end{aligned}$$

where the factor $a_{\overline{n}|i}$ (read "a angle n at i") represents the expression inside the brackets. Extensive tables have also been constructed giving values of $a_{\overline{n}|i}$ for different values of i and n (see Table 1, Appendix C).

Present Value of an Annuity

The present value P of an annuity of n payments of R dollars each, paid at the end of each investment period into an account that earns interest at the rate of i per period, is

$$P = R\left[\frac{1-(1+i)^{-n}}{i}\right] \tag{8}$$

EXAMPLE 2 Find the present value of an ordinary annuity of 24 payments of $100 each made monthly and earning interest at 9% per year compounded monthly.

SOLUTION ✓ Here, $R = 100$, $i = r/m = 0.09/12 = 0.0075$, and $n = 24$, so by Formula (8)

$$\begin{aligned} P &= \frac{100[1-(1.0075)^{-24}]}{0.0075} \\ &= 2188.92 \end{aligned}$$

or $2188.92. The same result may be obtained by using Table 1, Appendix C. Thus,

$$\begin{aligned} P &= 100a_{\overline{24}|0.0075} \\ &= 100(21.8892) \\ &= 2188.92 \end{aligned}$$

■■■■

APPLICATIONS

EXAMPLE 3 As a savings program toward their child's college education, parents decide to deposit $100 at the end of every month into a bank account paying interest at the rate of 6% per year compounded monthly. If the savings program began when the child was 6 years old, how much money would have accumulated by the time the child turns 18?

SOLUTION ✓ By the time the child turns 18, the parents would have made 144 deposits into the account. Thus, $n = 144$. Furthermore, we have $R = 100$, $r = 0.06$, and $m = 12$, so $i = 0.06/12 = 0.005$. Using Equation (6), we find that the amount of money that would have accumulated is given by

$$\begin{aligned} S &= \frac{100[(1.005)^{144}-1]}{0.005} \\ &\approx 21{,}015 \end{aligned}$$

or $21,015.

■■■■

EXAMPLE 4 After making a down payment of $2000 for an automobile, Murphy paid $200 per month for 36 months with interest charged at 12% per year compounded

monthly on the unpaid balance. What was the original cost of the car? What portion of Murphy's total car payments went toward interest charges?

SOLUTION ✓ The loan taken up by Murphy is given by the present value of the annuity

$$P = \frac{200[1 - (1.01)^{-36}]}{0.01} = 200a_{\overline{36}|0.01}$$
$$\approx 6021.50$$

or \$6021.50. Therefore, the original cost of the automobile is \$8021.50 (\$6021.50 plus the \$2000 down payment). The interest charges paid by Murphy are given by (36)(200) − 6021.50, or \$1178.50. ■■■■

One important application of annuities is in the area of tax planning. During the 1980s, Congress created many tax-sheltered retirement savings plans, such as Individual Retirement Accounts (IRAs), Keogh plans, and Simplified Employee Pension (SEP) plans. These plans are examples of annuities in which the individual is allowed to make contributions (which are often tax deductible) to an investment account. The amount of the contribution is limited by congressional legislation. The taxes on the contributions and/or the interest accumulated in these accounts are deferred until the money is withdrawn, ideally during retirement, when tax brackets should be lower. In the interim period, the individual has the benefit of tax-free growth on his or her investment.

Suppose, for example, you are eligible to make a fully deductible contribution to an IRA and you are in a marginal tax bracket of 28%. Additionally, suppose you receive a year-end bonus of \$2000 from your employer and have the option of depositing the \$2000 into either an IRA or a regular savings account, both accounts earning interest at an effective annual rate of 8% per year. If you choose to invest your bonus in a regular savings account, you will first have to pay taxes on the \$2000, leaving \$1440 to invest. At the end of 1 year, you will also have to pay taxes on the interest earned, leaving you with

$$\boxed{\text{Accumulated amount} - \text{Tax on interest} = \text{Net amount}}$$

$$1555.20 - 32.26 = 1522.94$$

or \$1522.94.

On the other hand, if you put the money into the IRA account, the entire sum will earn interest, and at the end of 1 year you will have (1.08)(\$2000), or \$2160, in your account. Of course, you will still have to pay taxes on this money when you withdraw it, but you will have gained the advantage of tax-free growth of the larger principal over the years. The disadvantage of this option is that if you withdraw the money before you reach the age of $59\frac{1}{2}$, you will be liable for taxes on both your contributions and the interest earned and you will also have to pay a 10% penalty.

REMARK In practice, the size of the contributions an individual might make to the various retirement plans might vary from year to year. Also, he or she might make the contributions at different payment periods. To simplify our discussion, we will consider examples in which fixed payments are made at regular intervals.

EXAMPLE 5 Caroline is planning to make the maximum contribution of $2000 on January 31 of each year into an IRA earning interest at an effective rate of 9% per year. After she makes her 25th payment on January 31 of the year following her retirement, how much will she have in her IRA?

The amount of money Caroline will have after her 25th payment into her account is found by using Equation (6) with $R = 2000$, $r = 0.09$, and $m = 1$, so $i = r/m = 0.09$ and $n = 25$. The required amount is given by

$$S = \frac{2000[(1.09)^{25} - 1]}{0.09}$$
$$\approx 169{,}401.79$$

or $169,401.79.

Tax-deferred annuities are another type of investment vehicle that allows an individual to build assets for retirement, college funds, or other future needs. The advantage gained in this type of investment is that the tax on the accumulated interest is deferred to a later date. Note that in this type of investment the contributions themselves are not tax deductible. At first glance the advantage thus gained may seem to be relatively inconsequential, but its true effect is illustrated by the next example.

EXAMPLE 6 Both Clark and Colby are salaried individuals, 45 years of age, who are saving for their retirement 20 years from now. Both Clark and Colby are also in the 28% marginal tax bracket. Clark makes a $1000 contribution annually on December 31 into a savings account earning an effective rate of 8% per year. At the same time, Colby makes a $1000 annual payment to an insurance company for a tax-deferred annuity. The annuity also earns interest at an effective rate of 8% per year. (Assume that both men remain in the same tax bracket throughout this period and disregard state income taxes.)

a. Calculate how much each man will have in his investment account at the end of 20 years.
b. Compute the interest earned on each account.
c. Show that even if the interest on Colby's investment were subjected to a tax of 28% upon withdrawal of his investment at the end of 20 years, the net accumulated amount of his investment would still be greater than that of Clark's.

SOLUTION ✓ **a.** Because Clark is in the 28% marginal tax bracket, the net yield for his investment is (0.72)(8), or 5.76%, per year.

Using Formula (6) with $R = 1000$, $r = 0.0576$, and $m = 1$, so that $i = 0.0576$ and $n = 20$, we see that Clark's investment will be worth

$$S = \frac{1000[(1 + 0.0576)^{20} - 1]}{0.0576}$$
$$\approx 35{,}850.49$$

or \$35,850.49 at his retirement.

Colby has a tax-sheltered investment with an effective yield of 8% per year. Using Formula (6) with $R = 1000$, $r = 0.08$, and $m = 1$, so that $i = 0.08$ and $n = 20$, we see that Colby's investment will be worth

$$S = \frac{1000[(1 + 0.08)^{20} - 1]}{0.08}$$
$$\approx 45{,}761.96$$

or \$45,761.96 at his retirement.

b. Each man will have paid 20(1000), or 20,000 dollars, into his account. Therefore, the total interest earned in Clark's account will be (35,850.49 − 20,000), or \$15,850.49, whereas the total interest earned in Colby's account will be (45,761.96 − 20,000), or \$25,761.96.

c. From part (b) we see that the total interest earned in Colby's account will be \$25,761.96. If it were taxed at 28%, he would still end up with (0.72)(25,761.96), or \$18,548.61. This is larger than the total interest of \$15,850.49 earned by Clark. ■■■■

SELF-CHECK EXERCISES 5.2

1. Phyliss Fletcher opened an IRA on January 31, 1990, with a contribution of \$2000. She plans to make a contribution of \$2000 thereafter on January 31 of each year until her retirement in the year 2009 (20 payments). If the account earns interest at the rate of 8%/year compounded yearly, how much will Phyliss have in her account when she retires?
2. The Denver Wildcatting Company has an immediate need for a loan. In an agreement worked out with its banker, Denver assigns its royalty income of \$4800/month for the next 3 years from certain oil properties to the bank, with the first payment due at the end of the first month. If the bank charges interest at the rate of 9%/year compounded monthly, what is the amount of the loan negotiated between the parties?

Solutions to Self-Check Exercises 5.2 can be found on page 307.

5.2 Exercises

A calculator is recommended for this exercise set.

In Exercises 1–6, find the amount (future value) of each ordinary annuity.

1. $1000 a year for 10 years at 10%/year compounded annually
2. $1500 per semiannual period for 8 years at 9%/year compounded semiannually
3. $1800 per quarter for 6 years at 8%/year compounded quarterly
4. $500 per semiannual period for 12 years at 11%/year compounded semiannually
5. $600 per quarter for 9 years at 12%/year compounded quarterly
6. $150 per month for 15 years at 10%/year compounded monthly

In Exercises 7–12, find the present value of each ordinary annuity.

7. $5000 a year for 8 years at 8%/year compounded annually
8. $1200 per semiannual period for 6 years at 10%/year compounded semiannually
9. $4000 a year for 5 years at 9%/year compounded yearly
10. $3000 per semiannual period for 6 years at 11%/year compounded semiannually
11. $800 per quarter for 7 years at 12%/year compounded quarterly
12. $150 per month for 10 years at 8%/year compounded monthly
13. IRAs If a merchant deposits $1500 annually at the end of each tax year in an IRA account paying interest at the rate of 8%/year compounded annually, how much will she have in her account at the end of 25 years?
14. SAVINGS ACCOUNTS If Jackson deposits $100 at the end of each month in a savings account earning interest at the rate of 8%/year compounded monthly, how much will he have on deposit in his savings account at the end of 6 years, assuming that he makes no withdrawals during that period?
15. SAVINGS ACCOUNTS Linda has joined a "Christmas Fund Club" at her bank. At the end of every month, December through October inclusive, she will make a deposit of $40 in her fund. If the money earns interest at the rate of 7%/year compounded monthly, how much will she have in her account on December 1 of the following year?
16. KEOGH ACCOUNTS Robin, who is self-employed, contributes $5000 a year into a Keogh account. How much will he have in the account after 25 years if the account earns interest at the rate of 8.5%/year compounded yearly?
17. RETIREMENT PLANNING As a fringe benefit for the past 12 years, Colin's employer has contributed $100 at the end of each month into an employee retirement account for Colin that pays interest at the rate of 7%/year compounded monthly. Colin has also contributed $2000 at the end of each of the last 8 years into an IRA that pays interest at the rate of 9%/year compounded yearly. How much does Colin have in his retirement fund at this time?
18. SAVINGS ACCOUNTS The Pirerras are planning to go to Europe 3 years from now and have agreed to set aside $150 each month for their trip. If they deposit this money at the end of each month into a savings account paying interest at the rate of 8%/year compounded monthly, how much money will be in their travel fund at the end of the third year?
19. AUTO LEASING The Betzes have leased an auto for 2 years at $450/month. If money is worth 9%/year compounded monthly, what is the equivalent cash payment (present value) of this annuity?
20. AUTO FINANCING Lupe made a down payment of $2000 toward the purchase of a new car. To pay the balance of the purchase price, she has secured a loan from her bank at the rate of 12%/year compounded monthly. Under the terms of her finance agreement, she is required to make payments of $210/month for 36 months. What is the cash price of the car?
21. INSTALLMENT PLANS R. G. Pierce Publishing Company sells encyclopedias under two payment plans: cash or installment. Under the installment plan, the customer pays $22/month over 3 years with interest charged on the balance at a rate of 18%/year compounded monthly. Find the cash price for a set of encyclopedias if it is equivalent to the price paid by a customer using the installment plan.

22. **Lottery Payouts** A state lottery commission pays the winner of the "Million Dollar" lottery 20 installments of \$50,000/year. The commission makes the first payment of \$50,000 immediately and the other $n = 19$ payments at the end of each of the next 19 years. Determine how much money the commission should have in the bank initially to guarantee the payments, assuming that the balance on deposit with the bank earns interest at the rate of 8%/year compounded yearly.

    Hint: Find the present value of an annuity.

23. **Purchasing a Home** The Johnsons have accumulated a nest egg of \$25,000 that they intend to use as a down payment toward the purchase of a new house. Because their present gross income has placed them in a relatively high tax bracket, they have decided to invest a minimum of \$800/month in monthly payments (to take advantage of their tax deductions) toward the purchase of their house. However, because of other financial obligations, their monthly payments should not exceed \$1000. If local mortgage rates are 9.5% per year compounded monthly for a conventional 30-year mortgage, what is the price range of houses that they should consider?

24. **Purchasing a Home** Refer to Exercise 23. If local mortgage rates were increased to 10%, how would this affect the price range of houses the Johnsons should consider?

25. **Purchasing a Home** Refer to Exercise 23. If the Johnsons decide to secure a 15-year mortgage instead of a 30-year mortgage, what is the price range of houses they should consider when the local mortgage rate for this type of loan is 9%?

In Exercises 26 and 27, determine whether the statement is true or false. If it is true, explain why it is true. If it is false, give an example to show why it is false.

26. The future value of an annuity can be found by adding together all the payments that are paid into the account.

27. If the future value of an annuity of n payments of R dollars each, paid at the end of each investment period into an account that earns interest at the rate of i per period, is S dollars, then,

$$R = \frac{iS}{(1+i)^n - 1}$$

Spreadsheet examples and exercises for this section that may be solved using the Microsoft® Excel program are given at the Brooks/Cole Web site:
http://www.brookscole.com/math/authors/tans/

Solutions to Self-Check Exercises 5.2

1. The amount Phyliss will have in her account when she retires may be found by using Formula (6) with $R = 2000$, $r = 0.08$, $m = 1$, so that $i = r = 0.08$ and $n = 20$. Thus,

$$S = \frac{2000[(1.08)^{20} - 1]}{0.08}$$
$$\approx 91{,}523.93$$

or \$91,523.93.

2. We want to find the present value of an ordinary annuity of 36 payments of \$4800 each made monthly and earning interest at 9%/year compounded monthly. Using Formula (8) with $R = 4800$, $m = 12$, so that $i = r/m = 0.09/12 = 0.0075$ and $n = (12)(3) = 36$, we find

$$P = \frac{4800[1 - (1.0075)^{-36}]}{0.0075} \approx 150{,}944.67$$

or \$150,944.67, the amount of the loan negotiated.

Finding the Amount of an Annuity

The following program, which we will call **FVAN**, can be used to calculate the future value S of an annuity of n payments of R dollars each earning interest at the rate of i per period [Formula (6), page 299]. When using this formula, recall that $i = r/m$ and $n = mt$, where r is the annual interest rate, m is the number of conversion periods per year, and t is the number of years in the term of the annuity.

```
: PROGRAM: FVAN
: Disp "R"
: Input R
: Disp "i"
: Input i
: Disp "N"
: Input N
: (R/i)((1 + i)^N - 1) → S
: Disp "AMOUNT IS"
: Disp S
```

REMARK Since the letter n is a reserved-name variable in some graphing calculators, we use N in lieu of n in the program **FVAN**.

EXAMPLE 1 Find the amount of an ordinary annuity of 36 quarterly payments of \$220 each that earns interest at 10% per year compounded quarterly.

SOLUTION Using the program **FVAN** and entering the values $R = 220$, $i = 0.1/4 = 0.025$, and $n = N = 36$, we find that $S = 12606.3107783$, or \$12,606.31.

Exercises

Use the program FVAN to solve Exercises 1–4.

1. Find the amount of an ordinary annuity of 20 payments of \$2500/quarter at $7\frac{1}{4}$%/year compounded quarterly.

2. Find the amount of an ordinary annuity of 24 payments of \$1790/quarter at $8\frac{3}{4}$%/year compounded quarterly.

3. Find the amount of an ordinary annuity of \$120/month for 5 years at $6\frac{3}{8}$%/year compounded monthly.

4. Find the amount of an ordinary annuity of \$225/month for 6 years at $7\frac{5}{8}$%/year compounded monthly.

5. Write a program to calculate the present value P of an annuity of n payments of R dollars each, paid at the end of each investment period into an account that earns interest at the rate of i per period [see Formula (8), page 302].

Use the program you wrote in Exercise 5 to solve Exercises 6–9.

6. Find the present value of an ordinary annuity of \$4500/semiannual period for 5 years earning interest at 9%/year compounded semiannually.

7. Find the present value of an ordinary annuity of \$2100/quarter for 7 years earning interest at $7\frac{1}{8}$%/year compounded quarterly.

8. Find the present value of an ordinary annuity of \$245/month for 6 years earning interest at $8\frac{3}{8}$%/year compounded monthly.

9. Find the present value of an ordinary annuity of \$185/month for 12 years earning interest at $6\frac{5}{8}$%/year compounded monthly.

5.3 Amortization and Sinking Funds

Amortization of Loans

The annuity formulas derived in Section 5.2 may be used to answer questions involving the amortization of certain types of installment loans. For example, in a typical housing loan, the mortgagor makes periodic payments toward reducing his indebtedness to the lender, who charges interest at a fixed rate on the unpaid portion of the debt. In practice, the borrower is required to repay the lender in periodic installments, usually of the same size and over a fixed term, so that the loan (principal plus interest charges) is amortized at the end of the term.

By thinking of the monthly loan repayments R as the payments in an annuity, we see that the original amount of the loan is given by P, the present value of the annuity. From Equation (8), Section 5.2, we have

$$P = R\left[\frac{1-(1+i)^{-n}}{i}\right] = Ra_{\overline{n}|i} \tag{9}$$

A question a financier might ask is: How much should the monthly installment be so that a loan will be amortized at the end of the term of the loan? To answer this question, we simply solve (9) for R in terms of P, obtaining

$$R = \frac{Pi}{1-(1+i)^{-n}} = \frac{P}{a_{\overline{n}|i}}$$

Amortization Formula

The periodic payment R on a loan of P dollars to be amortized over n periods with interest charged at the rate of i per period is

$$R = \frac{Pi}{1-(1+i)^{-n}} \tag{10}$$

EXAMPLE 1

A sum of \$50,000 is to be repaid over a 5-year period through equal installments made at the end of each year. If an interest rate of 8% per year is charged on the unpaid balance and interest calculations are made at the end of each year, determine the size of each installment so that the loan (principal plus interest charges) is amortized at the end of 5 years. Verify the result by displaying the amortization schedule.

SOLUTION ✓

Substituting $P = 50{,}000$, $i = r = 0.08$ (here, $m = 1$), and $n = 5$ into Formula (10), we obtain

$$R = \frac{(50{,}000)(0.08)}{1-(1.08)^{-5}} \approx 12{,}522.82$$

giving the required yearly installment as \$12,522.82.

Table 5.3

End of Period	Interest Charged	Repayment Made	Payment Toward Principal	Outstanding Principal
0	—	—	—	$50,000.00
1	$4,000.00	$12,522.82	$ 8,522.82	41,477.18
2	3,318.17	12,522.82	9,204.65	32,272.53
3	2,581.80	12,522.82	9,941.02	22,331.51
4	1,786.52	12,522.82	10,736.30	11,595.21
5	927.62	12,522.82	11,595.20	0.01

The amortization schedule is presented in Table 5.3. The outstanding principal at the end of 5 years is of course zero. (The figure of $.01 in Table 5.3 is the result of round-off errors.) Observe that initially the larger portion of the repayment goes toward payment of interest charges, but as time goes by the larger portion of the installment goes toward repayment of the principal.

Financing a Home

EXAMPLE 2 The Blakelys borrowed $120,000 from a bank to help finance the purchase of a house. The bank charges interest at a rate of 9% per year on the unpaid balance, with interest computations made at the end of each month. The Blakelys have agreed to repay the loan in equal monthly installments over 30 years. How much should each payment be if the loan is to be amortized at the end of the term?

SOLUTION ✓ Here, $P = 120{,}000$, $i = r/m = 0.09/12 = 0.0075$, and $n = (30)(12) = 360$. Using Formula (10) we find that the size of each monthly installment required is given by

$$R = \frac{(120{,}000)(0.0075)}{1 - (1.0075)^{-360}}$$
$$= 965.55$$

or $965.55.

EXAMPLE 3 Teresa and Raul purchased a house 10 years ago for $200,000. They made a down payment of 20% of the purchase price and secured a 30-year conventional home mortgage at 9% per year on the unpaid balance. The house is now worth $280,000. How much equity do Teresa and Raul have in their house now (after making 120 monthly payments)?

SOLUTION ✓ Since the down payment was 20%, we know that they secured a loan of 80% of $200,000, or $160,000. Furthermore, using Formula (10) with $P = 160{,}000$,

$i = r/m = 0.09/12 = 0.0075$, and $n = (30)(12) = 360$, we can determine their monthly installment to be

$$R = \frac{(160{,}000)(0.0075)}{1 - (1.0075)^{-360}}$$
$$\approx 1287.40$$

or \$1287.40.

After 120 monthly payments have been made, the outstanding principal is given by the sum of the present values of the remaining installments (that is, $360 - 120 = 240$ installments). But this sum is just the present value of an annuity with $n = 240$, $R = 1287.40$, and $i = 0.0075$. Using Formula (8), we find

$$P = 1{,}287.40\left[\frac{1 - (1 + 0.0075)^{-240}}{0.0075}\right]$$
$$\approx 143{,}088.01$$

or approximately \$143,088. Therefore, Teresa and Raul have an equity of $280{,}000 - 143{,}088 + 20{,}000$, or approximately \$156,912. ■■■■

Group Discussion and Exploring with Technology

1. Consider the amortization formula (10):

$$R = \frac{Pi}{1 - (1 + i)^{-n}}$$

Suppose you know the values of R, P, and n and you wish to determine i. Explain why you can accomplish this task by finding the point of intersection of the graphs of the functions

$$y_1 = R \qquad \text{and} \qquad y_2 = \frac{Pi}{1 - (1 + i)^{-n}}$$

2. Thalia knows that her monthly repayment on her 30-year conventional home loan of \$150,000 is \$1100.65 per month. Help Thalia determine the interest rate for her loan by verifying or executing the following steps:

a. Plot the graphs of

$$y_1 = 1100.65 \qquad \text{and} \qquad y_2 = \frac{150{,}000x}{1 - (1 + x)^{-360}}$$

using the viewing rectangle $[0, 0.01] \times [0, 1200]$.

b. Use the ISECT (intersection) function of the graphing utility to find the point of intersection of the graphs of part (a). Explain why this gives the value of i.

c. Compute r from the relationship $r = 12i$.

Group Discussion and Exploring with Technology

1. Suppose you secure a home mortgage loan of $P with an interest rate of r per year to be amortized over t years through monthly installments of $R. Show that after N installments your outstanding principal is given by

$$B(N) = P\left[\frac{(1+i)^n - (1+i)^N}{(1+i)^n - 1}\right] \qquad (0 \le N \le n)$$

Hint: $B(N) = R\left[\frac{1-(1+i)^{-n+N}}{i}\right]$. To see this, study Example 3, pages 311–312. Replace R using Formula (10).

2. Refer to Example 3, pages 311–312. Using the result of part 1 above, show that Teresa and Raul's outstanding balance after making N payments is

$$E(N) = \frac{160{,}000(1.0075^{360} - 1.0075^N)}{1.0075^{360} - 1} \qquad (0 \le N \le 360).$$

3. Using a graphing utility, plot the graph of

$$E(x) = \frac{160{,}000(1.0075^{360} - 1.0075^x)}{1.0075^{360} - 1}$$

using the viewing rectangle [0, 360] × [0, 160,000].

4. Referring to the graph in part 3, observe that the outstanding principal drops off slowly in the early years and accelerates quickly to zero toward the end of the loan. Can you explain why this is so?

5. How long does it take Teresa and Raul to repay one-half of the loan of $160,000?
Hint: See the Group Discussion and Exploring with Technology box following Example 3.

EXAMPLE 4 The Jacksons have determined that after making a down payment they could afford at most $1000 for a monthly house payment. The bank charges interest at the rate of 9.6% per year on the unpaid balance, with interest computations made at the end of each month. If the loan is to be amortized in equal monthly installments over 30 years, what is the maximum amount that the Jacksons can borrow from the bank?

SOLUTION ✓ Here, $i = r/m = 0.096/12 = 0.008$, $n = (30)(12) = 360$, $R = 1000$, and we are required to find P. From Equation (8), we have

$$P = \frac{R[1-(1+i)^{-n}]}{i}$$

Substituting the numerical values for R, n, and i into this expression for P, we obtain

$$P = \frac{1000[1-(1.008)^{-360}]}{0.008} \approx 117{,}902$$

Therefore, the Jacksons can borrow at most $117,902. ■■■■

SINKING FUNDS

Sinking funds are another important application of the annuity formulas. Simply stated, a **sinking fund** is a fund that is set up for a specific purpose at some future date. For example, an individual might establish a sinking fund for the purpose of discharging a debt at a future date. A corporation might establish a sinking fund in order to accumulate sufficient capital to replace equipment that is expected to be obsolete at some future date.

By thinking of the amount to be accumulated by a specific date in the future as the future value of an annuity [Equation (6), Section 5.2], we can answer questions about a large class of sinking fund problems.

EXAMPLE 5 The proprietor of Carson Hardware has decided to set up a sinking fund for the purpose of purchasing a computer in 2 years' time. It is expected that the computer will cost \$30,000. If the fund earns 10% interest per year compounded quarterly, determine the size of each (equal) quarterly installment the proprietor should pay into the fund. Verify the result by displaying the schedule.

SOLUTION ✓ The problem at hand is to find the size of each quarterly payment R of an annuity given that its future value is $S = 30{,}000$, the interest earned per conversion period is $i = r/m = 0.1/4 = 0.025$, and the number of payments is $n = (2)(4) = 8$. The formula for the annuity,

$$S = R\left[\frac{(1+i)^n - 1}{i}\right]$$

when solved for R yields

$$R = \frac{iS}{(1+i)^n - 1} \tag{11}$$

or, equivalently,

$$R = \frac{S}{s_{\overline{n}|i}}$$

Substituting the appropriate numerical values for i, S, and n into Equation (11), we obtain the desired quarterly payment

$$R = \frac{(0.025)(30{,}000)}{(1.025)^8 - 1} = 3434.02$$

or \$3434.02. Table 5.4 shows the required schedule.

Table 5.4

End of Period	Deposit Made	Interest Earned	Addition to Fund	Accumulated Amount in Fund
1	$3,434.02	0	$3,434.02	$ 3,434.02
2	3,434.02	$ 85.85	3,519.87	6,953.89
3	3,434.02	173.85	3,607.87	10,561.76
4	3,434.02	264.04	3,698.06	14,259.82
5	3,434.02	356.50	3,790.52	18,050.34
6	3,434.02	451.26	3,885.28	21,935.62
7	3,434.02	548.39	3,982.41	25,918.03
8	3,434.02	647.95	4,081.97	30,000.00

The formula derived in this last example is restated below.

Sinking Fund Payment

The periodic payment R required to accumulate a sum of S dollars over n periods with interest charged at the rate of i per period is

$$R = \frac{iS}{(1+i)^n - 1}$$

SELF-CHECK EXERCISES 5.3

1. The Mendozas wish to borrow $100,000 from a bank to help finance the purchase of a house. Their banker has offered the following plans for their consideration. In plan I, the Mendozas have 30 years to repay the loan in monthly installments with interest on the unpaid balance charged at 10.5%/year compounded monthly. In plan II, the loan is to be repaid in monthly installments over 15 years with interest on the unpaid balance charged at 9.75%/year compounded monthly.
   a. Find the monthly repayment for each plan.
   b. What is the difference in total payments made under each plan?
2. Harris, a self-employed individual who is 46 years old, is setting up a Defined-Benefit Keogh Plan for his retirement. If he wishes to have $250,000 in this Keogh account by age 65, what is the size of each yearly installment he will be required to make into a savings account earning interest at $8\frac{1}{4}$%/year? (Assume that Harris is eligible to make each of the 20 required contributions.)

Solutions to Self-Check Exercises 5.3 can be found on page 319.

Portfolio

JOHN DECKER

TITLE: Mortgage Counselor
INSTITUTION: A major Boston Bank

John Decker stresses that he and his colleagues "strive to grant loans. That's our job." But before allowing someone to file a formal application, Decker takes the person over several "prequalification hurdles" to gauge his or her ability to handle a mortgage.

To start, Decker relies on a two-tiered, debt-to-income ratio to see whether a person has sufficient gross monthly income to make payments. Under the first tier, the proposed monthly payment (principal and interest, property tax, homeowner's insurance, and, when applicable, a condo fee) cannot exceed 28% of an individual's gross monthly income. If Decker's initial calculations are positive, he then must determine the individual's ability to meet monthly mortgage payments while also repaying other debts, such as car and student loans, credit cards, alimony, and so on. These combined payments cannot exceed 36% of gross monthly income. A typical person with an $800 mortgage obligation and $550 in other payments would have to earn $3750 per month to clear these first two hurdles.

Contrary to popular belief, bankers *want* to lend money. "The idea is to grant mortgages," says Decker, which contribute substantially to a bank's profitability. But lending money means making sensible decisions about how much to lend as well as a person's ability to repay the loan.

Using a loan-to-value formula, banks might lend 80% of a property's value. In such cases, the applicant has to put 20% down to make the purchase. Or the bank may decide to lend up to 95% or as little as 75 percent of the appraised value.

Understandably, banks don't like to see bankruptcies, late payments, or liens on a personal credit history. Decker notes, however, that even this final hurdle doesn't necessarily mean failure in securing a mortgage. He works closely with each individual to overcome any stigma that might prompt a rejection.

Once Decker has put together a successful mortgage application, it is reviewed internally. Then, even though a mortgage is granted, it might come through at a slightly higher interest rate—10.4% instead of the expected 10% rate. Then he computes the change in monthly payments, for while this might seem like a small change, on a large mortgage it could be enough of a variable to affect the new customer's ability to make monthly payments. On a $100,000, 30-year, fixed-rate mortgage, payments will increase about $35 per month.

Using the debt-to-income ratio, Decker plugs the new variable into his formulas to determine whether a problem exists. Decker notes that such a small increase doesn't usually pose a significant problem. If the customer can't make the new payment, however, Decker explores alternatives until he finds a solution. For Decker, it comes down to this: "If there is any possible way to give a loan, we're going to make it work."

Decker's job is simple: to loan money to people who want to buy a home. The hard part is deciding whether applicants qualify for one of the bank's 40 different mortgage plans. Sifting through income figures, current indebtedness, and credit history helps Decker determine which individuals make the best mortgage candidates.

5.3 Exercises

A calculator is recommended for this exercise set.

In Exercises 1–8, find the periodic payment R required to amortize a loan of P dollars over n periods with interest earned at the rate of i per period.

1. $P = \$100{,}000$, $i = 0.08$, $n = 10$
2. $P = \$40{,}000$, $i = 0.015$, $n = 30$
3. $P = \$5000$, $i = 0.01$, $n = 12$
4. $P = \$16{,}000$, $i = 0.0075$, $n = 48$
5. $P = \$25{,}000$, $i = 0.0075$, $n = 48$
6. $P = \$80{,}000$, $i = 0.00875$, $n = 180$
7. $P = \$80{,}000$, $i = 0.00875$, $n = 360$
8. $P = \$100{,}000$, $i = 0.00875$, $n = 300$

In Exercises 9–14, find the periodic payment R required to accumulate a sum of S dollars over n periods with interest earned at the rate of i per period.

9. $S = \$20{,}000$, $i = 0.02$, $n = 12$
10. $S = \$40{,}000$, $i = 0.01$, $n = 36$
11. $S = \$100{,}000$, $i = 0.0075$, $n = 120$
12. $S = \$120{,}000$, $i = 0.0075$, $n = 180$
13. $S = \$250{,}000$, $i = 0.00875$, $n = 300$
14. $S = \$350{,}000$, $i = 0.00625$, $n = 120$
15. LOAN AMORTIZATION A sum of \$100,000 is to be repaid over a 10-year period through equal installments made at the end of each year. If an interest rate of 10%/year is charged on the unpaid balance and interest calculations are made at the end of each year, determine the size of each installment so that the loan (principal plus interest charges) is amortized at the end of 10 years.
16. LOAN AMORTIZATION What monthly payment is required to amortize a loan of \$30,000 over 10 years if interest at the rate of 12%/year is charged on the unpaid balance and interest calculations are made at the end of each month?
17. HOME MORTGAGES Complete the following table, which shows the monthly payments on a \$100,000, 30-year mortgage at the interest rates shown. Use this information to answer the following questions.

Amount of Mortgage	Interest Rate	Monthly Payment
\$100,000	7%	\$665.30
100,000	8	...
100,000	9	...
100,000	10	...
100,000	11	...
100,000	12	1028.61

a. What is the difference in monthly payments between a \$100,000, 30-year mortgage secured at 7%/year and one secured at 10%/year?
b. Use the table to calculate the monthly mortgage payments on a \$150,000 mortgage at 10%/year over 30 years and a \$50,000 mortgage at 10%/year over 30 years.

18. FINANCING A HOME The Flemings secured a bank loan of \$96,000 to help finance the purchase of a house. The bank charges interest at a rate of 9%/year on the unpaid balance, and interest computations are made at the end of each month. The Flemings have agreed to repay the loan in equal monthly installments over 25 years. What should be the size of each repayment if the loan is to be amortized at the end of the term?
19. FINANCING A CAR The price of a new car is \$16,000. Assume an individual makes a down payment of 25% toward the purchase of the car and secures financing for the balance at the rate of 10%/year compounded monthly.
a. What monthly payment will she be required to make if the car is financed over a period of 36 months? Over a period of 48 months?
b. What will the interest charges be if she elects the 36-month plan? The 48-month plan?
20. FINANCIAL ANALYSIS A group of private investors purchased a condominium complex for \$2 million. They made an initial down payment of 10% and obtained financing for the balance. If the loan is to be amortized over 15 years at an interest rate of 12%/year compounded quarterly, find the required quarterly payment.
21. FINANCING A HOME The Taylors have purchased a \$180,000 house. They made an initial down payment of \$20,000 and secured a mortgage with interest charged at the rate of 8%/year on the unpaid balance. Interest computations are made at the end of each month. If the loan is to be amortized over 30 years, what monthly

payment will the Taylors be required to make? What is their equity (disregarding appreciation) after 5 years? After 10 years? After 20 years?

22. **SINKING FUNDS** A city has $2.5 million worth of school bonds that are due in 20 years and has established a sinking fund to retire this debt. If the fund earns interest at the rate of 7%/year compounded annually, what amount must be deposited annually in this fund?

23. **SINKING FUNDS** The Lowell Corporation wishes to establish a sinking fund to retire a $200,000 debt that is due in 10 years. If the investment will earn interest at the rate of 9%/year compounded quarterly, find the amount of the quarterly deposit that must be made in order to accumulate the required sum.

24. **TRUST FUNDS** Carl is the beneficiary of a $20,000 trust fund set up for him by his grandparents. Under the terms of the trust, he is to receive the money over a 5-year period in equal installments at the end of each year. If the fund earns interest at the rate of 9%/year compounded annually, what amount will he receive each year?

25. **SINKING FUNDS** The management of the Gibraltar Brokerage Services, Inc., anticipates a capital expenditure of $20,000 in 3 years' time for the purpose of purchasing new fax machines and has decided to set up a sinking fund to finance this purchase. If the fund earns interest at the rate of 10%/year compounded quarterly, determine the size of each (equal) quarterly installment that should be deposited in the fund.

26. **KEOGH ACCOUNTS** Andrea, a self-employed individual, wishes to accumulate a retirement fund of $250,000. How much should she deposit each month into her Keogh account, which pays interest at the rate of 8.5%/year compounded monthly, to reach her goal upon retirement 25 years from now?

27. **IRAs** Martin has deposited $375 in his IRA at the end of each quarter for the past 20 years. His investment has earned interest at the rate of 8%/year compounded quarterly over this period. Now, at age 60, he is considering retirement. What quarterly payment will he receive over the next 15 years? (Assume that the money is earning interest at the same rate and payments are made at the end of each quarter.) If he continues working and makes quarterly payments of the same amount in his IRA account until age 65, what quarterly payment will he receive from his fund upon retirement over the following 10 years?

28. **FINANCING A CAR** Darla purchased a new car during a special sales promotion by the manufacturer. She secured a loan from the manufacturer in the amount of $16,000 at a rate of 7.9%/year compounded monthly. Her bank is now charging 11.5%/year compounded monthly for new car loans. Assuming that each loan would be amortized by 36 equal monthly installments, determine the amount of interest she would have paid at the end of 3 years for each loan. How much less will she have paid in interest payments over the life of the loan by borrowing from the manufacturer instead of her bank?

29. **FINANCING A HOME** The Sandersons are planning to refinance their home. The outstanding principal on their original loan is $100,000 and was to be amortized in 240 equal monthly installments at an interest rate of 10%/year compounded monthly. The new loan they expect to secure is to be amortized over the same period at an interest rate of 7.8%/year compounded monthly. How much less can they expect to pay over the life of the loan in interest payments by refinancing the loan at this time?

30. **FINANCING A HOME** After making a down payment of $25,000, the Meyers need to secure a loan of $140,000 to purchase a certain house. Their bank's current rate for 25-year home loans is 11%/year compounded monthly. The owner has offered to finance the loan at 9.8%/year compounded monthly. Assuming that both loans would be amortized over a 25-year period by 300 equal monthly installments, determine the difference in the amount of interest the Meyers would pay by choosing the seller's financing rather than their bank's.

31. **REFINANCING A HOME** The Martinez family is planning to refinance their home. The outstanding balance on their original loan is $150,000. Their finance company has offered them two options:

    Option A: A fixed-rate mortgage at an interest rate of 7.5%/year compounded monthly, payable over a 30-year period in 360 equal monthly installments.

    Option B: A fixed-rate mortgage at an interest rate of 7.25%/year compounded monthly, payable over a 15-year period in 180 equal monthly installments. (Assume that there are no additional finance charges.)

    a. Find the monthly payment required to amortize each of these loans over the life of the loan.
    b. How much interest would Mr. and Mrs. Martinez save if they chose the 15-year mortgage instead of the 30-year mortgage?

Spreadsheet examples and exercises for this section that may be solved using the Microsoft® Excel program are given at the Brooks/Cole Web site:
http://www.brookscole.com/math/authors/tans/

SOLUTIONS TO SELF-CHECK EXERCISES 5.3

1. a. We use Equation (10) in each instance. Under plan I,

$$P = 100{,}000, \qquad i = \frac{r}{m} = \frac{0.105}{12} = 0.00875, \qquad n = (30)(12) = 360$$

Therefore, the size of each monthly repayment under plan I is

$$R = \frac{100{,}000(0.00875)}{1 - (1.00875)^{-360}} \approx 914.74$$

or \$914.74.

Under plan II,

$$P = 100{,}000, \qquad i = \frac{r}{m} = \frac{0.0975}{12} = 0.008125, \qquad n = (15)(12) = 180$$

Therefore, the size of each monthly repayment under plan II is

$$R = \frac{100{,}000(0.008125)}{1 - (1.008125)^{-180}} \approx 1059.36$$

or \$1059.36.

b. Under plan I, the total amount of repayments will be

$$(360)(914.74) \quad \text{(Number of payments times the size of each installment)}$$
$$= 329{,}306.40$$

or \$329,306.40. Under plan II, the total amount of repayments will be

$$(180)(1059.36) = 190{,}684.80$$

or \$190,684.80. Therefore, the difference in payments is

$$329{,}306.40 - 190{,}684.80 = 138{,}621.60$$

or \$138,621.60.

2. We use Equation (11) with

$$S = 250{,}000$$
$$i = r = 0.0825 \quad \text{(Since } m = 1\text{)}$$
$$n = 20$$

giving the required size of each installment as

$$R = \frac{(0.0825)(250{,}000)}{(1.0825)^{20} - 1} \approx 5313.59$$

or \$5313.59.

AMORTIZING A LOAN

The following program, which we will call **AMORT,** enables us to calculate the periodic payment R on a loan of P dollars to be amortized over n periods with interest charged at the rate of i per period [Formula (10), page 310].

```
: PROGRAM: AMORT
: Disp "P"
: Input P
: Disp "i"
: Input i
: Disp "N"
: Input N
: P*i/(1 - (1 + i)^-N) → R
: Disp "R IS"
: Disp R
```

EXAMPLE 1 The Wongs are considering obtaining a preapproved 30-year loan of $120,000 to help finance the purchase of a house. The mortgage company charges interest at the rate of 8% per year on the unpaid balance, with interest computations made at the end of each month. What will be the monthly installments if the loan is amortized at the end of the term?

SOLUTION ✔ Using the **AMORT** program and entering the values $P = 120{,}000$, $i = r/m = 0.08/12$, and $n = N = (30)(12) = 360$, we find that $R = 880.517488654$, and so the monthly installment is $880.52.

Exercises

Use the program AMORT to solve Exercises 1–4.

1. Find the periodic payment required to amortize a loan of \$55,000 over 120 periods with interest earned at the rate of $6\frac{5}{8}$%/period.

2. Find the periodic payment required to amortize a loan of \$178,000 over 180 periods with interest earned at the rate of $7\frac{1}{8}$%/period.

3. Find the periodic payment required to amortize a loan of \$227,000 over 360 periods with interest earned at the rate of $8\frac{1}{8}$%/period.

4. Find the periodic payment required to amortize a loan of \$150,000 over 360 periods with interest earned at the rate of $7\frac{3}{8}$%/period.

5. Write a program to calculate the periodic payment R to accumulate a sum of S dollars over n periods with interest charged at the rate of i per period [see Formula (11), page 314].

Use the program you wrote in Exercise 5 to solve Exercises 6–9.

6. Find the periodic payment required to accumulate \$25,000 over 12 periods with interest earned at the rate of 2%/period.

7. Find the periodic payment required to accumulate \$50,000 over 36 periods with interest earned at the rate of $2\frac{1}{4}$%/period.

8. Find the periodic payment required to accumulate \$137,000 over 120 periods with interest earned at the rate of $\frac{3}{4}$%/period.

9. Find the periodic payment required to accumulate \$144,000 over 120 periods with interest earned at the rate of $\frac{5}{8}$%/period.

10. A loan of \$120,000 is to be repaid over a 10-year period through equal installments made at the end of each year. If an interest rate of 8.5%/year is charged on the unpaid balance and interest calculations are made at the end of each year, determine the size of each installment so that the loan is amortized at the end of 10 years. Verify the result by displaying the amortization schedule.

11. A loan of \$265,000 is to be repaid over an 8-year period through equal installments made at the end of each year. If an interest rate of 7.4%/year is charged on the unpaid balance and interest calculations are made at the end of each year, determine the size of each installment so that the loan is amortized at the end of 8 years. Verify the result by displaying the amortization schedule.

5.4 Arithmetic and Geometric Progressions (Optional)

ARITHMETIC PROGRESSIONS

An **arithmetic progression** is a sequence of numbers in which each term after the first is obtained by adding a constant d to the preceding term. The constant d is called the **common difference.** For example, the sequence

$$3, 6, 9, 12, \ldots$$

is an arithmetic progression with the common difference equal to 3.

Observe that an arithmetic progression is completely determined if the first term and the common difference are known. In fact, if

$$a_1, a_2, a_3, \ldots, a_n, \ldots$$

is an arithmetic progression with the first term given by a and common difference given by d, then by definition,

$$\begin{aligned}
a_1 &= a \\
a_2 &= a_1 + d = a + d \\
a_3 &= a_2 + d = (a + d) + d = a + 2d \\
a_4 &= a_3 + d = (a + 2d) + d = a + 3d \\
&\vdots \\
a_n &= a_{n-1} + d = a + (n - 2)d + d = a + (n - 1)d
\end{aligned}$$

Thus, we see that the nth term of an arithmetic progression with first term a and common difference d is given by

$$a_n = a + (n - 1)d \qquad \textbf{(12)}$$

nth Term of an Arithmetic Progression

The nth term of an arithmetic progression with first term a and common difference d is given by

$$a_n = a + (n - 1)d$$

EXAMPLE 1 Find the 12th term of the arithmetic progression

$$2, 7, 12, 17, 22, \ldots$$

SOLUTION ✓ The first term of the arithmetic progression is $a_1 = a = 2$, and the common difference is $d = 5$, so upon setting $n = 12$ in Equation (12), we find

$$a_{12} = 2 + (12 - 1)5 = 57$$

■■■■

EXAMPLE 2 Write the first five terms of an arithmetic progression whose 3rd and 11th terms are 21 and 85, respectively.

SOLUTION ✓ Using Equation (12), we obtain

$$\begin{aligned}
a_3 &= a + 2d = 21 \\
a_{11} &= a + 10d = 85
\end{aligned}$$

Subtracting the first equation from the second gives $8d = 64$, or $d = 8$. Substituting this value of d into the first equation yields $a + 16 = 21$, or $a = 5$. Thus, the required arithmetic progression is given by the sequence

$$5, 13, 21, 29, 37, \ldots$$

Let S_n denote the sum of the first n terms of an arithmetic progression with first term $a_1 = a$ and common difference d. Then,

$$S_n = a + (a + d) + (a + 2d) + \cdots + [a + (n - 1)d] \tag{13}$$

Rewriting the expression for S_n with the terms in reverse order gives

$$S_n = [a + (n - 1)d] + [a + (n - 2)d] + \cdots + (a + d) + a \tag{14}$$

Adding Equations (13) and (14), we obtain

$$\begin{aligned} 2S_n &= [2a + (n - 1)d] + [2a + (n - 1)d] \\ &\quad + \cdots + [2a + (n - 1)d] \\ &= n[2a + (n - 1)d] \\ S_n &= \frac{n}{2}[2a + (n - 1)d] \end{aligned}$$

Sum of Terms in an Arithmetic Progression

The sum of the first n terms of an arithmetic progression with first term a and common difference d is given by

$$S_n = \frac{n}{2}[2a + (n - 1)d] \tag{15}$$

EXAMPLE 3 Find the sum of the first 20 terms of the arithmetic progression of Example 1.

SOLUTION ✔ Letting $a = 2$, $d = 5$, and $n = 20$ in Equation (15), we obtain

$$S_{20} = \frac{20}{2}[2 \cdot 2 + 19 \cdot 5] = 990$$

EXAMPLE 4 The Madison Electric Company had sales of \$200,000 in its first year of operation. If the sales increased by \$30,000 per year thereafter, find Madison's sales in the fifth year and its total sales over the first 5 years of operation.

SOLUTION ✔ Madison's yearly sales follow an arithmetic progression, with the first term given by $a = 200{,}000$ and the common difference given by $d = 30{,}000$. The sales in the fifth year are found by using Equation (12) with $n = 5$. Thus,

$$a_5 = 200{,}000 + (5 - 1)30{,}000 = 320{,}000$$

or \$320,000.

Madison's total sales over the first 5 years of operation are found by using (15) with $n = 5$. Thus,

$$S_5 = \frac{5}{2}[2(200{,}000) + (5 - 1)30{,}000]$$
$$= 1{,}300{,}000$$

or $1,300,000. ■■■■

Geometric Progressions

A **geometric progression** is a sequence of numbers in which each term after the first is obtained by multiplying the preceding term by a constant r. The constant r is called the **common ratio.**

A geometric progression is completely determined if the first term and the common ratio are known. Thus, if

$$a_1, a_2, a_3, \ldots, a_n, \ldots$$

is a geometric progression with the first term given by a and common ratio given by r, then by definition,

$$\begin{aligned} a_1 &= a \\ a_2 &= a_1 r = ar \\ a_3 &= a_2 r = ar^2 \\ a_4 &= a_3 r = ar^3 \\ &\vdots \\ a_n &= a_{n-1} r = ar^{n-1} \end{aligned}$$

Thus, we see that the nth term of a geometric progression with first term a and common ratio r is given by

$$a_n = ar^{n-1} \tag{16}$$

nth Term of a Geometric Progression

The nth term of a geometric progression with first term a and common ratio r is given by

$$a_n = ar^{n-1}$$

EXAMPLE 5 Find the eighth term of a geometric progression whose first five terms are 162, 54, 18, 6, and 2.

SOLUTION ✓ The common ratio is found by taking the ratio of any term other than the first to the preceding term. Taking the ratio of the fourth term to the third term, for example, gives $r = 6/18 = 1/3$. To find the eighth term of the geometric progression, use Formula (16) with $a = 162$, $r = 1/3$, and $n = 8$, obtaining

$$a_8 = 162\left(\frac{1}{3}\right)^7$$
$$= \frac{2}{27}$$

■■■■

EXAMPLE 6 Find the tenth term of a geometric progression whose third term is 16 and whose seventh term is 1.

SOLUTION ✓ Using Equation (16) with $n = 3$ and $n = 7$, respectively, yields

$$a_3 = ar^2 = 16$$
$$a_7 = ar^6 = 1$$

Dividing a_7 by a_3 gives

$$\frac{ar^6}{ar^2} = \frac{1}{16}$$

from which we obtain $r^4 = 1/16$, or $r = 1/2$. Substituting this value of r into the expression for a_3, we obtain

$$a\left(\frac{1}{2}\right)^2 = 16 \qquad \text{or} \qquad a = 64$$

Finally, using (16) once again with $a = 64$, $r = 1/2$, and $n = 10$ gives

$$a_{10} = 64\left(\frac{1}{2}\right)^9 = \frac{1}{8}$$

■■■■

To find the sum of the first n terms of a geometric progression with the first term $a_1 = a$ and common ratio r, denote the required sum by S_n. Then,

$$S_n = a + ar + ar^2 + \cdots + ar^{n-2} + ar^{n-1} \tag{17}$$

Upon multiplying (17) by r, we obtain

$$rS_n = ar + ar^2 + ar^3 + \cdots + ar^{n-1} + ar^n \tag{18}$$

Subtracting (18) from (17) gives

$$S_n - rS_n = a - ar^n$$
$$(1 - r)S_n = a(1 - r^n)$$

If $r \neq 1$, we may divide both sides of the last equation by $(1 - r)$, obtaining

$$S_n = \frac{a(1 - r^n)}{(1 - r)}$$

If $r = 1$, then (17) gives

$$\begin{aligned} S_n &= a + a + a + \cdots + a \qquad (n \text{ terms}) \\ &= na \end{aligned}$$

Thus,

$$S_n = \begin{cases} \dfrac{a(1 - r^n)}{1 - r} & \text{if } r \neq 1 \\ na & \text{if } r = 1 \end{cases} \tag{19}$$

Sum of Terms in a Geometric Progression

The sum of the first n terms of a geometric progression with first term a and common ratio r is given by

$$S_n = \begin{cases} \dfrac{a(1-r^n)}{1-r} & \text{if } r \neq 1 \\ na & \text{if } r = 1 \end{cases}$$

EXAMPLE 7 Find the sum of the first six terms of the geometric progression

$$3, 6, 12, 24, \ldots$$

SOLUTION ✔ Here, $a = 3$ and $r = 6/3 = 2$, so Formula (19) gives

$$S_6 = \frac{3(1-2^6)}{1-2} = 189$$

■■■■

EXAMPLE 8 The Michaelson Land Development Company had sales of \$1 million in its first year of operation. If sales increased by 10% per year thereafter, find Michaelson's sales in the fifth year and its total sales over the first 5 years of operation.

SOLUTION ✔ Michaelson's yearly sales follow a geometric progression, with the first term given by $a = 1{,}000{,}000$ and the common ratio given by $r = 1.1$. The sales in the fifth year are found by using Formula (16) with $n = 5$. Thus,

$$a_5 = 1{,}000{,}000(1.1)^4 = 1{,}464{,}100$$

or \$1,464,100.

Michaelson's total sales over the first 5 years of operation are found by using Equation (19) with $n = 5$. Thus,

$$S_5 = \frac{1{,}000{,}000[1-(1.1)^5]}{1-1.1} = 6{,}105{,}100$$

or \$6,105,100.

■■■■

Double Declining-Balance Method of Depreciation

In Section 1.3 we discussed the straight-line, or linear, method of depreciating an asset. Linear depreciation assumes that the asset depreciates at a constant rate. For certain assets, such as machines, whose market values drop rapidly in the early years of usage and thereafter less rapidly, another method of depreciation called the **double declining-balance method** is often used. In practice, a business firm normally employs the double declining-balance method for depreciating such assets for a certain number of years and then switches over to the linear method.

To derive an expression for the book value of an asset being depreciated by the double declining-balance method, let C (in dollars) denote the original cost of the asset and let the asset be depreciated over N years. Using this method, the amount depreciated each year is $2/N$ times the value of the asset at the beginning of that year. Thus, the amount by which the asset is depreciated in its first year of use is given by $2C/N$, so if $V(1)$ denotes the book value of the asset at the end of the first year, then

$$V(1) = C - \frac{2C}{N} = C\left(1 - \frac{2}{N}\right)$$

Next, if $V(2)$ denotes the book value of the asset at the end of the second year, then a similar argument leads to

$$\begin{aligned} V(2) &= C\left(1 - \frac{2}{N}\right) - C\left(1 - \frac{2}{N}\right)\frac{2}{N} \\ &= C\left(1 - \frac{2}{N}\right)\left(1 - \frac{2}{N}\right) \\ &= C\left(1 - \frac{2}{N}\right)^2 \end{aligned}$$

Continuing, we find that if $V(n)$ denotes the book value of the asset at the end of n years, then the terms $C, V(1), V(2), \ldots, V(n)$ form a geometric progression with first term C and common ratio $(1 - 2/N)$. Consequently, the nth term, $V(n)$, is given by

$$V(n) = C\left(1 - \frac{2}{N}\right)^n \qquad (1 \le n \le N) \tag{20}$$

Also, if $D(n)$ denotes the amount by which the asset has been depreciated by the end of the nth year, then

$$\begin{aligned} D(n) &= C - C\left(1 - \frac{2}{N}\right)^n \\ &= C\left[1 - \left(1 - \frac{2}{N}\right)^n\right] \end{aligned} \tag{21}$$

EXAMPLE 9 A tractor purchased at a cost of \$60,000 is to be depreciated by the double declining-balance method over 10 years. What is the book value of the tractor at the end of 5 years? By what amount has the tractor been depreciated by the end of the fifth year?

SOLUTION ✓ We have $C = 60{,}000$ and $N = 10$. Thus, using Formula (20) with $n = 5$ gives the book value of the tractor at the end of 5 years as

$$\begin{aligned} V(5) &= 60{,}000\left(1 - \frac{2}{10}\right)^5 \\ &= 60{,}000\left(\frac{4}{5}\right)^5 = 19{,}660.80 \end{aligned}$$

or \$19,660.80.

The amount by which the tractor has been depreciated by the end of the fifth year is given by

$$60{,}000 - 19{,}660.80 = 40{,}339.20$$

or $40,339.20. You may verify the last result by using Equation (21) directly.

Exploring with Technology

A tractor purchased at a cost of $60,000 is to be depreciated over 10 years with a residual value of $0. Using the double declining-balance method, its value at the end of n years is $V_1(n) = 60{,}000(0.8)^n$ dollars. Using straight-line depreciation, its value at the end of n years is $V_2(n) = 60{,}000 - 6000n$. Use a graphing utility to sketch the graphs of V_1 and V_2 in the viewing rectangle $[0, 10] \times [0, 70{,}000]$. Comment on the relative merits of each method of depreciation.

Self-Check Exercises 5.4

1. Find the sum of the first five terms of the geometric progression with first term -24 and common ratio $-1/2$.
2. Office equipment purchased for $75,000 is to be depreciated by the double declining-balance method over 5 years. Find the book value at the end of 3 years.
3. Derive the formula for the future value of an annuity [Equation (6), Section 5.2].

Solutions to Self-Check Exercises 5.4 can be found on page 331.

5.4 Exercises

In Exercises 1–4, find the *n*th term of the arithmetic progression that has the given values of *a*, *d*, and *n*.

1. $a = 6, d = 3, n = 9$
2. $a = -5, d = 3, n = 7$
3. $a = -15, d = 3/2, n = 8$
4. $a = 1.2, d = 0.4, n = 98$
5. Find the first five terms of the arithmetic progression whose 4th and 11th terms are 30 and 107, respectively.
6. Find the first five terms of the arithmetic progression whose 7th and 23rd terms are -5 and -29, respectively.
7. Find the seventh term of the arithmetic progression: $x, x + y, x + 2y, \ldots$
8. Find the 11th term of the arithmetic progression: $a + b, 2a, 3a - b, \ldots$
9. Find the sum of the first 15 terms of the arithmetic progression: $4, 11, 18, \ldots$
10. Find the sum of the first 20 terms of the arithmetic progression: $5, -1, -7, \ldots$
11. Find the sum of the odd integers between 14 and 58.
12. Find the sum of the even integers between 21 and 99.

13. Find $f(1) + f(2) + f(3) + \cdots + f(20)$, given that $f(x) = 3x - 4$.

14. Find $g(1) + g(2) + g(3) + \cdots + g(50)$, given that $g(x) = 12 - 4x$.

15. Show that equation (15) can be written as $S_n = (n/2)(a + a_n)$, where a_n represents the last term of an arithmetic progression. Use this formula to find:
a. The sum of the first 11 terms of the arithmetic progression whose 1st and 11th terms are 3 and 47, respectively
b. The sum of the first 20 terms of the arithmetic progression whose 1st and 20th terms are 5 and −33, respectively

16. **SALES GROWTH** The Moderne Furniture Company had sales of $1,500,000 during its first year of operation. If the sales increased by $160,000/year thereafter, find Moderne's sales in the fifth year and its total sales over the first 5 years of operation.

17. **EXERCISE PROGRAM** As part of her fitness program, Karen has taken up jogging. If she jogs 1 mi the first day and increases her daily run by 1/4 mi every week, how long will it take her to reach her goal of 10 mi/day?

18. **COST OF DRILLING** A 100-ft oil well is to be drilled. The cost of drilling the first foot is $10, and the cost of drilling each additional foot is $4.50 more than that of the preceding foot. Find the cost of drilling the entire 100 ft.

19. **CONSUMER DECISIONS** Kunwoo wishes to go from the airport to his hotel, which is 25 mi away. The taxi rate is $1 for the first mile and 60 cents for each additional mile. The airport limousine also goes to his hotel and charges a flat rate of $7.50. How much money will the tourist save by taking the airport limousine?

20. **SALARY COMPARISONS** Markeeta, a recent college graduate, received two job offers. Company A offered her an initial salary of $28,800 with guaranteed annual increases of $1500/year for the first 5 years. Company B offered an initial salary of $30,400 with guaranteed annual increases of $1100 per year for the first 5 years.
a. Which company is offering a higher salary for the fifth year of employment?
b. Which company is offering more money for the first 5 years of employment?

21. **SUM-OF-THE-YEARS'-DIGITS METHOD OF DEPRECIATION** One of the methods that the Internal Revenue Service allows for computing depreciation of certain business property is the sum-of-the-years'-digits method. If a property valued at C dollars has an estimated useful life of N years and a salvage value of S dollars, then the amount of depreciation D_n allowed during the nth year is given by

$$D_n = (C - S)\frac{N - (n - 1)}{S_N} \qquad (0 \le n \le N)$$

where S_N is the sum of the first N positive integers representing the estimated useful life of the property. Thus,

$$S_N = 1 + 2 + \cdots + N = \frac{N(N + 1)}{2}$$

a. Verify that the sum of the arithmetic progression $S_N = 1 + 2 + \cdots + N$ is given by

$$\frac{N(N + 1)}{2}$$

b. If office furniture worth $6000 is to be depreciated by this method over $N = 10$ years and the salvage value of the furniture is $500, find the depreciation for the third year by computing D_3.

22. **SUM-OF-THE-YEARS'-DIGITS METHOD OF DEPRECIATION** Refer to Example 1, Section 1.3, where the amount of depreciation allowed for a printing machine, which has an estimated useful life of 5 years and an initial value of $100,000 (with no salvage value), was $20,000/year using the straight-line method of depreciation. Determine the amount of depreciation that would be allowed for the first year if the printing machine were depreciated using the sum-of-the-years'-digits method described in Exercise 21. Which method would result in a larger depreciation of the asset in its first year of use?

In Exercises 23–28, determine which of the given sequences are geometric progressions. For each geometric progression, find the seventh term and the sum of the first seven terms.

23. 4, 8, 16, 32, . . .

24. 1, −1/2, 1/4, −1/8, . . .

25. 1/2, −3/8, 1/4, −9/64, . . .

26. 0.004, 0.04, 0.4, 4, . . .

27. 243, 81, 27, 9, . . .

28. −1, 1, 3, 5, . . .

29. Find the 20th term and sum of the first 20 terms of the geometric progression −3, 3, −3, 3, . . .

30. Find the 23rd term in a geometric progression having the first term $a = 0.1$ and ratio $r = 2$.

A calculator is recommended for Exercises 31–42.

31. POPULATION GROWTH It has been projected that the population of a certain city in the Southwest will increase by 8% during each of the next 5 years. If the current population is 200,000, what is the expected population in 5 years?

32. SALES GROWTH The Metro Cable TV Company had sales of \$2,500,000 in its first year of operation. If thereafter the sales increased by 12% of the previous year, find the sales of the company in the fifth year and the total sales over the first 5 years of operation.

33. COLAs Suppose the cost-of-living index had increased by 9% during each of the past 6 years and that a member of the EUW Union had been guaranteed an annual increase equal to 2% above the cost-of-living index over that period. What would be the present salary of a union member whose salary 6 years ago was \$22,000?

34. SAVINGS PLANS The parents of a 9-year-old boy have agreed to deposit \$10 in their son's bank account on his 10th birthday and to double the size of their deposit every year thereafter until his 18th birthday.
    a. How much will they have to deposit on his 18th birthday?
    b. How much will they have deposited by his 18th birthday?

35. SALARY COMPARISONS Suppose an employee of the Stenton Printing Company whose current annual salary is \$28,000 has the option of taking an annual raise of 8%/year for the next 4 years or a fixed annual raise of \$1500/year. Which option would be more profitable to him considering his total earnings over the 4-year period?

36. BACTERIA GROWTH A culture of a certain bacteria is known to double in number every three hours. If the culture has an initial count of 20, what will be the population of the culture at the end of 24 hours?

37. TRUST FUNDS Sarah is the recipient of a trust fund that she will receive over a period of 6 years. Under the terms of the trust, she is to receive \$10,000 the first year and each succeeding annual payment is to be increased by 15%.
    a. How much will she receive during the sixth year?
    b. What is the total amount of the six payments she will receive?

In Exercises 38–40, find the book value of office equipment purchased at a cost *C* at the end of the *n*th year if it is to be depreciated by the double declining-balance method over 10 years. Assume a salvage value of \$0.

38. $C = \$20{,}000$, $n = 4$

39. $C = \$150{,}000$, $n = 8$

40. $C = \$80{,}000$, $n = 7$

41. DOUBLE DECLINING-BALANCE METHOD OF DEPRECIATION Restaurant equipment purchased at a cost of \$150,000 is to be depreciated by the double declining-balance method over 10 years. What is the book value of the equipment at the end of 6 years? By what amount has the equipment been depreciated at the end of the sixth year?

42. DOUBLE DECLINING-BALANCE METHOD OF DEPRECIATION Refer to Exercise 22. Recall that a printing machine that had an estimated useful life of 5 years and an initial value of \$100,000 (with no salvage value) was to be depreciated. At the end of the first year, using the straight-line method of depreciation, the amount of depreciation allowed was \$20,000, and when the sum-of-the-years'-digits method was used the depreciation was \$33,333. Determine the amount of depreciation that would be allowed for the first year if the printing machine were depreciated by the double declining-balance method. Which of these three methods would result in the largest depreciation of the printing machine at the end of its first year of use?

In Exercises 43 and 44, determine whether the statement is true or false. If it is true, explain why it is true. If it is false, give an example to show why it is false.

43. If $a_1, a_2, a_3, \ldots, a_n$ and $b_1, b_2, b_3, \ldots, b_n$ are arithmetic progressions, then $a_1 + b_1, a_2 + b_2, a_3 + b_3, \ldots, a_n + b_n$ is also an arithmetic progression.

44. If $a_1, a_2, a_3, \ldots, a_n$ and $b_1, b_2, b_3, \ldots, b_n$ are geometric progressions, then $a_1b_1, a_2b_2, a_3b_3, \ldots, a_nb_n$ is also a geometric progression.

Spreadsheet examples and exercises for this section that may be solved using the Microsoft® Excel program are given at the Brooks/Cole Web site:
http://www.brookscole.com/math/authors/tans/

SOLUTIONS TO SELF-CHECK EXERCISES 5.4

1. Use Equation (19) with $a = -24$ and $r = -1/2$, obtaining

$$S_5 = \frac{-24\left[1 - \left(-\frac{1}{2}\right)^5\right]}{1 - \left(-\frac{1}{2}\right)} = \frac{-24\left(1 + \frac{1}{32}\right)}{\frac{3}{2}} = -\frac{33}{2}$$

2. Use Equation (20) with $C = 75{,}000$, $N = 5$, and $n = 3$, giving the book value of the office equipment at the end of 3 years as

$$V(3) = 75{,}000\left(1 - \frac{2}{5}\right)^3 = 16{,}200$$

or \$16,200.

3. We have

$$S = R + R(1 + i) + R(1 + i)^2 + \cdots + R(1 + i)^{n-1}$$

Now, the sum on the right is easily seen to be the sum of the first n terms of a geometric progression with first term R and common ratio $(1 + i)$, so by virtue of Formula (19),

$$S = \frac{R[1 - (1 + i)^n]}{1 - (1 + i)} = R\left[\frac{(1 + i)^n - 1}{i}\right]$$

CHAPTER 5 Summary of Principal Formulas and Terms

Formulas

1. Simple interest — $A = P(1 + rt)$
2. Compound interest
   a. Accumulated amount — $A = P(1 + i)^n$
   b. Present value — $P = A(1 + i)^{-n}$
   c. Interest rate per compounding period — $i = r/m$
   d. Number of conversion periods — $n = mt$
3. Effective rate of interest — $r_{\text{eff}} = \left(1 + \frac{r}{m}\right)^m - 1$

4. Annuities

a. Future value $$S = R\left[\frac{(1+i)^n - 1}{i}\right]$$

b. Present value $$P = R\left[\frac{1-(1+i)^{-n}}{i}\right]$$

5. Amortization payment $$R = \frac{Pi}{1-(1+i)^{-n}}$$

6. Sinking fund payment $$R = \frac{iS}{(1+i)^n - 1}$$

Terms

simple interest	annuity
accumulated amount	ordinary annuity
compound interest	annuity certain
nominal rate (stated rate)	future value of an annuity
effective rate	present value of an annuity
present value	sinking fund
future value	

CHAPTER 5 REVIEW EXERCISES

1. Find the accumulated amount after 4 years if $5000 is invested at 10%/year compounded (a) annually, (b) semiannually, (c) quarterly, and (d) monthly.

2. Find the accumulated amount after 8 years if $12,000 is invested at 6.5%/year compounded (a) annually, (b) semiannually, (c) quarterly, and (d) monthly.

3. Find the effective rate of interest corresponding to a nominal rate of 12%/year compounded (a) annually, (b) semiannually, (c) quarterly, and (d) monthly.

4. Find the effective rate of interest corresponding to a nominal rate of 11.5%/year compounded (a) annually, (b) semiannually, (c) quarterly, and (d) monthly.

5. Find the present value of $41,413 due in 5 years at an interest rate of 6.5%/year compounded quarterly.

6. Find the present value of $64,540 due in 6 years at an interest rate of 8%/year compounded monthly.

7. Find the amount (future value) of an ordinary annuity of $150/quarter for 7 years at 8%/year compounded quarterly.

8. Find the future value of an ordinary annuity of $120/month for 10 years at 9%/year compounded monthly.

9. Find the present value of an ordinary annuity of 36 payments of $250 each made monthly and earning interest at 9%/year compounded monthly.

10. Find the present value of an ordinary annuity of 60 payments of $5000 each made quarterly and earning interest at 8%/year compounded quarterly.

11. Find the payment R needed to amortize a loan of $22,000 at 8.5%/year compounded monthly with 36 monthly installments over a period of 3 years.

12. Find the payment R needed to amortize a loan of $10,000 at 9.2%/year compounded monthly with 36 monthly installments over a period of 3 years.

13. Find the payment R needed to accumulate $18,000 with 48 monthly installments over a period of 4 years at an interest rate of 6%/year compounded monthly.

14. Find the payment R needed to accumulate $15,000 with

60 monthly installments over a period of 5 years at an interest rate of 7.2%/year compounded monthly.

15. Find the rate of interest per year compounded on a daily basis that is equivalent to 7.2%/year compounded monthly.

16. Find the rate of interest per year compounded on a daily basis that is equivalent to 9.6%/year compounded monthly.

17. The JCN Media Corporation had sales of $1,750,000 in the first year of operation. If the sales increased by 14%/year thereafter, find the company's sales in the fourth year and the total sales over the first 4 years of operation.

18. The manager of a money-market fund has invested $4.2 million in certificates of deposit that pay interest at the rate of 5.4%/year compounded quarterly over a period of 5 years. How much will the investment be worth at the end of 5 years?

19. Kim invested a sum of money 4 years ago in a savings account that has since paid interest at the rate of 6.5%/year compounded monthly. Her investment is now worth $19,440.31. How much did she originally invest?

20. Juan invested $24,000 in a mutual fund 5 years ago. Today his investment is worth $34,616. Find the effective annual rate of return on his investment over the 5-year period.

21. The Blakes have decided to start a monthly savings program in order to provide for their son's college education. How much should they deposit at the end of each month in a savings account earning interest at the rate of 8%/year compounded monthly so that at the end of the tenth year the accumulated amount will be $40,000?

22. Mai Lee has contributed $200 at the end of each month into her company's employee retirement account for the past 10 years. Her employer has matched her contribution each month. If the account has earned interest at the rate of 8%/year compounded monthly over the 10-year period, determine how much Mai Lee now has in her retirement account.

23. Maria has leased an auto for 4 years at $300/month. If money is worth 5%/year compounded monthly, what is the equivalent cash payment (present value) of this annuity? (Assume that the payments are made at the end of each month.)

24. Peggy made a down payment of $400 toward the purchase of new furniture. To pay the balance of the purchase price, she has secured a loan from her bank at 12%/year compounded monthly. Under the terms of her finance agreement, she is required to make payments of $75.32 at the end of each month for 24 months. What was the purchase price of the furniture?

25. The Turners have purchased a house for $150,000. They made an initial down payment of $30,000 and secured a mortgage with interest charged at the rate of 9%/year on the unpaid balance. (Interest computations are made at the end of each month.) Assume the loan is amortized over 30 years.
a. What monthly payment will the Turners be required to make?
b. What will be their total interest payment?
c. What will be their equity (disregard depreciation) after 10 years?

26. Refer to Exercise 25. If the loan is amortized over 15 years,
a. What monthly payment will the Turners be required to make?
b. What will be their total interest payment?
c. What will be their equity (disregard depreciation) after 10 years?

27. The management of a corporation anticipates a capital expenditure of $500,000 in 5 years for the purpose of purchasing replacement machinery. To finance this purchase, a sinking fund that earns interest at the rate of 10%/year compounded quarterly will be set up. Determine the amount of each (equal) quarterly installment that should be deposited in the fund. (Assume that the payments are made at the end of each quarter.)

28. The management of a condominium association anticipates a capital expenditure of $120,000 in 2 years for the purpose of painting the exterior of the condominium. To pay for this maintenance, a sinking fund will be set up that will earn interest at the rate of 5.8%/year compounded monthly. Determine the amount of each (equal) monthly installment the association will be required to deposit into the fund at the end of each month for the next 2 years.

29. The outstanding balance on Bill's credit card account is $3200. The bank issuing the credit card is charging 18.6%/year compounded monthly. If Bill decides to pay off this balance in equal monthly installments at the end of each month for the next 18 months, how much will be his monthly payment?

30. Refer to Exercise 29. What is the effective rate of interest the bank is charging Bill?

6 SETS AND COUNTING

We often deal with well-defined collections of objects called *sets*. In this chapter we see how sets can be combined algebraically to yield other sets. We also look at some techniques for determining the number of elements in a set and for determining the number of ways the elements of a set can be arranged or combined. These techniques enable us to solve many practical problems, as you will see throughout the chapter.

What are the investment options? An investor has decided to purchase shares of stock from a recommended list of aerospace, energy development, and electronics companies. In Example 5, page 357, we will determine how many ways the investor may select a group of three companies from the list.

6.1 Sets and Set Operations

SET TERMINOLOGY AND NOTATION

We often deal with collections of different kinds of objects. For example, in conducting a study of the distribution of the weights of newborn infants, we might consider the collection of all infants born in the Massachusetts General Hospital during 2000. In a study of the fuel consumption of compact cars, we might be interested in the collection of compact cars manufactured by General Motors in the 2000 model year. Such collections are examples of *sets*. More specifically, a **set** is a well-defined collection of objects. Thus, a set is not just any collection of objects, but it must be well defined in the sense that if we are given an object, then we should be able to determine whether or not it belongs to the collection.

The objects of a set are called the **elements,** or ***members,*** **of a set** and are usually denoted by lowercase letters $a, b, c, \ldots$; the sets themselves are usually denoted by uppercase letters $A, B, C, \ldots$. The elements of a set may be displayed by listing each element between braces. For example, using **roster notation,** the set A consisting of the first three letters of the English alphabet is written

$$A = \{a, b, c\}$$

The set B of all letters of the alphabet may be written

$$B = \{a, b, c, \ldots, z\}$$

Another kind of set notation commonly used is **set-builder notation.** Here, a rule is given that describes the definite property or properties an object x must satisfy to qualify for membership in the set. Using this notation, the set B is written as

$$B = \{x \mid x \text{ is a letter of the English alphabet}\}$$

and is read "B is the set of all elements x such that x is a letter of the English alphabet."

If a is an element of a set A, we write $a \in A$ and read "a belongs to A" or "a is an element of A." If, however, the element a does not belong to the set A, then we write $a \notin A$ and read "a does not belong to A." For example, if $A = \{1, 2, 3, 4, 5\}$, then $3 \in A$ but $6 \notin A$.

Group Discussion

1. Let A denote the collection of all the days in August 2000 in which the average daily temperature in San Francisco was approximately 75°F. Is A a set? Explain your answer.

2. Let B denote the collection of all the days in August 2000 in which the average daily temperature in San Francisco was between 73.5°F and 81.2°F, inclusive. Is B a set? Explain your answer.

Set Equality

> Two sets A and B are **equal**, written $A = B$, if and only if they have exactly the same elements.

EXAMPLE 1 Let A, B, and C be the sets

$$A = \{a, e, i, o, u\}$$
$$B = \{a, i, o, e, u\}$$
$$C = \{a, e, i, o\}$$

Then, $A = B$ since they both contain exactly the same elements. Note that the order in which the elements are displayed is immaterial. Also, $A \neq C$ since $u \in A$ but $u \notin C$. Similarly, we conclude that $B \neq C$. ■■■■

Subset

> If every element of a set A is also an element of a set B, then we say that A is a **subset** of B and write $A \subseteq B$.

By this definition, two sets A and B are equal if and only if (1) $A \subseteq B$ and (2) $B \subseteq A$. You may verify this (see Exercise 66).

EXAMPLE 2 Referring to Example 1, we find that $C \subseteq B$ since every element of C is also an element of B. Also, if D is the set

$$D = \{a, e, i, o, x\}$$

then D is not a subset of A, written $D \not\subseteq A$, since $x \in D$ but $x \notin A$. Observe that $A \not\subseteq D$ as well since $u \in A$ but $u \notin D$. ■■■■

If A and B are sets such that $A \subseteq B$ but $A \neq B$, then we say that A is a **proper subset** of B. In other words, a set A is a proper subset of a set B, written $A \subset B$, if (1) $A \subseteq B$ and (2) there exists at least one element in B that is not in A. The latter condition states that the set A is properly "smaller" than the set B.

EXAMPLE 3 Let $A = \{1, 2, 3, 4, 5, 6\}$ and $B = \{2, 4, 6\}$. Then, B is a proper subset of A since (1) $B \subseteq A$, which is easily verified, and (2) there exists at least one element in A that is not in B—for example, the element 1. ■■■■

Notice that when we are referring to sets and subsets we use the symbols $\subset$, $\subseteq$, $\supset$, and $\supseteq$ to express the idea of "containment." However, when we wish to show that an element is contained in a set, we use the symbol $\in$. Thus, in Example 3, we would write $1 \in A$ and *not* $\{1\} \in A$.

Empty Set

The set that contains no elements is called the **empty set** and is denoted by $\emptyset$.

The empty set, $\emptyset$, is a subset of every set. To see this, observe that $\emptyset$ has no elements and, therefore, contains no element that is not also in A.

EXAMPLE 4 List all subsets of the set $A = \{a, b, c\}$.

SOLUTION ✓ There is one subset consisting of no elements—namely, the empty set, $\emptyset$. Next, observe that there are three subsets consisting of one element,

$$\{a\}, \quad \{b\}, \quad \text{and} \quad \{c\}$$

three subsets consisting of two elements,

$$\{a, b\}, \quad \{a, c\}, \quad \text{and} \quad \{b, c\}$$

and one subset consisting of three elements, the set A itself. Therefore, the subsets of A are

$$\emptyset, \quad \{a\}, \quad \{b\}, \quad \{c\}, \quad \{a, b\}, \quad \{a, c\}, \quad \{b, c\}, \quad \{a, b, c\}$$

■■■■

In contrast with the empty set, we have, on the other extreme, the notion of a largest, or *universal,* set. A **universal set** is the set of all elements of interest in a particular discussion. It is the largest in the sense that all sets considered in the discussion of the problem are subsets of the universal set. Of course, different universal sets are associated with different problems, as shown in Example 5.

EXAMPLE 5 **a.** If the problem at hand is to determine the ratio of female to male students in a college, then a logical choice of a universal set is the set consisting of the whole student body of the college.
b. If the problem is to determine the ratio of female to male students in the business department of the college in part (a), then the set of all students in the business department may be chosen as the universal set. ■■■■

A visual representation of sets is realized through the use of **Venn diagrams,** which are of considerable help in understanding the concepts introduced earlier, as well as in solving problems involving sets. The universal set U is represented by a rectangle, and subsets of U are represented by regions lying inside the rectangle.

EXAMPLE 6 Use Venn diagrams to illustrate the following statements:

a. The sets A and B are equal.
b. The set A is a proper subset of the set B.
c. The sets A and B are not subsets of each other.

SOLUTION ✓ The respective Venn diagrams are shown in Figure 6.1a–c.

FIGURE 6.1

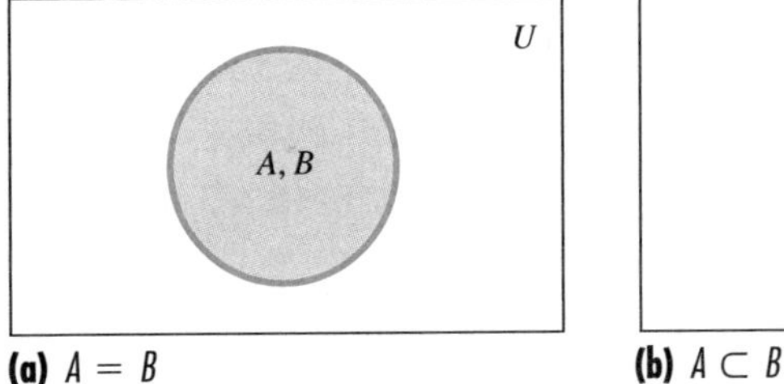

(a) $A = B$

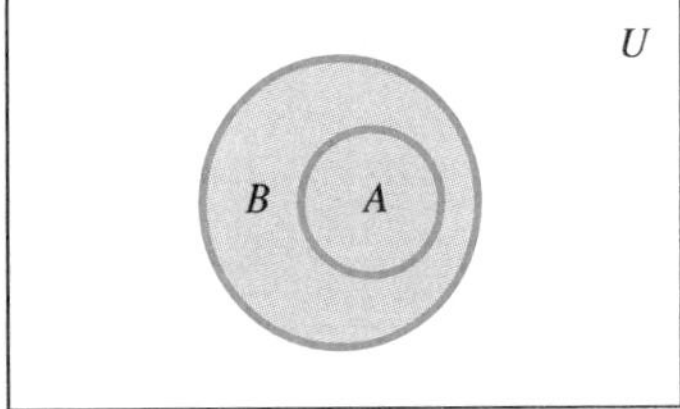

(b) $A \subset B$

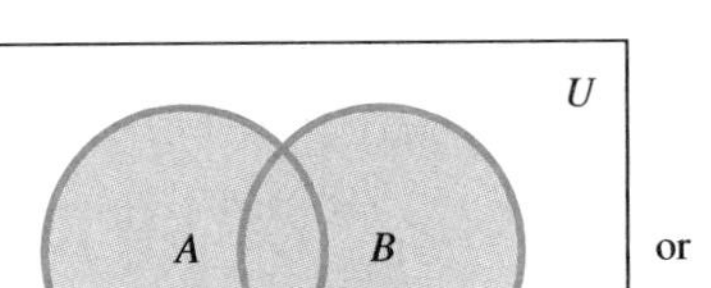

or

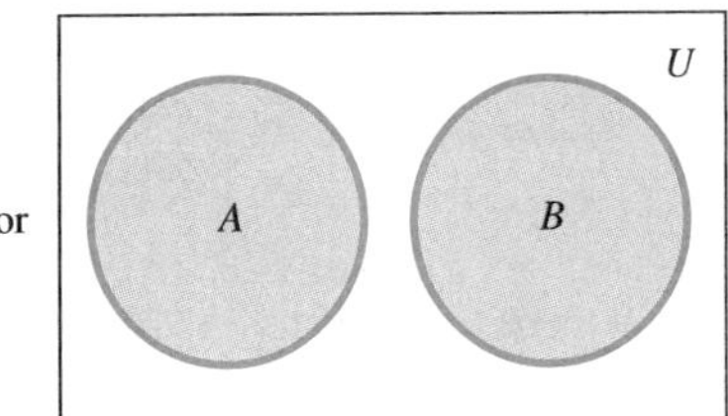

(c) $A \not\subset B$ and $B \not\subset A$

SET OPERATIONS

Having introduced the concept of a set, our next task is to consider operations on sets—that is, to consider ways in which sets may be combined to yield other sets. These operations enable us to combine sets in much the same way the operations of addition and multiplication enable us to combine numbers to obtain other numbers. In what follows, all sets are assumed to be subsets of a given universal set U.

Set Union

Let A and B be sets. The **union** of A and B, written $A \cup B$, is the set of all elements that belong to either A or B or both.

$$A \cup B = \{x \mid x \in A \quad \text{or} \quad x \in B \quad \text{or} \quad \text{both}\}$$

The shaded portion of the Venn diagram (Figure 6.2) depicts the set $A \cup B$.

FIGURE 6.2 Set union

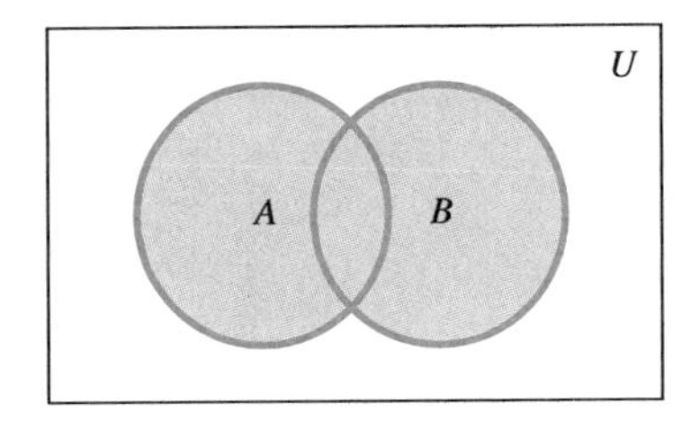

$A \cup B$

EXAMPLE 7 If $A = \{a, b, c\}$ and $B = \{a, c, d\}$, then $A \cup B = \{a, b, c, d\}$. ■■■■

Set Intersection

Let A and B be sets. The set of elements in common with the sets A and B, written $A \cap B$, is called the **intersection** of A and B.

$$A \cap B = \{x \mid x \in A \quad \text{and} \quad x \in B\}$$

The shaded portion of the Venn diagram (Figure 6.3) depicts the set $A \cap B$.

EXAMPLE 8 Let $A = \{a, b, c\}$ and $B = \{a, c, d\}$. Then, $A \cap B = \{a, c\}$. (Compare this result with Example 7.) ■■■■

EXAMPLE 9 Let $A = \{1, 3, 5, 7, 9\}$ and $B = \{2, 4, 6, 8, 10\}$. Then, $A \cap B = \varnothing$. ■■■■

The two sets of Example 9 have null intersection. In general, the sets A and B are said to be **disjoint** if they have no elements in common—that is, if $A \cap B = \varnothing$.

EXAMPLE 10 If U is the set of all students in the classroom and $M = \{x \in U \mid x \text{ is male}\}$ and $F = \{x \in U \mid x \text{ is female}\}$, then $F \cap M = \varnothing$, and F and M are disjoint. ■■■■

Complement of a Set

If U is a universal set and A is a subset of U, then the set of all elements in U that are not in A is called the **complement** of A and is denoted A^c.

$$A^c = \{x \mid x \in U, x \notin A\}$$

The shaded portion of the Venn diagram (Figure 6.4) shows the set A^c.

FIGURE 6.3
Set intersection

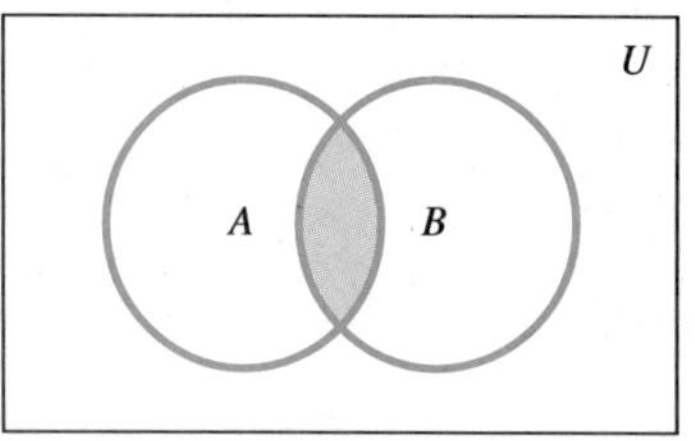

FIGURE 6.4
Set complementation

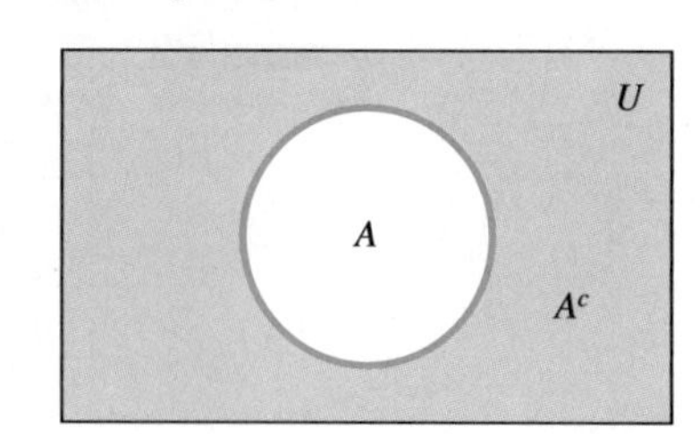

Group Discussion

Let A, B, and C be nonempty subsets of a set U.

1. Suppose $A \cap B \neq \emptyset$, $A \cap C \neq \emptyset$, and $B \cap C \neq \emptyset$. Can you conclude that $A \cap B \cap C \neq \emptyset$? Explain your answer with an example.
2. Suppose $A \cap B \cap C \neq \emptyset$. Can you conclude that $A \cap B \neq \emptyset$, $A \cap C \neq \emptyset$, and $B \cap C \neq \emptyset$ simultaneously? Explain your answer.

EXAMPLE 11

Let $U = \{1, 2, 3, 4, 5, 6, 7, 8, 9, 10\}$ and $A = \{2, 4, 6, 8, 10\}$. Then, $A^c = \{1, 3, 5, 7, 9\}$.

The following rules hold for the operation of complementation. See whether you can verify them.

Set Complementation

If U is a universal set and A is a subset of U, then

a. $U^c = \emptyset$ **b.** $\emptyset^c = U$ **c.** $(A^c)^c = A$
d. $A \cup A^c = U$ **e.** $A \cap A^c = \emptyset$

The following rules govern the operations on sets.

Set Operations

Let U be a universal set. If A, B, and C are arbitrary subsets of U, then

$A \cup B = B \cup A$	*Commutative law for union*
$A \cap B = B \cap A$	*Commutative law for intersection*
$A \cup (B \cup C) = (A \cup B) \cup C$	*Associative law for union*
$A \cap (B \cap C) = (A \cap B) \cap C$	*Associative law for intersection*
$A \cup (B \cap C) = (A \cup B) \cap (A \cup C)$	*Distributive law for union*
$A \cap (B \cup C) = (A \cap B) \cup (A \cap C)$	*Distributive law for intersection*

Two additional rules, referred to as De Morgan's laws, govern the operations on sets.

De Morgan's Laws

Let A and B be sets. Then,

$$(A \cup B)^c = A^c \cap B^c \quad (1)$$

$$(A \cap B)^c = A^c \cup B^c \quad (2)$$

Equation (1) states that the complement of the union of two sets is equal to the intersection of their complements. Equation (2) states that the complement of the intersection of two sets is equal to the union of their complements.

We will not prove De Morgan's laws here, but the plausibility of (2) is illustrated in the following example.

EXAMPLE 12 Using Venn diagrams, show that

$$(A \cap B)^c = A^c \cup B^c$$

SOLUTION ✓ $(A \cap B)^c$ is the set of elements in U but not in $A \cap B$ and is thus the shaded region shown in Figure 6.5. Next, A^c and B^c are shown in Figure 6.6a–b. Their union, $A^c \cup B^c$, is easily seen to be equivalent to $(A \cap B)^c$ by referring once again to Figure 6.5.

FIGURE 6.5
$(A \cap B)^c$

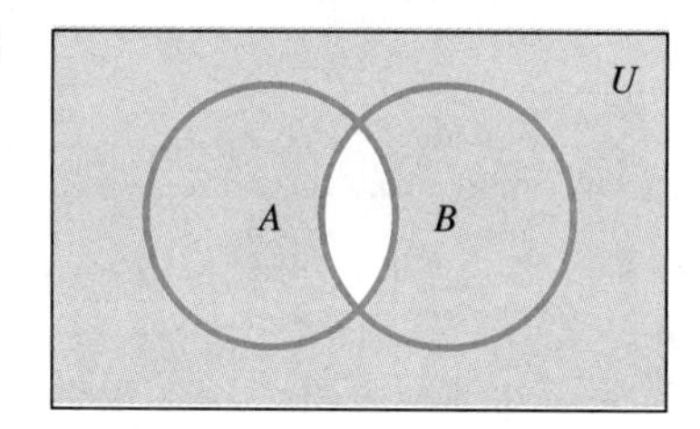

FIGURE 6.6
$A^c \cup B^c$ is the set obtained by joining (a) and (b).

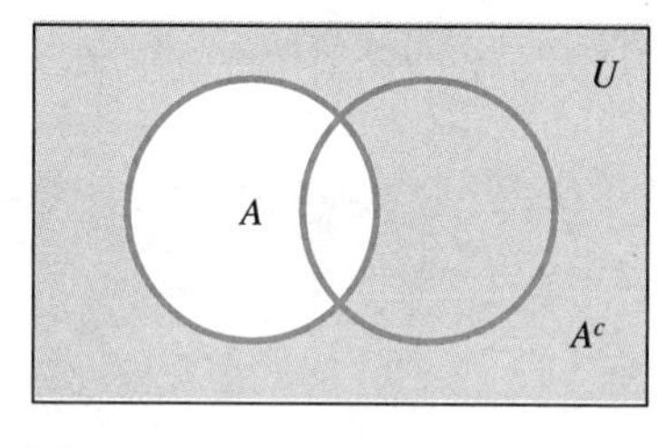

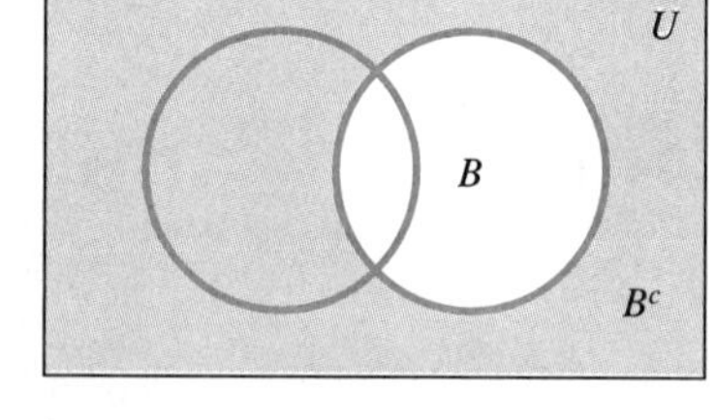

(a) (b)

■■■■

EXAMPLE 13 Let $U = \{1, 2, 3, 4, 5, 6, 7, 8, 9, 10\}$, $A = \{1, 2, 4, 8, 9\}$, and $B = \{3, 4, 5, 6, 8\}$. Verify by direct computation that $(A \cup B)^c = A^c \cap B^c$.

SOLUTION ✓ $A \cup B = \{1, 2, 3, 4, 5, 6, 8, 9\}$, so $(A \cup B)^c = \{7, 10\}$. However, $A^c = \{3, 5, 6, 7, 10\}$ and $B^c = \{1, 2, 7, 9, 10\}$, so $A^c \cap B^c = \{7, 10\}$. The required result follows. ■■■■

APPLICATION

EXAMPLE 14 Let U denote the set of all cars in a dealer's lot and

$$A = \{x \in U \mid x \text{ is equipped with automatic transmission}\}$$
$$B = \{x \in U \mid x \text{ is equipped with air conditioning}\}$$
$$C = \{x \in U \mid x \text{ is equipped with side air bags}\}$$

Find an expression in terms of A, B, and C for each of the following sets:

a. The set of cars with at least one of the given options
b. The set of cars with exactly one of the given options
c. The set of cars with automatic transmission and side air bags but no air conditioning

SOLUTION ✓

a. The set of cars with at least one of the given options is $A \cup B \cup C$ (Figure 6.7a).
b. The set of cars with automatic transmission only is given by $A \cap B^c \cap C^c$. Similarly, we find that the set of cars with air conditioning only is given by $B \cap C^c \cap A^c$, whereas the set of cars with side air bags only is given by $C \cap A^c \cap B^c$. Thus, the set of cars with exactly one of the given options is $(A \cap B^c \cap C^c) \cup (B \cap C^c \cap A^c) \cup (C \cap A^c \cap B^c)$ (Figure 6.7b).
c. The set of cars with automatic transmission and side air bags but no air conditioning is given by $A \cap C \cap B^c$ (Figure 6.7c).

FIGURE 6.7

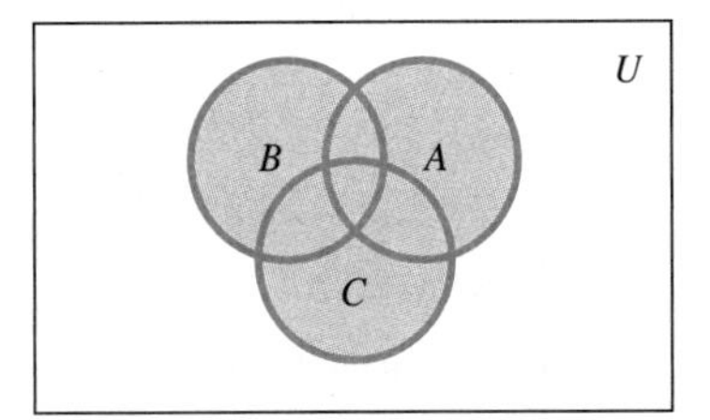

(a) The set of cars with at least one option

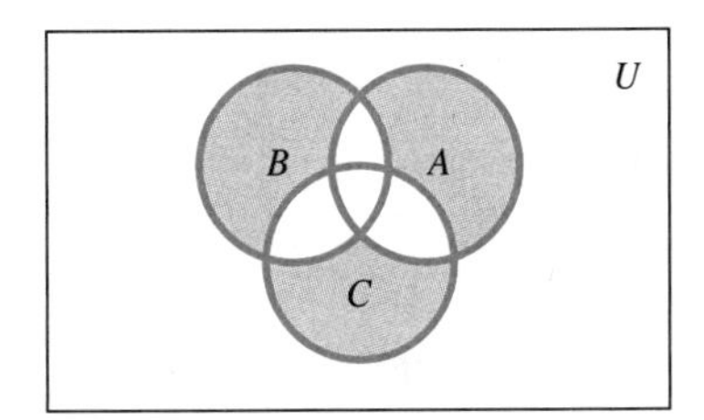

(b) The set of cars with exactly one option

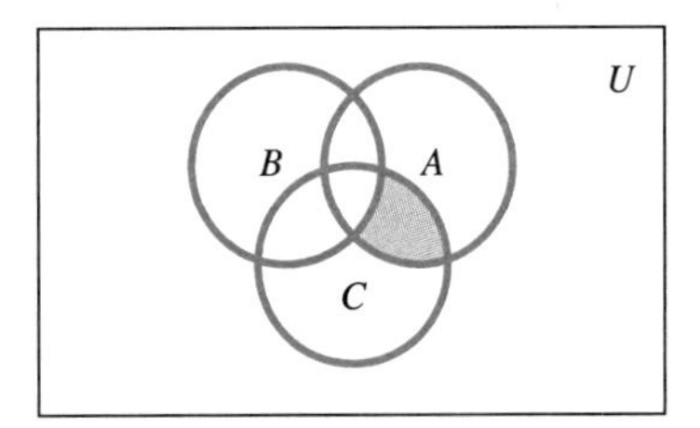

(c) The set of cars with automatic transmission and side air bags but no air conditioning

SELF-CHECK EXERCISES 6.1

1. Let $U = \{1, 2, 3, 4, 5, 6, 7\}$, $A = \{1, 2, 3\}$, $B = \{3, 4, 5, 6\}$, and $C = \{2, 3, 4\}$. Find the following sets:
a. A^c **b.** $A \cup B$ **c.** $B \cap C$
d. $(A \cup B) \cap C$ **e.** $(A \cap B) \cup C$ **f.** $A^c \cap (B \cup C)^c$

2. Let U denote the set of all members of the House of Representatives. Let

$$D = \{x \in U \mid x \text{ is a Democrat}\}$$
$$R = \{x \in U \mid x \text{ is a Republican}\}$$
$$F = \{x \in U \mid x \text{ is a female}\}$$
$$L = \{x \in U \mid x \text{ is a lawyer by training}\}$$

Describe each of the following sets in words.
a. $D \cap F$ **b.** $F^c \cap R$ **c.** $D \cap F \cap L^c$

Solutions to Self-Check Exercises 6.1 can be found on page 347.

6.1 Exercises

In Exercises 1–4, write the given set in set-builder notation.

1. The set of gold medalists in the 2000 Summer Olympic Games

2. The set of football teams in the NFL

3. $\{3, 4, 5, 6, 7\}$

4. $\{1, 3, 5, 7, 9, 11, \ldots, 39\}$

In Exercises 5–8, list the elements of the given set in roster notation.

5. $\{x \mid x$ is a digit in the number 352,646$\}$

6. $\{x \mid x$ is a letter in the word *HIPPOPOTAMUS*$\}$

7. $\{x \mid 2 - x = 4; x,$ an integer$\}$

8. $\{x \mid 2 - x = 4; x,$ a fraction$\}$

In Exercises 9–14, state whether the given statements are true or false.

9. **a.** $\{a, b, c\} = \{c, a, b\}$ **b.** $A \in A$

10. **a.** $\varnothing \in A$ **b.** $A \subset A$

11. **a.** $0 \in \varnothing$ **b.** $0 = \varnothing$

12. **a.** $\{\varnothing\} = \varnothing$ **b.** $\{a, b\} \in \{a, b, c\}$

13. $\{$Chevrolet, Pontiac, Buick$\} \subset \{x \mid x$ is a division of General Motors$\}$

14. $\{x \mid x$ is a silver medalist in the 2000 Summer Olympic Games$\} = \varnothing$

In Exercises 15 and 16, let $A = \{1, 2, 3, 4, 5\}$. Determine whether the given statements are true or false.

15. **a.** $2 \in A$ **b.** $A \subseteq \{2, 4, 6\}$

16. **a.** $0 \in A$ **b.** $\{1, 3, 5\} \in A$

17. Let $A = \{1, 2, 3\}$. Which of the following sets are equal to A?
 a. $\{2, 1, 3\}$ **b.** $\{3, 2, 1\}$ **c.** $\{0, 1, 2, 3\}$

18. Let $A = \{a, e, l, t, r\}$. Which of the following sets are equal to A?
 a. $\{x \mid x$ is a letter of the word *later*$\}$
 b. $\{x \mid x$ is a letter of the word *latter*$\}$
 c. $\{x \mid x$ is a letter of the word *relate*$\}$

19. List all subsets of the following sets:
 a. $\{1, 2\}$ **b.** $\{1, 2, 3\}$ **c.** $\{1, 2, 3, 4\}$

20. List all subsets of the set $A = \{$IBM, U.S. Steel, Union Carbide, Boeing$\}$. Which of these are proper subsets of A?

In Exercises 21–24, find the smallest possible set (that is, the set with the least number of elements) that contains the given sets as subsets.

21. $\{1, 2\}, \{1, 3, 4\}, \{4, 6, 8, 10\}$

22. $\{1, 2, 4\}, \{a, b\}$

23. $\{$Jill, John, Jack$\}$, $\{$Susan, Sharon$\}$

24. $\{$GM, Ford, Chrysler$\}$, $\{$Daimler-Benz, Volkswagen$\}$, $\{$Toyota, Nissan$\}$

25. Use Venn diagrams to represent the following relationships:
 a. $A \subset B$ and $B \subset C$
 b. $A \subset U$ and $B \subset U$, where A and B have no elements in common
 c. The sets A, B, and C are equal.

26. Let U denote the set of all students who applied for admission to the freshman class at Faber College for the upcoming academic year and let

$$A = \{x \in U \mid x \text{ is a successful applicant}\}$$
$$B = \{x \in U \mid x \text{ is a female student who enrolled in the freshman class}\}$$
$$C = \{x \in U \mid x \text{ is a male student who enrolled in the freshman class}\}$$

 a. Use Venn diagrams to represent the sets U, A, B, and C.
 b. Determine whether the following statements are true or false.
 i. $A \subseteq B$ **ii.** $B \subset A$ **iii.** $C \subset B$

In Exercises 27 and 28, shade the portion of the accompanying figure that represents each of the given sets.

27. **a.** $A \cap B^c$
 b. $A^c \cap B$

28. **a.** $A^c \cap B^c$
 b. $(A \cup B)^c$

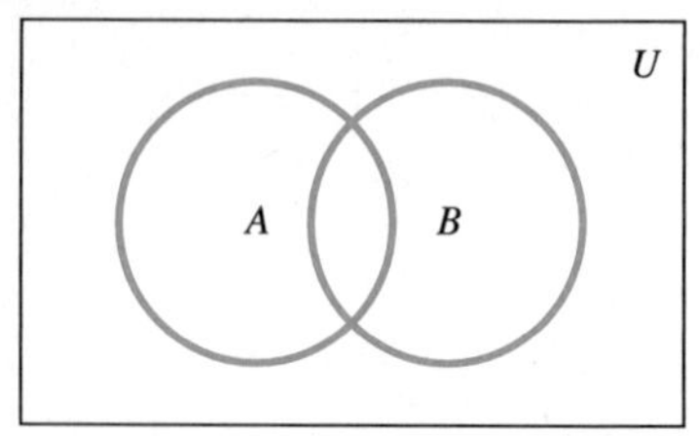

In Exercises 29–32, shade the portion of the accompanying figure that represents each of the given sets.

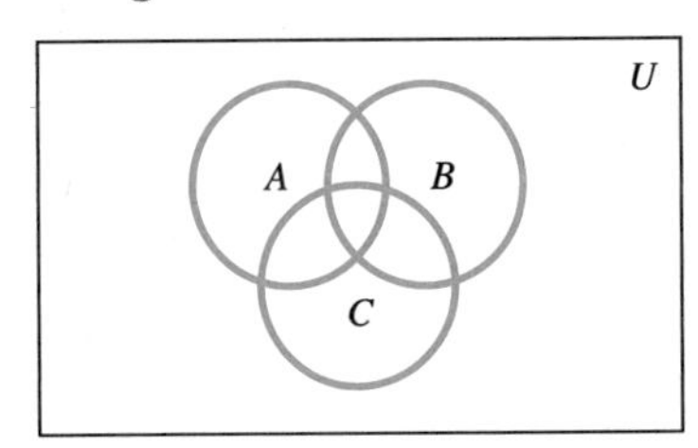

29. a. $A \cup B \cup C$ b. $A \cap B \cap C$

30. a. $A \cap B \cap C^c$ b. $A^c \cap B \cap C$

31. a. $A^c \cap B^c \cap C^c$ b. $(A \cup B)^c \cap C$

32. a. $A \cup (B \cap C)^c$ b. $(A \cup B \cup C)^c$

In Exercises 33–36, let $U = \{1, 2, 3, 4, 5, 6, 7, 8, 9, 10\}$, $A = \{1, 3, 5, 7, 9\}$, $B = \{2, 4, 6, 8, 10\}$, and $C = \{1, 2, 4, 5, 8, 9\}$. Find each of the given sets.

33. a. A^c b. $B \cup C$ c. $C \cup C^c$

34. a. $C \cap C^c$ b. $(A \cap C)^c$ c. $A \cup (B \cap C)$

35. a. $(A \cap B) \cup C$ b. $(A \cup B \cup C)^c$
c. $(A \cap B \cap C)^c$

36. a. $A^c \cap (B \cap C^c)$ b. $(A \cup B^c) \cup (B \cap C^c)$
c. $(A \cup B)^c \cap C^c$

In Exercises 37 and 38, determine whether the given pairs of sets are disjoint.

37. a. $\{1, 2, 3, 4\}$, $\{4, 5, 6, 7\}$
b. $\{a, c, e, g\}$, $\{b, d, f\}$

38. a. $\varnothing$, $\{1, 3, 5\}$ b. $\{0, 1, 3, 4\}$, $\{0, 2, 5, 7\}$

In Exercises 39–42, let U denote the set of all employees at the Universal Life Insurance Company. Let

$$T = \{x \in U \mid x \text{ drinks tea}\}$$
$$C = \{x \in U \mid x \text{ drinks coffee}\}$$

Describe each of the given sets in words.

39. a. T^c b. C^c

40. a. $T \cup C$ b. $T \cap C$

41. a. $T \cap C^c$ b. $T^c \cap C$

42. a. $T^c \cap C^c$ b. $(T \cup C)^c$

In Exercises 43–46, let U denote the set of all employees in a hospital. Let

$$N = \{x \in U \mid x \text{ is a nurse}\}$$
$$D = \{x \in U \mid x \text{ is a doctor}\}$$
$$A = \{x \in U \mid x \text{ is an administrator}\}$$
$$M = \{x \in U \mid x \text{ is a male}\}$$
$$F = \{x \in U \mid x \text{ is a female}\}$$

Describe each of the given sets in words.

43. a. D^c b. N^c

44. a. $N \cup D$ b. $N \cap M$

45. a. $D \cap M^c$ b. $D \cap A$

46. a. $N \cap F$ b. $(D \cup N)^c$

In Exercises 47 and 48, let U denote the set of all senators in Congress. Let

$$D = \{x \in U \mid x \text{ is a Democrat}\}$$
$$R = \{x \in U \mid x \text{ is a Republican}\}$$
$$F = \{x \in U \mid x \text{ is a female}\}$$
$$L = \{x \in U \mid x \text{ is a lawyer}\}$$

Write the set that represents each of the given statements.

47. a. The set of all Democrats who are female
b. The set of all Republicans who are male and are not lawyers

48. a. The set of all Democrats who are female or are lawyers
b. The set of all senators who are not Democrats or are lawyers

In Exercises 49 and 50, let U denote the set of all students in the business college of a certain university. Let

$$A = \{x \in U \mid x \text{ had taken a course in accounting}\}$$
$$B = \{x \in U \mid x \text{ had taken a course in economics}\}$$
$$C = \{x \in U \mid x \text{ had taken a course in marketing}\}$$

Write the set that represents each of the given statements.

49. **a.** The set of students who have not had a course in Economics
b. The set of students who have had courses in Accounting and Economics
c. The set of students who have had courses in Accounting and Economics but not Marketing

50. **a.** The set of students who have had courses in Economics but not courses in Accounting or Marketing
b. The set of students who have had at least one of the three courses
c. The set of students who have had all three courses

In Exercises 51 and 52, refer to the following diagram where U is the set of all tourists surveyed over a 1-week period in London and

$A = \{x \in U \mid x$ **has taken the underground (subway)**$\}$
$B = \{x \in U \mid x$ **has taken a cab**$\}$
$C = \{x \in U \mid x$ **has taken a bus**$\}$

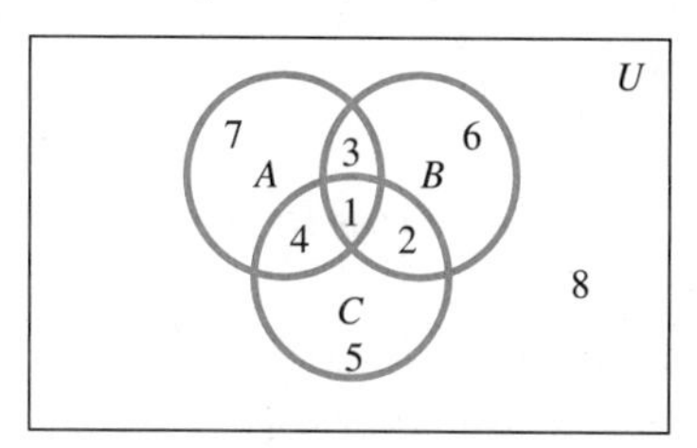

Express the indicated regions in set notation and in words.

51. **a.** Region 1
b. Regions 1 and 4 together
c. Regions 4, 5, 7, and 8 together

52. **a.** Region 3
b. Regions 4 and 6 together
c. Regions 5, 6, and 7 together

In Exercises 53–58, use Venn diagrams to illustrate each of the given statements.

53. $A \subset A \cup B$; $B \subset A \cup B$

54. $A \cap B \subset A$; $A \cap B \subset B$

55. $A \cup (B \cup C) = (A \cup B) \cup C$

56. $A \cap (B \cap C) = (A \cap B) \cap C$

57. $A \cap (B \cup C) = (A \cap B) \cup (A \cap C)$

58. $(A \cup B)^c = A^c \cap B^c$

In Exercises 59 and 60, let $U = \{1, 2, 3, 4, 5, 6, 7, 8, 9, 10\}$, $A = \{1, 3, 5, 7, 9\}$, $B = \{1, 2, 4, 7, 8\}$, and $C = \{2, 4, 6, 8\}$. Verify by direct computation each of the given equations.

59. **a.** $A \cup (B \cup C) = (A \cup B) \cup C$
b. $A \cap (B \cap C) = (A \cap B) \cap C$

60. **a.** $A \cap (B \cup C) = (A \cap B) \cup (A \cap C)$
b. $(A \cup B)^c = A^c \cap B^c$

In Exercises 61–64, refer to the accompanying figure and find the points that belong to each of the given sets.

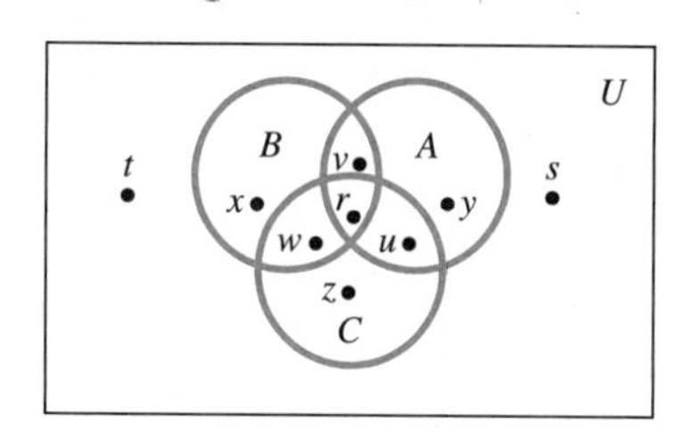

61. **a.** $A \cup B$ **b.** $A \cap B$

62. **a.** $A \cap (B \cup C)$ **b.** $(B \cap C)^c$

63. **a.** $(B \cup C)^c$ **b.** A^c

64. **a.** $(A \cap B) \cap C^c$ **b.** $(A \cup B \cup C)^c$

65. Suppose $A \subset B$ and $B \subset C$, where A and B are any two sets. What conclusion can be drawn regarding the sets A and C?

66. Verify the assertion that two sets A and B are equal if and only if (1) $A \subseteq B$ and (2) $B \subseteq A$.

In Exercises 67–72, determine whether the statement is true or false. If it is true, explain why it is true. If it is false, give an example to show why it is false.

67. A set is never a subset of itself.

68. A proper subset of a set is itself a subset of the set, but not vice versa.

69. If $A \cup B = \varnothing$, then $A = \varnothing$ and $B = \varnothing$.

70. If $A \cap B = \varnothing$, then $A = \varnothing$ or $B = \varnothing$ or both A and B are empty.

71. $(A \cup A^c)^c = \varnothing$

72. If $A \subseteq B$, then $A \cap B = A$.

SOLUTIONS TO SELF-CHECK EXERCISES 6.1

1. a. A^c is the set of all elements in U but not in A. Therefore,

$$A^c = \{4, 5, 6, 7\}$$

b. $A \cup B$ consists of all elements in A and/or B. So,

$$A \cup B = \{1, 2, 3, 4, 5, 6\}$$

c. $B \cap C$ is the set of all elements in both B and C. Therefore,

$$B \cap C = \{3, 4\}$$

d. Using the result from part (b), we find

$$\begin{aligned}(A \cup B) \cap C &= \{1, 2, 3, 4, 5, 6\} \cap \{2, 3, 4\} \\ &= \{2, 3, 4\}\end{aligned}$$

e. First, we compute

$$A \cap B = \{3\}$$

Next, since $(A \cap B) \cup C$ is the set of all elements in $(A \cap B)$ and/or C, we conclude that

$$\begin{aligned}(A \cap B) \cup C &= \{3\} \cup \{2, 3, 4\} \\ &= \{2, 3, 4\}\end{aligned}$$

f. From part (a), we have $A^c = \{4, 5, 6, 7\}$. Next, we compute

$$\begin{aligned}B \cup C &= \{3, 4, 5, 6\} \cup \{2, 3, 4\} \\ &= \{2, 3, 4, 5, 6\}\end{aligned}$$

from which we deduce that

$$(B \cup C)^c = \{1, 7\} \qquad \text{(The set of elements in } U \text{ but not in } B \cup C\text{)}$$

Finally, using these results, we obtain

$$A^c \cap (B \cup C)^c = \{4, 5, 6, 7\} \cap \{1, 7\} = \{7\}$$

2. a. $D \cap F$ denotes the set of all elements in both D and F. Since an element in D is a Democrat and an element in F is a female representative, we see that $D \cap F$ is the set of all female Democrats in the House of Representatives.
b. Since F^c is the set of male representatives and R is the set of Republicans, we see that $F^c \cap R$ is the set of male Republicans in the House of Representatives.
c. L^c is the set of representatives who are not lawyers by training. Therefore, $D \cap F \cap L^c$ is the set of female Democratic representatives who are not lawyers by training.

6.2 The Number of Elements in a Finite Set

COUNTING THE ELEMENTS IN A SET

The solution to some problems in mathematics calls for finding the number of elements in a set. Such problems are called **counting problems** and constitute a field of study known as **combinatorics.** Our study of combinatorics is restricted to the results that will be required for our work in probability later on.

The number of elements in a finite set is determined by simply counting the elements in the set. If A is a set, then $n(A)$ denotes the number of elements in A. For example, if

$$A = \{1, 2, 3, \ldots, 20\}, \qquad B = \{a, b\}, \qquad C = \{8\}$$

then $n(A) = 20$, $n(B) = 2$, and $n(C) = 1$.

The empty set has no elements in it, so $n(\varnothing) = 0$. Another result that is easily seen to be true is the following: If A and B are disjoint sets, then

$$n(A \cup B) = n(A) + n(B) \tag{3}$$

EXAMPLE 1

If $A = \{a, c, d\}$ and $B = \{b, e, f, g\}$, then $n(A) = 3$ and $n(B) = 4$, so $n(A) + n(B) = 7$. However, $A \cup B = \{a, b, c, d, e, f, g\}$ and $n(A \cup B) = 7$. Thus, Equation (3) holds true in this case. Note that $A \cap B = \varnothing$.

In the general case, A and B need not be disjoint, which leads us to the formula

$$\boxed{n(A \cup B) = n(A) + n(B) - n(A \cap B)} \tag{4}$$

FIGURE 6.8
$n(A \cup B) = x + y + z$

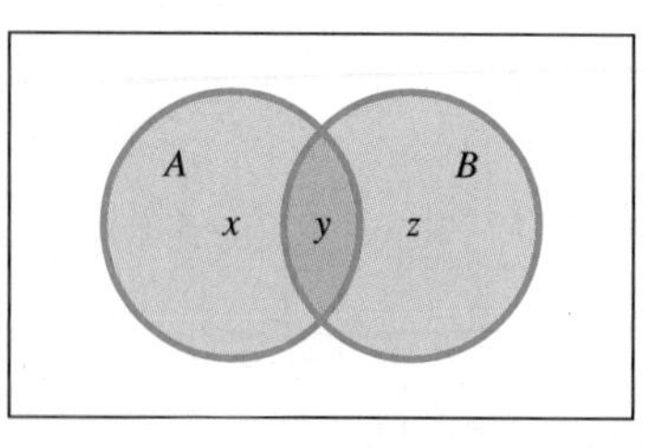

To see this, we observe that the set $A \cup B$ may be viewed as the union of three mutually disjoint sets with x, y, and z elements, respectively (Figure 6.8). This figure shows that

$$n(A \cup B) = x + y + z$$

Also,

$$n(A) = x + y \qquad \text{and} \qquad n(B) = y + z$$

so

$$\begin{aligned} n(A) + n(B) &= (x + y) + (y + z) \\ &= (x + y + z) + y \\ &= n(A \cup B) + n(A \cap B) \qquad [n(A \cap B) = y] \end{aligned}$$

Thus, solving for $n(A \cup B)$, we obtain

$$n(A \cup B) = n(A) + n(B) - n(A \cap B)$$

which is the desired result.

EXAMPLE 2

Let $A = \{a, b, c, d, e\}$ and $B = \{b, d, f, h\}$. Verify Equation (4) directly.

SOLUTION ✔

$$A \cup B = \{a, b, c, d, e, f, h\} \quad \text{so} \quad n(A \cup B) = 7$$
$$A \cap B = \{b, d\} \quad \text{so} \quad n(A \cap B) = 2$$

Furthermore,

$$n(A) = 5 \quad \text{and} \quad n(B) = 4$$

so

$$n(A) + n(B) - n(A \cap B) = 5 + 4 - 2 = 7 = n(A \cup B)$$

■■■■

APPLICATIONS

EXAMPLE 3

In a survey of 100 coffee drinkers, it was found that 70 take sugar, 60 take cream, and 50 take both sugar and cream with their coffee. How many coffee drinkers take sugar or cream with their coffee?

SOLUTION ✔

Let U denote the set of 100 coffee drinkers surveyed and let

$$A = \{x \in U \mid x \text{ takes sugar}\}$$
$$B = \{x \in U \mid x \text{ takes cream}\}$$

Then, $n(A) = 70$, $n(B) = 60$, and $n(A \cap B) = 50$. The set of coffee drinkers who take sugar or cream with their coffee is given by $A \cup B$. Using (4), we find

$$n(A \cup B) = n(A) + n(B) - n(A \cap B)$$
$$= 70 + 60 - 50 = 80$$

Thus, 80 out of the 100 coffee drinkers surveyed take cream or sugar with their coffee.

■■■■

An equation similar to (4) may be derived for the case that involves any finite number of finite sets. For example, a relationship involving the number of elements in the sets A, B, and C is given by

$$n(A \cup B \cup C) = n(A) + n(B) + n(C) - n(A \cap B) - n(A \cap C) - n(B \cap C) + n(A \cap B \cap C) \quad (5)$$

Group Discussion

Prove Formula (5) using an argument similar to that used to prove Formula (4). Another proof is outlined in Exercise 35 on page 354.

As useful as equations such as (5) are, in practice it is often easier to attack a problem directly with the aid of Venn diagrams, as shown by the following example.

EXAMPLE 4 A leading cosmetics manufacturer advertises its products in three magazines: *Cosmopolitan, McCalls,* and the *Ladies Home Journal.* A survey of 500 customers by the manufacturer reveals the following information:

180 learned of its products from *Cosmopolitan*
200 learned of its products from *McCalls*
192 learned of its products from the *Ladies Home Journal*
84 learned of its products from *Cosmopolitan* and *McCalls*
52 learned of its products from *Cosmopolitan* and the *Ladies Home Journal*
64 learned of its products from *McCalls* and the *Ladies Home Journal*
38 learned of its products from all three magazines

How many of the customers saw the manufacturer's advertisement in:

a. At least one magazine?
b. Exactly one magazine?

SOLUTION ✓ Let U denote the set of all customers surveyed and let

$$C = \{x \in U \mid x \text{ learned of the products from } \textit{Cosmopolitan}\}$$
$$M = \{x \in U \mid x \text{ learned of the products from } \textit{McCalls}\}$$
$$L = \{x \in U \mid x \text{ learned of the products from the } \textit{Ladies Home Journal}\}$$

The result that 38 customers learned of the products from all three magazines translates into $n(C \cap M \cap L) = 38$ (Figure 6.9a). Next, the result that 64 learned of the products from *McCalls* and the *Ladies Home Journal* translates into $n(M \cap L) = 64$. This leaves

$$64 - 38 = 26$$

who learned of the products from only *McCalls* and the *Ladies Home Journal* (Figure 6.9b). Similarly, $n(C \cap L) = 52$, so

$$52 - 38 = 14$$

learned of the products from only *Cosmopolitan* and the *Ladies Home Journal,* and $n(C \cap M) = 84$, so

$$84 - 38 = 46$$

learned of the products from only *Cosmopolitan* and *McCalls.* These numbers appear in the appropriate regions in Figure 6.9b.

FIGURE 6.9

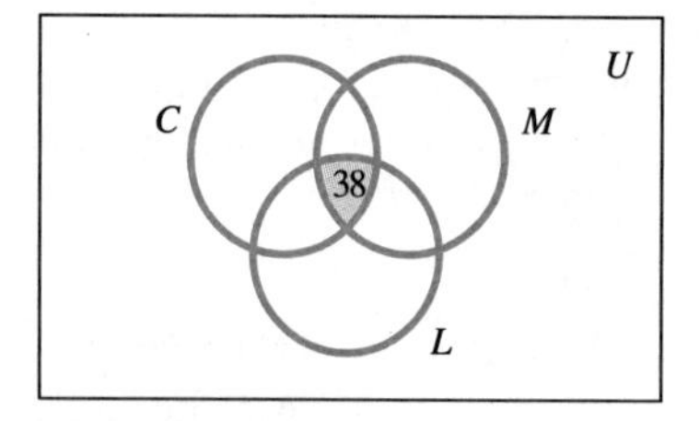

(a) All three magazines

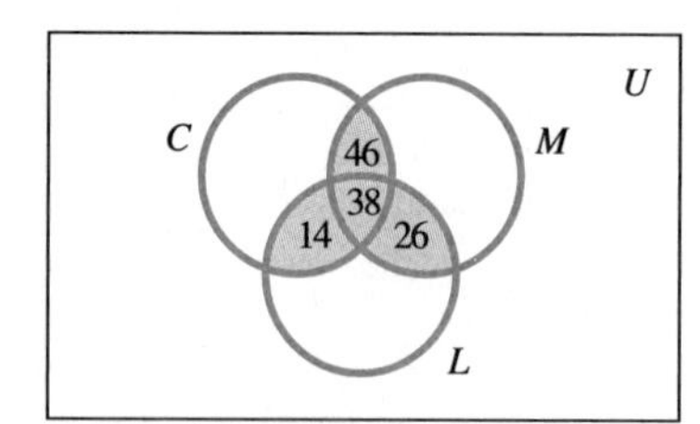

(b) Two or more magazines

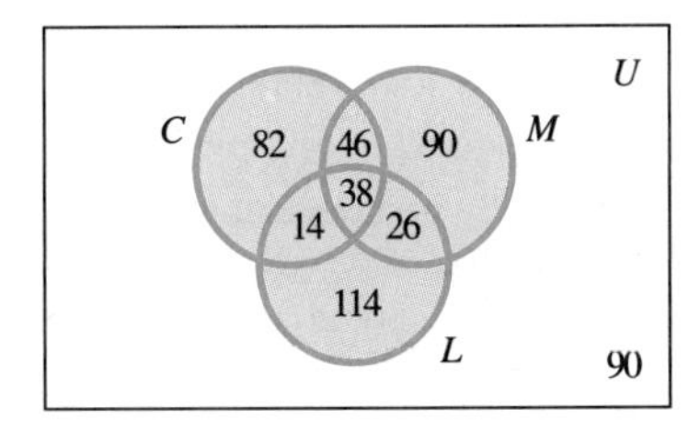

FIGURE 6.10
At least one magazine

Continuing, we have $n(L) = 192$, so the number who learned of the products from the *Ladies Home Journal* only is given by

$$192 - 14 - 38 - 26 = 114$$

(Figure 6.10). Similarly, $n(M) = 200$, so

$$200 - 46 - 38 - 26 = 90$$

learned of the products from only *McCalls,* and $n(C) = 180$, so

$$180 - 14 - 38 - 46 = 82$$

learned of the products from only *Cosmopolitan.* Finally,

$$500 - (90 + 26 + 114 + 14 + 82 + 46 + 38) = 90$$

learned of the products from other sources.

We are now in a position to answer questions (a) and (b).

a. Referring to Figure 6.10, we see that the number of customers who learned of the products from at least one magazine is given by

$$n(C \cup M \cup L) = 90 + 26 + 114 + 14 + 82 + 46 + 38 = 410$$

b. The number of customers who learned of the products from exactly one magazine (Figure 6.11) is given by

$$n(L \cap C^c \cap M^c) + n(M \cap C^c \cap L^c) + n(C \cap L^c \cap M^c)$$
$$= 114 + 90 + 82 = 286$$

FIGURE 6.11
Exactly one magazine

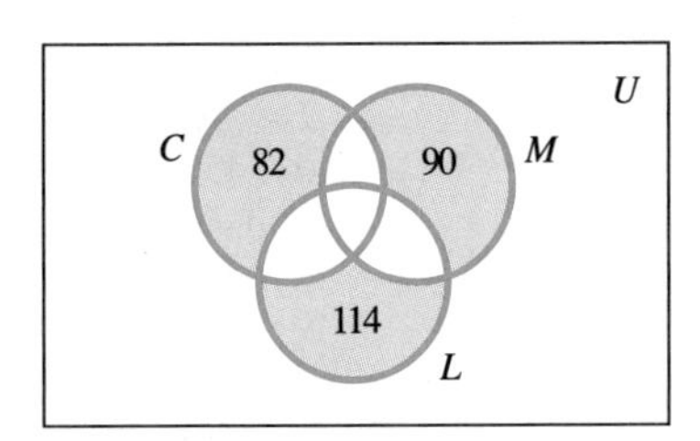

SELF-CHECK EXERCISES 6.2

1. Let A and B be subsets of a universal set U and suppose $n(U) = 100$, $n(A) = 60$, $n(B) = 40$, and $n(A \cap B) = 20$. Compute:
   a. $n(A \cup B)$ **b.** $n(A \cap B^c)$ **c.** $n(A^c \cap B)$
2. In a recent survey of 1000 readers of *Video Magazine,* it was found that 900 own at least one videocassette recorder (VCR) in the VHS format, 240 own at least one VCR in the S-VHS format, and 160 own VCRs in both formats. How many of the readers surveyed own VCRs in the VHS format only? How many of the readers surveyed do not own a VCR in either format?

Solutions to Self-Check Exercises 6.2 can be found on page 354.

6.2 Exercises

In Exercises 1 and 2, verify the equation

$$n(A \cup B) = n(A) + n(B)$$

for the given disjoint sets.

1. $A = \{a, e, i, o, u\}$ and $B = \{g, h, k, l, m\}$
2. $A = \{x \mid x \text{ is a whole number between 0 and 4}\}$
   $B = \{x \mid x \text{ is a negative integer greater than } -4\}$
3. Let $A = \{2, 4, 6, 8\}$ and $B = \{6, 7, 8, 9, 10\}$. Compute:
   a. $n(A)$ **b.** $n(B)$
   c. $n(A \cup B)$ **d.** $n(A \cap B)$
4. Verify directly that $n(A \cup B) = n(A) + n(B) - n(A \cap B)$ for the sets in Exercise 3.
5. Let $A = \{a, e, i, o, u\}$ and $B = \{b, d, e, o, u\}$. Verify by direct computation that $n(A \cup B) = n(A) + n(B) - n(A \cap B)$.
6. If $n(A) = 15$, $n(A \cap B) = 5$, and $n(A \cup B) = 30$, what is $n(B)$?
7. If $n(A) = 10$, $n(A \cup B) = 15$, and $n(B) = 8$, what is $n(A \cap B)$?

In Exercises 8 and 9, let A and B be subsets of a universal set U and suppose $n(U) = 200$, $n(A) = 100$, $n(B) = 80$, and $n(A \cap B) = 40$. Compute:

8. **a.** $n(A \cup B)$ **b.** $n(A^c)$
   c. $n(A \cap B^c)$
9. **a.** $n(A^c \cap B)$ **b.** $n(B^c)$
   c. $n(A^c \cap B^c)$
10. Find $n(A \cup B)$ given that $n(A) = 6$, $n(B) = 10$, and $n(A \cap B) = 3$.
11. If $n(B) = 6$, $n(A \cup B) = 14$, and $n(A \cap B) = 3$, find $n(A)$.
12. If $n(A) = 4$, $n(B) = 5$, and $n(A \cup B) = 9$, find $n(A \cap B)$.
13. If $n(A) = 16$, $n(B) = 16$, $n(C) = 14$, $n(A \cap B) = 6$, $n(A \cap C) = 5$, $n(B \cap C) = 6$, and $n(A \cup B \cup C) = 31$, find $n(A \cap B \cap C)$.
14. If $n(A) = 12$, $n(B) = 12$, $n(A \cap B) = 5$, $n(A \cap C) = 5$, $n(B \cap C) = 4$, $n(A \cap B \cap C) = 2$, and $n(A \cup B \cup C) = 25$, find $n(C)$.
15. A survey of 1000 subscribers to the *Los Angeles Times* revealed that 900 people subscribe to the daily morning edition and 500 subscribe to both the daily and the Sunday editions. How many subscribe to the Sunday edition? How many subscribe to the Sunday edition only?
16. Of 100 clock radios sold recently in a department store, 70 had FM circuitry, and 90 had AM circuitry. How many radios had both FM and AM circuitry? How many could receive FM transmission only? How many could receive AM transmission only?
17. On a certain day, the Wilton County Jail had 190 prisoners. Of these, 130 were accused of felonies, and 121 were accused of misdemeanors. How many prisoners were accused of both a felony and a misdemeanor?
18. CONSUMER SURVEYS In a recent survey of 200 members of a local sports club, 100 members indicated that they plan to attend the next Summer Olympic Games, 60 indicated that they plan to attend the next Winter Olympic Games, and 40 indicated that they plan to attend both games. How many members of the club plan to attend:
    a. At least one of the two games?
    b. Exactly one of the games?
    c. The Summer Olympic Games only?
    d. None of the games?
19. CONSUMER SURVEYS In a survey of 120 consumers conducted in a shopping mall, 80 consumers indicated that they buy brand A of a certain product, 68 buy brand B, and 42 buy both brands. How many consumers participating in the survey buy:
    a. At least one of these brands?
    b. Exactly one of these brands?
    c. Only brand A?
    d. None of these brands?
20. COMMUTER TRENDS Of 50 employees of a store located in downtown Boston, 18 people take the subway to work, 12 take the bus, and 7 take both the subway and the bus. How many employees:
    a. Take the subway or the bus to work?
    b. Take only the bus to work?
    c. Take either the bus or the subway to work?
    d. Get to work by some other means?
21. INVESTING In a poll conducted among 200 active investors, it was found that 120 use discount brokers, 126 use full-service brokers, and 64 use both discount and full-service brokers. How many investors:
    a. Use at least one kind of broker?
    b. Use exactly one kind of broker?
    c. Use only discount brokers?
    d. Don't use a broker?

In Exercises 22–25, let A, B, and C be subsets of a universal set U and suppose $n(U) = 100$, $n(A) = 28$, $n(B) = 30$, $n(C) = 34$, $n(A \cap B) = 8$, $n(A \cap C) = 10$, $n(B \cap C) = 15$, and $n(A \cap B \cap C) = 5$. Compute:

22. **a.** $n(A \cup B \cup C)$ **b.** $n(A^c \cap B \cap C)$

23. **a.** $n[A \cap (B \cup C)]$ **b.** $n[A \cap (B \cup C)^c]$

24. **a.** $n(A^c \cap B^c \cap C^c)$ **b.** $n[A^c \cap (B \cup C)]$

25. **a.** $n[A \cup (B \cap C)]$ **b.** $n(A^c \cap B^c \cap C^c)^c$

26. STUDENT DROPOUT RATE Data released by the Department of Education regarding the rate (percentage) of ninth-grade students that don't graduate showed that out of 50 states,

12 states had an increase in the dropout rate during the past 2 yr.

15 states had a dropout rate of at least 30% during the past 2 yr.

21 states had an increase in the dropout rate and/or a dropout rate of at least 30% during the past 2 yr.

a. How many states had both a dropout rate of at least 30% and an increase in the dropout rate over the 2-yr period?
b. How many states had a dropout rate that was less than 30% but that had increased over the 2-yr period?

27. ECONOMIC SURVEYS A survey of the opinions of 10 leading economists in a certain country showed that because oil prices were expected to drop in that country over the next 12 months,

7 had lowered their estimate of the consumer inflation rate.

8 had raised their estimate of the gross national product growth rate.

2 had lowered their estimate of the consumer inflation rate but had not raised their estimate of the gross national product growth rate.

How many economists had both lowered their estimate of the consumer inflation rate and raised their estimate of the gross national product growth rate for that period?

28. SAT SCORES Results of a Department of Education survey of SAT test scores in 22 states showed that

10 states had an average composite test score of at least 900 during the past 3 yr.

15 states had an increase of at least 10 points in the average composite score during the past 3 yr.

8 states had both an average composite SAT score of at least 900 and an increase in the average composite score of at least 10 points during the past 3 yr.

a. How many of the 22 states had composite scores less than 900 and showed an increase of at least 10 points over the 3-yr period?
b. How many of the 22 states had composite scores of at least 900 and did not show an increase of at least 10 points over the 3-yr period?

29. STUDENT READING HABITS A survey of 100 college students who frequent the reading lounge of a university revealed the following results:

40 read *Time.*

30 read *Newsweek.*

25 read *U.S. News & World Report.*

15 read *Time* and *Newsweek.*

12 read *Time* and *U.S. News & World Report.*

10 read *Newsweek* and *U.S. News & World Report.*

4 read all three magazines.

How many of the students surveyed read:
a. At least one magazine?
b. Exactly one magazine?
c. Exactly two magazines?
d. None of these magazines?

30. STUDENT SURVEYS To help plan the number of meals to be prepared in a college cafeteria, a survey was conducted, and the following data were obtained:

130 students ate breakfast.

180 students ate lunch.

275 students ate dinner.

68 students ate breakfast and lunch.

112 students ate breakfast and dinner.

90 students ate lunch and dinner.

58 students ate all three meals.

How many of the students:
a. Ate at least one meal in the cafeteria?
b. Ate exactly one meal in the cafeteria?
c. Ate only dinner in the cafeteria?
d. Ate exactly two meals in the cafeteria?

31. CONSUMER SURVEYS The 120 consumers of Exercise 19 were also asked about their buying preferences concerning another product that is sold in the market under three labels. The results were:

12 buy only those sold under label A.

25 buy only those sold under label B.

26 buy only those sold under label C.

15 buy only those sold under labels A and B.

10 buy only those sold under labels A and C.

12 buy only those sold under labels B and C.

8 buy the product sold under all three labels.

How many of the consumers surveyed buy the product sold under:

a. At least one of the three labels?
b. Labels A and B but not C?
c. Label A?
d. None of these labels?

In Exercises 32–34, determine whether the statement is true or false. If it is true, explain why it is true. If it is false, give an example to show why it is false.

32. If $A \cap B \neq \varnothing$, then $n(A \cup B) \neq n(A) + n(B)$.

33. If $A \subseteq B$, then $n(B) = n(A) + n(A^c \cap B)$.

34. If $n(A \cup B) = n(A) + n(B)$, then $A \cap B = \varnothing$.

35. Derive Equation (5).
Hint: Equation (4) may be written as $n(D \cup E) = n(D) + n(E) - n(D \cap E)$. Now, put $D = A \cup B$ and $E = C$. Use (4) again if necessary.

SOLUTIONS TO SELF-CHECK EXERCISES 6.2

1. Refer to the following Venn diagram.

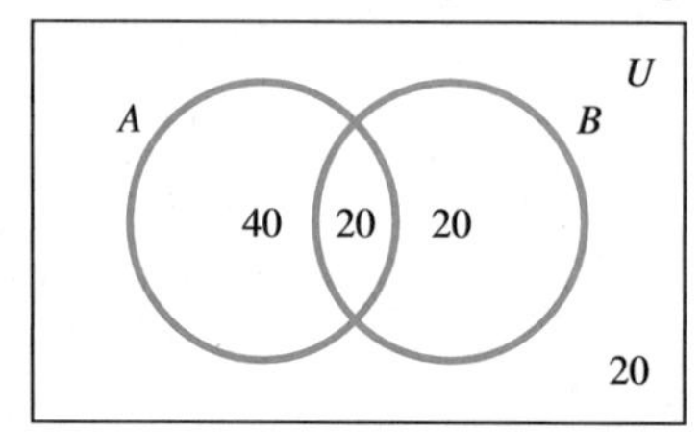

Using this result, we see that
a. $n(A \cup B) = 40 + 20 + 20 = 80$
b. $n(A \cap B^c) = 40$
c. $n(A^c \cap B) = 20$

2. Let U denote the set of all readers surveyed and let

$$A = \{x \in U \mid x \text{ owns at least one VCR in the VHS format}\}$$
$$B = \{x \in U \mid x \text{ owns at least one VCR in the S-VHS format}\}$$

Then, the result that 160 of the readers own VCRs in both formats gives $n(A \cap B) = 160$. Also, $n(A) = 900$ and $n(B) = 240$. Using this information, we obtain the following Venn diagram:

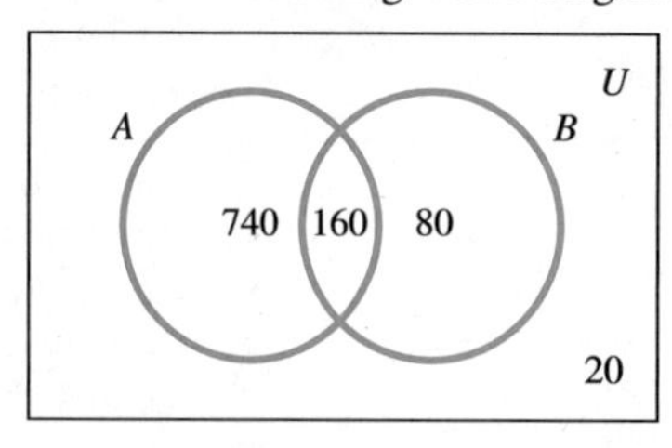

From the Venn diagram we see that the number of readers who own VCRs in the VHS format only is given by

$$n(A \cap B^c) = 740$$

The number of readers who do not own a VCR in either format is given by

$$n(A^c \cap B^c) = 20$$

6.3 The Multiplication Principle

THE FUNDAMENTAL PRINCIPLE OF COUNTING

The solution of certain problems requires more sophisticated counting techniques than those developed in the previous section. We look at some such techniques in this and the following section. We begin by stating a fundamental principle of counting called the **multiplication principle.**

The Multiplication Principle

> Suppose there are m ways of performing a task T_1 and n ways of performing a task T_2. Then, there are mn ways of performing the task T_1 followed by the task T_2.

EXAMPLE 1 Three trunk roads connect town A and town B, and two trunk roads connect town B and town C.

a. Use the multiplication principle to find the number of ways a journey from town A to town C via town B may be completed.
b. Verify part (a) directly by exhibiting all possible routes.

SOLUTION ✓ **a.** Since there are three ways of performing the first task (going from town A to town B) followed by two ways of performing the second task (going from town B to town C), the multiplication principle says that there are $3 \cdot 2$, or 6, ways to complete a journey from town A to town C via town B.
b. Label the trunk roads connecting town A and town B with the Roman numerals I, II, and III and the trunk roads connecting town B and town C with the lowercase letters a and b. A schematic of this is shown in Figure 6.12. Then the routes from town A to town C via town B may be exhibited

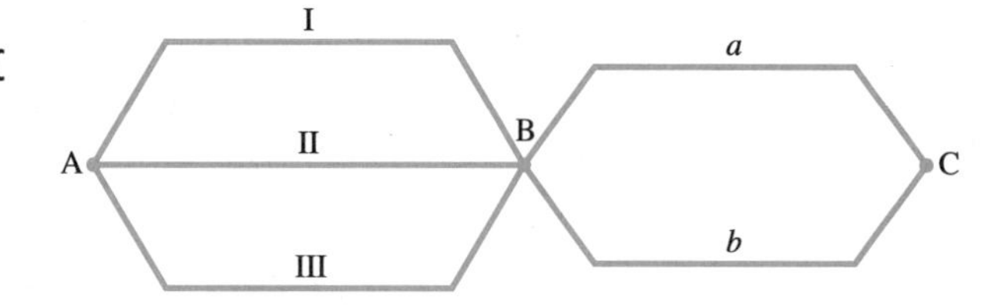

FIGURE 6.12
Roads from towns A to C

with the aid of a **tree diagram** (Figure 6.13). If we follow all of the branches from the initial point A to the right-hand edge of the tree, we obtain the six routes represented by the six ordered pairs

$$(\text{I}, a), \quad (\text{I}, b), \quad (\text{II}, a), \quad (\text{II}, b), \quad (\text{III}, a), \quad (\text{III}, b)$$

where (I, a) means that the journey from town A to town B is made on trunk road I with the rest of the journey from town B to town C to be completed on trunk road a, and so forth.

FIGURE 6.13
Tree diagram displaying the possible routes from town A to town C

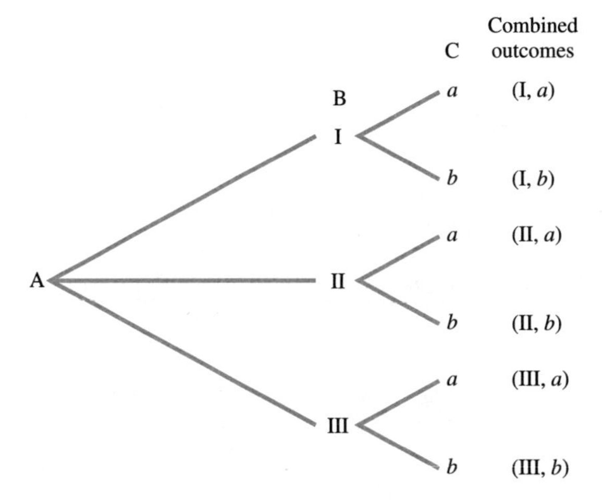

■■■■

Group Discussion

One way of gauging the performance of an airline is to track the arrival times of its flights. Suppose we denote by E, O, and L, a flight that arrives early, on time, or late, respectively.

1. Use a tree diagram to exhibit the possible outcomes when you track two successive flights of the airline. How many outcomes are there?
2. How many outcomes are there if you track three successive flights? Justify your answer.

EXAMPLE 2 Diners at Angelo's Spaghetti Bar may select their entree from 6 varieties of pasta and 28 choices of sauce. How many such combinations are there that consist of 1 variety of pasta and 1 kind of sauce?

SOLUTION ✓ There are 6 ways of choosing a pasta followed by 28 ways of choosing a sauce, so by the multiplication principle, there are $6 \cdot 28$, or 168, combinations of this pasta dish. ■■■■

The multiplication principle may be easily extended, which leads to the **generalized multiplication principle.**

Generalized Multiplication Principle

Suppose a task T_1 can be performed in N_1 ways, a task T_2 can be performed in N_2 ways, . . . , and, finally, a task T_n can be performed in N_n ways. Then, the number of ways of performing the tasks $T_1, T_2, \ldots, T_n$ in succession is given by the product

$$N_1 N_2 \cdots N_n$$

We now illustrate the application of the generalized multiplication principle to several diverse situations.

EXAMPLE 3

A coin is tossed three times, and the sequence of heads and tails is recorded.

a. Use the generalized multiplication principle to determine the number of outcomes of this activity.
b. Exhibit all the sequences by means of a tree diagram.

SOLUTION ✓

a. The coin may land in two ways. Therefore, in three tosses the number of outcomes (sequences) is given by $2 \cdot 2 \cdot 2$, or 8.
b. Let H and T denote the outcomes "a head" and "a tail," respectively. Then the required sequences may be obtained as shown in Figure 6.14, giving the sequence as HHH, HHT, HTH, HTT, THH, THT, TTH, and TTT.

FIGURE 6.14
Tree diagram displaying possible outcomes of three consecutive coin tosses

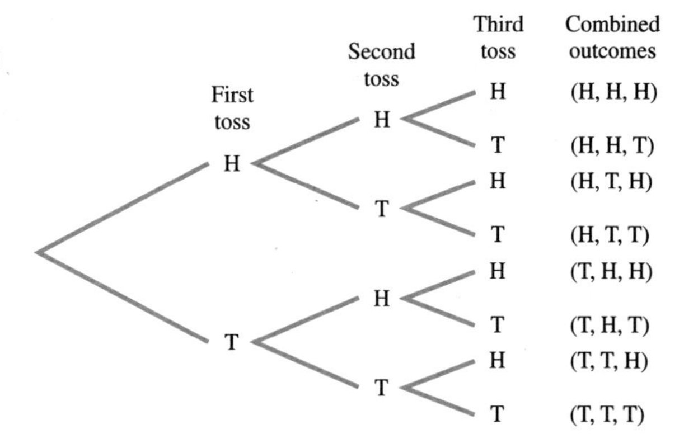

■■■■

APPLICATIONS

EXAMPLE 4

A combination lock is unlocked by dialing a sequence of numbers, first to the left, then to the right, and to the left again. If there are ten digits on the dial, determine the number of possible combinations.

SOLUTION ✓

There are ten choices for the first number, followed by ten for the second and ten for the third, so by the generalized multiplication principle, there are $10 \cdot 10 \cdot 10$, or 1000, possible combinations. ■■■■

EXAMPLE 5

An investor has decided to purchase shares in the stock of three companies: one engaged in aerospace activities, one involved in energy development, and one involved in electronics. After some research, the account executive of a brokerage firm has recommended that the investor consider stock from five aerospace companies, three energy development companies, and four electronics companies. In how many ways may the investor select the group of three companies from the executive's list?

SOLUTION ✓

The investor has five choices for selecting an aerospace company, three choices for selecting an energy development company, and four choices for selecting an electronics company. Therefore, by the generalized multiplication principle, there are $5 \cdot 3 \cdot 4$, or 60, ways in which she can select a group of three companies, one from each industry group. ■■■■

Portfolio

JOHN L. HIGGINS

TITLE: Concierge
INSTITUTION: RVM&G Property Managers

John Higgins's responsibility is to provide building services that "assist the tenants in their day-to-day work." Although not the official building manager, he oversees outside contractors, requesting bids and outlining the scope of the work to be done. He ensures that all such work is within budget and performed correctly. Since the building doesn't have a security guard, Higgins also monitors lobby traffic.

If a major problem arises, tenants feel comfortable approaching Higgins, knowing he'll do his best to resolve it. He stresses that "positive tenant relations" are critical to the building's success. In Higgins's book, "the tenant is always right."

Each month Higgins chronicles building operations in a report to the owner, listing the status of capital improvements, ongoing maintenance, significant tenant complaints, and any variances in the building's accounts. To determine a variance, he has to balance each account monthly. If an account's funds are too high or too low, Higgins has to explain the fluctuation. Math at this level requires simple arithmetic skills, but the calculations become more complicated when it comes to computing a tenant's base rent and future increases.

First, the building's operating costs—utility bills, staff salaries, and vendor fees (cleaning, exterminating, window washing, and so on)—have to be calculated. For example, if the costs add up to $420,000 per year, Higgins divides the total by the building's 70,000 square feet to calculate a general operating expense of $6 per square foot. Dividing the annual real estate tax of $280,000 by the square footage yields another $4 per square foot.

If the owner agrees to modify a new tenant's offices, construction costs—which can run into tens of thousands of dollars—have to be factored in. The construction costs are then amortized like a mortgage over the life of a typical 5-year lease. The principal and interest are calculated so that the tenant pays a fixed amount each month until construction costs are repaid.

Miscellaneous costs (broker, legal, and architectural fees) and the owner's expected profit margin round out the tenant's base for the first year. The base is multiplied by the number of square feet a tenant occupies to determine monthly rent. As operating costs and real estate taxes increase in subsequent years, Higgins has to refigure each tenant's monthly rent using the original base as his starting point.

As the concierge for a small, ten-story office building in Boston, Higgins supervises daily operations. From his post in the lobby, he resolves tenant problems, oversees building maintenance and security, and coordinates the budget. Part building manager, lobby security guard, and friend, Higgins brings an upbeat quality to his job, joking with tenants as they enter and leave.

EXAMPLE 6 Tom is planning to leave for New York City from Washington, D.C., on Monday morning and has decided that he will either fly or take the train. There are five flights and two trains departing for New York City from Washington that morning. When he returns on Sunday afternoon, Tom plans to either fly or hitch a ride with a friend. There are two flights departing from New York City to Washington that afternoon. In how many ways can Tom complete this round trip?

SOLUTION ✓ There are seven ways Tom can go from Washington, D.C., to New York City (five by plane and two by train). On the return trip, Tom can travel in three ways (two by plane and one by car). Therefore, by the multiplication principle, Tom can complete the round trip in $7 \cdot 3$, or 21, ways. ■■■■

Self-Check Exercises 6.3

1. Encore Travel, Inc., offers a "Theater Week in London" package originating from New York City. There is a choice of eight flights departing from New York City per week, a choice of five hotel accommodations, and a choice of one complimentary ticket to one of eight shows. How many such travel packages can one choose from?
2. The Café Napolean offers a dinner special on Wednesdays consisting of a choice of two entrées (Beef Bourguignon and Chicken Basquaise); one dinner salad; one French roll; a choice of three vegetables; a choice of a carafe of Burgundy, Rosé, or Chablis wine; a choice of coffee or tea; and a choice of six French pastries for dessert. How many combinations of dinner specials are there?

Solutions to Self-Check Exercises 6.3 can be found on page 361.

6.3 Exercises

1. Rental Rates Lynbrook West, an apartment complex financed by the State Housing Finance Agency, consists of one-, two-, three-, and four-bedroom units. The rental rate for each type of unit—low, moderate, or market—is determined by the income of the tenant. How many different rates are there?

2. Commuter Passes Five different types of monthly commuter passes are offered by a city's local transit authority for three different groups of passengers: youths, adults, and senior citizens. How many different kinds of passes must be printed each month?

3. Blackjack In the game of blackjack, a 2-card hand consisting of an ace and either a face card or a 10 is called a "blackjack." If a standard 52-card deck is used, determine how many blackjack hands can be dealt.

4. Coin Tosses A coin is tossed four times and the sequence of heads and tails is recorded.
   a. Use the generalized multiplication principle to determine the number of outcomes of this activity.
   b. Exhibit all the sequences by means of a tree diagram.

5. Wardrobe Selection A female executive selecting her wardrobe purchased two blazers, four blouses, and three skirts in coordinating colors. How many ensembles consisting of a blazer, a blouse, and a skirt can she create from this collection?

6. Commuter Options Four commuter trains and three express buses depart from city A to city B in the morning, and three commuter trains and three express buses operate on the return trip in the evening. In how many ways can a commuter from city A to city B complete a daily round trip via bus and/or train?

7. **Psychology Experiments** A psychologist has constructed the following maze for use in an experiment. The maze is constructed so that a rat must pass through a series of one-way doors. How many different paths are there from start to finish?

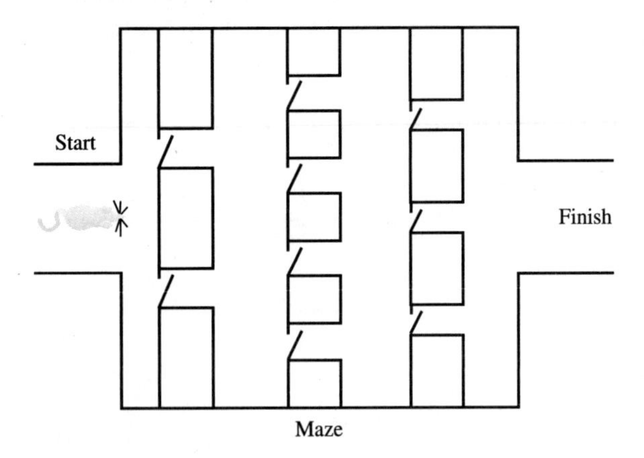

Maze

8. **Union Bargaining Issues** In a survey conducted by a union, members were asked to rate the importance of the following issues: (1) job security, (2) increased fringe benefits, and (3) improved working conditions. Five different responses were allowed for each issue. Among completed surveys, how many different responses to this survey were possible?

9. **Health-Care Plan Options** A new state employee is offered a choice of ten basic health plans, three dental plans, and two vision care plans. How many different health-care plans are there to choose from if one plan is selected from each category?

10. **Code Words** How many three-letter code words can be constructed from the first ten letters of the Greek alphabet if no repetitions are allowed?

11. **Social Security Numbers** A Social Security number has nine digits. How many Social Security numbers are possible?

12. **Serial Numbers** Computers manufactured by a certain company have a serial number consisting of a letter of the alphabet followed by a four-digit number. If all the serial numbers of this type have been used, how many sets have already been manufactured?

13. **Computer Dating** A computer dating service uses the results of its compatibility survey for arranging dates. The survey consists of 50 questions, each having five possible answers. How many different responses are possible if every question is answered?

14. **Automobile Selection** An automobile manufacturer has three different subcompact cars in the line. Customers selecting one of these cars have a choice of three engine sizes, four body styles, and three color schemes. How many different selections can a customer make?

15. **Menu Selections** Two soups, five entrées, and three desserts are listed on the "Special" menu at the Neptune Restaurant. How many different selections consisting of one soup, one entrée, and one dessert can a customer choose from this menu?

16. **Television-Viewing Polls** An opinion poll is to be conducted among cable TV viewers. Six multiple-choice questions, each with four possible answers, will be asked. In how many different ways can a viewer complete the poll if exactly one response is given to each question?

17. **ATM Cards** To gain access to his account, a customer using an automatic teller machine (ATM) must enter a four-digit code. If repetition of the same four digits is not allowed (for example, 5555), how many possible combinations are there?

18. **Political Polls** An opinion poll was conducted by the Morris Polling Group. Respondents were classified according to their sex (M or F), political affiliation (D, I, R), and the region of the country in which they reside (NW, W, C, S, E, NE).
a. Use the generalized multiplication principle to determine the number of possible classifications.
b. Construct a tree diagram to exhibit all possible classifications of females.

19. **License Plate Numbers** Over the years the state of California has used different combinations of letters of the alphabet and digits on its automobile license plates.
a. At one time, license plates were issued that consisted of three letters followed by three digits. How many different license plates can be issued under this arrangement?
b. Later on, license plates were issued that consisted of three digits followed by three letters. How many different license plates can be issued under this arrangement?

20. **License Plate Numbers** In recent years the state of California issued license plates using a combination of one letter of the alphabet followed by three digits, followed by another three letters of the alphabet. How many different license plates can be issued using this configuration?

21. **Exams** An exam consists of ten true-or-false questions. Assuming that every question is answered, in how many different ways can a student complete the exam? In how many ways may the exam be completed if a penalty is

imposed for each incorrect answer, so that a student may leave some questions unanswered?

22. **Warranty Numbers** A warranty identification number for a certain product consists of a letter of the alphabet followed by a five-digit number. How many possible identification numbers are there if the first digit of the five-digit number must be nonzero?

23. **Lotteries** In a state lottery there are 15 finalists eligible for the Big Money Draw. In how many ways can the first, second, and third prizes be awarded if no ticket holder may win more than one prize?

24. **Telephone Numbers**
a. How many seven-digit telephone numbers are possible if the first digit must be nonzero?
b. How many international direct-dialing numbers are possible if each number consists of a three-digit area code (the first digit of which must be nonzero) and a number of the type described in part (a)?

25. **Slot Machines** A "lucky dollar" is one of the nine symbols printed on each reel of a slot machine with three reels. A player receives one of various payouts whenever one or more "lucky dollars" appear in the window of the machine. Find the number of winning combinations for which the machine gives a payoff.
Hint: (a) Compute the number of ways in which the nine symbols on the first, second, and third wheels can appear in the window slot and (b) compute the number of ways in which the eight symbols other than the "lucky dollar" can appear in the window slot. The difference $(a - b)$ is the number of ways in which the "lucky dollar" can appear in the window slot. Why?

In Exercises 26 and 27, determine whether the statement is true or false. If it is true, explain why it is true. If it is false, give an example to show why it is false.

26. The number of three-digit numbers that can be formed from the digits 1, 2, 3, and 4 if the number must be odd is 32.

27. If there are six toppings available, then the number of different pizzas that can be made is 2^5, or 32, different pizzas.

Solutions to Self-Check Exercises 6.3

1. A tourist has a choice of eight flights, five hotel accommodations, and eight tickets. By the generalized multiplication principle, there are $8 \cdot 5 \cdot 8$, or 320, travel packages.
2. There is a choice of two entrées, one dinner salad, one French roll, a choice of three vegetables, a choice of three wines, a choice of two nonalcoholic beverages, and a choice of six pastries. Therefore, by the generalized multiplication principle, there are $2 \cdot 1 \cdot 1 \cdot 3 \cdot 3 \cdot 2 \cdot 6$, or 216, combinations of dinner specials.

6.4 Permutations and Combinations

Permutations

In this section we apply the generalized multiplication principle to the solution of two types of counting problems. Both types involve determining the number of ways the elements of a set may be arranged, and both play an important role in the solution of problems in probability.

We begin by considering the permutations of a set. Specifically, given a set of distinct objects, a **permutation** of the set is an arrangement of these objects in a *definite order*. To see why the order in which objects are arranged is important in certain practical situations, suppose the winning number for the first prize in a raffle is 9237. Then, the number 2973, although it contains the same digits as the winning number, cannot be a first-prize winner (Figure 6.15). Here, the four objects—the numbers 9, 2, 3, and 7—are arranged in a

FIGURE 6.15
The same digits appear on each ticket, but the order of the digits is different.

different order; one arrangement is associated with the winning number for the first prize, and the other is not.

EXAMPLE 1 Let $A = \{a, b, c\}$.

a. Find the number of permutations of A.
b. List all the permutations of A with the aid of a tree diagram.

SOLUTION ✓ **a.** Each permutation of A consists of a sequence of the three letters a, b, c. Therefore, we may think of such a sequence as being constructed by filling in each of the three blanks

$$\underline{\quad} \ \underline{\quad} \ \underline{\quad}$$

with one of the three letters. Now, there are three ways in which we may fill the first blank—we may choose a, b, or c. Having selected a letter for the first blank, there are two letters left for the second blank. Finally, there is but one way left to fill the third blank. Schematically, we have

$$\underline{\ 3\ } \ \underline{\ 2\ } \ \underline{\ 1\ }$$

Invoking the generalized multiplication principle, we conclude that there are $3 \cdot 2 \cdot 1$, or 6, permutations of the set A.
b. The tree diagram associated with this problem appears in Figure 6.16, and the six permutations of A are *abc*, *acb*, *bac*, *bca*, *cab*, and *cba*.

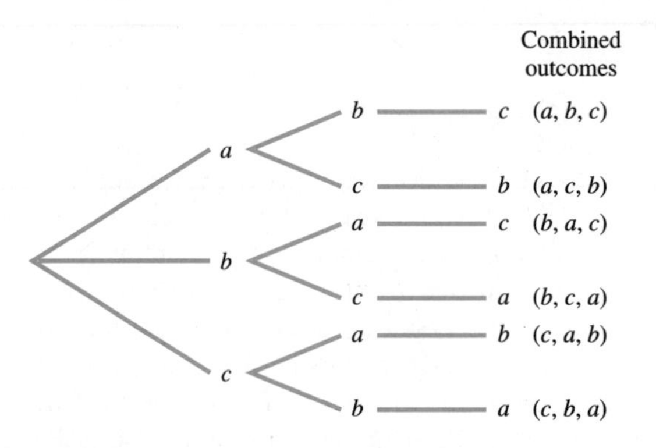

FIGURE 6.16
Permutations of three objects

REMARK Notice that, when the possible outcomes are listed in the tree diagram in Example 1, order is taken into account. Thus, (a, b, c) and (a, c, b) are two different arrangements.

EXAMPLE 2 Find the number of ways a baseball team consisting of nine people can arrange themselves in a line for a group picture.

SOLUTION ✓ We want to determine the number of permutations of the nine members of the baseball team. Each permutation in this situation consists of an arrangement of the nine team members in a line. The nine positions can be represented by nine blanks. Thus,

	___	___	___	___	___	___	___	___	___
Position	1	2	3	4	5	6	7	8	9

There are nine ways to choose from among the nine players to fill the first position. When that position is filled, eight players are left, which gives us eight ways to fill the second position. Proceeding in a similar manner, we find that there are seven ways to fill the third position, and so on. Schematically, we have

Number of ways to fill each position	9	8	7	6	5	4	3	2	1

Invoking the generalized multiplication principle, we conclude that there are $9 \cdot 8 \cdot 7 \cdot 6 \cdot 5 \cdot 4 \cdot 3 \cdot 2 \cdot 1$, or 362,880, ways the baseball team can be arranged for the picture. ■■■■

Whenever we are asked to determine the number of ways the objects of a set can be arranged in a line, order is important. For example, if we take a picture of two baseball players, A and B, then the two players can line up for the picture in two ways, AB or BA, and the two pictures will be different.

Pursuing the same line of argument used in solving the problems in the last two examples, we can derive an expression for the number of ways of permuting a set A of n distinct objects taken n at a time. In fact, each permutation may be viewed as being obtained by filling each of n blanks with one and only one element from the set. There are n ways of filling the first blank, followed by $(n - 1)$ ways of filling the second blank, and so on, so by the generalized multiplication principle there are

$$n(n-1)(n-2) \cdot \cdots \cdot 3 \cdot 2 \cdot 1$$

ways of permuting the elements of the set A.

Before stating this result formally, let's introduce a notation that will enable us to write in a compact form many of the expressions that follow. We use the symbol $n!$ (read "**n-factorial**") to denote the product of the first n natural numbers.

n-Factorial

For any natural number n,

$$n! = n(n-1)(n-2) \cdot \cdots \cdot 3 \cdot 2 \cdot 1$$
$$0! = 1$$

For example,

$$\begin{aligned}
1! &= 1 \\
2! &= 2 \cdot 1 = 2 \\
3! &= 3 \cdot 2 \cdot 1 = 6 \\
4! &= 4 \cdot 3 \cdot 2 \cdot 1 = 24 \\
5! &= 5 \cdot 4 \cdot 3 \cdot 2 \cdot 1 = 120 \\
&\vdots \\
10! &= 10 \cdot 9 \cdot 8 \cdot 7 \cdot 6 \cdot 5 \cdot 4 \cdot 3 \cdot 2 \cdot 1 = 3{,}628{,}800
\end{aligned}$$

Using this notation, we may express *the number of permutations of n distinct objects taken n at a time, $P(n, n)$, as*

$$P(n, n) = n!$$

In many situations we are interested in determining the number of ways of permuting n distinct objects taken r at a time, where $r \leq n$. To derive a formula for computing the number of ways of permuting a set consisting of n distinct objects taken r at a time, we observe that each such permutation may be viewed as being obtained by filling each of r blanks with precisely one element from the set. Now there are n ways of filling the first blank, followed by $(n - 1)$ ways of filling the second blank, and so on. Finally, there are $(n - r + 1)$ ways of filling the rth blank. We may represent this argument schematically:

Number of ways	n	$n-1$	$n-2$	. . .	$n-r+1$
Position	1st	2nd	3rd		rth

Using the generalized multiplication principle, we conclude that *the number of ways of permuting n distinct objects taken r at a time, $P(n, r)$, is given by*

$$P(n, r) = \underbrace{n(n-1)(n-2)\cdots(n-r+1)}_{r \text{ factors}}$$

Since

$$\begin{aligned}
&n(n-1)(n-2)\cdots(n-r+1) \\
&\quad = [n(n-1)(n-2)\cdots(n-r+1)] \cdot \underbrace{\frac{[(n-r)(n-r-1)\cdot\cdots\cdot 3 \cdot 2 \cdot 1]}{[(n-r)(n-r-1)\cdot\cdots\cdot 3 \cdot 2 \cdot 1]}}_{\text{Here we are multiplying by 1.}} \\
&\quad = \frac{[n(n-1)(n-2)\cdots(n-r+1)][(n-r)(n-r-1)\cdot\cdots\cdot 3 \cdot 2 \cdot 1]}{(n-r)(n-r-1)\cdot\cdots\cdot 3 \cdot 2 \cdot 1} \\
&\quad = \frac{n!}{(n-r)!}
\end{aligned}$$

we have the following formula:

Permutations of *n* Distinct Objects

The number of *permutations* of n distinct objects taken r at a time is

$$P(n, r) = \frac{n!}{(n-r)!} \tag{6}$$

REMARK When $r = n$, Equation (6) reduces to

$$P(n, n) = \frac{n!}{0!} = \frac{n!}{1} = n! \qquad \text{(Note that } 0! = 1.)$$

In other words, the number of permutations of a set of n distinct objects, taken all together, is $n!$. ■■■

EXAMPLE 3

Compute (a) $P(4, 4)$ and (b) $P(4, 2)$ and interpret your results.

SOLUTION ✓

a. $P(4, 4) = \frac{4!}{(4-4)!} = \frac{4!}{0!} = \frac{4!}{1} = \frac{4 \cdot 3 \cdot 2 \cdot 1}{1} = 24$ (Recall that $0! = 1$.)

This gives the number of permutations of four objects taken four at a time.

b. $P(4, 2) = \frac{4!}{(4-2)!} = \frac{4!}{2!} = \frac{4 \cdot 3 \cdot 2 \cdot 1}{2 \cdot 1} = 12$

This is the number of permutations of four objects taken two at a time. ■■■■

EXAMPLE 4

Let $A = \{a, b, c, d\}$.

a. Use Equation (6) to compute the number of permutations of the set A taken two at a time.
b. Display the permutations of part (a) with the aid of a tree diagram.

SOLUTION ✓

a. Here, $n = 4$ and $r = 2$, so the required number of permutations is given by

$$P(4, 2) = \frac{4!}{(4-2)!} = \frac{4!}{2!} = \frac{4 \cdot 3 \cdot 2 \cdot 1}{2 \cdot 1} = 4 \cdot 3$$
$$= 12$$

b. The tree diagram associated with the problem is shown in Figure 6.17, and the permutations of A taken two at a time are

ab, *ac*, *ad*, *ba*, *bc*, *bd*,
ca, *cb*, *cd*, *da*, *db*, *dc*

FIGURE 6.17
Permutations of four objects taken two at a time

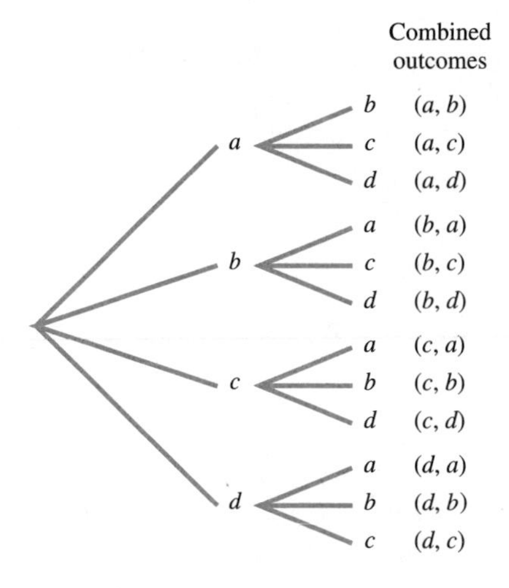

■■■■

EXAMPLE 5 Find the number of ways a chairman, a vice-chairman, a secretary, and a treasurer can be chosen from a committee of eight members.

SOLUTION ✔ The problem is equivalent to finding the number of permutations of eight distinct objects taken four at a time. Therefore, there are

$$P(8, 4) = \frac{8!}{(8-4)!} = \frac{8!}{4!} = 8 \cdot 7 \cdot 6 \cdot 5 = 1680$$

ways of choosing the four officials from the committee of eight members.

■■■■

The permutations considered thus far have been those involving sets of *distinct* objects. In many situations we are interested in finding the number of permutations of a set of objects in which not all of the objects are distinct.

Permutations of *n* Objects, Not All Distinct

Given a set of n objects in which n_1 objects are alike and of one kind, n_2 objects are alike and of another kind, . . . , and, finally, n_r objects are alike and of yet another kind so that

$$n_1 + n_2 + \cdots + n_r = n$$

then the number of permutations of these n objects taken n at a time is given by

$$\frac{n!}{n_1!n_2! \cdots n_r!} \quad \textbf{(7)}$$

To establish Equation (7), let's denote the number of such permutations by x. Now, if we *think* of the n_1 objects as being distinct, then they may be permuted in $n_1!$ ways. Similarly, if we *think* of the n_2 objects as being distinct, then they may be permuted in $n_2!$ ways, and so on. Therefore, if we *think* of the n objects as being distinct, then, by the generalized multiplication principle, there are $x \cdot n_1! \cdot n_2! \cdot \cdots \cdot n_r!$ permutations of these objects. But, the number of permutations of a set of n distinct objects taken n at a time is just equal to $n!$. Therefore, we have

$$x(n_1! \cdot n_2! \cdot \cdots \cdot n_r!) = n!$$

from which we deduce that

$$x = \frac{n!}{n_1!n_2! \cdots n_r!}$$

EXAMPLE 6 Find the number of permutations that can be formed from all the letters in the word *ATLANTA*.

SOLUTION ✓ There are seven objects (letters) involved, so $n = 7$. However, three of them are alike and of one kind (the three *A*s), two of them are alike and of another kind (the two *T*s), so that in this case, $n_1 = 3$, $n_2 = 2$, $n_3 = 1$, and $n_4 = 1$. Therefore, using Formula (7), there are

$$\frac{7!}{3!2!1!1!} = \frac{7 \cdot 6 \cdot 5 \cdot 4 \cdot 3 \cdot 2 \cdot 1}{3 \cdot 2 \cdot 1 \cdot 2 \cdot 1} = 420$$

required permutations. ■■■■

EXAMPLE 7 Weaver and Kline, a stock brokerage firm, has received nine inquiries regarding new accounts. In how many ways can these inquiries be directed to three of the firm's account executives if each account executive is to handle three inquiries?

SOLUTION ✓ If we think of the nine inquiries as being slots arranged in a row with inquiry 1 on the left and inquiry 9 on the right, then the problem can be thought of as one of filling each slot with a business card from an account executive. Then nine business cards would be used, of which three are alike and of one kind, three are alike and of another kind, and three are alike and of yet another kind. Thus, using (7) with $n = 9$, $n_1 = n_2 = n_3 = 3$, there are

$$\frac{9!}{3!3!3!} = \frac{9 \cdot 8 \cdot 7 \cdot 6 \cdot 5 \cdot 4 \cdot 3 \cdot 2 \cdot 1}{3 \cdot 2 \cdot 1 \cdot 3 \cdot 2 \cdot 1 \cdot 3 \cdot 2 \cdot 1} = 1680$$

ways of assigning the inquiries. ■■■■

Combinations

Up to now, we have dealt with permutations of a set—that is, with arrangements of the objects of the set in which the *order* of the elements is taken into consideration. In many situations one is interested in determining the number of ways of selecting r objects from a set of n objects without any regard to the order in which the objects are selected. Such a subset is called a **combination.**

For example, if one is interested in knowing the number of 5-card poker hands that can be dealt from a standard deck of 52 cards, then the order in which the poker hand is dealt is unimportant (Figure 6.18). In this situation, we are interested in determining the number of combinations of 5 cards (objects) selected from a deck (set) of 52 cards (objects). (We will solve this problem in Example 10.)

FIGURE 6.18

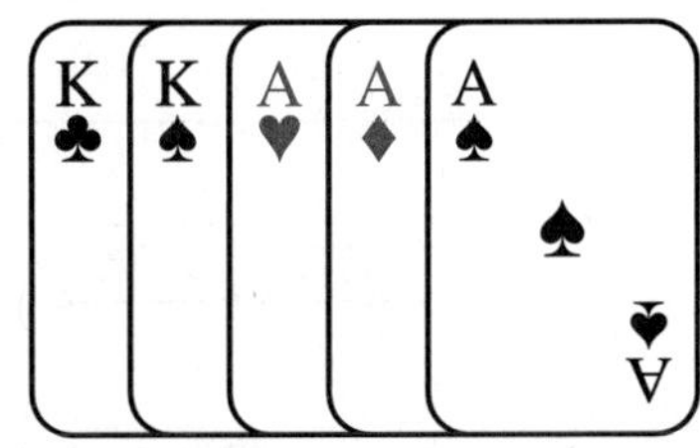

To derive a formula for determining the number of combinations of n objects taken r at a time, written

$$C(n, r) \quad \text{or} \quad \binom{n}{r}$$

we observe that each of the $C(n, r)$ combinations of r objects can be permuted in $r!$ ways (Figure 6.19).

FIGURE 6.19

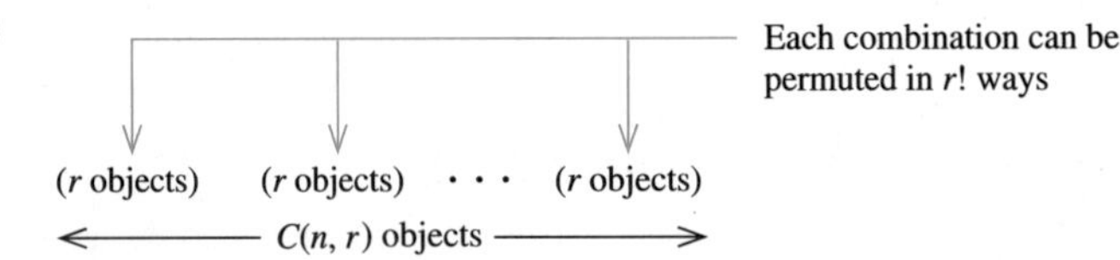

Thus, by the multiplication principle, the product $r!C(n, r)$ gives the number of permutations of n objects taken r at a time; that is,

$$r!C(n, r) = P(n, r)$$

from which we find

$$C(n, r) = \frac{P(n, r)}{r!}$$

or, using Equation (6),

$$C(n, r) = \frac{n!}{r!(n - r)!}$$

Combinations of *n* Objects

The number of combinations of n distinct objects taken r at a time is given by

$$C(n, r) = \frac{n!}{r!(n - r)!} \qquad (\text{where } r \leq n) \qquad \textbf{(8)}$$

EXAMPLE 8

Compute (a) $C(4, 4)$ and (b) $C(4, 2)$ and interpret the results.

SOLUTION ✓

a. $C(4,4) = \dfrac{4!}{4!(4-4)!} = \dfrac{4!}{4!0!} = 1$ (Recall that $0! = 1$.)

This gives 1 as the number of combinations of four distinct objects taken four at a time.

b. $C(4,2) = \dfrac{4!}{2!(4-2)!} = \dfrac{4!}{2!2!} = \dfrac{4 \cdot 3}{2} = 6$

This gives 6 as the number of combinations of four distinct objects taken two at a time.

APPLICATIONS

EXAMPLE 9

A Senate investigation subcommittee of four members is to be selected from a Senate committee of ten members. Determine the number of ways this can be done.

SOLUTION ✓

Since the order in which the members of the subcommittee are selected is unimportant, the number of ways of choosing the subcommittee is given by $C(10, 4)$, the number of combinations of ten objects taken four at a time. Therefore, there are

$$C(10,4) = \frac{10!}{4!(10-4)!} = \frac{10!}{4!6!} = \frac{10 \cdot 9 \cdot 8 \cdot 7}{4 \cdot 3 \cdot 2 \cdot 1} = 210$$

ways of choosing such a subcommittee.

REMARK Remember, a combination is a selection of objects *without* regard to order. Thus, in Example 9, we used a combination formula rather than a permutation formula to solve the problem because the order of selection was not important; that is, it did not matter whether a member of the subcommittee was selected first, second, third, or fourth.

EXAMPLE 10

How many poker hands of 5 cards can be dealt from a standard deck of 52 cards?

SOLUTION ✓

The order in which the 5 cards are dealt is not important. The number of ways of dealing a poker hand of 5 cards from a standard deck of 52 cards is given by $C(52, 5)$, the number of combinations of 52 objects taken five at a time. Therefore, there are

$$\begin{aligned} C(52, 5) &= \frac{52!}{5!(52-5)!} = \frac{52!}{5!47!} \\ &= \frac{52 \cdot 51 \cdot 50 \cdot 49 \cdot 48}{5 \cdot 4 \cdot 3 \cdot 2 \cdot 1} \\ &= 2{,}598{,}960 \end{aligned}$$

ways of dealing such a poker hand.

The next several examples show that solving a counting problem often involves the repeated application of Equation (6) and/or (8), possibly in conjunction with the multiplication principle.

EXAMPLE 11 The members of a string quartet composed of two violinists, a violist, and a cellist are to be selected from a group of six violinists, three violists, and two cellists, respectively.

a. In how many ways can the string quartet be formed?
b. In how many ways can the string quartet be formed if one of the violinists is to be designated as the first violinist and the other is to be designated as the second violinist?

SOLUTION ✓ **a.** Since the order in which each musician is selected is not important, we use combinations. The violinists may be selected in $C(6, 2)$, or 15, ways; the violist may be selected in $C(3, 1)$, or 3, ways; and the cellist may be selected in $C(2, 1)$, or 2, ways. By the multiplication principle, there are $15 \cdot 3 \cdot 2$, or 90, ways of forming the string quartet.
b. The order in which the violinists are selected is important here. Consequently, the number of ways of selecting the violinists is given by $P(6, 2)$, or 30, ways. The number of ways of selecting the violist and the cellist are, of course, 3 and 2, respectively. Therefore, the number of ways in which the string quartet may be formed is given by $30 \cdot 3 \cdot 2$, or 180, ways.

REMARK The solution of Example 11 involves both a permutation and a combination. When we select two violinists from six violinists, order is not important, and we use a combination formula to solve the problem. However, when one of the violinists is designated as a first violinist, order is important, and we use a permutation formula to solve the problem.

EXAMPLE 12 Refer to Example 5, page 357. Suppose the investor has decided to purchase shares in the stocks of two aerospace companies, two energy development companies, and two electronics companies. In how many ways may the investor select the group of six companies for the investment from the recommended list of five aerospace companies, three energy development companies, and four electronics companies?

SOLUTION ✓ There are $C(5, 2)$ ways in which the investor may select the aerospace companies, $C(3, 2)$ ways in which she may select the companies involved in energy development, and $C(4, 2)$ ways in which she may select the electronics companies as investments. By the generalized multiplication principle, there are

$$C(5,2)C(3,2)C(4,2) = \frac{5!}{2!3!} \cdot \frac{3!}{2!1!} \cdot \frac{4!}{2!2!}$$
$$= \frac{5 \cdot 4}{2} \cdot 3 \cdot \frac{4 \cdot 3}{2} = 180$$

ways of selecting the group of six companies for her investment. ■■■■

EXAMPLE 13 The Futurists, a rock group, are planning a concert tour with performances to be given in five cities: San Francisco, Los Angeles, San Diego, Denver, and Las Vegas. In how many ways can they arrange their itinerary if:

a. There are no restrictions?
b. The three performances in California must be given consecutively?

SOLUTION ✓ **a.** The order is important here, and we see that there are

$$P(5, 5) = 5! = 120$$

ways of arranging their itinerary.
b. First, note that there are $P(3, 3)$ ways of choosing between performing in California and in the two cities outside that state. Next, there are $P(3, 3)$ ways of arranging their itinerary in the three cities in California. Therefore, by the multiplication principle, there are

$$P(3,3) \cdot P(3,3) = \frac{3!}{(3-3)!} \cdot \frac{3!}{(3-3)!} = (6)(6) = 36$$

ways of arranging their itinerary. ■■■■

EXAMPLE 14 The U.N. Security Council consists of 5 permanent members and 10 nonpermanent members. Decisions made by the council require nine votes for passage. However, any permanent member may veto a measure and thus block its passage. In how many ways can a measure be passed if all 15 members of the Council vote (no abstentions)?

SOLUTION ✓ If a measure is to be passed, then all 5 permanent members must vote for passage of that measure. This can be done in $C(5, 5)$, or 1, way.

Next, observe that since nine votes are required for passage of a measure, *at least* 4 of the 10 nonpermanent members must also vote for its passage. To determine the number of ways this can be done, notice that there are $C(10, 4)$ ways in which exactly 4 of the nonpermanent members can vote for passage of a measure, $C(10, 5)$ ways in which exactly 5 of them can vote for passage of a measure, and so on. Finally, there are $C(10, 10)$ ways in which all 10 nonpermanent members can vote for passage of a measure. Therefore, there are

$$C(10, 4) + C(10, 5) + \cdots + C(10, 10)$$

ways in which at least 4 of the 10 nonpermanent members can vote for a measure. So, by the multiplication principle, there are

$$\begin{aligned} &C(5,5) \cdot [C(10,4) + C(10,5) + \cdots + C(10,10)] \\ &= (1)\left[\frac{10!}{4!6!} + \frac{10!}{5!5!} + \cdots + \frac{10!}{10!0!}\right] \\ &= (1)(210 + 252 + 210 + 120 + 45 + 10 + 1) = 848 \end{aligned}$$

ways a measure can be passed.

Self-Check Exercises 6.4

1. Evaluate: **a.** 5! **b.** $C(7, 4)$ **c.** $P(6, 2)$
2. A space shuttle crew consists of a shuttle commander, a pilot, three engineers, a scientist, and a civilian. The shuttle commander and pilot are to be chosen from 8 candidates, the three engineers from 12 candidates, the scientist from 5 candidates, and the civilian from 2 candidates. How many such space shuttle crews can be formed?

Solutions to Self-Check Exercises 6.4 can be found on page 378.

6.4 Exercises

In Exercises 1–22, evaluate the given expression.

1. 3(5!) **2.** 2(7!) **3.** $\frac{5!}{2!3!}$

4. $\frac{6!}{4!2!}$ **5.** $P(5, 5)$ **6.** $P(6, 6)$

7. $P(5, 2)$ **8.** $P(5, 3)$ **9.** $P(n, 1)$

10. $P(k, 2)$ **11.** $C(6, 6)$ **12.** $C(8, 8)$

13. $C(7, 4)$ **14.** $C(9, 3)$ **15.** $C(5, 0)$

16. $C(6, 5)$ **17.** $C(9, 6)$ **18.** $C(10, 3)$

19. $C(n, 2)$ **20.** $C(7, r)$ **21.** $P(n, n - 2)$

22. $C(n, n - 2)$

In Exercises 23–30, classify each problem according to whether it involves a permutation or a combination.

23. In how many ways can the letters of the word *GLACIER* be arranged?

24. In the eighth-grade dance class, there are 10 girls and 14 boys. In how many ways can these students be paired off to form dance couples consisting of 1 boy and 1 girl?

25. As part of a quality-control program, 3 record-o-phones are selected at random for testing from each 100 phones produced by the manufacturer. In how many ways can this test batch be chosen?

26. How many three-digit numbers can be formed using the numerals in the set $\{3, 2, 7, 9\}$ if repetition is not allowed?

27. In how many ways can nine different books be arranged on a shelf?

28. A member of a book club wishes to purchase two books from a selection of eight books recommended for a certain month. In how many ways can she choose them?

29. How many five-card poker hands can be dealt consisting of three queens and a pair?

30. Four couples are to be seated at an oval dining table. If each couple must be seated together, how many possible seating arrangements are there?

31. How many four-letter permutations can be formed from the first four letters of the alphabet?

32. How many three-letter permutations can be formed from the first five letters of the alphabet?

33. In how many ways can four students be seated in a row of four seats?

34. In how many ways can five people line up at a checkout counter in a supermarket?

35. How many different batting orders can be formed for a nine-member baseball team?

36. In how many ways can the names of six candidates for political office be listed on a ballot?

37. In how many ways can a member of a hiring committee select 3 of 12 job applicants for further consideration?

38. In how many ways can an investor select four mutual funds for his investment portfolio from a recommended list of eight mutual funds?

39. Find the number of distinguishable permutations that can be formed from the letters of the word *ANTARCTICA*.

40. Find the number of distinguishable permutations that can be formed from the letters of the word *PHILIPPINES*.

41. MANAGEMENT DECISIONS In how many ways can a supermarket chain select 3 out of 12 possible sites for the construction of new supermarkets?

42. BOOK SELECTIONS A student is given a reading list of ten books from which he must select two for an outside reading requirement. In how many ways can he make his selections?

43. QUALITY CONTROL In how many ways can a quality-control engineer select a sample of 3 transistors for testing from a batch of 100 transistors?

44. STUDY GROUPS A group of five students studying for a bar exam had formed a study group. Each member of the group will be responsible for preparing a study outline for one of five courses. In how many different ways can the five courses be assigned to the members of the group?

45. TELEVISION PROGRAMMING In how many ways can a television-programming director schedule six different commercials in the six time slots allocated to commercials during a 1-hr program?

46. WAITING LINES Seven people arrive at the ticket counter of a cinema at the same time. In how many ways can they line up to purchase their tickets?

47. MANAGEMENT DECISIONS Weaver and Kline, a stock brokerage firm, has received six inquiries regarding new accounts. In how many ways can these inquiries be directed to its 12 account executives if each executive handles no more than one inquiry?

48. CAR POOLS A company car that has a seating capacity of six is to be used by six employees who have formed a car pool. If only four of these employees can drive, how many possible seating arrangements are there for the group?

49. BOOK DISPLAYS At a college library exhibition of faculty publications, three mathematics books, four social science books, and three biology books will be displayed on a shelf. (Assume that none of the books is alike.)
a. In how many ways can the ten books be arranged on the shelf?
b. In how many ways can the ten books be arranged on the shelf if books on the same subject matter are placed together?

50. SEATING In how many ways can four married couples attending a concert be seated in a row of eight seats if:
a. There are no restrictions?
b. Each married couple is seated together?
c. The members of each sex are seated together?

51. NEWSPAPER ADVERTISEMENTS Four items from five different departments of the Metro Department Store will be featured in a one-page newspaper advertisement as shown in the following diagram:

Advertisement

1	2	3	4
5	6	7	8
9	10	11	12
13	14	15	16
17	18	19	20

a. In how many different ways can the 20 featured items be arranged on the page?
b. If items from the same department must be in the same row, how many arrangements are possible?

52. **MANAGEMENT DECISIONS** The C & J Realty Company has received 12 inquiries from prospective home buyers. In how many ways can the inquiries be directed to four of the firm's real estate agents if each agent handles three inquiries?

53. **SPORTS** In the women's tennis tournament at Wimbledon, two finalists, A and B, are competing for the title, which will be awarded to the first player to win two sets. In how many different ways can the match be completed?

54. **SPORTS** In the men's tennis tournament at Wimbledon, two finalists, A and B, are competing for the title, which will be awarded to the first player to win three sets. In how many different ways can the match be completed?

55. **U.N. VOTING** Refer to Example 14. In how many ways can a measure be passed if two particular permanent and two particular nonpermanent members of the Council abstain from voting?

56. **JURY SELECTION** In how many different ways can a panel of 12 jurors and 2 alternate jurors be chosen from a group of 30 prospective jurors?

57. **TEACHING ASSISTANTSHIPS** Twelve graduate students have applied for three available teaching assistantships. In how many ways can the assistantships be awarded among these applicants if:
a. No preference is given to any student?
b. One particular student must be awarded an assistantship?
c. The group of applicants includes seven men and five women and it is stipulated that at least one woman must be awarded an assistantship?

58. **EXAMS** A student taking an examination is required to answer 10 out of 15 questions.
a. In how many ways can the 10 questions be selected?
b. In how many ways can the 10 questions be selected if exactly 2 of the first 3 questions must be answered?

59. **CONTRACT BIDDING** The U.B.S. Television Company is considering bids submitted by seven different firms for three different contracts. In how many ways can the contracts be awarded among these firms if no firm is to receive more than two contracts?

60. **SENATE COMMITTEES** In how many ways can a subcommittee of four be chosen from a Senate committee of five Democrats and four Republicans if:
a. All members are eligible?
b. The subcommittee must consist of two Republicans and two Democrats?

61. **COURSE SELECTION** A student planning her curriculum for the upcoming year must select one of five business courses, one of three mathematics courses, two of six elective courses, and either one of four history courses or one of three social science courses. How many different curricula are available for her consideration?

62. **PERSONNEL SELECTION** The J.C.L. Computer Company has five vacancies in its executive trainee program. In how many ways can the company select five trainees from a group of ten female and ten male applicants if the vacancies:
a. May be filled by any combination of men and women?
b. Must be filled by two men and three women?

63. **DRIVERS' TESTS** The State Motor Vehicle Department requires learners to pass a written test on the motor vehicle laws of the state. The exam consists of ten true-or-false questions, of which eight must be answered correctly to qualify for a permit. In how many different ways can a learner who answers all the questions on the exam qualify for a permit?

64. **QUALITY CONTROL** The Goodman Tire Company has 32 tires of a particular size and grade in stock, 2 of which are defective. If a set of 4 tires is to be selected,
a. How many different selections can be made?
b. How many different selections can be made that do not include any defective tires?

A list of poker hands ranked in order from the highest to the lowest is shown in the following table, along with a description and example of each hand. Use the table to answer Exercises 65–70.

Hand	Description	Example
Straight flush	5 cards in sequence in the same suit	A ♥ 2 ♥ 3 ♥ 4 ♥ 5 ♥
Four of a kind	4 cards of the same rank and any other card	K ♥ K ♦ K ♠ K ♣ 2 ♥
Full house	3 of a kind and a pair	3 ♥ 3 ♦ 3 ♣ 7 ♥ 7 ♦
Flush	5 cards of the same suit that are not all in sequence	5 ♥ 6 ♥ 9 ♥ J ♥ K ♥
Straight	5 cards in sequence but not all of the same suit	10 ♥ J ♦ Q ♣ K ♠ A ♥
Three of a kind	3 cards of the same rank and 2 unmatched cards	K ♥ K ♦ K ♠ 2 ♥ 4 ♦
Two pair	2 cards of the same rank and 2 cards of any other rank with an unmatched card	K ♥ K ♦ 2 ♥ 2 ♠ 4 ♣
One pair	2 cards of the same rank and 3 unmatched cards	K ♥ K ♦ 5 ♥ 2 ♠ 4 ♥

If a 5-card poker hand is dealt from a well-shuffled deck of 52 cards, how many different hands consist of the following:

65. A straight flush? (Note that an ace may be played as either a high or a low card in a straight sequence—that is, A, 2, 3, 4, 5 or 10, J, Q, K, A. Hence, there are ten possible sequences for a straight in one suit.)

66. A straight (but not a straight flush)?

67. A flush (but not a straight flush)?

68. Four of a kind?

69. A full house?

70. Two pairs?

71. **Bus Routing** The following is a schematic diagram of a city's street system between the points A and B. The City Transit Authority is in the process of selecting a route from A to B along which to provide bus service. If the company's intention is to keep the route as short as possible, how many routes must be considered?

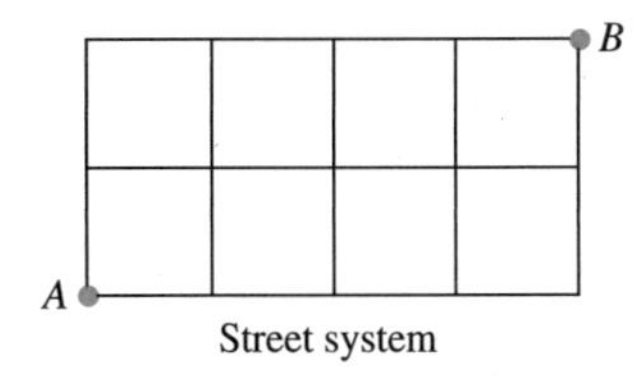

Street system

72. **Sports** In the World Series, one National League team and one American League team compete for the coveted title that is awarded to the first team to win four games. In how many different ways can the series of seven games be completed?

73. **Voting Quorums** A quorum (minimum) of 6 voting members is required at all meetings of the Curtis Townhomes Owners Association. If there is a total of 12 voting members in the group, find the number of ways this quorum can be formed.

74. **Circular Permutations** Suppose n distinct objects are arranged in a circle. Show that the number of (different) circular arrangements of the n objects is $(n - 1)!$

Hint: Consider the arrangement of the five letters A, B, C, D, and E in the accompanying figure. The permutations $ABCDE$, $BCDEA$, $CDEAB$, $DEABC$, and $EABCD$ are not distinguishable. Generalize this observation to the case of n objects.

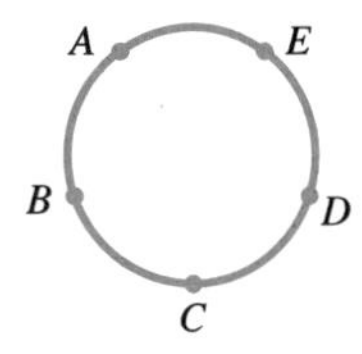

75. Refer to Exercise 74. In how many ways can five TV commentators be seated at a round table for a discussion?

Using Technology

Evaluating $n!$, $P(n, r)$, and $C(n, r)$

A graphing calculator can be used to calculate factorials, permutations, and combinations with relative ease. A graphing calculator is, therefore, an indispensable tool in solving counting problems involving large numbers of objects. Here we use the **nPr** (permutation) and **nCr** (combination) functions of a graphing calculator.

EXAMPLE 1 Use a graphing calculator to find (a) 12!, (b) $P(52, 5)$, and (c) $C(38, 10)$.

SOLUTION ✓

a. Using the factorial function, we find that $12! = 479{,}001{,}600$.

b. Using the **nPr** function, we have

$$P(52, 5) = 52 \textbf{ nPr } 5 = 311{,}875{,}200$$

c. Using the **nCr** function, we obtain

$$C(38, 10) = 38 \textbf{ nCr } 10 = 472{,}733{,}756$$

■■■■

Exercises

In Exercises 1–10, evaluate the expression.

1. $15!$ **2.** $20!$

3. $4(18!)$ **4.** $\frac{30!}{18!}$

5. $P(52, 7)$ **6.** $P(24, 8)$

7. $C(52, 7)$ **8.** $C(26, 8)$

9. $P(10, 4) \cdot C(12, 6)$

10. $P(20, 5) \cdot C(9, 3) \cdot C(8, 4)$

11. A mathematics professor uses a computerized test bank to prepare her final exam. If 25 different problems are available for the first three exam questions, 40 different problems available for the next five questions, and 30 different problems available for the last two questions, how many different ten-question exams can she set? (Assume that the order of the questions within each group is not important.)

12. The S & S Brokerage Company has received 100 inquiries from prospective clients. In how many ways can the inquiries be directed to five of the firm's brokers if each broker handles 20 inquiries?

76. Refer to Exercise 74. In how many ways can four men and four women be seated at a round table at a dinner party if each guest is seated next to members of the opposite sex?

77. At the end of Section 3.3, we mentioned that a linear programming problem in three variables and eight constraints may have up to 56 feasible corner points and that the determination of these feasible corner points calls for the solution of 56 3×3 systems of linear equations. Prove this assertion.

78. Refer to Exercise 77. Show that a linear programming problem in five variables and 15 constraints may have up to 3003 feasible corner points. This assertion was also made at the end of Section 3.3.

In Exercises 79–82, determine whether the statement is true or false. If it is true, explain why it is true. If it is false, give an example to show why it is false.

79. The number of permutations of n distinct objects taken all together is $n!$

80. $P(n, r) = r!C(n, r)$

81. The number of combinations of n objects taken $n - r$ at a time is the same as the number taken r at a time.

82. If a set of n objects consists of r elements of one kind and $n - r$ elements of another kind, then the number of permutations of the n objects, taken all together is $P(n, r)$.

SOLUTIONS TO SELF-CHECK EXERCISES 6.4

1. **a.** $5! = 5 \cdot 4 \cdot 3 \cdot 2 \cdot 1 = 120$

   b. $C(7, 4) = \dfrac{7!}{4!3!} = \dfrac{7 \cdot 6 \cdot 5}{3 \cdot 2 \cdot 1} = 35$

   c. $P(6, 2) = \dfrac{6!}{4!} = 6 \cdot 5 = 30$

2. There are $P(8, 2)$ ways of picking the shuttle commander and pilot (the order *is* important here), $C(12, 3)$ ways of picking the engineers (the order is not important here), $C(5, 1)$ ways of picking the scientist, and $C(2, 1)$ ways of picking the civilian. By the multiplication principle, there are

$$\begin{aligned} &P(8,2) \cdot C(12,3) \cdot C(5,1) \cdot C(2,1) \\ &= \frac{8!}{6!} \cdot \frac{12!}{9!3!} \cdot \frac{5!}{4!1!} \cdot \frac{2!}{1!1!} \\ &= \frac{(8)(7)(12)(11)(10)(5)(2)}{(3)(2)} \\ &= 123{,}200 \end{aligned}$$

ways a crew can be selected.

CHAPTER 6 Summary of Principal Formulas and Terms

Formulas

1. Commutative laws $\quad A \cup B = B \cup A$
 $A \cap B = B \cap A$

2. Associative laws $\quad A \cup (B \cup C) = (A \cup B) \cup C$
 $A \cap (B \cap C) = (A \cap B) \cap C$

3. Distributive laws
$A \cup (B \cap C) = (A \cup B) \cap (A \cup C)$
$A \cap (B \cup C) = (A \cap B) \cup (A \cap C)$

4. De Morgan's laws
$(A \cup B)^c = A^c \cap B^c$
$(A \cap B)^c = A^c \cup B^c$

5. Number of elements in the union of two finite sets
$n(A \cup B) = n(A) + n(B) - n(A \cap B)$

6. Permutation of n distinct objects, taken r at a time
$P(n, r) = \dfrac{n!}{(n-r)!}$

7. Permutation of n objects, not all distinct, taken r at a time
$\dfrac{n!}{n_1!n_2! \cdots n_r!}$

8. Combination of n distinct objects, taken r at a time
$C(n, r) = \dfrac{n!}{r!(n-r)!}$

Terms

set
element of a set
roster notation
set-builder notation
set equality
subset
empty set
universal set
Venn diagram
set union
set intersection
set complementation
multiplication principle
generalized multiplication principle
permutation
n-factorial
combination

CHAPTER 6 REVIEW EXERCISES

In Exercises 1–4, list the elements of the given set in roster notation.

1. $\{x \mid 3x - 2 = 7; x, \text{ an integer}\}$
2. $\{x \mid x$ is a letter of the word *TALLAHASSEE*$\}$
3. The set whose elements are the even numbers between 3 and 11
4. $\{x \mid (x - 3)(x + 4) = 0; x, \text{ a negative integer}\}$

Let A = {*a*, *c*, *e*, *r*}. In Exercises 5–8, determine whether the given set is equal to A.

5. $\{r, e, c, a\}$
6. $\{x \mid x$ is a letter of the word *career*$\}$
7. $\{x \mid x$ is a letter of the word *racer*$\}$
8. $\{x \mid x$ is a letter of the word *cares*$\}$

In Exercises 9–12, shade the portion of the accompanying figure that represents the given set.

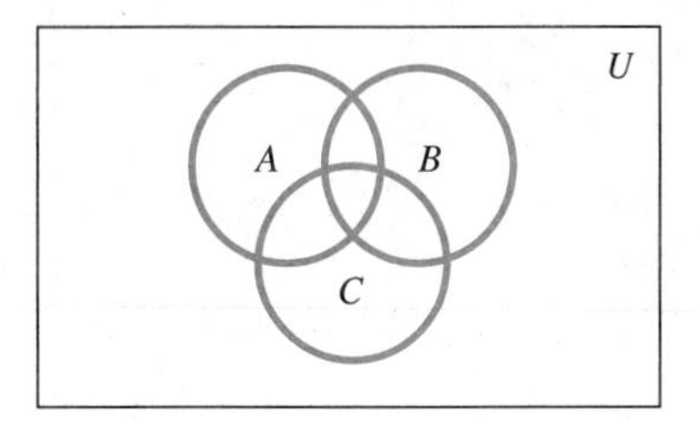

9. $A \cup (B \cap C)$

10. $(A \cap B \cap C)^c$

11. $A^c \cap B^c \cap C^c$

12. $A^c \cap (B^c \cup C^c)$

Let $U = \{a, b, c, d, e\}$, $A = \{a, b\}$, $B = \{b, c, d\}$, and $C = \{a, d, e\}$. In Exercises 13–16, verify the given equation by direct computation.

13. $A \cup (B \cup C) = (A \cup B) \cup C$

14. $A \cap (B \cap C) = (A \cap B) \cap C$

15. $A \cap (B \cup C) = (A \cap B) \cup (A \cap C)$

16. $A \cup (B \cap C) = (A \cup B) \cap (A \cup C)$

Let U = {all participants in a consumer-behavior survey conducted by a national polling group}

A = {consumers who avoided buying a product because it is not recyclable}

B = {consumers who used cloth rather than disposable diapers}

C = {consumers who boycotted a company's products because of their record on the environment}

D = {consumers who voluntarily recycled their garbage}

In Exercises 17–20, describe each of the given sets in words.

17. $A \cap C$ 18. $A \cup D$

19. $B^c \cap D$ 20. $C^c \cup D^c$

Let A and B be subsets of a universal set U and suppose $n(U) = 350$, $n(A) = 120$, $n(B) = 80$, and $n(A \cap B) = 50$. In Exercises 21–26, find the number of elements in each of the given sets.

21. $n(A \cup B)$ 22. $n(A^c)$

23. $n(B^c)$ 24. $n(A^c \cap B)$

25. $n(A \cap B^c)$ 26. $n(A^c \cap B^c)$

In Exercises 27–30, evaluate each of the given quantities.

27. $C(20, 18)$ 28. $P(9, 7)$

29. $C(5, 3) \cdot P(4, 2)$ 30. $4 \cdot P(5, 3) \cdot C(7, 4)$

31. A comparison of five major credit cards showed that

Three offered cash advances.

Three offered extended payments for *all* goods and services puchased.

Two required an annual fee of less than $35.

Two offered both cash advances and extended payments.

One offered extended payments and had an annual fee less than $35.

No card had an annual fee less than $35 and offered both cash advances and extended payments.

How many cards had an annual fee less than $35 and offered cash advances? (Assume that every card had at least one of the three mentioned features.)

32. The Department of Foreign Languages of a liberal arts college conducted a survey of its recent graduates to determine the foreign language courses they had taken while undergraduates at the college. Of the 480 graduates

200 had at least 1 yr of Spanish.

178 had at least 1 yr of French.

140 had at least 1 yr of German.

33 had at least 1 yr of Spanish and French.

24 had at least 1 yr of Spanish and German.

18 had at least 1 yr of French and German.

3 had at least 1 yr of all three languages.

How many of the graduates had:

a. At least 1 yr of at least one of the three languages?
b. At least 1 yr of exactly one of the three languages?
c. Less than 1 yr of any of the three languages?

33. In how many ways can six different compact discs be arranged on a shelf?

34. In how many ways can three pictures be selected from a group of six different pictures?

35. In an election being held by the Associated Students Organization, there are six candidates for president, four for vice-president, five for secretary, and six for treasurer. How many different possible outcomes are there for this election?

36. How many three-digit numbers can be formed from the numerals in the set {1, 2, 3, 4, 5} if:
a. Repetition of digits is not allowed?
b. Repetition of digits is allowed?

37. Find the number of distinguishable permutations that can be formed from the letters of each word.
a. *CINCINNATI* **b.** *HONOLULU*

38. From a standard 52-card deck, how many 5-card poker hands can be dealt consisting of:
a. Five clubs? **b.** Three kings and one pair?

39. In how many ways can seven students be assigned seats in a row containing seven desks if:
a. There are no restrictions?
b. Two of the students must not be seated next to each other?

40. There are eight seniors and six juniors in the Math Club at Jefferson High School. In how many ways can a math team consisting of four seniors and two juniors be selected from the members of the Math Club?

41. A sample of 4 balls is to be selected at random from an urn containing 15 balls numbered 1 to 15. If 6 balls are green, 5 are white, and 4 are black:
a. How many different samples can be selected?
b. How many samples can be selected that contain at least 1 white ball?

42. From a shipment of 60 transistors, 5 of which are defective, a sample of 4 transistors is selected at random.
a. In how many different ways can the sample be selected?
b. How many samples contain 3 defective transistors?
c. How many samples do not contain any defective transistors?

7 PROBABILITY

The systematic study of probability began in the seventeenth century, when certain aristocrats wanted to discover superior strategies to use in the gaming rooms of Europe. Some of the best mathematicians of the period were engaged in this pursuit. Since then, probability has evolved in virtually every sphere of human endeavor in which an element of uncertainty is present.

We begin by introducing some of the basic terminology used in the study of the subject. Then, in Section 7.2, we give the technical meaning of the term *probability.* The rest of this chapter is devoted to the development of techniques for computing the probabilities of the occurrence of certain events.

Where did the defective picture tube come from? Picture tubes for the Pulsar 19-inch color television sets are manufactured in three locations and then shipped to the main plant of the Vista Vision Corporation for final assembly. Each location produces a certain number of the picture tubes with different degrees of reliability. In Example 1, page 442, we will determine the likelihood that a defective picture tube is manufactured in a particular location.

7.1 Experiments, Sample Spaces, and Events

TERMINOLOGY

A number of specialized terms are used in the study of probability. We begin by defining the term *experiment.*

Experiment

An **experiment** is an activity with observable results.

The results of the experiment are called the **outcomes** of the experiment. Three examples of experiments are the following:

- Tossing a coin and observing whether it falls "heads" or "tails"
- Casting a die and observing which of the numbers 1, 2, 3, 4, 5, or 6 shows up
- Testing a spark plug from a batch of 100 spark plugs and observing whether or not it is defective

In our discussion of experiments, we use the following terms:

Sample Point, Sample Space, and Event

Sample point: an outcome of an experiment
Sample space: the set consisting of all possible sample points of an experiment
Event: a subset of a sample space of an experiment

The sample space of an experiment is a universal set whose elements are precisely the outcomes, or the sample points, of the experiment; the events of the experiment are the subsets of the universal set. A sample space associated with an experiment that has a finite number of possible outcomes (sample points) is called a **finite sample space.**

Since the events of an experiment are subsets of a universal set (the sample space of the experiment), we may use the results for set theory given in Chapter 6 to help us study probability. The event B is said to **occur** in a trial of an experiment whenever B contains the observed outcome. We begin by explaining the roles played by the empty set and a universal set when viewed as events associated with an experiment. The empty set, $\varnothing$, is called the *impossible event;* it cannot occur since the $\varnothing$ has no elements (outcomes). Next, the universal set S is referred to as the *certain event;* it must occur since S contains all the outcomes of the experiment.

This terminology is illustrated in the next several examples.

EXAMPLE 1 Describe the sample space associated with the experiment of tossing a coin and observing whether it falls "heads" or "tails." What are the events of this experiment?

SOLUTION ✓ The two outcomes are "heads" and "tails," and the required sample space is given by $S = \{H, T\}$, where H denotes the outcome "heads" and T denotes the outcome "tails." The events of the experiment, the subsets of S, are

$$\varnothing, \quad \{H\}, \quad \{T\}, \quad S$$

Note that we have included the impossible event, $\varnothing$, and the certain event, S.

■■■■

Since the events of an experiment are subsets of the sample space of the experiment, we may talk about the union and intersection of any two events; we can also consider the complement of an event with respect to the sample space.

Union of Two Events

The **union** of the two events E and F is the event $E \cup F$.

Thus, the event $E \cup F$ contains the set of outcomes of E and/or F.

Intersection of Two Events

The **intersection** of the two events E and F is the event $E \cap F$.

Thus, the event $E \cap F$ contains the set of outcomes of E and F.

Complement of an Event

The **complement** of an event E is the event E^c.

Thus, the event E^c is the set containing all the outcomes in the sample space S that are not in E.

Venn diagrams depicting the union, intersection, and complementation of events are shown in Figure 7.1.

FIGURE 7.1

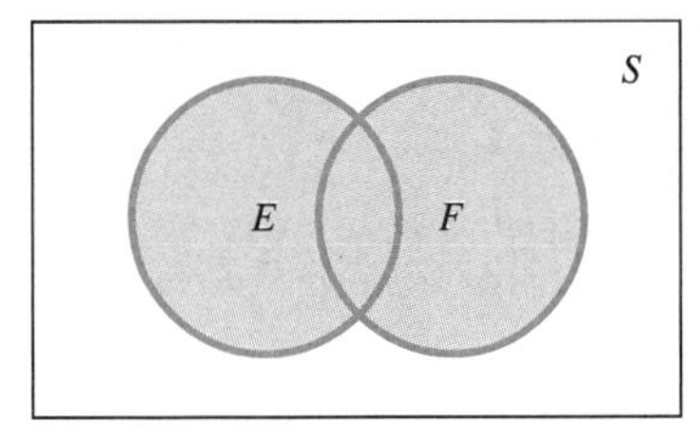

$E \cup F$

(a) The union of two events

S

E F

$E \cap F$

(b) The intersection of two events

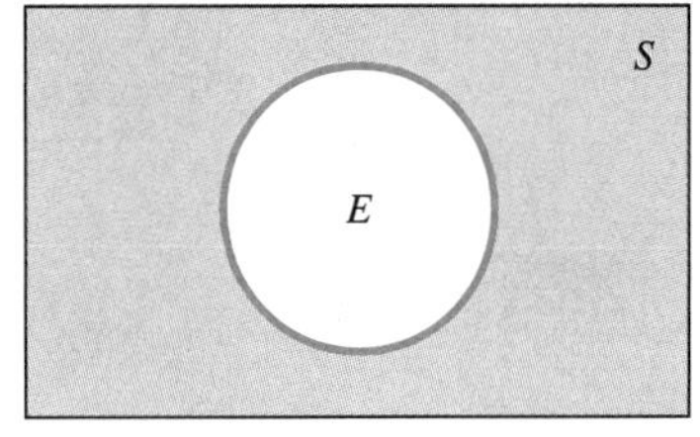

E^c

(c) The complement of the event E

These concepts are illustrated in the following example.

EXAMPLE 2

Consider the experiment of casting a die and observing the number that falls uppermost. Let $S = \{1, 2, 3, 4, 5, 6\}$ denote the sample space of the experiment and $E = \{2, 4, 6\}$ and $F = \{1, 3\}$ be events of this experiment. Compute (a) $E \cup F$, (b) $E \cap F$, and (c) F^c. Interpret your results.

SOLUTION ✓

a. $E \cup F = \{1, 2, 3, 4, 6\}$ and is the event that the outcome of the experiment is a 1, a 2, a 3, a 4, or a 6.
b. $E \cap F = \varnothing$ is the impossible event; the number appearing uppermost when a die is cast cannot be both even and odd at the same time.
c. $F^c = \{2, 4, 5, 6\}$ is precisely the event that the event F does not occur.

■■■■

If two events cannot occur at the same time, they are said to be mutually exclusive. Using set notation, we have the following definition.

Mutually Exclusive Events

E and F are **mutually exclusive** if $E \cap F = \varnothing$.

As before, we may use Venn diagrams to illustrate these events. In this case the two mutually exclusive events are depicted as two nonintersecting circles (Figure 7.2).

FIGURE 7.2
Mutually exclusive events

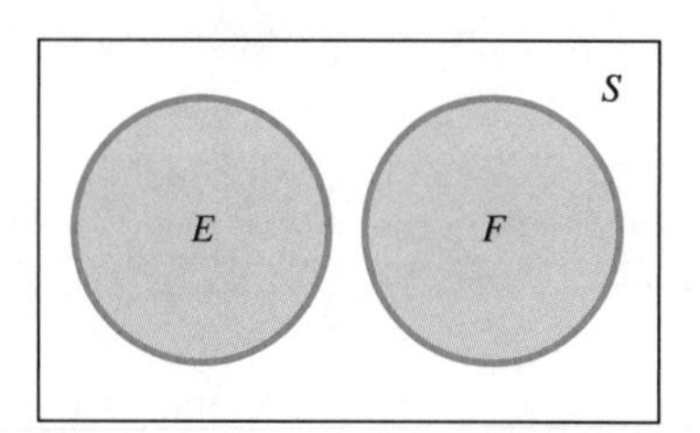

Group Discussion

1. Suppose E and F are two complementary events. Must E and F be mutually exclusive? Explain your answer.

2. Suppose E and F are mutually exclusive events. Must E and F be complementary? Explain your answer.

EXAMPLE 3

An experiment consists of tossing a coin three times and observing the resulting sequence of "heads" and "tails."

a. Describe the sample space S of the experiment.
b. Determine the event E that exactly two heads appear.
c. Determine the event F that at least one head appears.

SOLUTION ✓

a. The sample points may be obtained with the aid of a tree diagram (Figure 7.3).

FIGURE 7.3

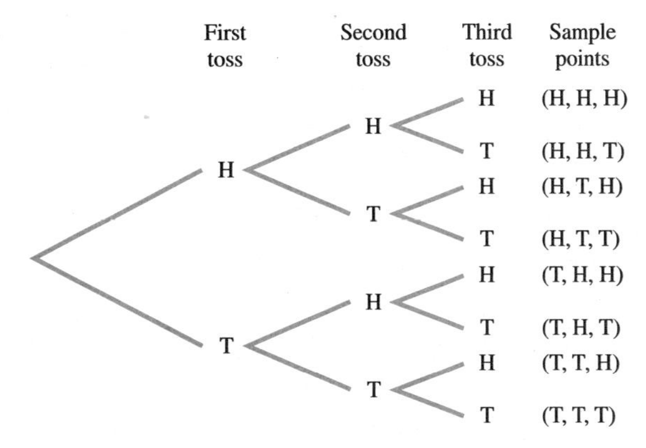

The required sample space S is given by

$$S = \{HHH, HHT, HTH, HTT, THH, THT, TTH, TTT\}$$

b. By scanning the sample space S obtained in part (a), we see that the outcomes in which exactly two heads appear are given by the event

$$E = \{HHT, HTH, THH\}$$

c. Proceeding as in part (b), we find

$$F = \{HHH, HHT, HTH, HTT, THH, THT, TTH\}$$

EXAMPLE 4

An experiment consists of casting a pair of dice and observing the number that falls uppermost on each die.

a. Describe an appropriate sample space S for this experiment.
b. Determine the events $E_2, E_3, E_4, \ldots, E_{12}$ that the sum of the numbers falling uppermost is 2, 3, 4, . . . , 12, respectively.

SOLUTION ✓

a. We may represent each outcome of the experiment by an ordered pair of numbers, the first representing the number that appears uppermost on the first die and the second representing the number that appears uppermost on the second die. To distinguish between the two dice, think of the first die as being red and the second as being green. Since there are six possible outcomes for each die, the multiplication principle implies that there are $6 \cdot 6$, or 36, elements in the sample space:

$$\begin{aligned} S = \{&(1,1), (1,2), (1,3), (1,4), (1,5), (1,6), \\ &(2,1), (2,2), (2,3), (2,4), (2,5), (2,6), \\ &(3,1), (3,2), (3,3), (3,4), (3,5), (3,6), \\ &(4,1), (4,2), (4,3), (4,4), (4,5), (4,6), \\ &(5,1), (5,2), (5,3), (5,4), (5,5), (5,6), \\ &(6,1), (6,2), (6,3), (6,4), (6,5), (6,6)\} \end{aligned}$$

b. With the aid of the results of part (a), we obtain the required list of events, shown in Table 7.1.

Table 7.1

Sum of Uppermost Numbers	Event
2	$E_2 = \{(1, 1)\}$
3	$E_3 = \{(1, 2), (2, 1)\}$
4	$E_4 = \{(1, 3), (2, 2), (3, 1)\}$
5	$E_5 = \{(1, 4), (2, 3), (3, 2), (4, 1)\}$
6	$E_6 = \{(1, 5), (2, 4), (3, 3), (4, 2), (5, 1)\}$
7	$E_7 = \{(1, 6), (2, 5), (3, 4), (4, 3), (5, 2), (6, 1)\}$
8	$E_8 = \{(2, 6), (3, 5), (4, 4), (5, 3), (6, 2)\}$
9	$E_9 = \{(3, 6), (4, 5), (5, 4), (6, 3)\}$
10	$E_{10} = \{(4, 6), (5, 5), (6, 4)\}$
11	$E_{11} = \{(5, 6), (6, 5)\}$
12	$E_{12} = \{(6, 6)\}$

APPLICATIONS

EXAMPLE 5 The manager of a local cinema records the number of patrons attending a first-run movie at the 1 P.M. screening. The theater has a seating capacity of 500.

a. What is an appropriate sample space for this experiment?
b. Describe the event E that fewer than 50 people attend the screening.
c. Describe the event F that the theater is more than half full at the screening.

SOLUTION ✓ **a.** The number of patrons at the screening (the outcome) could run from 0 to 500. Therefore, a sample space for this experiment is

$$S = \{0, 1, 2, 3, \ldots, 500\}$$

b. $E = \{0, 1, 2, 3, \ldots, 49\}$
c. $F = \{251, 252, 253, \ldots, 500\}$

EXAMPLE 6 An experiment consists of studying the composition of a three-child family in which the children were born at different times.

a. Describe an appropriate sample space S for this experiment.
b. Describe the event E that there are two girls and a boy in the family.
c. Describe the event F that the oldest child is a girl.
d. Describe the event G that the oldest child is a girl and the youngest child is a boy.

SOLUTION ✓

a. The sample points of the experiment may be obtained with the aid of the tree diagram shown in Figure 7.4, where b denotes a boy and g denotes a girl.

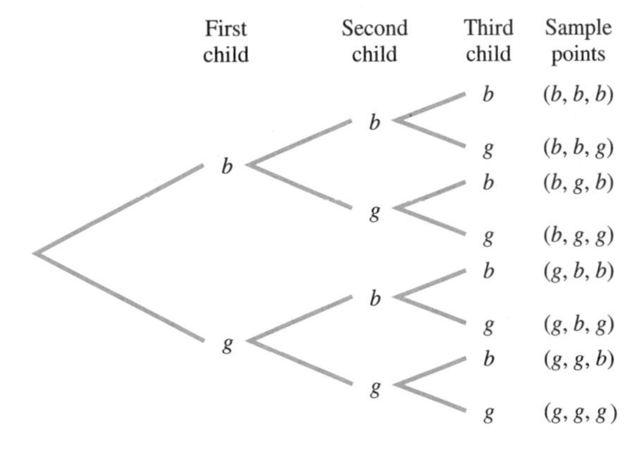

FIGURE 7.4
Tree diagram for three-child families

We see from the tree diagram that the required sample space is given by

$$S = \{(b, b, b), (b, b, g), (b, g, b), (b, g, g), (g, b, b), (g, b, g), (g, g, b), (g, g, g)\}$$

Using the tree diagram, we find that

b. $E = \{(b, g, g), (g, b, g), (g, g, b)\}$
c. $F = \{(g, b, b), (g, b, g), (g, g, b), (g, g, g)\}$
d. $G = \{(g, b, b), (g, g, b)\}$

Group Discussion
Think of an experiment.

1. Describe the sample point(s) and sample space of the experiment.
2. Construct two events, E and F, of the experiment.
3. Find the union and intersection of E and F and the complement of E.
4. Are E and F mutually exclusive? Explain your answer.

The next example shows that sample spaces may be infinite.

EXAMPLE 7

The Ever-Brite Battery Company is developing a high-amperage, high-capacity battery as a source for powering electric cars. The battery is tested by installing it in a prototype electric car and running the car with a fully charged battery on a test track at a constant speed of 55 mph until the car runs out of power. The distance covered by the car is then observed.

a. What is the sample space for this experiment?
b. Describe the event E that the driving range under test conditions is less than 150 miles.

c. Describe the event F that the driving range is between 200 and 250 miles, inclusive.

SOLUTION ✓ **a.** Since the distance d covered by the car in any run may be given by any nonnegative number, the sample space S is given by

$$S = \{d \mid d \geq 0\}$$

b. The event E is given by

$$E = \{d \mid d < 150\}$$

c. The event F is given by

$$F = \{d \mid 200 \leq d \leq 250\}$$

Self-Check Exercises 7.1

1. A sample of three apples taken from Cavallero's Fruit Stand are examined to determine whether they are good or rotten.
   a. What is an appropriate sample space for this experiment?
   b. Describe the event E that exactly one of the apples picked is rotten.
   c. Describe the event F that the first apple picked is rotten.
2. Refer to Self-Check Exercise 1.
   a. Find $E \cup F$.
   b. Find $E \cap F$.
   c. Find F^c.
   d. Are the events E and F mutually exclusive?

Solutions to Self-Check Exercises 7.1 can be found on page 393.

7.1 Exercises

In Exercises 1–6, let $S = \{a, b, c, d, e, f\}$ be a sample space of an experiment and let $E = \{a, b\}$, $F = \{a, d, f\}$, and $G = \{b, c, e\}$ be events of this experiment.

1. Find the events $E \cup F$ and $E \cap F$.
2. Find the events $F \cup G$ and $F \cap G$.
3. Find the events F^c and $E \cap G^c$.
4. Find the events E^c and $F^c \cap G$.
5. Are the events E and F mutually exclusive?
6. Are the events $E \cup F$ and $E \cap F^c$ mutually exclusive?

In Exercises 7–12, let $S = \{1, 2, 3, 4, 5, 6\}$, $E = \{2, 4, 6\}$, $F = \{1, 3, 5\}$, and $G = \{5, 6\}$.

7. Find the event $E \cup F \cup G$.
8. Find the event $E \cap F \cap G$.
9. Find the event $(E \cup F \cup G)^c$.
10. Find the event $(E \cap F \cap G)^c$.
11. Are the events E and F mutually exclusive?
12. Are the events F and G mutually exclusive?

In Exercises 13–18, let S be any sample space and E, F, and G be any three events associated with the experiment. Describe the given events using the symbols $\cup$, $\cap$, and ^{c}.

13. The event that E and/or F occurs

14. The event that both E and F occur

15. The event that G does not occur

16. The event that E but not F occurs

17. The event that none of the events E, F, and G occurs

18. The event that E occurs but neither of the events F or G occurs

19. Let $S = \{a, b, c\}$ be a sample space of an experiment with outcomes a, b, and c. List all the events of this experiment.

20. Let $S = \{1, 2, 3\}$ be a sample space associated with an experiment.
a. List all events of this experiment.
b. How many subsets of S contain the number 3?
c. How many subsets of S contain either the number 2 or the number 3?

21. An experiment consists of selecting a card from a standard deck of playing cards and noting whether it is black (B) or red (R).
a. Describe an appropriate sample space for this experiment.
b. What are the events of this experiment?

22. An experiment consists of selecting a letter at random from the letters in the word *MASSACHUSETTS* and observing the outcomes.
a. What is an appropriate sample space for this experiment?
b. Describe the event "the letter selected is a vowel."

23. An experiment consists of tossing a coin and casting a die and observing the outcomes.
a. Describe an appropriate sample space for this experiment.
b. Describe the event "a head is tossed and an even number is cast."

24. An experiment consists of spinning the hand of the numbered disc shown in the following figure and observing the region in which the pointer stops. (If the needle stops on a line, the result is discounted and the needle is spun again.)

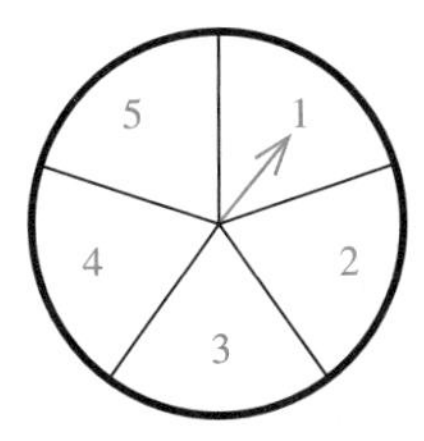

a. What is an appropriate sample space for this experiment?
b. Describe the event "the spinner points to the number 2."
c. Describe the event "the spinner points to an odd number."

25. **QUALITY CONTROL** A sample of three transistors taken from a local electronics store was examined to determine whether the transistors were defective (d) or nondefective (n). What is an appropriate sample space for this experiment?

26. **BLOOD TYPING** Human blood is classified by the presence or absence of three main antigens (A, B, and Rh). When a blood specimen is typed, the presence of the A and/or B antigen is indicated by listing the letter A and/or the letter B. If neither the A nor B antigen is present, the letter O is used. The presence or absence of the Rh antigen is indicated by the symbols + or −, respectively. Thus, if a blood specimen is classified as AB^+, it contains the A and the B antigens as well as the Rh antigen. Similarly, O^- blood contains none of the three antigens. Using this information, determine the sample space corresponding to the different blood groups.

27. **GAME SHOWS** In a television game show, the winner is asked to select three prizes from five different prizes, A, B, C, D, and E.
a. Describe a sample space of possible outcomes (order is not important).
b. How many points are there in the sample space corresponding to a selection that includes A?
c. How many points are there in the sample space corresponding to a selection that includes A and B?
d. How many points are there in the sample space corresponding to a selection that includes either A or B?

28. **AUTOMATIC TELLERS** The manager of a local bank observes how long it takes a customer to complete his transactions at the automatic bank teller.
a. Describe an appropriate sample space for this experiment.
b. Describe the event that it takes a customer between 2 and 3 min to complete his transactions at the automatic bank teller.

29. **COMMON STOCKS** Robin purchased shares of a machine tool company and shares of an airline company. Let E be the event that the shares of the machine tool company increase in value over the next 6 mo, and let F be the event that the shares of the airline company increase in value over the next 6 mo. Using the symbols $\cup$, $\cap$, and ^{c}, describe the following events.
a. The shares in the machine tool company do not increase in value.
b. The shares in both the machine tool company and the airline company do not increase in value.
c. The shares of at least one of the two companies increase in value.
d. The shares of only one of the two companies increase in value.

30. **CUSTOMER SERVICE SURVEYS** The customer service department of the Universal Instruments Company, manufacturer of the Galaxy home computer, conducted a survey among customers who had returned their purchase registration cards. Purchasers of its deluxe model home computer were asked to report the length of time (t) in days before service was required.
a. Describe a sample space corresponding to this survey.
b. Describe the event E that a home computer required service before a period of 90 days had elapsed.
c. Describe the event F that a home computer did not require service before a period of 1 yr had elapsed.

31. **ASSEMBLY-TIME STUDIES** A time study was conducted by the production manager of the Vista Vision Corporation to determine the length of time in minutes required by an assembly worker to complete a certain task during the assembly of its Pulsar color television sets.
a. Describe a sample space corresponding to this time study.
b. Describe the event E that an assembly worker took 2 min or less to complete the task.
c. Describe the event F that an assembly worker took more than 2 min to complete the task.

32. **POLITICAL POLLS** An opinion poll is conducted among a state's electorate to determine the relationship between their income levels and their stands on a proposition aimed at reducing state income taxes. Voters are classified as belonging to either the low-, middle-, or upper-income group. They are asked whether they favor, oppose, or are undecided about the proposition. Let the letters L, M, and U represent the low-, middle-, and upper-income groups, respectively, and let the letters f, o, and u represent the responses—favor, oppose, and undecided, respectively.
a. Describe a sample space corresponding to this poll.
b. Describe the event E_1 that a respondent favors the proposition.
c. Describe the event E_2 that a respondent opposes the proposition and does not belong to the low-income group.
d. Describe the event E_3 that a respondent does not favor the proposition and does not belong to the upper-income group.

33. **QUALITY CONTROL** As part of a quality-control procedure, an inspector at Bristol Farms randomly selects ten eggs from each consignment of eggs he receives and records the number of broken eggs.
a. What is an appropriate sample space for this experiment?
b. Describe the event E that at most three eggs are broken.
c. Describe the event F that at least five eggs are broken.

34. **POLITICAL POLLS** In the opinion poll of Exercise 32, the voters were also asked to indicate their political affiliations—Democrat, Republican, or Independent. As before, let the letters L, M, and U represent the low-, middle-, and upper-income groups, respectively, and let the letters D, R, and I represent Democrat, Republican, and Independent, respectively.
a. Describe a sample space corresponding to this poll.
b. Describe the event E_1 that a respondent is a Democrat.
c. Describe the event E_2 that a respondent belongs to the upper-income group and is a Republican.
d. Describe the event E_3 that a respondent belongs to the middle-income group and is not a Democrat.

35. **SHUTTLE BUS USAGE** A certain airport hotel operates a shuttle bus service between the hotel and the airport. The maximum capacity of a bus is 20 passengers. On alternate trips of the shuttle bus over a period of 1 wk, the hotel manager kept a record of the number of passengers arriving at the hotel in each bus.
a. What is an appropriate sample space for this experiment?
b. Describe the event E that a shuttle bus carried fewer than ten passengers.
c. Describe the event F that a shuttle bus arrived with a full capacity.

36. **SPORTS** Eight players, A, B, C, D, E, F, G, and H, are competing in a series of elimination matches of a tennis tournament in which the winner of each preliminary

match will advance to the semifinals and the winner of the semifinals will advance to the finals. An outline of the scheduled matches follows. Describe a sample space listing the possible participants in the finals.

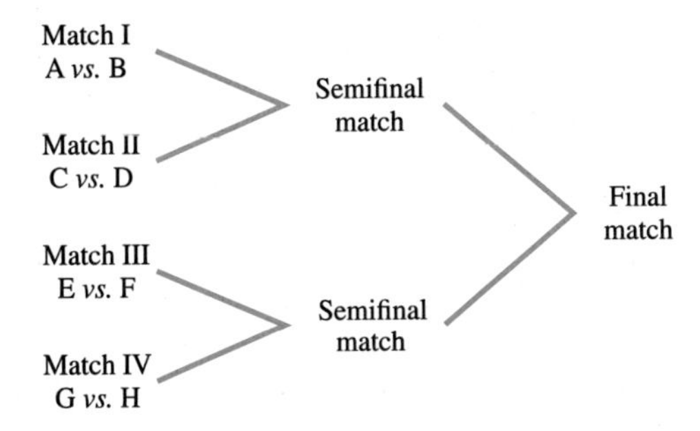

37. Let S be a sample space for an experiment. Show that if E is any event of an experiment, then E and E^c are mutually exclusive.

38. Let S be a sample space for an experiment and let E and F be events of this experiment. Show that the events $E \cup F$ and $E^c \cap F^c$ are mutually exclusive.
Hint: Use De Morgan's Law.

39. Let S be a sample space of an experiment with n outcomes. Determine the number of events of this experiment.

In Exercises 40–41, determine whether the statement is true or false. If it is true, explain why it is true. If it is false, give an example to show why it is false.

40. If E and F are mutually exclusive and E and G are mutually exclusive, then F and G are mutually exclusive.

41. The numbers 1, 2, and 3 are written separately on three pieces of paper. These slips of paper are then placed in a bowl. If you draw two slips from the bowl, one at a time, without replacement, then the sample space for this experiment consists of six elements.

SOLUTIONS TO SELF-CHECK EXERCISES 7.1

1. a. Let g denote a good apple and r a rotten apple. Thus, the required sample points may be obtained with the aid of a tree diagram (compare with Example 3). The required sample space is given by

$$S = \{ggg, ggr, grg, grr, rgg, rgr, rrg, rrr\}$$

b. By scanning the sample space S obtained in part (a), we identify the outcomes in which exactly one apple is rotten. We find

$$E = \{ggr, grg, rgg\}$$

c. Proceeding as in part (b), we find

$$F = \{rgg, rgr, rrg, rrr\}$$

2. Using the results of Self-Check Exercise 1, we find

a. $E \cup F = \{ggr, grg, rgg, rgr, rrg, rrr\}$

b. $E \cap F = \{rgg\}$

c. F^c is the set of outcomes in S but not in F. Thus,

$$F^c = \{ggg, ggr, grg, grr\}$$

d. Since $E \cap F \neq \varnothing$, we conclude that E and F are not mutually exclusive.

7.2 Definition of Probability

Finding the Probability of an Event

Let's return to the coin-tossing experiment. The sample space of this experiment is given by $S = \{H, T\}$, where the sample points H and T correspond to the two possible outcomes, a *head* and a *tail*. If the coin is *unbiased,* then there is *one chance out of two* of obtaining a head (or a tail) and we say that the *probability* of tossing a head (tail) is 1/2, abbreviated

$$P(\text{H}) = \frac{1}{2} \quad \text{and} \quad P(\text{T}) = \frac{1}{2}$$

An alternative method of obtaining the values of $P(\text{H})$ and $P(\text{T})$ is based on continued experimentation and does not depend on the assumption that the two outcomes are equally likely. Table 7.2 summarizes the results of such an exercise.

Table 7.2 As the Number of Trials Increases, the Relative Frequency Approaches .5

Number of Tosses, n	Number of Heads, m	Relative Frequency of Heads, m/n
10	4	.4000
100	58	.5800
1,000	492	.4920
10,000	5,034	.5034
20,000	10,024	.5012
40,000	20,032	.5008

Observe that the relative frequencies (column 3) differ considerably when the number of trials is small, but as the number of trials becomes very large, the relative frequency approaches the number .5. This result suggests that we assign to $P(\text{H})$ the value 1/2, as before.

More generally, consider an experiment that may be repeated over and over again under independent and similar conditions. Suppose that in n trials an event E occurs m times. We call the ratio m/n the **relative frequency** of the event E after n repetitions. If this relative frequency approaches some value $P(E)$ as n becomes larger and larger, then $P(E)$ is called the **empirical probability** of E. Thus, the probability $P(E)$ of an event occurring is a measure of the proportion of the time that the event E will occur in the long run. Observe that this method of computing the probability of a head occurring is effective even in the case when a biased coin is used in the experiment. The relative frequency distribution is often referred to as an observed or **empirical probability distribution.**

The **probability of an event** is a number that lies between 0 and 1. In general, the larger the probability of an event, the more likely the event will occur. Thus, an event with a probability of .8 is more likely to occur than an event with a probability of .6. An event with a probability of 1/2, or .5, has a "fifty-fifty," or equal, chance of occurring.

Now suppose we are given an experiment and wish to determine the probabilities associated with certain events of the experiment. This problem could be solved by computing $P(E)$ directly for each event E of interest. However, in practice, the number of events that we may be interested in is usually quite large, so this approach is not satisfactory.

The following approach is particularly suitable when the sample space of an experiment is finite.* Let S be a finite sample space with n outcomes; that is,

$$S = \{s_1, s_2, s_3, \ldots, s_n\}$$

Then, the events

$$\{s_1\}, \{s_2\}, \{s_3\}, \ldots, \{s_n\}$$

which consist of exactly one point, are called **elementary,** or **simple, events** of the experiment. They are elementary in the sense that any (nonempty) event of the experiment may be obtained by taking a finite union of suitable elementary events. The simple events of an experiment are also **mutually exclusive;** that is, given any two simple events of the experiment, only one can occur.

By assigning probabilities to each of the simple events, we obtain the results shown in Table 7.3. This table is called a **probability distribution** for the experiment. The function P, which assigns a probability to each of the simple events, is called a **probability function.**

Table 7.3 A Probability Distribution

Simple Event	Probability*
$\{s_1\}$	$P(s_1)$
$\{s_2\}$	$P(s_2)$
$\{s_3\}$	$P(s_3)$
⋮	⋮
$\{s_n\}$	$P(s_n)$

* For simplicity, we use the notation $P(s_i)$ instead of the technically more correct $P(\{s_i\})$.

The numbers $P(s_1), P(s_2), \ldots, P(s_n)$ have the following properties:

1. $0 \leq P(s_i) \leq 1 \quad (i = 1, 2, \ldots, n)$
2. $P(s_1) + P(s_2) + \cdots + P(s_n) = 1$
3. $P(\{s_i\} \cup \{s_j\}) = P(s_i) + P(s_j) \quad (i \neq j) \quad (i = 1, 2, \ldots, n; j = 1, 2, \ldots, n)$

The first property simply states that the probability of a simple event must be between 0 and 1 inclusive. The second property states that the sum of the probabilities of all simple events of the sample space is 1. This follows from the fact that the event S is certain to occur. The third property states that the probability of the union of two mutually exclusive events is given by the sum of their probabilities.

As we saw earlier, there is no unique method for assigning probabilities to the simple events of an experiment. In practice, the methods used to determine these probabilities may range from theoretical considerations of the problem on the one extreme to the reliance on "educated guesses" on the other.

Sample spaces in which the outcomes are equally likely are called **uniform sample spaces.** Assigning probabilities to the simple events in these spaces is relatively easy.

* For the remainder of the chapter we assume that all sample spaces are finite.

Probability of an Event in a Uniform Sample Space

If

$$S = \{s_1, s_2, \ldots, s_n\}$$

is the sample space for an experiment in which the outcomes are equally likely, then we assign the probabilities

$$P(s_1) = P(s_2) = \cdots = P(s_n) = \frac{1}{n}$$

to each of the simple events $s_1, s_2, \ldots, s_n$.

EXAMPLE 1 A fair die is cast, and the number that falls uppermost is observed. Determine the probability distribution for the experiment.

SOLUTION ✓ The sample space for the experiment is $S = \{1, 2, 3, 4, 5, 6\}$, and the simple events are accordingly given by the sets {1}, {2}, {3}, {4}, {5}, and {6}. Since the die is assumed to be fair, the six outcomes are equally likely. We therefore assign a probability of 1/6 to each of the simple events and obtain the probability distribution shown in Table 7.4.

Table 7.4 A Probability Distribution

Simple Event	Probability
{1}	$\frac{1}{6}$
{2}	$\frac{1}{6}$
{3}	$\frac{1}{6}$
{4}	$\frac{1}{6}$
{5}	$\frac{1}{6}$
{6}	$\frac{1}{6}$

Group Discussion

You suspect that a die is biased.

1. Describe a method you might use to prove your assertion.
2. How would you assign the probability to each outcome 1 through 6 of the experiment of casting the die and observing which number lands uppermost?

The next example shows how the *relative frequency* interpretation of probability lends itself to the computation of probabilities.

EXAMPLE 2

Refer to Example 7, Section 7.1. The data shown in Table 7.5 were obtained in tests involving 200 test runs. Each run was made with a fully charged battery.

Table 7.5 Data Obtained During 200 Test Runs of an Electric Car

Distance Covered in Miles, x	Frequency of Occurrence
$0 \leq x \leq 50$	4
$50 < x \leq 100$	10
$100 < x \leq 150$	30
$150 < x \leq 200$	100
$200 < x \leq 250$	40
$250 < x$	16

a. Describe an appropriate sample space for this experiment.
b. Find the empirical probability distribution for this experiment.

SOLUTION ✔

a. Let s_1 denote the outcome that the distance covered by the car does not exceed 50 miles; let s_2 denote the outcome that the distance covered by the car is greater than 50 miles but does not exceed 100 miles, and so on. Finally, let s_6 denote the outcome that the distance covered by the car is greater than 250 miles. Then, the required sample space is given by

$$S = \{s_1, s_2, s_3, s_4, s_5, s_6\}$$

b. To compute the empirical probability distribution for the experiment, we turn to the relative frequency interpretation of probability. Accepting the inaccuracies inherent in a relatively small number of trials (200 runs), we take the probability of s_1 occurring as

$$P(s_1) = \frac{\text{Number of trials in which } s_1 \text{ occurs}}{\text{Total number of trials}}$$

$$= \frac{4}{200} = .02$$

Table 7.6 A Probability Distribution

Simple Event	Probability
$\{s_1\}$	.02
$\{s_2\}$	.05
$\{s_3\}$	.15
$\{s_4\}$	.50
$\{s_5\}$	.20
$\{s_6\}$	.08

In a similar manner, we assign probabilities to the other simple events, obtaining the probability distribution shown in Table 7.6. ■■■■

We are now in a position to give a procedure for computing the probability $P(E)$ of an arbitrary event E of an experiment.

Finding the Probability of an Event E

1. Determine a sample space S associated with the experiment.
2. Assign probabilities to the simple events of S.
3. If $E = \{s_1, s_2, s_3, \ldots, s_n\}$ where $\{s_1\}, \{s_2\}, \{s_3\}, \ldots, \{s_n\}$ are simple events, then

$$P(E) = P(s_1) + P(s_2) + P(s_3) + \cdots + P(s_n)$$

If E is the empty set, $\varnothing$, then $P(E) = 0$.

The principle stated in step 3 is called the **addition principle** and is a consequence of Property 3 of the probability function (page 395). This principle allows us to find the probabilities of all other events once the probabilities of the simple events are known.

The addition rule applies *only* to the addition of probabilities of simple events.

APPLICATIONS

EXAMPLE 3 A pair of fair dice is cast.

a. Calculate the probability that the two dice show the same number.
b. Calculate the probability that the sum of the numbers of the two dice is 6.

SOLUTION ✓ From the results of Example 4, page 388, we see that the sample space S of the experiment consists of 36 outcomes:

$$S = \{(1, 1), (1, 2), \ldots, (6, 5), (6, 6)\}$$

Since both dice are fair, each of the 36 outcomes is equally likely. Accordingly, we assign the probability of 1/36 to each simple event. We are now in a position to answer the questions posed.

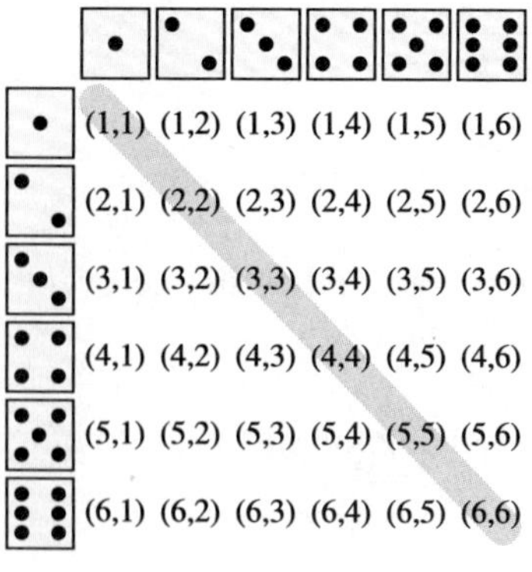

a. The event that the two dice show the same number is given by

$$E = \{(1, 1), (2, 2), (3, 3), (4, 4), (5, 5), (6, 6)\}$$

FIGURE 7.5
The event that the two dice show the same number

	(1,1)	(1,2)	(1,3)	(1,4)	(1,5)	(1,6)
	(2,1)	(2,2)	(2,3)	(2,4)	(2,5)	(2,6)
	(3,1)	(3,2)	(3,3)	(3,4)	(3,5)	(3,6)
	(4,1)	(4,2)	(4,3)	(4,4)	(4,5)	(4,6)
	(5,1)	(5,2)	(5,3)	(5,4)	(5,5)	(5,6)
	(6,1)	(6,2)	(6,3)	(6,4)	(6,5)	(6,6)

(Figure 7.5). Therefore, by the addition principle, the probability that the two dice show the same number is given by

$$\begin{aligned} P(E) &= P[(1, 1)] + P[(2, 2)] + \cdots + P[(6, 6)] \\ &= \frac{1}{36} + \frac{1}{36} + \cdots + \frac{1}{36} \qquad \text{(Six terms)} \\ &= \frac{1}{6} \end{aligned}$$

b. The event that the sum of the numbers of the two dice is 6 is given by

$$E_6 = \{(1, 5), (2, 4), (3, 3), (4, 2), (5, 1)\}$$

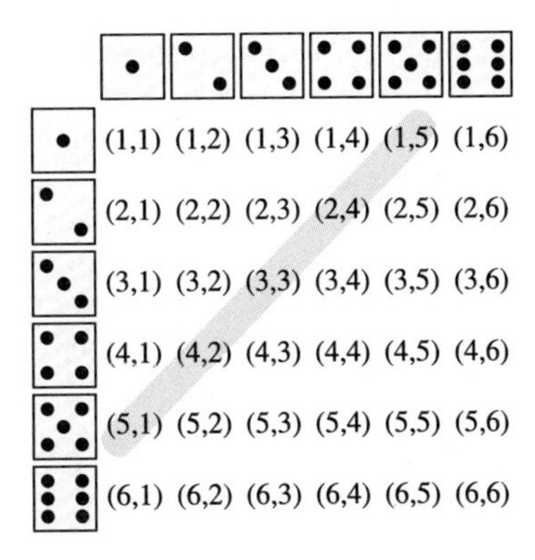

FIGURE 7.6
The event that the sum of the numbers on the two dice is 6

(Figure 7.6). Therefore, the probability that the sum of the numbers on the two dice is 6 is given by

$$\begin{aligned} P(E_6) &= P[(1,5)] + P[(2,4)] + P[(3,3)] + P[(4,2)] + P[(5,1)] \\ &= \frac{1}{36} + \frac{1}{36} + \cdots + \frac{1}{36} \qquad \text{(Five terms)} \\ &= \frac{5}{36} \end{aligned}$$

EXAMPLE 4 Consider the experiment by the Ever-Brite Company in Example 2. What is the probability that the prototype car will travel more than 150 miles on a fully charged battery?

SOLUTION ✓ Using the results of Example 2, we see that the event that the car will travel more than 150 miles on a fully charged battery is given by $E = \{s_4, s_5, s_6\}$. Therefore, the probability that the car will travel more than 150 miles on one charge is given by

$$P(E) = P(s_4) + P(s_5) + P(s_6)$$

or, using the probability distribution for the experiment obtained in Example 2,

$$P(E) = .50 + .20 + .08 = .78$$

SELF-CHECK EXERCISES 7.2

1. A biased die was cast repeatedly, and the results of the experiment are summarized in the following table:

Outcome	1	2	3	4	5	6
Frequency of Occurrence	142	173	158	175	162	190

Using the relative frequency interpretation of probability, find the empirical probability distribution for this experiment.

2. In an experiment conducted to study the effectiveness of an eye-level third brake light in the prevention of rear-end collisions, 250 of the 500 highway patrol cars of a certain state were equipped with such lights. At the end of the 1-yr trial period, the records revealed that of those equipped with a third brake light, there were 14 incidents of rear-end collision. There were 22 such incidents involving the cars not equipped with the accessory. Based on these data, what is the probability that a highway patrol car equipped with a third brake light will be rear-ended within a 1-yr period? What is the probability that a car not so equipped will be rear-ended within a 1-yr period?

Solutions to Self-Check Exercises 7.2 can be found on page 404.

7.2 Exercises

In Exercises 1–8, list the simple events associated with each of the given experiments.

1. A nickel and a dime are tossed, and the result of heads or tails is recorded for each coin.

2. A card is selected at random from a standard 52-card deck, and its suit—hearts (h), diamonds (d), spades (s), or clubs (c)—is recorded.

3. OPINION POLLS An opinion poll is conducted among a group of registered voters. Their political affiliation, Democrat (D), Republican (R), or Independent (I), and their sex, male (m) or female (f), are recorded.

4. QUALITY CONTROL As part of a quality-control procedure, eight circuit boards are checked, and the number of defectives is recorded.

5. MOVIE ATTENDANCE In a survey conducted to determine whether movie attendance is increasing (i), decreasing (d), or holding steady (s) among various sectors of the population, participants are classified as follows:

   Group 1: those aged 10–19
   Group 2: those aged 20–29
   Group 3: those aged 30–39
   Group 4: those aged 40–49
   Group 5: those 50 and over

   The response of each participant and his or her age group are recorded.

6. DURABLE GOODS ORDERS Data concerning durable goods orders are obtained each month by an economist. A record is kept for a 1-yr period of any increase (i), decrease (d), or unchanged movement (u) in the number of durable goods orders for each month as compared with the number of such orders in the same month in the previous year.

7. BLOOD TYPES Blood tests are given as a part of the admission procedure at the Monterey Garden Community Hospital. The blood type of each patient (A, B, AB, or O) and the presence or absence of the Rh factor in each patient's blood (Rh^+ or Rh^-) are recorded.

8. METEOROLOGY A meteorologist preparing a weather map classifies the expected average temperature in each of five neighboring states for the upcoming week as follows:
   a. More than 10° below average
   b. Normal to 10° below average
   c. Higher than normal to 10° above average
   d. More than 10° above average
   Using each state's abbreviation and the categories—(a), (b), (c), and (d)—the meteorologist records these data.

9. GRADE DISTRIBUTIONS The grade distribution for a certain class is shown in the following table. Find the probability distribution associated with these data.

Grade	A	B	C	D	F
Frequency of Occurrence	4	10	18	6	2

10. BLOOD TYPES The percentage of the general population that has each blood type is shown in the following table:

Blood Type	A	B	AB	O
Percentage of Population	41%	12%	3%	44%

Determine the probability distribution associated with these data.

11. **TRAFFIC SURVEYS** The number of cars entering a tunnel leading to an airport in a major city over a period of 200 peak hours was observed and the following data were obtained:

Number of Cars, x	Frequency of Occurrence
$0 < x \le 200$	15
$200 < x \le 400$	20
$400 < x \le 600$	35
$600 < x \le 800$	70
$800 < x \le 1000$	45
$x > 1000$	15

a. Describe an appropriate sample space for this experiment.
b. Find the empirical probability distribution for this experiment.

12. **PRODUCT SURVEYS** The accompanying data were obtained from a survey of 1500 Americans who were asked: How safe are American-made consumer products?

Rating	A (very safe)	B (somewhat safe)
Number of Respondents	285	915

Rating	C (not too safe)	D (not safe at all)
Number of Respondents	225	30

Rating	E (don't know)
Number of Respondents	45

Determine the empirical probability distribution associated with these data.

13. **POLITICAL VIEWS** In a poll conducted among 2000 college freshmen to ascertain the political views of college students, the accompanying data were obtained:

Political Views	A (far left)	B (liberal)
Number of Respondents	52	398

Political Views	C (middle-of-the-road)
Number of Respondents	1140

Political Views	D (conservative)	E (far right)
Number of Respondents	386	24

Determine the empirical probability distribution associated with these data.

14. **SERVICE-UTILIZATION STUDIES** The Metro Telephone Company of Belmont compiled the accompanying information during a service-utilization study pertaining to the number of customers using their Dial-the-Time service from 7 A.M. to 9 A.M. on a certain weekday morning. Using these data, find the empirical probability distribution associated with the experiment.

Number of Calls Received/Minute	Frequency of Occurrence
10	6
11	15
12	12
13	3
14	12
15	36
16	24
17	0
18	6
19	6

15. **ASSEMBLY-TIME STUDIES** The results of a time study conducted by the production manager of the Ace Novelty Company are shown in the accompanying table, where the number of space action-figures produced each quar-

ter hour during an 8-hr workday has been tabulated. Find the empirical probability distribution associated with this experiment.

Number of Figures Produced (in dozens)	Frequency of Occurrence
30	4
31	0
32	6
33	8
34	6
35	4
36	4

16. **ELECTRONIC MAIL SERVICES** The number of subscribers to five leading electronic mail services is shown in the accompanying table:

Company	A	B	C
Subscribers	300,000	200,000	120,000

Company	D	E
Subscribers	80,000	60,000

Find the empirical probability distribution associated with these data.

17. **CORRECTIVE LENS USE** According to Mediamark Research, Inc., 84 million out of 179 million adults in the United States correct their vision by using prescription eyeglasses, bifocals, or contact lenses. (Some respondents use more than one type.) What is the probability that an adult selected at random from the adult population uses corrective lenses?

18. **CORRECTIONAL SUPERVISION** A study conducted by the Corrections Department of a certain state revealed that 163,605 people out of a total adult population of 1,778,314 were under correctional supervision (on probation, parole, or in jail). What is the probability that a person selected at random from the adult population in that state is under correctional supervision?

19. **LIGHTNING DEATHS** According to data obtained from the National Weather Service, 376 of the 439 people killed by lightning in the United States between 1985 and 1992 were men. (Job and recreational habits of men make them more vulnerable to lightning.) Assuming that this trend holds in the future, what is the probability that a person killed by lightning:
   a. Is a male?
   b. Is a female?

20. **QUALITY CONTROL** One light bulb is selected at random from a lot of 120 light bulbs, of which 5% are defective. What is the probability that the light bulb selected is defective?

21. **EFFORTS TO STOP SHOPLIFTING** According to a survey of 176 retailers, 46% of them use electronic tags as protection against shoplifting and employee theft. If one of these retailers is selected at random, what is the probability that he or she uses electronic tags as antitheft devices?

22. If a ball is selected at random from an urn containing three red balls, two white balls, and five blue balls, what is the probability that it will be a white ball?

23. If 1 card is drawn at random from a standard 52-card deck, what is the probability that the card drawn is:
   a. A diamond? **b.** A black card?
   c. An ace?

24. A pair of fair dice is cast. What is the probability that:
   a. The sum of the numbers shown uppermost is less than 5?
   b. At least one 6 is cast?

25. **TRAFFIC LIGHTS** What is the probability of arriving at a traffic light when it is red if the red signal is flashed for 30 sec, the yellow signal for 5 sec, and the green signal for 45 sec?

26. **ROULETTE** What is the probability that a roulette ball will come to rest on an even number other than 0 or 00? (Assume that there are 38 equally likely outcomes consisting of the numbers 1–36, 0, and 00.)

27. Refer to Exercise 9. What is the probability that a student selected at random from this class received a passing grade (D or better)?

28. Refer to Exercise 11. What is the probability that more than 600 cars will enter the airport tunnel during a peak hour?

29. **DISPOSITION OF CRIMINAL CASES** Of the 98 first-degree murder cases from 1990 through the first half of 1992 in the Suffolk superior court, 9 cases were thrown out of the system, 62 cases were plea-bargained, and 27 cases

went to trial. What is the probability that a case selected at random:

a. Was settled through plea bargaining?
b. Went to trial?

30. SWEEPSTAKES One hundred thousand entries have been received in a sweepstakes sponsored by the Gemini Paper Products Company. If 1 grand prize, 5 first prizes, 25 second prizes, and 500 third prizes are to be awarded, what is the probability that a person who has submitted one entry will win:

a. The grand prize?
b. A prize?

31. A pair of fair dice is cast and the sum of the two numbers falling uppermost observed. The probability of obtaining a sum of 2 is the same as that of obtaining a 7 since there is only one way of getting a 2—namely, by each die showing a 1; and there is only one way of obtaining a 7—namely, by one die showing a 3 and the other die showing a 4. What is wrong with this argument?

In Exercises 32–35, determine whether the given experiment has a sample space with equally likely outcomes.

32. A loaded die is cast, and the number appearing uppermost on the die is recorded.

33. Two fair dice are cast, and the sum of the numbers appearing uppermost is recorded.

34. A ball is selected at random from an urn containing six black balls and six red balls, and the color of the ball is recorded.

35. A weighted coin is thrown, and the outcome of heads or tails is recorded.

36. Let $S = \{s_1, s_2, s_3, s_4, s_5\}$ be the sample space associated with an experiment having the following probability distribution:

Outcome	$\{s_1\}$	$\{s_2\}$	$\{s_3\}$	$\{s_4\}$	$\{s_5\}$
Probability	$\frac{1}{14}$	$\frac{3}{14}$	$\frac{6}{14}$	$\frac{2}{14}$	$\frac{2}{14}$

Find the probability of the event:

a. $A = \{s_1, s_2, s_4\}$ **b.** $B = \{s_1, s_5\}$
c. $C = S$

37. Let $S = \{s_1, s_2, s_3, s_4, s_5, s_6\}$ be the sample space associated with an experiment having the following probability distribution:

Outcome	$\{s_1\}$	$\{s_2\}$	$\{s_3\}$	$\{s_4\}$	$\{s_5\}$	$\{s_6\}$
Probability	$\frac{1}{12}$	$\frac{1}{4}$	$\frac{1}{12}$	$\frac{1}{6}$	$\frac{1}{3}$	$\frac{1}{12}$

Find the probability of the event:

a. $A = \{s_1, s_3\}$ **b.** $B = \{s_2, s_4, s_5, s_6\}$
c. $C = S$

38. POLITICAL POLLS An opinion poll was conducted among a group of registered voters in a certain state concerning a proposition aimed at limiting state and local taxes. Results of the poll indicated that 35% of the voters favored the proposition, 32% were against it, and the remaining group were undecided. If the results of the poll are assumed to be representative of the opinions of the state's electorate, what is the probability that a registered voter selected at random from the electorate:

a. Favors the proposition?
b. Is undecided about the proposition?

39. Consider the composition of a three-child family in which the children were born at different times. Assume that a girl is as likely as a boy at each birth. What is the probability that:

a. There are two girls and a boy in the family?
b. The oldest child is a girl?
c. The oldest child is a girl and the youngest child is a boy?

40. AIRFONE USAGE The number of planes in the fleets of five leading airlines that contain Airfones is shown in the accompanying table:

Airline	No. of Planes with Airfones	Size of Fleet
A	50	295
B	40	325
C	31	167
D	29	50
E	25	248

a. If a plane is selected at random from airline A, what is the probability that it contains an Airfone?
b. If a plane is selected at random from the entire fleet of the five airlines, what is the probability that it contains an Airfone?

41. AIRLINE SAFETY In an attempt to study the leading causes of airline crashes, the following data were compiled from records of airline crashes from 1959 to 1994 (excluding sabotage and military action).

Primary Factor	No. of Accidents
Flight crew	327
Airplane	49
Maintenance	14
Weather	22
Airport/air traffic control	19
Miscellaneous/other	15

Assume that you have just learned of an airline crash and that the data give a good indication of the causes of airline crashes, in general. Give an estimate of the probability that the primary cause of the crash was due to pilot error or bad weather.
Source: National Transportation Safety Board

In Exercises 42 and 43, determine whether the statement is true or false. If it is true, explain why it is true. If it is false, give an example to show why it is false.

42. If $S = \{s_1, s_2, \ldots, s_n\}$ is a uniform sample space with n outcomes, then $0 \leq P(s_1) + P(s_2) + \cdots + P(s_n) \leq 1$.

43. Let $S = \{s_1, s_2, \ldots, s_n\}$ be a uniform sample space for an experiment. If $n \geq 5$ and $E = \{s_1, s_2, s_5\}$, then $P(E) = 3/n$.

SOLUTIONS TO SELF-CHECK EXERCISES 7.2

1. $$P(1) = \frac{\text{Number of trials in which a 1 appears uppermost}}{\text{Total number of trials}} = \frac{142}{1000} = .142$$

Similarly, we compute $P(2), \ldots, P(6)$, obtaining the following probability distribution.

Outcome	1	2	3	4	5	6
Probability	.142	.173	.158	.175	.162	.190

2. The probability that a highway patrol car equipped with a third brake light will be rear-ended within a 1-yr period is given by

$$\frac{\text{Number of rear-end collisions involving cars equipped with a third brake light}}{\text{Total number of such cars}} = \frac{14}{250} = .056$$

The probability that a highway patrol car not equipped with a third brake light will be rear-ended within a 1-yr period is given by

$$\frac{\text{Number of rear-end collisions involving cars not equipped with a third brake light}}{\text{Total number of such cars}} = \frac{22}{250} = .088$$

7.3 Rules of Probability

PROPERTIES OF THE PROBABILITY FUNCTION AND THEIR APPLICATIONS

In this section we examine some of the properties of the probability function and look at the role they play in solving certain problems. We begin by looking at the generalization of the three properties of the probability function, which were stated for simple events in the last section. Let S be a sample space of an experiment and suppose E and F are events of the experiment. We have:

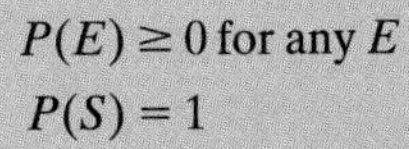

Property 1 $P(E) \geq 0$ for any E

Property 2 $P(S) = 1$

Property 3 If E and F are mutually exclusive (that is, only one of them can occur, or equivalently, $E \cap F = \varnothing$), then

$$P(E \cup F) = P(E) + P(F)$$

(Figure 7.7).

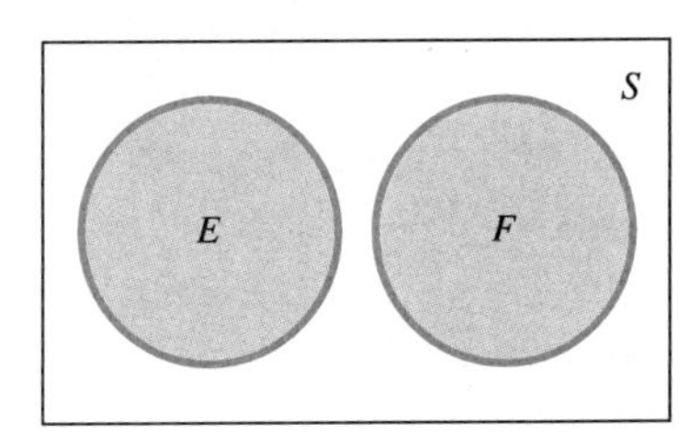

FIGURE 7.7
If E and F are mutually exclusive events, then $P(E \cup F) = P(E) + P(F)$.

Property 3 may be easily extended to the case involving any finite number of mutually exclusive events. Thus, if $E_1, E_2, \ldots, E_n$ are mutually exclusive events, then

$$P(E_1 \cup E_2 \cup \cdots \cup E_n) = P(E_1) + P(E_2) + \cdots + P(E_n)$$

EXAMPLE 1 The superintendent of a metropolitan school district has estimated the probabilities associated with the SAT verbal scores of students from that district. The results are shown in Table 7.7.

Table 7.7 Probability Distribution

Score, x	Probability
$x > 700$	.01
$600 < x \leq 700$	.07
$500 < x \leq 600$	.19
$400 < x \leq 500$	.23
$300 < x \leq 400$	.31
$x \leq 300$	.19

If a student is selected at random, what is the probability that his or her SAT verbal score will be:

a. More than 400?
b. Less than or equal to 500?
c. Greater than 400 but less than or equal to 600?

SOLUTION ✓

Let A, B, C, D, E, and F denote, respectively, the event that the score is greater than 700, greater than 600 but less than or equal to 700, greater than 500 but less than or equal to 600, and so forth. Then, these events are mutually exclusive. Therefore,

a. The probability that the student's score will be more than 400 is given by

$$\begin{aligned} P(D \cup C \cup B \cup A) &= P(D) + P(C) + P(B) + P(A) \\ &= .23 + .19 + .07 + .01 \\ &= .5 \end{aligned}$$

b. The probability that the student's score will be less than or equal to 500 is given by

$$\begin{aligned} P(D \cup E \cup F) &= P(D) + P(E) + P(F) \\ &= .23 + .31 + .19 \\ &= .73 \end{aligned}$$

c. The probability that the student's score will be greater than 400 but less than or equal to 600 is given by

$$\begin{aligned} P(C \cup D) &= P(C) + P(D) \\ &= .19 + .23 \\ &= .42 \end{aligned}$$

■■■■

Property 3 holds if and only if E and F are mutually exclusive. In the general case, we have the following rule:

Property 4 Addition Rule

If E and F are any two events of an experiment, then

$$P(E \cup F) = P(E) + P(F) - P(E \cap F)$$

(Figure 7.8).

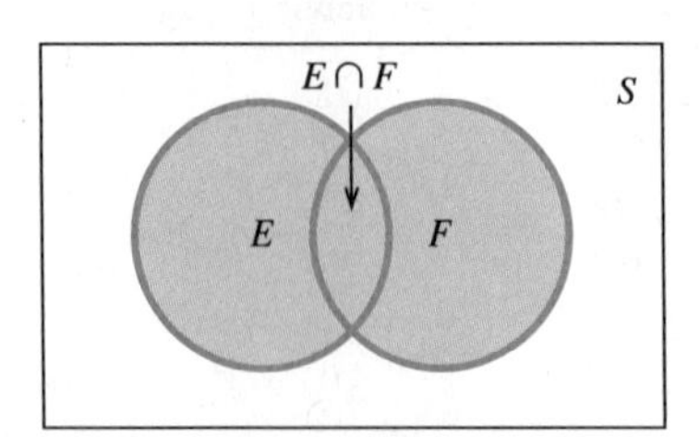

FIGURE 7.8
$P(E \cup F) = P(E) + P(F) - P(E \cap F)$

To derive the property for uniform sample spaces, we use Equation (4), Section 6.2, to see that

$$n(E \cup F) = n(E) + n(F) - n(E \cap F)$$

where E and F are events of an experiment with sample space S. Dividing both sides of this equation by $n(S)$, we obtain

$$\frac{n(E \cup F)}{n(S)} = \frac{n(E)}{n(S)} + \frac{n(F)}{n(S)} - \frac{n(E \cap F)}{n(S)}$$

Recalling the definition of the probability of an event then leads to

$$P(E \cup F) = P(E) + P(F) - P(E \cap F)$$

as we wish to show.

REMARK Observe that when E and F are mutually exclusive—that is, when $E \cap F = \varnothing$—then the equation of Property 4 reduces to that of Property 3. In other words, if E and F are mutually exclusive events, then

$$P(E \cup F) = P(E) + P(F)$$

If E and F are not mutually exclusive events, then

$$P(E \cup F) = P(E) + P(F) - P(E \cap F)$$

■■■

EXAMPLE 2 A card is drawn from a well-shuffled deck of 52 playing cards. What is the probability that it is an ace or a spade?

SOLUTION ✓ Let E denote the event that the card drawn is an ace and let F denote the event that the card drawn is a spade. Then,

$$P(E) = \frac{4}{52} \quad \text{and} \quad P(F) = \frac{13}{52}$$

Furthermore, E and F are not mutually exclusive events. In fact, $E \cap F$ is the event that the card drawn is an ace of spades. Consequently,

$$P(E \cap F) = \frac{1}{52}$$

FIGURE 7.9
$P(E \cup F) = P(E) + P(F) - P(E \cap F)$

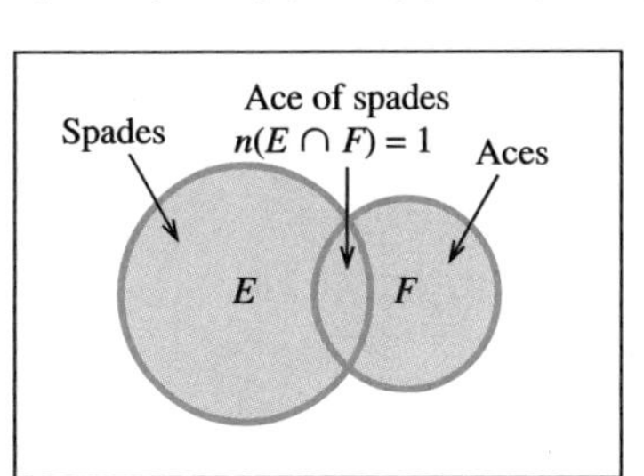

The event that a card drawn is an ace or a spade is $E \cup F$, with probability given by

$$\begin{aligned} P(E \cup F) &= P(E) + P(F) - P(E \cap F) \\ &= \frac{4}{52} + \frac{13}{52} - \frac{1}{52} = \frac{16}{52} = \frac{4}{13} \end{aligned}$$

(Figure 7.9). This result, of course, can be obtained by arguing that 16 of the 52 cards are either spades or aces of other suits. ■■■■

Group Discussion

Let E, F, and G be any three events of an experiment. Use Formula (5) of Section 6.2 to show that

$$P(E \cup F \cup G) = P(E) + P(F) + P(G) - P(E \cap F) - P(E \cap G) - P(F \cap G) + P(E \cap F \cap G)$$

If E, F, and G are mutually exclusive, what is $P(E \cup F \cup G)$?

EXAMPLE 3 The quality-control department of the Vista Vision Corporation, manufacturer of the Pulsar 19-inch color television set, has determined from records obtained from the company's service centers that 3% of the sets sold experience video problems, 1% experience audio problems, and 0.1% experience both video as well as audio problems before the expiration of the 90-day warranty. Find the probability that a set purchased by a consumer will experience video or audio problems before the warranty expires.

SOLUTION ✓ Let E denote the event that a set purchased will experience video problems within 90 days and let F denote the event that a set purchased will experience audio problems within 90 days. Then,

$$P(E) = .03, \qquad P(F) = .01, \qquad P(E \cap F) = .001$$

FIGURE 7.10
$P(E \cup F) = P(E) + P(F) - P(E \cap F)$

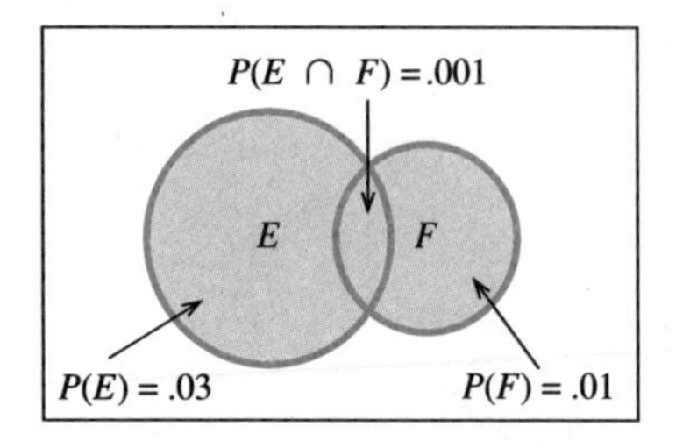

The event that a set purchased will experience video problems or audio problems before the warranty expires is $E \cup F$, and the probability of this event is given by

$$\begin{aligned} P(E \cup F) &= P(E) + P(F) - P(E \cap F) \\ &= .03 + .01 - .001 \\ &= .039 \end{aligned}$$

(Figure 7.10).

Another property of the probability function that is of considerable aid in computing probability follows.

Property 5 Rule of Complements

If E is an event of an experiment and E^c denotes the complement of E, then

$$P(E^c) = 1 - P(E)$$

Property 5 is an immediate consequence of Properties 2 and 3. Indeed, we have $E \cup E^c = S$ and $E \cap E^c = \varnothing$, so

$$1 = P(S) = P(E \cup E^c) = P(E) + P(E^c)$$

and, therefore,

$$P(E^c) = 1 - P(E)$$

EXAMPLE 4 Refer to Example 3. What is the probability that a Pulsar 19-inch color television set bought by a consumer will not experience video or audio difficulties before the warranty expires?

SOLUTION ✓ Let E denote the event that a set bought by a consumer will experience video or audio difficulties before the warranty expires. Then, the event that the set will not experience either problem before the warranty expires is given by E^c, with probability

$$\begin{aligned} P(E^c) &= 1 - P(E) \\ &= 1 - .039 \\ &= .961 \end{aligned}$$

Computations Involving the Rules of Probability

We close this section by looking at two additional examples that illustrate the rules of probability.

EXAMPLE 5 Let E and F be two mutually exclusive events and suppose $P(E) = .1$ and $P(F) = .6$. Compute:

a. $P(E \cap F)$ **b.** $P(E \cup F)$ **c.** $P(E^c)$
d. $P(E^c \cap F^c)$ **e.** $P(E^c \cup F^c)$

SOLUTION ✓

a. Since the events E and F are mutually exclusive—that is, $E \cap F = \varnothing$—we have $P(E \cap F) = 0$.

b.
$$\begin{aligned} P(E \cup F) &= P(E) + P(F) \qquad \text{(Since } E \text{ and } F \text{ are mutually exclusive)} \\ &= .1 + .6 \\ &= .7 \end{aligned}$$

c.
$$\begin{aligned} P(E^c) &= 1 - P(E) \qquad \text{(Property 5)} \\ &= 1 - .1 \\ &= .9 \end{aligned}$$

d. Observe that, by De Morgan's law, $E^c \cap F^c = (E \cup F)^c$, so

$$\begin{aligned} P(E^c \cap F^c) &= P[(E \cup F)^c] \qquad \text{(See Figure 7.11.)} \\ &= 1 - P(E \cup F) \qquad \text{(Property 5)} \\ &= 1 - .7 \qquad \text{[Using the result of part (b)]} \\ &= .3 \end{aligned}$$

FIGURE 7.11
$P(E^c \cap F^c) = P[(E \cup F)^c]$

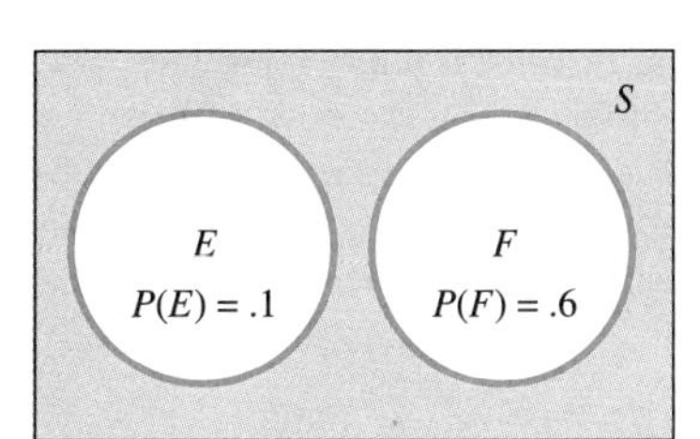

e. Again, using De Morgan's law, we find

$$\begin{aligned} P(E^c \cup F^c) &= P[(E \cap F)^c] \\ &= 1 - P(E \cap F) \\ &= 1 - 0 \qquad \text{[Using the result of part (a)]} \\ &= 1 \end{aligned}$$

EXAMPLE 6 Let E and F be two events of an experiment with sample space S. Suppose $P(E) = .2$, $P(F) = .1$, and $P(E \cap F) = .05$. Compute:

a. $P(E \cup F)$ **b.** $P(E^c \cap F^c)$
c. $P(E^c \cap F)$ (*Hint:* Draw a Venn diagram.)

SOLUTION ✓ **a.**

$$\begin{aligned} P(E \cup F) &= P(E) + P(F) - P(E \cap F) \quad \text{(Property 4)} \\ &= .2 + .1 - .05 \\ &= .25 \end{aligned}$$

b. Using De Morgan's law, we have

$$\begin{aligned} P(E^c \cap F^c) &= P[(E \cup F)^c] \\ &= 1 - P(E \cup F) \quad \text{(Property 5)} \\ &= 1 - .25 \quad \text{[Using the result of part (a)]} \\ &= .75 \end{aligned}$$

c. From the Venn diagram describing the relationship between E, F, and S (Figure 7.12), we have

$$P(E^c \cap F) = .05 \quad \text{(The shaded subset is the event } E^c \cap F.)$$

FIGURE 7.12
$P(E^c \cap F)$: the probability that the event F, but not the event E, will occur

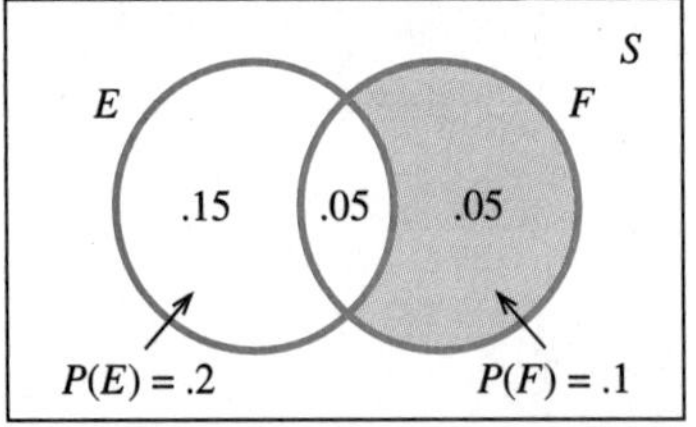

This result may also be obtained by using the relationship

$$\begin{aligned} P(E^c \cap F) &= P(F) - P(E \cap F) \\ &= .1 - .05 \\ &= .05 \end{aligned}$$

as before. ■■■■

SELF-CHECK EXERCISES 7.3

1. Let E and F be events of an experiment with sample space S. Suppose $P(E) = .4$, $P(F) = .5$, and $P(E \cap F) = .1$. Compute:
a. $P(E \cup F)$ **b.** $P(E \cap F^c)$
2. Susan Garcia wishes to sell or lease a condominium through a realty company. The realtor estimates that the probability of finding a buyer within a month of the date the property is listed for sale or lease is .3, the probability of finding a lessee is .8, and the probability of finding both a buyer and a lessee is .1. Determine the probability that the property will be sold or leased within 1 mo from the date the property is listed for sale or lease.

Solutions to Self-Check Exercises 7.3 can be found on page 414.

7.3 Exercises

A pair of dice is cast, and the number that appears uppermost on each die is observed. In Exercises 1–6, refer to this experiment and find the probability of the given event.

1. The sum of the numbers is an even number.
2. The sum of the numbers is either 7 or 11.
3. A pair of 1s is thrown.
4. A double is thrown.
5. One die shows a 6 and the other a number less than 3.
6. The sum of the numbers is at least 4.

An experiment consists of selecting a card at random from a 52-card deck. In Exercises 7–12, refer to this experiment and find the probability of the given event.

7. A king of diamonds is drawn.
8. A diamond or a king is drawn.
9. A face card is drawn.
10. A red face card is drawn.
11. An ace is not drawn.
12. A black face card is not drawn.
13. Five hundred people have purchased raffle tickets. What is the probability that a person holding one ticket will win the first prize? What is the probability that he or she will not win the first prize?
14. TV HOUSEHOLDS The results of a recent television survey of American TV households revealed that 87 out of every 100 TV households have at least one remote control. What is the probability that a randomly selected TV household does not have at least one remote control?

In Exercises 15–24, explain why the given statement is incorrect.

15. The sample space associated with an experiment is given by $S = \{a, b, c\}$, where $P(a) = .3$, $P(b) = .4$, and $P(c) = .4$.
16. The probability that a bus will arrive late at the Civic Center is .35, and the probability that it will be on time or early is .60.
17. A person participates in a weekly office pool in which he has one chance in ten of winning the purse. If he participates for 5 wk in succession, the probability of winning at least one purse is 5/10.
18. The probability that a certain stock will increase in value over a period of 1 wk is .6. Therefore, the probability that the stock will decrease in value is .4.
19. A red die and a green die are tossed. The probability that a 6 will appear uppermost on the red die is 1/6, and the probability that a 1 will appear uppermost on the green die is 1/6. Hence, the probability that the red die will show a 6 or the green die will show a 1 is 1/6 + 1/6.
20. Joanne, a high school senior, has applied for admission to four colleges, A, B, C, and D. She has estimated that the probability that she will be accepted for admission by A, B, C, and D is .5, .3, .1, and .08, respectively. Thus, the probability that she will be accepted for admission by at least one college is $P(A) + P(B) + P(C) + P(D) = .5 + .3 + .1 + .08 = .98$.
21. The sample space associated with an experiment is given by $S = \{a, b, c, d, e\}$. The events $E = \{a, b\}$ and $F = \{c, d\}$ are mutually exclusive. Hence, the events E^c and F^c are mutually exclusive.
22. A 5-card poker hand is dealt from a 52-card deck. Let A denote the event that a flush is dealt and let B be the event that a straight is dealt. Then the events A and B are mutually exclusive.
23. Mark Owens, an optician, estimates that the probability that a customer coming into his store will purchase one or more pairs of glasses but not contact lenses is .40, and the probability that he will purchase one or more pairs of contact lenses but not glasses is .25. Hence, Owens concludes that the probability that a customer coming into his store will purchase neither a pair of glasses nor a pair of contact lenses is .35.
24. There are eight grades in the Garfield Elementary School. If a student is selected at random from the school, then the probability that the student is in the first grade is 1/8.
25. Let E and F be two events that are mutually exclusive and suppose $P(E) = .2$ and $P(F) = .5$. Compute:
    a. $P(E \cap F)$ b. $P(E \cup F)$
    c. $P(E^c)$ d. $P(E^c \cap F^c)$

26. Let E and F be two events of an experiment with sample space S. Suppose $P(E) = .6$, $P(F) = .4$, and $P(E \cap F) = .2$. Compute:

a. $P(E \cup F)$ **b.** $P(E^c)$
c. $P(F^c)$ **d.** $P(E^c \cap F)$

27. Let $S = \{s_1, s_2, s_3, s_4\}$ be the sample space associated with an experiment having the probability distribution shown in the accompanying table. If $A = \{s_1, s_2\}$ and $B = \{s_1, s_3\}$, find:

a. $P(A)$, $P(B)$ **b.** $P(A^c)$, $P(B^c)$
c. $P(A \cap B)$ **d.** $P(A \cup B)$

Outcome	Probability
$\{s_1\}$	$\frac{1}{8}$
$\{s_2\}$	$\frac{3}{8}$
$\{s_3\}$	$\frac{1}{4}$
$\{s_4\}$	$\frac{1}{4}$

28. Let $S = \{s_1, s_2, s_3, s_4, s_5, s_6\}$ be the sample space associated with an experiment having the probability distribution shown in the accompanying table. If $A = \{s_1, s_2\}$ and $B = \{s_1, s_5, s_6\}$, find:

a. $P(A)$, $P(B)$ **b.** $P(A^c)$, $P(B^c)$
c. $P(A \cap B)$ **d.** $P(A \cup B)$
e. $P(A^c \cap B^c)$ **f.** $P(A^c \cup B^c)$

Outcome	Probability
$\{s_1\}$	$\frac{1}{3}$
$\{s_2\}$	$\frac{1}{8}$
$\{s_3\}$	$\frac{1}{6}$
$\{s_4\}$	$\frac{1}{6}$
$\{s_5\}$	$\frac{1}{12}$
$\{s_6\}$	$\frac{1}{8}$

29. TEACHER ATTITUDES In a survey of 2140 teachers in a certain metropolitan area, conducted by a nonprofit organization regarding teacher attitudes, the following data were obtained:

900 said that lack of parental support is a problem.

890 said that abused or neglected children are problems.

680 said that malnutrition or students in poor health is a problem.

120 said that lack of parental support and abused or neglected children are problems.

110 said that lack of parental support and malnutrition or poor health are problems.

140 said that abused or neglected children and malnutrition or poor health are problems.

40 said that lack of parental support, abuse or neglect, and malnutrition or poor health are problems.

What is the probability that a teacher selected at random from this group said that lack of parental support is the only problem hampering a student's schooling?

Hint: Draw a Venn diagram.

30. COURSE ENROLLMENTS Among 500 freshmen pursuing a business degree at a university, 320 are enrolled in an Economics course, 225 are enrolled in a Mathematics course, and 140 are enrolled in both an Economics and a Mathematics course. What is the probability that a freshman selected at random from this group is enrolled in:

a. An Economics and/or a Mathematics course?
b. Exactly one of these two courses?
c. Neither an Economics course nor a Mathematics course?

31. CONSUMER SURVEYS A leading manufacturer of kitchen appliances advertised its products in two magazines: *Good Housekeeping* and the *Ladies Home Journal*. A survey of 500 customers revealed that 140 learned of its products from *Good Housekeeping*, 130 learned of its products from the *Ladies Home Journal*, and 80 learned of its products from both magazines. What is the probability that a person selected at random from this group saw the manufacturer's advertisement in:

a. Both magazines?
b. At least one of the two magazines?
c. Exactly one magazine?

32. STUDY HABITS Students at a certain university were asked to state how many hours per week they spent studying in the library. Results of the survey revealed the following information:

Time Spent, x	Percentage of Students
$0 \le x \le 1$	32.3
$1 < x \le 4$	40.7
$4 < x \le 10$	16.5
$x > 10$	10.5

Find the probability distribution associated with these data. What is the probability that a student selected at random at the university studied in the library:
a. More than 4 hr/wk?
b. No more than 10 hr/wk?

33. **ASSEMBLY-TIME STUDIES** A time study was conducted by the production manager of the Universal Instruments Company to determine how much time it took an assembly worker to complete a certain task during the assembly of its Galaxy home computers. Results of the study indicated that 20% of the workers were able to complete the task in less than 3 min, 60% of the workers were able to complete the task in 4 min or less, and 10% of the workers required more than 5 min to complete the task. If an assembly-line worker is selected at random from this group, what is the probability that:
a. He or she will be able to complete the task in 5 min or less?
b. He or she will not be able to complete the task within 4 min?
c. The time taken for the worker to complete the task will be between 3 and 4 min (inclusive)?

34. **PLANS TO KEEP CARS** In a survey conducted to see how long Americans keep their cars, 2000 automobile owners were asked how long they plan to keep their present cars. The results of the survey follow:

Number of Years Car Is Kept, x	Number of Respondents
$0 \le x < 1$	60
$1 \le x < 3$	440
$3 \le x < 5$	360
$5 \le x < 7$	340
$7 \le x < 10$	240
$x \ge 10$	560

Find the probability distribution associated with these data. What is the probability that an automobile owner selected at random from those surveyed plans to keep his or her present car:
a. Less than five years?
b. Three or more years?

35. **GUN-CONTROL LAWS** A poll was conducted among 250 residents of a certain city regarding tougher gun-control laws. The results of the poll are shown in the table:

	Own Only a Handgun	Own Only a Rifle	Own a Handgun and a Rifle	Own Neither	Total
Favor Tougher Laws	0	12	0	138	150
Oppose Tougher Laws	58	5	25	0	88
No Opinion	0	0	0	12	12
Total	58	17	25	150	250

If one of the participants in this poll is selected at random, what is the probability that he or she:
a. Favors tougher gun-control laws?
b. Owns a handgun?
c. Owns a handgun but not a rifle?
d. Favors tougher gun-control laws and does not own a handgun?

36. **RISK OF AN AIRPLANE CRASH** According to a study conducted by the National Transportation Safety Board, of Western-built commercial jets involved in crashes from 1988 to 1998, the percentages of airplane crashes that occur at each stage of flight are as follows:

Phase	Percentage
On ground, taxiing	4
During takeoff	10
Climbing to cruise altitude	19
En route	5
Descent and approach	31
Landing	31

If one of the doomed flights in the period 1988–1998 is picked at random, what is the probability that it crashed:
a. While taxiing on the ground or while en route?
b. During takeoff or landing?
If the study is indicative of airplane crashes in general, when is the risk of a plane crash the highest?
Source: National Transportation and Safety Board

37. Suppose the probability that Bill can solve a problem is p_1 and the probability that Mike can solve it is p_2. Show that the probability that Bill and Mike working independently can solve the problem is $p_1 + p_2 - p_1 p_2$.

38. Fifty raffle tickets are numbered 1 through 50, and one of them is drawn at random. What is the probability that the number is a multiple of 5 or 7? Consider the following "solution": Since 10 tickets bear numbers that are multiples of 5 and 7 tickets bear numbers that are multiples of 7, we conclude that the required probability is

$$\frac{10}{50} + \frac{7}{50} = \frac{17}{50}$$

What is wrong with this argument? What is the correct answer?

In Exercises 39–42, determine whether the statement is true or false. If it is true, explain why it is true. If it is false, give an example to show why it is false.

39. If A is a subset of B and $P(B) = 0$, then $P(A) = 0$.

40. If A is a subset of B, then $P(A) \leq P(B)$.

41. If $E_1, E_2, \ldots, E_n$ are events of an experiment, then $P(E_1 \cup E_2 \cup \cdots \cup E_n) = P(E_1) + P(E_2) + \cdots + P(E_n)$.

42. If E is an event of an experiment, then $P(E) + P(E^c) = 1$.

SOLUTIONS TO SELF-CHECK EXERCISES 7.3

1. a. Using Property 4, we find

$$\begin{aligned} P(E \cup F) &= P(E) + P(F) - P(E \cap F) \\ &= .4 + .5 - .1 \\ &= .8 \end{aligned}$$

b. From the accompanying Venn diagram, in which the subset $E \cap F^c$ is shaded, we see that

$$P(E \cap F^c) = .3$$

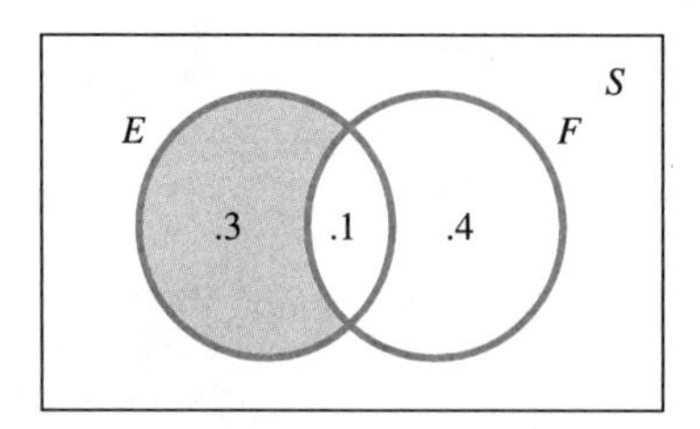

The result may also be obtained by using the relationship

$$\begin{aligned} P(E \cap F^c) &= P(E) - P(E \cap F) \\ &= .4 - .1 = .3 \end{aligned}$$

2. Let E denote the event that the property will be sold within 1 mo of the date it is listed for sale or lease and let F denote the event that the property will be leased within the same time period. Then,

$$P(E) = .3, \qquad P(F) = .8, \qquad P(E \cap F) = .1$$

The probability of the event that the property will be sold or leased within 1 mo of the date it is listed for sale or lease is given by

$$\begin{aligned} P(E \cup F) &= P(E) + P(F) - P(E \cap F) \\ &= .3 + .8 - .1 = 1 \end{aligned}$$

—that is, a certainty.

7.4 Use of Counting Techniques in Probability

Further Applications of Counting Techniques

As we have seen many times before, a problem in which the underlying sample space has a small number of elements may be solved by first determining all such sample points. For problems involving sample spaces with a large number of sample points, however, this approach is neither practical nor desirable.

In this section we see how the counting techniques studied in Chapter 6 may be employed to help us solve problems in which the associated sample spaces contain large numbers of sample points. In particular, we restrict our attention to the study of uniform sample spaces—that is, sample spaces in which the outcomes are equally likely. For such spaces we have the following result:

Computing the Probability of an Event in a Uniform Sample Space

Let S be a uniform sample space and let E be any event. Then,

$$P(E) = \frac{\text{Number of favorable outcomes in } E}{\text{Number of possible outcomes in } S} = \frac{n(E)}{n(S)} \tag{1}$$

EXAMPLE 1 An unbiased coin is tossed six times. What is the probability that the coin will land heads:

a. Exactly three times?
b. At most three times?
c. On the first and the last toss?

SOLUTION ✓ **a.** Each outcome of the experiment may be represented as a sequence of heads and tails. Using the generalized multiplication principle, we see that the number of outcomes of this experiment is given by 2^6, or 64. Let E denote the event that the coin lands heads exactly three times. Since there are $C(6, 3)$ ways this can occur, we see that the required probability is

$$P(E) = \frac{n(E)}{n(S)} = \frac{C(6,3)}{64} = \frac{\frac{6!}{3!3!}}{64} \qquad (S\text{, sample space of the experiment})$$

$$= \frac{\frac{6 \cdot 5 \cdot 4}{3 \cdot 2}}{64} = \frac{20}{64} = \frac{5}{16} = .3125$$

b. Let F denote the event that the coin lands heads at most three times. Then $n(F)$ is given by the sum of the number of ways the coin lands heads zero times (no heads!), the number of ways it lands heads exactly once, the number of ways it lands heads exactly twice, and the number of ways it lands heads exactly three times. That is,

$$\begin{aligned} n(F) &= C(6,0) + C(6,1) + C(6,2) + C(6,3) \\ &= \frac{6!}{0!6!} + \frac{6!}{1!5!} + \frac{6!}{2!4!} + \frac{6!}{3!3!} \\ &= 1 + 6 + \frac{(6)(5)}{2} + \frac{(6)(5)(4)}{(3)(2)} = 42 \end{aligned}$$

Therefore, the required probability is

$$P(F) = \frac{n(F)}{n(S)} = \frac{42}{64} = \frac{21}{32} \approx .66$$

c. Let F denote the event that the coin lands heads on the first and the last toss. Then $n(F) = 1 \cdot 2 \cdot 2 \cdot 2 \cdot 2 \cdot 1 = 2^4$, so the probability that this event occurs is

$$\begin{aligned} P(F) &= \frac{2^4}{2^6} \\ &= \frac{1}{2^2} \\ &= \frac{1}{4} \end{aligned}$$

■■■■

EXAMPLE 2

Two cards are selected at random from a well-shuffled pack of 52 playing cards. What is the probability that:

a. They are both aces?
b. Neither of them is an ace?

SOLUTION ✓

a. The experiment consists of selecting 2 cards from a pack of 52 playing cards. Since the order in which the cards are selected is immaterial, the sample points are combinations of 52 cards taken 2 at a time. Now, there are $C(52, 2)$ ways of selecting 52 cards taken 2 at a time, so the number of elements in the sample space S is given by $C(52, 2)$. Next, we observe that there are $C(4, 2)$ ways of selecting 2 aces from the 4 in the deck. Therefore, if E denotes the event that the cards selected are both aces, then

$$\begin{aligned} P(E) &= \frac{n(E)}{n(S)} \\ &= \frac{C(4,2)}{C(52,2)} = \frac{\frac{4!}{2!2!}}{\frac{52!}{2!50!}} \\ &= \frac{1}{221} \end{aligned}$$

b. Let F denote the event that neither of the two cards selected is an ace. Since there are $C(48, 2)$ ways of selecting two cards, neither of which is an ace, we find that

$$P(F) = \frac{n(F)}{n(S)} = \frac{C(48,2)}{C(52,2)} = \frac{\frac{48!}{2!46!}}{\frac{52!}{2!50!}} = \frac{48 \cdot 47}{2} \cdot \frac{2}{52 \cdot 51}$$

$$= \frac{188}{221}$$

■■■■

EXAMPLE 3

A bin in the hi-fi department of Building 20, a bargain outlet, contains 100 blank cassette tapes, of which 10 are known to be defective. If a customer selects 6 of these cassette tapes, determine the probability that:

a. Two of them are defective.
b. At least 1 of them is defective.

SOLUTION ✓

a. There are $C(100, 6)$ ways of selecting a set of 6 cassette tapes from the 100, and this gives $n(S)$, the number of outcomes in the sample space associated with the experiment. Next, we observe that there are $C(10, 2)$ ways of selecting a set of 2 defective cassette tapes from the 10 defective cassette tapes and $C(90, 4)$ ways of selecting a set of 4 nondefective cassette tapes from the 90 nondefective cassette tapes (Figure 7.13). Thus, by the multiplication principle,

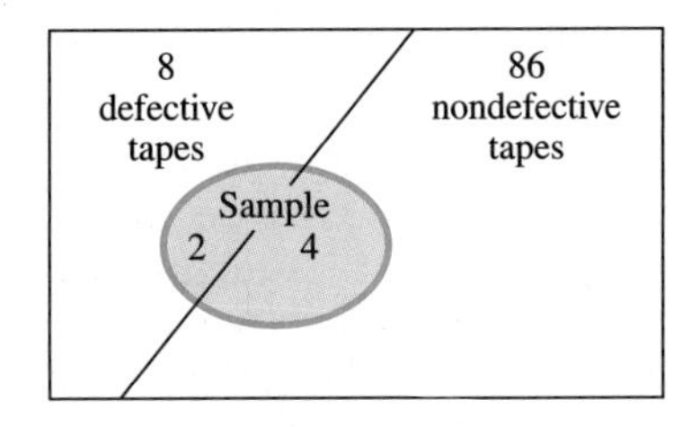

FIGURE 7.13
A sample of 6 tapes selected from 90 nondefective tapes and 10 defective tapes

there are $C(10, 2) \cdot C(90, 4)$ ways of selecting 2 defective and 4 nondefective cassette tapes. Therefore, the probability of selecting 6 cassette tapes, of which 2 are defective, is given by

$$\frac{C(10,2) \cdot C(90,4)}{C(100,6)} = \frac{\frac{10!}{2!8!}\frac{90!}{4!86!}}{\frac{100!}{6!94!}}$$

$$= \frac{10 \cdot 9}{2} \cdot \frac{90 \cdot 89 \cdot 88 \cdot 87}{4 \cdot 3 \cdot 2 \cdot 1} \cdot \frac{6 \cdot 5 \cdot 4 \cdot 3 \cdot 2 \cdot 1}{100 \cdot 99 \cdot 98 \cdot 97 \cdot 96 \cdot 95}$$

$$\approx .096$$

b. Let E denote the event that none of the cassette tapes selected is defective. Then E^c gives the event that at least 1 of the cassette tapes is defective. But, by the rule of complements,

$$P(E^c) = 1 - P(E)$$

To compute $P(E)$, we observe that there are $C(90, 6)$ ways of selecting a set of 6 cassette tapes that are nondefective. Therefore,

$$P(E) = \frac{C(90, 6)}{C(100, 6)}$$

$$\begin{aligned} P(E^c) &= 1 - \frac{C(90, 6)}{C(100, 6)} \\ &= 1 - \frac{\frac{90!}{6!84!}}{\frac{100!}{6!94!}} \\ &= 1 - \frac{90 \cdot 89 \cdot 88 \cdot 87 \cdot 86 \cdot 85}{6 \cdot 5 \cdot 4 \cdot 3 \cdot 2 \cdot 1} \cdot \frac{6 \cdot 5 \cdot 4 \cdot 3 \cdot 2 \cdot 1}{100 \cdot 99 \cdot 98 \cdot 97 \cdot 96 \cdot 95} \\ &\approx .48 \end{aligned}$$

THE BIRTHDAY PROBLEM

EXAMPLE 4

A group of five people is selected at random. What is the probability that at least two of them have the same birthday?

SOLUTION ✓

For simplicity we assume that none of the five people was born on February 29 of a leap year. Since the five people were selected at random, we may also assume that each of them is equally likely to have any of the 365 days of a year as his or her birthday. If we let A, B, C, D, and E represent the five people, then an outcome of the experiment may be represented by (a, b, c, d, e), where the numbers a, b, c, d, and e give the birthdays of A, B, C, D, and E, respectively.

We first observe that since there are 365 possibilities for each of the dates a, b, c, d, and e, the multiplication principle implies that there are

$$\underset{a}{\boxed{365}} \cdot \underset{b}{\boxed{365}} \cdot \underset{c}{\boxed{365}} \cdot \underset{d}{\boxed{365}} \cdot \underset{e}{\boxed{365}}$$

or 365^5 outcomes of the experiment. Therefore,

$$n(S) = 365^5$$

where S denotes the sample space of the experiment.

Next, let E denote the event that two or more of the five people have the same birthday. It is now necessary to compute $P(E)$. However, a direct computation of $P(E)$ is relatively difficult. It is much easier to compute $P(E^c)$, where E^c is the event that no two of the five people have the same birthday, and then use the relation

$$P(E) = 1 - P(E^c)$$

To compute $P(E^c)$, observe that there are 365 ways (corresponding to the 365 dates) on which A's birthday can occur, followed by 364 ways on which B's birthday could occur if B were not to have the same birthday as A, and so on. Therefore, by the generalized multiplication principle,

$$n(E^c) = \underset{\text{A's birthday}}{365} \cdot \underset{\text{B's birthday}}{364} \cdot \underset{\text{C's birthday}}{363} \cdot \underset{\text{D's birthday}}{362} \cdot \underset{\text{E's birthday}}{361}$$

Thus,

$$\begin{aligned} P(E^c) &= \frac{n(E^c)}{n(S)} \\ &= \frac{365 \cdot 364 \cdot 363 \cdot 362 \cdot 361}{365^5} \\ P(E) &= 1 - P(E^c) \\ &= 1 - \frac{365 \cdot 364 \cdot 363 \cdot 362 \cdot 361}{365^5} \\ &\approx .027 \end{aligned}$$

■■■■

We can extend the result obtained in Example 4 to the general case involving r people. In fact, if E denotes the event that at least two of the r people have the same birthday, an argument similar to that used in Example 4 leads to the result

$$P(E) = 1 - \frac{365 \cdot 364 \cdot 363 \cdot \cdots \cdot (365 - r + 1)}{365^r}$$

By letting r take on the values 5, 10, 15, 20, . . . , 50, in turn, we obtain the probabilities that at least 2 of 5, 10, 15, 20, . . . , 50 people, respectively, have the same birthday. These results are summarized in Table 7.8.

Table 7.8 Probability That at Least Two People in a Randomly Selected Group of r People Have the Same Birthday

r	5	10	15	20	22	23	25	30	40	50
$P(E)$	.027	.117	.253	.411	.476	.507	.569	.706	.891	.970

The results show that in a group of 23 randomly selected people the chances are greater than 50% that at least 2 of them will have the same birthday. In a group of 50 people, it is an excellent bet that at least 2 people in the group will have the same birthday.

Group Discussion

During an episode of the *Tonight Show,* Johnny Carson related "The Birthday Problem" to the audience—noting that, in a group of 50 or more people, probabilists have calculated that the probability of at least 2 people having the same birthday is very high. To illustrate this point, he proceeded to conduct his own experiment. A person selected at random from the audience was asked to state his birthday. Carson then asked if anyone in the audience had the same birthday. The response was negative. He repeated the experiment. Once again, the response was negative. These results, observed Carson, were contrary to expectations. In a later episode of the show, Carson explained why this experiment had been improperly conducted. Explain why Carson failed to illustrate the point he was trying to make in the earlier episode.

Self-Check Exercises 7.4

1. Four balls are selected at random without replacement from an urn containing 10 white balls and 8 red balls. What is the probability that all the chosen balls are white?
2. A box contains 20 microchips, of which 4 are substandard. If 2 of the chips are taken from the box, what is the probability that they are both substandard?

Solutions to Self-Check Exercises 7.4 can be found on page 423.

7.4 Exercises

A calculator is recommended for this exercise set.

An unbiased coin is tossed five times. In Exercises 1–4, find the probability of the given event.

1. The coin lands heads all five times.
2. The coin lands heads exactly once.
3. The coin lands heads at least once.
4. The coin lands heads more than once.

Two cards are selected at random without replacement from a well-shuffled deck of 52 playing cards. In Exercises 5–8, find the probability of the given event.

5. A pair is drawn.
6. A pair is not drawn.
7. Two black cards are drawn.
8. Two cards of the same suit are drawn.

Four balls are selected at random without replacement from an urn containing three white balls and five blue balls. In Exercises 9–12, find the probability of the given event.

9. Two of the balls are white, and two are blue.

10. All of the balls are blue.

11. Exactly three of the balls are blue.

12. Two or three of the balls are white.

Assume that the probability of a boy being born is the same as the probability of a girl being born. In Exercises 13–16, find the probability that a family with three children will have the given composition.

13. Two boys and one girl

14. At least one girl

15. No girls

16. The two oldest children are girls.

17. An exam consists of ten true-or-false questions. If a student guesses at every answer, what is the probability that he or she will answer exactly six questions correctly?

18. PERSONNEL SELECTION Jacobs & Johnson, Inc., an accounting firm, employs 14 accountants, of whom 8 are CPAs. If a delegation of 3 accountants is randomly selected from the firm to attend a conference, what is the probability that 3 CPAs will be selected?

19. QUALITY CONTROL Two light bulbs are selected at random from a lot of 24, of which 4 are defective. What is the probability that:
a. Both of the light bulbs are defective?
b. At least 1 of the light bulbs is defective?

20. A customer at Cavallaro's Fruit Stand picks a sample of 3 oranges at random from a crate containing 60 oranges, of which 4 are rotten. What is the probability that the sample contains 1 or more rotten oranges?

21. QUALITY CONTROL A shelf in the Metro Department Store contains 80 colored ink cartridges for a popular ink-jet printer. Six of the cartridges are defective. If a customer selects 2 of these cartridges at random from the shelf, what is the probability that:
a. Both are defective?
b. At least 1 is defective?

22. QUALITY CONTROL Electronic baseball games manufactured by Tempco Electronics are shipped in lots of 24. Before shipping, a quality-control inspector randomly selects a sample of 8 from each lot for testing. If the sample contains any defective games, the entire lot is rejected. What is the probability that a lot containing exactly 2 defective games will still be shipped?

23. PERSONNEL SELECTION The City Transit Authority plans to hire 12 new bus drivers. From a group of 100 qualified applicants, of which 60 are men and 40 are women, 12 names are to be selected by lot. Suppose that Mary and John Lewis are among the 100 qualified applicants.
a. What is the probability that Mary's name will be selected? That both Mary's and John's names will be selected?
b. If it is stipulated that an equal number of men and women are to be selected (6 men from the group of 60 men and 6 women from the group of 40 women), what is the probability that Mary's name will be selected? That Mary's and John's names will be selected?

24. PUBLIC HOUSING The City Housing Authority has received 50 applications from qualified applicants for eight low-income apartments. Three of the apartments are on the north side of town, and five are on the south side. If the apartments are to be assigned by means of a lottery, what is the probability that:
a. A specific qualified applicant will be selected for one of these apartments?
b. Two specific qualified applicants will be selected for apartments on the same side of town?

25. A student studying for a vocabulary test knows the meanings of 12 words from a list of 20 words. If the test contains 10 words from the study list, what is the probability that at least 8 of the words on the test are words that the student knows?

26. DRIVERS' TESTS Four different written driving tests are administered by the City Motor Vehicle Department. One of these four tests is selected at random for each applicant for a driver's license. If a group consisting of two women and three men apply for a license, what is the probability that:

a. Exactly two of the five will take the same test?
b. The two women will take the same test?

27. **BRAND SELECTION** A druggist wishes to select three brands of aspirin to sell in his store. He has five major brands to choose from: brands A, B, C, D, and E. If he selects the three brands at random, what is the probability that he will select:
a. Brand B? b. Brands B and C?
c. At least one of the two brands B and C?

28. **BLACKJACK** In the game of blackjack, a 2-card hand consisting of an ace and a face card or a 10 is called a blackjack.
a. If a player is dealt 2 cards from a standard deck of 52 well-shuffled cards, what is the probability that the player will receive a blackjack?
b. If a player is dealt 2 cards from 2 well-shuffled standard decks, what is the probability that the player will receive a blackjack?

29. **SLOT MACHINES** Refer to Exercise 25, Section 6.3, where the "lucky dollar" slot machine was described. What is the probability that the three "lucky dollar" symbols will appear in the window of the slot machine?

30. **ROULETTE** In 1959 a world record was set for the longest run on an ungaffed (fair) roulette wheel at the El San Juan Hotel in Puerto Rico. The number 10 appeared six times in a row. What is the probability of the occurrence of this event? (Assume that there are 38 equally likely outcomes consisting of the numbers 1–36, 0, and 00.)

In "The Numbers Game," a state lottery, four numbers are drawn with replacement from an urn containing the digits 0–9, inclusive. In Exercises 31–34, find the probability of a ticket holder having the indicated winning ticket.

31. All four digits in exact order (the grand prize)

32. Two specified, consecutive digits in exact order (the first two digits, the middle two digits, or the last two digits)

33. One specified digit in exact order (the first, second, third, or fourth digit)

34. All four digits in any order (including the other winning tickets)

A list of poker hands ranked in order from the highest to the lowest is shown in the accompanying table along with a description and example of each hand. Use the table to answer Exercises 35–40.

Hand	Description	Example
Straight flush	5 cards in sequence in the same suit	A ♥ 2 ♥ 3 ♥ 4 ♥ 5 ♥
Four of a kind	4 cards of the same rank and any other card	K ♥ K ♦ K ♠ K ♣ 2 ♥
Full house	3 of a kind and a pair	3 ♥ 3 ♦ 3 ♣ 7 ♥ 7 ♦
Flush	5 cards of the same suit that are not all in sequence	5 ♥ 6 ♥ 9 ♥ J ♥ K ♥
Straight	5 cards in sequence but not all of the same suit	10 ♥ J ♦ Q ♣ K ♠ A ♥
Three of a kind	3 cards of the same rank and 2 unmatched cards	K ♥ K ♦ K ♠ 2 ♥ 4 ♦
Two pair	2 cards of the same rank and 2 cards of any other rank with an unmatched card	K ♥ K ♦ 2 ♥ 2 ♠ 4 ♣
One pair	2 cards of the same rank and 3 unmatched cards	K ♥ K ♦ 5 ♥ 2 ♠ 4 ♥

If a 5-card poker hand is dealt from a well-shuffled deck of 52 cards, what is the probability of being dealt the given hand?

35. A straight flush. (Note that an ace may be played as either a high or low card in a straight sequence—that is, A, 2, 3, 4, 5 or 10, J, Q, K, A. Hence, there are ten possible sequences for a straight in one suit.)

36. A straight (but not a straight flush)

37. A flush (but not a straight flush)

38. Four of a kind

39. A full house

40. Two pairs

41. **ZODIAC SIGNS** There are 12 signs of the Zodiac: Aries, Taurus, Gemini, Cancer, Leo, Virgo, Libra, Scorpio, Sagittarius, Capricorn, Aquarius, and Pisces. Each sign corresponds to a different calendar period of approximately 1 mo. Assuming that a person is just as likely to be born under one sign as another, what is the probability that in a group of five people at least two of them:
  a. Have the same sign?
  b. Were born under the sign of Aries?

42. **BIRTHDAY PROBLEM** What is the probability that at least two justices of the U.S. Supreme Court have the same birthday?

43. **BIRTHDAY PROBLEM** Fifty people are selected at random. What is the probability that none of the people in this group have the same birthday?

44. There were 41 different presidents of the United States from 1789 through 2000. What is the probability that at least two of them had the same birthday? Compare your calculation with the facts by checking an almanac or some other source.

SOLUTIONS TO SELF-CHECK EXERCISES 7.4

1. The probability that all 4 balls selected are white is given by

$$\frac{\begin{array}{c}\text{The number of ways of selecting 4 white}\\ \text{balls from the 10 in the urn}\end{array}}{\begin{array}{c}\text{The number of ways of selecting any}\\ \text{4 balls from the 18 balls in the urn}\end{array}}$$

$$= \frac{C(10,4)}{C(18,4)}$$

$$= \frac{\frac{10!}{4!6!}}{\frac{18!}{4!14!}}$$

$$= \frac{10 \cdot 9 \cdot 8 \cdot 7}{4 \cdot 3 \cdot 2} \cdot \frac{4 \cdot 3 \cdot 2}{18 \cdot 17 \cdot 16 \cdot 15}$$

$$= .069$$

2. The probability that both chips are substandard is given by

$$\frac{\begin{array}{c}\text{The number of ways of choosing any}\\ \text{2 of the 4 substandard chips}\end{array}}{\begin{array}{c}\text{The number of ways of choosing any}\\ \text{2 of the 20 chips}\end{array}}$$

$$= \frac{C(4,2)}{C(20,2)}$$

$$= \frac{\frac{4!}{2!2!}}{\frac{20!}{2!18!}}$$

$$= \frac{4 \cdot 3}{2} \cdot \frac{2}{20 \cdot 19}$$

$$= .032$$

7.5 Conditional Probability and Independent Events

CONDITIONAL PROBABILITY

Three cities, A, B, and C, are vying to play host to the Summer Olympic Games in the year 2000. If each city has the same chance of winning the right to host the Games, then the probability of city A hosting the Games is 1/3. Suppose city B then decides to pull out of contention because of fiscal problems. Then it would seem that city A's chances of playing host will increase. In fact, if each of the two remaining cities have equal chances of winning, then the probability of city A playing host to the Games is 1/2.

In general, the probability of an event is affected by the occurrence of other events and/or by the knowledge of information relevant to the event. Basically, the injection of conditions into a problem modifies the underlying sample space of the original problem. This in turn leads to a change in the probability of the event.

EXAMPLE 1

Two cards are drawn without replacement from a well-shuffled deck of 52 playing cards.

a. What is the probability that the first card drawn is an ace?
b. What is the probability that the second card drawn is an ace given that the first card drawn was not an ace?
c. What is the probability that the second card drawn is an ace given that the first card drawn was an ace?

SOLUTION ✓

a. The sample space here consists of 52 equally likely outcomes, 4 of which are aces. Therefore, the probability that the first card drawn is an ace is 4/52.

b. Having drawn the first card, there are 51 cards left in the deck. In other words, for the second phase of the experiment, we are working in a *reduced* sample space. If the first card drawn was not an ace, then this modified sample space of 51 points contains 4 "favorable" outcomes (the 4 aces), so the probability that the second card drawn is an ace is given by 4/51.

c. If the first card drawn was an ace, then there are 3 aces left in the deck of 51 playing cards, so the probability that the second card drawn is an ace is given by 3/51. ■■■■

Observe that in Example 1 the occurrence of the first event reduces the size of the original sample space. The information concerning the first card drawn also leads us to the consideration of modified sample spaces: In part (b) the deck contained 4 aces, and in part (c) the deck contained 3 aces.

The probability found in part (b) or (c) of Example 1 is known as a *conditional probability,* since it is the probability of an event occurring given that another event has already occurred. For example, in part (b) we computed the probability of the event that the second card drawn is an ace, given the event that the first card drawn was not an ace. In general, given two events A and B of an experiment, one may, under certain circumstances, compute the probability of the event B given that the event A has already occurred. This probability, denoted by $P(B \mid A)$, is called the **conditional probability of B given A.**

A formula for computing the conditional probability of B given A may be discovered with the aid of a Venn diagram. Consider an experiment with a uniform sample space S and suppose A and B are two events of the experiment (Figure 7.14).

FIGURE 7.14

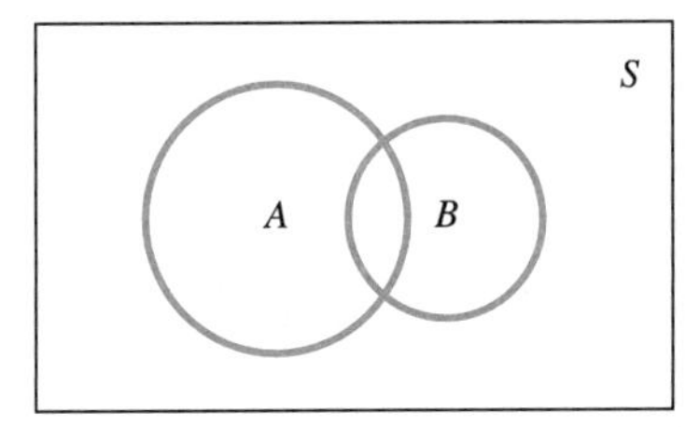

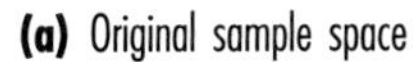

(a) Original sample space

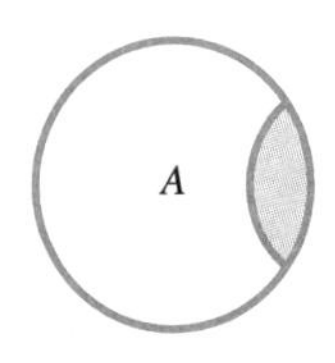

(b) Reduced sample space A. The shaded area is $A \cap B$.

The condition that the event A has occurred tells us that the possible outcomes of the experiment in the second phase are restricted to those outcomes (elements) in the set A. In other words, we may work with the reduced sample space A instead of the original sample space S in the experiment. Next, we observe that, with respect to the reduced sample space A, the outcomes favorable to the event B are precisely those elements in the set $A \cap B$. Consequently, the conditional probability of B given A is

$$P(B \mid A) = \frac{\text{Number of elements in } A \cap B}{\text{Number of elements in } A}$$

$$= \frac{n(A \cap B)}{n(A)} \qquad [n(A) \neq 0]$$

Dividing the numerator and the denominator by $n(S)$, the number of elements in S, we have

$$P(B \mid A) = \frac{\dfrac{n(A \cap B)}{n(S)}}{\dfrac{n(A)}{n(S)}}$$

which is equivalent to the following formula:

Conditional Probability of an Event

If A and B are events in an experiment and $P(A) \neq 0$, then the conditional probability that the event B will occur given that the event A has already occurred is

$$P(B \mid A) = \frac{P(A \cap B)}{P(A)} \tag{2}$$

EXAMPLE 2

A pair of fair dice is cast. What is the probability that the sum of the numbers falling uppermost is 7 if it is known that one of the numbers is a 5?

SOLUTION ✓

Let A denote the event that the sum of the numbers falling uppermost is 7 and let B denote the event that one of the numbers is a 5. From the results of Example 4, Section 7.1, we find that

$$A = \{(6,1), (5,2), (4,3), (3,4), (2,5), (1,6)\}$$
$$B = \{(5,1), (5,2), (5,3), (5,4), (5,5), (5,6), (1,5), (2,5), (3,5), (4,5), (6,5)\}$$

so that

$$A \cap B = \{(5,2), (2,5)\}$$

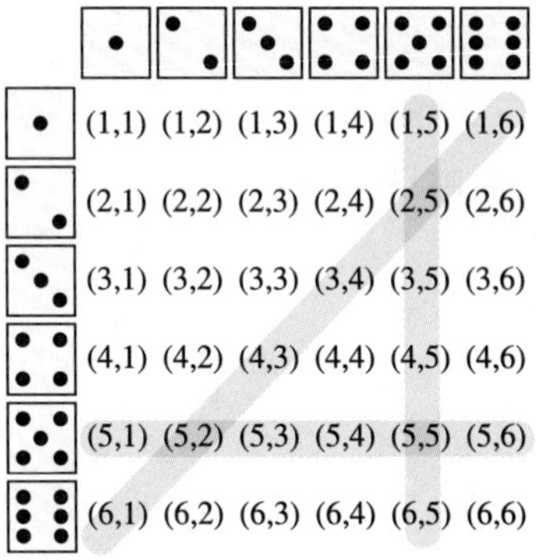

(Figure 7.15). Since the dice are fair, each outcome of the experiment is equally likely; therefore,

$$P(A \cap B) = \frac{2}{36} \quad \text{and} \quad P(B) = \frac{11}{36} \quad \text{[Recall that } n(S) = 36.\text{]}$$

FIGURE 7.15
$A \cap B = \{(5, 2), (2, 5)\}$

	1	2	3	4	5	6
1	(1,1)	(1,2)	(1,3)	(1,4)	(1,5)	(1,6)
2	(2,1)	(2,2)	(2,3)	(2,4)	(2,5)	(2,6)
3	(3,1)	(3,2)	(3,3)	(3,4)	(3,5)	(3,6)
4	(4,1)	(4,2)	(4,3)	(4,4)	(4,5)	(4,6)
5	(5,1)	(5,2)	(5,3)	(5,4)	(5,5)	(5,6)
6	(6,1)	(6,2)	(6,3)	(6,4)	(6,5)	(6,6)

Thus, the probability that the sum of the numbers falling uppermost is 7 given that one of the numbers is a 5 is, by virtue of Equation (2),

$$P(A \mid B) = \frac{\frac{2}{36}}{\frac{11}{36}} = \frac{2}{11}$$

■■■■

EXAMPLE 3

In a test recently conducted by the U.S. Army, it was found that of 1000 new recruits, 600 men and 400 women, 50 of the men and 4 of the women were red-green color-blind. Given that a recruit selected at random from this group is red-green color-blind, what is the probability that the recruit is a male?

SOLUTION ✓

Let C denote the event that a randomly selected subject is red-green color-blind and let M denote the event that the subject is a male recruit. Since 54 out of the 1000 subjects are color-blind, we may take

$$P(C) = \frac{54}{1000} = .054$$

Therefore, by Equation (2), the probability that a subject is male given that the subject is red-green color-blind is

$$P(M \mid C) = \frac{P(M \cap C)}{P(C)} = \frac{.05}{.054} = \frac{25}{27}$$

■■■■

Group Discussion

Let A and B be events in an experiment and suppose $P(A) \neq 0$. Suppose that in n trials the event A occurs m times, the event B occurs k times, and the events A and B occur together l times.

1. Explain why it makes good sense to call the ratio l/m the conditional relative frequency of the event B given the event A.
2. Show that the relative frequencies l/m, m/n, and l/n satisfy the equation

$$\frac{l}{m} = \frac{\frac{l}{n}}{\frac{m}{n}}$$

3. Explain why the result of part (2) suggests that Formula (2)

$$P(B \mid A) = \frac{P(A \cap B)}{P(A)} \qquad [P(A) \neq 0]$$

is plausible.

In certain problems, the probability of an event B occurring given that A has occurred, written $P(B \mid A)$, is known, and we wish to find the probability of A *and* B occurring. The solution to such a problem is facilitated by the use of the following formula:

Product Rule

$$P(A \cap B) = P(A) \cdot P(B \mid A) \qquad [\text{if } P(A) \neq 0] \qquad (3)$$

This formula is obtained from (2) by multiplying both sides of the equation by $P(A)$. We illustrate the use of the product rule in the next several examples.

EXAMPLE 4

There are 300 seniors in Jefferson High School, of which 140 are males. It is known that 80% of the males and 60% of the females have their driver's license. If a student is selected at random from this senior class, what is the probability that the student is:

a. A male and has a driver's license?
b. A female who does not have a driver's license?

SOLUTION ✓

a. Let M denote the event that the student is a male and let D denote the event that the student has a driver's license. Then,

$$P(M) = \frac{140}{300} \qquad \text{and} \qquad P(D \mid M) = .8$$

Now, the event that the student selected at random is a male and has a driver's license is $M \cap D$, and, by the product rule, the probability of this event occurring is given by

$$P(M \cap D) = P(M) \cdot P(D \mid M)$$
$$= \left(\frac{140}{300}\right)(.8) = \frac{28}{75}$$

b. Let F denote the event that the student is a female and let D be as before. Then D^c is the event that the student does not have a driver's license. We have

$$P(F) = \frac{160}{300} \quad \text{and} \quad P(D^c \mid F) = 1 - .6 = .4$$

Note that we have used the rule of complements in the computation of $P(D^c \mid F)$. Now, the event that the student selected at random is a female and does not have a driver's license is $F \cap D^c$, so by the product rule, the probability of this event occurring is given by

$$P(F \cap D^c) = P(F) \cdot P(D^c \mid F)$$
$$= \left(\frac{160}{300}\right)(.4) = \frac{16}{75}$$

■■■■

EXAMPLE 5 Two cards are drawn without replacement from a well-shuffled deck of 52 playing cards. What is the probability that the first card drawn is an ace and the second card drawn is a face card?

SOLUTION ✓ Let A denote the event that the first card drawn is an ace and let F denote the event that the second card drawn is a face card. Then $P(A) = 4/52$. After drawing the first card, there are 51 cards left in the deck, of which 12 are face cards. Therefore, the probability of drawing a face card given that the first card drawn was an ace is given by

$$P(F \mid A) = \frac{12}{51}$$

By the product rule, the probability that the first card drawn is an ace and the second card drawn is a face card is given by

$$P(A \cap F) = P(A) \cdot P(F \mid A)$$
$$= \left(\frac{4}{52}\right)\left(\frac{12}{51}\right) = \frac{4}{221}$$

■■■■

Group Discussion

The product rule can be extended to the case involving three or more events. For example, if A, B, and C are three events in an experiment, then it can be shown that

$$P(A \cap B \cap C) = P(A) \cdot P(B \mid A) \cdot P(C \mid A \cap B)$$

1. Explain the formula in words.
2. Suppose 3 cards are drawn without replacement from a well-shuffled deck of 52 playing cards. Use the given formula to find the probability that the 3 cards are aces.

The product rule may be generalized to the case involving any finite number of events. For example, in the case involving the three events E, F, and G, it may be shown that

$$P(E \cap F \cap G) = P(E) \cdot P(F \mid E) \cdot P(G \mid E \cap F) \quad \textbf{(4)}$$

More on Tree Diagrams

Formula (4) and its generalizations may be used to help us solve problems that involve finite stochastic processes. More specifically, a **finite stochastic process** is an experiment consisting of a finite number of stages in which the outcomes and associated probabilities of each stage depend on the outcomes and associated probabilities of the preceding stages.

We can use tree diagrams to help us solve problems involving finite stochastic processes. Consider, for example, the experiment consisting of drawing 2 cards without replacement from a well-shuffled deck of 52 playing cards. What is the probability that the second card drawn is a face card?

We may think of this experiment as a stochastic process with two stages. The events associated with the first stage are F, that the card drawn is a face card, and F^c, that the card drawn is not a face card. Since there are 12 face cards, we have

$$P(F) = \frac{12}{52} \quad \text{and} \quad P(F^c) = 1 - \frac{12}{52} = \frac{40}{52}$$

FIGURE 7.16
F is the probability that a face card is drawn.

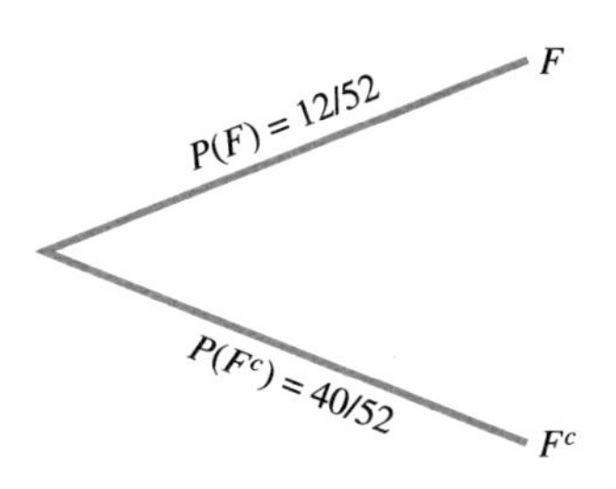

The outcomes of this trial, together with the associated probabilities, may be represented along two "branches" of a tree diagram, as shown in Figure 7.16.

In the second trial, we again have two events: G, that the card drawn is a face card, and G^c, that the card drawn is not a face card. But the outcome of the second trial depends on the outcome of the first trial. For example, if the first card drawn was a face card, then the event G that the second card drawn is a face card has probability given by the *conditional probability*

$P(G \mid F)$. Since the occurrence of a face card in the first draw leaves 11 face cards in a deck of 51 cards for the second draw, we see that

$$P(G \mid F) = \frac{11}{51}$$ (The probability of drawing a face card given that a face card has already been drawn)

Similarly, the occurrence of a face card in the first draw leaves 40 that are other than face cards in a deck of 51 cards for the second draw. Therefore, the probability of drawing other than a face card in the second draw given that the first card drawn is a face card is

$$P(G^c \mid F) = \frac{40}{51}$$

Using these results, we extend the tree diagram of Figure 7.16 by displaying another two branches of the tree growing from its upper branch (Figure 7.17).

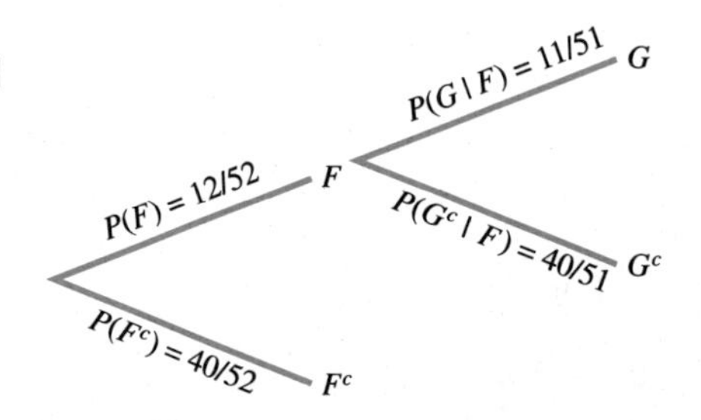

FIGURE 7.17
G is the probability that the second card drawn is a face card.

To complete the tree diagram, we compute $P(G \mid F^c)$ and $P(G^c \mid F^c)$, the conditional probabilities that the second card drawn is a face card and other than a face card, respectively, given that the first card drawn is not a face card. We find that

$$P(G \mid F^c) = \frac{12}{51} \quad \text{and} \quad P(G^c \mid F^c) = \frac{39}{51}$$

This leads to the completion of the tree diagram, shown in Figure 7.18, where the branches of the tree that lead to the two outcomes of interest have been highlighted.

Having constructed the tree diagram associated with the problem, we are now in a position to answer the question posed earlier—namely, "What is the probability of the second card being a face card?" Observe that Figure 7.18 shows the two ways in which a face card may result in the second draw—namely, the two Gs on the extreme right of the diagram.

Now, by the product rule, the probability that the second card drawn is a face card and the first card drawn is a face card (this is represented by the upper branch) is

$$P(G \cap F) = P(F) \cdot P(G \mid F)$$

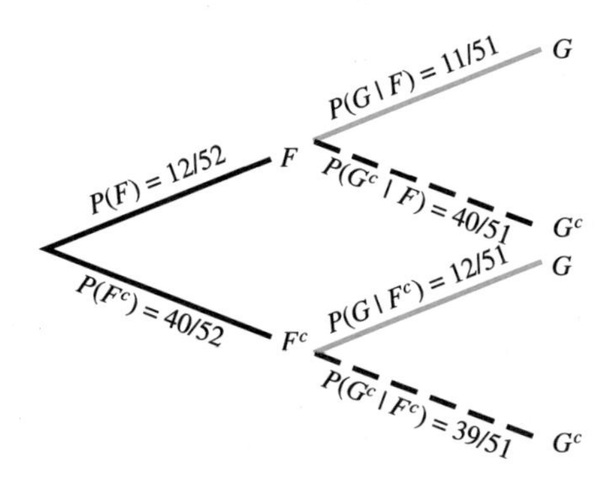

FIGURE 7.18
Tree diagram showing the two trials of the experiment

Similarly, the probability that the second card drawn is a face card and the first card drawn is other than a face card (this corresponds to the other branch) is

$$P(G \cap F^c) = P(F^c) \cdot P(G \mid F^c)$$

Observe that each of these probabilities is obtained by taking the *product of the probabilities appearing on the respective branch.* Since $G \cap F$ and $G \cap F^c$ are mutually exclusive events (why?), the probability that the second card drawn is a face card is given by

$$P(G \cap F) + P(G \cap F^c) = P(F) \cdot P(G \mid F) + P(F^c) \cdot P(G \mid F^c)$$

or, upon replacing the probabilities on the right of the expression by their numerical values,

$$\begin{aligned} P(G \cap F) + P(G \cap F^c) &= \left(\frac{12}{52}\right)\left(\frac{11}{51}\right) + \left(\frac{40}{52}\right)\left(\frac{12}{51}\right) \\ &= \frac{3}{13} \end{aligned}$$

EXAMPLE 6

The picture tubes for the Pulsar 19-inch color television sets are manufactured in three locations and then shipped to the main plant of the Vista Vision Corporation for final assembly. Plants A, B, and C supply 50%, 30%, and 20%, respectively, of the picture tubes used by the company. The quality-control department of the company has determined that 1% of the picture tubes produced by plant A are defective, whereas 2% of the picture tubes produced by plants B and C are defective. What is the probability that a randomly selected Pulsar 19-inch color television set will have a defective picture tube?

SOLUTION ✓

Let A, B, and C denote the events that the set chosen has a picture tube manufactured in plant A, plant B, and plant C, respectively. Also, let D denote the event that a set has a defective picture tube. Using the given information, we draw the tree diagram shown in Figure 7.19.

FIGURE 7.19
Tree diagram showing the probabilities of producing defective picture tubes at each plant

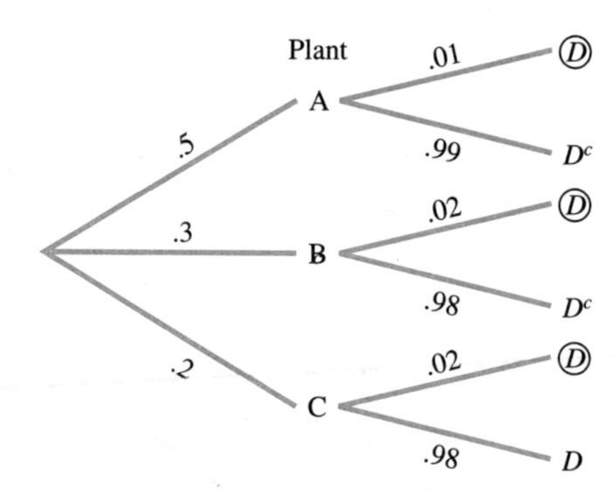

(The events that result in a set with a defective picture tube being selected are circled.) Taking the product of the probabilities along each branch leading to such an event and adding them yields the probability that a set chosen at random has a defective picture tube. Thus, the required probability is given by

$$\begin{aligned}(.5)(.01) + (.3)(.02) + (.2)(.02) &= .005 + .006 + .004 \\ &= .015\end{aligned}$$

■■■■

EXAMPLE 7

A box contains eight 9-volt transistor batteries, of which two are known to be defective. The batteries are selected one at a time without replacement and tested until a nondefective one is found. What is the probability that the number of batteries tested is (a) one, (b) two, and (c) three?

SOLUTION ✓

We may view this experiment as a multistage process with up to three stages. In the first stage, a battery is selected with a probability of 6/8 of its being nondefective and a probability of 2/8 of its being defective. If the battery selected is good, the experiment is terminated. Otherwise, a second battery is selected with probabilities of 6/7 and 1/7, respectively, of its being nondefective and defective. If the second battery selected is good, the experiment is terminated. Otherwise, a third battery is selected with probabilities of 1 and 0, respectively, of its being nondefective and defective. The tree diagram associated with this experiment is shown in Figure 7.20, where N denotes the event that the battery selected is nondefective and D denotes the event that the battery selected is defective.

FIGURE 7.20
In this experiment, batteries are selected until a nondefective one is found.

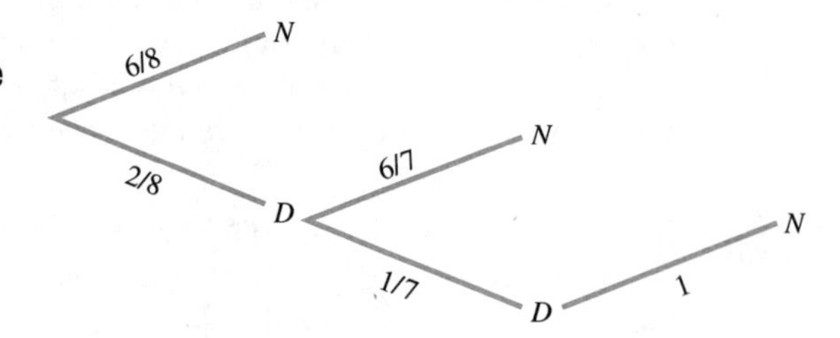

With the aid of the tree diagram we see that (a) the probability that only one battery is selected is 6/8, (b) the probability that two batteries are selected is (2/8)(6/7), or 3/14, and (c) the probability that three batteries are selected is (2/8)(1/7)(1) = 1/28. ■■■■

INDEPENDENT EVENTS

Let's return to the experiment of drawing 2 cards in succession without replacement from a well-shuffled deck of 52 playing cards considered in Example 5. Let E denote the event that the first card drawn is not a face card and let F denote the event that the second card drawn is a face card. It is intuitively clear that the events E and F are *not* independent of each other since whether or not the first card drawn is a face card affects the likelihood that the second card drawn is a face card.

Next, let's consider the experiment of tossing a coin twice and observing the outcomes: If H denotes the event that the first toss produces "heads" and T denotes the event that the second toss produces "tails," then it is intuitively clear that H and T *are* independent of each other since the outcome of the first toss does not affect the outcome of the second.

In general, two events A and B are **independent** if the outcome of one does not affect the outcome of the other. Thus, if A and B are independent events, then

$$P(A \mid B) = P(A) \qquad \text{or} \qquad P(B \mid A) = P(B)$$

Using the product rule, we can find a simple test to determine the independence of two events. Suppose A and B are independent and $P(A) \neq 0$ and $P(B) \neq 0$. Then,

$$P(B \mid A) = P(B)$$

Thus, by the product rule, we have

$$P(A \cap B) = P(A) \cdot P(B \mid A) = P(A) \cdot P(B)$$

Conversely, if this equation holds, then it can be seen that $P(B \mid A) = P(B)$; that is, A and B are independent. Accordingly, we have the following test for the independence of two events.

Test for the Independence of Two Events

Two events A and B are independent if and only if

$$P(A \cap B) = P(A) \cdot P(B) \tag{5}$$

Do not confuse *independent* events with *mutually exclusive* events. The former pertains to how the occurrence of one event affects the occurrence of another event, whereas the latter pertains to the question of whether the events can occur at the same time.

EXAMPLE 8

Consider the experiment consisting of tossing a fair coin twice and observing the outcomes. Show that the event of "heads" in the first toss and "tails" in the second toss are independent events.

SOLUTION ✓

Let A denote the event that the outcome of the first toss is a *head* and let B denote the event that the outcome of the second toss is a *tail.* The sample space of the experiment is

$$S = \{(\mathrm{H}, \mathrm{H}), (\mathrm{H}, \mathrm{T}), (\mathrm{T}, \mathrm{H}), (\mathrm{T}, \mathrm{T})\}$$
$$A = \{(\mathrm{H}, \mathrm{H}), (\mathrm{H}, \mathrm{T})\}$$
$$B = \{(\mathrm{H}, \mathrm{T}), (\mathrm{T}, \mathrm{T})\}$$

so that

$$A \cap B = \{(\mathrm{H}, \mathrm{T})\}$$

Next, we compute

$$P(A \cap B) = \frac{1}{4}, \qquad P(A) = \frac{1}{2}, \qquad P(B) = \frac{1}{2}$$

and observe that Equation (5) is satisfied in this case so that A and B are independent events, as we set out to show. ■■■■

EXAMPLE 9

A survey conducted by an independent agency for the National Lung Society found that of 2000 women, 680 were heavy smokers and 50 had emphysema. Of those who had emphysema, 42 were also heavy smokers. Using the data in this survey, determine whether the events "being a heavy smoker" and "having emphysema" are independent events.

SOLUTION ✓

Let A denote the event that a woman is a heavy smoker and let B denote the event that a woman has emphysema. Then, the probability that a woman is a heavy smoker and has emphysema is given by

$$P(A \cap B) = \frac{42}{2000} = .021$$

Next,

$$P(A) = \frac{680}{2000} = .34 \qquad \text{and} \qquad P(B) = \frac{50}{2000} = .025$$

so that

$$P(A) \cdot P(B) = (.34)(.025) = .0085$$

Since $P(A \cap B) \neq P(A) \cdot P(B)$, we conclude that A and B are not independent events. ■■■■

Group Discussion

Let E and F be independent events in a sample space S. Are E^c and F^c independent?

The solution of many practical problems involves more than two independent events. In such cases we use the following result.

Independence of More Than Two Events

If $E_1, E_2, \ldots, E_n$ are independent events, then

$$P(E_1 \cap E_2 \cap \cdots \cap E_n) = P(E_1) \cdot P(E_2) \cdot \cdots \cdot P(E_n) \qquad (6)$$

Formula (6) states that the probability of the simultaneous occurrence of n independent events is equal to the product of the probabilities of the n events.

It is important to note that the mere requirement that the n events $E_1, E_2, \ldots, E_n$ satisfy (6) is not sufficient to guarantee that the n events are indeed independent. However, a criterion does exist for determining the independence of n events and may be found in more advanced texts on probability.

EXAMPLE 10

It is known that the three events A, B, and C are independent and $P(A) = .2$, $P(B) = .4$, and $P(C) = .5$. Compute:

a. $P(A \cap B)$ **b.** $P(A \cap B \cap C)$

SOLUTION ✓

Using Formulas (5) and (6), we find

a. $P(A \cap B) = P(A) \cdot P(B)$
$= (.2)(.4) = .08$

b. $P(A \cap B \cap C) = P(A) \cdot P(B) \cdot P(C)$
$= (.2)(.4)(.5) = .04$

EXAMPLE 11

The Acrosonic model F loudspeaker system has four loudspeaker components: a woofer, a midrange, a tweeter, and an electrical crossover. The quality-control manager of Acrosonic has determined that on the average 1% of the woofers, 0.8% of the midranges, and 0.5% of the tweeters are defective, while 1.5% of the electrical crossovers are defective. Determine the probability that an Acrosonic model F loudspeaker system selected at random coming off the assembly line and before final inspection is not defective. Assume that the defects in the manufacturing of the components are unrelated.

SOLUTION ✓

Let A, B, C, and D denote, respectively, the events that the woofer, the midrange, the tweeter, and the electrical crossover are defective. Then,

$$P(A) = .01, \quad P(B) = .008, \quad P(C) = .005, \quad P(D) = .015$$

and the probabilities of the corresponding complementary events are

$$P(A^c) = .99, \quad P(B^c) = .992, \quad P(C^c) = .995, \quad P(D^c) = .985$$

The event that a loudspeaker system selected at random is not defective is given by $A^c \cap B^c \cap C^c \cap D^c$, and since the events A, B, C, and D (and therefore also A^c, B^c, C^c, and D^c) are assumed to be independent, we find that the required probability is given by

$$\begin{aligned} P(A^c \cap B^c \cap C^c \cap D^c) &= P(A^c) \cdot P(B^c) \cdot P(C^c) \cdot P(D^c) \\ &= (.99)(.992)(.995)(.985) \\ &\approx .96 \end{aligned}$$

SELF-CHECK EXERCISES 7.5

1. Let A and B be events in a sample space S such that $P(A) = .4$, $P(B) = .8$, and $P(A \cap B) = .3$. Find:
   a. $P(A \mid B)$ **b.** $P(B \mid A)$
2. Three friends—Alice, Betty, and Cathy—are unmarried and are 30, 35, and 40 years old, respectively. While they were having tea one afternoon, Alice recalled reading an article by Yale sociologist Neil Bennett in which he concluded that women who are still unmarried at age 30 have only a 20% chance of marrying, those who are still unmarried at 35 have a 5.4% chance, and those who are still unmarried at 40 have a 1.3% chance. Betty wondered what the probability was that all three of them would eventually "tie the knot." Cathy, a statistician, pulled out her pocket calculator and answered Betty's question. What was her answer?

Solutions to Self-Check Exercises 7.5 can be found on page 440.

7.5 Exercises

1. Let A and B be events in a sample space S such that $P(A) = .6$, $P(B) = .5$, and $P(A \cap B) = .2$. Find:
   a. $P(A \mid B)$ **b.** $P(B \mid A)$
2. Let A and B be two events in a sample space S such that $P(A) = .4$, $P(B) = .6$, and $P(A \cap B) = .3$. Find:
   a. $P(A \mid B)$ **b.** $P(B \mid A)$
3. Let A and B be two events in a sample space S such that $P(A) = .6$ and $P(B \mid A) = .5$. Find $P(A \cap B)$.
4. Let A and B be the events described in Exercise 1. Find:
   a. $P(A \mid B^c)$ **b.** $P(B \mid A^c)$
   Hint: $(A \cap B^c) \cup (A \cap B) = A$.

In Exercises 5–8, determine whether the given events A and B are independent.

5. $P(A) = .3$, $P(B) = .6$, $P(A \cap B) = .18$
6. $P(A) = .6$, $P(B) = .8$, $P(A \cap B) = .2$
7. $P(A) = .5$, $P(B) = .7$, $P(A \cup B) = .85$
8. $P(A^c) = .3$, $P(B^c) = .4$, $P(A \cap B) = .42$
9. If A and B are independent events and $P(A) = .4$ and $P(B) = .6$, find:
   a. $P(A \cap B)$ **b.** $P(A \cup B)$
10. If A and B are independent events and $P(A) = .35$ and $P(B) = .45$, find:
    a. $P(A \cap B)$ **b.** $P(A \cup B)$
11. The accompanying tree diagram represents an experiment consisting of two trials:

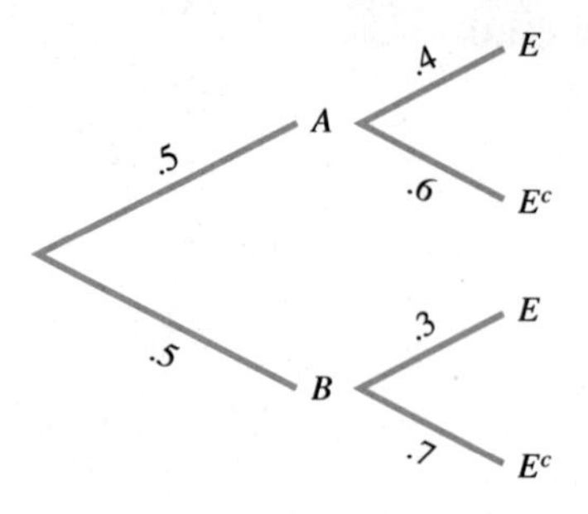

Use the diagram to find:

a. $P(A)$ **b.** $P(E \mid A)$
c. $P(A \cap E)$ **d.** $P(E)$
e. Does $P(A \cap E) = P(A) \cdot P(E)$?
f. Are A and E independent events?

12. The accompanying tree diagram represents an experiment consisting of two trials. Use the diagram to find:

a. $P(A)$ **b.** $P(E \mid A)$
c. $P(A \cap E)$ **d.** $P(E)$
e. Does $P(A \cap E) = P(A) \cdot P(E)$?
f. Are A and E independent events?

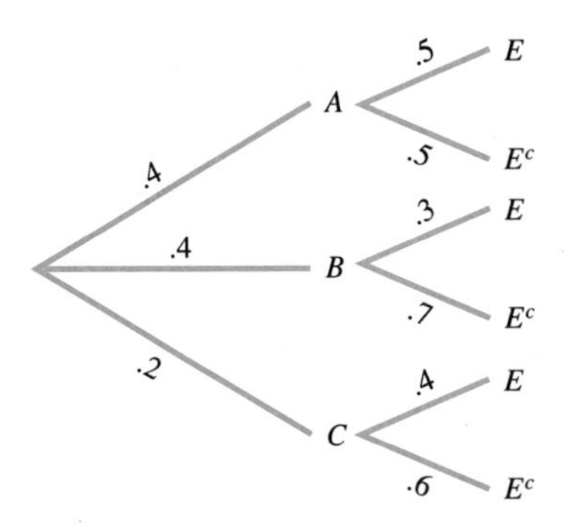

13. An experiment consists of two trials. The outcomes of the first trial are A and B with probabilities of occurring of .4 and .6. There are also two outcomes, C and D, in the second trial with probabilities of .3 and .7. Draw a tree diagram representing this experiment. Use this diagram to find:

a. $P(A)$ **b.** $P(C \mid A)$
c. $P(A \cap C)$ **d.** $P(C)$
e. Does $P(A \cap C) = P(A) \cdot P(C)$?
f. Are A and C independent events?

14. An experiment consists of two trials. The outcomes of the first trial are A, B, and C, with probabilities of occurring of .2, .5, and .3, respectively. The outcomes of the second trial are E and F, with probabilities of occurring of .6 and .4. Draw a tree diagram representing this experiment. Use this diagram to find:

a. $P(B)$ **b.** $P(F \mid B)$
c. $P(B \cap F)$ **d.** $P(F)$
e. Does $P(B \cap F) = P(B) \cdot P(F)$?
f. Are B and F independent events?

15. A pair of fair dice is cast. What is the probability that the sum of the numbers falling uppermost is less than 9, if it is known that one of the numbers is a 6?

16. A pair of fair dice is cast. What is the probability that the number landing uppermost on the first die is a 4, if it is known that the sum of the numbers falling uppermost is 7?

17. A pair of fair dice is cast. Let E denote the event that the number landing uppermost on the first die is a 3 and let F denote the event that the sum of the numbers falling uppermost is 7. Determine whether E and F are independent events.

18. A pair of fair dice is cast. Let E denote the event that the number landing uppermost on the first die is a 3 and let F denote the event that the sum of the numbers landing uppermost is 6. Determine whether E and F are independent events.

19. PRODUCT RELIABILITY The probability that a battery will last 10 hr or more is .80, and the probability that it will last 15 hr or more is .15. Given that a battery has lasted 10 hr, find the probability that it will last 15 hr or more.

20. Two cards are drawn without replacement from a well-shuffled deck of 52 playing cards.

a. What is the probability that the first card drawn is a heart?
b. What is the probability that the second card drawn is a heart given that the first card drawn was not a heart?
c. What is the probability that the second card drawn is a heart given that the first card drawn was a heart?

21. Five black balls and four white balls are placed in an urn. Two balls are then drawn in succession. What is the probability that the second ball drawn is a white ball if:

a. The second ball is drawn without replacing the first?
b. The first ball is replaced before the second is drawn?

22. AUDITING TAX RETURNS A tax specialist has estimated the probability that a tax return selected at random will be audited is .02. Furthermore, he estimates the probability that an audited return will result in additional assessments being levied on the taxpayer is .60. What is the probability that a tax return selected at random will result in additional assessments being levied on the taxpayer?

23. STUDENT ENROLLMENT At a certain medical school, 1/7 of the students are from a minority group. Of those students who belong to a minority group, 1/3 are Black.

a. What is the probability that a student selected at random from this medical school is Black?
b. What is the probability that a student selected at random from this medical school is Black if it is known that the student is a member of a minority group?

24. **EDUCATIONAL LEVEL OF VOTERS** In a survey of 1000 eligible voters selected at random, it was found that 80 had a college degree. Additionally, it was found that 80% of those who had a college degree voted in the last presidential election, whereas 55% of the people who did not have a college degree voted in the last presidential election. Assuming that the poll is representative of all eligible voters, find the probability that an eligible voter selected at random:
a. Had a college degree and voted in the last presidential election.
b. Did not have a college degree and did not vote in the last presidential election.
c. Voted in the last presidential election.
d. Did not vote in the last presidential election.

25. Three cards are drawn without replacement from a well-shuffled deck of 52 playing cards. What is the probability that the third card drawn is a diamond?

26. A coin is tossed three times. What is the probability that the coin will land heads:
a. At least twice?
b. On the second toss given that heads were thrown on the first toss?
c. On the third toss given that tails were thrown on the first toss?

27. In a three-child family, what is the probability that all three children are girls given that one of the children is a girl? (Assume that the probability of a boy being born is the same as the probability of a girl being born.)

28. **QUALITY CONTROL** An automobile manufacturer obtains the microprocessors used to regulate fuel consumption in its automobiles from three microelectronic firms: A, B, and C. The quality-control department of the company has determined that 1% of the microprocessors produced by firm A are defective, 2% of those produced by firm B are defective, and 1.5% of those produced by firm C are defective. Firms A, B, and C supply 45%, 25%, and 30%, respectively, of the microprocessors used by the company. What is the probability that a randomly selected automobile manufactured by the company will have a defective microprocessor?

29. **CAR THEFT** Figures obtained from a city's police department seem to indicate that, of all motor vehicles reported as stolen, 64% were stolen by professionals whereas 36% were stolen by amateurs (primarily for joy rides). Of those vehicles presumed stolen by professionals, 24% were recovered within 48 hr, 16% were recovered after 48 hr, and 60% were never recovered. Of those vehicles presumed stolen by amateurs, 38% were recovered within 48 hr, 58% were recovered after 48 hr, and 4% were never recovered.
a. Draw a tree diagram representing these data.
b. What is the probability that a vehicle stolen by a professional in this city will be recovered within 48 hours?
c. What is the probability that a vehicle stolen in this city will never be recovered?

30. **HOUSING LOANS** The chief loan officer of the La Crosse Home Mortgage Company summarized the housing loans extended by the company in 1993 according to type and term of the loan. Her list shows that 70% of the loans were fixed-rate mortgages (F), 25% were adjustable-rate mortgages (A), and 5% belong to some other category (O) (mostly second trust-deed loans and loans extended under the graduated payment plan). Of the fixed-rate mortgages, 80% were 30-yr loans and 20% were 15-yr loans; of the adjustable-rate mortgages, 40% were 30-yr loans and 60% were 15-yr loans; finally, of the other loans extended, 30% were 20-yr loans, 60% were 10-yr loans, and 10% were for a term of 5 yr or less.
a. Draw a tree diagram representing this experiment.
b. What is the probability that a home loan extended by La Crosse has an adjustable rate and is for a term of 15 yr?
c. What is the probability that a home loan extended by La Crosse is for a term of 15 yr?

31. **COLLEGE ADMISSIONS** The admissions office of a private university released the following admission data for the preceding academic year: From a pool of 3900 male applicants, 40% were accepted by the university, and of these, 40% subsequently enrolled. Additionally, from a pool of 3600 female applicants, 45% were accepted by the university, and of these, 40% subsequently enrolled. What is the probability that:
a. A male applicant will be accepted by and subsequently will enroll in the university?
b. A student who applies for admissions will be accepted by the university?
c. A student who applies for admission will be accepted by the university and subsequently will enroll?

32. **QUALITY CONTROL** Suppose a box contains two defective Christmas tree lights that have been inadvertently mixed with eight nondefective lights. If the lights are selected one at a time without replacement and tested until both defective lights are found, what is the probability that both defective lights will be found after three trials?

33. **QUALITY CONTROL** It is estimated that 0.80% of a large consignment of eggs in a certain supermarket is broken.
a. What is the probability that a customer who randomly selects a dozen of these eggs receives at least one broken egg?
b. What is the probability that a customer who selects these eggs at random will have to check three cartons before finding a carton without any broken eggs? (Each carton contains a dozen eggs.)

34. **STUDENT FINANCIAL AID** The accompanying data were obtained from the financial aid office of a certain university:

	Receiving Financial Aid	Not Receiving Financial Aid	Total
Undergraduates	4,222	3,898	8,120
Graduates	1,879	731	2,610
Total	6,101	4,629	10,730

Let A be the event that a student selected at random from this university is an undergraduate student and let B be the event that a student selected at random is receiving financial aid.
a. Find each of the following probabilities: $P(A)$, $P(B)$, $P(A \cap B)$, $P(B \mid A)$, and $P(B \mid A^c)$.
b. Are the events A and B independent events?

35. **EMPLOYEE EDUCATION AND INCOME** The personnel department of the Franklin National Life Insurance Company compiled the accompanying data regarding the income and education of its employees:

	Income \$40,000 or Below	Income Above \$40,000
Noncollege Graduate	2040	840
College Graduate	400	720

Let A be the event that a randomly chosen employee has a college degree and B the event that the chosen employee's income is more than \$40,000.
a. Find each of the following probabilities: $P(A)$, $P(B)$, $P(A \cap B)$, $P(B \mid A)$, and $P(B \mid A^c)$.
b. Are the events A and B independent events?

36. Two cards are drawn without replacement from a well-shuffled deck of 52 cards. Let A be the event that the first card drawn is a heart and let B be the event that the second card drawn is a red card. Show that the events A and B are dependent events.

37. **MEDICAL RESEARCH** A nationwide survey conducted by the National Cancer Society revealed the following information: Of 10,000 people surveyed, 3200 were "heavy coffee drinkers" and 160 had cancer of the pancreas. Of those who had cancer of the pancreas, 132 were heavy coffee drinkers. Using the data in this survey, determine whether the events "being a heavy coffee drinker" and "having cancer of the pancreas" are independent events.

38. **MAIL DELIVERY** Suppose the probability that your mail will be delivered before 2 P.M. on a delivery day is .90. What is the probability that your mail will be delivered before 2 P.M. for:
a. Two consecutive delivery days?
b. Three consecutive delivery days?

39. **RELIABILITY OF SECURITY SYSTEMS** Before being allowed to enter a maximum-security area at a military installation, a person must pass three identification tests: a voice-pattern test, a fingerprint test, and a handwriting test. If the reliability of the first test is 97%, the reliability of the second test is 98.5%, and that of the third is 98.5%, what is the probability that this security system will allow an improperly identified person to enter the maximum-security area?

40. **RELIABILITY OF A HOME THEATER SYSTEM** In a home theater system, the probability that the video component needs repair within 1 yr is .01, the probability that the electronic components need repair within 1 yr is .005, and the probability that the audio component needs repair within 1 yr is .001. Assuming the probabilities are independent, find the probability that:
a. At least one component will need repair within 1 yr.
b. Exactly one component will need repair within 1 yr.

41. **PROBABILITY OF TRANSPLANT REJECTION** The independent probabilities that the three patients who are scheduled to receive kidney transplants at General Hospital will suffer rejection are 1/2, 1/3, and 1/10. Find the probability that:
a. At least one patient will suffer rejection.
b. Exactly two patients will suffer rejection.

42. **QUALITY CONTROL** Copykwik, Inc., has four photocopy machines A, B, C, and D. The probability that a given machine will break down on a particular day is

$$P(A) = \frac{1}{50}, \quad P(B) = \frac{1}{60}, \quad P(C) = \frac{1}{75}, \quad P(D) = \frac{1}{40}$$

Assuming independence, what is the probability on a particular day that:
a. All four machines will break down?
b. None of the machines will break down?
c. Exactly one machine will break down?

43. **Product Reliability** The proprietor of Cunningham's Hardware Store has decided to install floodlights on the premises as a measure against vandalism and theft. If the probability is .01 that a certain brand of floodlight will burn out within a year, find the minimum number of floodlights that must be installed to ensure that the probability that at least one of them will remain functional within the year is at least .99999. (Assume that the floodlights operate independently.)

44. Let E be any event in a sample space S.
 a. Are E and S independent? Explain your answer.
 b. Are E and $\varnothing$ independent? Explain your answer.

45. Suppose the probability that an event will occur in one trial is p. Show that the probability that the event will occur at least once in n independent trials is $1 - (1 - p)^n$.

46. Suppose A and B are mutually exclusive events and $P(A \cup B) \neq 0$. What is $P(A \mid A \cup B)$?

In Exercises 47–49, determine whether the statement is true or false. If it is true, explain why it is true. If it is false, give an example to show why it is false.

47. If A and B are mutually exclusive and $P(B) \neq 0$, then $P(A \mid B) = 0$.

48. If A is an event of an experiment, then $P(A \mid A^c) \neq 0$.

49. If A and B are mutually exclusive and $P(A \cup B) \neq 0$, then

$$P(A \mid A \cup B) = \frac{P(A)}{P(A) + P(B)}$$

Solutions to Self-Check Exercises 7.5

1. a. $P(A \mid B) = \dfrac{P(A \cap B)}{P(B)} = \dfrac{.3}{.8} = \dfrac{3}{8}$

 b. $P(B \mid A) = \dfrac{P(A \cap B)}{P(A)} = \dfrac{.3}{.4} = \dfrac{3}{4}$

2. Let A, B, and C denote the events that three women who are still unmarried at ages 30, 35, and 40 *will* marry eventually. Then, it seems reasonable to assume that these events are independent, with $P(A) = .20$, $P(B) = .054$, and $P(C) = .013$. Based on these figures, the probability that all three women will be married eventually is given by

$$\begin{aligned} P(A)P(B)P(C) &= (.2)(.054)(.013) \\ &= .00014 \end{aligned}$$

7.6 Bayes' Theorem

A Posteriori Probabilities

Suppose three machines, A, B, and C, produce similar engine components. Machine A produces 45% of the total components, machine B produces 30%, and machine C, 25%. For the usual production schedule, 6% of the components produced by machine A do not meet established specifications; for machine B and machine C, the corresponding figures are 4% and 3%. One component is selected at random from the total output and is found to be defective. What is the probability that the component selected was produced by machine A?

FIGURE 7.21
D is the event that a defective component is produced by machine A, machine B, or machine C.

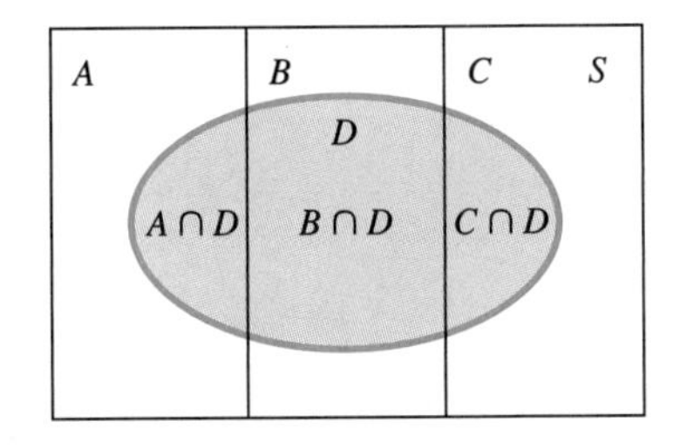

The answer to this question is found by calculating the probability *after* the outcomes of the experiment have been observed. Such probabilities are called **a posteriori probabilities** as opposed to **a priori probabilities**—probabilities that give the likelihood that an event *will* occur, the subject of the last several sections.

Returning to the example under consideration, we need to determine the a posteriori probability for the event that the component selected was produced by machine A. To this end, let A, B, and C denote the event that a component is produced by machine A, machine B, and machine C, respectively. We may represent this experiment with a Venn diagram (Figure 7.21).

The three mutually exclusive events A, B, and C form a **partition** of the sample space S. That is, aside from being mutually exclusive, their union is precisely S. The event D that a component is defective is the shaded area. Again referring to Figure 7.21, we see that

1. The event D may be expressed as

$$D = (A \cap D) \cup (B \cap D) \cup (C \cap D)$$

2. The event that a component is defective and is produced by machine A is given by $A \cap D$.

Thus, the a posteriori probability that a defective component selected was produced by machine A is given by

$$P(A \mid D) = \frac{n(A \cap D)}{n(D)}$$

Upon dividing both the numerator and the denominator by $n(S)$ and observing that the events $A \cap D$, $B \cap D$, and $C \cap D$ are mutually exclusive, we obtain

$$\begin{aligned} P(A \mid D) &= \frac{P(A \cap D)}{P(D)} \\ &= \frac{P(A \cap D)}{P(A \cap D) + P(B \cap D) + P(C \cap D)} \end{aligned} \tag{7}$$

Next, using the product rule, we may express

$$\begin{aligned} P(A \cap D) &= P(A) \cdot P(D \mid A) \\ P(B \cap D) &= P(B) \cdot P(D \mid B) \\ P(C \cap D) &= P(C) \cdot P(D \mid C) \end{aligned}$$

so that Equation (7) may be expressed in the form

$$P(A \mid D) = \frac{P(A) \cdot P(D \mid A)}{P(A) \cdot P(D \mid A) + P(B) \cdot P(D \mid B) + P(C) \cdot P(D \mid C)} \tag{8}$$

which is a special case of a result known as **Bayes' theorem.**

Observe that the expression on the right of (8) involves the probabilities $P(A)$, $P(B)$, $P(C)$ and the conditional probabilities $P(D \mid A)$, $P(D \mid B)$, and $P(D \mid C)$, all of which may be calculated in the usual fashion. In fact, by

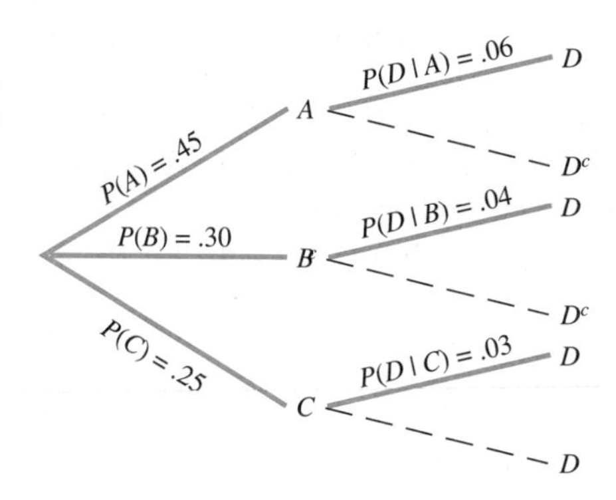

FIGURE 7.22
A tree diagram displaying the probabilities that a defective component is produced by machine A, machine B, or machine C

displaying these quantities on a tree diagram, we obtain Figure 7.22. We may compute the required probability by substituting the relevant quantities into (8), or we may make use of the following device:

$$P(A \mid D) = \frac{\text{Product of probabilities along the limb through } A}{\text{Sum of products of the probabilities along each limb terminating at } D}$$

In either case, we obtain

$$\begin{aligned} P(A \mid D) &= \frac{(.45)(.06)}{(.45)(.06) + (.3)(.04) + (.25)(.03)} \\ &= .58 \end{aligned}$$

Before looking at any further examples, let's state the general form of Bayes' theorem.

BAYES' THEOREM

Let $A_1, A_2, \ldots, A_n$ be a partition of a sample space S and let E be an event of the experiment such that $P(E) \neq 0$. Then the a posteriori probability $P(A_i \mid E)$ $(1 \leq i \leq n)$ is given by

$$P(A_i \mid E) = \frac{P(A_i) \cdot P(E \mid A_i)}{P(A_1) \cdot P(E \mid A_1) + P(A_2) \cdot P(E \mid A_2) + \cdots + P(A_n) \cdot P(E \mid A_n)} \tag{9}$$

APPLICATIONS

EXAMPLE 1

The picture tubes for the Pulsar 19-inch color television sets are manufactured in three locations and then shipped to the main plant of the Vista Vision Corporation for final assembly. Plants A, B, and C supply 50%, 30%, and 20%, respectively, of the picture tubes used by Vista Vision. The quality-control department of the company has determined that 1% of the picture tubes produced by plant A are defective, whereas 2% of the picture tubes produced by plants B and C are defective. If a Pulsar 19-inch color television set is selected at random and the picture tube is found to be defective, what is the probability that the picture tube was manufactured in plant C? Compare with Example 6, page 431.

SOLUTION ✓

Let A, B, and C denote the event that the set chosen has a picture tube manufactured in plant A, plant B, and plant C, respectively. Also, let D denote the event that a set has a defective picture tube. Using the given information,

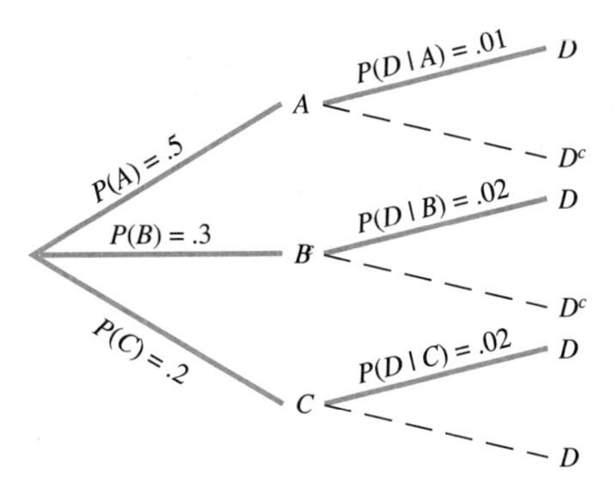

FIGURE 7.23
$P(C \mid D) = \dfrac{\text{Product of probabilities of branches to } D \text{ through } C}{\text{Sum of product of probabilities of branches leading to } D}$

we may draw the tree diagram shown in Figure 7.23. Next, using formula (9), we find that the required a posteriori probability is given by

$$
\begin{aligned}
P(C \mid D) &= \frac{P(C) \cdot P(D \mid C)}{P(A) \cdot P(D \mid A) + P(B) \cdot P(D \mid B) + P(C) \cdot P(D \mid C)} \\
&= \frac{(.2)(.02)}{(.5)(.01) + (.3)(.02) + (.2)(.02)} \\
&\approx .27
\end{aligned}
$$

■■■■

EXAMPLE 2

A study was conducted in a large metropolitan area to determine the annual incomes of married couples in which the husbands were the sole providers and of those in which the husbands and wives were both employed. Table 7.9 gives the results of this study.

Table 7.9

Annual Family Income, $	Percentage of Married Couples	Percentage of Income Group with Both Spouses Working
125,000 and over	4	65
100,000–124,999	10	73
75,000–99,999	21	68
50,000–74,999	24	63
30,000–49,999	30	43
Under 30,000	11	28

a. What is the probability that a couple selected at random from this area has two incomes?
b. If a randomly chosen couple has two incomes, what is the probability that the annual income of this couple is over \$125,000?
c. If a randomly chosen couple has two incomes, what is the probability that the annual income of this couple is greater than \$49,999?

SOLUTION ✓ Let A denote the event that the annual income of the couple is \$125,000 and over; let B denote the event that the annual income is between \$100,000 and \$124,999; let C denote the event that the annual income is between \$75,000 and \$99,999; and so on. Finally, let F denote the event that the annual income is less than \$30,000 and let T denote the event that both spouses work. The probabilities of the occurrence of these events are displayed in Figure 7.24.

FIGURE 7.24

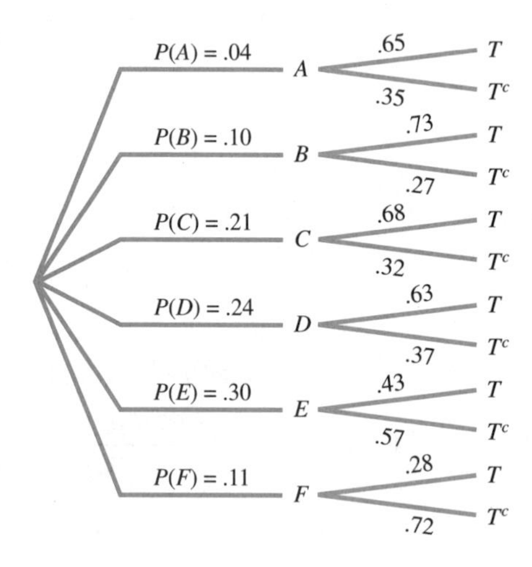

a. The probability that a couple selected at random from this group has two incomes is given by

$$\begin{aligned} P(T) &= P(A) \cdot P(T \mid A) + P(B) \cdot P(T \mid B) + P(C) \cdot P(T \mid C) \\ &\quad + P(D) \cdot P(T \mid D) + P(E) \cdot P(T \mid E) + P(F) \cdot P(T \mid F) \\ &= (.04)(.65) + (.10)(.73) + (.21)(.68) + (.24)(.63) \\ &\quad + (.30)(.43) + (.11)(.28) \\ &= .5528 \end{aligned}$$

b. Using the results of part (a) and Bayes' theorem, we find that the probability that a randomly chosen couple has an annual income over \$125,000 given that both spouses are working is

$$P(A \mid T) = \frac{P(A) \cdot P(T \mid A)}{P(T)} = \frac{(.04)(.65)}{.5528}$$
$$= .047$$

c. The probability that a randomly chosen couple has an annual income greater than $49,999 given that both spouses are working is

$$\begin{aligned} &P(A \mid T) + P(B \mid T) + P(C \mid T) + P(D \mid T) \\ &= \frac{P(A) \cdot P(T \mid A) + P(B) \cdot P(T \mid B) + P(C) \cdot P(T \mid C) + P(D) \cdot P(T \mid D)}{P(T)} \\ &= \frac{(.04)(.65) + (.1)(.73) + (.21)(.68) + (.24)(.63)}{.5528} \\ &= .711 \end{aligned}$$

SELF-CHECK EXERCISES 7.6

1. The accompanying tree diagram represents a two-stage experiment. Use the diagram to find $P(B \mid D)$.

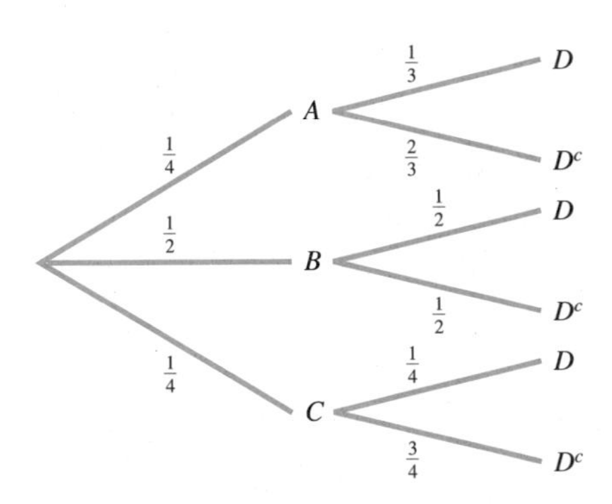

2. In a recent presidential election, it was estimated that the probability that the Republican candidate would be elected was 3/5 and therefore the probability that the Democratic candidate would be elected was 2/5 (the two Independent candidates were given little chance of being elected). It was also estimated that if the Republican candidate were elected, then the probability that research for a new manned bomber would continue was 4/5. But if the Democratic candidate were successful, then the probability that the research would continue was 3/10. Research was terminated shortly after the successful presidential candidate took office. What is the probability that the Republican candidate won that election?

Solutions to Self-Check Exercises 7.6 can be found on page 450.

7.6 Exercises

In Exercises 1–3, refer to the accompanying Venn diagram. An experiment in which the three mutually exclusive events *A*, *B*, and *C* form a partition of the uniform sample space *S* is depicted in the diagram.

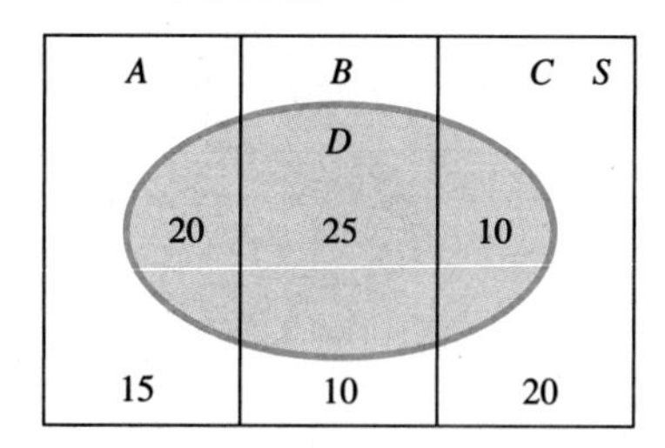

1. Draw a tree diagram using the information given in the Venn diagram illustrating the probabilities of the events A, B, C, and D.

2. Find: **a.** $P(D)$ **b.** $P(A \mid D)$

3. Find: **a.** $P(D^c)$ **b.** $P(B \mid D^c)$

In Exercises 4–6, refer to the accompanying Venn diagram. An experiment in which the three mutually exclusive events *A*, *B*, and *C* form a partition of the uniform sample space *S* is depicted in the diagram.

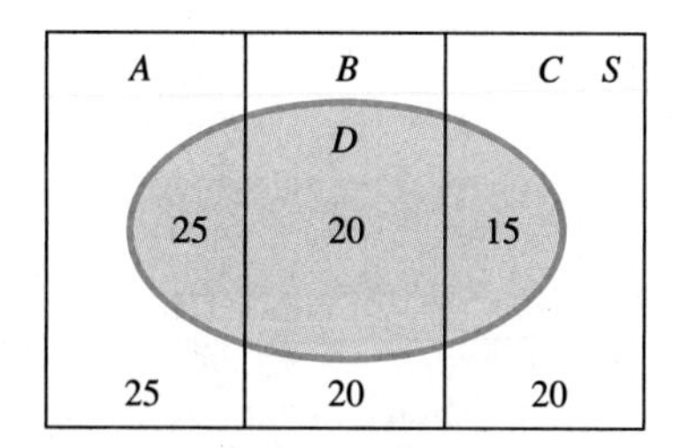

4. Draw a tree diagram using the information given in the Venn diagram illustrating the probabilities of the events A, B, C, and D.

5. Find: **a.** $P(D)$ **b.** $P(B \mid D)$

6. Find: **a.** $P(D^c)$ **b.** $P(B \mid D^c)$

7. The accompanying tree diagram represents a two-stage experiment. Use the diagram to find:
a. $P(A) \cdot P(D \mid A)$ **b.** $P(B) \cdot P(D \mid B)$
c. $P(A \mid D)$

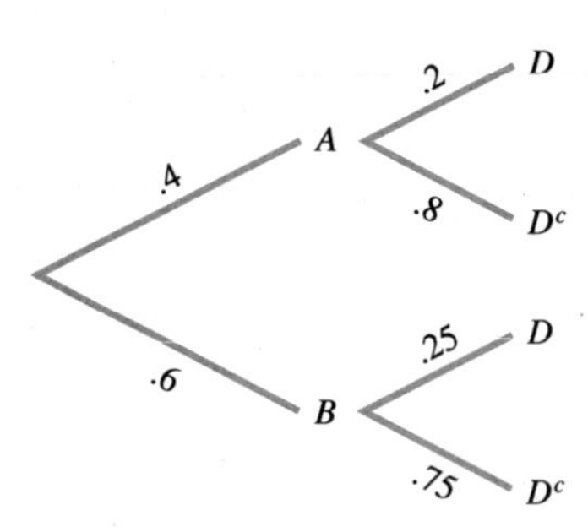

8. The accompanying tree diagram represents a two-stage experiment. Use the diagram to find:
a. $P(A) \cdot P(D \mid A)$ **b.** $P(B) \cdot P(D \mid B)$
c. $P(A \mid D)$

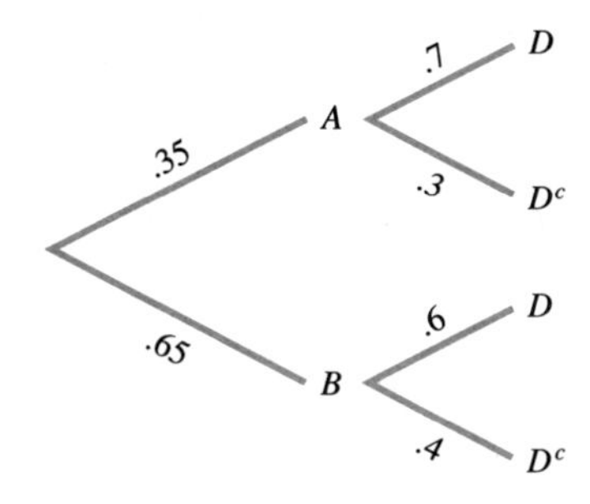

9. The accompanying tree diagram represents a two-stage experiment. Use the diagram to find:
a. $P(A) \cdot P(D \mid A)$ **b.** $P(B) \cdot P(D \mid B)$
c. $P(C) \cdot P(D \mid C)$ **d.** $P(A \mid D)$

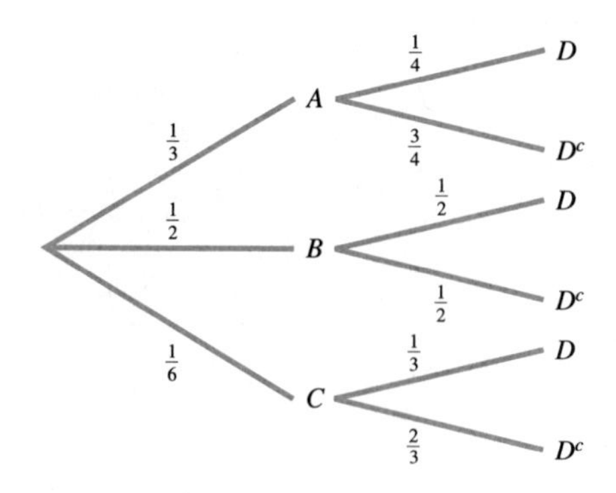

10. The accompanying tree diagram represents a two-stage experiment. Use this diagram to find:
a. $P(A \cap D)$ **b.** $P(B \cap D)$
c. $P(C \cap D)$ **d.** $P(D)$
e. Verify:

$$P(A \mid D) = \frac{P(A \cap D)}{P(D)} = \frac{P(A) \cdot P(D \mid A)}{P(A) \cdot P(D \mid A) + P(B) \cdot P(D \mid B) + P(C) \cdot P(D \mid C)}$$

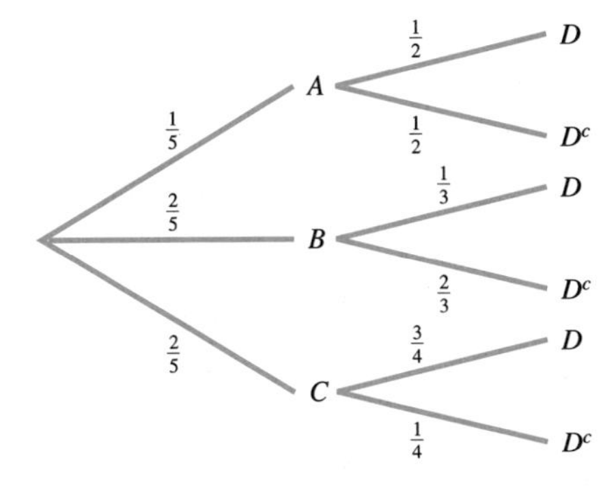

In Exercises 11–14, refer to the following experiment: Two cards are drawn in succession without replacement from a standard deck of 52 cards.

11. What is the probability that the first card is a heart given that the second card is a heart?

12. What is the probability that the first card is a heart given that the second card is a diamond?

13. What is the probability that the first card is a jack given that the second card is an ace?

14. What is the probability that the first card is a face card given that the second card is an ace?

In Exercises 15–18, refer to the following experiment: Urn A contains four white and six black balls. Urn B contains three white and five black balls. A ball is drawn from urn A and then transferred to urn B. A ball is then drawn from urn B.

15. Represent the probabilities associated with this two-stage experiment in the form of a tree diagram.

16. What is the probability that the transferred ball was white given that the second ball drawn was white?

17. What is the probability that the transferred ball was black given that the second ball drawn was white?

18. What is the probability that the transferred ball was black given that the second ball drawn was black?

19. POLITICS The 1992 U.S. Senate was composed of 57 Democrats and 43 Republicans. Sixty-six percent of the Democrats served in the military, whereas 64% of the Republicans had seen military service. If a senator selected at random had served in the military, what is the probability that he was Republican? *Note:* No congresswoman had served in the military.

20. QUALITY CONTROL Jansen Electronics has four machines that produce an identical component for use in its videocassette players. The proportion of the components produced by each machine and the probability of that component being defective are shown in the accompanying table. What is the probability that a component selected at random:
a. Is defective
b. Was produced by machine I, given it is defective?
c. Was produced by machine II, given it is defective?

Machine	Proportion of Components Produced	Probability of Defective Component
I	.15	.04
II	.30	.02
III	.35	.02
IV	.20	.03

21. An experiment consists of randomly selecting one of three coins, tossing it, and observing the outcome—heads or tails. The first coin is a two-headed coin, the second is a biased coin such that P(H) = .75, and the third is a fair coin.
a. What is the probability that the coin that is tossed will show heads?
b. If the coin selected shows heads, what is the probability that this coin is the fair coin?

A calculator is recommended for the remainder of this exercise set.

22. RELIABILITY OF MEDICAL TESTS A medical test has been designed to detect the presence of a certain disease. Among those who have the disease, the probability that the disease will be detected by the test is .95. However, the probability that the test will erroneously indicate the presence of the disease in those who do not actually

have it is .04. It is estimated that 4% of the population who take this test have the disease.

a. If the test administered to an individual is positive, what is the probability that the person actually has the disease?

b. If an individual takes the test twice and both times the test is positive, what is the probability that the person actually has the disease?

23. **RELIABILITY OF MEDICAL TESTS** Refer to Exercise 22. Suppose 20% of the people who were referred to a clinic for the test did in fact have the disease. If the test administered to an individual from this group is positive, what is the probability that the person actually has the disease?

24. **QUALITY CONTROL** A desk lamp produced by the Luminar Company was found to be defective. The company has three factories where the lamps are manufactured. The percentage of the total number of desk lamps produced by each factory and the probability that a lamp manufactured by that factory is defective are shown in the accompanying table. What is the probability that the defective lamp was manufactured in factory III?

Factory	Percentage of Total Production	Probability of Defective Component
I	.35	.015
II	.35	.01
III	.30	.02

25. **AUTO-ACCIDENT RATES** An insurance company has compiled the accompanying data relating the age of drivers and the accident rate (the probability of being involved in an accident during a 1-yr period) for drivers within that group:

Age Group	Percentage of Insured Drivers	Accident Rate
Under 25	.16	.055
25–44	.40	.025
45–64	.30	.02
65 and over	.14	.04

a. What is the probability that an insured driver will be involved in an accident during a particular 1-yr period?

b. What is the probability that an insured driver who is involved in an accident is under 25?

26. **SEAT-BELT COMPLIANCE** Data compiled by the Highway Patrol Department regarding the use of seat belts by drivers in a certain area after the passage of a compulsory seat-belt law are shown in the accompanying table:

Drivers	Percentage of Drivers in Group	Percentage of Group Stopped for Moving Violation
Group I (using seat belts)	.64	.002
Group II (not using seat belts)	.36	.005

If a driver in that area is stopped for a moving violation, what is the probability that he or she:

a. Will have a seat belt on?

b. Will not have a seat belt on?

27. **MEDICAL RESEARCH** Based on data obtained from the National Institute of Dental Research, it has been determined that 42% of 12-year-olds have never had a cavity, 34% of 13-year-olds have never had a cavity, and 28% of 14-year-olds have never had a cavity. If a child is selected at random from a group of 24 junior high school students comprising six 12-year-olds, eight 13-year-olds, and ten 14-year-olds and this child does not have a cavity, what is the probability that this child is 14 years old?

28. **VOTING PATTERNS** In a recent senatorial election, 50% of the voters in a certain district were registered as Democrats, 35% were registered as Republicans, and 15% were registered as Independents. The incumbent Democratic senator was reelected over her Republican and Independent opponents. Exit polls indicated that she gained 75% of the Democratic vote, 25% of the Republican vote, and 30% of the Independent vote. Assuming that the exit poll is accurate, what is the probability that a vote for the incumbent was cast by a registered Republican?

29. **CRIME RATES** Data compiled by the Department of Justice on the number of people arrested for serious crimes (murder, forcible rape, robbery, and so on) in 1988 revealed that 89% were male and 11% were female. Of the males, 30% were under 18, whereas 27% of the females arrested were under 18.

a. What is the probability that a person arrested for a serious crime in 1988 was under 18?

b. If a person arrested for a serious crime in 1988 was known to be under 18, what is the probability that the person is female?

30. **OPINION POLLS** A poll was conducted among 500 registered voters in a certain area regarding their position on a national lottery to raise revenue for the government.

The results of the poll are shown in the accompanying table:

Sex	Percentage of Voters Polled	Percentage Favoring Lottery	Percentage Not Favoring Lottery	Percentage Expressing No Opinion
Male	.51	.62	.32	.06
Female	.49	.68	.28	.04

What is the probability that a registered voter who:
a. Favored a national lottery was a woman?
b. Expressed no opinion regarding the lottery was a woman?

31. Customer Surveys The sales department of the Thompson Drug Company released the accompanying data concerning the sales of a certain pain reliever manufactured by the company:

Pain Reliever	Percentage of Drug Sold	Percentage of Group Sold in Extra-Strength Dosage
Group I (capsule form)	.57	.38
Group II (tablet form)	.43	.31

If a customer purchased the extra-strength dosage of this drug, what is the probability that it was in capsule form?

32. Selection of Supreme Court Judges In a past presidential election, it was estimated that the probability that the Republican candidate would be elected was 3/5, and therefore the probability that the Democratic candidate would be elected was 2/5 (the two Independent candidates were given little chance of being elected). It was also estimated that if the Republican candidate were elected, the probabilities that a conservative, moderate, or liberal judge would be appointed to the Supreme Court (one retirement was expected during the presidential term) were 1/2, 1/3, and 1/6, respectively. If the Democratic candidate were elected, the probabilities that a conservative, moderate, or liberal judge would be appointed to the Supreme Court would be 1/8, 3/8, and 1/2, respectively. A conservative judge was appointed to the Supreme Court during the presidential term. What is the probability that the Democratic candidate was elected?

33. Personnel Selection Applicants for temporary office work at the Carter Temporary Help Agency who have successfully completed a typing test are then placed in suitable positions by Nancy Dwyer and Darla Newberg. Employers who hire temporary help through the agency return a card indicating satisfaction or dissatisfaction with the work performance of those hired. From past experience it is known that 80% of the employees placed by Nancy are rated as satisfactory, whereas 70% of those placed by Darla are rated as satisfactory. Darla places 55% of the temporary office help at the agency and Nancy the remaining 45%. If a Carter office worker is rated unsatisfactory, what is the probability that he or she was placed by Darla?

34. College Majors The Office of Admissions and Records of a large western university released the accompanying information concerning the contemplated majors of its freshman class:

Major	Percentage of Freshmen Choosing This Major	Percentage of Females	Percentage of Males
Business	.24	.38	.62
Humanities	.08	.60	.40
Education	.08	.66	.34
Social science	.07	.58	.42
Natural sciences	.09	.52	.48
Other	.44	.48	.52

a. What is the probability that a student selected at random from the freshman class is a female?
b. What is the probability that a business student selected at random from the freshman class is a male?
c. What is the probability that a female student selected at random from the freshman class is majoring in business?

35. Medical Diagnoses A study was conducted among a certain group of union members whose health insurance policies required second opinions prior to surgery. Of those members whose doctors advised them to have surgery, 20% were informed by a second doctor that no surgery was needed. Of these, 70% took the second doctor's opinion and did not go through with the surgery. Of the members who were advised to have surgery by both doctors, 95% went through with the surgery. What is the probability that a union member who had surgery was advised to do so by a second doctor?

36. **AGE DISTRIBUTION OF RENTERS** A study conducted by the Metro Housing Agency in a midwestern city revealed the accompanying information concerning the age distribution of renters within the city:

Age	Percentage of Adult Population	Percentage of Group Who Are Renters
21–44	.51	.58
45–64	.31	.45
65 and over	.18	.60

a. What is the probability that an adult selected at random from this population is a renter?
b. If a renter is selected at random, what is the probability that he or she is in the 21–44 age bracket?
c. If a renter is selected at random, what is the probability that he or she is 45 years of age or older?

37. **THE SOCIAL LADDER** The following table summarizes the results of a poll conducted with 1154 adults by the New York Times/CBS News.

Annual Household Income, $	Percentage of Respondents Within That Income Range	Percentage of Respondents Who Call Themselves		
		Rich	Middle Class	Poor
Less than 15,000	11.2	0	24	76
15,000–29,999	18.6	3	60	37
30,000–49,999	24.5	0	86	14
50,000–74,999	21.9	2	90	8
75,000 and higher	23.8	5	91	4

a. What is the probability that a respondent chosen at random calls himself or herself middle class?
b. If a randomly chosen respondent calls himself or herself middle class, what is the probability that the annual household income of that individual is between \$30,000 and \$49,999, inclusive?
c. If a randomly chosen respondent calls himself or herself middle class, what is the probability that the individual's income is less than or equal to \$29,999 or greater than or equal to \$50,000?

Source: New York Times/CBS News, *Wall Street Journal Almanac*

SOLUTIONS TO SELF-CHECK EXERCISES 7.6

1. By Bayes' theorem, we have, using the probabilities given in the tree diagram,

$$P(B \mid D) = \frac{P(B)P(D \mid B)}{P(A)P(D \mid A) + P(B)P(D \mid B) + P(C)P(D \mid C)}$$

$$= \frac{\left(\frac{1}{2}\right)\left(\frac{1}{2}\right)}{\left(\frac{1}{4}\right)\left(\frac{1}{3}\right) + \left(\frac{1}{2}\right)\left(\frac{1}{2}\right) + \left(\frac{1}{4}\right)\left(\frac{1}{4}\right)}$$

$$= \frac{12}{19}$$

2. Let R and D, respectively, denote the event that the Republican and the Democratic candidate won the presidential election. Then, $P(R) = 3/5$ and $P(D) = 2/5$. Also, let C denote the event that research for the new manned bomber would continue. These data may be exhibited as in the accompanying tree diagram:

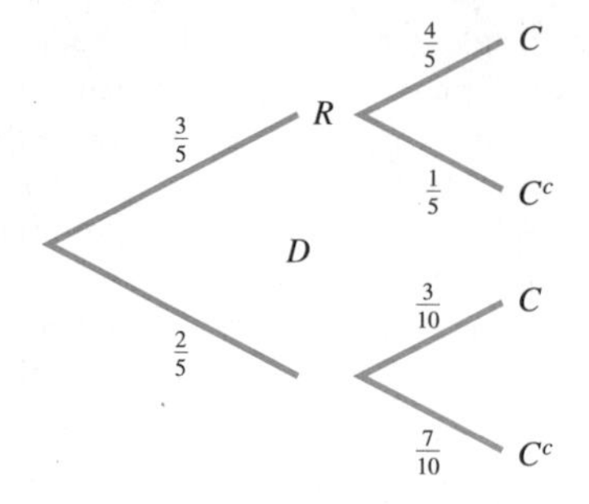

Using Bayes' theorem, we find that the probability that the Republican candidate had won the election is given by

$$P(R \mid C^c) = \frac{P(R)P(C^c \mid R)}{P(R)P(C^c \mid R) + P(D)P(C^c \mid D)}$$

$$= \frac{\left(\frac{3}{5}\right)\left(\frac{1}{5}\right)}{\left(\frac{3}{5}\right)\left(\frac{1}{5}\right) + \left(\frac{2}{5}\right)\left(\frac{7}{10}\right)} = \frac{3}{10}$$

CHAPTER 7 Summary of Principal Formulas and Terms

Formulas

1. Probability of an event in a uniform sample space — $P(E) = \dfrac{n(E)}{n(S)}$
2. Probability of the union of two mutually exclusive events — $P(E \cup F) = P(E) + P(F)$
3. Addition rule — $P(E \cup F) = P(E) + P(F) - P(E \cap F)$
4. Rule of complements — $P(E^c) = 1 - P(E)$
5. Conditional probability — $P(B \mid A) = \dfrac{P(A \cap B)}{P(A)}$
6. Product rule — $P(A \cap B) = P(A) \cdot P(B \mid A)$
7. Test for independence — $P(A \cap B) = P(A) \cdot P(B)$

Terms

experiment
sample point
sample space
event
finite sample space
union of two events
intersection of two events
complement of an event
mutually exclusive events
relative frequency
empirical probability
probability of an event
elementary (simple) event
probability distribution
probability function
uniform sample space
addition principle
conditional probability
finite stochastic process
independent events

CHAPTER 7 REVIEW EXERCISES

1. Let E and F be two mutually exclusive events and suppose $P(E) = .4$ and $P(F) = .2$. Compute:
   a. $P(E \cap F)$ **b.** $P(E \cup F)$
   c. $P(E^c)$ **d.** $P(E^c \cap F^c)$
   e. $P(E^c \cup F^c)$

2. Let E and F be two events of an experiment with sample space S. Suppose $P(E) = .3$, $P(F) = .2$, and $P(E \cap F) = .15$. Compute:
   a. $P(E \cup F)$ **b.** $P(E^c \cap F^c)$
   c. $P(E^c \cap F)$

3. Suppose a die is loaded and it was determined that the probability distribution associated with the experiment of casting the die and observing which number falls uppermost is given by

Simple Event	Probability
{1}	.20
{2}	.12
{3}	.16
{4}	.18
{5}	.15
{6}	.19

   a. What is the probability of the number being even?
   b. What is the probability of the number being either a 1 or a 6?
   c. What is the probability of the number being less than 4?

4. An urn contains six red, five black, and four green balls. If two balls are selected at random without replacement from the urn, what is the probability that a red ball and a black ball will be selected?

5. The quality-control department of the Starr Communications Company, the manufacturer of video-game cartridges, has determined from records that 1.5% of the cartridges sold have video defects, 0.8% have audio defects, and 0.4% have both audio and video defects. What is the probability that a cartridge purchased by a customer:
   a. Will have a video or audio defect?
   b. Will not have a video or audio defect?

6. Let E and F be two events and suppose $P(E) = .35$, $P(F) = .55$, and $P(E \cup F) = .70$. Find $P(E \mid F)$.

The accompanying tree diagram represents an experiment consisting of two trials. In Exercises 7–11, use the diagram to find the given probability.

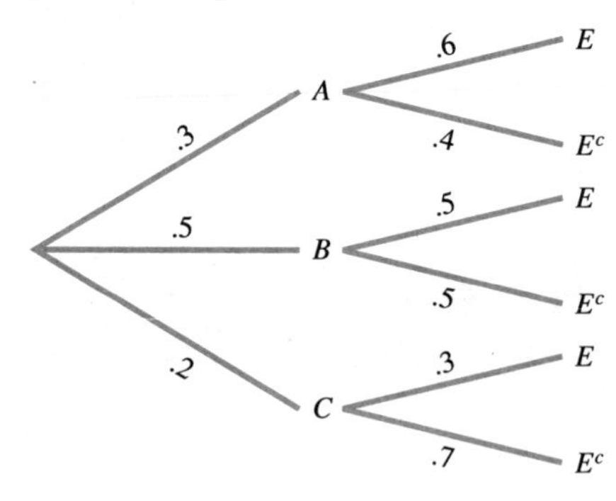

7. $P(A \cap E)$ 8. $P(B \cap E)$

9. $P(C \cap E)$ 10. $P(A \mid E)$

11. $P(E)$

12. An experiment consists of tossing a fair coin three times and observing the outcomes. Let A be the event that at least one head is thrown and B the event that at most two tails are thrown.
    a. Find $P(A)$.
    b. Find $P(B)$.
    c. Are A and B independent events?

13. In a group of 20 ballpoint pens on a shelf in the stationery department of the Metro Department Store, 2 are known to be defective. If a customer selects 3 of these pens, what is the probability that:
    a. At least 1 is defective?
    b. No more than 1 is defective?

14. Five people are selected at random. What is the probability that none of the people in this group were born on the same day of the week?

15. A pair of fair dice is cast. What is the probability that the sum of the numbers falling uppermost is 8 if it is known that the two numbers are different?

Three cards are drawn at random without replacement from a standard deck of 52 playing cards. In Exercises 16–20, find the probability of each of the given events.

16. All three cards are aces.

17. All three cards are face cards.

18. The second and third cards are red.

19. The second card is black, given that the first card was red.

20. The second card is a club, given that the first card was black.

21. Of 320 male and 280 female employees at the home office of the Gibraltar Insurance Company, 160 of the men and 190 of the women are on flex-time (flexible working hours). Given that an employee selected at random from this group is on flex-time, what is the probability that the employee is a male?

22. In a manufacturing plant, three machines, A, B, and C, produce 40%, 35%, and 25%, respectively, of the total production. The company's quality-control department has determined that 1% of the items produced by machine A, 1.5% of the items produced by machine B, and 2% of the items produced by machine C are defective. If an item is selected at random and found to be defective, what is the probability that it was produced by machine B?

23. Applicants who wish to be admitted to a certain professional school in a large university are required to take a screening test that was devised by an educational testing service. From past results, the testing service has estimated that 70% of all applicants are eligible for admission and that 92% of those who are eligible for admission pass the exam, whereas 12% of those who are ineligible for admission pass the exam. Using these results, what is the probability that an applicant for admission:

a. Passed the exam?

b. Passed the exam but was actually ineligible?

8 PROBABILITY DISTRIBUTIONS AND STATISTICS

- 8.1 Distributions of Random Variables
- 8.2 Expected Value
- 8.3 Variance and Standard Deviation
- 8.4 The Binomial Distribution
- 8.5 The Normal Distribution
- 8.6 Applications of the Normal Distribution

Statistics is that branch of mathematics concerned with the collection, analysis, and interpretation of data. In Sections 8.1–8.3 of this chapter, we take a look at descriptive statistics; here, our interest lies in the description and presentation of data in the form of tables and graphs. In the rest of the chapter, we briefly examine inductive statistics, and we see how mathematical tools such as those developed in Chapter 7 may be used in conjunction with these data to help us draw certain conclusions and make forecasts.

Where do we go from here? Some students in the top 10% of this senior class will further their education at one of the campuses of the State University system. In Example 3, page 519, we will determine the minimum grade point average a senior needs to be eligible for admission to a state university.

8.1 Distributions of Random Variables

Random Variables

In many situations it is desirable to assign numerical values to the outcomes of an experiment. For example, if an experiment consists of casting a die and observing the face that lands uppermost, then it is natural to assign the numbers 1, 2, 3, 4, 5, and 6, respectively, to the outcomes *one, two, three, four, five,* and *six* of the experiment. If we let X denote the outcome of the experiment, then X assumes one of the numbers. Because the values assumed by X depend on the outcomes of a chance experiment, the outcome X is referred to as a random variable.

Random Variable

A **random variable** is a rule that assigns a number to each outcome of a chance experiment.

More precisely, a random variable is a function with domain given by the set of outcomes of a chance experiment and range contained in the set of real numbers.

EXAMPLE 1 A coin is tossed three times. Let the random variable X denote the number of heads that occur in the three tosses.

a. List the outcomes of the experiment; that is, find the domain of the function X.

b. Find the value assigned to each outcome of the experiment by the random variable X.

c. Find the event comprising the outcomes to which a value of 2 has been assigned by X. This event is written $(X = 2)$ and is the event comprising the outcomes in which two heads occur.

SOLUTION ✓

a. From the results of Example 3, Section 7.1 (page 386), we see that the set of outcomes of the experiment is given by the sample space

$$S = \{\text{HHH, HHT, HTH, THH, HTT, THT, TTH, TTT}\}$$

b. The outcomes of the experiment are displayed in the first column of Table 8.1. The corresponding value assigned to each such outcome by the random variable X (the number of heads) appears in the second column.

Table 8.1 No. of Heads in Three Coin Tosses

Outcome	Value of X
HHH	3
HHT	2
HTH	2
THH	2
HTT	1
THT	1
TTH	1
TTT	0

c. With the aid of Table 8.1, we see that the event $(X = 2)$ is given by the set

$$\{\text{HHT, HTH, THH}\}$$

■■■■

EXAMPLE 2

A coin is tossed repeatedly until a head occurs. Let the random variable Y denote the number of coin tosses in the experiment. What are the values of Y?

SOLUTION ✓

The outcomes of the experiment make up the infinite set

$$S = \{\text{H, TH, TTH, TTTH, TTTTH}, \ldots\}$$

These outcomes of the experiment are displayed in the first column of Table 8.2. The corresponding values assumed by the random variable Y (the number of tosses) appear in the second column.

Table 8.2 No. of Coin Tosses Before Heads Appears

Outcome	Value of Y
H	1
TH	2
TTH	3
TTTH	4
TTTTH	5
⋮	⋮

■■■■

EXAMPLE 3

A disposable flashlight is turned on until its battery runs out. Let the random variable Z denote the length (in hours) of the life of the battery. What values may Z assume?

SOLUTION ✓

The values assumed by Z may be any nonnegative real numbers; that is, the possible values of Z comprise the interval $0 \leq Z < \infty$. ■■■■

One advantage of working with random variables rather than working directly with the outcomes of an experiment is that random variables are functions that may be added, subtracted, and multiplied. Because of this, results developed in the field of algebra and other areas of mathematics may be used freely to help us solve problems in the fields of probability and statistics.

A random variable is classified into three categories depending on the set of values it assumes. A random variable is called **finite discrete** if it assumes only finitely many values. For example, the random variable X of Example 1 is finite discrete since it may assume values only from the finite set $\{0, 1, 2, 3\}$ of numbers. Next, a random variable is said to be **infinite discrete** if it takes on infinitely many values, which may be arranged in a sequence. For example, the random variable Y of Example 2 is infinite discrete since it assumes values from the set $\{1, 2, 3, 4, 5, \ldots\}$, which has been arranged in the form of an infinite sequence. Finally, a random variable is called **continuous** if the values it may assume comprise an interval of real numbers. For example, the random variable Z of Example 3 is continuous since the values it may assume comprise the interval of nonnegative real numbers. For the remainder of this section *all random variables will be assumed to be finite discrete.*

PROBABILITY DISTRIBUTIONS OF RANDOM VARIABLES

In Section 7.2 we learned how to construct the probability distribution for an experiment. There, the probability distribution took the form of a table that gave the probabilities associated with the outcomes of an experiment. Since the random variable associated with an experiment is related to the outcomes of the experiment, it is clear that we should be able to construct a probability distribution associated with the *random variable* rather than one associated with the outcomes of the experiment. Such a distribution is called the **probability distribution of a random variable** and may be given in the form of a formula or displayed in a table that gives the distinct (numerical) values of the random variable X and the probabilities associated with these values. Thus, if $x_1, x_2, \ldots, x_n$ are the values assumed by the random variable X with associated probabilities $P(X = x_1), P(X = x_2), \ldots, P(X = x_n)$, respectively, then the required probability distribution of the random variable X, where $p_i = P(X = x_i)$, $i = 1, 2, \ldots, n$, may be expressed in the form of a table:

A Probability Distribution for the Random Variable X

x	$P(X = x)$
x_1	p_1
x_2	p_2
x_3	p_3
$\vdots$	$\vdots$
x_n	p_n

The next several examples illustrate the construction of probability distributions.

EXAMPLE 4 Find the probability distribution of the random variable associated with the experiment of Example 1.

SOLUTION ✓ From the results of Example 1, we see that the values assumed by the random variable X are 0, 1, 2, and 3, corresponding to the events of 0, 1, 2, and 3 heads occurring, respectively. Referring to Table 8.1 once again, we see that the outcome associated with the event $(X = 0)$ is given by the set {TTT}.

Table 8.3 A Probability Distribution

x	$P(X = x)$
0	$\frac{1}{8}$
1	$\frac{3}{8}$
2	$\frac{3}{8}$
3	$\frac{1}{8}$

Consequently, the probability associated with the random variable X when it assumes the value 0 is given by

$$P(X = 0) = \frac{1}{8} \qquad \text{[Note that } n(S) = 8.]$$

Next, observe that the event $(X = 1)$ is given by the set {HTT, THT, TTH}, so

$$P(X = 1) = \frac{3}{8}$$

In a similar manner we may compute $P(X = 2)$ and $P(X = 3)$, which gives the probability distribution shown in Table 8.3. ■■■■

EXAMPLE 5 Let X denote the random variable that gives the sum of the faces that fall uppermost when two fair dice are cast. Find the probability distribution of X.

SOLUTION ✓ The values assumed by the random variable X are 2, 3, 4, . . . , 12, corresponding to the events E_2, E_3, E_4, . . . , E_{12} (see Example 4, Section 7.1). Next, the probabilities associated with the random variable X when X assumes the values 2, 3, 4, . . . , 12 are precisely the probabilities $P(E_2)$, $P(E_3)$, . . . , $P(E_{12})$, respectively, and may be computed in much the same way as the solution to Example 3, Section 7.2. Thus,

$$P(X = 2) = P(E_2) = \frac{1}{36}$$

$$P(X = 3) = P(E_3) = \frac{2}{36}$$

and so on. The required probability distribution of X is given in Table 8.4.

Table 8.4 A Probability Distribution for the Random Variable That Gives the Sum of the Faces of Two Dice

x	2	3	4	5	6	7	8	9	10	11	12
$P(X = x)$	1/36	2/36	3/36	4/36	5/36	6/36	5/36	4/36	3/36	2/36	1/36

■■■■

EXAMPLE 6 The following data give the number of cars observed waiting in line at the beginning of 2-minute intervals between 3 and 5 P.M. on a certain Friday at the drive-in teller of the Westwood Savings Bank and the corresponding frequency of occurrence. Find the probability distribution of the random variable X, where X denotes the number of cars observed waiting in line.

Number of Cars	0	1	2	3	4	5	6	7	8
Frequency of Occurrence	2	9	16	12	8	6	4	2	1

SOLUTION ✓

Dividing each number in the last row of the given table by 60 (the sum of these numbers) gives the respective probabilities (here, we use the relative frequency interpretation of probability) associated with the random variable X when X assumes the values 0, 1, 2, . . . , 8. For example,

$$P(X = 0) = \frac{2}{60} \approx .03$$

$$P(X = 1) = \frac{9}{60} = .15$$

and so on. The resulting probability distribution is shown in Table 8.5.

Table 8.5 A Probability Distribution

x	$P(X = x)$
0	.03
1	.15
2	.27
3	.20
4	.13
5	.10
6	.07
7	.03
8	.02

HISTOGRAMS

A probability distribution of a random variable may be exhibited graphically by means of a **histogram.** To construct a histogram of a particular probability distribution, first locate the values of the random variable on the number line. Then, above each such number, erect a rectangle with width 1 and height equal to the probability associated with that value of the random variable. For example, the histogram of the probability distribution appearing in Table 8.3 is shown in Figure 8.1. The histograms of the probability distributions of Examples 5 and 6 are constructed in a similar manner and are displayed in Figures 8.2 and 8.3, respectively.

Observe that in each histogram, the area of a rectangle associated with a value of a random variable X gives precisely the probability associated with the value of X. This follows because each such rectangle, by construction, has width 1 and height corresponding to the probability associated with the value of the random variable. Another consequence arising from the method of

FIGURE 8.1
Histogram showing the probability distribution for the number of heads occurring in three coin tosses

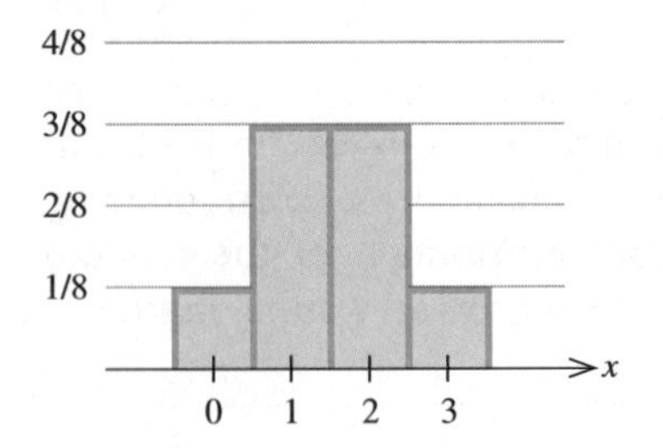

FIGURE 8.2
Histogram showing the probability distribution for the sum of the faces of two dice

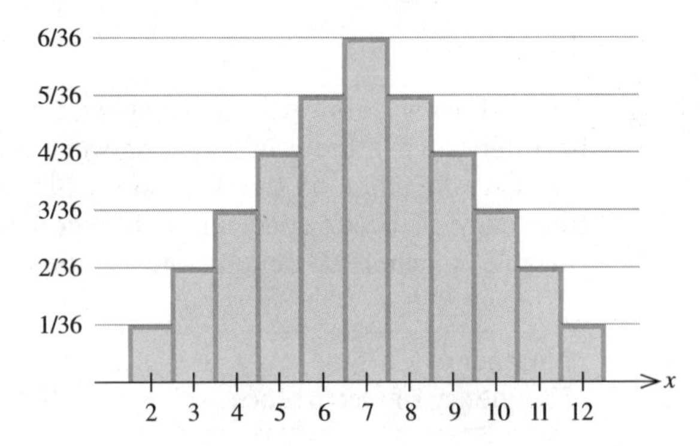

FIGURE 8.3
Histogram showing the probability distribution for the number of cars waiting in line

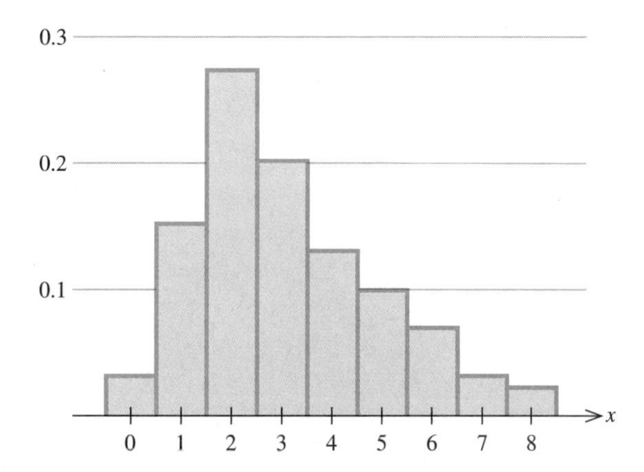

construction of a histogram is that *the probability associated with more than one value of the random variable X is given by the sum of the areas of the rectangles associated with those values of X.* For example, in the coin-tossing experiment of Example 1, the event of obtaining at least two heads, which corresponds to the event $(X = 2)$ or $(X = 3)$, is given by

$$P(X = 2) + P(X = 3)$$

and may be obtained from the histogram depicted in Figure 8.1 by adding the areas associated with the values 2 and 3, respectively, of the random variable X. We obtain

$$P(X = 2) + P(X = 3) = (1)\left(\frac{3}{8}\right) + (1)\left(\frac{1}{8}\right) = \frac{1}{2}$$

This result provides us with a method of computing the probabilities of events directly from the knowledge of a histogram of the probability distribution of the random variable associated with the experiment.

EXAMPLE 7

Suppose the probability distribution of a random variable X is represented by the histogram shown in Figure 8.4. Identify that part of the histogram whose area gives the probability $P(10 \leq X \leq 20)$. Do not evaluate the result.

FIGURE 8.4

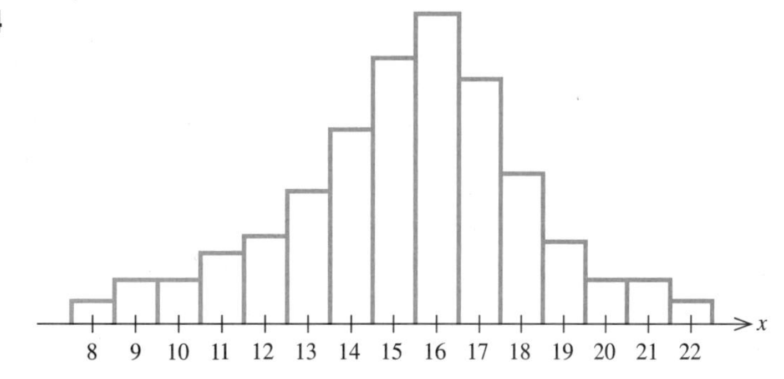

SOLUTION ✓ The event $(10 \le X \le 20)$ is the event comprising outcomes related to the values 10, 11, 12, . . . , 20 of the random variable X. The probability of this event $P(10 \le X \le 20)$ is therefore given by the shaded area of the histogram in Figure 8.5.

FIGURE 8.5
$P(10 \le X \le 20)$

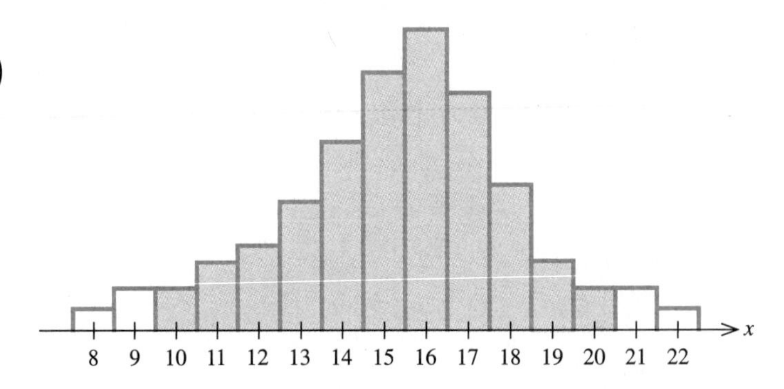

SELF-CHECK EXERCISES 8.1

1. Three balls are selected at random without replacement from an urn containing four black balls and five white balls. Let the random variable X denote the number of black balls drawn.
   a. List the outcomes of the experiment.
   b. Find the value assigned to each outcome of the experiment by the random variable X.
   c. Find the event comprising the outcomes to which a value of 2 has been assigned by X.
2. The following data, extracted from the records of the Dover Public Library, give the number of books borrowed by the library's members over a 1-mo period:

Number of Books	0	1	2	3	4	5	6	7	8
Frequency of Occurrence	780	300	412	205	98	54	57	30	6

   a. Find the probability distribution of the random variable X, where X denotes the number of books checked out over a 1-mo period by a randomly chosen member.
   b. Draw the histogram representing this probability distribution.

Solutions to Self-Check Exercises 8.1 can be found on page 467.

8.1 Exercises

1. Three balls are selected at random without replacement from an urn containing four green balls and six red balls. Let the random variable X denote the number of green balls drawn.
   a. List the outcomes of the experiment.
   b. Find the value assigned to each outcome of the experiment by the random variable X.
   c. Find the event comprising the outcomes to which a value of 3 has been assigned by X.

2. A coin is tossed four times. Let the random variable X denote the number of tails that occur.
   a. List the outcomes of the experiment.
   b. Find the value assigned to each outcome of the experiment by the random variable X.
   c. Find the event comprising the outcomes to which a value of 2 has been assigned by X.

3. A die is cast repeatedly until a 6 falls uppermost. Let the random variable X denote the number of times the die is cast. What are the values that X may assume?

4. Cards are selected one at a time without replacement from a well-shuffled deck of 52 cards until an ace is drawn. Let X denote the random variable that gives the number of cards drawn. What values may X assume?

5. Let X denote the random variable that gives the sum of the faces that fall uppermost when two fair dice are cast. Find $P(X = 7)$.

6. Two cards are drawn from a well-shuffled deck of 52 playing cards. Let X denote the number of aces drawn. Find $P(X = 2)$.

In Exercises 7–12, give the range of values that the random variable X may assume and classify the random variable as finite discrete, infinite discrete, or continuous.

7. X = The number of times a die is thrown until a 2 appears

8. X = The number of defective watches in a sample of eight watches

9. X = The distance a commuter travels to work

10. X = The number of hours a child watches television on a given day

11. X = The number of times an accountant takes the CPA examination before passing

12. X = The number of boys in a four-child family

13. The probability distribution of the random variable X is shown in the accompanying table:

x	−10	−5	0	5
$P(X = x)$	.20	.15	.05	.1

x	10	15	20
$P(X = x)$	.25	.1	.15

Find:
a. $P(X = -10)$ **b.** $P(X \geq 5)$
c. $P(-5 \leq X \leq 5)$ **d.** $P(X \leq 20)$

14. The probability distribution of the random variable X is shown in the accompanying table.

x	−5	−3	−2
$P(X = x)$	.17	.13	.33

x	0	2	3
$P(X = x)$	.16	.11	.10

Find:
a. $P(X \leq 0)$ **b.** $P(X \leq -3)$
c. $P(-2 \leq X \leq 2)$

15. Suppose a probability distribution of a random variable X is represented by the accompanying histogram. Shade that part of the histogram whose area gives the probability $P(17 \leq X \leq 20)$.

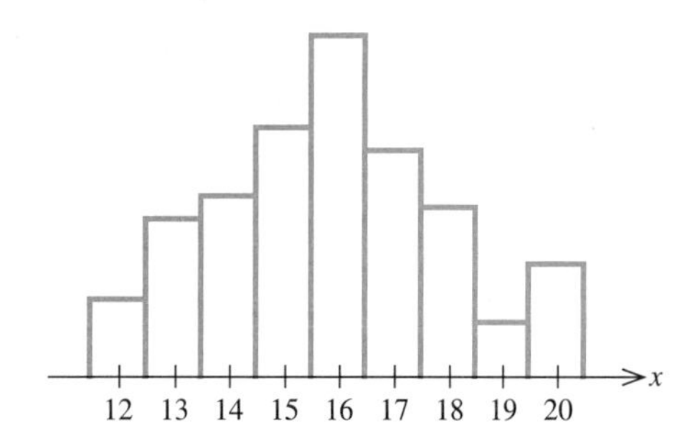

(continued on p. 466)

Using Technology

GRAPHING A HISTOGRAM

The graphing calculator can be used to plot the histogram for a given set of data, as illustrated by the following example.

EXAMPLE 1 A survey of 90,000 households conducted in 1995 revealed the following percentage of women who wear the given shoe size.

Shoe Size	<5	5–5½	6–6½	7–7½	8–8½	9–9½	10–10½	>10
Women, %	1	5	15	27	29	14	7	2

Source: Footwear Market Insights survey

a. Plot a histogram for the given data.
b. What is the percentage of women in the survey who wear size 7–7½ or 8–8½ shoes?

SOLUTION ✓ **a.** Let X denote the random variable taking on the values 1 through 8, where 1 corresponds to a shoe size less than 5, 2 corresponds to a shoe size of 5–5½, and so on. Entering the values of X as $x_1 = 1$, $x_2 = 2$, . . . , $x_8 = 8$ and the corresponding values of Y as $y_1 = 1$, $y_2 = 5$, . . . , $y_8 = 2$, and then using the **DRAW** function from the Statistics menu, we draw the histogram shown in Figure T1.

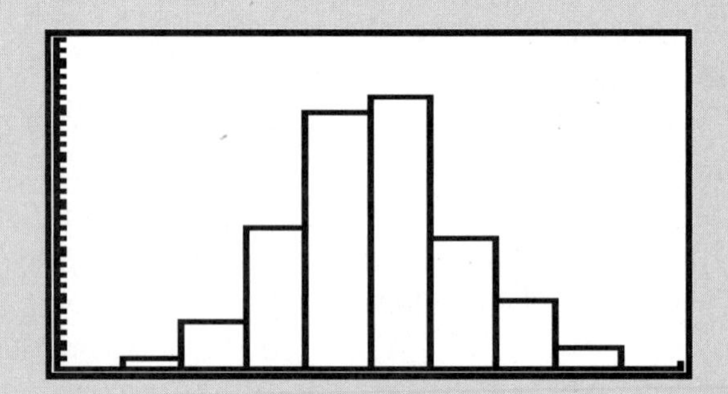

FIGURE T1
The histogram for the given data, using the viewing rectangle $[0, 9] \times [0, 35]$

b. The probability that a woman participating in the survey wears size 7–7½ or 8–8½ shoes is given by

$$P(X = 4) + P(X = 5) = .27 + .29 = .56$$

which tells us that 56% of the women wear either size 7–7½ or size 8–8½ shoes.

Exercises

1. Graph the histogram associated with the data given in Table 8.1, page 456. Compare your graph with that given in Figure 8.1, page 460.
2. Graph the histogram associated with the data given in Exercise 18, page 466.
3. Graph the histogram associated with the data given in Exercise 19, page 466.
4. Graph the histogram associated with the data given in Exercise 21, page 466.

16. **EXAMS** An examination consisting of ten true-or-false questions was taken by a class of 100 students. The probability distribution of the random variable X, where X denotes the number of questions answered correctly by a randomly chosen student, is represented by the accompanying histogram. The rectangle with base centered on the number 8 is missing. What should be the height of this rectangle?

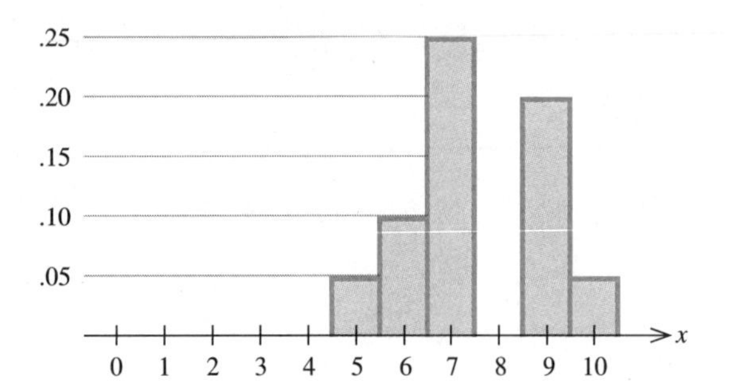

17. Two dice are cast. Let the random variable X denote the number that falls uppermost on the first die and let Y denote the number that falls uppermost on the second die.
a. Find the probability distributions of X and Y.
b. Find the probability distribution of $X + Y$.

18. **DISTRIBUTION OF FAMILIES BY SIZE** A survey was conducted by the Public Housing Authority in a certain community among 1000 families to determine the distribution of families by size. The results follow:

Family Size	2	3	4	5
Frequency of Occurrence	350	200	245	125

Family Size	6	7	8
Frequency of Occurrence	66	10	4

a. Find the probability distribution of the random variable X, where X denotes the number of persons in a randomly chosen family.
b. Draw the histogram corresponding to the probability distribution found in part (a).

19. **WAITING LINES** The accompanying data were obtained in a study conducted by the manager of the Sav-More Supermarket. In this study the number of customers waiting in line at the express checkout at the beginning of each 3-min interval between 9 A.M. and 12 noon on Saturday was observed.

Number of Customers	0	1	2	3	4
Frequency of Occurrence	1	4	2	7	14

Number of Customers	5	6	7	8	9	10
Frequency of Occurrence	8	10	6	3	4	1

a. Find the probability distribution of the random variable X, where X denotes the number of customers observed waiting in line.
b. Draw the histogram representing this probability distribution.

20. **MONEY-MARKET RATES** The rates paid by 30 financial institutions on a certain day for money-market deposit accounts are shown in the accompanying table:

Rate, %	6	6.25	6.55	6.56
Number of Institutions	1	7	7	1

Rate, %	6.58	6.60	6.65	6.85
Number of Institutions	1	8	3	2

Let the random variable X denote the interest paid by a randomly chosen financial institution on its money-market deposit accounts and find the probability distribution associated with these data.

21. **TELEVISION PILOTS** After the private screening of a new television pilot, audience members were asked to rate the new show on a scale of 1 to 10 (10 being the highest rating). From a group of 140 people, the accompanying responses were obtained:

Rating	1	2	3	4	5
Frequency of Occurrence	1	4	3	11	23

Rating	6	7	8	9	10
Frequency of Occurrence	21	28	29	16	4

Let the random variable X denote the rating given to the show by a randomly chosen audience member. Find the probability distribution associated with these data.

In Exercises 22 and 23, determine whether the statement is true or false. If it is true, explain why it is true. If it is false, give an example to show why it is false.

22. Suppose X is a finite discrete random variable assuming the values $x_1, x_2, \ldots, x_n$ and associated probabilities $p_1, p_2, \ldots, p_n$, then $p_1 + p_2 + \cdots + p_n = 1$.

23. The area of a histogram associated with a probability distribution is a number between 0 and 1.

SOLUTIONS TO SELF-CHECK EXERCISES 8.1

1. **a.** Using the accompanying tree diagram, we see that the outcomes of the experiment are

$$S = \{(B, B, B), (B, B, W), (B, W, B), (B, W, W), (W, B, B), (W, B, W), (W, W, B), (W, W, W)\}$$

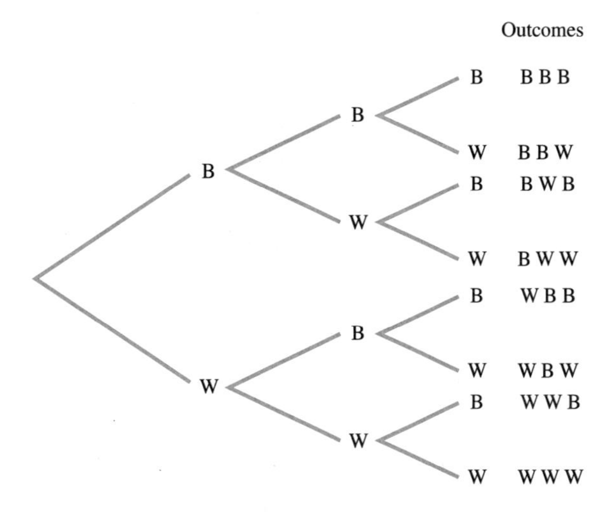

b. Using the results of part (a), we obtain the values assigned to the outcomes of the experiment as follows:

Outcome	BBB	BBW	BWB	BWW
Value	3	2	2	1

Outcome	WBB	WBW	WWB	WWW
Value	2	1	1	0

c. The required event is {BBW, BWB, WBB}.

2. a. We divide each number in the bottom row of the given table by 1942 (the sum of these numbers) to obtain the probabilities associated with the random variable X when X takes on the values 0, 1, 2, 3, 4, 5, 6, 7, and 8. For example,

$$P(X=0)=\frac{780}{1942}\approx .402$$

$$P(X=1)=\frac{300}{1942}\approx .154$$

The required probability distribution and histogram follow:

x	0	1	2	3	4
$P(X = x)$	.402	.154	.212	.106	.050

x	5	6	7	8
$P(X = x)$	.028	.029	.015	.003

b.

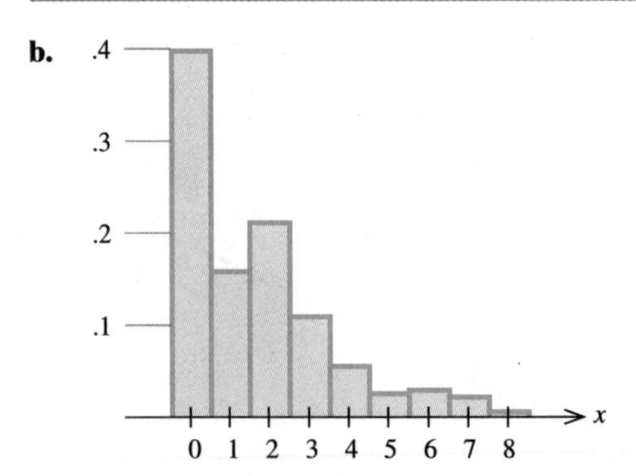

8.2 Expected Value

MEAN

The average value of a set of numbers is a familiar notion to most people. For example, to compute the average of the four numbers

$$12, \quad 16, \quad 23, \quad 37$$

we simply add these numbers and divide the resulting sum by 4, giving the required average as

$$\frac{12+16+23+37}{4}=\frac{88}{4}=22$$

In general, we have the following definition:

Average, or Mean

The **average,** or **mean,** of the n numbers

$$x_1, x_2, \ldots, x_n$$

is $\bar{x}$ (read "x bar"), where

$$\bar{x} = \frac{x_1 + x_2 + \cdots + x_n}{n}$$

EXAMPLE 1 Refer to Example 6, Section 8.1. Find the average number of cars waiting in line at the bank's drive-in teller at the beginning of each 2-minute interval during the period in question.

SOLUTION ✓ The number of cars, together with its corresponding frequency of occurrence, are reproduced in Table 8.6. Observe that the number 0 (of cars) occurs twice, the number 1 occurs 9 times, and so on. There are altogether

$$2 + 9 + 16 + 12 + 8 + 6 + 4 + 2 + 1 = 60$$

Table 8.6

Number of Cars	0	1	2	3	4	5	6	7	8
Frequency of Occurrence	2	9	16	12	8	6	4	2	1

numbers to be averaged. Therefore, the required average is given by

$$\frac{0\cdot 2 + 1\cdot 9 + 2\cdot 16 + 3\cdot 12 + 4\cdot 8 + 5\cdot 6 + 6\cdot 4 + 7\cdot 2 + 8\cdot 1}{60} \approx 3.1 \quad \textbf{(1)}$$

or approximately 3.1 cars. ■■■■

Expected Value

Let's reconsider the expression in Equation (1) that gives the average of the frequency distribution shown in Table 8.6. Dividing each term by the denominator, the expression may be rewritten in the form

$$0\cdot\left(\frac{2}{60}\right) + 1\cdot\left(\frac{9}{60}\right) + 2\cdot\left(\frac{16}{60}\right) + 3\cdot\left(\frac{12}{60}\right) + 4\cdot\left(\frac{8}{60}\right) + 5\cdot\left(\frac{6}{60}\right)$$
$$+ 6\cdot\left(\frac{4}{60}\right) + 7\cdot\left(\frac{2}{60}\right) + 8\cdot\left(\frac{1}{60}\right)$$

Observe that each term in the sum is a product of two factors; the first factor is the value assumed by the random variable X, where X denotes the number of cars waiting in line, and the second factor is just the probability associated with that value of the random variable. This observation suggests the following general method for calculating the expected value (that is, the average, or mean) of a random variable X that assumes a finite number of values from the knowledge of its probability distribution.

Expected Value of a Random Variable X

Let X denote a random variable that assumes the values $x_1, x_2, \ldots, x_n$ with associated probabilities $p_1, p_2, \ldots, p_n$, respectively. Then the **expected value** of X, $E(X)$, is given by

$$E(X) = x_1p_1 + x_2p_2 + \cdots + x_np_n \tag{2}$$

REMARK The numbers $x_1, x_2, \ldots, x_n$ may be positive, zero, or negative. For example, such a number will be positive if it represents a profit and negative if it represents a loss.

EXAMPLE 2 Re-solve Example 1 by using the probability distribution associated with the experiment, reproduced in Table 8.7.

Table 8.7 A Probability Distribution

x	$P(X = x)$
0	.03
1	.15
2	.27
3	.20
4	.13
5	.10
6	.07
7	.03
8	.02

SOLUTION ✔ Let X denote the number of cars waiting in line. Then, the average number of cars waiting in line is given by the expected value of X—that is, by

$$\begin{aligned} E(X) &= (0)(.03) + (1)(.15) + (2)(.27) + (3)(.20) + (4)(.13) \\ &\quad + (5)(.10) + (6)(.07) + (7)(.03) + (8)(.02) \\ &= 3.1 \text{ cars} \end{aligned}$$

which agrees with the earlier result.

The expected value of a random variable X is a measure of the central tendency of the probability distribution associated with X. In repeated trials of an experiment with random variable X, the average of the observed values of X gets closer and closer to the expected value of X as the number of trials gets larger and larger. Geometrically, the expected value of a random variable X has the following simple interpretation: If a laminate is made of the histo-

FIGURE 8.6
Expected value of a random variable X

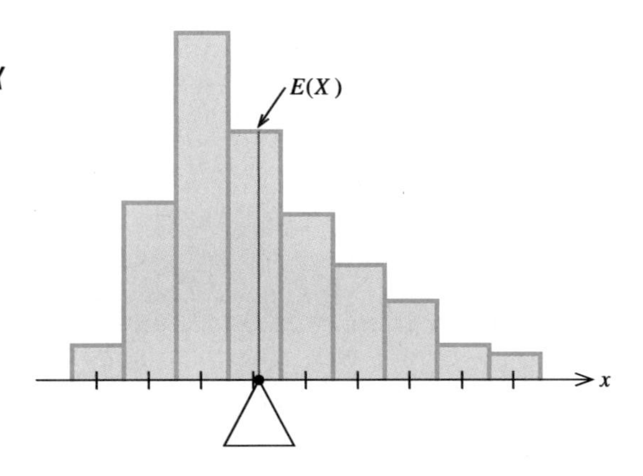

gram of a probability distribution associated with a random variable X, then the expected value of X corresponds to the point on the base of the laminate at which the latter will balance perfectly when the point is directly over a fulcrum (Figure 8.6).

EXAMPLE 3

Let X denote the random variable that gives the sum of the faces that fall uppermost when two fair dice are cast. Find the expected value, $E(X)$, of X.

SOLUTION ✓

The probability distribution of X, reproduced in Table 8.8, was found in Example 5, Section 8.1. Using this result, we find

$$\begin{aligned} E(X) &= 2\left(\frac{1}{36}\right) + 3\left(\frac{2}{36}\right) + 4\left(\frac{3}{36}\right) + 5\left(\frac{4}{36}\right) + 6\left(\frac{5}{36}\right) + 7\left(\frac{6}{36}\right) \\ &\quad + 8\left(\frac{5}{36}\right) + 9\left(\frac{4}{36}\right) + 10\left(\frac{3}{36}\right) + 11\left(\frac{2}{36}\right) + 12\left(\frac{1}{36}\right) \\ &= 7 \end{aligned}$$

Table 8.8 Probability Distribution

x	$P(X = x)$
2	$\frac{1}{36}$
3	$\frac{2}{36}$
4	$\frac{3}{36}$
5	$\frac{4}{36}$
6	$\frac{5}{36}$
7	$\frac{6}{36}$
8	$\frac{5}{36}$
9	$\frac{4}{36}$
10	$\frac{3}{36}$
11	$\frac{2}{36}$
12	$\frac{1}{36}$

FIGURE 8.7
Histogram showing the probability distribution for the sum of the faces of two dice

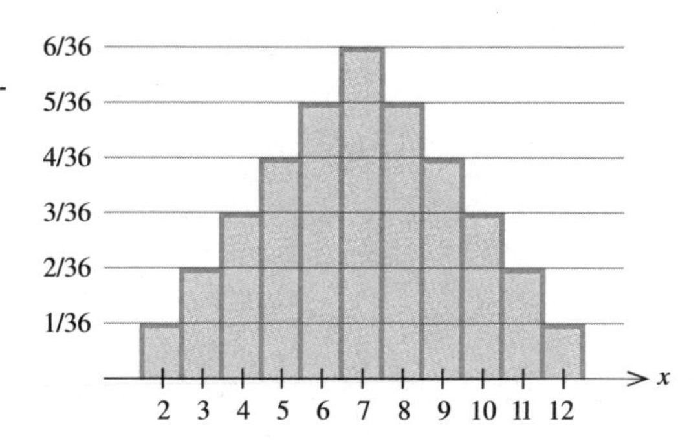

Note that, because of the symmetry of the histogram of the probability distribution with respect to the vertical line $x = 7$, the result could have been obtained by merely inspecting Figure 8.7.

APPLICATIONS

The next example shows how we can use the concept of expected value to help us make the best investment decision.

EXAMPLE 4

A group of private investors intends to purchase one of two motels currently being offered for sale in a certain city. The terms of sale of the two motels are similar, although the Regina Inn has 52 rooms and is in a slightly better location than the Merlin Motor Lodge, which has 60 rooms. Records obtained for each motel reveal that the occupancy rates, with corresponding probabilities, during the May–September tourist season are as shown in the accompanying tables:

Regina Inn

Occupancy Rate	.80	.85	.90	.95	1.00
Probability	.19	.22	.31	.23	.05

Merlin Motor Lodge

Occupancy Rate	.75	.80	.85	.90	.95	1.00
Probability	.35	.21	.18	.15	.09	.02

The average profit per day for each occupied room at the Regina Inn is \$10, whereas the average profit per day for each occupied room at the Merlin Motor Lodge is \$9.

a. Find the average number of rooms occupied per day at each motel.

b. If the investors' objective is to purchase the motel that generates the higher daily profit, which motel should they purchase? (Compare the expected daily profit of the two motels.)

SOLUTION ✓

a. Let X denote the occupancy rate at the Regina Inn. Then the average daily occupancy rate at the Regina Inn is given by the expected value of X—that is, by

$$\begin{aligned} E(X) &= (.80)(.19) + (.85)(.22) + (.90)(.31) \\ &\quad + (.95)(.23) + (1.00)(.05) \\ &= .8865 \end{aligned}$$

The average number of rooms occupied per day at the Regina is

$$(.8865)(52) \approx 46.1$$

or approximately 46.1 rooms. Similarly, letting Y denote the occupancy rate at the Merlin Motor Lodge, we have

$$\begin{aligned} E(Y) &= (.75)(.35) + (.80)(.21) + (.85)(.18) + (.90)(.15) \\ &\quad + (.95)(.09) + (1.00)(.02) \\ &= .8240 \end{aligned}$$

The average number of rooms occupied per day at the Merlin is

$$(.8240)(60) \approx 49.4$$

or approximately 49.4 rooms.

b. The expected daily profit at the Regina is given by

$$46.1(10) = 461$$

or \$461. The expected daily profit at the Merlin is given by

$$(49.4)(9) \approx 445$$

or \$445. From these results we conclude that the investors should purchase the Regina Inn, which is expected to yield a higher daily profit. ■■■■

EXAMPLE 5

The Island Club is holding a fund-raising raffle. Ten thousand tickets have been sold for \$2 each. There will be a first prize of \$3000, 3 second prizes of \$1000 each, 5 third prizes of \$500 each, and 20 consolation prizes of \$100 each. Letting X denote the net winnings (that is, winnings less the cost of the ticket) associated with the tickets, find $E(X)$. Interpret your results.

SOLUTION ✓

The values assumed by X are $(0 - 2)$, $(100 - 2)$, $(500 - 2)$, $(1000 - 2)$, and $(3000 - 2)$—that is, -2, 98, 498, 998, and 2998—which correspond, respectively, to the value of a losing ticket, a consolation prize, a third prize, and so on. The probability distribution of X may be calculated in the usual manner and appears in Table 8.9. Using the table, we find

Table 8.9 Probability Distribution for a Raffle

x	$P(X = x)$
−2	.9971
98	.0020
498	.0005
998	.0003
2998	.0001

$$\begin{aligned} E(X) &= (-2)(.9971) + 98(.0020) + 498(.0005) \\ &\quad + 998(.0003) + 2998(.0001) \\ &= -0.95 \end{aligned}$$

This expected value gives the long-run average loss (negative gain) of a holder of one ticket; that is, if one participated in such a raffle by purchasing one ticket each time, in the long run, one may expect to lose, on the average, 95 cents per raffle. ■■■■

EXAMPLE 6

In the game of roulette as played in Las Vegas casinos, the wheel is divided into 38 compartments numbered 1 through 36, 0, and 00. One-half of the numbers 1 through 36 are red, the other half black, and 0 and 00 are green (Figure 8.8). Of the many types of bets that may be placed, one type involves betting on the outcome of the color of the winning number. For example, one may place a certain sum of money on *red*. If the winning number is red, one wins an amount equal to the bet placed and loses the bet otherwise. Find the expected value of the winnings on a $1 bet placed on *red*.

FIGURE 8.8
Roulette wheel

SOLUTION ✓

Let X be a random variable whose values are 1 and -1, which correspond to a win and a loss. The probabilities associated with the values 1 and -1 are 18/38 and 20/38, respectively. Therefore, the expected value is given by

$$E(X) = 1\left(\frac{18}{38}\right) + (-1)\left(\frac{20}{38}\right) = -\frac{2}{38}$$
$$\approx -0.053$$

Thus, if one places a $1 bet on *red* over and over again, one may expect to lose, on the average, approximately 5 cents per bet in the long run. ■■■■

Examples 5 and 6 illustrate games that are not "fair." Of course, most participants in such games are aware of this fact and participate in them for other reasons. In a fair game, neither party has an advantage, a condition that translates into the condition that $E(X) = 0$, where X takes on the values of a player's winnings.

EXAMPLE 7

Mike and Bill play a card game with a standard deck of 52 cards. Mike selects a card from a well-shuffled deck and receives A dollars from Bill if the card selected is a diamond; otherwise, Mike pays Bill a dollar. Determine the value of A if the game is to be fair.

SOLUTION ✓

Let X denote a random variable whose values are associated with Mike's winnings. Then X takes on the value A with probability $P(X = A) = 1/4$ (since there are 13 diamonds in the deck) if Mike wins and takes on the value -1 with probability $P(X = -1) = 3/4$ if Mike loses. Since the game is to be a fair one, the expected value $E(X)$ of Mike's winnings must be equal to zero, that is,

$$E(X) = A\left(\frac{1}{4}\right) + (-1)\left(\frac{3}{4}\right) = 0$$

Solving this equation for A gives $A = 3$. Thus, the card game will be fair if Bill makes a \$3 payoff for a winning bet of \$1 placed by Mike. ■■■■

ODDS

In everyday parlance the probability of the occurrence of an event is often stated in terms of the *odds in favor of* (or *odds against*) the occurrence of the event. For example, one often hears statements such as "the odds that the Dodgers will win the World Series this season are 7 to 5" and "the odds that it will not rain tomorrow are 3 to 2." We will return to these examples later. But first, let us look at a definition that ties together these two concepts.

Odds in Favor Of and Odds Against

If $P(E)$ is the probability of an event E occurring, then

1. The **odds in favor of E** occurring are

$$\frac{P(E)}{1-P(E)} = \frac{P(E)}{P(E^c)} \qquad [P(E) \neq 1] \qquad \textbf{(3a)}$$

2. The **odds against E** occurring are

$$\frac{1-P(E)}{P(E)} = \frac{P(E^c)}{P(E)} \qquad [P(E) \neq 0] \qquad \textbf{(3b)}$$

REMARKS

1. Notice that the odds in favor of the occurrence of an event are given by the ratio of the probability of the event occurring to the probability of the event not occurring. The odds against the occurrence of an event are given by the reciprocal of the odds in favor of the occurrence of the event.
2. Whenever possible, odds are expressed as ratios of whole numbers. If the odds in favor of E are a/b, we say the odds in favor of E are a to b. If the odds against E occurring are b/a, we say the odds against E are b to a.

■■■

EXAMPLE 8 Find the odds in favor of winning a bet on *red* in American roulette. What are the odds against winning a bet on *red*?

SOLUTION ✓ The probability of winning a bet here—the probability that the ball lands in a red compartment—is given by $P = 18/38$. Therefore, using Formula (3a), we see that the odds in favor of winning a bet on *red* are

$$\frac{P(E)}{1-P(E)} = \frac{\frac{18}{38}}{1-\frac{18}{38}} \qquad (E\text{, event of winning a bet on } red)$$

$$= \frac{\frac{18}{38}}{\frac{38-18}{38}}$$

$$= \frac{18}{38} \cdot \frac{38}{20}$$

$$= \frac{18}{20} = \frac{9}{10}$$

or 9 to 10. Next, using (3b), we see that the odds against winning a bet on *red* are 10/9, or 10 to 9. ■■■■

Now, suppose the odds in favor of the occurrence of an event are a to b. Then, (3a) gives

$$\frac{a}{b} = \frac{P(E)}{1-P(E)}$$

$$a[1 - P(E)] = bP(E) \qquad \text{(Cross-multiplying)}$$

$$a - aP(E) = bP(E)$$

$$a = (a + b)P(E)$$

$$P(E) = \frac{a}{a+b}$$

which leads us to the following result.

Probability of an Event (Given the Odds)

If the odds in favor of an event E occurring are a to b, then the probability of E occurring is

$$P(E) = \frac{a}{a+b} \qquad \textbf{(4)}$$

Formula (4) is often used to determine subjective probabilities, as the next example shows.

EXAMPLE 9 Consider each of the following statements.

a. "The odds that the Dodgers will win the World Series this season are 7 to 5."
b. "The odds that it will not rain tomorrow are 3 to 2."

Express each of these odds as a probability of the event occurring.

SOLUTION ✓

a. Using Formula (4) with $a = 7$ and $b = 5$ gives the required probability as

$$\frac{7}{7+5} = \frac{7}{12} \approx .5833$$

b. Here, the event is that it will not rain tomorrow. Using (4) with $a = 3$ and $b = 2$, we conclude that the probability that it will not rain tomorrow is

$$\frac{3}{3+2} = \frac{3}{5} = .6$$

In concluding this section we should mention that in addition to the mean there are two other measures of central tendency of a set of numerical data: the median and the mode of a set of numbers. The **median** is the middle value in a set of data arranged in increasing or decreasing order (when there is an odd number of entries). If there is an even number of entries, the median is the mean of the two middle numbers. The **mode** is the value that occurs most frequently in a set of data. (See Exercises 34 and 35.)

Group Discussion

In the movie *Casino,* the executive of the Tangiers Casino, Sam Rothstein (Robert DeNiro), fired the manager of the slot machines in the casino after three gamblers hit three "million dollar" jackpots in a span of 20 minutes. Rothstein claimed that it was a scam and that somebody had gotten into those machines to set the wheels. He was especially annoyed at the slot machine manager's assertion that there was no way to determine this. According to Rothstein the odds of hitting a jackpot in a four-wheel machine is 1 in $1\frac{1}{2}$ million, and the probability of hitting three jackpots in a row is "in the billions." "It cannot happen! It will not happen!" To see why Mr. Rothstein was so indignant, find the odds of hitting the jackpots in three of the machines in quick succession and comment on the likelihood of this happening.

SELF-CHECK EXERCISES 8.2

1. Find the expected value of a random variable X having the following probability distribution:

x	-4	-3	-1	0	1	2
$P(X = x)$	.10	.20	.25	.10	.25	.10

2. The developer of the Shoreline Condominiums has provided the following estimate of the probability that 20, 25, 30, 35, 40, 45, or 50 of the townhouses will be sold within the first month they are offered for sale:

Number of Units	20	25	30	35	40	45	50
Probability	.05	.10	.30	.25	.15	.10	.05

How many townhouses can the developer expect to sell within the first month they are put on the market?

Solutions to Self-Check Exercises 8.2 can be found on page 483.

8.2 Exercises

A calculator is recommended for this exercise set.

1. During the first year at a university that uses a 4-point grading system, a freshman took ten 3-credit courses and received two As, three Bs, four Cs, and one D.
   a. Compute this student's grade point average.
   b. Let the random variable X denote the number of points corresponding to a given letter grade. Find the probability distribution of the random variable X and compute $E(X)$, the expected value of X.

2. Records kept by the chief dietitian at the university cafeteria over a 30-wk period show the following weekly consumption of milk (in gallons):

Milk	200	205	210	215	220	225	230	235	240
Number of Weeks	3	4	6	5	4	3	2	2	1

   a. Find the average number of gallons of milk consumed per week in the cafeteria.
   b. Let the random variable X denote the number of gallons of milk consumed in a week at the cafeteria. Find the probability distribution of the random variable X and compute $E(X)$, the expected value of X.

3. Find the expected value of a random variable X having the following probability distribution:

x	-5	-1	0	1	5	8
$P(X = x)$	.12	.16	.28	.22	.12	.1

4. Find the expected value of a random variable X having the following probability distribution:

x	0	1	2	3	4	5
$P(X = x)$	$\frac{1}{8}$	$\frac{1}{4}$	$\frac{3}{16}$	$\frac{1}{4}$	$\frac{1}{16}$	$\frac{1}{8}$

5. The daily earnings X of an employee who works on a commission basis are given by the following probability distribution. Find the employee's expected earnings.

x (in \$)	0	25	50	75	100	125	150
$P(X = x)$	.07	.12	.17	.14	.28	.18	.04

6. In a four-child family, what is the expected number of boys? (Assume that the probability of a boy being born is the same as the probability of a girl being born.)

7. Based on past experience, the manager of the Video-Rama Store has compiled the following table, which gives the probabilities that a customer who enters the Video-Rama Store will buy 0, 1, 2, 3, or 4 videocassettes.

How many videocassettes can a customer entering this store be expected to buy?

Number of Video-cassettes	0	1	2	3	4
Probability	.42	.36	.14	.05	.03

8. If a sample of three batteries is selected from a lot of ten, of which two are defective, what is the expected number of defective batteries?

9. AUTO ACCIDENTS The number of accidents that occur at a certain intersection known as "Five Corners" on a Friday afternoon between the hours of 3 P.M. and 6 P.M., along with the corresponding probabilities, are shown in the following table. Find the expected number of accidents during the period in question.

Number of Accidents	0	1	2	3	4
Probability	.935	.03	.02	.01	.005

10. EXPECTED DEMAND The owner of a newsstand in a college community estimates the weekly demand for a certain magazine as follows:

Quantity Demanded	10	11	12
Probability	.05	.15	.25

Quantity Demanded	13	14	15
Probability	.30	.20	.05

Find the number of issues of the magazine that the newsstand owner can expect to sell per week.

11. EXPECTED PRODUCT RELIABILITY A bank has two automatic tellers at its main office and two at each of its three branches. The number of machines that break down on a given day, along with the corresponding probabilities, are shown in the following table.

Number of Machines That Break Down	0	1	2	3	4
Probability	.43	.19	.12	.09	.04

Number of Machines That Break Down	5	6	7	8
Probability	.03	.03	.02	.05

Find the expected number of machines that will break down on a given day.

12. LOTTERIES In a lottery, 5000 tickets are sold for $1 each. One first prize of $2000, 1 second prize of $500, 3 third prizes of $100, and 10 consolation prizes of $25 are to be awarded. What are the expected net earnings of a person who buys one ticket?

13. LIFE INSURANCE PREMIUMS A man wishes to purchase a 5-yr term-life insurance policy that will pay the beneficiary $20,000 in the event that the man's death occurs during the next 5 yr. Using life insurance tables, he determines that the probability that he will live another 5 yr is .96. What is the minimum amount that he can expect to pay for his premium?
Hint: The minimum premium occurs when the insurance company's expected profit is zero.

14. EXPECTED GAIN A woman purchased a $10,000, 1-yr term-life insurance policy for $130. Assuming that the probability that she will live another year is .992, find the company's expected gain.

15. LIFE INSURANCE POLICIES As a fringe benefit, Dennis Taylor receives a $25,000 life insurance policy from his employer. The probability that Dennis will live another year is .9935. If he purchases the same coverage for himself, what is the minimum amount that he can expect to pay for the policy?

16. EXPECTED PROFIT A buyer for Discount Fashions, an outlet for women's apparel, is considering buying a batch of clothing for $64,000. She estimates that the company will be able to sell it for $80,000, $75,000, or $70,000 with probabilities of .30, .60, and .10, respectively. Based on these estimates, what will be the company's expected gross profit?

17. INVESTMENT ANALYSIS The proprietor of Midland Construction Company has to decide between two projects. He estimates that the first project will yield a profit of $180,000 with a probability of .7 or a profit of $150,000 with a probability of .3; the second project will yield a profit of $220,000 with a probability of .6 or a profit of $80,000 with a probability of .4. Which project should the proprietor choose if he wants to maximize his expected profit?

18. CABLE TELEVISION The management of Multi-Vision, Inc., a cable TV company, intends to submit a bid for the cable television rights in one of two cities, A or B. If

the company obtains the rights to city A, the probability of which is .2, the estimated profit over the next 10 yr is $10 million; if the company obtains the rights to city B, the probability of which is .3, the estimated profit over the next 10 yr is $7 million. The cost of submitting a bid for rights in city A is $250,000 and that of city B is $200,000. By comparing the expected profits for each venture, determine whether the company should bid for the rights in city A or city B.

19. **EXPECTED AUTO SALES** Roger Hunt intends to purchase one of two car dealerships currently for sale in a certain city. Records obtained from each of the two dealers reveal that their weekly volume of sales, with corresponding probabilities, are as follows:

Dahl Motors

Number of Cars Sold per Week	5	6	7	8
Probability	.05	.09	.14	.24

Number of Cars Sold per Week	9	10	11	12
Probability	.18	.14	.11	.05

Farthington Auto Sales

Number of Cars Sold per Week	5	6	7
Probability	.08	.21	.31

Number of Cars Sold per Week	8	9	10
Probability	.24	.10	.06

The average profit per car at Dahl Motors is $362, and the average profit per car at Farthington Auto Sales is $436.

a. Find the average number of cars sold per week at each dealership.

b. If Roger's objective is to purchase the dealership that generates the higher weekly profit, which dealership should he purchase? (Compare the expected weekly profit for each dealership.)

20. **EXPECTED HOME SALES** Sally Leonard, a real estate broker, is relocating in a large metropolitan area where she has received job offers from realty company A and realty company B. The number of houses she expects to sell in a year at each firm and the associated probabilities are shown in the following tables:

Company A

Number of Houses Sold	12	13	14	15	16
Probability	.02	.03	.05	.07	.07

Number of Houses Sold	17	18	19	20
Probability	.16	.17	.13	.11

Number of Houses Sold	21	22	23	24
Probability	.09	.06	.03	.01

Company B

Number of Houses Sold	6	7	8	9	10
Probability	.01	.04	.07	.06	.11

Number of Houses Sold	11	12	13	14
Probability	.12	.19	.17	.13

Number of Houses Sold	15	16	17	18
Probability	.04	.03	.02	.01

The average price of a house in the locale of company A is $104,000, whereas the average price of a house in the locale of company B is $177,000. If Sally will receive a 3% commission on sales at both companies, which job offer should she accept to maximize her expected yearly commission?

(continued on p. 482)

Portfolio

LILLI MEISELMAN

TITLE: Buyer

Lilli Meiselman's job is challenging. She visits New York at least twice a month to buy clothes from a number of different manufacturers. She works with the store's in-house advertising agency, planning advertising as well as approving copy. Meiselman notes that the vast majority of the ads appear in local newspapers, the store's principal advertising medium.

For all the demands on her time, Meiselman enjoys her work. "It's gratifying to see things I bought for the stores being sold," she says. Although Meiselman decides how much to spend on particular items and which styles, sizes, and colors to select, her decisions are the end result of detailed plans that guide her buying decisions.

Based on the previous year's sales, Meiselman will "work with a departmental planner" to develop a seasonal merchandise plan (in retail there are only two seasons, fall and spring, each covering a 6-month time span).

As part of that merchandising plan, Meiselman must determine how much inventory is required per month. If she needs $600,000 worth of inventory at the beginning of June, she subtracts projected sales of $400,000 from the May inventory of $800,000, leaving a balance of $400,000. To meet her June inventory goal of $600,000, she has to increase her inventory by an additional $200,000.

The finances have to be coordinated with actual quantities of suits, dresses, blouses, outerwear, and so on. Using a sales-to-stock ratio, Meiselman estimates that perhaps half of a particular item will sell in any given month. To meet her financial goals, she needs sufficient quantity on hand to sell. For example, Meiselman would need 4000 raincoats in April retailing at $99 each to reach a $200,000 sales goal if the stores' sales were expected to reach a 50% sales-to-stock ratio.

Sales volume is the key to success in any discount business, whether clothing or home appliances. Meiselman's merchandise is marked up or down depending on that volume. Her goal is a 45% markup on all merchandise. If a line of suits doesn't sell, she may mark it down, achieving only a 35% markup over the wholesale price. To balance out the loss, she marks up another line of items so that her *average* markup hits 45%.

With its high overhead, the chain has to generate sufficient sales volume and profits to stay in business. Meiselman's fashion choices have helped make that goal a continuing reality.

Buying women's clothes for a chain of 50 discount clothing stores is a demanding job, requiring Meiselman to be part fashion arbiter and part accountant. She must balance her fashion choices against a bottom line that has to show a profit. To make her job even more difficult, Meiselman's selections are judged by women from northern New England to the midwestern states. What sells in one area may not sell in another. An incorrect choice can be a costly mistake.

21. **WEATHER PREDICTIONS** Suppose the probability that it will rain tomorrow is .3.
 a. What are the odds that it will rain tomorrow?
 b. What are the odds that it will not rain tomorrow?

22. **ROULETTE** In American roulette, as described in Example 6, a player may bet on a split (two adjacent numbers). In this case, if the player bets $1 and either number comes up, the player wins $17 and gets his $1 back. If neither comes up, he loses his $1 bet. Find the expected value of the winnings on a $1 bet placed on a split.

23. **ROULETTE** If a player placed a $1 bet on *red* and a $1 bet on *black* in a single play in American roulette, what would be the expected value of his winnings?

24. **ROULETTE** In European roulette the wheel is divided into 37 compartments numbered 1 through 36 and 0. (In American roulette there are 38 compartments numbered 1 through 36, 0, and 00.) Find the expected value of the winnings on a $1 bet placed on *red* in European roulette.

25. The probability of an event E occurring is .8. What are the odds in favor of E occurring? What are the odds against E occurring?

26. The probability of an event E not occurring is .6. What are the odds in favor of E occurring? What are the odds against E occurring?

27. The odds in favor of an event E occurring are 9 to 7. What is the probability of E occurring?

28. The odds against an event E occurring are 2 to 3. What is the probability of E not occurring?

29. **ODDS** Carmen, a computer sales representative, feels that the odds are 8 to 5 that she will clinch the sale of a minicomputer to a certain company. What is the (subjective) probability that Carmen will make the sale?

30. **SPORTS** Steffi feels that the odds in favor of her winning her tennis match tomorrow are 7 to 5. What is the (subjective) probability that she will win her match tomorrow?

31. **SPORTS** If a sports forecaster states that the odds of a certain boxer winning a match are 4 to 3, what is the probability that the boxer will win the match?

32. **ODDS** Bob, the proprietor of Midland Lumber, feels that the odds in favor of a business deal going through are 9 to 5. What is the (subjective) probability that this deal will not materialize?

33. **ROULETTE**
 a. Show that for any number c,

$$E(cX) = cE(X)$$

 b. Use this result to find the expected loss if a gambler bets $300 on red in a single play in American roulette.
 Hint: Use the results of Example 6.

34. **EXAM SCORES** In an examination given to a class of 20 students, the following test scores were obtained:

40	45	50	50	55	60	60	75	75	80
80	85	85	85	85	90	90	95	95	100

 a. Find the mean, or average, score, the mode, and the median score.
 b. Which of these three measures of central tendency do you think is the least representative of the set of scores?

35. **WAGE RATES** The frequency distribution of the hourly wage rates (in dollars) among blue-collar workers in a certain factory is given in the following table. Find the mean (or average) wage rate, the mode, and the median wage rate of these workers.

Wage Rate	10.70	10.80	10.90	11.00	11.10	11.20
Frequency	60	90	75	120	60	45

In Exercises 36 and 37, determine whether the statement is true or false. If it is true, explain why it is true. If it is false, give an example to show why it is false.

36. A game between two persons is fair if the expected value to both persons is zero.

37. If the odds in favor of an event E occurring are a to b, then the probability of E^c occurring is $b/(a + b)$.

Spreadsheet examples and exercises for this section that may be solved using the Microsoft® Excel program are given at the Brooks/Cole Web site:
http://www.brookscole.com/math/authors/tans/

SOLUTIONS TO SELF-CHECK EXERCISES 8.2

1. $E(X) = (-4)(.10) + (-3)(.20) + (-1)(.25) + (0)(.10) + (1)(.25) + (2)(.10) = -0.8$

2. Let X denote the number of townhouses that will be sold within 1 mo of being put on the market. Then, the number of townhouses the developer expects to sell within 1 mo is given by the expected value of X—that is, by

$$\begin{aligned} E(X) &= 20(.05) + 25(.10) + 30(.30) + 35(.25) \\ &\quad + 40(.15) + 45(.10) + 50(.05) \\ &= 34.25 \end{aligned}$$

or 34 townhouses.

8.3 Variance and Standard Deviation

VARIANCE

FIGURE 8.9
The histograms of two probability distributions

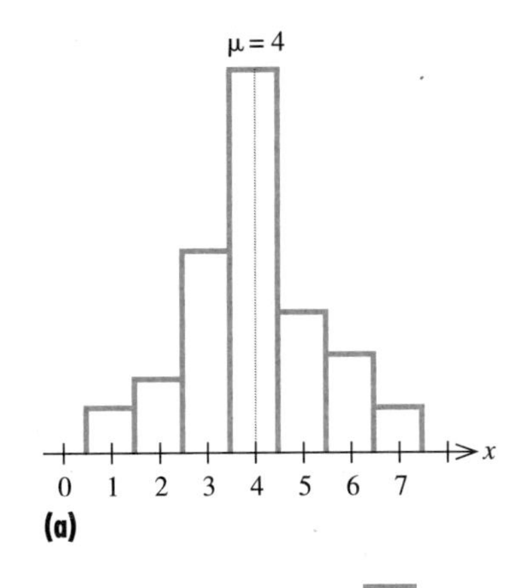

(a)

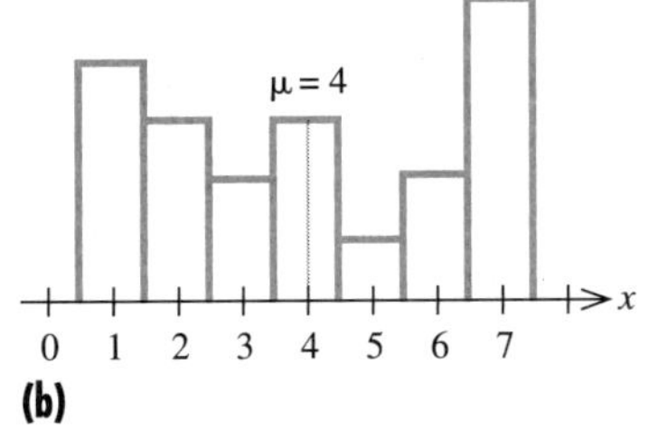

(b)

The mean, or expected value, of a random variable enables us to express an important property of the probability distribution associated with the random variable in terms of a single number. But the knowledge of the location, or central tendency, of a probability distribution alone is usually not enough to give a reasonably accurate picture of the probability distribution. Consider, for example, the two probability distributions whose histograms appear in Figure 8.9. Both distributions have the same expected value, or mean, $\mu = 4$ (the Greek letter μ is read "mu"). Note that the probability distribution with the histogram shown in Figure 8.9a is closely concentrated about its mean μ, whereas the one with the histogram shown in Figure 8.9b is widely dispersed or spread about its mean.

As another example, suppose David Horowitz, host of the popular television show *The Consumer Advocate,* decides to demonstrate the accuracy of the weights of two popular brands of potato chips. Ten packages of potato chips of each brand are selected at random and weighed carefully. The results are as follows:

Weight in Ounces										
Brand A	16.1	16	15.8	16	15.9	16.1	15.9	16	16	16.2
Brand B	16.3	15.7	15.8	16.2	15.9	16.1	15.7	16.2	16	16.1

In Example 3 we will verify that the mean weights for each of the two brands is 16 ounces. However, a cursory examination of the data now shows that the weights of the brand B packages exhibit much greater dispersion about the mean than those of brand A.

One measure of the degree of dispersion, or spread, of a probability distribution about its mean is given by the variance of the random variable associated with the probability distribution. A probability distribution with a small spread about its mean will have a small variance, whereas one with a larger spread will have a larger variance. Thus, the variance of the random variable associated with the probability distribution whose histogram appears in Figure 8.9a is smaller than the variance of the random variable associated with the probability distribution whose histogram is shown in Figure 8.9b (see Example 1). Also, as we will see in Example 3, the variance of the random variable associated with the weights of the brand A potato chips is smaller than that of the random variable associated with the weights of the brand B potato chips. (This observation was made earlier.)

We now define the variance of a random variable.

Variance of a Random Variable X

Suppose a random variable has the probability distribution

x	x_1	x_2	x_3	$\cdots$	x_n
$P(X = x)$	p_1	p_2	p_3	$\cdots$	p_n

and expected value

$$E(X) = \mu$$

Then the **variance** of the random variable X is

$$\text{Var}(X) = p_1(x_1 - \mu)^2 + p_2(x_2 - \mu)^2 + \cdots + p_n(x_n - \mu)^2 \qquad (5)$$

Let's look a little closer at Equation (5). First, note that the numbers

$$x_1 - \mu, x_2 - \mu, \ldots, x_n - \mu \qquad (6)$$

measure the **deviations** of $x_1, x_2, \ldots, x_n$ from μ, respectively. Thus, the numbers

$$(x_1 - \mu)^2, (x_2 - \mu)^2, \ldots, (x_n - \mu)^2 \qquad (7)$$

measure the squares of the deviations of $x_1, x_2, \ldots, x_n$ from μ, respectively. Next, by multiplying each of the numbers in (7) by the probability associated with each value of the random variable X, the numbers are weighted accordingly so that their sum is a measure of the variance of X about its mean. An attempt to define the variance of a random variable about its mean in a similar manner using the deviations in (6) rather than their squares would not be fruitful since some of the deviations may be positive whereas others may be negative and (because of cancellations) the sum will not give a satisfactory measure of the variance of the random variable.

EXAMPLE 1 Find the variance of the random variable X and of the random variable Y whose probability distributions are shown in the following table. These are the probability distributions associated with the histograms shown in Figure 8.9a–b.

x	$P(X = x)$	y	$P(Y = y)$
1	.05	1	.2
2	.075	2	.15
3	.2	3	.1
4	.375	4	.15
5	.15	5	.05
6	.1	6	.1
7	.05	7	.25

SOLUTION ✓ The mean of the random variable X is given by

$$\begin{aligned}\mu_X &= (1)(.05) + (2)(.075) + (3)(.2) + (4)(.375) + (5)(.15) \\ &\quad + (6)(.1) + (7)(.05) \\ &= 4\end{aligned}$$

Therefore, using Equation (5) and the data from the probability distribution of X, we find that the variance of X is given by

$$\begin{aligned}\text{Var}(X) &= (.05)(1 - 4)^2 + (.075)(2 - 4)^2 + (.2)(3 - 4)^2 \\ &\quad + (.375)(4 - 4)^2 + (.15)(5 - 4)^2 \\ &\quad + (.1)(6 - 4)^2 + (.05)(7 - 4)^2 \\ &= 1.95\end{aligned}$$

Next, we find that the mean of the random variable Y is given by

$$\begin{aligned}\mu_Y &= (1)(.2) + (2)(.15) + (3)(.1) + (4)(.15) + (5)(.05) \\ &\quad + (6)(.1) + (7)(.25) \\ &= 4\end{aligned}$$

and so the variance of Y is given by

$$\begin{aligned}\text{Var}(Y) &= (.2)(1 - 4)^2 + (.15)(2 - 4)^2 + (.1)(3 - 4)^2 \\ &\quad + (.15)(4 - 4)^2 + (.05)(5 - 4)^2 \\ &\quad + (.1)(6 - 4)^2 + (.25)(7 - 4)^2 \\ &= 5.2\end{aligned}$$

Note that Var(X) is smaller than Var(Y), which confirms the earlier observations about the spread, or dispersion, of the probability distribution of X and Y, respectively. ■■■■

STANDARD DEVIATION

Because Equation (5), which gives the variance of the random variable X, involves the squares of the deviations, the unit of measurement of $\text{Var}(X)$ is the square of the unit of measurement of the values of X. For example, if the values assumed by the random variable X are measured in units of a gram, then $\text{Var}(X)$ will be measured in units involving the *square* of a gram. To remedy this situation, one normally works with the square root of $\text{Var}(X)$ rather than $\text{Var}(X)$ itself. The former is called the standard deviation of X.

Standard Deviation of a Random Variable X

The **standard deviation** of a random variable X, σ (pronounced "sigma"), is defined by

$$\begin{aligned}\sigma &= \sqrt{\text{Var}(X)} \\ &= \sqrt{p_1(x_1-\mu)^2 + p_2(x_2-\mu)^2 + \cdots + p_n(x_n-\mu)^2}\end{aligned} \qquad (8)$$

where $x_1, x_2, \ldots, x_n$ denote the values assumed by the random variable X and $p_1 = P(X = x_1)$, $p_2 = P(X = x_2)$, $\ldots$, $p_n = P(X = x_n)$.

EXAMPLE 2

Find the standard deviations of the random variables X and Y of Example 1.

SOLUTION ✓

From the results of Example 1, we have $\text{Var}(X) = 1.95$ and $\text{Var}(Y) = 5.2$. Taking their respective square roots, we have

$$\begin{aligned}\sigma_X &= \sqrt{1.95} \\ &\approx 1.40 \\ \sigma_Y &= \sqrt{5.2} \\ &\approx 2.28\end{aligned}$$

■■■■

EXAMPLE 3

Let X and Y denote the random variables whose values are the weights of the brand A and brand B potato chips, respectively (see page 483). Compute the means and standard deviations of X and Y and interpret your results.

SOLUTION ✓

The probability distributions of X and Y may be computed from the given data as follows:

Brand A

x	Relative Frequency of Occurrence	$P(X = x)$
15.8	1	.1
15.9	2	.2
16.0	4	.4
16.1	2	.2
16.2	1	.1

Brand B

y	Relative Frequency of Occurrence	$P(Y = y)$
15.7	2	.2
15.8	1	.1
15.9	1	.1
16.0	1	.1
16.1	2	.2
16.2	2	.2
16.3	1	.1

The means of X and Y are given by

$$\begin{aligned}\mu_X &= (.1)(15.8) + (.2)(15.9) + (.4)(16.0) + (.2)(16.1) \\ &\quad + (.1)(16.2) \\ &= 16 \\ \mu_Y &= (.2)(15.7) + (.1)(15.8) + (.1)(15.9) + (.1)(16.0) \\ &\quad + (.2)(16.1) + (.2)(16.2) + (.1)(16.3) \\ &= 16\end{aligned}$$

Therefore,

$$\begin{aligned}\text{Var}(X) &= (.1)(15.8 - 16)^2 + (.2)(15.9 - 16)^2 + (.4)(16 - 16)^2 \\ &\quad + (.2)(16.1 - 16)^2 + (.1)(16.2 - 16)^2 \\ &= 0.012 \\ \text{Var}(Y) &= (.2)(15.7 - 16)^2 + (.1)(15.8 - 16)^2 + (.1)(15.9 - 16)^2 \\ &\quad + (.1)(16 - 16)^2 + (.2)(16.1 - 16)^2 + (.2)(16.2 - 16)^2 \\ &\quad + (.1)(16.3 - 16)^2 \\ &= 0.042\end{aligned}$$

so that the required standard deviations are

$$\begin{aligned}\sigma_X &= \sqrt{\text{Var}(X)} \\ &= \sqrt{0.012} \\ &\approx 0.11 \\ \sigma_Y &= \sqrt{\text{Var}(Y)} \\ &= \sqrt{0.042} \\ &\approx 0.20\end{aligned}$$

The mean of X and that of Y are both equal to 16. Therefore, the average weight of a package of potato chips of either brand is 16 ounces. However, the standard deviation of Y is greater than that of X. This tells us that the weights of the packages of brand B potato chips are more widely dispersed about the common mean of 16 than are those of brand A. ■■■■

Group Discussion

A useful alternative formula for the variance is

$$\sigma^2 = E(X^2) - \mu^2$$

where $E(X)$ is the expected value of X.

1. Establish the validity of the formula.
2. Use the formula to verify the calculations in Example 3.

Group Discussion

Suppose the mean weight of m packages of brand A potato chips is μ_1 and the standard deviation from the mean of their weight distribution is σ_1. Also suppose the mean weight of n packages of brand B potato chips is μ_2 and the standard deviation from the mean of their weight distribution is σ_2.

1. Show that the mean of the weights of packages of brand A and brand B combined is

$$\mu = \frac{m\mu_1 + n\mu_2}{m + n}$$

2. If $\mu_1 = \mu_2$, show that the standard deviation from the mean of the combined-weight distribution is

$$\sigma = \left[\frac{m\sigma_1^2 + n\sigma_2^2}{m + n}\right]^{1/2}$$

3. Refer to Example 3, page 486. Using the results of parts 1 and 2, find the mean and the standard deviation of the combined-weight distribution.

CHEBYCHEV'S INEQUALITY

The standard deviation of a random variable X may be used in statistical estimations. For example, the following result, derived by the Russian mathematician P. L. Chebychev (1821–1894), gives the proportion of the values of X lying within k standard deviations of the expected value of X.

Chebychev's Inequality

Let X be a random variable with expected value μ and standard deviation σ. Then, the probability that a randomly chosen outcome of the experiment lies between $\mu - k\sigma$ and $\mu + k\sigma$ is at least $1 - (1/k^2)$; that is,

$$P(\mu - k\sigma \leq X \leq \mu + k\sigma) \geq 1 - \frac{1}{k^2} \tag{9}$$

To shed some light on this result, let's take $k = 2$ in Inequality (9) and compute

$$P(\mu - 2\sigma \leq X \leq \mu + 2\sigma) \geq 1 - \frac{1}{2^2} = 1 - \frac{1}{4} = .75$$

This tells us that at least 75% of the outcomes of the experiment lie within 2 standard deviations of the mean (Figure 8.10). Taking $k = 3$ in Formula (9), we have

FIGURE 8.10
At least 75% of the outcomes fall within this interval

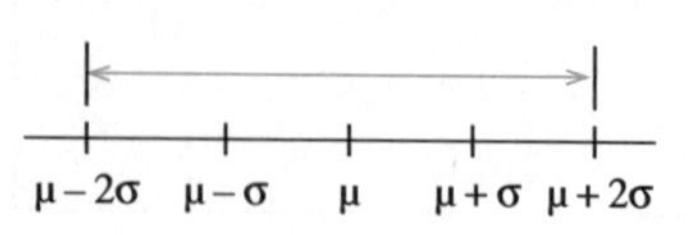

$$P(\mu - 3\sigma \leq X \leq \mu + 3\sigma) \geq 1 - \frac{1}{3^2} = 1 - \frac{1}{9} = \frac{8}{9} \approx .89$$

This tells us that at least 89% of the outcomes of the experiment lie within 3 standard deviations of the mean (Figure 8.11).

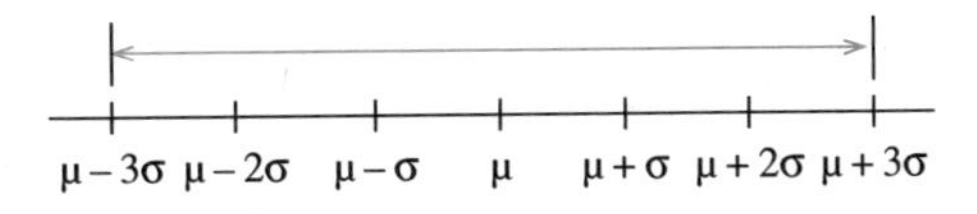

FIGURE 8.11
At least 89% of the outcomes fall within this interval

EXAMPLE 4

A probability distribution has a mean of 10 and a standard deviation of 1.5. Use Chebychev's inequality to estimate the probability that an outcome of the experiment lies between 7 and 13.

SOLUTION ✓

Here, $\mu = 10$ and $\sigma = 1.5$. Next, to determine the value of k, note that $\mu - k\sigma = 7$ and $\mu + k\sigma = 13$. Substituting the appropriate values for μ and σ, we find $k = 2$. Using Chebychev's Inequality (9), we see that the probability that an outcome of the experiment lies between 7 and 13 is given by

$$P(7 \leq X \leq 13) \geq 1 - \left(\frac{1}{2^2}\right)$$
$$= \frac{3}{4}$$

—that is, at least 75%. ■■■■

REMARK The results of Example 4 tell us that at least 75% of the outcomes of the experiment lie between $10 - 2\sigma$ and $10 + 2\sigma$—that is, between 7 and 13. ■■■

EXAMPLE 5

The Great Northwest Lumber Company employs 400 workers in its mills. It has been estimated that X, the random variable measuring the number of mill workers who have industrial accidents during a 1-year period, is distributed with a mean of 40 and a standard deviation of 6. Using Chebychev's Inequality (9), estimate the probability that the number of workers who will have an industrial accident over a 1-year period is between 30 and 50, inclusive.

SOLUTION ✓

Here, $\mu = 40$ and $\sigma = 6$. We wish to estimate $P(30 \leq X \leq 50)$. To use Chebychev's Inequality (9), we first determine the value of k from the equation

$$\mu - k\sigma = 30 \qquad \text{or} \qquad \mu + k\sigma = 50$$

Since $\mu = 40$ and $\sigma = 6$ in this case, we see that k satisfies

$$40 - 6k = 30 \qquad \text{and} \qquad 40 + 6k = 50$$

from which we deduce that $k = 5/3$. Thus, the probability that the number of mill workers who will have an industrial accident during a 1-year period is between 30 and 50 is given by

$$P(30 \leq X \leq 50) \geq 1 - \frac{1}{(5/3)^2}$$
$$= \frac{16}{25}$$

—that is, at least 64%. ■■■■

SELF-CHECK EXERCISES 8.3

1. Compute the mean, variance, and standard deviation of the random variable X with probability distribution as follows:

x	-4	-3	-1	0	2	5
$P(X = x)$	.1	.1	.2	.3	.1	.2

2. James recorded the following travel times (the length of time in minutes it took him to drive to work) on 10 consecutive days:

55 50 52 48 50 52 46 48 50 51

Calculate the mean and standard deviation of the random variable X associated with these data.

Solutions to Self-Check Exercises 8.3 can be found on page 493.

8.3 Exercises

A calculator is recommended for this exercise set.

In Exercises 1–6, the probability distribution of a random variable X is given. Compute the mean, variance, and standard deviation of X.

1.

x	1	2	3	4
$P(X = x)$	.4	.3	.2	.1

2.

x	-4	-2	0	2	4
$P(X = x)$	.1	.2	.3	.1	.3

3.

x	-2	-1	0	1	2
$P(X = x)$	1/16	4/16	6/16	4/16	1/16

4.

x	10	11	12	13	14	15
$P(X = x)$	1/8	2/8	1/8	2/8	1/8	1/8

5.

x	430	480	520	565	580
$P(X = x)$	.1	.2	.4	.2	.1

6.

x	-198	-195	-193	-188	-185
$P(X = x)$	.15	.30	.10	.25	.20

7. The following histograms represent the probability distributions of the random variables X and Y. Determine by inspection which probability distribution has the larger variance.

a.

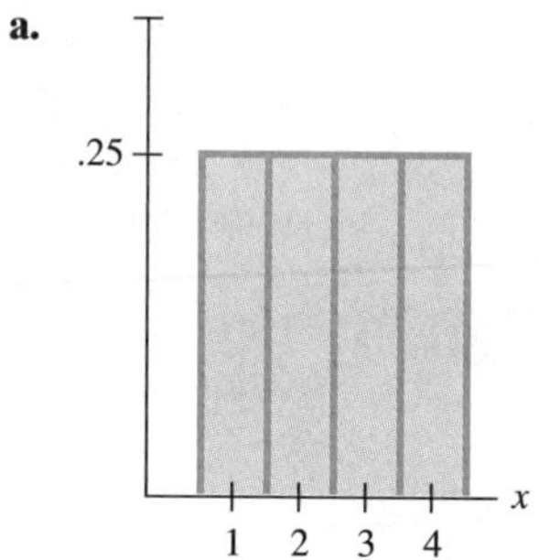

b.

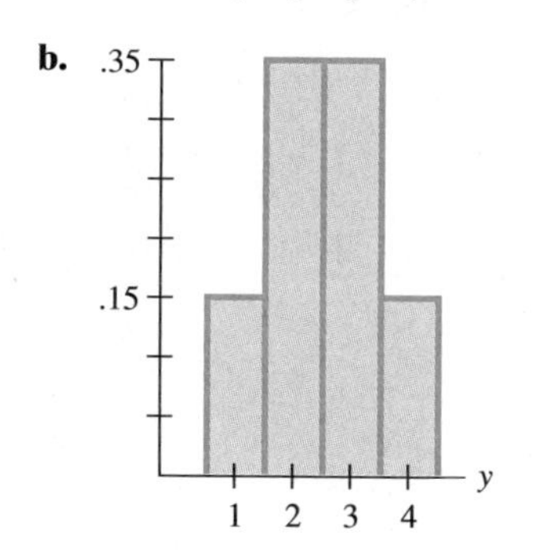

8. The following histograms represent the probability distributions of the random variables X and Y.

a.
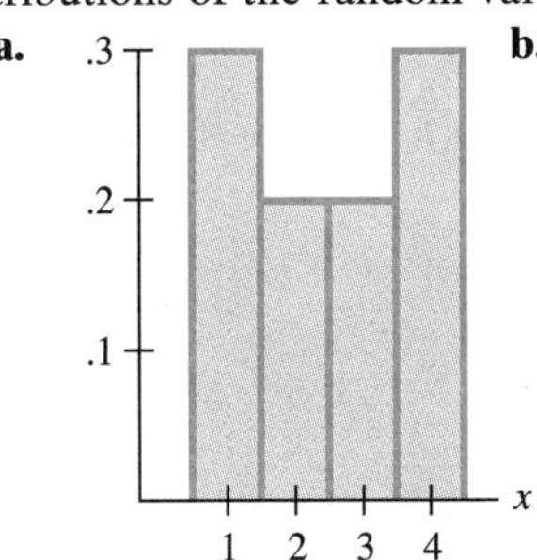

b.
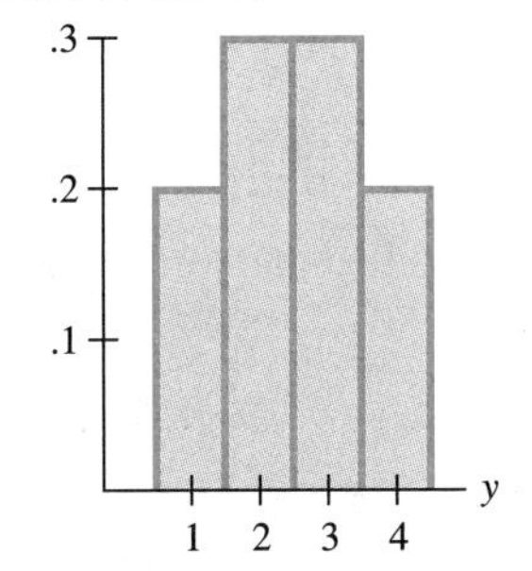

Determine by inspection which probability distribution has the larger variance.

In Exercises 9 and 10, find the variance of the probability distribution for the given histogram.

9.
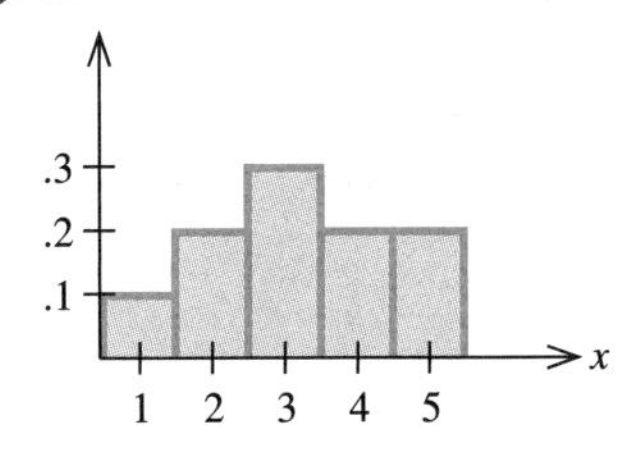

10.
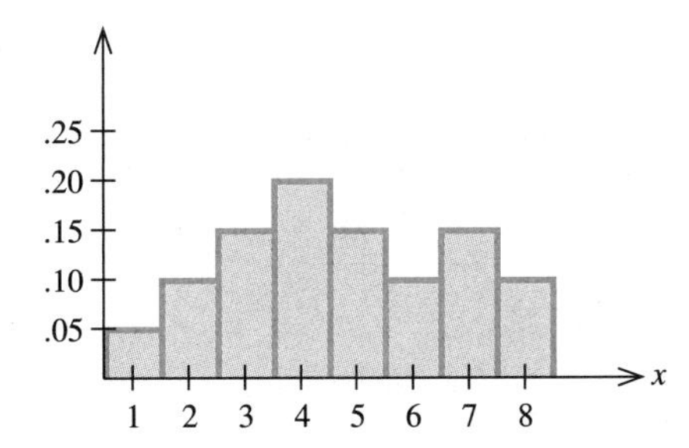

11. An experiment consists of casting an eight-sided die (numbered 1 through 8) and observing the number that appears uppermost. Find the mean and variance of this experiment.

12. DRIVING AGE REQUIREMENTS The minimum age requirement for a regular driver's license differs from state to state. The frequency distribution for this age requirement in the 50 states is given in the following table.

Minimum Age	15	16	17	18	19	21
Frequency of Occurrence	1	15	4	28	1	1

a. Describe a random variable X that is associated with these data.
b. Find the probability distribution for the random variable X.
c. Compute the mean, variance, and standard deviation of X.

13. BIRTH RATES The birth rates in the United States for the years 1981–1990 are given in the following table. (The birth rate is the number of live births/1000 population.)

Year	1981	1982	1983	1984
Birth Rate	15.9	15.5	15.5	15.7

Year	1985	1986	1987
Birth Rate	15.7	15.6	15.7

Year	1988	1989	1990
Birth Rate	15.9	16.2	16.7

a. Describe a random variable X that is associated with these data.
b. Find the probability distribution for the random variable X.
c. Compute the mean, variance, and standard deviation of X.

Source: The World Almanac

14. INVESTMENT ANALYSIS Paul Hunt is considering two business ventures. The anticipated returns (in thousands of dollars) of each venture are described by the following probability distributions:

Venture A

Earnings	Probability
−20	.3
40	.4
50	.3

Venture B

Earnings	Probability
−15	.2
30	.5
40	.3

a. Compute the mean and variance for each venture.
b. Which investment would provide Paul with the higher expected return (the greater mean)?
c. In which investment would the element of risk be less (that is, which probability distribution has the smaller variance)?

15. INVESTMENT ANALYSIS Rosa Walters is considering investing \$10,000 in two mutual funds. The anticipated returns from price appreciation and dividends (in hundreds of dollars) are described by the following probability distributions:

Mutual Fund A

Returns	Probability
−4	.2
8	.5
10	.3

Mutual Fund B

Returns	Probability
−2	.2
6	.4
8	.4

a. Compute the mean and variance associated with the returns for each mutual fund.
b. Which investment would provide Rosa with the higher expected return (the greater mean)?
c. In which investment would the element of risk be less (that is, which probability distribution has the smaller variance)?

16. The distribution of the number of chocolate chips (x) in a cookie is shown in the following table. Find the mean and the variance of the number of chocolate chips in a cookie.

x	0	1	2
$P(X = x)$	.01	.03	.05

x	3	4	5
$P(X = x)$	.11	.13	.24

x	6	7	8
$P(X = x)$	.22	.16	.05

17. Formula (5) can also be expressed in the form

$$\text{Var}(X) = (p_1x_1^2 + p_2x_2^2 + \cdots + p_nx_n^2) - \mu^2$$

Find the variance of the distribution of Exercise 1, using this formula.

18. Find the variance of the distribution of Exercise 16, using the formula

$$\text{Var}(X) = (p_1x_1^2 + p_2x_2^2 + \cdots + p_nx_n^2) - \mu^2$$

19. HOUSING PRICES A survey was conducted by the market research department of the National Real Estate Company among 500 prospective buyers in a large metropolitan area to determine the maximum price a prospective buyer would be willing to pay for a house. From the data collected, the distribution that follows was obtained. Compute the mean, variance, and standard deviation of the maximum price (in thousands of dollars) that these buyers were willing to pay for a house.

Maximum Price Considered, x	$P(X = x)$
180	$\frac{10}{500}$
190	$\frac{20}{500}$
200	$\frac{75}{500}$
210	$\frac{85}{500}$
220	$\frac{70}{500}$
250	$\frac{90}{500}$
280	$\frac{90}{500}$
300	$\frac{55}{500}$
350	$\frac{5}{500}$

20. A probability distribution has a mean of 20 and a standard deviation of 3. Use Chebychev's inequality to estimate the probability that an outcome of the experiment lies between:
a. 15 and 25. **b.** 10 and 30.

21. A probability distribution has a mean of 42 and a standard deviation of 2. Use Chebychev's inequality to estimate the probability that an outcome of the experiment lies between:
a. 38 and 46. **b.** 32 and 52.

22. A probability distribution has a mean of 50 and a standard deviation of 1.4. Use Chebychev's inequality to find the value of c that guarantees that the probability is at least 96% that an outcome of the experiment lies between $50 - c$ and $50 + c$.

23. Suppose X is a random variable with mean μ and standard deviation σ. If a large number of trials is observed, at least what percentage of these values is expected to lie between $\mu - 2\sigma$ and $\mu + 2\sigma$?

24. PRODUCT RELIABILITY A Christmas tree light has an expected life of 200 hr and a standard deviation of 2 hr.
a. Estimate the probability that one of these Christmas tree lights will last between 190 and 210 hr.
b. Suppose 150,000 of these Christmas tree lights are used by a large city as part of its Christmas decorations. Estimate the number of lights that will require replacement between 180 and 220 hr of use.

25. PRODUCT RELIABILITY The expected lifetime of the deluxe model hair dryer produced by the Roland Electric Company has a mean life of 24 mo and a standard deviation of 3 mo. Find the probability that one of these hair dryers will last between 20 and 28 mo.

26. STARTING SALARIES The mean annual starting salary of a new graduate in a certain profession is $42,000 with a standard deviation of $500. What is the probability that the starting salary of a new graduate in this profession will be between $40,000 and $44,000?

27. QUALITY CONTROL Sugar packaged by a certain machine has a mean weight of 5 lb and a standard deviation of 0.02 lb. For what values of c can the manufacturer of the machinery claim that the sugar packaged by this machine has a weight between $5 - c$ and $5 + c$ lb with probability at least 96%?

In Exercises 28 and 29, determine whether the statement is true or false. If it is true, explain why it is true. If it is false, give an example to show why it is false.

28. Both the variance and the standard deviation of a random variable measure the spread of a probability distribution.

29. Chebychev's inequality is useless when $k \leq 1$.

SOLUTIONS TO SELF-CHECK EXERCISES 8.3

1. The mean of the random variable X is

$$\mu = (-4)(.1) + (-3)(.1) + (-1)(.2) + (0)(.3) + (2)(.1) + (5)(.2) = 0.3$$

The variance of X is

$$\text{Var}(X) = (.1)(-4 - 0.3)^2 + (.1)(-3 - 0.3)^2 + (.2)(-1 - 0.3)^2 + (.3)(0 - 0.3)^2 + (.1)(2 - 0.3)^2 + (.2)(5 - 0.3)^2 = 8.01$$

The standard deviation of X is

$$\sigma = \sqrt{\text{Var}(X)} = \sqrt{8.01} \approx 2.83$$

2. We first compute the probability distribution of X from the given data as follows:

x	Relative Frequency of Occurrence	$P(X = x)$
46	1	.1
48	2	.2
50	3	.3
51	1	.1
52	2	.2
55	1	.1

(continued on p. 496)

Using Technology

FINDING THE MEAN AND STANDARD DEVIATION

The calculation of the mean and standard deviation of a random variable is facilitated by the use of a graphing calculator.

EXAMPLE 1

A survey conducted in 1995 of the Fortune 1000 companies revealed the following age distribution of the company directors:

Age	20–25	25–30	30–35	35–40	40–45	45–50	50–55
Number of Directors	1	6	28	104	277	607	1142

Age	55–60	60–65	65–70	70–75	75–80	80–85	85–90
Number of Directors	1413	1424	494	159	62	31	5

Source: Directorship

a. Plot a histogram for the given data.
b. Find the mean age and the standard deviation of the company directors.

SOLUTION ✓

a. Let X denote the random variable taking on the values 1 through 14, where 1 corresponds to the age bracket 20–25, 2 corresponds to the age bracket 25–30, and so on. Entering the values of X as $x_1 = 1, x_2 = 2, \ldots, x_{14} = 14$ and the corresponding values of Y as $y_1 = 1, y_2 = 6, \ldots, y_{14} = 5$, and then using the DRAW function from the Statistics menu of a graphing calculator, we obtain the histogram shown in Figure T1.

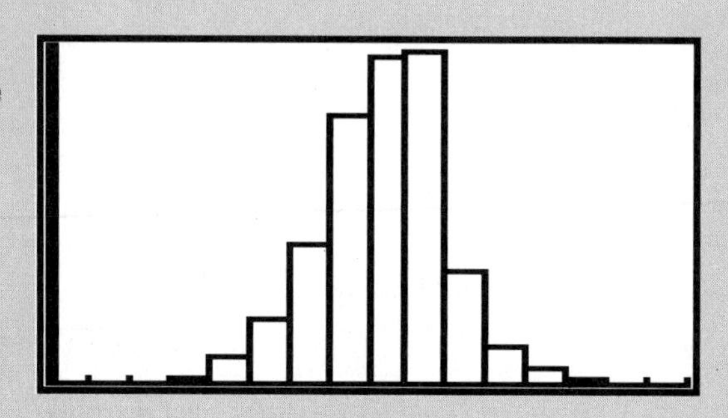

FIGURE T1
The histogram for the given data using the viewing rectangle $[0, 16] \times [0, 1500]$

b. Using the appropriate function from the Statistics menu, we find that $\bar{x} = 7.9193$ and $\sigma x = 1.6378$; that is, the mean of X is $\mu \approx 7.9$, and the standard deviation is $\sigma \approx 1.6$. Thus, the average age of the directors is in the 55- to 60-year-old bracket. ■■■■

Exercises

1. **a.** Graph the histogram associated with the random variable X in Example 1, page 485.
b. Find the mean and the standard deviation for these data.

2. **a.** Graph the histogram associated with the random variable Y in Example 1, page 485.
b. Find the mean and the standard deviation for these data.

3. **a.** Graph the histogram associated with the data given in Exercise 12, page 491.
b. Find the mean and the standard deviation for these data.

4. **a.** Graph the histogram associated with the data given in Exercise 16, page 492.
b. Find the mean and the standard deviation for these data.

5. A sugar refiner uses a machine to pack sugar in 5-lb cartons. To check the machine's accuracy, cartons are selected at random and weighed. The results follow:

4.98	5.02	4.96	4.97	5.03
4.96	4.98	5.01	5.02	5.06
4.97	5.04	5.04	5.01	4.99
4.98	5.04	5.01	5.03	5.05
4.96	4.97	5.02	5.04	4.97
5.03	5.01	5.00	5.01	4.98

a. Describe a random variable X that is associated with these data.
b. Find the probability distribution for the random variable X.
c. Compute the mean and standard deviation of X.

6. The scores of 25 students in a mathematics examination follow:

90	85	74	92	68	94	66
87	85	70	72	68	73	72
69	66	58	70	74	88	90
98	71	75	68			

a. Describe a random variable X that is associated with these data.
b. Find the probability distribution for the random variable X.
c. Compute the mean and standard deviation of X.

7. Heights of Females The following data, obtained from the records of the Westwood Health Club, give the heights (to the nearest inch) of 200 female members of the club.

Height	62	$62\frac{1}{2}$	63	$63\frac{1}{2}$	64	$64\frac{1}{2}$	65	$65\frac{1}{2}$	66
Frequency	2	3	4	8	11	20	32	30	18

Height	$66\frac{1}{2}$	67	$67\frac{1}{2}$	68	$68\frac{1}{2}$	69	$69\frac{1}{2}$	70	$70\frac{1}{2}$	71
Frequency	18	16	8	10	5	5	4	3	2	1

a. Plot a histogram for the given data.
b. Find the mean and the standard deviation (from the mean).

8. Age Distribution in a Town The following tables give the distribution of the ages of the residents of the town of Monroe under the age of 40:

Age (in years)	0–3	4–7	8–11	12–15	16–19
Number (in hundreds)	30	42	50	60	50

Age (in years)	20–23	24–27	28–31	32–35	36–39
Number (in hundreds)	41	50	45	42	34

Let X denote the random variable taking on the values 1 through 10, where 1 corresponds to the range 0–3, . . . , and 10 corresponds to the range 36–39.
a. Plot a histogram for the given data.
b. Find the mean and the standard deviation.

The mean of X is

$$\begin{aligned}\mu &= (.1)(46) + (.2)(48) + (.3)(50) \\ &\quad + (.1)(51) + (.2)(52) + (.1)(55) \\ &= 50.2\end{aligned}$$

The variance of X is

$$\begin{aligned}\text{Var}(X) &= (.1)(46 - 50.2)^2 + (.2)(48 - 50.2)^2 \\ &\quad + (.3)(50 - 50.2)^2 + (.1)(51 - 50.2)^2 \\ &\quad + (.2)(52 - 50.2)^2 + (.1)(55 - 50.2)^2 \\ &= 5.76\end{aligned}$$

from which we deduce the standard deviation

$$\begin{aligned}\sigma &= \sqrt{5.76} \\ &= 2.4\end{aligned}$$

8.4 The Binomial Distribution

BERNOULLI TRIALS

An important class of experiments have (or may be viewed as having) two outcomes. For example, in a coin-tossing experiment, the two outcomes are *heads* and *tails.* In the card game played by Mike and Bill (Example 7, Section 8.2), one may view the selection of a diamond as a *win* (for Mike) and the selection of a card of another suit as a *loss* for Mike. For a third example, consider the experiment in which a person is inoculated with a flu vaccine. Here, the vaccine may be classified as being "effective" or "ineffective" with respect to that particular person.

In general, experiments with two outcomes are called **Bernoulli trials,** or **binomial trials.** It is standard practice to label one of the outcomes of a binomial trial a *success* and the other a *failure.* For example, in a coin-tossing experiment, the outcome *a head* may be called a success, in which case the outcome *a tail* is called a failure. Note that by using the terms *success* and *failure* in this way, we depart from their usual connotations.

A sequence of Bernoulli (binomial) trials is called a binomial experiment. More precisely, we have the following definition:

Binomial Experiment

A **binomial experiment** has the following properties:

1. The number of trials in the experiment is fixed.
2. There are two outcomes of the experiment: "success" and "failure."
3. The probability of success in each trial is the same.
4. The trials are independent of each other.

In a binomial experiment it is customary to denote the probability of a success by the letter p and the probability of a failure by the letter q. Because the event of a success and the event of a failure are complementary events, we have the relationship

$$p + q = 1$$

or, equivalently,

$$q = 1 - p$$

The properties of a binomial experiment are illustrated in the following example.

EXAMPLE 1 A fair die is cast four times. Compute the probability of obtaining exactly one 6 in the four throws.

SOLUTION ✔ There are four trials in this experiment. Each trial consists of casting the die once and observing the face that lands uppermost. We may view each trial as an experiment with two outcomes: a success (S) if the face that lands uppermost is a 6 and a failure (F) if it is any of the other five numbers. Letting p and q denote the probability of success and failure, respectively, of a single trial of the experiment, we find that

$$p = \frac{1}{6} \quad \text{and} \quad q = 1 - \frac{1}{6} = \frac{5}{6}$$

Furthermore, we may assume that the trials of this experiment are independent. Thus, we have a binomial experiment.

With the aid of the multiplication principle, we see that the experiment has 2^4, or 16, outcomes. We can obtain these outcomes by constructing the tree diagram associated with the experiment (see Table 8.10, where the outcomes are listed according to the number of successes). From the table, we see that the event of obtaining exactly one success in four trials is given by

$$E = \{\text{SFFF, FSFF, FFSF, FFFS}\}$$

with probability given by

$$P(E) = P(\text{SFFF}) + P(\text{FSFF}) + P(\text{FFSF}) + P(\text{FFFS}) \quad \textbf{(10)}$$

Table 8.10

0 Success	1 Success	2 Successes	3 Successes	4 Successes
FFFF	SFFF	SSFF	SSSF	SSSS
	FSFF	SFSF	SSFS	
	FFSF	SFFS	SFSS	
	FFFS	FSSF	FSSS	
		FSFS		
		FFSS		

Since the trials (throws) are independent, the terms on the right-hand side of Equation (10) may be computed as follows:

$$P(\text{SFFF}) = P(\text{S})P(\text{F})P(\text{F})P(\text{F}) = p \cdot q \cdot q \cdot q = pq^3$$
$$P(\text{FSFF}) = P(\text{F})P(\text{S})P(\text{F})P(\text{F}) = q \cdot p \cdot q \cdot q = pq^3$$
$$P(\text{FFSF}) = P(\text{F})P(\text{F})P(\text{S})P(\text{F}) = q \cdot q \cdot p \cdot q = pq^3$$
$$P(\text{FFFS}) = P(\text{F})P(\text{F})P(\text{F})P(\text{S}) = q \cdot q \cdot q \cdot p = pq^3$$

Therefore, upon substituting these values in (10), we obtain

$$\begin{aligned} P(E) &= pq^3 + pq^3 + pq^3 + pq^3 = 4pq^3 \\ &= 4\left(\frac{1}{6}\right)\left(\frac{5}{6}\right)^3 \approx .386 \end{aligned}$$

■■■■

Probabilities in Bernoulli Trials

Let's reexamine the computations performed in the last example. There it was found that the probability of obtaining exactly one success in a binomial experiment with four independent trials with probability of success in a single trial p is given by

$$P(E) = 4pq^3 \qquad \text{(where } q = 1 - p\text{)} \tag{11}$$

Observe that the coefficient 4 of pq^3 appearing in Equation (11) is precisely the number of outcomes of the experiment with exactly one success and three failures, the outcomes being

SFFF, FSFF, FFSF, FFFS

Another way of obtaining this coefficient is to think of the outcomes as arrangements of the letters S and F. Then, the number of ways of selecting one position for S from four possibilities is given by

$$\begin{aligned} C(4, 1) &= \frac{4!}{1!(4-1)!} \\ &= 4 \end{aligned}$$

Next, observe that, because the trials are independent, each of the four outcomes of the experiment has the same probability, given by

$$pq^3$$

where the exponents 1 and 3 of p and q, respectively, correspond to exactly one success and three failures in the trials that make up each outcome.

As a result of the foregoing discussion, we may write (11) as

$$P(E) = C(4, 1)pq^3 \tag{12}$$

We are also in a position to generalize this result. Suppose that in a binomial experiment the probability of success in any trial is p. What is the probability of obtaining exactly x successes in n independent trials? We start by counting the number of outcomes of the experiment, each of which has exactly x successes. Now, one such outcome involves x successive successes followed by $(n - x)$ failures—that is,

$$\underbrace{\text{SS}\cdots\text{S}}_{x}\ \underbrace{\text{FF}\cdots\text{F}}_{n-x} \tag{13}$$

The other outcomes, each of which has exactly x successes, are obtained by rearranging the Ss (x of them) and Fs ($n - x$ of them). But there are $C(n, x)$ ways of arranging these letters. Next, arguing as in Example 1, we see that each such outcome has probability given by

$$p^x q^{n-x}$$

For example, for the outcome (13), we find

$$\begin{aligned} P(\underbrace{\text{SS}\cdots\text{S}}_{x}\ \underbrace{\text{FF}\cdots\text{F}}_{(n-x)}) &= \underbrace{P(\text{S})P(\text{S})\cdots P(\text{S})}_{x}\ \underbrace{P(\text{F})P(\text{F})\cdots P(\text{F})}_{(n-x)} \\ &= \underbrace{pp\cdots p}_{x}\ \underbrace{qq\cdots q}_{n-x} \\ &= p^x q^{n-x} \end{aligned}$$

Let's state this important result formally:

Computation of Probabilities in Bernoulli Trials

> In a binomial experiment in which the probability of success in any trial is p, the probability of exactly x successes in n independent trials is given by
>
> $$C(n, x)p^x q^{n-x}$$

If we let X be the random variable that gives the number of successes in a binomial experiment, then the probability of exactly x successes in n independent trials may be written

$$P(X = x) = C(n, x)p^x q^{n-x} \qquad (x = 0, 1, 2, \ldots, n) \tag{14}$$

The random variable X is called a **binomial random variable,** and the probability distribution of X is called a **binomial distribution.**

EXAMPLE 2

A fair die is cast five times. If a 1 or a 6 lands uppermost in a trial, then the throw is considered a success. Otherwise, the throw is considered a failure.

a. Find the probability of obtaining exactly 0, 1, 2, 3, 4, and 5 successes, respectively, in this experiment.

b. Using the results obtained in the solution to part (a), construct the binomial distribution for this experiment and draw the histogram associated with it.

SOLUTION ✓

a. This is a binomial experiment with X, the binomial random variable, taking on each of the values 0, 1, 2, 3, 4, and 5 corresponding to exactly 0, 1, 2, 3, 4, and 5 successes, respectively, in five trials. Since the die is fair, the probability of a 1 or a 6 landing uppermost in any trial is given by $p = 2/6 = 1/3$, from which it also follows that $q = 1 - p = 2/3$. Finally, $n = 5$ since there are five trials (throws of the die) in this experiment. Using Equation (14), we find that the required probabilities are

$$P(X=0) = C(5,0)\left(\frac{1}{3}\right)^0\left(\frac{2}{3}\right)^5 = \frac{5!}{0!5!}\cdot 1\cdot\frac{32}{243} \approx .132$$

$$P(X=1) = C(5,1)\left(\frac{1}{3}\right)^1\left(\frac{2}{3}\right)^4 = \frac{5!}{1!4!}\cdot\frac{16}{243} \approx .329$$

$$P(X=2) = C(5,2)\left(\frac{1}{3}\right)^2\left(\frac{2}{3}\right)^3 = \frac{5!}{2!3!}\cdot\frac{8}{243} \approx .329$$

$$P(X=3) = C(5,3)\left(\frac{1}{3}\right)^3\left(\frac{2}{3}\right)^2 = \frac{5!}{3!2!}\cdot\frac{4}{243} \approx .165$$

$$P(X=4) = C(5,4)\left(\frac{1}{3}\right)^4\left(\frac{2}{3}\right)^1 = \frac{5!}{4!1!}\cdot\frac{2}{243} \approx .041$$

$$P(X=5) = C(5,5)\left(\frac{1}{3}\right)^5\left(\frac{2}{3}\right)^0 = \frac{5!}{5!0!}\cdot\frac{1}{243} \approx .004$$

Table 8.11 A Probability Distribution

x	$P(X = x)$
0	.132
1	.329
2	.329
3	.165
4	.041
5	.004

b. Using these results, we find the required binomial distribution associated with this experiment given in Table 8.11. Next, we use this table to construct the histogram associated with the probability distribution (Figure 8.12).

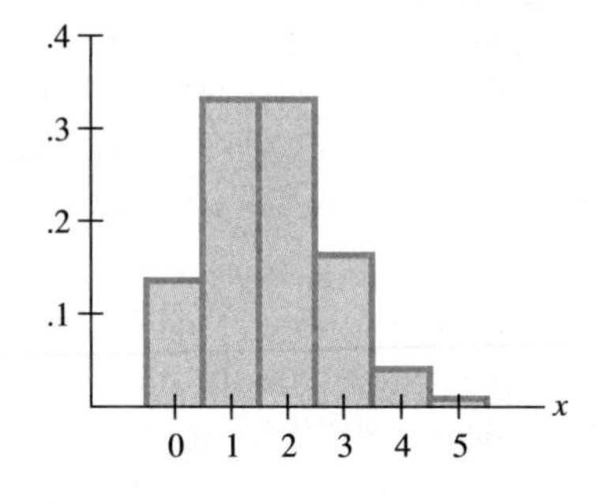

FIGURE 8.12
The probability of the number of successes in five throws

■■■■

EXAMPLE 3

A fair die is cast five times. If a 1 or a 6 lands uppermost in a trial, then the throw is considered a success. Use the results from Example 2 to answer the following questions:

a. What is the probability of obtaining 0 or 1 success in the experiment?
b. What is the probability of obtaining at least 1 success in the experiment?

SOLUTION ✓

Interpreting the probability associated with the random variable X, when X assumes the value $X = a$, as the area of the rectangle centered about $X = a$ (Figure 8.12), or otherwise, we find that:

a. The probability of obtaining 0 or 1 success in the experiment is given by

$$P(X = 0) + P(X = 1) = .132 + .329 = .461$$

b. The probability of obtaining at least 1 success in the experiment is given by

$$\begin{aligned} &P(X=1)+P(X=2)+P(X=3)+P(X=4)+P(X=5) \\ &= .329 + .329 + .165 + .041 + .004 \\ &= .868 \end{aligned}$$

Group Discussion

Consider the equation

$$P(X = x) = C(n, x)p^x q^{n-x}$$

for the binomial distribution.

1. Construct the histogram with $n = 5$ and $p = .2$; the histogram with $n = 5$ and $p = .5$; and the histogram with $n = 5$ and $p = .8$.

2. Comment on the shape of the histograms and give an interpretation.

The following formulas (which we state without proof) will be useful in solving problems involving binomial experiments.

Mean, Variance, and Standard Deviation of a Random Variable X

If X is a binomial random variable associated with a binomial experiment consisting of n trials with probability of success p and probability of failure q, then the **mean** (expected value), **variance,** and **standard deviation** of X are

$$\mu = E(X) = np \qquad \textbf{(15a)}$$

$$\text{Var}(X) = npq \qquad \textbf{(15b)}$$

$$\sigma_X = \sqrt{npq} \qquad \textbf{(15c)}$$

EXAMPLE 4 For the experiment in Examples 2 and 3, compute the mean, the variance, and the standard deviation of X, (a) using Formulas (15a), (15b), and (15c) and (b) using the definition of each term (Sections 8.2 and 8.3).

SOLUTION ✓ **a.** We use (15a), (15b), and (15c), with $p = 1/3$, $q = 2/3$, and $n = 5$, obtaining

$$\begin{aligned} \mu &= E(X) = (5)\left(\frac{1}{3}\right) = \frac{5}{3} \\ &\approx 1.67 \\ \text{Var}(X) &= (5)\left(\frac{1}{3}\right)\left(\frac{2}{3}\right) = \frac{10}{9} \\ &\approx 1.11 \\ \sigma_X &= \sqrt{\text{Var}(X)} = \sqrt{1.11} \\ &\approx 1.05 \end{aligned}$$

We leave it to you to interpret the results.

b. Using the definition of expected value (Section 8.2) and the values of the probability distribution shown in Table 8.11, we find

$$\begin{aligned}\mu = E(X) &= (0)(.132) + (1)(.329) + (2)(.329)\\ &\quad + (3)(.165) + (4)(.041) + (5)(.004)\\ &\approx 1.67\end{aligned}$$

which agrees with the result obtained in part (a). Next, using the definition of variance and the fact that $\mu = 1.67$, we find

$$\begin{aligned}\text{Var}(X) &= (.132)(-1.67)^2 + (.329)(-0.67)^2 + (.329)(0.33)^2\\ &\quad + (.165)(1.33)^2 + (.041)(2.33)^2 + (.004)(3.33)^2\\ &\approx 1.11\\ \sigma_X &\approx \sqrt{1.11}\\ &\approx 1.05\end{aligned}$$

which again agrees with the earlier results. ■■■■

APPLICATIONS

We close this section by looking at several examples involving binomial experiments. In working through these examples, you may use a calculator, or you may consult Table II, Appendix C.

EXAMPLE 5

A division of the Solaron Corporation manufactures photovoltaic cells to use in the company's solar energy converters. It is estimated that 5% of the cells manufactured are defective. If a random sample of 20 is selected from a large lot of cells manufactured by the company, what is the probability that it will contain at most 2 defective cells?

SOLUTION ✓

We may view this as a binomial experiment. To see this, first note that a fixed number of trials ($n = 20$) correspond to the selection of exactly 20 photovoltaic cells. Second, observe that there are exactly two outcomes in the experiment, defective ("success") and nondefective ("failure"). Third, the probability of success in each trial is .05 ($p = .05$) and the probability of failure in each trial is .95 ($q = .95$). This assumption is justified by virtue of the fact that the lot from which the cells are selected is "large," so the removal of a few cells will not appreciably affect the percentage of defective cells in the lot in each successive trial. Finally, the trials are independent of each other, once again because of the lot size.

Letting X denote the number of defective cells, we find that the probability of finding at most 2 defective cells in the sample of 20 is given by

$$\begin{aligned}&P(X = 0) + P(X = 1) + P(X = 2)\\ &\quad = C(20, 0)(.05)^0(.95)^{20} + C(20, 1)(.05)^1(.95)^{19}\\ &\qquad + C(20, 2)(.05)^2(.95)^{18}\\ &\quad \approx .3585 + .3774 + .1887\\ &\quad = .9246\end{aligned}$$

Thus, for lots of photovoltaic cells manufactured by Solaron, approximately 92% of the samples will have at most 2 defective cells; equivalently, approximately 8% of the samples will contain more than 2 defective cells. ■■■■

EXAMPLE 6 The probability that a heart transplant performed at the Medical Center is successful (that is, the patient survives 1 year or more after undergoing such an operation) is .7. Of six patients who have recently undergone such an operation, what is the probability that 1 year from now:

a. None of the heart recipients will be alive?
b. Exactly three will be alive?
c. At least three will be alive?
d. All will be alive?

SOLUTION ✓ Here, $n = 6$, $p = .7$, and $q = .3$. Let X denote the number of successful operations. Then,

a. The probability that no heart recipients will be alive after 1 year is given by

$$\begin{aligned} P(X=0) &= C(6,0)(.7)^0(.3)^6 \\ &= \frac{6!}{0!6!} \cdot 1 \cdot (.3)^6 \\ &\approx .0007 \end{aligned}$$

b. The probability that exactly three will be alive after 1 year is given by

$$\begin{aligned} P(X=3) &= C(6,3)(.7)^3(.3)^3 \\ &= \frac{6!}{3!3!}(.7)^3(.3)^3 \\ &\approx .19 \end{aligned}$$

c. The probability that at least three will be alive after 1 year is given by

$$\begin{aligned} &P(X=3) + P(X=4) + P(X=5) + P(X=6) \\ &= C(6,3)(.7)^3(.3)^3 + C(6,4)(.7)^4(.3)^2 \\ &\quad + C(6,5)(.7)^5(.3)^1 + C(6,6)(.7)^6(.3)^0 \\ &= \frac{6!}{3!3!}(.7)^3(.3)^3 + \frac{6!}{4!2!}(.7)^4(.3)^2 + \frac{6!}{5!1!}(.7)^5(.3)^1 \\ &\quad + \frac{6!}{6!0!}(.7)^6 \cdot 1 \\ &\approx .93 \end{aligned}$$

d. The probability that all will be alive after 1 year is given by

$$\begin{aligned} P(X=6) &= C(6,6)(.7)^6(.3)^0 = \frac{6!}{6!0!}(.7)^6 \\ &\approx .12 \end{aligned}$$

■■■■

EXAMPLE 7 The P.A.R. Bearings Company manufactures ball bearings packaged in lots of 100 each. The company's quality-control department has determined that 2% of the ball bearings manufactured do not meet the specifications imposed by a buyer. Find the average number of ball bearings per package that fail to meet with the specification imposed by the buyer.

SOLUTION ✓ The experiment under consideration is binomial. The average number of ball bearings per package that fail to meet with the specifications is therefore given by the expected value of the associated binomial random variable. Using (15a), we find that

$$\mu = E(X) = np = (100)(.02) = 2$$

substandard ball bearings in a package of 100. ■■■■

SELF-CHECK EXERCISES 8.4

1. A binomial experiment consists of four independent trials. The probability of success in each trial is .2.
   a. Find the probability of obtaining exactly 0, 1, 2, 3, and 4 successes, respectively, in this experiment.
   b. Construct the binomial distribution and draw the histogram associated with this experiment.
   c. Compute the mean and the standard deviation of the random variable associated with this experiment.
2. A recent survey shows that 60% of the households in a large metropolitan area have microwave ovens. If ten households are selected at random, what is the probability that five or fewer of these households have microwave ovens?

Solutions to Self-Check Exercises 8.4 can be found on page 507.

8.4 Exercises

A calculator is recommended for this exercise set.

In Exercises 1–6, determine whether the given experiment is a binomial experiment. Justify your answer.

1. Casting a fair die three times and observing the number of times a 6 is thrown
2. Casting a fair die and observing the number of times the die is thrown until a 6 appears uppermost
3. Casting a fair die three times and observing the number that appears uppermost
4. A card is selected from a deck of 52 cards, and its color is observed. A second card is then drawn (without replacement), and its color is observed.
5. Recording the number of accidents that occur at a given intersection on 4 clear days and 1 rainy day
6. Recording the number of hits a baseball player, whose batting average is .325, gets after being up to bat five times

In Exercises 7–10, find $C(n, x)p^xq^{n-x}$ for the given values of n, x, and p.

7. $n = 4, x = 2, p = 1/3$
8. $n = 6, x = 4, p = 1/4$
9. $n = 5, x = 3, p = .2$
10. $n = 6, x = 5, p = .4$

In Exercises 11–16, use the formula $C(n, x)p^x q^{n-x}$ to determine the probability of the given event.

11. The probability of exactly no successes in five trials of a binomial experiment in which $p = 1/3$

12. The probability of exactly three successes in six trials of a binomial experiment in which $p = 1/2$

13. The probability of at least three successes in six trials of a binomial experiment in which $p = 1/2$

14. The probability of no successful outcomes in six trials of a binomial experiment in which $p = 1/3$

15. The probability of no failures in five trials of a binomial experiment in which $p = 1/3$

16. The probability of at least one failure in five trials of a binomial experiment in which $p = 1/3$

17. A fair die is cast four times. Calculate the probability of obtaining exactly two 6s.

18. Let X be the number of successes in five independent trials of a binomial experiment in which the probability of success is $p = 2/5$. Find:
a. $P(X = 4)$ **b.** $P(2 \leq X \leq 4)$

19. A binomial experiment consists of five independent trials. The probability of success in each trial is .4.
a. Find the probability of obtaining exactly 0, 1, 2, 3, 4, and 5 successes, respectively, in this experiment.
b. Construct the binomial distribution and draw the histogram associated with this experiment.
c. Compute the mean and the standard deviation of the random variable associated with this experiment.

20. Let the random variable X denote the number of girls in a five-child family. If the probability of a female birth is .5,
a. Find the probability of 0, 1, 2, 3, 4, and 5 girls in a five-child family.
b. Construct the binomial distribution and draw the histogram associated with this experiment.
c. Compute the mean and the standard deviation of the random variable X.

21. The probability that a fuse produced by a certain manufacturing process will be defective is 1/50. Is it correct to infer from this statement that there is at most 1 defective fuse in each lot of 50 produced by this process? Justify your answer.

22. SPORTS If the probability that a certain tennis player will serve an ace is 1/4, what is the probability that he will serve exactly two aces out of five serves?

23. CUSTOMER SERVICES Mayco, a mail-order department store, has six telephone lines available for customers who wish to place their orders. If the probability that during business hours any one of the six telephone lines is engaged is 1/4, find the probability that when a customer calls to place an order all six lines will be in use.

24. SALES PREDICTIONS From experience, the manager of Kramer's Book Mart knows that 40% of the people who are browsing in the store will make a purchase. What is the probability that among ten people who are browsing in the store, at least three will make a purchase?

25. ADVERTISEMENTS An advertisement for brand A chicken noodle soup claims that 60% of all consumers prefer brand A over brand B, the chief competitor's product. To test this claim, David Horowitz, host of *The Consumer Advocate,* selected ten people at random from the audience. After tasting both soups, each person was asked to state his or her preference. Assuming the company's claim is correct, find the probability that:
a. The company's claim was supported by the experiment; that is, six or more people stated a preference for brand A.
b. The company's claim was not supported by the experiment; that is, fewer than six people stated a preference for brand A.

26. RESTAURANT VIOLATIONS OF THE HEALTH CODE Suppose that 30% of the restaurants in a certain part of a town are in violation of the health code. If a health inspector randomly selects five of the restaurants for inspection, what is the probability that:
a. None of the restaurants are in violation of the health code?
b. Just one of the restaurants is in violation of the health code?
c. At least two of the restaurants are in violation of the health code?

27. VIOLATIONS OF THE BUILDING CODE Suppose that one-third of the new buildings in a town are in violation of the building code. If a building inspector inspects five of the buildings, find the probability that:
a. The first three buildings will pass the inspection and the remaining two will fail the inspection.
b. Just three of the buildings will pass inspection.

28. VOTERS In a certain congressional district, it is known that 40% of the registered voters classify themselves as conservatives. If ten registered voters are selected at random from this district, what is the probability that four of them will be conservatives?

29. **Blood Types** It is estimated that one-third of the general population has blood type A^+. If a sample of nine people is selected at random, what is the probability that:
a. Exactly three of them have blood type A^+?
b. At most three of them have blood type A^+?

30. **Exams** A biology quiz consists of eight multiple-choice questions. Five must be answered correctly to receive a passing grade. If each question has five possible answers, of which only one is correct, what is the probability that a student who guesses at random on each question will pass the examination?

31. **Exams** A psychology quiz consists of ten true-or-false questions. If a student knows the correct answer to six of the questions but determines the answers to the remaining questions by flipping a coin, what is the probability that she will obtain a score of at least 90%?

32. **Quality Control** The probability that a videodisc player produced by the VCA Television Company is defective is estimated to be .02. If a sample of ten sets is selected at random, what is the probability that the sample contains:
a. No defectives?
b. At most two defectives?

33. **Quality Control** As part of its quality-control program, the video cartridges produced by the Starr Communications Company are subjected to a final inspection before shipment. A sample of six cartridges is selected at random from each lot of cartridges produced, and the lot is rejected if the sample contains one or more defective cartridges. If 1.5% of the cartridges produced by Starr is defective, find the probability that a shipment will be accepted.

34. **Robot Reliability** An automobile manufacturing company uses ten industrial robots as welders on its assembly line. On a given working day, the probability that a robot will be inoperative is .05. What is the probability that on a given working day
a. Exactly two robots are inoperative?
b. More than two robots are inoperative?

35. **Engine Failures** The probability that an airplane engine will fail in a transcontinental flight is .001. Assuming that engine failures are independent of each other, what is the probability that, on a certain transcontinental flight, a four-engine plane will experience:
a. Exactly one engine failure?
b. Exactly two engine failures?
c. More than two engine failures? (*Note:* In this event, the airplane will crash!)

36. **Quality Control** The manager of Toy World has decided to accept a shipment of electronic games if none of a random sample of 20 is found to be defective.
a. What is the probability that he will accept the shipment if 10% of the electronic games is defective?
b. What is the probability that he will accept the shipment if 5% of the electronic games is defective?

37. **Quality Control** Refer to Exercise 36. If the manager's criterion for accepting shipment is that there be no more than 1 defective electronic game in a random sample of 20, what is the probability that he will accept the shipment if 10% of the electronic games is defective?

38. **Quality Control** Refer to Exercise 36. If the manager of the store changes his sample size to 10 and decides to accept shipment if none of the games is defective, what is the probability that he will accept the shipment if 10% of the games is defective?

39. How many times must a person toss a coin if the chances of obtaining at least one head are 99% or better?

40. **Drug Testing** A new drug has been found to be effective in treating 75% of the people afflicted by a certain disease. If the drug is administered to 500 people who have this disease, what are the mean and the standard deviation of the number of people for whom the drug can be expected to be effective?

41. **College Graduates** At a certain university the probability that an entering freshman will graduate within 4 yr is .6. From an incoming class of 2000 freshmen, find:
a. The expected number of students who will graduate within 4 yr.
b. The standard deviation of the number of students who will graduate within 4 yr.

In Exercises 42–44, determine whether the statement is true or false. If it is true, explain why it is true. If it is false, give an example to show why it is false.

42. In a binomial experiment, the outcomes of the experiment may be any finite number.

43. In a binomial experiment with $n = 3$, $P(X = 1 \text{ or } 2) = 3pq$.

44. If the probability that a batter gets a hit is 1/4, then the batter is sure to get a hit if she bats four times.

Spreadsheet examples and exercises for this section that may be solved using the Microsoft® Excel program are given at the Brooks/Cole Web site:
http://www.brookscole.com/math/authors/tans/

SOLUTIONS TO SELF-CHECK EXERCISES 8.4

1. a. We use Formula (14) with $n = 4$, $p = .2$, and $q = 1 - .2 = .8$, obtaining

$$\begin{aligned}
P(X = 0) &= C(4, 0)(.2)^0(.8)^4 \\
&= \frac{4!}{0!4!} \cdot 1 \cdot (.8)^4 \approx .410 \\
P(X = 1) &= C(4, 1)(.2)^1(.8)^3 \\
&= \frac{4!}{1!3!}(.2)(.8)^3 \approx .410 \\
P(X = 2) &= C(4, 2)(.2)^2(.8)^2 \\
&= \frac{4!}{2!2!}(.2)^2(.8)^2 \approx .154 \\
P(X = 3) &= C(4, 3)(.2)^3(.8)^1 \\
&= \frac{4!}{3!1!}(.2)^3(.8) \approx .026 \\
P(X = 4) &= C(4, 4)(.2)^4(.8)^0 \\
&= \frac{4!}{4!0!}(.2)^4 \cdot 1 \approx .002
\end{aligned}$$

b. The required binomial distribution and histogram are as follows:

x	0	1	2	3	4
$P(X = x)$	.410	.410	.154	.026	.002

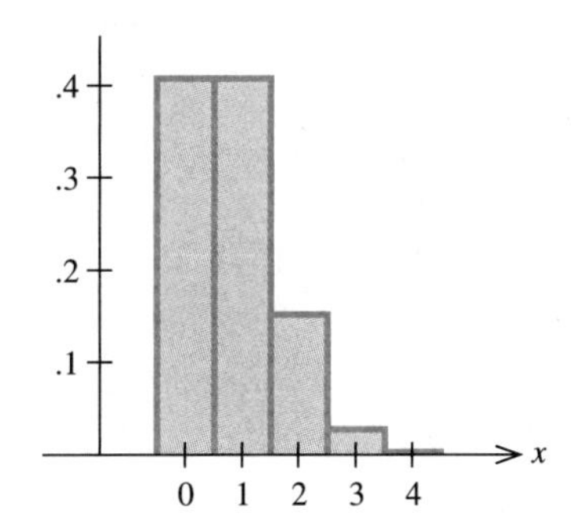

c. The mean is

$$\begin{aligned}
\mu = E(X) = np &= (4)(.2) \\
&= 0.8
\end{aligned}$$

and the standard deviation is

$$\begin{aligned}
\sigma = \sqrt{npq} &= \sqrt{(4)(.2)(.8)} \\
&= 0.8
\end{aligned}$$

2. This is a binomial experiment with $n = 10$, $p = .6$, and $q = .4$. Let X denote the number of households that have microwave ovens. Then, the probability that five or fewer households have microwave ovens is given by

$$\begin{aligned}
&P(X=0) + P(X=1) + P(X=2) + P(X=3) \\
&\quad + P(X=4) + P(X=5) \\
&\quad = C(10,0)(.6)^0(.4)^{10} + C(10,1)(.6)^1(.4)^9 \\
&\qquad + C(10,2)(.6)^2(.4)^8 + C(10,3)(.6)^3(.4)^7 \\
&\qquad + C(10,4)(.6)^4(.4)^6 + C(10,5)(.6)^5(.4)^5 \\
&\quad \approx 0 + .002 + .011 + .042 + .111 + .201 \\
&\quad \approx .37
\end{aligned}$$

8.5 The Normal Distribution

Probability Density Functions

The probability distributions discussed in the preceding sections were all associated with finite random variables—that is, random variables that take on finitely many values. Such probability distributions are referred to as *finite probability distributions.* In this section, we consider probability distributions associated with a continuous random variable—that is, a random variable that may take on any value lying in an interval of real numbers. Such probability distributions are called **continuous probability distributions.**

Unlike a finite probability distribution, which may be exhibited in the form of a table, a continuous probability distribution is defined by a function f whose domain coincides with the interval of values taken on by the random variable associated with the experiment. Such a function f is called the **probability density function** associated with the probability distribution and has the following properties:

1. $f(x)$ is nonnegative for all values of x.
2. The area of the region between the graph of f and the x-axis is equal to 1 (Figure 8.13).

Now suppose we are given a continuous probability distribution defined by a probability density function f. Then, the probability that the random

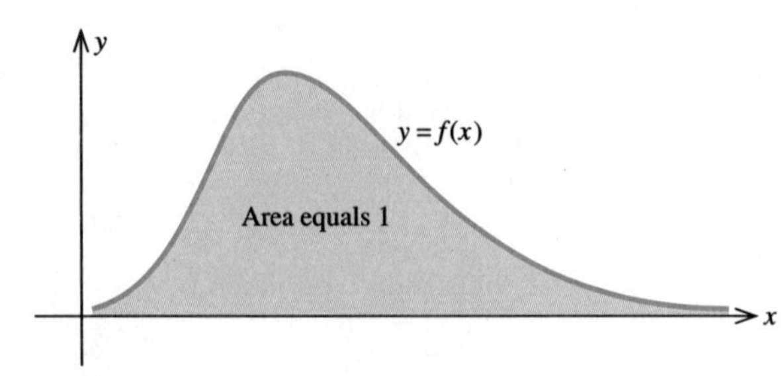

FIGURE 8.13
A probability density function

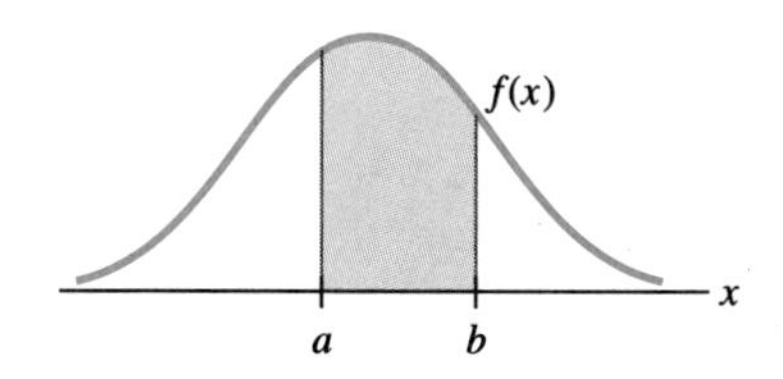

FIGURE 8.14
$P(a < X < b)$ is given by the area of the shaded region.

variable X assumes a value in an interval $a < x < b$ is given by the area of the region between the graph of f and the x-axis from $x = a$ to $x = b$ (Figure 8.14). We denote the value of this probability by $P(a < X < b)$.* Observe that Property 2 of the probability density function states that the probability that a continuous random variable takes on a value lying in its range is 1, a certainty, which is expected. Note the analogy between the areas under the probability density curves and the histograms associated with finite probability distributions (see Section 8.1).

Normal Distributions

The mean μ and the standard deviation σ of a continuous probability distribution have roughly the same meanings as the mean and standard deviation of a finite probability distribution. Thus, the mean of a continuous probability distribution is a measure of the central tendency of the probability distribution, and the standard deviation of the probability distribution measures its spread about its mean. Both of these numbers will play an important role in the following discussion.

For the remainder of this section, we will discuss a special class of continuous probability distributions known as **normal distributions.** The normal distribution is without doubt the most important of all the probability distributions. Many phenomena, such as the heights of people in a given population, the weights of newborn infants, the IQs of college students, the actual weights of 16-ounce packages of cereals, and so on, have probability distributions that are normal. The normal distribution also provides us with an accurate approximation to the distributions of many random variables associated with random-sampling problems. In fact, in the next section we will see how a normal distribution may be used to approximate a binomial distribution under certain conditions.

The graph of a normal distribution, which is bell shaped, is called a **normal curve** (Figure 8.15).

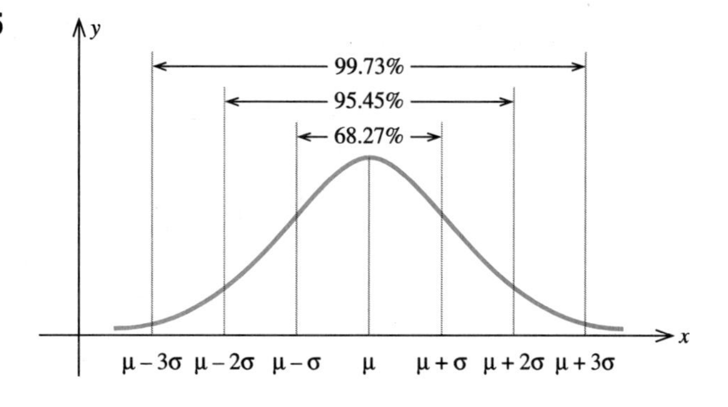

FIGURE 8.15
A normal curve

* Because the area under one point of the graph of f is equal to zero, we see that $P(a < X < b) = P(a < X \leq b) = P(a \leq X < b) = P(a \leq X \leq b)$.

The normal curve (and therefore the corresponding normal distribution) is completely determined by its mean μ and standard deviation σ. In fact, the normal curve has the following characteristics, described in terms of these two parameters:*

1. The curve has a peak at $x = \mu$.
2. The curve is symmetric with respect to the vertical line $x = \mu$.
3. The curve always lies above the x-axis but approaches the x-axis as x extends indefinitely in either direction.
4. The area under the curve is 1.
5. For any normal curve, 68.27% of the area under the curve lies within 1 standard deviation of the mean (that is, between $\mu - \sigma$ and $\mu + \sigma$), 95.45% of the area lies within 2 standard deviations of the mean, and 99.73% of the area lies within 3 standard deviations of the mean.

Figure 8.16 shows two normal curves with different means μ_1 and μ_2 but the same deviation. Next, Figure 8.17 shows two normal curves with the same mean but different standard deviations σ_1 and σ_2. (Which number is smaller?)

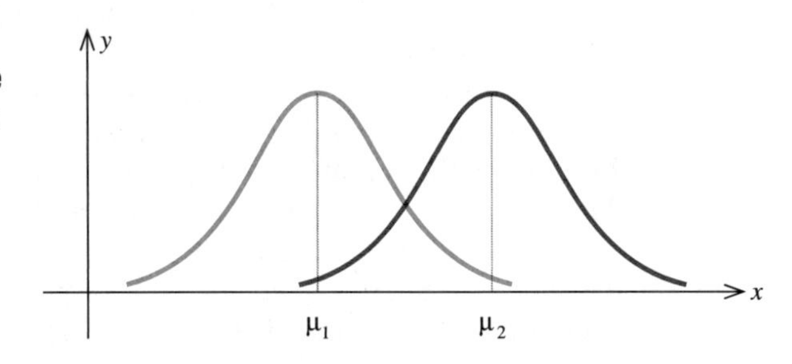

FIGURE 8.16
Two normal curves that have the same standard deviation but different means

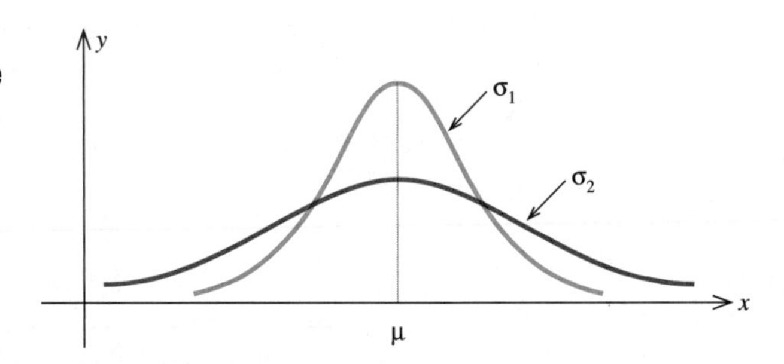

FIGURE 8.17
Two normal curves that have the same mean but different standard deviations

In general, the mean μ of a normal distribution determines where the center of the curve is located, whereas the standard deviation σ of a normal distribution determines the sharpness (or flatness) of the curve.

As this discussion reveals, there are infinitely many normal curves corresponding to different choices of the parameters μ and σ, which characterize

* The probability density function associated with this normal curve is given by

$$y = \frac{1}{\sigma\sqrt{2\pi}} e^{-(1/2)[(x-\mu)/\sigma]^2}$$

but the direct use of this formula will not be required in our discussion of the normal distribution.

Exploring with Technology

Consider the probability density function

$$f(x) = \frac{1}{\sqrt{2\pi}} e^{-x^2/2}$$

which is the formula given in the footnote on page 510 with $\mu = 0$ and $\sigma = 1$.

1. Use a graphing utility to plot the graph of f, using the viewing rectangle $[-4, 4] \times [0, 0.5]$.
2. Use the numerical integration function of a graphing calculator to find the area of the region under the graph of f on the intervals $[-1, 1]$, $[-2, 2]$, and $[-3, 3]$ and thus verify Property 5 of normal distributions for the special case where $\mu = 0$ and $\sigma = 1$.

such curves. Fortunately, any normal curve may be transformed into any other normal curve (as we will see later), so in the study of normal curves it suffices to single out one such particular curve for special attention. The normal curve with mean $\mu = 0$ and standard deviation $\sigma = 1$ is called the **standard normal curve.** The corresponding distribution is called the **standard normal distribution.** The random variable itself is called the **standard normal variable** and is commonly denoted by Z.

Computations of Probabilities Associated with Normal Distributions

Areas under the standard normal curve have been extensively computed and tabulated. Table 3, Appendix C, gives the areas of the regions under the standard normal curve to the left of the number z; these areas correspond, of course, to probabilities of the form $P(Z < z)$ or $P(Z \leq z)$. The next several examples illustrate the use of this table in computations involving the probabilities associated with the standard normal variable.

EXAMPLE 1 Let Z be the standard normal variable. By first making a sketch of the appropriate region under the standard normal curve, find the values of:

a. $P(Z < 1.24)$ **b.** $P(Z > 0.5)$ **c.** $P(0.24 < Z < 1.48)$
d. $P(-1.65 < Z < 2.02)$

SOLUTION ✓ **a.** The region under the standard normal curve associated with the probability $P(Z < 1.24)$ is shown in Figure 8.18 on page 512. To find the area of the required region using Table 3, Appendix C, we first locate the number 1.2 in the column and the number 0.04 in the row, both headed by z, and read off the number 0.8925 appearing in the body of the table. Thus,

$$P(Z < 1.24) = .8925$$

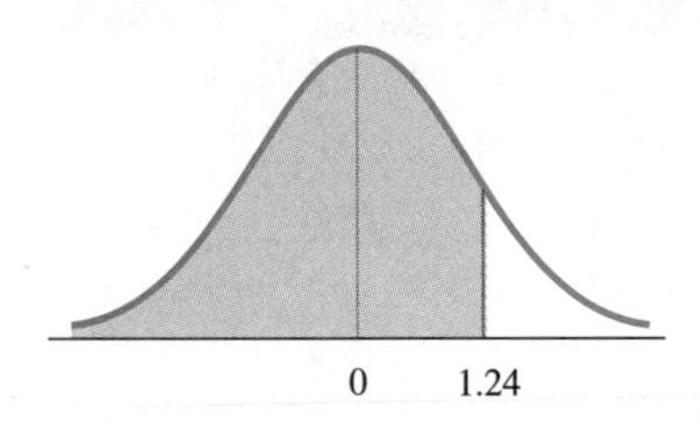

FIGURE 8.18
$P(Z < 1.24)$

b. The region under the standard normal curve associated with the probability $P(Z > 0.5)$ is shown in Figure 8.19a. Observe, however, that the required area is, by virtue of the symmetry of the standard normal curve, equal to the shaded area shown in Figure 8.19b. Thus,

$$\begin{aligned} P(Z > 0.5) &= P(Z < -0.5) \\ &= .3085 \end{aligned}$$

FIGURE 8.19

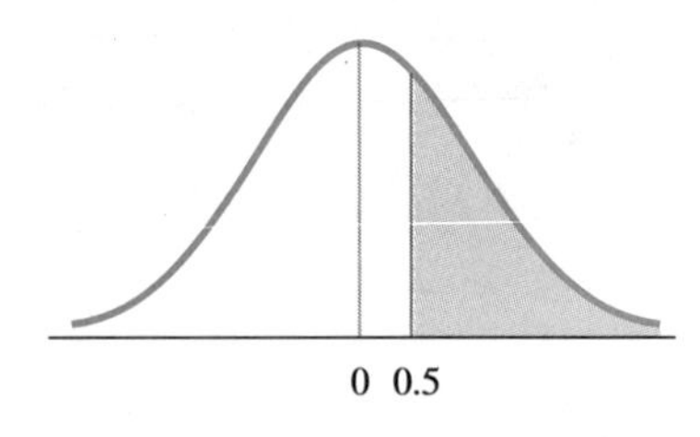

(a) $P(Z > 0.5)$

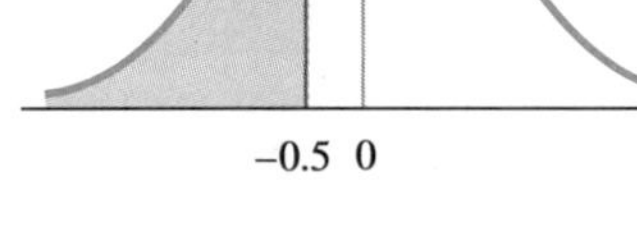

(b) $P(Z < -0.5)$

c. The probability $P(0.24 < Z < 1.48)$ is equal to the shaded area shown in Figure 8.20. But this area is obtained by subtracting the area under the curve to the left of $z = 0.24$ from the area under the curve to the left of $z = 1.48$; that is,

$$\begin{aligned} P(0.24 < Z < 1.48) &= P(Z < 1.48) - P(Z < 0.24) \\ &= .9306 - .5948 \\ &= .3358 \end{aligned}$$

d. The probability $P(-1.65 < Z < 2.02)$ is given by the shaded area shown in Figure 8.21. We have

$$\begin{aligned} P(-1.65 < Z < 2.02) &= P(Z < 2.02) - P(Z < -1.65) \\ &= .9783 - .0495 \\ &= .9288 \end{aligned}$$

FIGURE 8.20
$P(0.24 < Z < 1.48)$

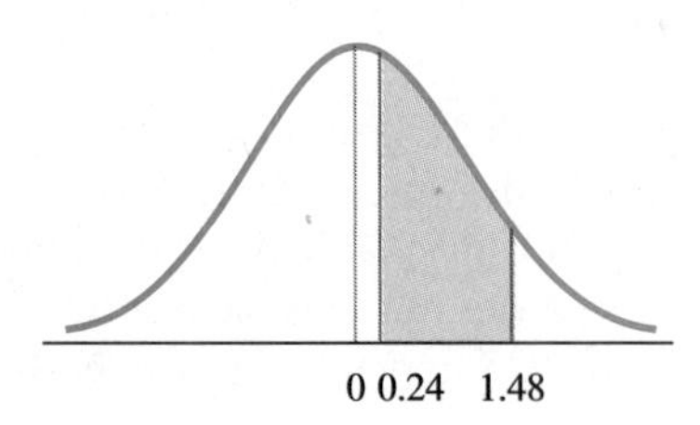

FIGURE 8.21
$P(-1.65 < Z < 2.02)$

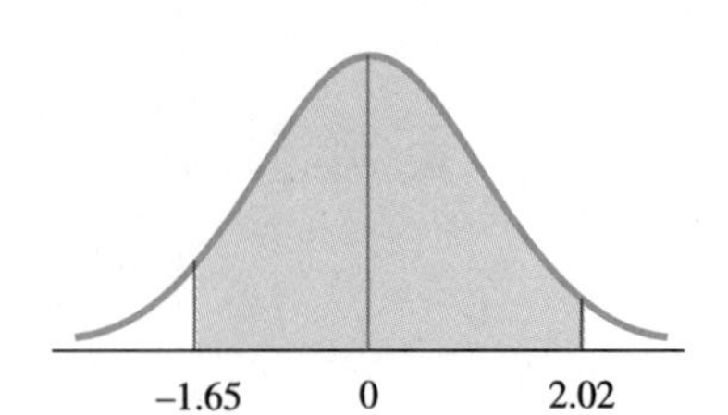

EXAMPLE 2 Let Z be the standard normal variable. Find the value of z if z satisfies:

a. $P(Z < z) = .9474$ **b.** $P(Z > z) = .9115$
c. $P(-z < Z < z) = .7888$

SOLUTION ✓

a. Refer to Figure 8.22. We want the value of Z such that the area of the region under the standard normal curve and to the left of $Z = z$ is .9474. Locating the number .9474 in Table 3, Appendix C, and reading back, we find that $z = 1.62$.

FIGURE 8.22
$P(Z < z) = .9474$

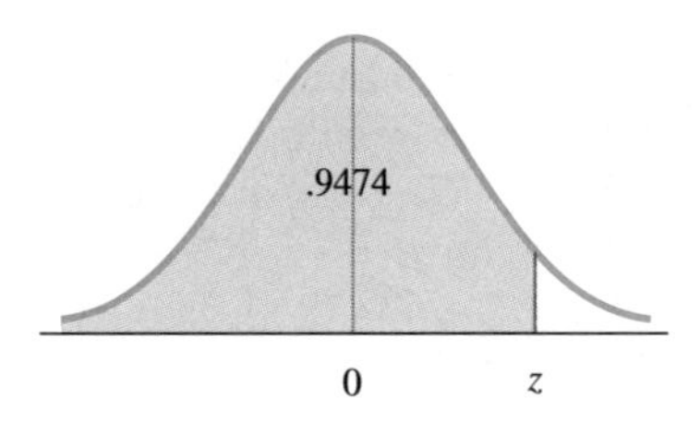

b. Since $P(Z > z)$, or equivalently, the area of the region to the right of z is greater than 0.5, z must be negative (Figure 8.23). Therefore $-z$ is positive. Furthermore, the area of the region to the right of z is the same as the area of the region to the left of $-z$. Therefore,

$$\begin{aligned} P(Z > z) &= P(Z < -z) \\ &= .9115 \end{aligned}$$

Looking up the table, we find $-z = 1.35$, so $z = -1.35$.

FIGURE 8.23
$P(Z > z) = .9115$

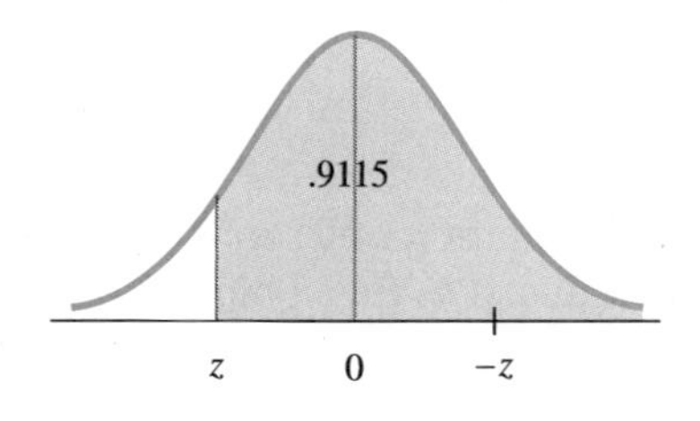

c. The region associated with $P(-z < Z < z)$ is shown in Figure 8.24. Observe that by symmetry the area of this region is just double that of the area of the region between $Z = 0$ and $Z = z$; that is,

$$P(-z < Z < z) = 2P(0 < Z < z)$$

$$P(0 < Z < z) = P(Z < z) - \frac{1}{2}$$

(Figure 8.25). Therefore,

$$\frac{1}{2}P(-z < Z < z) = P(Z < z) - \frac{1}{2}$$

or, solving for $P(Z < z)$,

$$\begin{aligned} P(Z < z) &= \frac{1}{2} + \frac{1}{2}P(-z < Z < z) \\ &= \frac{1}{2}(1 + .7888) \\ &= .8944 \end{aligned}$$

Consulting the table, we find $z = 1.25$.

FIGURE 8.24
$P(-z < Z < z) = .7888$

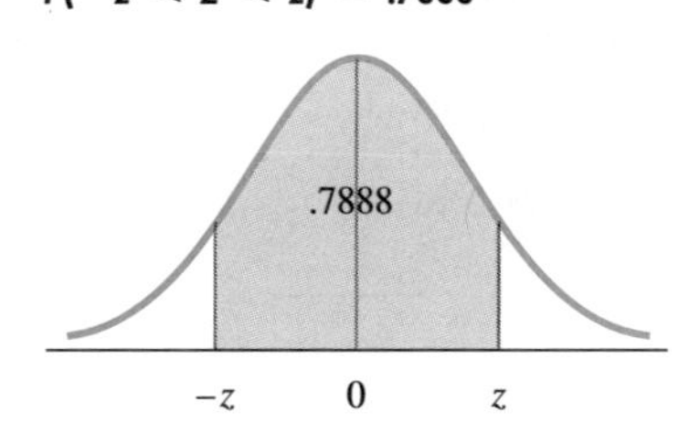

FIGURE 8.25

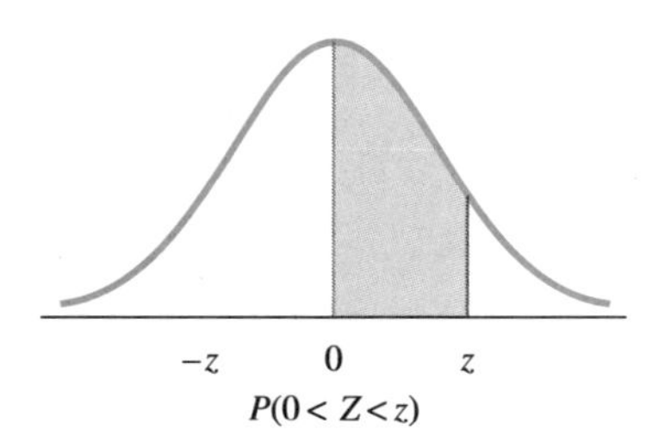

=

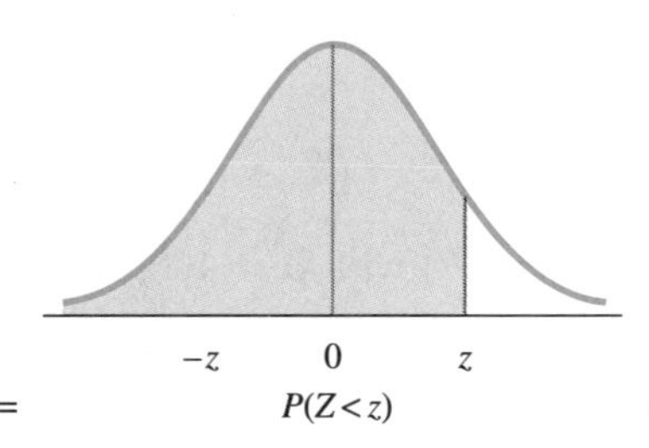

−

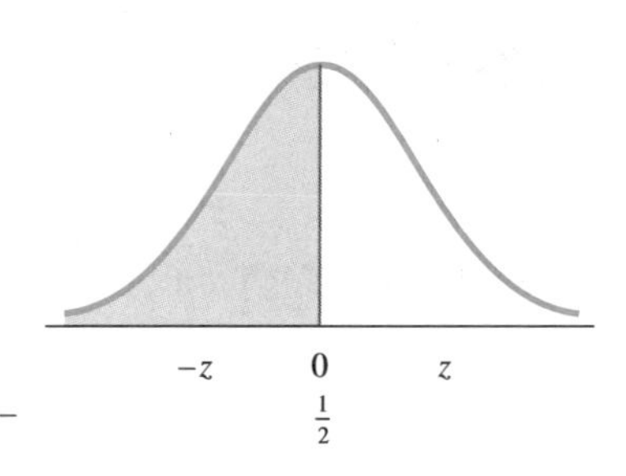

■■■■

We now turn our attention to the computation of probabilities associated with normal distributions whose means and standard deviations are not necessarily equal to 0 and 1, respectively. As mentioned earlier, any normal curve may be transformed into the standard normal curve. In particular, it may be shown that if X is a normal random variable with mean μ and standard deviation σ, then it can be transformed into the standard normal random variable Z by means of the substitution

$$Z = \frac{X - \mu}{\sigma}$$

The area of the region under the normal curve (with random variable X) between $x = a$ and $x = b$ is *equal* to the area of the region under the standard normal curve between $z = (a - \mu)/\sigma$ and $z = (b - \mu)/\sigma$. In terms of probabilities associated with these distributions, we have

$$P(a < X < b) = P\left(\frac{a - \mu}{\sigma} < Z < \frac{b - \mu}{\sigma}\right) \tag{16}$$

(Figure 8.26). Similarly, we have

$$P(X < b) = P\left(Z < \frac{b - \mu}{\sigma}\right) \tag{17}$$

$$P(X > a) = P\left(Z > \frac{a - \mu}{\sigma}\right) \tag{18}$$

Thus, with the help of Equations (16)–(18), computations of probabilities associated with any normal distribution may be reduced to the computations of areas of regions under the standard normal curve.

FIGURE 8.26

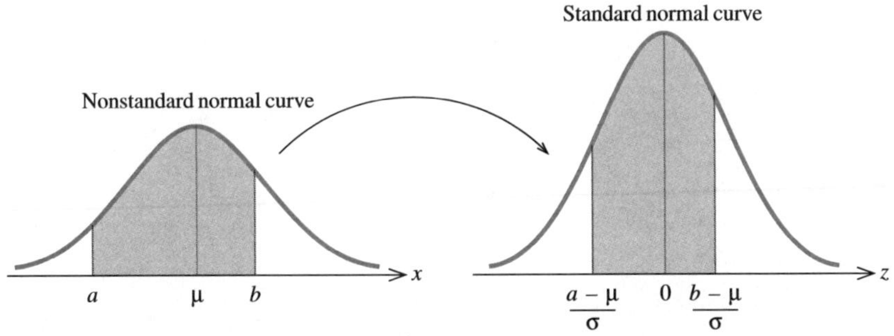

Area under the curve between a and b = Area under the curve between $\frac{a-\mu}{\sigma}$ and $\frac{b-\mu}{\sigma}$.

EXAMPLE 3 Suppose X is a normal random variable with $\mu = 100$ and $\sigma = 20$. Find the values of:

a. $P(X < 120)$ **b.** $P(X > 70)$ **c.** $P(75 < X < 110)$

SOLUTION ✓ **a.** Using Formula (17) with $\mu = 100$, $\sigma = 20$, and $b = 120$, we have

$$\begin{aligned} P(X < 120) &= P\left(Z < \frac{120 - 100}{20}\right) \\ &= P(Z < 1) = .8413 \qquad \text{(Using the table of values of } Z\text{)} \end{aligned}$$

b. Using Formula (18) with $\mu = 100$, $\sigma = 20$, and $a = 70$, we have

$$\begin{aligned} P(X > 70) & \\ &= P\left(Z > \frac{70 - 100}{20}\right) \\ &= P(Z > -1.5) = P(Z < 1.5) = .9332 \end{aligned}$$

(Using the table of values of Z)

c. Using Formula (16) with $\mu = 100$, $\sigma = 20$, $a = 75$, and $b = 110$, we have

$$\begin{aligned} P(75 < X < 110) & \\ &= P\left(\frac{75 - 100}{20} < Z < \frac{110 - 100}{20}\right) \\ &= P(-1.25 < Z < 0.5) \\ &= P(Z < 0.5) - P(Z < -1.25) \\ &= .6915 - .1056 = .5859 \end{aligned}$$

(See Figure 8.27.)

(Using the table of values of Z) ■■■■

FIGURE 8.27

−1.25 0 0.5

SELF-CHECK EXERCISES 8.5

1. Let Z be a standard normal variable.
   a. Find the value of $P(-1.2 < Z < 2.1)$ by first making a sketch of the appropriate region under the standard normal curve.
   b. Find the value of z if z satisfies $P(-z < Z < z) = .8764$.
2. Let X be a normal random variable with $\mu = 80$ and $\sigma = 10$. Find the values of:
   a. $P(X < 100)$ **b.** $P(X > 60)$ **c.** $P(70 < X < 90)$

Solutions to Self-Check Exercises 8.5 can be found on page 516.

8.5 Exercises

In Exercises 1–6, find the value of the probability of the standard normal variable Z corresponding to the shaded area under the standard normal curve.

1. $P(Z < 1.45)$

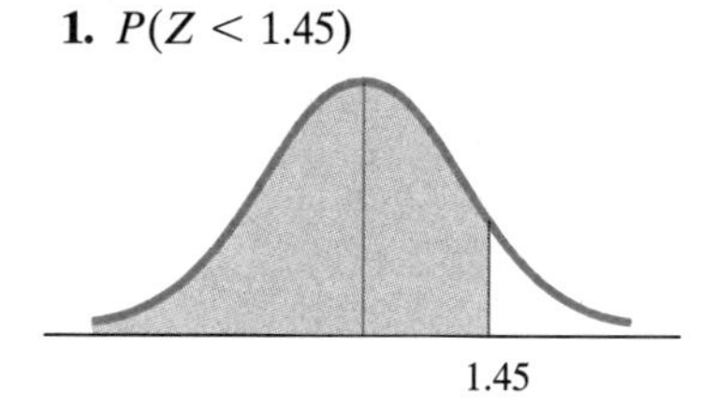

2. $P(Z > 1.11)$

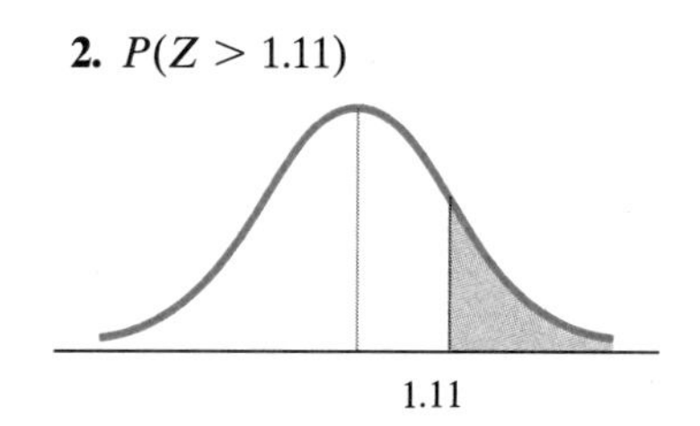

3. $P(Z < -1.75)$

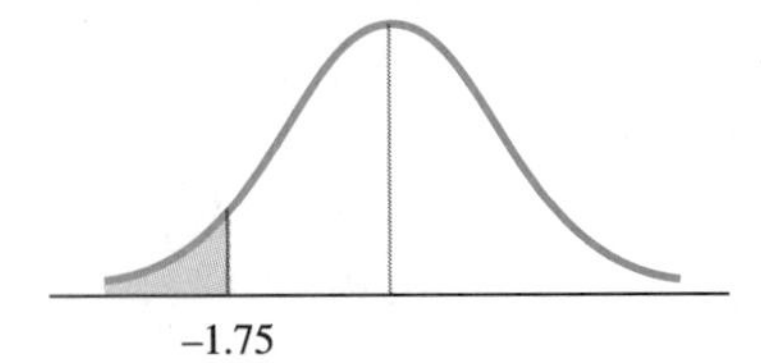

4. $P(0.3 < Z < 1.83)$

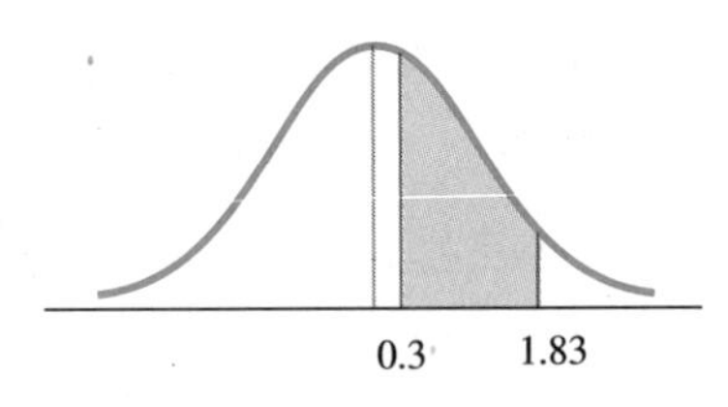

5. $P(-1.32 < Z < 1.74)$

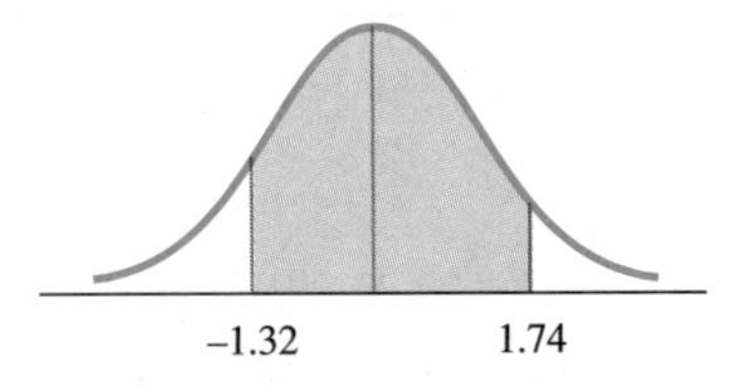

6. $P(-2.35 < Z < -0.51)$

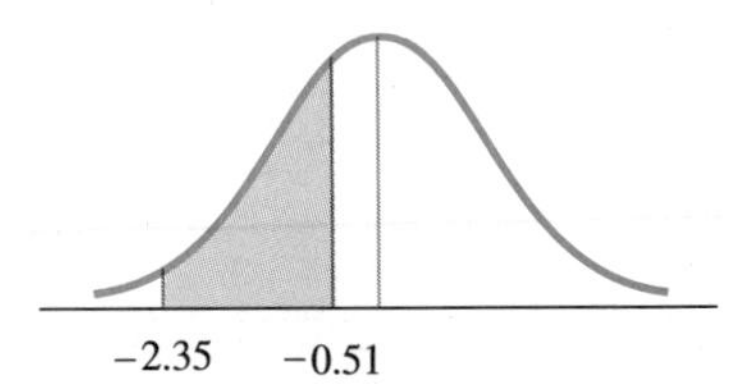

In Exercises 7–14, (a) make a sketch of the area under the standard normal curve corresponding to the given probability and (b) find the value of the probability of the standard normal variable Z corresponding to this area.

7. $P(Z < 1.37)$
8. $P(Z > 2.24)$
9. $P(Z < -0.65)$
10. $P(0.45 < Z < 1.75)$
11. $P(Z > -1.25)$
12. $P(-1.48 < Z < 1.54)$
13. $P(0.68 < Z < 2.02)$
14. $P(-1.41 < Z < -0.24)$

15. Let Z be the standard normal variable. Find the values of z if z satisfies:
   a. $P(Z < z) = .8907$
   b. $P(Z < z) = .2090$

16. Let Z be the standard normal variable. Find the values of z if z satisfies:
   a. $P(Z > z) = .9678$
   b. $P(-z < Z < z) = .8354$

17. Let Z be the standard normal variable. Find the values of z if z satisfies:
   a. $P(Z > -z) = .9713$
   b. $P(Z < -z) = .9713$

18. Suppose X is a normal random variable with $\mu = 380$ and $\sigma = 20$. Find the value of:
   a. $P(X < 405)$ **b.** $P(400 < X < 430)$
   c. $P(X > 400)$

19. Suppose X is a normal random variable with $\mu = 50$ and $\sigma = 5$. Find the value of:
   a. $P(X < 60)$ **b.** $P(X > 43)$
   c. $P(46 < X < 58)$

20. Suppose X is a normal random variable with $\mu = 500$ and $\sigma = 75$. Find the value of:
   a. $P(X < 750)$ **b.** $P(X > 350)$
   c. $P(400 < X < 600)$

SOLUTIONS TO SELF-CHECK EXERCISES 8.5

1. **a.** The probability $P(-1.2 < Z < 2.1)$ is given by the shaded area in the accompanying figure:

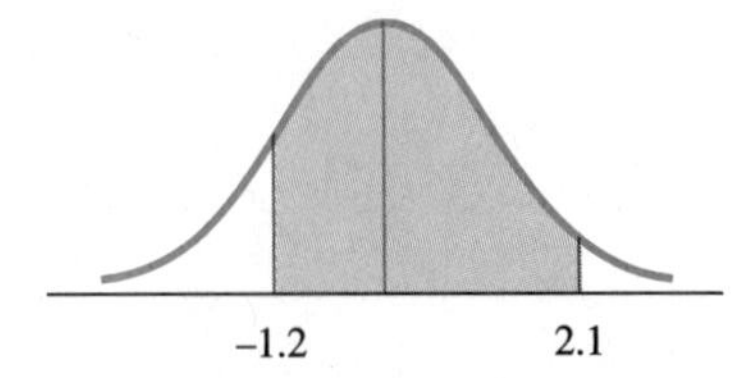

We have

$$\begin{aligned} P(-1.2 < Z < 2.1) &= P(Z < 2.1) - P(Z < -1.2) \\ &= .9821 - .1151 \\ &= .867 \end{aligned}$$

b. The region associated with $P(-z < Z < z)$ is shown in the accompanying figure:

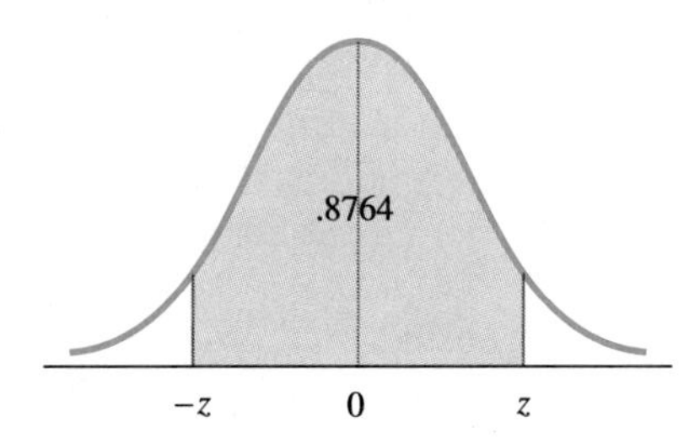

Observe that we have the following relationship:

$$P(Z < z) = \frac{1}{2}[1 + P(-z < Z < z)]$$

(see Example 2c). With $P(-z < Z < z) = .8764$, we find

$$\begin{aligned} P(Z < z) &= \frac{1}{2}(1 + .8764) \\ &= .9382 \end{aligned}$$

Consulting the table, we find $z = 1.54$.

2. Using the transformation (16) and the table of values of Z, we have

a. $$\begin{aligned} P(X < 100) &= P\left(Z < \frac{100 - 80}{10}\right) \\ &= P(Z < 2) \\ &= .9772 \end{aligned}$$

b. $$\begin{aligned} P(X > 60) &= P\left(Z > \frac{60 - 80}{10}\right) \\ &= P(Z > -2) \\ &= P(Z < 2) \\ &= .9772 \end{aligned}$$

c. $$\begin{aligned} P(70 < X < 90) &= P\left(\frac{70 - 80}{10} < Z < \frac{90 - 80}{10}\right) \\ &= P(-1 < Z < 1) \\ &= P(Z < 1) - P(Z < -1) \\ &= .8413 - .1587 \\ &= .6826 \end{aligned}$$

8.6 Applications of the Normal Distribution

Applications Involving Normal Random Variables

In this section we look at some applications involving the normal distribution.

EXAMPLE 1 The medical records of infants delivered at the Kaiser Memorial Hospital show that the infants' birth weights in pounds are normally distributed with a mean of 7.4 and a standard deviation of 1.2. Find the probability that an infant selected at random from among those delivered at the hospital weighed more than 9.2 pounds at birth.

SOLUTION ✓ Let X be the normal random variable denoting the birth weights of infants delivered at the hospital. Then, the probability that an infant selected at random has a birth weight of more than 9.2 pounds is given by $P(X > 9.2)$. To compute $P(X > 9.2)$ we use Formula (18), Section 8.5, with $\mu = 7.4$, $\sigma = 1.2$, and $a = 9.2$. We find

$$\begin{aligned} P(X > 9.2) &= P\left(Z > \frac{9.2 - 7.4}{1.2}\right) \qquad \left[P(X > a) = P\left(Z > \frac{a-\mu}{\sigma}\right)\right] \\ &= P(Z > 1.5) \\ &= P(Z < -1.5) \\ &= .0668 \end{aligned}$$

Thus, the probability that an infant delivered at the hospital weighs more than 9.2 pounds is .0668. ■■■■

EXAMPLE 2 The Idaho Natural Produce Corporation ships potatoes to its distributors in bags whose weights are normally distributed with a mean weight of 50 pounds and standard deviation of 0.5 pound. If a bag of potatoes is selected at random from a shipment, what is the probability that it weighs:

a. More than 51 pounds? **b.** Less than 49 pounds?
c. Between 49 and 51 pounds?

SOLUTION ✓ Let X denote the weight of potatoes packed by the company. Then, the mean and standard deviation of X are $\mu = 50$ and $\sigma = 0.5$, respectively.

a. The probability that a bag selected at random weighs more than 51 pounds is given by

$$\begin{aligned} P(X > 51) &= P\left(Z > \frac{51 - 50}{0.5}\right) \qquad \left[P(X > a) = P\left(Z > \frac{a-\mu}{\sigma}\right)\right] \\ &= P(Z > 2) \\ &= P(Z < -2) \\ &= .0228 \end{aligned}$$

b. The probability that a bag selected at random weighs less than 49 pounds is given by

$$P(X < 49) = P\left(Z < \frac{49 - 50}{0.5}\right) \qquad \left[P(X < b) = P\left(Z < \frac{b - \mu}{\sigma}\right)\right]$$
$$= P(Z < -2)$$
$$= .0228$$

c. The probability that a bag selected at random weighs between 49 and 51 pounds is given by

$$P(49 < X < 51) \qquad \left[P(a < X < b) = P\left(\frac{a - \mu}{\sigma} < Z < \frac{b - \mu}{\sigma}\right)\right]$$
$$= P\left(\frac{49 - 50}{0.5} < Z < \frac{51 - 50}{0.5}\right)$$
$$= P(-2 < Z < 2)$$
$$= P(Z < 2) - P(Z < -2)$$
$$= .9772 - .0228$$
$$= .9544$$

■■■■

EXAMPLE 3 The grade point average (GPA) of the senior class of Jefferson High School is normally distributed with a mean of 2.7 and a standard deviation of 0.4 point. If a senior in the top 10% of his or her class is eligible for admission to any of the nine campuses of the State University system, what is the minimum GPA that a senior should have to ensure eligibility for admission to the State University system?

SOLUTION ✓ Let X denote the GPA of a randomly selected senior at Jefferson High School and let x denote the minimum GPA to ensure his or her eligibility for admission to the university. Since only the top 10% is eligible for admission, x must satisfy the equation

$$P(X \geq x) = .1$$

Using Formula (18), Section 8.5, with $\mu = 2.7$ and $\sigma = 0.4$, we find

$$P(X \geq x) = P\left(Z \geq \frac{x - 2.7}{0.4}\right) = .1 \qquad \left[P(X > a) = P\left(Z > \frac{a - \mu}{\sigma}\right)\right]$$

But, this is equivalent to the equation

$$P\left(Z \leq \frac{x - 2.7}{0.4}\right) = .9 \qquad \text{(Why?)}$$

Consulting Table 3, Appendix C, we find

$$\frac{x - 2.7}{0.4} = 1.28$$

Upon solving for x, we obtain

$$x = (1.28)(0.4) + 2.7$$
$$\approx 3.2$$

Thus, to ensure eligibility for admission to one of the nine campuses of the State University system, a senior at Jefferson High School should have a minimum 3.2 GPA. ■■■■

APPROXIMATING BINOMIAL DISTRIBUTIONS

As mentioned in the last section, one important application of the normal distribution is that it provides us with an accurate approximation of other continuous probability distributions. We now show how a binomial distribution may be approximated by a suitable normal distribution. This technique leads to a convenient and simple solution to certain problems involving binomial probabilities.

Recall that a binomial distribution is a probability distribution of the form

$$P(X = x) = C(n, x)p^x q^{n-x} \qquad (x = 0, 1, 2, \ldots, n) \tag{19}$$

(See Section 8.4.) For small values of n, the arithmetic computations of the binomial probabilities may be done with relative ease. However, if n is large, then the work involved becomes prodigious, even when tables of $P(X = x)$ are available. For example, if $n = 50$, $p = .3$, and $q = .7$, then the probability of ten or more successes is given by

$$P(X \geq 10) = P(X = 10) + P(X = 11) + \cdots + P(X = 50)$$
$$= \frac{50!}{10!40!}(.3)^{10}(.7)^{40} + \frac{50!}{11!39!}(.3)^{11}(.7)^{39} + \cdots + \frac{50!}{50!0!}(.3)^{50}(.7)^{0}$$

Table 8.12 A Probability Distribution

x	$P(X = x)$
0	.0000
1	.0000
2	.0002
3	.0011
4	.0046
5	.0148
6	.0370
7	.0739
8	.1201
9	.1602
10	.1762
11	.1602
12	.1201
⋮	⋮
20	.0000

To see how the normal distribution helps us in such situations, let's consider a coin-tossing experiment. Suppose a fair coin is tossed 20 times and we wish to compute the probability of obtaining ten or more heads. The solution to this problem may be obtained of course by computing

$$P(X \geq 10) = P(X = 10) + P(X = 11) + \cdots + P(X = 20)$$

The inconvenience of this approach for solving the problem at hand has already been pointed out. As an alternative solution, let's begin by interpreting the solution in terms of finding the area of suitable rectangles of the histogram for the distribution associated with the problem. Using Formula (19) we compute the probability of obtaining exactly x heads in 20 coin tosses. The results lead to the binomial distribution displayed in Table 8.12.

Using the data from the table, we next construct the histogram for the distribution (Figure 8.28). The probability of obtaining ten or more heads in 20 coin tosses is equal to the sum of the areas of the shaded rectangles of the histogram of the binomial distribution shown in Figure 8.29.

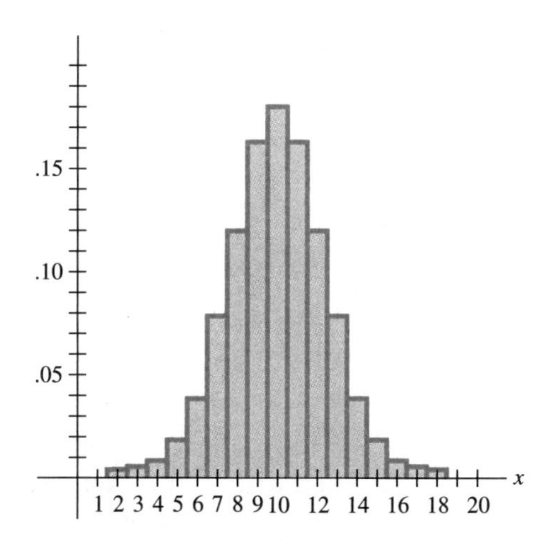

FIGURE 8.28
Histogram showing the probability of obtaining x heads in 20 coin tosses

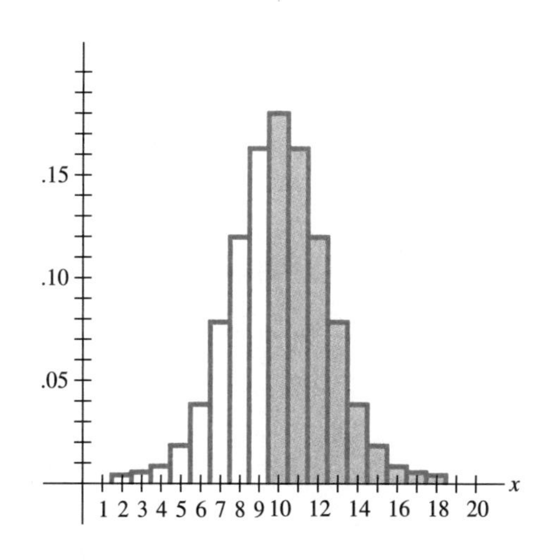

FIGURE 8.29
The shaded area gives the probability of obtaining ten or more heads in 20 coin tosses

Next, observe that the shape of the histogram suggests that the binomial distribution under consideration may be approximated by a suitable normal distribution. Since the mean and standard deviation of the binomial distribution are given by

$$\begin{aligned}\mu &= np \\ &= (20)(.5) = 10 \\ \sigma &= \sqrt{npq} \\ &= \sqrt{(20)(.5)(.5)} \\ &= 2.24\end{aligned}$$

respectively (see Section 8.4), the natural choice of a normal curve for this purpose is one with a mean of 10 and standard deviation of 2.24. Figure 8.30 shows such a normal curve superimposed on the histogram of the binomial distribution.

FIGURE 8.30
Normal curve superimposed on the histogram for a binomial distribution

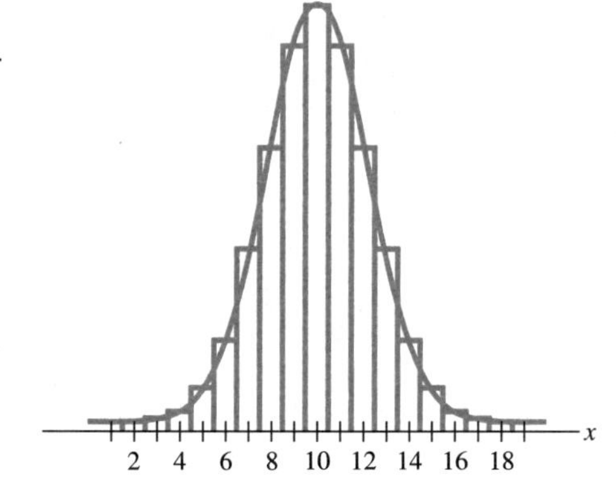

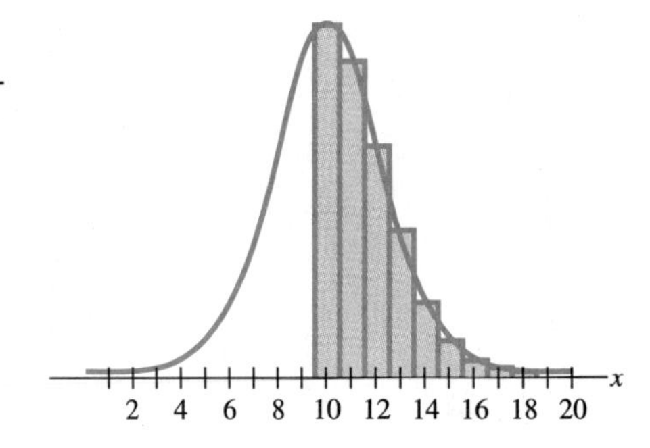

FIGURE 8.31
$P(X \geq 10)$ is approximated by the area under the normal curve

The good fit suggests that the sum of the areas of the rectangles representing $P(X \geq 10)$, the probability of obtaining ten or more heads in 20 coin tosses, may be approximated by the area of an appropriate region under the normal curve. To determine this region, let's note that the base of the portion of the histogram representing the required probability extends from $x = 9.5$ on, since the base of the leftmost rectangle is centered at $x = 10$ and the base of each rectangle has length 1 (Figure 8.31). Therefore, the required region under the normal curve should also have $x \geq 9.5$. Letting Y denote the continuous normal variable, we have

$$
\begin{aligned}
P(X \geq 10) &\approx P(Y \geq 9.5) \\
&= P(Y > 9.5) \\
&= P\left(Z > \frac{9.5 - 10}{2.24}\right) \qquad \left[P(X > a) = P\left(Z > \frac{a - \mu}{\sigma}\right)\right] \\
&= P(Z > -0.22) \\
&= P(Z < 0.22) \\
&= .5871 \qquad \text{(Using the table of values of } Z\text{)}
\end{aligned}
$$

The exact value of $P(X \geq 10)$ may be found by computing

$$P(X = 10) + P(X = 11) + \cdots + P(X = 20)$$

in the usual fashion and is equal to .5881. Thus, the normal distribution with suitably chosen mean and standard deviation does provide us with a good approximation of the binomial distribution.

In the general case, the following result, which is a special case of the *central limit theorem,* guarantees the accuracy of the approximation of a binomial distribution by a normal distribution under certain conditions.

THEOREM

Suppose we are given a binomial distribution associated with a binomial experiment involving n trials, each with a probability of success p and probability of failure q. Then, if n is large and p is not close to 0 or 1, the binomial distribution may be approximated by a normal distribution with

$$\mu = np \qquad \text{and} \qquad \sigma = \sqrt{npq}$$

REMARK It can be shown that if both np and nq are greater than 5, then the error resulting from this approximation is negligible.

Applications Involving Binomial Random Variables

EXAMPLE 4 An automobile manufacturer receives the microprocessors used to regulate fuel consumption in its automobiles in shipments of 1000 each from a certain supplier. It has been estimated that, on the average, 1% of the microprocessors manufactured by the supplier are defective. Determine the probability that more than 20 of the microprocessors in a single shipment are defective.

SOLUTION ✓ Let X denote the number of defective microprocessors in a single shipment. Then, X has a binomial distribution with $n = 1000$, $p = .01$, and $q = .99$, so

$$\begin{aligned}\mu &= (1000)(.01) = 10\\ \sigma &= \sqrt{(1000)(.01)(.99)}\\ &\approx 3.15\end{aligned}$$

Approximating the binomial distribution by a normal distribution with a mean of 10 and a standard deviation of 3.15, we find that the probability that more than 20 microprocessors in a shipment are defective is given by

$$\begin{aligned}P(X > 20) &\approx P(Y > 20.5) && \text{(Where } Y \text{ denotes the normal random variable)}\\ &= P\left(Z > \frac{20.5 - 10}{3.15}\right) && \left[P(X > a) = P\left(Z > \frac{a-\mu}{\sigma}\right)\right]\\ &= P(Z > 3.33)\\ &= P(Z < -3.33)\\ &= .0004\end{aligned}$$

In other words, approximately 0.04% of the shipments containing 1000 microprocessors each will contain more than 20 defective units.

EXAMPLE 5 The probability that a heart transplant performed at the Medical Center is successful (that is, the patient survives 1 year or more after undergoing the surgery) is .7. Of 100 patients who have undergone such an operation, what is the probability that:

a. Fewer than 75 will survive 1 year or more after the operation?
b. Between 80 and 90, inclusive, will survive 1 year or more after the operation?

SOLUTION ✓ Let X denote the number of patients who survive 1 year or more after undergoing a heart transplant at the Medical Center. Then, X is a binomial random variable. Also, $n = 100$, $p = .7$, and $q = .3$, so

$$\begin{aligned}\mu &= (100)(.7) = 70\\ \sigma &= \sqrt{(100)(.7)(.3)}\\ &\approx 4.58\end{aligned}$$

Approximating the binomial distribution by a normal distribution with a mean of 70 and a standard deviation of 4.58, we find, upon letting Y denote the associated normal random variable:

a. The probability that fewer than 75 patients will survive 1 year or more is given by

$$\begin{aligned} P(X<75) &\approx P(Y<74.5) && \text{(Why?)} \\ &= P\left(Z<\frac{74.5-70}{4.58}\right) && \left[P(X<b)=P\left(Z<\frac{b-\mu}{\sigma}\right)\right] \\ &= P(Z<0.98) \\ &= .8365 \end{aligned}$$

b. The probability that between 80 and 90, inclusive, of the patients will survive 1 year or more is given by

$$\begin{aligned} &P(80 \le X \le 90) \\ &\quad\approx P(79.5<Y<90.5) && \left[P(a<X<b)\right. \\ &\quad= P\left(\frac{79.5-70}{4.58}<Z<\frac{90.5-70}{4.58}\right) && \left.= P\left(\frac{a-\mu}{\sigma}<Z<\frac{b-\mu}{\sigma}\right)\right] \\ &\quad= P(2.07<Z<4.48) \\ &\quad= P(Z<4.48)-P(Z<2.07) \\ &\quad= 1-.9808 && [\textit{Note: } P(Z<4.48)\approx 1.] \\ &\quad= .0192 \end{aligned}$$

■■■■

SELF-CHECK EXERCISES 8.6

1. The serum cholesterol levels in milligrams/decaliter (mg/dL) in a current Mediterranean population are found to be normally distributed with a mean of 160 and a standard deviation of 50. Scientists at the National Heart, Lung, and Blood Institute consider this pattern ideal for a minimal risk of heart attacks. Find the percentage of the population having blood cholesterol levels between 160 and 180 mg/dL.
2. It has been estimated that 4% of the luggage manufactured by The Luggage Company fails to meet the standards established by the company and is sold as "seconds" to discount and outlet stores. If 500 bags are produced, what is the probability that more than 30 will be classified as "seconds"?

Solutions to Self-Check Exercises 8.6 can be found on page 527.

8.6 Exercises

1. **Medical Records** The medical records of infants delivered at the Kaiser Memorial Hospital show that the infants' lengths at birth (in inches) are normally distributed with a mean of 20 and a standard deviation of 2.6. Find the probability that an infant selected at random from among those delivered at the hospital measures:
   a. More than 22 in. **b.** Less than 18 in.
   c. Between 19 and 21 in.

2. **Factory Workers' Wages** According to the data released by the Chamber of Commerce of a certain city, the weekly wages of factory workers are normally distributed with a mean of \$500 and a standard deviation of \$50. What is the probability that a worker selected at random from the city makes a weekly wage:
   a. Of less than \$500? **b.** Of more than \$660?
   c. Between \$450 and \$550?

3. **Product Reliability** The TKK Products Corporation manufactures electric light bulbs in the 50-, 60-, 75-, and 100-watt range. Laboratory tests show that the lives of these light bulbs are normally distributed with a mean of 750 hr and a standard deviation of 75 hr. What is the probability that a TKK light bulb selected at random will burn:
   a. For more than 900 hr?
   b. For less than 600 hr?
   c. Between 750 and 900 hr?
   d. Between 600 and 800 hr?

4. **Education** On the average, a student takes 100 words/minute midway through an advanced court reporting course at the American Institute of Court Reporting. Assuming that the dictation speeds of the students are normally distributed and that the standard deviation is 20 words/minute, what is the probability that a student randomly selected from the course can take dictation at a speed:
   a. Of more than 120 words/minute?
   b. Between 80 and 120 words/minute?
   c. Of less than 80 words/minute?

5. **IQs** The IQs of students at the Wilson Elementary School were measured recently and found to be normally distributed with a mean of 100 and a standard deviation of 15. What is the probability that a student selected at random will have an IQ:
   a. Of 140 or higher? **b.** Of 120 or higher?
   c. Between 100 and 120? **d.** Of 90 or less?

6. **Product Reliability** The tread lives of the Super Titan radial tires under normal driving conditions are normally distributed with a mean of 40,000 mi and a standard deviation of 2000 mi. What is the probability that a tire selected at random will have a tread life of more than 35,000 mi? If four new tires are installed in a car and they experience even wear, determine the probability that all four tires still have useful tread lives after 35,000 mi of driving.

7. **Female Factory Workers' Wages** According to data released by the Chamber of Commerce of a certain city, the weekly wages (in dollars) of female factory workers are normally distributed with a mean of 475 and a standard deviation of 50. Find the probability that a female factory worker selected at random from the city makes a weekly wage of \$450 to \$550.

8. **Civil Service Exams** To be eligible for further consideration, applicants for certain Civil Service positions must first pass a written qualifying examination on which a score of 70 or more must be obtained. In a recent examination it was found that the scores were normally distributed with a mean of 60 points and a standard deviation of 10 points. Determine the percentage of applicants who passed the written qualifying examination.

9. **Warranties** The general manager of the Service Department of the MCA Television Company has estimated that the time that elapses between the dates of purchase and the dates on which the 19-in. sets manufactured by the company first require service is normally distributed with a mean of 22 mo and a standard deviation of 4 mo. If the company gives a 1-yr warranty on parts and labor for these sets, determine the percentage of sets manufactured and sold by the company that may require service before the warranty period runs out.

10. **Grade Distributions** The scores on an economics examination are normally distributed with a mean of 72 and a standard deviation of 16. If the instructor assigns a grade of A to 10% of the class, what is the lowest score a student may have and still obtain an A?

11. **Grade Distributions** The scores on a sociology examination are normally distributed with a mean of 70 and a standard deviation of 10. If the instructor assigns As to 15%, Bs to 25%, Cs to 40%, Ds to 15%, and Fs to 5% of the class, find the cutoff points for these grades.

In Exercises 12–22, use the appropriate normal distributions to approximate the resulting binomial distributions.

12. A fair coin is tossed 20 times. What is the probability of obtaining:
    a. Fewer than 8 heads? **b.** More than 6 heads?
    c. Between 6 and 10 heads inclusive?

13. A coin is weighted so that the probability of obtaining a head in a single toss is .4. If the coin is tossed 25 times, what is the probability of obtaining:
    a. Fewer than 10 heads?
    b. Between 10 and 12 heads, inclusive?
    c. More than 15 heads?

14. SPORTS A basketball player has a 75% chance of making a free throw. What is the probability of his making 100 or more free throws in 120 trials?

15. SPORTS A marksman's chance of hitting a target with each of his shots is 60%. If he fires 30 shots, what is the probability of his hitting the target:
    a. At least 20 times? **b.** Fewer than 10 times?
    c. Between 15 and 20 times, inclusive?

16. QUALITY CONTROL The P.A.R. Bearings Company is the principal supplier of ball bearings for the Sperry Gyroscope Company. It has been determined that 6% of the ball bearings shipped are rejected because they fail to meet tolerance requirements. What is the probability that a shipment of 200 ball bearings contains more than 10 rejects?

17. QUALITY CONTROL The manager of C & R Clothiers, a major manufacturer of men's shirts, has determined that 3% of C & R's shirts do not meet with company standards and are sold as "seconds" to discount and outlet stores. What is the probability that in a day's production of 200 dozen shirts, less than 10 dozen will be classified as "seconds"?

18. INDUSTRIAL ACCIDENTS The Colorado Mining and Mineral Company has 800 employees engaged in its mining operations. It has been estimated that the probability of a worker meeting with an accident during a 1-yr period is .1. What is the probability that more than 70 workers will meet with an accident during the 1-yr period?

19. DRUG TESTING An experiment was conducted to test the effectiveness of a new drug in treating a certain disease. The drug was administered to 50 mice that had been previously exposed to the disease. It was found that 35 mice subsequently recovered from the disease. It was determined that the natural recovery rate from the disease is 0.5.
    a. Determine the probability that 35 or more of the mice not treated with the drug would recover from the disease.
    b. Using the results obtained in part (a), comment on the effectiveness of the drug in the treatment of the disease.

20. LOAN DELINQUENCIES The manager of the Madison Finance Company has estimated that, because of a recession year, 5% of its 400 loan accounts will be delinquent. If the manager's estimate is correct, what is the probability that 25 or more of the accounts will be delinquent?

21. CRUISE SHIP BOOKINGS Because of late cancellations, Neptune Lines, an operator of cruise ships, has a policy of accepting more reservations than there are accommodations available. From experience, 8% of the bookings for the 90-day around-the-world cruise on the S.S. *Drion,* which has accommodations for 2000 passengers, are subsequently canceled. If the management of Neptune Lines has decided, for public relations reasons, that a person who has made a reservation should have a probability of .99 of obtaining accommodation on the ship, determine the largest number of reservations that should be taken for a cruise on the S.S. *Drion.*

22. THEATER BOOKINGS The Preview Showcase, a research firm, screens pilots of new TV shows before a randomly selected audience and then solicits their opinions of the shows. Based on past experience, 20% of those who get complimentary tickets are "no-shows." The theater has a seating capacity of 500. Management has decided, for public relations reasons, that a person who has been solicited for a screening should have a probability of .99 of being seated. How many tickets should the company send out to prospective viewers for each screening?

Spreadsheet examples and exercises for this section that may be solved using the Microsoft® Excel program are given at the Brooks/Cole Web site: http://www.brookscole.com/math/authors/tans/

SOLUTIONS TO SELF-CHECK EXERCISES 8.6

1. Let X be the normal random variable denoting the serum cholesterol levels in mg/dL in the current Mediterranean population under consideration. Then, the percentage of the population having blood cholesterol levels between 160 and 180 mg/dL is given by $P(160 < X < 180)$. To compute $P(160 < X < 180)$, we use Formula (16), Section 8.5, with $\mu = 160$, $\sigma = 50$, $a = 160$, and $b = 180$. We find

$$\begin{aligned} P(160 < X < 180) &= P\left(\frac{160 - 160}{50} < Z < \frac{180 - 160}{50}\right) \\ &= P(0 < Z < 0.4) \\ &= P(Z < 0.4) - P(Z < 0) \\ &= .6554 - .5000 \\ &= .1554 \end{aligned}$$

Thus, approximately 15.5% of the population has blood cholesterol levels between 160 and 180 mg/dL.

2. Let X denote the number of substandard bags in the production. Then, X has a binomial distribution with $n = 500$, $p = .04$, $q = .96$, so

$$\mu = (500)(.04) = 20$$
$$\sigma = \sqrt{(500)(.04)(.96)} = 4.38$$

Approximating the binomial distribution by a normal distribution with a mean of 20 and standard deviation of 4.38, we find that the probability that more than 30 bags in the production of 500 will be substandard is given by

$$\begin{aligned} P(X > 30) &\approx P(Y > 30.5) \qquad \text{(Where } Y \text{ denotes the normal random variable)} \\ &= P\left(Z > \frac{30.5 - 20}{4.38}\right) \\ &= P(Z > 2.40) \\ &= P(Z < -2.40) \\ &= .0082 \end{aligned}$$

or approximately 0.8%.

CHAPTER 8 Summary of Principal Formulas and Terms

Formulas

1. Mean of n numbers $\qquad \bar{x} = \dfrac{x_1 + x_2 + \cdots + x_n}{n}$

2. Expected value $\qquad E(X) = x_1p_1 + x_2p_2 + \cdots + x_np_n$

3. Odds in favor of E occurring $\qquad \dfrac{P(E)}{P(E^c)}$

4. Odds against E occurring $\quad \dfrac{P(E^c)}{P(E)}$

5. Probability of an event occurring given the odds $\quad \dfrac{a}{a+b}$

6. Variance of a random variable $\quad \text{Var}(X) = p_1(x_1 - \mu)^2 + p_2(x_2 - \mu)^2 + \cdots + p_n(x_n - \mu)^2$

7. Standard deviation of a random variable $\quad \sigma = \sqrt{\text{Var}(X)}$

8. Chebychev's inequality $\quad P(\mu - k\sigma \le X \le \mu + k\sigma) \ge 1 - \dfrac{1}{k^2}$

9. Probability of x successes in n Bernoulli trials $\quad C(n, x)p^x q^{n-x}$

10. Binomial random variable:
    - Mean $\quad \mu = E(X) = np$
    - Variance $\quad \text{Var}(X) = npq$
    - Standard deviation $\quad \sigma_x = \sqrt{npq}$

Terms

random variable
finite discrete random variable
infinite discrete random variable
continuous random variable
probability distribution of a random variable
histogram
average (mean)
expected value
variance
standard deviation
Bernoulli (binomial) trial
binomial experiment
binomial random variable
binomial distribution
probability density function
normal distribution

CHAPTER 8 REVIEW EXERCISES

1. Three balls are selected at random without replacement from an urn containing three white balls and four blue balls. Let the random variable X denote the number of blue balls drawn.
   a. List the outcomes of this experiment.
   b. Find the value assigned to each outcome of this experiment by the random variable X.
   c. Find the probability distribution of the random variable associated with this experiment.
   d. Draw the histogram representing this distribution.
2. A man purchased a \$25,000, 1-yr term-life insurance policy for \$375. Assuming that the probability that he will live for another year is .989, find the company's expected gain.

3. The probability distribution of a random variable X is shown in the accompanying table:

x	$P(X = x)$
0	.1
1	.1
2	.2
3	.3
4	.2
5	.1

a. Compute $P(1 \leq X \leq 4)$.
b. Compute the mean and standard deviation of X.

4. A binomial experiment consists of four trials in which the probability of success in any one trial is 2/5.
a. Construct the probability distribution for the experiment.
b. Compute the mean and standard deviation of the probability distribution.

In Exercises 5–8, let Z be the standard normal variable. Make a rough sketch of the appropriate region under the standard normal curve and find the given probability.

5. $P(Z < 0.5)$ **6.** $P(Z < -0.75)$

7. $P(-0.75 < Z < 0.5)$ **8.** $P(-0.42 < Z < 0.66)$

In Exercises 9–12, let Z be the standard normal variable. Find z if z satisfies the given value.

9. $P(Z < z) = .9922$ **10.** $P(Z < z) = .1469$

11. $P(Z > z) = .9788$

12. $P(-z < Z < z) = .8444$

In Exercises 13–16, let X be a normal random variable with $\mu = 10$ and $\sigma = 2$. Find the value of the given probability.

13. $P(X < 11)$ **14.** $P(X > 8)$

15. $P(7 < X < 9)$ **16.** $P(6.5 < X < 11.5)$

17. If the probability that a bowler will bowl a strike is .7, what is the probability that he will get exactly two strikes in four attempts? At least two strikes in four attempts?

18. The heights of 4000 women who participated in a recent survey were found to be normally distributed with a mean of 64.5 in. and a standard deviation of 2.5 in. What percentage of these women have heights of 67 in. or greater?

19. Refer to Exercise 18. Use Chebychev's inequality to estimate the probability that the height of a woman who participated in the survey will fall within 2 standard deviations of the mean—that is, that her height will be between 59.5 and 69.5 in.

20. The proprietor of a hardware store will accept a shipment of ceramic wall tiles if no more than 2 of a random sample of 20 are found to be defective. What is the probability that he will accept shipment if exactly 10% of the tiles in a certain shipment is defective?

21. An experimental drug has been found to be effective in treating 15% of the people afflicted by a certain disease. If the drug is administered to 800 people who have this disease, what are the mean and standard deviation of the number of people for whom the drug can be expected to be effective?

22. The Dayton Iron Works Company manufactures steel rods to a specification of 1-in. diameter. These rods are accepted by the buyer if they fall within the tolerance limits of 0.995 and 1.005. Assuming that the diameter of the rods is normally distributed about a mean of 1 in. and a standard deviation of 0.002 in., estimate the percentage of rods that will be rejected by the buyer.

23. A coin is biased so that the probability of it landing heads is .6. If the coin is tossed 100 times, what is the probability that heads will appear more than 50 times in the 100 tosses?

24. A division of the Solaron Corporation manufactures photovoltaic cells for use in the company's solar energy converters. It is estimated that 5% of the cells manufactured is defective. In a batch of 200 cells manufactured by the company, what is the probability that it will contain at most 20 defective units?

9 MARKOV CHAINS AND THE THEORY OF GAMES

In this chapter we look at two important applications of mathematics that are based primarily on matrix theory and the theory of probability. Both of these applications, *Markov chains* and the *theory of games*, are relatively recent developments in the field of mathematics and have wide applications in many practical areas.

Will the flowers be red? A certain species of plant produces red, pink, or white flowers, depending on its genetic makeup. In Example 4, page 563, we will show that if the offspring of two plants are crossed successively with plants of a certain genetic makeup only, then in the long run all the flowers produced by the plants will be red.

9.1 Markov Chains

TRANSITIONAL PROBABILITIES

A finite stochastic process, you may recall, is an experiment consisting of a finite number of stages in which the outcomes and associated probabilities at each stage depend on the outcomes and associated probabilities of the *preceding stages.* In this chapter we are concerned with a special class of stochastic processes—namely, those in which the probabilities associated with the outcomes at any stage of the experiment depend only on the outcomes of the *preceding stage.* Such a process is called a **Markov process,** or a **Markov chain,** named after the Russian mathematician A. A. Markov (1856–1922).

The outcome at any stage of the experiment in a Markov process is called the **state** of the experiment. In particular, the outcome at the current stage of the experiment is called the **current state** of the process. Here is a typical problem involving a Markov chain:

Starting from one state of a process (the current state), determine the probability that the process will be at a particular state at some future time.

EXAMPLE 1

An analyst at Weaver and Kline, a stock brokerage firm, observes that the closing price of the preferred stock of an airline company over a short span of time depends only on its previous closing price. At the end of each trading day, he makes a note of the stock's performance for that day, recording the closing price as "higher," unchanged," or "lower" according to whether the stock closes higher, unchanged, or lower than the previous day's closing price. This sequence of observations may be viewed as a Markov chain. ■■■■

The transition from one state to another in a Markov chain may be studied with the aid of tree diagrams, as in the next example.

EXAMPLE 2

Refer to Example 1. If on a certain day the stock's closing price is higher than that of the previous day, then the probability that it closes higher, unchanged, or lower on the next trading day is .2, .3, and .5, respectively. Next, if the stock's closing price is unchanged from the previous day, then the probability that it closes higher, unchanged, or lower on the next trading day is .5, .2, and .3, respectively. Finally, if the stock's closing price is lower than that of the previous day, then the probability that it closes higher, unchanged, or lower on the next trading day is .4, .4, and .2, respectively. With the aid of tree diagrams, describe the transition between states and the probabilities associated with these transitions.

SOLUTION ✓

The Markov chain being described has three states: higher, unchanged, and lower. If the current state is higher, then the transition to the other states from this state may be displayed by constructing a tree diagram in which the associated probabilities are shown on the appropriate limbs (Figure 9.1). Tree

FIGURE 9.1
Tree diagrams showing transition probabilities between states

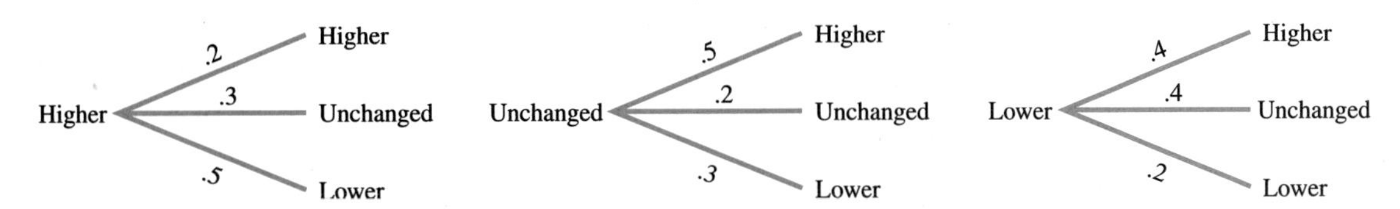

diagrams describing the transition from each of the other two possible current states, unchanged and lower, to the other states are constructed in a similar manner. ■■■■

The probabilities encountered in this example are called **transition probabilities** because they are associated with the transition from one state to the next in the Markov process. These transition probabilities may be conveniently represented in the form of a matrix. Suppose for simplicity that we have a Markov chain with three possible outcomes at each stage of the experiment. Let's refer to these outcomes as state 1, state 2, and state 3. Then the transition probabilities associated with the transition from state 1 to each of the states 1, 2, and 3 in the next phase of the experiment are precisely the respective conditional probabilities that the outcome is state 1, state 2, and state 3 *given* that the outcome state 1 has occurred. In short, the desired transition probabilities are $P(\text{state } 1 \mid \text{state } 1)$, $P(\text{state } 2 \mid \text{state } 1)$, and $P(\text{state } 3 \mid \text{state } 1)$, respectively. Thus, we write

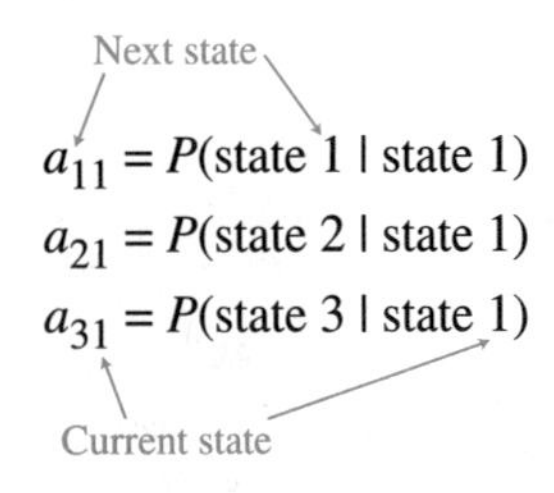

$$a_{11} = P(\text{state } 1 \mid \text{state } 1)$$
$$a_{21} = P(\text{state } 2 \mid \text{state } 1)$$
$$a_{31} = P(\text{state } 3 \mid \text{state } 1)$$

Note that the first subscript in this notation refers to the state in the next stage of the experiment, and the second subscript refers to the current state. Using a tree diagram, we have the following representation:

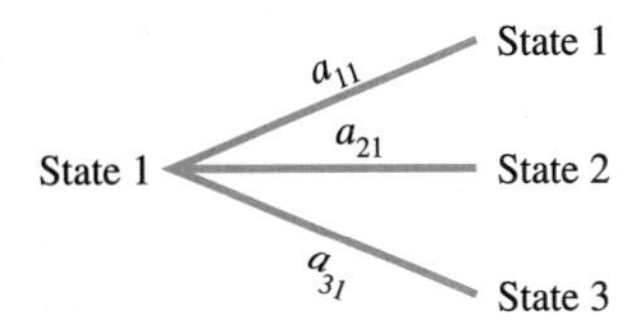

Similarly, the transition probabilities associated with the transition from state 2 and state 3 to each of the states 1, 2, and 3 are

$$\begin{aligned} a_{12} &= P(\text{state } 1 \mid \text{state } 2) & & & a_{13} &= P(\text{state } 1 \mid \text{state } 3) \\ a_{22} &= P(\text{state } 2 \mid \text{state } 2) & &\text{and} & a_{23} &= P(\text{state } 2 \mid \text{state } 3) \\ a_{32} &= P(\text{state } 3 \mid \text{state } 2) & & & a_{33} &= P(\text{state } 3 \mid \text{state } 3) \end{aligned}$$

These observations lead to the following matrix representation of the transition probabilities:

		Current state		
		State 1	State 2	State 3
Next state	State 1	a_{11}	a_{12}	a_{13}
	State 2	a_{21}	a_{22}	a_{23}
	State 3	a_{31}	a_{32}	a_{33}

EXAMPLE 3 Use a matrix to represent the transition probabilities obtained in Example 2.

SOLUTION ✓ There are three states at each stage of the Markov chain under consideration. Letting state 1, state 2, and state 3 denote the states "higher," "unchanged," and "lower," respectively, we find that

$$a_{11} = .2, \qquad a_{21} = .3, \qquad a_{31} = .5$$

and so on, so the required matrix representation is given by

$$T = \begin{bmatrix} .2 & .5 & .4 \\ .3 & .2 & .4 \\ .5 & .3 & .2 \end{bmatrix}$$

■■■■

The matrix obtained in Example 3 is a transition matrix. In the general case, we have the following definition:

Transition Matrix

A **transition matrix** associated with a Markov chain with n states is an $n \times n$ matrix T with entries a_{ij} $(1 \le i \le n;\ 1 \le j \le n)$

$$T = \text{Next state} \begin{array}{c} \\ \text{State } 1 \\ \text{State } 2 \\ \vdots \\ \text{State } i \\ \vdots \\ \text{State } n \end{array} \overset{\text{Current state}}{\begin{array}{c} \begin{array}{ccccccc} \text{State } 1 & \text{State } 2 & \cdots & \text{State } j & \cdots & \text{State } n \end{array} \\ \begin{bmatrix} a_{11} & a_{12} & \cdots & a_{1j} & \cdots & a_{1n} \\ a_{21} & a_{22} & \cdots & a_{2j} & \cdots & a_{2n} \\ \vdots & \vdots & & \vdots & & \vdots \\ a_{i1} & a_{i2} & \cdots & a_{ij} & \cdots & a_{in} \\ \vdots & \vdots & & \vdots & & \vdots \\ a_{n1} & a_{n2} & \cdots & a_{nj} & \cdots & a_{nn} \end{bmatrix} \end{array}}$$

having the following properties:

1. $a_{ij} \ge 0$ for all i and j.
2. The sum of the entries in each column of T is 1.

Since $a_{ij} = P(\text{state } i \mid \text{state } j)$ is the probability of the occurrence of an event, it must be nonnegative, and this is precisely what Property 1 implies. Property 2 follows from the fact that the transition from any one of the current states must terminate in one of the n states in the next stage of the experiment. Any square matrix satisfying properties 1 and 2 is referred to as a **stochastic matrix.**

Group Discussion

Let

$$A = \begin{bmatrix} p & q \\ 1-p & 1-q \end{bmatrix} \quad \text{and} \quad B = \begin{bmatrix} r & s \\ 1-r & 1-s \end{bmatrix}$$

be two 2×2 stochastic matrices, where $0 \le p \le 1, 0 \le q \le 1, 0 \le r \le 1$, and $0 \le s \le 1$.

1. Show that AB is a 2×2 stochastic matrix.
2. Use the result of part (a) to explain why $A^2, A^3, \ldots, A^n$, where n is a positive integer, are also 2×2 stochastic matrices.

One advantage in representing the transition probabilities in the form of a matrix is that we may use the results from matrix theory to help us solve problems involving Markov processes, as we will see in the next several sections.

Next, for simplicity, let's consider the following Markov process where each stage of the experiment has precisely two possible states.

EXAMPLE 4 Because of the continued successful implementation of an urban renewal program, it is expected that each year 3% of the population currently residing in the city will move to the suburbs and 6% of the population currently residing in the suburbs will move into the city. At present, 65% of the total population of the metropolitan area lives in the city itself, while the remaining 35% lives in the suburbs. Assuming that the total population of the metropolitan area remains constant, what will be the distribution of the population one year from now?

SOLUTION ✔ This problem may be solved with the aid of a tree diagram and the techniques of Chapter 7. The required tree diagram describing this process is shown in Figure 9.2. Using the method of Section 7.5, we find the probability that a person selected at random will be a city dweller one year from now is given by

$$(.65)(.97) + (.35)(.06) = .6515$$

In a similar manner, we find that the probability that a person selected at random will reside in the suburbs one year from now is given by

$$(.65)(.03) + (.35)(.94) = .3485$$

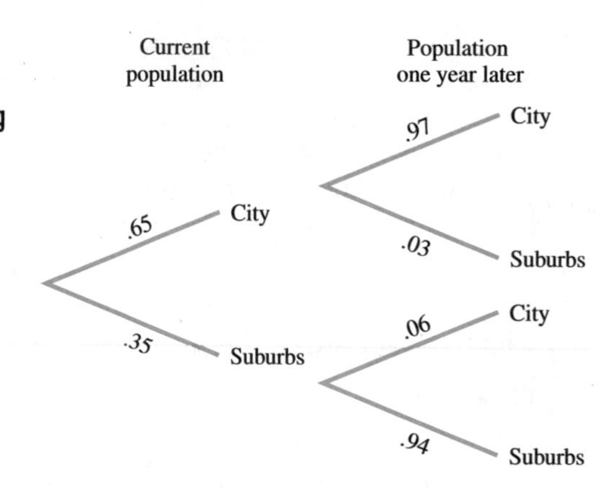

FIGURE 9.2
Tree diagram showing a Markov process with two states: living in the city and living in the suburbs

Thus, the population of the area one year from now may be expected to be distributed as follows: 65.15% living in the city and 34.85% residing in the suburbs. ■■■■

Let us reexamine the solution to this problem. As noted earlier, the process under consideration may be viewed as a Markov chain with two possible states at each stage of the experiment: "living in the city" (state 1) and "living in the suburbs" (state 2). The transition matrix associated with this Markov chain is

$$T = \begin{array}{c} \text{State 1} \\ \text{State 2} \end{array} \overset{\begin{array}{cc} \text{State 1} & \text{State 2} \end{array}}{\begin{bmatrix} .97 & .06 \\ .03 & .94 \end{bmatrix}} \qquad \text{(Transition matrix)}$$

Next, observe that the initial (current) probability distribution of the population may be summarized in the form of a column vector of dimension 2 (that is, a 2×1 matrix). Thus,

$$X_0 = \begin{array}{c} \text{State 1} \\ \text{State 2} \end{array} \begin{bmatrix} .65 \\ .35 \end{bmatrix} \qquad \text{(Initial-state matrix)}$$

Using the results of Example 4, we may write the population distribution one year later as

$$X_1 = \begin{array}{c} \text{State 1} \\ \text{State 2} \end{array} \begin{bmatrix} .6515 \\ .3485 \end{bmatrix} \qquad \text{(Distribution after one year)}$$

You may now verify that

$$TX_0 = \begin{bmatrix} .97 & .06 \\ .03 & .94 \end{bmatrix} \begin{bmatrix} .65 \\ .35 \end{bmatrix} = \begin{bmatrix} .6515 \\ .3485 \end{bmatrix} = X_1$$

so this problem may be solved using matrix multiplication.

EXAMPLE 5 Refer to Example 4. What is the population distribution of the city after two years? After three years?

SOLUTION ✓ Let X_2 be the column vector representing the probability population distribution of the metropolitan area after two years. We may view X_1, the vector representing the probability population distribution of the metropolitan area after one year, as representing the "initial" probability distribution in this part of our calculation. Thus,

$$X_2 = TX_1 = \begin{bmatrix} .97 & .06 \\ .03 & .94 \end{bmatrix} \begin{bmatrix} .6515 \\ .3485 \end{bmatrix} = \begin{bmatrix} .6529 \\ .3471 \end{bmatrix}$$

The vector representing the probability distribution of the metropolitan area after three years is given by

$$X_3 = TX_2 = \begin{bmatrix} .97 & .06 \\ .03 & .94 \end{bmatrix} \begin{bmatrix} .6529 \\ .3471 \end{bmatrix} = \begin{bmatrix} .6541 \\ .3459 \end{bmatrix}$$

That is, after three years, the population will be distributed as follows: 65.41% will live in the city and 34.59% will live in the suburbs. ■■■■

Distribution Vectors

Observe that, in the foregoing computations, we have $X_1 = TX_0$, $X_2 = TX_1 = T^2X_0$, and $X_3 = TX_2 = T^3X_0$. These results are easily generalized. To see this, suppose we have a Markov process in which there are n possible states at each stage of the experiment. Suppose further that the probability of the system being in state 1, state 2, . . . , state n, initially, is given by p_1, $p_2, \ldots, p_n$, respectively. This distribution may be represented as an n-dimensional vector

$$X_0 = \begin{bmatrix} p_1 \\ p_2 \\ \vdots \\ p_n \end{bmatrix}$$

called a **distribution vector.** If T represents the $n \times n$ transition matrix associated with the Markov process, then the probability distribution of the system after m observations is given by

$$X_m = T^m X_0 \qquad \textbf{(1)}$$

EXAMPLE 6 To keep track of the location of its cabs, the Zephyr Cab Company has divided a town into three zones: zone I, zone II, and zone III. Zephyr's management has determined from company records that of the passengers picked up in zone I, 60% are discharged in the same zone, 30% are discharged in zone II, and 10% are discharged in zone III. Of those picked up in zone II, 40% are

discharged in zone I, 30% are discharged in zone II, and 30% are discharged in zone III. Of those picked up in zone III, 30% are discharged in zone I, 30% are discharged in zone II, and 40% are discharged in zone III. Suppose that at the beginning of the day 80% of the cabs are in zone I, 15% are in zone II, and 5% are in zone III, and that a taxi without a passenger will cruise within the zone it is currently in until a pickup is made.

a. Find the transition matrix for the Markov chain that describes the successive locations of a cab.
b. What is the distribution of the cabs after all of them have made one pickup and discharge?
c. What is the distribution of the cabs after all of them have made two pickups and discharges?

SOLUTION ✓ Let zone I, zone II, and zone III correspond to state 1, state 2, and state 3 of the Markov chain.

a. The required transition matrix is given by

$$T = \begin{bmatrix} .6 & .4 & .3 \\ .3 & .3 & .3 \\ .1 & .3 & .4 \end{bmatrix}$$

b. The initial distribution vector associated with the problem is

$$X_0 = \begin{bmatrix} .8 \\ .15 \\ .05 \end{bmatrix}$$

If X_1 denotes the distribution vector after one observation—that is, after all the cabs have made one pickup and discharge—then

$$\begin{aligned} X_1 &= TX_0 \\ &= \begin{bmatrix} .6 & .4 & .3 \\ .3 & .3 & .3 \\ .1 & .3 & .4 \end{bmatrix} \begin{bmatrix} .8 \\ .15 \\ .05 \end{bmatrix} = \begin{bmatrix} .555 \\ .3 \\ .145 \end{bmatrix} \end{aligned}$$

That is, 55.5% of the cabs are in zone I, 30% are in zone II, and 14.5% are in zone III.

c. Let X_2 denote the distribution vector after all the cabs have made two pickups and discharges. Then

$$\begin{aligned} X_2 &= TX_1 \\ &= \begin{bmatrix} .6 & .4 & .3 \\ .3 & .3 & .3 \\ .1 & .3 & .4 \end{bmatrix} \begin{bmatrix} .555 \\ .3 \\ .145 \end{bmatrix} = \begin{bmatrix} .4965 \\ .3 \\ .2035 \end{bmatrix} \end{aligned}$$

That is, 49.65% of the cabs are in zone I, 30% are in zone II, and 20.35% are in zone III. You should verify that the same result may be obtained by computing T^2X_0. ■■■■

REMARK In this simplified model, we do not take into consideration variable demand and variable delivery time.

SELF-CHECK EXERCISES 9.1

1. Three supermarkets serve a certain section of a city. During the upcoming year, supermarket A is expected to retain 80% of its customers, lose 5% of its customers to supermarket B, and lose 15% to supermarket C. Supermarket B is expected to retain 90% of its customers and lose 5% of its customers to each of supermarkets A and C. Supermarket C is expected to retain 75% of its customers, lose 10% to supermarket A, and lose 15% to supermarket B. Construct the transition matrix for the Markov chain that describes the change in the market share of the three supermarkets.
2. Refer to Self-Check Exercise 1. Currently the market shares of supermarket A, supermarket B, and supermarket C are 0.4, 0.3, and 0.3, respectively.
   a. Find the initial distribution vector for this Markov chain.
   b. What share of the market will be held by each supermarket after 1 year? Assuming that the trend continues, what will the market share be after 2 years?

Solutions to Self-Check Exercises 9.1 can be found on page 544.

9.1 Exercises

In Exercises 1–10, determine which of the given matrices are stochastic.

1. $\begin{bmatrix} .4 & .7 \\ .6 & .3 \end{bmatrix}$

2. $\begin{bmatrix} .8 & .2 \\ .3 & .7 \end{bmatrix}$

3. $\begin{bmatrix} \frac{1}{4} & \frac{1}{8} \\ \frac{3}{4} & \frac{7}{8} \end{bmatrix}$

4. $\begin{bmatrix} \frac{1}{3} & 0 & \frac{1}{2} \\ \frac{1}{2} & 1 & 0 \\ \frac{1}{4} & 0 & \frac{1}{2} \end{bmatrix}$

5. $\begin{bmatrix} .3 & .2 & .4 \\ .4 & .7 & .3 \\ .3 & .1 & .2 \end{bmatrix}$

6. $\begin{bmatrix} \frac{1}{3} & \frac{1}{4} & \frac{1}{2} \\ \frac{1}{3} & 0 & -\frac{1}{2} \\ \frac{1}{4} & \frac{3}{4} & \frac{1}{2} \end{bmatrix}$

7. $\begin{bmatrix} .1 & .4 & .3 \\ .7 & .2 & .1 \\ .2 & .4 & .6 \end{bmatrix}$

8. $\begin{bmatrix} 1 & 0 & 0 \\ 0 & 0 & 1 \\ 0 & 1 & 0 \end{bmatrix}$

9. $\begin{bmatrix} .2 & .3 \\ .3 & .1 \\ .5 & .6 \end{bmatrix}$

10. $\begin{bmatrix} .5 & .2 & .3 \\ .2 & .3 & .2 \\ .3 & .4 & .1 \\ 0 & .1 & .4 \end{bmatrix}$

11. The transition matrix for a Markov process is given by

$$T = \text{State}\ \begin{matrix} \\ 1 \\ 2 \end{matrix} \begin{matrix} \text{State} \\ \begin{matrix} 1 & 2 \end{matrix} \\ \begin{bmatrix} .3 & .6 \\ .7 & .4 \end{bmatrix} \end{matrix}$$

   a. What does the entry $a_{11} = .3$ represent?
   b. Given that the outcome state 1 has occurred, what is the probability that the next outcome of the experiment will be state 2?
   c. If the initial-state distribution vector is given by

$$X_0 = \begin{matrix} \text{State 1} \\ \text{State 2} \end{matrix} \begin{bmatrix} .4 \\ .6 \end{bmatrix}$$

   find TX_0, the probability distribution of the system after one observation.

12. The transition matrix for a Markov process is given by

$$T = \text{State} \begin{matrix} 1 \\ 2 \end{matrix} \overset{\text{State} \atop 1 \quad 2}{\begin{bmatrix} \frac{1}{6} & \frac{2}{3} \\ \frac{5}{6} & \frac{1}{3} \end{bmatrix}}$$

a. What does the entry $a_{22} = \frac{1}{3}$ represent?
b. Given that the outcome state 1 has occurred, what is the probability that the next outcome of the experiment will be state 2?
c. If the initial-state distribution vector is given by

$$X_0 = \text{State} \begin{matrix} 1 \\ 2 \end{matrix} \begin{bmatrix} \frac{1}{4} \\ \frac{3}{4} \end{bmatrix}$$

find TX_0, the probability distribution of the system after one observation.

13. The transition matrix for a Markov process is given by

$$T = \text{State} \begin{matrix} 1 \\ 2 \end{matrix} \overset{\text{State} \atop 1 \quad 2}{\begin{bmatrix} .6 & .2 \\ .4 & .8 \end{bmatrix}}$$

and the initial-state distribution vector is given by

$$X_0 = \text{State} \begin{matrix} 1 \\ 2 \end{matrix} \begin{bmatrix} .5 \\ .5 \end{bmatrix}$$

Find TX_0 and interpret your result with the aid of a tree diagram.

14. The transition matrix for a Markov process is given by

$$X_0 = \text{State} \begin{matrix} 1 \\ 2 \end{matrix} \overset{\text{State} \atop 1 \quad 2}{\begin{bmatrix} \frac{1}{2} & \frac{3}{4} \\ \frac{1}{2} & \frac{1}{4} \end{bmatrix}}$$

and the initial-state distribution vector is given by

$$X_0 = \text{State} \begin{matrix} 1 \\ 2 \end{matrix} \begin{bmatrix} \frac{1}{3} \\ \frac{2}{3} \end{bmatrix}$$

Find TX_0 and interpret your result with the aid of a tree diagram.

In Exercises 15–18, find X_2 (the probability distribution of the system after two observations) for the given distribution vector X_0 and the given transition matrix T.

15. $X_0 = \begin{bmatrix} .6 \\ .4 \end{bmatrix}$, $T = \begin{bmatrix} .4 & .8 \\ .6 & .2 \end{bmatrix}$

16. $X_0 = \begin{bmatrix} \frac{1}{2} \\ \frac{1}{2} \\ 0 \end{bmatrix}$, $T = \begin{bmatrix} \frac{1}{2} & \frac{1}{3} & \frac{1}{2} \\ 0 & \frac{1}{3} & \frac{1}{4} \\ \frac{1}{2} & \frac{1}{3} & \frac{1}{4} \end{bmatrix}$

17. $X_0 = \begin{bmatrix} \frac{1}{4} \\ \frac{1}{2} \\ \frac{1}{4} \end{bmatrix}$, $T = \begin{bmatrix} \frac{1}{4} & \frac{1}{4} & \frac{1}{2} \\ \frac{1}{4} & \frac{1}{2} & \frac{1}{2} \\ \frac{1}{2} & \frac{1}{4} & 0 \end{bmatrix}$

18. $X_0 = \begin{bmatrix} .25 \\ .40 \\ .35 \end{bmatrix}$, $T = \begin{bmatrix} .1 & .1 & .3 \\ .8 & .7 & .2 \\ .1 & .2 & .5 \end{bmatrix}$

19. Psychology Experiments A psychologist conducts an experiment in which a mouse is placed in a T-maze, where it has a choice at the T-junction of turning left and receiving a reward (cheese) or turning right and receiving a mild electric shock (see accompanying figure). At the end of each trial, a record is kept of the mouse's response. It is observed that the mouse is as likely to turn left (state 1) as right (state 2) during the first trial. In subsequent trials, however, the observation is made that if the mouse had turned left in the previous trial, then on the next trial the probability that it will turn left is .8, whereas the probability that it will turn right is .2. If the

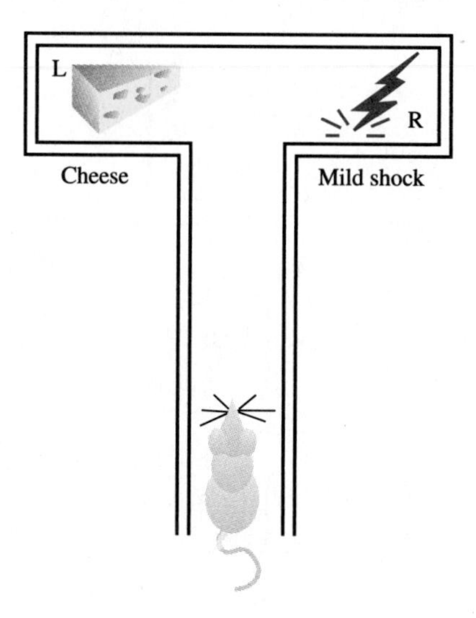

mouse had turned right in the previous trial, then the probability that it will turn right on the next trial is .1, whereas the probability that it will turn left is .9.

a. Using a tree diagram, describe the transitions between states and the probabilities associated with these transitions.

b. Represent the transition probabilities obtained in part (a) in terms of a matrix.

c. What is the initial-state probability vector?

d. Use the results of parts (b) and (c) to find the probability that a mouse will turn left on the second trial.

20. **SMALL-TOWN REVIVAL** At the beginning of 1990, the population of a certain state was 55.4% rural and 44.6% urban. Based on past trends, it is expected that 10% of the population currently residing in the rural areas will move into the urban areas, while 17% of the population currently residing in the urban areas will move into the rural areas in the next decade. What was the population distribution in that state at the beginning of the year 2000?

21. **POLITICAL POLLS** The Morris Polling Group conducted a poll 6 mo before an election in a state in which a Democrat and a Republican were running for governor and found that 60% of the voters intended to vote for the Republican and 40% intended to vote for the Democrat. In a poll conducted 3 mo later, it was found that 70% of those who had earlier stated a preference for the Republican candidate still maintained that preference, whereas 30% of these voters now preferred the Democratic candidate. Of those who had earlier stated a preference for the Democrat, 80% still maintained that preference, whereas 20% now preferred the Republican candidate.

a. If the election were held at this time, who would win?

b. Assuming that this trend continues, which candidate is expected to win the election?

22. **COMMUTER TRENDS** Within a large metropolitan area, 20% of the commuters currently use the public transportation system, whereas the remaining 80% commute via automobile. The city has recently revitalized and expanded its public transportation system. It is expected that 6 mo from now 30% of those who are now commuting to work via automobile will switch to public transportation, and 70% will continue to commute via automobile. At the same time, it is expected that 20% of those now using public transportation will commute via automobile and 80% will continue to use public transportation.

a. Construct the transition matrix for the Markov chain that describes the change in the mode of transportation used by these commuters.

b. Find the initial distribution vector for this Markov chain.

c. What percentage of the commuters are expected to use public transportation 6 mo from now?

23. Refer to Example 6. If the initial distribution vector for the location of the taxis is

$$X_0 = \begin{matrix} \text{Zone I} \\ \text{Zone II} \\ \text{Zone III} \end{matrix} \begin{bmatrix} .6 \\ .2 \\ .2 \end{bmatrix}$$

what will be the distribution after all of them have made one pickup and discharge?

24. **URBAN–SUBURBAN POPULATION FLOW** Refer to Example 4. If the initial probability distribution is

$$X_0 = \begin{matrix} \text{City} \\ \text{Suburb} \end{matrix} \begin{bmatrix} .80 \\ .20 \end{bmatrix}$$

what will be the population distribution of the city after 1 yr? After 2 yr?

25. **MARKET SHARE** At a certain university, three bookstores—the University Bookstore, the Campus Bookstore, and the Book Mart—currently serve the university community. From a survey conducted at the beginning of the fall quarter, it was found that the University Bookstore and the Campus Bookstore each had 40% of the market, whereas the Book Mart had 20% of the market. Each quarter the University Bookstore retains 80% of its customers but loses 10% to the Campus Bookstore and 10% to the Book Mart. The Campus Bookstore retains 75% of its customers but loses 10% to the University Bookstore and 15% to the Book Mart. The Book Mart retains 90% of its customers but loses 5% to the University Bookstore and 5% to the Campus Bookstore. If these trends continue, what percentage of the market will each store have at the beginning of the second quarter? The third quarter?

26. **MARKET SHARE OF AUTO MANUFACTURERS** In a study of the domestic market share of the three major automobile manufacturers A, B, and C in a certain country, it was found that their current market shares were 60%, 30%, and 10%, respectively. Furthermore, it was found that of the customers who bought a car manufactured by A, 75% would again buy a car manufactured by A, 15%

(continued on p. 544)

FINDING DISTRIBUTION VECTORS

Since the computation of the probability distribution of a system after a certain number of observations involves matrix multiplication, the graphing calculator may be used to facilitate the work.

EXAMPLE 1 Consider the problem posed in Example 6, page 537, where

$$T = \begin{bmatrix} .6 & .4 & .3 \\ .3 & .3 & .3 \\ .1 & .3 & .4 \end{bmatrix} \quad \text{and} \quad X_0 = \begin{bmatrix} .8 \\ .15 \\ .05 \end{bmatrix}$$

Verify that

$$X_2 = \begin{bmatrix} .4965 \\ .3 \\ .2035 \end{bmatrix}$$

as obtained in that example.

SOLUTION ✔ First, we enter into the calculator the matrix X_0 as the matrix A and the matrix T as the matrix B. Then, performing the indicated multiplication, we find that

$$\text{B^2 * A} = \begin{bmatrix} .4965 \\ .3 \\ .2035 \end{bmatrix}$$

That is,

$$X_2 = T^2 X_0 = \begin{bmatrix} .4965 \\ .3 \\ .2035 \end{bmatrix}$$

as was to be shown. ■■■■

Exercises

In Exercises 1–2, use a graphing calculator to find X_5 [the probability distribution of the system after five observations] for the given distribution vector X_0 and the given transition matrix T.

1. $X_0 = \begin{bmatrix} .2 \\ .3 \\ .2 \\ .1 \\ .2 \end{bmatrix} \qquad T = \begin{bmatrix} .2 & .2 & .3 & .2 & .1 \\ .1 & .2 & .1 & .2 & .1 \\ .3 & .4 & .1 & .3 & .3 \\ .2 & .1 & .2 & .2 & .2 \\ .2 & .1 & .3 & .1 & .3 \end{bmatrix}$

2. $X_0 = \begin{bmatrix} .1 \\ .2 \\ .2 \\ .3 \\ .2 \end{bmatrix} \qquad T = \begin{bmatrix} .3 & .2 & .1 & .3 & .1 \\ .2 & .1 & .2 & .1 & .2 \\ .1 & .2 & .3 & .2 & .2 \\ .1 & .3 & .2 & .3 & .2 \\ .3 & .2 & .2 & .1 & .3 \end{bmatrix}$

3. Refer to Exercise 26 on page 541. Using the same data, determine the market share that will be held by each manufacturer five model years after the study began.

would buy a car manufactured by B, and 10% would buy a car manufactured by C. Of the customers who bought a car manufactured by B, 90% would again buy a car manufactured by B, whereas 5% each would buy cars manufactured by A and C, respectively. Finally, of the customers who bought a car manufactured by C, 85% would again buy a car manufactured by C, 5% would buy a car manufactured by A, and 10% would buy a car manufactured by B. Assuming that these sentiments reflect the buying habits of customers in the future, determine the market share that will be held by each manufacturer after the next two model years.

27. **College Majors** Records compiled by the Admissions Office at a state university indicating the percentage of students that change their major each year are shown in the following transition matrix. Of the freshmen now at the university, 30% have chosen their major field in business, 30% in the humanities, 20% in education, and 20% in the natural sciences and other fields. Assuming that this trend continues, find the percentage of these students that will be majoring in each of the given areas in their senior year.

Hint: Find T^3X_0.

	Bus.	Hum.	Educ.	Nat. sci. and others
Business	.80	.10	.20	.10
Humanities	.10	.70	.10	.05
Education	.05	.10	.60	.05
Nat. sci. and others	.05	.10	.10	.80

28. **Homeowners' Choice of Energy** A study conducted by the Urban Energy Commission in a large metropolitan area indicates the probabilities that homeowners within the area will use certain heating fuels or solar energy during the next 10 years as the major source of heat for their homes. The transition matrix representing the transition probabilities from one state to another is

	Elec.	Gas	Oil	Solar
Electricity	.70	0	0	0
Natural gas	.15	.90	.20	.05
Fuel oil	.05	.02	.75	0
Solar energy	.10	.08	.05	.95

Among homeowners within the area, 20% currently use electricity, 35% use natural gas, 40% use oil, and 5% use solar energy as the major source of heat for their homes. What is the expected distribution of the homeowners that will be using each type of heating fuel or solar energy within the next decade?

In Exercises 29 and 30, determine whether the statement is true or false. If it is true, explain why it is true. If it is false, give an example to show why it is false.

29. A Markov chain is a process in which the outcomes at any stage of the experiment depend on the outcomes of the preceding stages.

30. The sum of the entries in each column of a transition matrix must not exceed 1.

Solutions to Self-Check Exercises 9.1

1. The required transition matrix is

$$T = \begin{bmatrix} .80 & .05 & .10 \\ .05 & .90 & .15 \\ .15 & .05 & .75 \end{bmatrix}$$

2. **a.** The initial distribution vector is

$$X_0 = \begin{bmatrix} .4 \\ .3 \\ .3 \end{bmatrix}$$

b. The vector representing the market share of each supermarket after 1 yr is

$$X_1 = TX_0$$
$$= \begin{bmatrix} .80 & .05 & .10 \\ .05 & .90 & .15 \\ .15 & .05 & .75 \end{bmatrix} \begin{bmatrix} .4 \\ .3 \\ .3 \end{bmatrix} = \begin{bmatrix} .365 \\ .335 \\ .3 \end{bmatrix}$$

That is, after 1 yr supermarket A will command a 36.5% market share, supermarket B will have a 33.5% share, and supermarket C will have a 30% market share.

The vector representing the market share of the supermarkets after 2 yr is

$$X_2 = TX_1$$
$$= \begin{bmatrix} .80 & .05 & .10 \\ .05 & .90 & .15 \\ .15 & .05 & .75 \end{bmatrix} \begin{bmatrix} .365 \\ .335 \\ .3 \end{bmatrix} = \begin{bmatrix} .3388 \\ .3648 \\ .2965 \end{bmatrix}$$

That is, 2 yr later the market shares of supermarkets A, B, and C will be 33.88%, 36.48%, and 29.65%, respectively.

9.2 Regular Markov Chains

STEADY-STATE DISTRIBUTION VECTORS

In the last section we derived a formula for computing the likelihood that a physical system will be in any one of the possible states associated with each stage of a Markov process describing the system. In this section we use this formula to help us investigate the long-term trends of certain Markov processes.

EXAMPLE 1

A survey conducted by the National Commission on the Educational Status of Women reveals that 70% of the daughters of women who have completed 2 or more years of college have also completed 2 or more years of college, whereas 20% of the daughters of women who have had less than 2 years of college have completed 2 or more years of college. If this trend continues, determine, in the long run, the percentage of women in the population who will have completed at least 2 years of college given that currently only 20% of the women have completed at least 2 years of college.

SOLUTION ✓

This problem may be viewed as a Markov process with two possible states: "completed 2 or more years of college" (state 1) and "completed less than 2 years of college" (state 2). The transition matrix associated with this Markov chain is given by

$$T = \begin{bmatrix} .7 & .2 \\ .3 & .8 \end{bmatrix}$$

The initial distribution vector is given by

$$X_0 = \begin{bmatrix} .2 \\ .8 \end{bmatrix}$$

To study the long-term trend pertaining to this particular aspect of the educational status of women, let's compute $X_1, X_2, \ldots$, the distribution vectors associated with the Markov process under consideration. These vectors give the percentage of women with 2 or more years of college and that of women with less than 2 years of college after one generation, after two generations, and so on. With the aid of Formula (1), Section 9.1, we find (to four decimal places)

After one generation
$$X_1 = TX_0 = \begin{bmatrix} .7 & .2 \\ .3 & .8 \end{bmatrix} \begin{bmatrix} .2 \\ .8 \end{bmatrix} = \begin{bmatrix} .3 \\ .7 \end{bmatrix}$$

After two generations
$$X_2 = TX_1 = \begin{bmatrix} .7 & .2 \\ .3 & .8 \end{bmatrix} \begin{bmatrix} .3 \\ .7 \end{bmatrix} = \begin{bmatrix} .35 \\ .65 \end{bmatrix}$$

After three generations
$$X_3 = TX_2 = \begin{bmatrix} .7 & .2 \\ .3 & .8 \end{bmatrix} \begin{bmatrix} .35 \\ .65 \end{bmatrix} = \begin{bmatrix} .375 \\ .625 \end{bmatrix}$$

Proceeding further, we obtain the following sequence of vectors:

$$X_4 = \begin{bmatrix} .3875 \\ .6125 \end{bmatrix}$$

$$X_5 = \begin{bmatrix} .3938 \\ .6062 \end{bmatrix}$$

$$X_6 = \begin{bmatrix} .3969 \\ .6031 \end{bmatrix}$$

$$X_7 = \begin{bmatrix} .3984 \\ .6016 \end{bmatrix}$$

$$X_8 = \begin{bmatrix} .3992 \\ .6008 \end{bmatrix}$$

$$X_9 = \begin{bmatrix} .3996 \\ .6004 \end{bmatrix}$$

After ten generations
$$X_{10} = \begin{bmatrix} .3998 \\ .6002 \end{bmatrix}$$

From the results of these computations, we see that as m increases, the probability distribution vector X_m approaches the probability distribution vector

$$\begin{bmatrix} .4 \\ .6 \end{bmatrix} \quad \text{or} \quad \begin{bmatrix} \frac{2}{5} \\ \frac{3}{5} \end{bmatrix}$$

Such a vector is called the **limiting,** or **steady-state, distribution vector** for the system. We interpret these results in the following way: Initially, 20% of the women in the population have completed 2 or more years of college, whereas

80% have completed less than 2 years of college. After one generation, the former has increased to 30% of the population, and the latter has dropped to 70% of the population. The trend continues, and eventually, 40% of all women in future generations will have completed 2 or more years of college, whereas 60% will have completed less than 2 years of college. ■■■■

To explain the foregoing result, let's analyze Formula (1), Section 9.1, more closely. Now, the initial distribution vector X_0 is a constant; that is, it remains fixed throughout our computation of $X_1, X_2, \ldots$. It appears reasonable, therefore, to conjecture that this phenomenon is a result of the behavior of the powers, T^m, of the transition matrix T. Pursuing this line of investigation, we compute

$$T^2 = \begin{bmatrix} .7 & .2 \\ .3 & .8 \end{bmatrix} \begin{bmatrix} .7 & .2 \\ .3 & .8 \end{bmatrix} = \begin{bmatrix} .55 & .3 \\ .45 & .7 \end{bmatrix}$$

$$T^3 = \begin{bmatrix} .7 & .2 \\ .3 & .8 \end{bmatrix} \begin{bmatrix} .55 & .3 \\ .45 & .7 \end{bmatrix} = \begin{bmatrix} .475 & .35 \\ .525 & .65 \end{bmatrix}$$

Proceeding further, we obtain the following sequence of matrices:

$$T^4 = \begin{bmatrix} .4375 & .375 \\ .5625 & .625 \end{bmatrix} \qquad T^5 = \begin{bmatrix} .4188 & .3875 \\ .5813 & .6125 \end{bmatrix}$$

$$T^6 = \begin{bmatrix} .4094 & .3938 \\ .5906 & .6062 \end{bmatrix} \qquad T^7 = \begin{bmatrix} .4047 & .3969 \\ .5953 & .6031 \end{bmatrix}$$

$$T^8 = \begin{bmatrix} .4023 & .3984 \\ .5977 & .6016 \end{bmatrix} \qquad T^9 = \begin{bmatrix} .4012 & .3992 \\ .5988 & .6008 \end{bmatrix}$$

$$T^{10} = \begin{bmatrix} .4006 & .3996 \\ .5994 & .6004 \end{bmatrix} \qquad T^{11} = \begin{bmatrix} .4003 & .3998 \\ .5997 & .6002 \end{bmatrix}$$

These results show that the powers T^m of the transition matrix T tend toward a fixed matrix as m gets larger and larger. In this case, the "limiting matrix" is the matrix

$$L = \begin{bmatrix} .40 & .40 \\ .60 & .60 \end{bmatrix} \quad \text{or} \quad \begin{bmatrix} \frac{2}{5} & \frac{2}{5} \\ \frac{3}{5} & \frac{3}{5} \end{bmatrix}$$

Such a matrix is called the **steady-state matrix** for the system. Thus, as suspected, the long-term behavior of a Markov process such as the one in this example depends on the behavior of the limiting matrix of the powers of the transition matrix—the steady-state matrix for the system. In view of this, the long-term (steady-state) distribution vector for this problem may be found by computing the product

$$LX_0 = \begin{bmatrix} .40 & .40 \\ .60 & .60 \end{bmatrix} \begin{bmatrix} .2 \\ .8 \end{bmatrix} = \begin{bmatrix} .40 \\ .60 \end{bmatrix}$$

which agrees with the result obtained earlier.

Next, since the transition matrix T in this situation seems to have a stabilizing effect over the long term, we are led to wonder whether the steady state would be reached regardless of the initial state of the system. To answer this question, suppose the initial distribution vector is

$$X_0 = \begin{bmatrix} p \\ 1-p \end{bmatrix}$$

Then, as before, the steady-state distribution vector is given by

$$LX_0 = \begin{bmatrix} .40 & .40 \\ .60 & .60 \end{bmatrix} \begin{bmatrix} p \\ 1-p \end{bmatrix} = \begin{bmatrix} .40 \\ .60 \end{bmatrix}$$

Thus, the steady state is reached regardless of the initial state of the system!

REGULAR MARKOV CHAINS

The transition matrix T of Example 1 has several important properties, which we emphasized in the foregoing discussion. First, the sequence $T, T^2, T^3, \ldots$ approaches a steady-state matrix in which the rows of the limiting matrix are all equal and all entries are positive. A matrix T having this property is given the special name regular Markov chain.

Regular Markov Chain

A stochastic matrix T is a **regular Markov chain** if the sequence

$$T, T^2, T^3, \ldots$$

approaches a steady-state matrix in which the rows of the limiting matrix are all equal and all the entries are positive.

It can be shown that *a stochastic matrix T is regular if and only if some power of T has entries that are all positive.* Second, as in the case of Example 1, a Markov chain with a regular transition matrix has a steady-state distribution vector whose elements coincide with those of a row (since they are all the same) of the steady-state matrix; thus, this steady-state distribution vector is always reached regardless of the initial distribution vector.

We will return to computations involving regular Markov chains, but for the moment let's see how one may determine whether a given matrix is indeed regular.

EXAMPLE 2 Determine which of the following matrices are regular:

a. $\begin{bmatrix} .7 & .2 \\ .3 & .8 \end{bmatrix}$ **b.** $\begin{bmatrix} .4 & 1 \\ .6 & 0 \end{bmatrix}$ **c.** $\begin{bmatrix} 0 & 1 \\ 1 & 0 \end{bmatrix}$

SOLUTION ✓

a. Since all the entries of the matrix are positive, the given matrix is regular. Note that this is the transition matrix of Example 1.

b. In this case, one of the entries of the given matrix is equal to zero. Let's compute

$$\begin{bmatrix} .4 & 1 \\ .6 & 0 \end{bmatrix}^2 = \begin{bmatrix} .4 & 1 \\ .6 & 0 \end{bmatrix}\begin{bmatrix} .4 & 1 \\ .6 & 0 \end{bmatrix} = \begin{bmatrix} .76 & .4 \\ .24 & .6 \end{bmatrix}$$

↑ All entries are positive.

Since the second power of the matrix has entries that are all positive, we conclude that the given matrix is in fact regular.

c. Denote the given matrix by A. Then,

$$A = \begin{bmatrix} 0 & 1 \\ 1 & 0 \end{bmatrix}$$

$$A^2 = \begin{bmatrix} 0 & 1 \\ 1 & 0 \end{bmatrix}\begin{bmatrix} 0 & 1 \\ 1 & 0 \end{bmatrix} = \begin{bmatrix} 1 & 0 \\ 0 & 1 \end{bmatrix}$$

$$A^3 = \begin{bmatrix} 0 & 1 \\ 1 & 0 \end{bmatrix}\begin{bmatrix} 1 & 0 \\ 0 & 1 \end{bmatrix} = \begin{bmatrix} 0 & 1 \\ 1 & 0 \end{bmatrix}$$

(Not all entries are positive.)

Observe that $A^3 = A$. It therefore follows that $A^4 = A^2$, $A^5 = A$, and so on. In other words, any power of A must coincide with either A or A^2. Since not all entries of A and A^2 are positive, the same is true of any power of A. We conclude accordingly that the given matrix is not regular. ■■■■

Group Discussion

Find the set of all 2×2 stochastic matrices with elements that are either 0 or 1.

We now return to the study of regular Markov chains. In Example 1 we found the steady-state distribution vector associated with a regular Markov chain by studying the limiting behavior of a sequence of distribution vectors. Alternatively, as pointed out in the subsequent discussion, the steady-state distribution vector may also be obtained by first determining the steady-state matrix associated with the regular Markov chain.

Fortunately, there is a relatively simple procedure for finding the steady-state distribution vector associated with a regular Markov process. It does not involve the rather tedious computations required to obtain the sequences in Example 1. The procedure follows.

Finding the Steady-State Distribution Vector

Let T be a regular stochastic matrix. Then the steady-state distribution vector X may be found by solving the vector equation

$$TX = X$$

together with the condition that the sum of the elements of the vector X be equal to 1.

A justification of the foregoing procedure is given in Exercise 29.

EXAMPLE 3 Find the steady-state distribution vector for the regular Markov chain whose transition matrix is

$$T = \begin{bmatrix} .7 & .2 \\ .3 & .8 \end{bmatrix} \qquad \text{(See Example 1.)}$$

SOLUTION ✓ Let

$$X = \begin{bmatrix} x \\ y \end{bmatrix}$$

be the steady-state distribution vector associated with the Markov process, where the numbers x and y are to be determined. The condition $TX = X$ translates into the matrix equation

$$\begin{bmatrix} .7 & .2 \\ .3 & .8 \end{bmatrix} \begin{bmatrix} x \\ y \end{bmatrix} = \begin{bmatrix} x \\ y \end{bmatrix}$$

or, equivalently, the system of linear equations

$$\begin{aligned} 0.7x + 0.2y &= x \\ 0.3x + 0.8y &= y \end{aligned}$$

But each of the equations that make up this system of equations is equivalent to the single equation

$$0.3x - 0.2y = 0 \qquad \begin{array}{l} (0.7x - x + 0.2y = 0) \\ (0.3x + 0.8y - y = 0) \end{array}$$

Next, the condition that the sum of the elements of X add up to 1 gives

$$x + y = 1$$

Thus, the fulfillment of the two conditions simultaneously implies that x and y are the solutions of the system

$$\begin{aligned} 0.3x - 0.2y &= 0 \\ x + \quad y &= 1 \end{aligned}$$

Solving the first equation for x, we obtain

$$x = \frac{2}{3}y$$

which, upon substitution into the second, yields

$$\frac{2}{3}y + y = 1$$

$$y = \frac{3}{5}$$

Thus, $x = 2/5$, and the required steady-state distribution vector is given by

$$X = \begin{bmatrix} \frac{2}{5} \\ \frac{3}{5} \end{bmatrix}$$

which agrees with the result obtained earlier. ■■■■

APPLICATION

EXAMPLE 4 In Example 6, Section 9.1, we showed that the transition matrix that described the movement of taxis from zone to zone was given by the regular stochastic matrix

$$T = \begin{bmatrix} .6 & .4 & .3 \\ .3 & .3 & .3 \\ .1 & .3 & .4 \end{bmatrix}$$

Use this information to determine the long-term distribution of the taxis in the three zones.

SOLUTION ✓ Let

$$X = \begin{bmatrix} x \\ y \\ z \end{bmatrix}$$

be the steady-state distribution vector associated with the Markov process under consideration, where x, y, and z are to be determined. The condition $TX = X$ translates into the matrix equation

$$\begin{bmatrix} .6 & .4 & .3 \\ .3 & .3 & .3 \\ .1 & .3 & .4 \end{bmatrix} \begin{bmatrix} x \\ y \\ z \end{bmatrix} = \begin{bmatrix} x \\ y \\ z \end{bmatrix}$$

or, equivalently, the system of linear equations

$$\begin{aligned} 0.6x + 0.4y + 0.3z &= x \\ 0.3x + 0.3y + 0.3z &= y \\ 0.1x + 0.3y + 0.4z &= z \end{aligned}$$

This system simplifies into

$$\begin{aligned} 4x - 4y - 3z &= 0 \\ 3x - 7y + 3z &= 0 \\ x + 3y - 6z &= 0 \end{aligned}$$

Since $x + y + z = 1$ as well, we are required to solve the system

$$\begin{aligned} x + y + z &= 1 \\ 4x - 4y - 3z &= 0 \\ 3x - 7y + 3z &= 0 \\ x + 3y - 6z &= 0 \end{aligned}$$

Using the Gauss–Jordan elimination procedure of Chapter 2, we find that

$$x = \frac{33}{70}, \quad y = \frac{3}{10}, \quad \text{and} \quad z = \frac{8}{35}$$

or $x \approx 0.47$, $y = 0.30$, and $z \approx 0.23$. Thus, in the long run, 47% of the taxis will be in zone I, 30% in zone II, and 23% in zone III.

SELF-CHECK EXERCISES 9.2

1. Find the steady-state distribution vector for the regular Markov chain whose transition matrix is

$$T = \begin{bmatrix} .5 & .8 \\ .5 & .2 \end{bmatrix}$$

2. Three supermarkets serve a certain section of a city. During the year, supermarket A is expected to retain 80% of its customers, lose 5% of its customers to supermarket B, and lose 15% to supermarket C. Supermarket B is expected to retain 90% of its customers and lose 5% to each of supermarket A and supermarket C. Supermarket C is expected to retain 75% of its customers, lose 10% to supermarket A, and lose 15% to supermarket B. If these trends continue, what will the market share of each supermarket be in the long run?

Solutions to Self-Check Exercises 9.2 can be found on page 555.

9.2 Exercises

In Exercises 1–8, determine which of the given matrices are regular.

1. $\begin{bmatrix} \frac{2}{5} & \frac{3}{4} \\ \frac{3}{5} & \frac{1}{4} \end{bmatrix}$

2. $\begin{bmatrix} 0 & .3 \\ 1 & .7 \end{bmatrix}$

3. $\begin{bmatrix} 1 & .8 \\ 0 & .2 \end{bmatrix}$

4. $\begin{bmatrix} \frac{1}{3} & 0 \\ \frac{2}{3} & 1 \end{bmatrix}$

5. $\begin{bmatrix} \frac{1}{2} & \frac{3}{4} & 0 \\ \frac{1}{2} & 0 & \frac{1}{2} \\ 0 & \frac{1}{4} & \frac{1}{2} \end{bmatrix}$

6. $\begin{bmatrix} 1 & .3 & .1 \\ 0 & .4 & .8 \\ 0 & .3 & .1 \end{bmatrix}$

7. $\begin{bmatrix} .7 & .2 & .3 \\ .3 & .8 & .3 \\ 0 & 0 & .4 \end{bmatrix}$

8. $\begin{bmatrix} 0 & 0 & \frac{1}{4} \\ 1 & 0 & 0 \\ 0 & 1 & \frac{3}{4} \end{bmatrix}$

In Exercises 9–16, find the steady-state vector for the given transition matrix.

9. $\begin{bmatrix} \frac{1}{3} & \frac{1}{4} \\ \frac{2}{3} & \frac{3}{4} \end{bmatrix}$

10. $\begin{bmatrix} \frac{4}{5} & \frac{3}{5} \\ \frac{1}{5} & \frac{2}{5} \end{bmatrix}$

11. $\begin{bmatrix} .5 & .2 \\ .5 & .8 \end{bmatrix}$

12. $\begin{bmatrix} .9 & 1 \\ .1 & 0 \end{bmatrix}$

13. $\begin{bmatrix} 0 & \frac{1}{8} & 1 \\ 1 & \frac{5}{8} & 0 \\ 0 & \frac{1}{4} & 0 \end{bmatrix}$

14. $\begin{bmatrix} .6 & .3 & 0 \\ .4 & .4 & .6 \\ 0 & .3 & .4 \end{bmatrix}$

15. $\begin{bmatrix} .2 & 0 & .3 \\ 0 & .6 & .4 \\ .8 & .4 & .3 \end{bmatrix}$

16. $\begin{bmatrix} .1 & .2 & .3 \\ .1 & .2 & .3 \\ .8 & .6 & .4 \end{bmatrix}$

17. **PSYCHOLOGY EXPERIMENTS** A psychologist conducts an experiment in which a mouse is placed in a T-maze, where it has a choice at the T-junction of turning left and receiving a reward (cheese) or turning right and receiving a mild shock. At the end of each trial a record is kept of the mouse's response. It is observed that the mouse is as likely to turn left (state 1) as right (state 2) during the first trial. In subsequent trials, however, the observation is made that if the mouse had turned left in the previous trial, then the probability that it will turn left in the next trial is .8, whereas the probability that it will turn right is .2. If the mouse had turned right in the previous trial, then the probability that it will turn right in the next trial is .1, whereas the probability that it will turn left is .9. In the long run, what percentage of the time will the mouse turn left at the T-junction?

18. **COMMUTER TRENDS** Within a large metropolitan area, 20% of the commuters currently use the public transportation system, whereas the remaining 80% commute via automobile. The city has recently revitalized and expanded its public transportation system. It is expected that six months from now 30% of those who are now commuting to work via automobile will switch to public transportation, and 70% will continue to commute via automobile. At the same time, it is expected that 20% of those now using public transportation will commute via automobile, and 80% will continue to use public transportation. In the long run, what percentage of the commuters will be using public transportation?

19. **ONE- AND TWO-INCOME FAMILIES** From data compiled over a 10-year period by Manpower, Inc., in a statewide study of married couples in which at least one spouse was working, the following transition matrix was constructed. It gives the transitional probabilities for one and two wage earners among married couples.

		Current State	
		1 Wage Earner	2 Wage Earners
Next State	1 Wage Earner	.72	.12
	2 Wage Earners	.28	.88

At the present time, 48% of the married couples (in which at least one spouse is working) have one wage earner and 52% have two wage earners. Assuming that this trend continues, what will be the distributon of one- and two-wage earner families among married couples in this area 10 years from now? Over the long run?

20. **PROFESSIONAL WOMEN** From data compiled over a 5-year period by *Women's Daily* in a study of the number of women in the professions, the following transition matrix was constructed. It gives the transitional probabilities for the number of men and women in the professions.

		Current State	
		Men	Women
Next State	Men	.95	.04
	Women	.05	.96

As of the beginning of 1986, 52.9% of professional jobs were held by men. If this trend continues, what percentage of professional jobs will be held by women in the long run?

21. **BUYING TRENDS OF HOME BUYERS** From data collected by the Association of Realtors of a certain city, the following transition matrix was obtained. The matrix describes the buying pattern of home buyers who buy single-family homes (S) or condominiums (C).

		Current State	
		S	C
Next State	S	.85	.35
	C	.15	.65

Currently, 80% of the homeowners live in single-family homes, whereas 20% live in condominiums. If this trend continues, what will be the percentage of homeowners in this city who will own single-family homes and condominiums 2 yr from now? In the long run?

22. **HOMEOWNERS' CHOICE OF ENERGY** A study conducted by the Urban Energy Commission in a large metropolitan area indicates the probabilities that homeowners within the area will use certain heating fuels or solar energy during the next 10 yr as the major source of heat for their homes. The following transition matrix represents the transition probabilities from one state to another:

$$\begin{array}{l} \\ \text{Electricity} \\ \text{Natural gas} \\ \text{Fuel oil} \\ \text{Solar energy} \end{array} \begin{array}{c} \begin{array}{cccc} \text{Elec.} & \text{Gas} & \text{Oil} & \text{Solar} \end{array} \\ \begin{bmatrix} .70 & 0 & .10 & 0 \\ .15 & .90 & .10 & .05 \\ .05 & .02 & .75 & .05 \\ .10 & .08 & .05 & .90 \end{bmatrix} \end{array}$$

Among the homeowners within the area, 20% currently use electricity, 35% use natural gas, 40% use oil, and 5% use solar energy as their major source of heat for their homes. In the long run, what percentage of homeowners within the area will be using solar energy as their major source of heating fuel?

23. **Network News Viewership** A television poll was conducted among regular viewers of the national news in a certain region where the three national networks share the same time slot for the evening news. Results of the poll indicate that 30% of the viewers watch the ABC evening news, 40% watch the CBS evening news, and 30% watch the NBC evening news. Furthermore, it was found that of those viewers who watched the ABC evening news during 1 week, 80% would again watch the ABC evening news during the next week, 10% would watch the CBS news, and 10% would watch the NBC news. Of those viewers who watched the CBS evening news during 1 week, 85% would again watch the CBS evening news during the next week, 10% would watch the ABC news, and 5% would watch the NBC news. Of those viewers who watched the NBC evening news during 1 week, 85% would again watch the NBC news during the next week, 10% would watch ABC, and 5% would watch CBS.
   a. What share of the audience consisting of regular viewers of the national news will each network command after 2 weeks?
   b. In the long run, what share of the audience will each network command?

24. **Network News Viewership** Refer to Exercise 23. If the initial distribution vector is

$$X_0 = \begin{matrix} \text{ABC} \\ \text{CBS} \\ \text{NBC} \end{matrix} \begin{bmatrix} .40 \\ .40 \\ .20 \end{bmatrix}$$

what share of the audience will each network command in the long run?

25. **Genetics** In a certain species of roses, a plant with genotype (genetic makeup) AA has red flowers, a plant with genotype Aa has pink flowers, and a plant with genotype aa has white flowers, where A is the dominant gene and a is the recessive gene for color. If a plant with one genotype is crossed with another plant, then the color of the offspring's flowers is determined by the genotype of the parent plants. If a plant of each genotype is crossed with a pink-flowered plant, then the transition matrix used to determine the color of the offspring's flowers is given by

$$\begin{array}{ll} & \qquad\qquad\quad \text{Parent} \\ & \qquad\quad \text{Red} \quad \text{Pink} \quad \text{White} \\ \text{Offspring} & \begin{array}{l} \text{Red } (AA) \\ \text{Pink } (Aa) \text{ or } (aA) \\ \text{White } (aa) \end{array} \begin{bmatrix} \frac{1}{2} & \frac{1}{4} & 0 \\ \frac{1}{2} & \frac{1}{2} & \frac{1}{2} \\ 0 & \frac{1}{4} & \frac{1}{2} \end{bmatrix} \end{array}$$

If the offspring of each generation are crossed only with pink-flowered plants, in the long run what percentage of the plants will have red flowers? Pink flowers? White flowers?

26. **Market Share of Auto Manufacturers** In a study of the domestic market share of the three major automobile manufacturers A, B, and C in a certain country, it was found that of the customers who bought a car manufactured by A, 75% would again buy a car manufactured by A, 15% would buy a car manufactured by B, and 10% would buy a car manufactured by C. Of the customers who bought a car manufactured by B, 90% would again buy a car manufactured by B, whereas 5% each would buy cars manufactured by A and C, respectively. Finally, of the customers who bought a car manufactured by C, 85% would again buy a car manufactured by C, 5% would buy a car manufactured by A, and 10% would buy a car manufactured by B. Assuming that these sentiments reflect the buying habits of customers in the future model years, determine the market share that will be held by each manufacturer in the long run.

In Exercises 27 and 28, determine whether the statement is true or false. If it is true, explain why it is true. If it is false, give an example to show why it is false.

27. A stochastic matrix T is a regular Markov chain if the powers of T approach a fixed matrix whose rows are all equal.

28. To find the steady-state distribution vector X, we solve the system

$$\begin{cases} TX = X \\ x_1 + x_2 + \cdots + x_n = 1 \end{cases}$$

where T is the regular stochastic matrix associated with the Markov process and

$$X = \begin{bmatrix} x_1 \\ x_2 \\ \vdots \\ x_n \end{bmatrix}$$

29. Let T be a regular stochastic matrix. Show that the steady-state distribution vector X may be found by solving the vector equation $TX = X$ together with the condition that the sum of the elements of X be equal to 1.

Hint: Take the initial distribution vector to be X, the steady-state distribution vector. Then, when n is large, $X \approx T^n X$. (Why?) Multiply both sides of the last equation by T (on the left) and consider the resulting equation when n is large.

SOLUTIONS TO SELF-CHECK EXERCISES 9.2

1. Let

$$X = \begin{bmatrix} x \\ y \end{bmatrix}$$

be the steady-state distribution vector associated with the Markov process, where the numbers x and y are to be determined. The condition $TX = X$ translates into the matrix equation

$$\begin{bmatrix} .5 & .8 \\ .5 & .2 \end{bmatrix} \begin{bmatrix} x \\ y \end{bmatrix} = \begin{bmatrix} x \\ y \end{bmatrix}$$

which is equivalent to the system of linear equations

$$\begin{aligned} 0.5x + 0.8y &= x \\ 0.5x + 0.2y &= y \end{aligned}$$

Each equation in the system is equivalent to the equation

$$0.5x - 0.8y = 0$$

Next, the condition that the sum of the elements of X adds up to 1 gives

$$x + y = 1$$

Thus, the fulfillment of the two conditions simultaneously implies that x and y are the solutions of the system

$$\begin{aligned} 0.5x - 0.8y &= 0 \\ x + \quad y &= 1 \end{aligned}$$

Solving the first equation for x, we obtain

$$x = \frac{8}{5}y$$

which, upon substitution into the second, yields

$$\frac{8}{5}y + y = 1$$

$$y = \frac{5}{13}$$

(continued on p. 558)

Using Technology

FINDING THE LONG-TERM DISTRIBUTION VECTOR

The problem of finding the long-term distribution vector for a regular Markov chain ultimately rests on the problem of solving a system of linear equations. As such, the **rref** or equivalent function of a graphing calculator proves indispensable, as the following example shows.

EXAMPLE 1 Find the steady-state distribution vector for the regular Markov chain whose transition matrix is

$$T = \begin{bmatrix} .4 & .2 & .1 \\ .3 & .4 & .5 \\ .3 & .4 & .4 \end{bmatrix}$$

SOLUTION ✓ Let

$$\begin{bmatrix} x \\ y \\ z \end{bmatrix}$$

be the steady-state distribution vector, where x, y, and z are to be determined. The condition $TX = X$ translates into the matrix equation

$$\begin{bmatrix} .4 & .2 & .1 \\ .3 & .4 & .5 \\ .3 & .4 & .4 \end{bmatrix} \begin{bmatrix} x \\ y \\ z \end{bmatrix} = \begin{bmatrix} x \\ y \\ z \end{bmatrix}$$

or, equivalently, the system of linear equations

$$\begin{aligned} 0.4x + 0.2y + 0.1z &= x \\ 0.3x + 0.4y + 0.5z &= y \\ 0.3x + 0.4y + 0.4z &= z \end{aligned}$$

Since $x + y + z = 1$, we are required to solve the system

$$\begin{aligned} -0.6x + 0.2y + 0.1z &= 0 \\ 0.3x - 0.6y + 0.5z &= 0 \\ 0.3x + 0.4y - 0.6z &= 0 \\ x + \quad y + \quad z &= 1 \end{aligned}$$

Entering this system into the graphing calculator as the augmented matrix

$$A = \left[\begin{array}{ccc|c} -.6 & .2 & .1 & 0 \\ .3 & -.6 & .5 & 0 \\ .3 & .4 & -.6 & 0 \\ 1 & 1 & 1 & 1 \end{array}\right]$$

and then using the **rref** function, we obtain the equivalent system (to two decimal places)

$$\left[\begin{array}{ccc|c} 1 & 0 & 0 & .20 \\ 0 & 1 & 0 & .42 \\ 0 & 0 & 1 & .38 \\ 0 & 0 & 0 & 0 \end{array}\right]$$

Therefore, $x \approx 0.20$, $y \approx 0.42$, and $z \approx 0.38$, and so the required steady-state distribution vector is

$$\begin{bmatrix} .20 \\ .42 \\ .38 \end{bmatrix}$$

■■■■

Exercises

In Exercises 1 and 2, use a graphing calculator to find the steady-state vector for the given matrix *T*.

1. $$T = \begin{bmatrix} .2 & .2 & .3 & .2 & .1 \\ .1 & .2 & .1 & .2 & .1 \\ .3 & .4 & .1 & .3 & .3 \\ .2 & .1 & .2 & .2 & .2 \\ .2 & .1 & .3 & .1 & .3 \end{bmatrix}$$

2. $$T = \begin{bmatrix} .3 & .2 & .1 & .3 & .1 \\ .2 & .1 & .2 & .1 & .2 \\ .1 & .2 & .3 & .2 & .2 \\ .1 & .3 & .2 & .3 & .2 \\ .3 & .2 & .2 & .1 & .3 \end{bmatrix}$$

3. Use a graphing calculator to verify that the steady-state vector for Example 4, page 551, is

$$X = \begin{bmatrix} .47 \\ .30 \\ .23 \end{bmatrix}$$

Therefore, $x = 8/13$, and the required steady-state distribution vector is

$$\begin{bmatrix} \frac{8}{13} \\ \frac{5}{13} \end{bmatrix}$$

2. The transition matrix for the Markov process under consideration is

$$T = \begin{bmatrix} .80 & .05 & .10 \\ .05 & .90 & .15 \\ .15 & .05 & .75 \end{bmatrix}$$

Now, let

$$X = \begin{bmatrix} x \\ y \\ z \end{bmatrix}$$

be the steady-state distribution vector associated with the Markov process under consideration, where x, y, and z are to be determined. The condition $TX = X$ is

$$\begin{bmatrix} .80 & .05 & .10 \\ .05 & .90 & .15 \\ .15 & .05 & .75 \end{bmatrix} \begin{bmatrix} x \\ y \\ z \end{bmatrix} = \begin{bmatrix} x \\ y \\ z \end{bmatrix}$$

or, equivalently, the system of linear equations

$$\begin{aligned} 0.80x + 0.05y + 0.10z &= x \\ 0.05x + 0.90y + 0.15z &= y \\ 0.15x + 0.05y + 0.75z &= z \end{aligned}$$

This system simplifies into

$$\begin{aligned} 4x - y - 2z &= 0 \\ x - 2y + 3z &= 0 \\ 3x + y - 5z &= 0 \end{aligned}$$

Since $x + y + z = 1$ as well, we are required to solve the system

$$\begin{aligned} 4x - y - 2z &= 0 \\ x - 2y + 3z &= 0 \\ 3x + y - 5z &= 0 \\ x + y + z &= 1 \end{aligned}$$

Using the Gauss–Jordan elimination procedure, we find

$$x = \frac{1}{4}, \quad y = \frac{1}{2}, \quad z = \frac{1}{4}$$

Therefore, in the long run, supermarkets A and C will each have one-quarter of the customers, and supermarket B will have half the customers.

9.3 Absorbing Markov Chains

ABSORBING MARKOV CHAINS

In this section we investigate the long-term trends of a certain class of Markov chains that involve transition matrices that are not regular. In particular, we study Markov chains in which the transition matrices, known as absorbing stochastic matrices, have the special properties to be described presently.

Consider the stochastic matrix

$$\begin{bmatrix} 1 & 0 & .2 & 0 \\ 0 & 1 & .3 & 1 \\ 0 & 0 & .5 & 0 \\ 0 & 0 & 0 & 0 \end{bmatrix}$$

associated with a Markov process. Interpreting it in the usual fashion, we see that after one observation, the probability is 1 (a certainty) that an object previously in state 1 will remain in state 1. Similarly, we see that an object previously in state 2 must remain in state 2. Next, we find that an object previously in state 3 has a probability of .2 of going to state 1, a probability of .3 of going to state 2, a probability of .5 of remaining in state 3, and no chance of going to state 4. Finally, an object previously in state 4 must, after one observation, end up in state 2.

This stochastic matrix exhibits certain special characteristics. First, as observed earlier, an object in state 1 or state 2 must stay in state 1 or state 2, respectively. Such states are called absorbing states. In general, an **absorbing state** is one from which it is impossible for an object to leave. To identify the absorbing states of a stochastic matrix, we simply examine each column of the matrix. If column i has a 1 in the a_{ii} position (that is, on the main diagonal of the matrix) and zeros elsewhere in that column, then and only then is state i an absorbing state.

Second, observe that states 3 and 4, although not absorbing states, have the property that an object in each of these states has a possibility of going to an absorbing state. For example, an object currently in state 3 has a probability of .2 of ending up in state 1, an absorbing state, and an object in state 4 must end up in state 2, also an absorbing state, after one transition.

Absorbing Stochastic Matrix

An **absorbing stochastic matrix** has the following properties:

1. There is at least one absorbing state.
2. It is possible to go from any nonabsorbing state to an absorbing state in one or more stages.

A Markov chain is said to be an **absorbing Markov chain** if the transition matrix associated with the process is an absorbing stochastic matrix.

EXAMPLE 1 Determine whether the following matrices are absorbing stochastic matrices.

a. $\begin{bmatrix} .7 & 0 & .1 & 0 \\ 0 & 1 & .5 & 0 \\ .3 & 0 & .2 & 0 \\ 0 & 0 & .2 & 1 \end{bmatrix}$ **b.** $\begin{bmatrix} 1 & 0 & 0 & 0 \\ 0 & 1 & 0 & 0 \\ 0 & 0 & .5 & .4 \\ 0 & 0 & .5 & .6 \end{bmatrix}$

SOLUTION ✓

a. States 2 and 4 are both absorbing states. Furthermore, even though state 1 is not an absorbing state, there is a possibility (with probability .3) that an object may go from this state to state 3. State 3 itself is nonabsorbing, but an object in that state has a probability of .5 of going to the absorbing state 2 and a probability of .2 of going to the absorbing state 4. Thus, the given matrix is an absorbing stochastic matrix.

b. States 1 and 2 are absorbing states. However, it is impossible for an object to go from the nonabsorbing states 3 and 4 to either or both of the absorbing states. Thus, the given matrix is not an absorbing stochastic matrix.

■■■■

Given an absorbing stochastic matrix, it is always possible, by suitably reordering the states if necessary, to rewrite it so that the absorbing states appear first. Then, the resulting matrix can be partitioned into four sub matrices,

$$\begin{array}{cc} \text{Absorbing} & \text{Nonabsorbing} \end{array}$$
$$\left[\begin{array}{c|c} I & S \\ \hline O & R \end{array}\right]$$

where I is an identity matrix whose order is determined by the number of absorbing states and O is a zero matrix. The submatrices R and S correspond to the nonabsorbing states. As an example, the absorbing stochastic matrix of Example 1(a)

$$\begin{array}{c} \\ 1 \\ 2 \\ 3 \\ 4 \end{array}\begin{array}{c} \begin{array}{cccc} 1 & 2 & 3 & 4 \end{array} \\ \begin{bmatrix} .7 & 0 & .1 & 0 \\ 0 & 1 & .5 & 0 \\ .3 & 0 & .2 & 0 \\ 0 & 0 & .2 & 1 \end{bmatrix} \end{array} \quad \text{may be written as} \quad \begin{array}{c} \\ 4 \\ 2 \\ 1 \\ 3 \end{array}\begin{array}{c} \begin{array}{cccc} 4 & 2 & 1 & 3 \end{array} \\ \left[\begin{array}{cc|cc} 1 & 0 & 0 & .2 \\ 0 & 1 & 0 & .5 \\ \hline 0 & 0 & .7 & .1 \\ 0 & 0 & .3 & .2 \end{array}\right] \end{array}$$

upon reordering the states as indicated.

APPLICATIONS

EXAMPLE 2
Gambler's Ruin

John has decided to risk \$2 in the following game of chance. He places a \$1 bet on each repeated play of the game in which the probability of his winning \$1 is .4, and he continues to play until he has accumulated a total of \$3 or he has lost all of his money. Write the transition matrix for the related absorbing Markov chain.

SOLUTION ✓ There are four states in this Markov chain, which correspond to John accumulating a total of \$0, \$1, \$2, and \$3. Since the first and last states listed are absorbing states, we will list these states first, resulting in the transition matrix

$$\begin{array}{c} \\ \\ \$0 \\ \$3 \\ \$1 \\ \$2 \end{array} \begin{array}{c} \overbrace{\quad\quad\quad}^{\text{Absorbing}} \; \overbrace{\quad\quad\quad}^{\text{Nonabsorbing}} \\ \begin{array}{cccc} \$0 & \$3 & \$1 & \$2 \end{array} \\ \begin{bmatrix} 1 & 0 & .6 & 0 \\ 0 & 1 & 0 & .4 \\ 0 & 0 & 0 & .6 \\ 0 & 0 & .4 & 0 \end{bmatrix} \end{array}$$

which is constructed as follows: Since the state "\$0" is an absorbing state, we see that $a_{11} = 1$, $a_{21} = a_{31} = a_{41} = 0$. Similarly, the state "\$3" is an absorbing state, so $a_{22} = 1$, and $a_{12} = a_{32} = a_{42} = 0$. To construct the column corresponding to the nonabsorbing state "\$1," we note that there is a probability of .6 (John loses) in going from an accumulated amount of \$1 to \$0, so $a_{13} = .6$; $a_{23} = a_{33} = 0$ because it is not feasible to go from an accumulated amount of \$1 to either an accumulated amount of \$3 or \$1 in one transition (play). Finally, there is a probability of .4 (John wins) in going from an accumulated amount of \$1 to an accumulated amount of \$2, so $a_{43} = .4$. The last column of the transition matrix is constructed by reasoning in a similar manner. ■■■■

The following question arises in connection with the last example: If John continues to play the game as originally planned, what is the probability that he will depart from the game victorious—that is, leave with an accumulated amount of \$3?

To answer this question, we have to look at the long-term trend of the relevant Markov chain. Taking a cue from our work in the last section, we may compute the powers of the transition matrix associated with the Markov chain. Just as in the case of regular stochastic matrices, it turns out that the powers of an absorbing stochastic matrix approach a steady-state matrix. However, instead of demonstrating this, we use the following result, which we state without proof, for computing the steady-state matrix:

Finding the Steady-State Matrix for an Absorbing Stochastic Matrix

Suppose an absorbing stochastic matrix A has been partitioned into submatrices

$$A = \left[\begin{array}{c|c} I & S \\ \hline O & R \end{array}\right]$$

Then the *steady-state matrix* of A is given by

$$\left[\begin{array}{c|c} I & S(I-R)^{-1} \\ \hline O & O \end{array}\right]$$

where the order of the identity matrix appearing in the expression $(I - R)^{-1}$ is chosen to have the same order as R.

EXAMPLE 3
Gambler's Ruin (Continued)

Refer to Example 2. If John continues to play the game until he has accumulated a sum of \$3 or until he has lost all his money, what is the probability that he will accumulate \$3?

SOLUTION ✓ The transition matrix associated with the Markov process is (see Example 2)

$$A = \left[\begin{array}{cc|cc} 1 & 0 & .6 & 0 \\ 0 & 1 & 0 & .4 \\ \hline 0 & 0 & 0 & .6 \\ 0 & 0 & .4 & 0 \end{array}\right]$$

We need to find the steady-state matrix of A. In this case,

$$R = \begin{bmatrix} 0 & .6 \\ .4 & 0 \end{bmatrix} \quad \text{and} \quad S = \begin{bmatrix} .6 & 0 \\ 0 & .4 \end{bmatrix}$$

so

$$I - R = \begin{bmatrix} 1 & 0 \\ 0 & 1 \end{bmatrix} - \begin{bmatrix} 0 & .6 \\ .4 & 0 \end{bmatrix} = \begin{bmatrix} 1 & -.6 \\ -.4 & 1 \end{bmatrix}$$

Using the formula in Section 2.6 for finding the inverse of a 2×2 matrix, we find that

$$(I - R)^{-1} = \begin{bmatrix} 1.32 & .79 \\ .53 & 1.32 \end{bmatrix}$$

and so

$$S(I - R)^{-1} = \begin{bmatrix} .6 & 0 \\ 0 & .4 \end{bmatrix} \begin{bmatrix} 1.32 & .79 \\ .53 & 1.32 \end{bmatrix} = \begin{bmatrix} .79 & .47 \\ .21 & .53 \end{bmatrix}$$

Therefore, the required steady-state matrix of A is given by

$$\left[\begin{array}{c|c} I & S(I - R^{-1}) \\ \hline O & O \end{array}\right] = \begin{array}{c} \\ \$0 \\ \$3 \\ \$1 \\ \$2 \end{array} \begin{array}{c} \begin{array}{cccc} \$0 & \$3 & \$1 & \$2 \end{array} \\ \left[\begin{array}{cc|cc} 1 & 0 & .79 & .47 \\ 0 & 1 & .21 & .53 \\ \hline 0 & 0 & 0 & 0 \\ 0 & 0 & 0 & 0 \end{array}\right] \end{array}$$

From this result we see that starting with \$2, the probability is .53 that John will leave the game with an accumulated amount of \$3—that is, that he wins \$1.

■■■■

Our last example shows an application of Markov chains in the field of genetics.

Group Discussion
Consider the stochastic matrix

$$A = \begin{bmatrix} 1 & 0 & a \\ 0 & 1 & b \\ 0 & 0 & 1 - a - b \end{bmatrix}$$

where a and b satisfy $0 < a < 1$, $0 < b < 1$, and $0 < a + b < 1$.

1. Find the steady-state matrix.
2. What is the probability that state 3 will be absorbed in state 2?

EXAMPLE 4 In a certain species of flowers, a plant of genotype (genetic makeup) AA has red flowers, a plant of genotype Aa has pink flowers, and a plant of genotype aa has white flowers, where A is the dominant gene and a is the recessive gene for color. If a plant of one genotype is crossed with another plant, then the color of the offspring's flowers is determined by the genotype of the parent plants. If the offspring are crossed successively with plants of genotype AA only, show that in the long run all the flowers produced by the plants will be red.

SOLUTION ✓ First, let's construct the transition matrix associated with the resulting Markov chain. In crossing a plant of genotype AA with another of the same genotype AA, the offspring will inherit one dominant gene from each parent and thus will have genotype AA. Therefore, the probabilities of the offspring being genotype AA, Aa, and aa are 1, 0, and 0, respectively.

Next, in crossing a plant of genotype AA with one of genotype Aa, the probability of the offspring having genotype AA (inheriting an A gene from the first parent and an A from the second) is 1/2; the probability of the offspring having genotype Aa (inheriting an A gene from the first parent and an a gene from the second parent) is 1/2; finally, the probability of the offspring being of genotype aa is 0 since this is clearly impossible.

A similar argument shows that in crossing a plant of genotype AA with one of genotype aa, the probabilities of the offspring having genotype AA, Aa, and aa are 0, 1, and 0, respectively.

The required transition matrix is thus given by

Absorbing state ↓

$$T = \begin{array}{c} \\ AA \\ Aa \\ aa \end{array} \begin{array}{c} \begin{array}{ccc} AA & Aa & aa \end{array} \\ \begin{bmatrix} 1 & \frac{1}{2} & 0 \\ 0 & \frac{1}{2} & 1 \\ 0 & 0 & 0 \end{bmatrix} \end{array}$$

Observe that the state AA is an absorbing state. Furthermore, it is possible to go from each of the other two nonabsorbing states to the absorbing state AA. Thus, the Markov chain is an absorbing Markov chain. To determine the long-term effects of this experiment, let's compute the steady-state matrix of T. Partitioning T in the usual manner, we find

$$T = \left[\begin{array}{c|cc} 1 & \frac{1}{2} & 0 \\ \hline 0 & \frac{1}{2} & 1 \\ 0 & 0 & 0 \end{array}\right]$$

so that

$$R = \begin{bmatrix} \frac{1}{2} & 1 \\ 0 & 0 \end{bmatrix} \quad \text{and} \quad S = \begin{bmatrix} \frac{1}{2} & 0 \end{bmatrix}$$

Next, we compute

$$I - R = \begin{bmatrix} 1 & 0 \\ 0 & 1 \end{bmatrix} - \begin{bmatrix} \frac{1}{2} & 1 \\ 0 & 0 \end{bmatrix} = \begin{bmatrix} \frac{1}{2} & -1 \\ 0 & 1 \end{bmatrix}$$

and, using the formula for finding the inverse of a 2×2 matrix in Section 2.6,

$$(I - R)^{-1} = \begin{bmatrix} 2 & 2 \\ 0 & 1 \end{bmatrix}$$

Thus,

$$S(I - R)^{-1} = [\tfrac{1}{2} \quad 0]\begin{bmatrix} 2 & 2 \\ 0 & 1 \end{bmatrix} = [1 \quad 1]$$

Therefore, the steady-state matrix of T is given by

$$\left[\begin{array}{c|c} I & S(I-R)^{-1} \\ \hline O & O \end{array}\right] = \begin{array}{c} \\ AA \\ Aa \\ aa \end{array} \begin{array}{c} \begin{array}{ccc} AA & Aa & aa \end{array} \\ \left[\begin{array}{c|cc} 1 & 1 & 1 \\ \hline 0 & 0 & 0 \\ 0 & 0 & 0 \end{array}\right] \end{array}$$

Interpreting the steady-state matrix of T, we see that the long-term result of crossing the offspring with plants of genotype AA only leads to the absorbing state AA. In other words, such a procedure will result in the production of plants that will bear only red flowers, as we set out to demonstrate. ■■■■

SELF-CHECK EXERCISES 9.3

1. Let

$$T = \begin{bmatrix} .2 & 0 & 0 \\ .3 & 1 & .6 \\ .5 & 0 & .4 \end{bmatrix}$$

a. Show that T is an absorbing stochastic matrix.
b. Rewrite T so that the absorbing states appear first, partition the resulting matrix, and identify the submatrices R and S.
c. Compute the steady-state matrix of T.

2. There is a trend toward increased use of computer-aided transcription (CAT) and electronic recording (ER) as alternatives to manual transcription (MT) of court proceedings by court stenographers in a certain state. Suppose the following stochastic matrix is the transition matrix associated with the Markov process:

$$T = \begin{array}{c} \\ \text{CAT} \\ \text{ER} \\ \text{MT} \end{array} \begin{array}{c} \begin{array}{ccc} \text{CAT} & \text{ER} & \text{MT} \end{array} \\ \begin{bmatrix} 1 & .3 & .2 \\ 0 & .6 & .3 \\ 0 & .1 & .5 \end{bmatrix} \end{array}$$

Determine the probability that a court now using electronic recording or manual transcribing of its proceedings will eventually change to computer-aided transcription.

Solutions to Self-Check Exercises 9.3 can be found on page 566.

9.3 Exercises

In Exercises 1–8, determine whether the given matrix is an absorbing stochastic matrix.

1. $\begin{bmatrix} \frac{2}{5} & 0 \\ \frac{3}{5} & 1 \end{bmatrix}$

2. $\begin{bmatrix} 1 & 0 \\ 0 & 1 \end{bmatrix}$

3. $\begin{bmatrix} 1 & .5 & 0 \\ 0 & 0 & 1 \\ 0 & .5 & 0 \end{bmatrix}$

4. $\begin{bmatrix} 1 & 0 & 0 \\ 0 & .7 & .2 \\ 0 & .3 & .8 \end{bmatrix}$

5. $\begin{bmatrix} \frac{1}{8} & 0 & 0 \\ \frac{1}{4} & 1 & 0 \\ \frac{5}{8} & 0 & 1 \end{bmatrix}$

6. $\begin{bmatrix} 1 & 0 & 0 & 0 \\ 0 & \frac{5}{8} & 0 & \frac{1}{6} \\ 0 & \frac{1}{8} & 1 & 0 \\ 0 & \frac{1}{4} & 0 & \frac{5}{6} \end{bmatrix}$

7. $\begin{bmatrix} 1 & 0 & .3 & 0 \\ 0 & 1 & .2 & 0 \\ 0 & 0 & .1 & .5 \\ 0 & 0 & .4 & .5 \end{bmatrix}$

8. $\begin{bmatrix} 1 & 0 & 0 & 0 \\ 0 & 1 & 0 & 0 \\ 0 & 0 & .2 & .6 \\ 0 & 0 & .8 & .4 \end{bmatrix}$

In Exercises 9–14, rewrite each of the given absorbing stochastic matrices so that the absorbing states appear first, partition the resulting matrix, and identify the submatrices *R* and *S*.

9. $\begin{bmatrix} .6 & 0 \\ .4 & 1 \end{bmatrix}$

10. $\begin{bmatrix} \frac{1}{4} & 0 & 0 \\ \frac{1}{4} & 1 & 0 \\ \frac{1}{2} & 0 & 1 \end{bmatrix}$

11. $\begin{bmatrix} 0 & .2 & 0 \\ .5 & .4 & 0 \\ .5 & .4 & 1 \end{bmatrix}$

12. $\begin{bmatrix} .5 & 0 & .3 \\ 0 & 1 & .1 \\ .5 & 0 & .6 \end{bmatrix}$

13. $\begin{bmatrix} .4 & .2 & 0 & 0 \\ .2 & .3 & 0 & 0 \\ 0 & .3 & 1 & 0 \\ .4 & .2 & 0 & 1 \end{bmatrix}$

14. $\begin{bmatrix} .1 & 0 & 0 & 0 \\ .2 & 1 & 0 & .2 \\ .3 & 0 & 1 & 0 \\ .4 & 0 & 0 & .8 \end{bmatrix}$

In Exercises 15–24, compute the steady-state matrix of each of the given stochastic matrices.

15. $\begin{bmatrix} .55 & 0 \\ .45 & 1 \end{bmatrix}$

16. $\begin{bmatrix} \frac{3}{5} & 0 \\ \frac{2}{5} & 1 \end{bmatrix}$

17. $\begin{bmatrix} 1 & .2 & .3 \\ 0 & .4 & .2 \\ 0 & .4 & .5 \end{bmatrix}$

18. $\begin{bmatrix} \frac{1}{5} & 0 & 0 \\ 0 & 1 & \frac{3}{8} \\ \frac{4}{5} & 0 & \frac{5}{8} \end{bmatrix}$

19. $\begin{bmatrix} \frac{1}{2} & 0 & \frac{1}{3} & 0 \\ \frac{1}{2} & 1 & 0 & 0 \\ 0 & 0 & \frac{2}{3} & 0 \\ 0 & 0 & 0 & 1 \end{bmatrix}$

20. $\begin{bmatrix} 1 & \frac{1}{8} & \frac{1}{3} & 0 \\ 0 & \frac{1}{8} & 0 & 0 \\ 0 & \frac{1}{4} & \frac{2}{3} & 0 \\ 0 & \frac{1}{2} & 0 & 1 \end{bmatrix}$

21. $\begin{bmatrix} 1 & 0 & \frac{1}{4} & \frac{1}{3} \\ 0 & 1 & \frac{1}{4} & \frac{1}{3} \\ 0 & 0 & \frac{1}{2} & 0 \\ 0 & 0 & 0 & \frac{1}{3} \end{bmatrix}$

22. $\begin{bmatrix} 1 & 0 & .2 & .1 \\ 0 & 1 & .4 & .2 \\ 0 & 0 & 0 & .4 \\ 0 & 0 & .4 & .3 \end{bmatrix}$

23. $\begin{bmatrix} 1 & 0 & 0 & .2 & .1 \\ 0 & 1 & 0 & .1 & .2 \\ 0 & 0 & 1 & .3 & .1 \\ 0 & 0 & 0 & .2 & .2 \\ 0 & 0 & 0 & .2 & .4 \end{bmatrix}$

24. $\begin{bmatrix} 1 & 0 & \frac{1}{4} & \frac{1}{3} & 0 \\ 0 & 1 & 0 & \frac{1}{3} & \frac{1}{2} \\ 0 & 0 & \frac{1}{4} & \frac{1}{3} & 0 \\ 0 & 0 & \frac{1}{2} & 0 & \frac{1}{2} \\ 0 & 0 & 0 & 0 & 0 \end{bmatrix}$

25. GASOLINE CONSUMPTION As more and more old cars are taken off the road and replaced by late models that use unleaded fuel, the consumption of leaded gasoline will continue to drop. Suppose the transition matrix

$$A = \begin{array}{c} \\ \text{L} \\ \text{UL} \end{array} \begin{array}{c} \begin{array}{cc} \text{L} & \text{UL} \end{array} \\ \begin{bmatrix} .80 & 0 \\ .20 & 1 \end{bmatrix} \end{array}$$

describes this Markov process, where L denotes leaded gasoline and UL denotes unleaded gasoline.

a. Show that A is an absorbing stochastic matrix and rewrite it so that the absorbing state appears first. Partition the resulting matrix and identify the submatrices R and S.

b. Compute the steady-state matrix of A and interpret your results.

26. Diane has decided to play the following game of chance. She places a \$1 bet on each repeated play of the game in which the probability of her winning \$1 is .5. She has further decided to continue playing the game until she has either accumulated a total of \$3 or has lost all her money. What is the probability that Diane will eventually leave the game a winner if she started with a capital of \$1? Of \$2?

27. Refer to Exercise 26. Suppose Diane has decided to stop playing only after she has accumulated a sum of \$4 or has lost all her money. All other conditions being the same, what is the probability that Diane will leave the game a winner if she started with a capital of \$1? Of \$2? Of \$3?

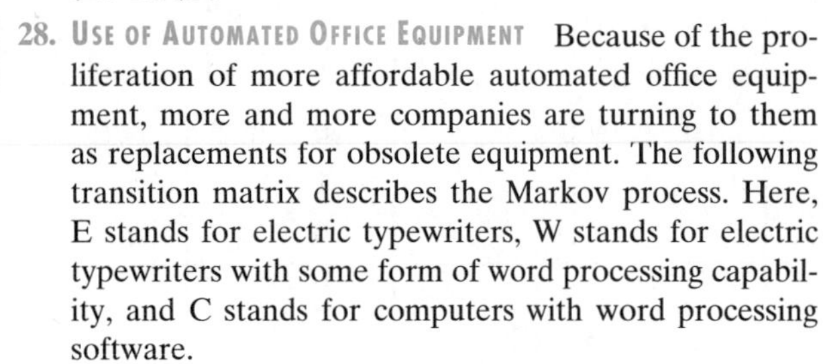

28. USE OF AUTOMATED OFFICE EQUIPMENT Because of the proliferation of more affordable automated office equipment, more and more companies are turning to them as replacements for obsolete equipment. The following transition matrix describes the Markov process. Here, E stands for electric typewriters, W stands for electric typewriters with some form of word processing capability, and C stands for computers with word processing software.

$$A = \begin{array}{c} \\ E \\ W \\ C \end{array} \begin{array}{c} \begin{array}{ccc} E & W & C \end{array} \\ \begin{bmatrix} .10 & 0 & 0 \\ .70 & .60 & 0 \\ .20 & .40 & 1 \end{bmatrix} \end{array}$$

a. Show that A is an absorbing stochastic matrix and rewrite it so that the absorbing state appears first. Partition the resulting matrix and identify the submatrices R and S.
b. Compute the steady-state matrix of A and interpret your results.

29. EDUCATION RECORDS The registrar of the Computronics Institute has compiled the following statistics on the progress of the school's students in their 2-yr computer programming course leading to an associate degree: Of beginning students in a particular year, 75% successfully complete their first year of study and move on to the second year, whereas 25% drop out of the program; of second-year students in a particular year, 90% go on to graduate at the end of the year, whereas 10% drop out of the program.
a. Construct the transition matrix associated with this Markov process.
b. Compute the steady-state matrix.
c. Determine the probability that a beginning student enrolled in the program will complete the course successfully.

30. EDUCATION RECORDS The registrar of a law school has compiled the following statistics on the progress of the school's students working toward the LLB degree: Of the first-year students in a particular year, 85% successfully complete their course of studies and move on to the second year, whereas 15% drop out of the program; of the second-year students in a particular year, 92% go on to the third year, whereas 8% drop out of the program; of the third-year students in a particular year, 98% go on to graduate at the end of the year, whereas 2% drop out of the program.
a. Construct the transition matrix associated with the Markov process.
b. Find the steady-state matrix.
c. Determine the probability that a beginning law student enrolled in the program will go on to graduate.

In Exercises 31 and 32, determine whether the statement is true or false. If it is true, explain why it is true. If it is false, give an example to show why it is false.

31. An absorbing stochastic matrix need not contain an absorbing state.

32. In partitioning an absorbing matrix into subdivisions,

$$A = \left[\begin{array}{c|c} I & S \\ \hline O & R \end{array}\right]$$

the identity matrix I is chosen to have the same order as R.

33. GENETICS Refer to Example 4. If the offspring are crossed successively with plants of genotype *aa* only, show that in the long run all the flowers produced by the plants will be white.

SOLUTIONS TO SELF-CHECK EXERCISES 9.3

1. a. State 2 is an absorbing state. States 1 and 3 are not absorbing, but each has a possibility (with probability .3 and .6) that an object may go from these states to state 2. Therefore, the matrix T is an absorbing stochastic matrix.
b. Denoting the states as indicated, we rewrite

$$\begin{array}{c} \\ 1 \\ 2 \\ 3 \end{array} \begin{array}{c} \begin{array}{ccc} 1 & 2 & 3 \end{array} \\ \begin{bmatrix} .2 & 0 & 0 \\ .3 & 1 & .6 \\ .5 & 0 & .4 \end{bmatrix} \end{array}$$

in the form

$$\begin{array}{c} \\ 2 \\ 3 \\ 1 \end{array}\begin{array}{c} \begin{array}{ccc} 2 & 3 & 1 \end{array} \\ \left[\begin{array}{c|cc} 1 & .6 & .3 \\ \hline 0 & .4 & .5 \\ 0 & 0 & .2 \end{array}\right] \end{array}$$

We see that

$$S = [.6 \quad .3] \qquad \text{and} \qquad R = \begin{bmatrix} .4 & .5 \\ 0 & .2 \end{bmatrix}$$

c. We compute

$$I - R = \begin{bmatrix} 1 & 0 \\ 0 & 1 \end{bmatrix} - \begin{bmatrix} .4 & .5 \\ 0 & .2 \end{bmatrix} = \begin{bmatrix} .6 & -.5 \\ 0 & .8 \end{bmatrix}$$

and, using the formula for finding the inverse of a 2×2 matrix in Section 2.6,

$$(I - R)^{-1} = \begin{bmatrix} 1.67 & 1.04 \\ 0 & 1.25 \end{bmatrix}$$

and so

$$S(I - R)^{-1} = [.6 \quad .3]\begin{bmatrix} 1.67 & 1.04 \\ 0 & 1.25 \end{bmatrix} = [1 \quad 1]$$

Therefore, the steady-state matrix of T is

$$\left[\begin{array}{c|cc} 1 & 1 & 1 \\ \hline 0 & 0 & 0 \\ 0 & 0 & 0 \end{array}\right]$$

2. We want to compute the steady-state matrix of T. Note that T is in the form

$$\left[\begin{array}{c|c} I & S \\ \hline O & R \end{array}\right]$$

where

$$S = [.3 \quad .2] \qquad \text{and} \qquad R = \begin{bmatrix} .6 & .3 \\ .1 & .5 \end{bmatrix}$$

We compute

$$I - R = \begin{bmatrix} 1 & 0 \\ 0 & 1 \end{bmatrix} - \begin{bmatrix} .6 & .3 \\ .1 & .5 \end{bmatrix} = \begin{bmatrix} .4 & -.3 \\ -.1 & .5 \end{bmatrix}$$

and, using the inverse formula in Section 2.6,

$$(I - R)^{-1} = \begin{bmatrix} 2.94 & 1.76 \\ 0.59 & 2.36 \end{bmatrix}$$

so

$$S(I - R)^{-1} = [.3 \quad .2]\begin{bmatrix} 2.94 & 1.76 \\ 0.59 & 2.36 \end{bmatrix} = [1 \quad 1]$$

Therefore, the steady-state matrix of T is

$$\begin{array}{c} \\ \text{CAT} \\ \text{ER} \\ \text{MT} \end{array} \begin{array}{c} \begin{array}{ccc} \text{CAT} & \text{ER} & \text{MT} \end{array} \\ \left[\begin{array}{c|cc} 1 & 1 & 1 \\ \hline 0 & 0 & 0 \\ 0 & 0 & 0 \end{array}\right] \end{array}$$

Interpreting the steady-state matrix of T, we see that in the long run all courts in this state will use computer-aided transcription.

9.4 Game Theory and Strictly Determined Games

The theory of games is a relatively new branch of mathematics and owes much of its development to John von Neumann (1903–1957), one of the mathematical giants of this century. Basically, the theory of games combines matrix methods with the theory of probability to determine the optimal strategies to be employed by two or more opponents involved in a competitive situation, with each opponent seeking to maximize his or her "gains," or, equivalently, to minimize his or her "losses." As such, the players may be poker players, managers of rival corporations seeking to extend their share of the market, campaign managers, or generals of opposing armies, to name a few.

For simplicity, we limit our discussion to games with two players. Such games are, naturally enough, called two-person games.

Two-Person Games

EXAMPLE 1 Richie and Chuck play a coin-matching game in which each player selects a side of a penny without prior knowledge of the other's choice. Then, upon a predetermined signal, both players disclose their choices simultaneously. Chuck agrees to pay Richie a sum of $3 if both choose heads; if Richie chooses heads and Chuck chooses tails, then Richie pays Chuck $6; if Richie chooses tails and Chuck chooses heads, then Chuck pays Richie $2; finally, if both Richie and Chuck choose tails, then Chuck pays Richie $1. In this game, the objective of each player is to discover a strategy that will ensure that his winnings are maximized (equivalently, that his losses are minimized). ■■■■

The coin-matching game is an example of a **zero-sum game**—that is, a game in which the payoff to one party results in an equal loss to the other. For such games, the sum of the payments made by both players at the end of each play adds up to zero.

To facilitate the analysis of the problem, we represent the given data in the form of a matrix.

$$\begin{array}{c c c} & & \text{C's moves} \\ & & \begin{array}{cc}\text{Heads} & \text{Tails}\end{array} \\ \text{R's moves} & \begin{array}{c}\text{Heads} \\ \text{Tails}\end{array} & \begin{bmatrix} 3 & -6 \\ 2 & 1 \end{bmatrix} \end{array}$$

Each row of the matrix corresponds to one of the two possible moves by Richie (referred to as the row player, R), whereas each column corresponds to one of the two possible moves by Chuck (referred to as the column player, C). Each entry in the matrix represents the payoff from C to R. For example, the entry $a_{11} = 3$ represents a \$3 payoff from Chuck to Richie (C to R) when Richie chooses to play row 1 (heads) and Chuck chooses to play column 1 (heads). On the other hand, the entry $a_{12} = -6$ represents (because it's negative) a \$6 payoff to C (from R) when R chooses to play row 1 (heads) and C chooses to play column 2 (tails). (Interpret the meaning of $a_{21} = 2$ and $a_{22} = 1$ for yourself.)

More generally, suppose we are given a two-person game with two players R and C. Furthermore, suppose that R has m possible moves $R_1, R_2, \ldots, R_m$ and that C has n possible moves $C_1, C_2, \ldots, C_n$. Then, we can represent the game in terms of an $m \times n$ matrix in which each row of the matrix represents one of the m possible moves of R and each column of the matrix represents one of the n possible moves of C:

$$\begin{array}{c c c} & & \text{C's moves} \\ & & \begin{array}{ccccccc} C_1 & C_2 & \cdots & C_j & \cdots & C_n \end{array} \\ \text{R's moves} & \begin{array}{c} R_1 \\ R_2 \\ \vdots \\ R_i \\ \vdots \\ R_m \end{array} & \begin{bmatrix} a_{11} & a_{12} & \cdots & a_{1j} & \cdots & a_{1n} \\ a_{21} & a_{22} & \cdots & a_{2j} & \cdots & a_{2n} \\ \vdots & \vdots & & \vdots & & \vdots \\ a_{i1} & a_{i2} & \cdots & a_{ij} & \cdots & a_{in} \\ \vdots & \vdots & & \vdots & & \vdots \\ a_{m1} & a_{m2} & \cdots & a_{mj} & \cdots & a_{mn} \end{bmatrix} \end{array}$$

The entry a_{ij} in the ith row and jth column of the (payoff) matrix represents the payoff to R when R chooses move R_i and C chooses move C_j. In this context, note that a payoff to R means, in actuality, a payoff to C in the event that the value of a_{ij} is negative.

EXAMPLE 2 The payoff matrix associated with a game is given by

$$\begin{array}{c c c} & & \text{C's moves} \\ & & \begin{array}{ccc} C_1 & C_2 & C_3 \end{array} \\ \text{R's moves} & \begin{array}{c} R_1 \\ R_2 \end{array} & \begin{bmatrix} 1 & -2 & 3 \\ 4 & -5 & -1 \end{bmatrix} \end{array}$$

Give an interpretation of this payoff matrix.

SOLUTION ✓ In this two-person game, player R has two possible moves, whereas player C has three possible moves. The payoffs are determined as follows: If R chooses R_1, then

R wins 1 unit if C chooses C_1.
R loses 2 units if C chooses C_2.
R wins 3 units if C chooses C_3.

If R chooses R_2, then

R wins 4 units if C chooses C_1.
R loses 5 units if C chooses C_2.
R loses 1 unit if C chooses C_3.

■■■■

Optimal Strategies

Let's return to the payoff matrix of Example 1 and see how it may be used to help us determine the "best" strategy for each of the two players R and C. For convenience, this matrix is reproduced here:

$$\text{R's moves} \begin{matrix} \\ R_1 \\ R_2 \end{matrix} \overset{\text{C's moves}}{\begin{matrix} C_1 & C_2 \\ \begin{bmatrix} 3 & -6 \\ 2 & 1 \end{bmatrix} \end{matrix}}$$

Let's first consider the game from R's point of view. Since the entries in the payoff matrix represent payoffs to him, his initial reaction might be to seek out the largest entry in the matrix and consider the row containing such an entry as a possible move. Thus, he is led to the consideration of R_1 as a possible move.

Let's examine this choice a little more closely. To be sure, R would realize the largest possible payoff to himself ($3) if C chose C_1; but if C chose C_2, then R would lose $6! Since R does not know beforehand what C's move will be, a more prudent approach on his part would be to assume that no matter what row he chooses, C will counter with a move (column) that will result in the smallest payoff to him. To maximize the payoff to himself under these circumstances, R would then select from among the moves (rows) the one in which the smallest payoff is as large as possible. This strategy for R, called, for obvious reasons, the **maximin strategy,** may be summarized as follows:

Maximin Strategy

1. For each row of the payoff matrix, find the smallest entry in that row.
2. Choose the row for which the entry found in step 1 is as large as possible. This row constitutes R's "best" move.

For the problem under consideration, we can organize our work as follows:

$$\begin{bmatrix} 3 & -6 \\ 2 & 1 \end{bmatrix} \qquad \begin{matrix} \text{Row minima} \\ -6 \\ \textcircled{1} \end{matrix} \leftarrow \text{Larger of the row minima}$$

From these results, it is seen that R's "best" move is row 2. By choosing this move R stands to win at least \$1.

Next, let's consider the game from C's point of view. His objective is to minimize the payoff to R. This is accomplished by choosing the column whose largest payoff is as small as possible. This strategy for C, called the **minimax strategy,** may be summarized as follows:

Minimax Strategy

1. For each column of the payoff matrix, find the largest entry in that column.
2. Choose the column for which the entry found in step 1 is as small as possible. This column constitutes C's "best" move.

We can organize the work involved in determining C's "best" move as follows:

$$\begin{matrix} & \begin{bmatrix} 3 & -6 \\ 2 & 1 \end{bmatrix} \\ \text{Column maxima} & 3 \quad \textcircled{1} \end{matrix}$$

↑ Smaller of the column maxima

From these results, we see that C's "best" move is column 2. By choosing this move, C stands to lose at most \$1.

EXAMPLE 3 For the game with the following payoff matrix, determine the maximin and minimax strategies for each player.

$$\begin{bmatrix} -3 & -2 & 4 \\ -2 & 0 & 3 \\ 6 & -1 & 1 \end{bmatrix}$$

SOLUTION ✓ We determine the minimum of each row and the maximum of each column of the payoff matrix and then display these numbers by circling the largest of the row minima and the smallest of the column maxima:

$$\begin{matrix} & \begin{bmatrix} -3 & -2 & 4 \\ -2 & 0 & 3 \\ 6 & -1 & 1 \end{bmatrix} & \begin{matrix} \text{Row minima} \\ -3 \\ -2 \\ \textcircled{-1} \end{matrix} \leftarrow \text{Largest of the row minima} \\ \text{Column maxima} & 6 \quad \textcircled{0} \quad 4 & \end{matrix}$$

↑ Smallest of the column maxima

From these results, we conclude that the maximin strategy (for the row player) is to play row 3, whereas the minimax strategy (for the column player) is to play column 2. ■■■■

EXAMPLE 4 Determine the maximin and minimax strategies for each player in a game whose payoff matrix is given by

$$\begin{bmatrix} 3 & 4 & -4 \\ 2 & -1 & -3 \end{bmatrix}$$

SOLUTION ✓ Proceeding as in the last example, we obtain the following:

$$\begin{array}{lccc|c} & & & & \text{Row minima} \\ & \left[\begin{matrix} 3 \\ 2 \end{matrix}\right. & \begin{matrix} 4 \\ -1 \end{matrix} & \left.\begin{matrix} -4 \\ -3 \end{matrix}\right] & \begin{matrix} -4 \\ \textcircled{-3} \end{matrix} \leftarrow \text{Larger of the row minima} \\ \text{Column maxima} & 3 & 4 & \textcircled{-3} & \\ & & & \uparrow & \\ & & & \text{Smallest of the column maxima} & \end{array}$$

from which we conclude that the maximin strategy for the row player is to play row 2, whereas the minimax strategy for the column player is to play column 3. ■■■■

In arriving at the maximin and minimax strategies for the respective players, we have assumed that both players always act rationally, with the knowledge that their opponents will always act rationally. This means that each player adopts a strategy of always making the same move and assumes that his opponent is always going to counter that move with a move that will maximize the payoff to the opponent. Thus, each player adopts the pure strategy of always making the move that will minimize the payoff his opponent can receive and thereby maximize the payoff to himself.

This raises the following question: Suppose a game is played repeatedly and one of the players realizes that the opponent is employing his maximin (or minimax) strategy. Can this knowledge be used to the player's advantage? To obtain a partial answer to this question, let's consider the game posed in Example 3. There, the minimax strategy for the column player is to play column 2. Suppose, in repeated plays of the game, a player consistently plays that column and this strategy becomes known to the row player. The row player may then change the strategy from playing row 3 (the maximin strategy) to playing row 2, thereby reducing losses from 1 unit to zero. Thus, at least for this game, the knowledge that a player is employing the maximin or minimax strategy can be used to the opponent's advantage.

There is, however, a class of games in which the knowledge that a player is using the maximin (or minimax) strategy proves of no help to the opponent. Consider, for example, the game of Example 4. There, the row player's maximin strategy is to play row 2, and the column player's minimax strategy is to play column 3. Suppose, in repeated plays of the game, R (the row player) has discovered that C (the column player) consistently chooses to play column 3 (the minimax strategy). Can this knowledge be used to R's advantage? Now, other than playing row 2 (the maximin strategy), R may choose to play row 1. But if R makes this choice, then he would lose 4 units instead of 3 units! Clearly, in this case, the knowledge that C is using the minimax strategy cannot be used to advantage. R's optimal (best) strategy is the maximin strategy.

Optimal Strategy

The **optimal strategy** in a game is the strategy that is most profitable to a particular player.

Next, suppose that in repeated plays of the game, C has discovered that R consistently plays row 2 (his optimal strategy). Can this knowledge be used to C's advantage? Another glance at the payoff matrix reveals that by playing column 1, C stands to lose 2 units, and by playing column 2, he stands to win 1 unit, as compared with winning 3 units by playing column 3, as called for by the minimax strategy. Thus, as in the case of R, C does not benefit from knowing his opponent's move. Furthermore, his optimal strategy coincides with the minimax strategy.

This game, in which the row player cannot benefit from knowing that his opponent is using the minimax strategy and the column player cannot benefit from knowing that his opponent is using the maximin strategy, is said to be strictly determined.

Strictly Determined Game

A **strictly determined game** is characterized by the following properties:

1. There is an entry in the payoff matrix that is *simultaneously* the smallest entry in its row and the largest entry in its column. This entry is called the **saddle point** for the game.
2. The optimal strategy for the row player is precisely the maximin strategy and is the row containing the saddle point. The optimal strategy for the column player is the minimax strategy and is the column containing the saddle point.

The saddle point of a strictly determined game is also referred to as the **value of the game.** If the value of a strictly determined game is positive, then the game favors the row player. If the value is negative, it favors the column player. If the value of the game is zero, the game is called a **fair game.**

Returning to the coin-matching game discussed earlier, we conclude that Richie's optimal strategy consists of playing row 2 repeatedly, whereas Chuck's optimal strategy consists of playing column 2 repeatedly. Furthermore, the value of the game is 1, implying that the game favors the row player, Richie.

EXAMPLE 5 A two-person, zero-sum game is defined by the payoff matrix

$$A = \begin{bmatrix} 1 & 2 & -3 \\ -1 & 2 & -2 \\ 2 & 3 & -4 \end{bmatrix}$$

a. Show that the game is strictly determined and find the saddle point(s) for the game.
b. What is the optimal strategy for each player?
c. What is the value of the game? Does the game favor one player over the other?

SOLUTION ✔ **a.** First, we determine the minimum of each row and the maximum of each column of the payoff matrix A and display these minima and maxima as follows:

$$\begin{array}{lccc} & \begin{bmatrix} 1 & 2 & -3 \\ -1 & 2 & \textcircled{-2} \\ 2 & 3 & -4 \end{bmatrix} & \begin{array}{c}\text{Row minima} \\ -3 \\ -2 \\ -4 \end{array} & \begin{array}{c} \\ \\ \leftarrow \text{Largest of the row minima} \\ \\ \end{array} \\ \text{Column maxima} & \begin{array}{ccc} 2 & 3 & -2 \end{array} & & \end{array}$$

↑ Smallest of the column maxima

From these results, we see that the circled entry, −2, is simultaneously the smallest entry in its row and the largest entry in its column. Therefore, the game is strictly determined, with the entry $a_{23} = -2$ as its saddle point.

b. From these results, we see that the optimal strategy for the row player is to make the move represented by the second row of the matrix, and the optimal strategy for the column player is to make the move represented by the third column.

c. The value of the game is −2, which implies that if both players adopt their best strategy, the column player will win 2 units in a play. Consequently, the game favors the column player. ■■■■

A game may have more than one saddle point, as the next example shows.

EXAMPLE 6 A two-person, zero-sum game is defined by the payoff matrix

$$A = \begin{bmatrix} 4 & 5 & 4 \\ -2 & -5 & -3 \\ 4 & 6 & 8 \end{bmatrix}$$

a. Show that the game is strictly determined and find the saddle points for the game.
b. Discuss the optimal strategies for the players.
c. Does the game favor one player over the other?

SOLUTION ✓

a. Proceeding as in the previous example, we obtain the following information:

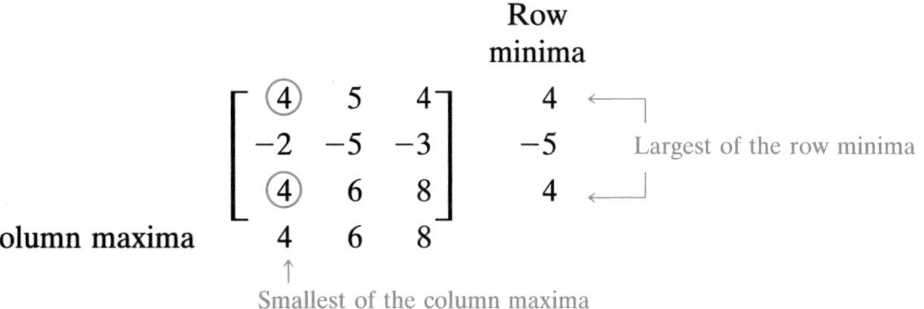

$$\begin{bmatrix} ④ & 5 & 4 \\ -2 & -5 & -3 \\ ④ & 6 & 8 \end{bmatrix} \quad \begin{matrix} 4 \\ -5 \\ 4 \end{matrix}$$

Column maxima 4 6 8

We see that each of the circled entries, 4, is simultaneously the smallest entry in the row and the largest entry in the column containing it. Therefore, the game is strictly determined, and in this case it has two saddle points: the entry $a_{11} = 4$ and the entry $a_{31} = 4$. In general, it can be shown that every saddle point of a payoff matrix must have the same value.

b. Since the game has two saddle points, both lying in the first column and in the first and third rows of the payoff matrix, we see that the row player's optimal strategy consists of playing either row 1 or row 3 consistently, whereas the column player's optimal strategy consists of playing column 1 repeatedly.

c. The value of the game is 4, which implies that it favors the row player.

■■■■

EXAMPLE 7

Two television subscription companies, UBS and Telerama, are planning to extend their operations to a certain city. Each has the option of making its services available to prospective subscribers with a special introductory subscription rate. It is estimated that if both UBS and Telerama offer the special subscription rate, each will get 50% of the market, whereas if UBS offers the special subscription rate and Telerama does not, UBS will get 70% of the market. If Telerama offers the special subscription rate and UBS does not, it is estimated that UBS will get 40% of the market. If both companies elect not to offer the special subscription rate, it is estimated that UBS will get 60% of the market.

a. Construct the payoff matrix for the game.
b. Show that the game is strictly determined.
c. Determine the optimal strategy for each company and find the value of the game.

SOLUTION ✓

a. The required payoff matrix is given by

		Telerama: Intro. rate	Telerama: Usual rate
UBS	Intro. rate	.50	.70
	Usual rate	.40	.60

b. The entry $a_{11} = .50$ is the smaller entry in its row and the larger entry in its column. Therefore, the entry $a_{11} = .50$ is a saddle point of the game and the game is strictly determined.

c. The entry $a_{11} = .50$ is the only saddle point of the game, so UBS's optimal strategy is to choose row 1, and Telerama's optimal strategy is to choose column 1. In other words, both companies should offer their potential customers their respective introductory subscription rates.

Group Discussion

A two-person, zero-sum game is defined by the payoff matrix

$$A = \begin{bmatrix} a & a \\ c & d \end{bmatrix}$$

1. Show that the game is strictly determined.
2. What can you say about the game with payoff matrix

$$B = \begin{bmatrix} a & b \\ c & c \end{bmatrix}$$

SELF-CHECK EXERCISES 9.4

1. A two-person, zero-sum game is defined by the payoff matrix

$$A = \begin{bmatrix} -2 & 1 & 3 \\ 3 & 2 & 2 \\ 2 & -1 & 4 \end{bmatrix}$$

a. Show that the game is strictly determined and find the saddle point(s) for the game.
b. What is the optimal strategy for each player?
c. What is the value of the game? Does the game favor one player over the other?

2. The management of Delta Corporation, a construction and development company, is deciding whether to go ahead with the construction of a large condominium complex. A financial analysis of the project indicates that if Delta goes ahead with the development and the home mortgage rate drops 1 point or more by next year, when the complex is expected to be completed, it will stand to make a profit of \$750,000. If Delta goes ahead with the development and the mortgage rate stays within 1 point of the current rate by next year, it will stand to make a profit of \$600,000. If Delta goes ahead with the development and the mortgage rate increases 1 point or more by next year, it will stand to make a profit of \$350,000. If Delta does not go ahead with the development and the mortgage rate drops 1 point or more by next year, it will stand to make a profit of \$400,000. If Delta does not go ahead with the development and the mortgage rate stays within 1 point of the current rate by next year, it will stand to make a profit of \$350,000. Finally, if Delta

does not go ahead with the development and the mortgage rate increases 1 point or more by next year, it stands to make $250,000.

a. Represent this information in the form of a payoff matrix.

b. Assuming that the home mortgage rate trend is volatile over the next year, determine whether or not Delta should go ahead with the project.

Solutions to Self-Check Exercises 9.4 can be found on page 579.

9.4 Exercises

In Exercises 1–8, determine the maximin and minimax strategies for each of the given two-person, zero-sum matrix games.

1. $\begin{bmatrix} 2 & 3 \\ 4 & 1 \end{bmatrix}$ **2.** $\begin{bmatrix} -1 & 3 \\ 2 & 5 \end{bmatrix}$

3. $\begin{bmatrix} 1 & 3 & 2 \\ 0 & -1 & 4 \end{bmatrix}$ **4.** $\begin{bmatrix} 1 & 4 & -2 \\ 4 & 6 & -3 \end{bmatrix}$

5. $\begin{bmatrix} 3 & 2 & 1 \\ 1 & -2 & 3 \\ 6 & 4 & 1 \end{bmatrix}$ **6.** $\begin{bmatrix} 1 & 4 \\ 2 & -2 \\ 3 & 0 \end{bmatrix}$

7. $\begin{bmatrix} 4 & 2 & 1 \\ 1 & 0 & -1 \\ 2 & 1 & 3 \end{bmatrix}$ **8.** $\begin{bmatrix} -1 & 1 & 2 \\ 3 & 1 & 1 \\ -1 & 1 & 2 \\ 3 & 2 & -1 \end{bmatrix}$

In Exercises 9–18, determine whether the given two-person, zero-sum matrix game is strictly determined. If a game is strictly determined,

a. Find the saddle point(s) of the game.

b. Find the optimal strategy for each player.

c. Find the value of the game.

d. Determine whether the game favors one player over the other.

9. $\begin{bmatrix} 2 & 3 \\ 1 & -4 \end{bmatrix}$ **10.** $\begin{bmatrix} 1 & 0 \\ 0 & -1 \end{bmatrix}$

11. $\begin{bmatrix} 1 & 3 & 2 \\ -1 & 4 & -6 \end{bmatrix}$ **12.** $\begin{bmatrix} 3 & 2 \\ -1 & -2 \\ 4 & 1 \end{bmatrix}$

13. $\begin{bmatrix} 1 & 3 & 4 & 2 \\ 0 & 2 & 6 & -4 \\ -1 & -3 & -2 & 1 \end{bmatrix}$ **14.** $\begin{bmatrix} 2 & 4 & 2 \\ 0 & 3 & 0 \\ -1 & -2 & 1 \end{bmatrix}$

15. $\begin{bmatrix} 1 & 2 \\ 0 & 3 \\ -1 & 2 \\ 2 & -2 \end{bmatrix}$ **16.** $\begin{bmatrix} -1 & 2 & 4 \\ 2 & 3 & 5 \\ 0 & 1 & -3 \\ -2 & 4 & -2 \end{bmatrix}$

17. $\begin{bmatrix} 1 & -1 & 3 & 2 \\ 1 & 0 & 2 & 2 \\ -2 & 2 & 3 & -1 \end{bmatrix}$

18. $\begin{bmatrix} 3 & -1 & 0 & -4 \\ 2 & 1 & 0 & 2 \\ -3 & 1 & -2 & 1 \\ -1 & -1 & -2 & 1 \end{bmatrix}$

19. Robin and Cathy play a game of matching fingers. On a predetermined signal, both players simultaneously extend 1, 2, or 3 fingers from a closed fist. If the sum of the number of fingers extended is even, then Robin receives an amount in dollars equal to that sum from Cathy. If the sum of the number of fingers extended is odd, then Cathy receives an amount in dollars equal to that sum from Robin.

a. Construct the payoff matrix for the game.

b. Find the maximin and the minimax strategies for Robin and Cathy, respectively.

c. Is the game strictly determined?

d. If the answer to part (c) is yes, what is the value of the game?

20. MANAGEMENT DECISIONS Brady's, a conventional department store, and Value-Mart, a discount department store, are each considering opening new stores at one of two possible sites: the Civic Center and North Shore Plaza. The strategies available to the management of each store are given in the following payoff matrix, where each entry represents the amounts (in hundreds of thousands of dollars) either gained or lost by one business from or to the other as a result of the sites selected.

		Value-Mart Center	Value-Mart Plaza
Brady's	Civic Center	2	−2
	North Shore Plaza	3	−4

a. Show that the game is strictly determined.
b. What is the value of the game?
c. Determine the best strategy for the management of each store (that is, determine the ideal locations for each store).

21. **Financial Analysis** The management of the Acrosonic Company is faced with the problem of deciding whether to expand the production of its line of electrostatic loudspeaker systems. It has been estimated that an expansion will result in an annual profit of $200,000 for Acrosonic if the general economic climate is good. On the other hand, an expansion during a period of economic recession will cut its annual profit to $120,000. As an alternative, Acrosonic may hold the production of its electrostatic loudspeaker systems at the current level and expand its line of conventional loudspeaker systems. In this event, the company is expected to make a profit of $50,000 in an expanding economy (because many potential customers will be expected to buy electrostatic loudspeaker systems from other competitors) and a profit of $150,000 in a recessionary economy.
a. Construct the payoff matrix for this game.
Hint: The row player is the management of the company and the column player is the economy.

b. Should management recommend expanding the company's line of electrostatic loudspeaker systems?

22. **Financial Analysis** The proprietor of the Belvedere Restaurant is faced with the problem of deciding whether to expand his restaurant facilities now or to wait until some future date to do so. If he expands the facilities now and the economy experiences a period of growth during the coming year, he will make a net profit of $442,000; if he expands now and a period of zero growth follows, then he will make a net profit of $40,000; and if he expands now and an economic recession follows, he will suffer a net loss of $108,000. If he does not expand the restaurant now and the economy experiences a period of growth during the coming year, he will make a net profit of $280,000; if he does not expand now and a period of zero growth follows, he will make a net profit of $190,000. Finally, if he does not expand now and an economic recession follows, he will make a net profit of $100,000.
a. Represent this information in the form of a payoff matrix.
b. Assuming that the state of the economy during the coming year is uncertain, determine whether the owner of the restaurant should expand his facilities at this time.

23. **Market Share** Two barber shops, Roland's Barber Shop and Charley's Barber Shop, are both located in the business district of a certain town. Roland estimates that if he raises the price of a haircut by $1, he will increase his market share by 3% if Charley raises his price by the same amount; he will decrease his market share by 1% if Charley holds his price at the same level; and he will decrease his market share by 3% if Charley lowers his price by $1. If Roland keeps his price the same, he will increase his market share by 2% if Charley raises his price by $1; he will keep the same market share if Charley holds the price at the same level; and he will decrease his market share by 2% if Charley lowers his price by $1. Finally, if Roland lowers the price he charges by $1, his market share will increase by 5% if Charley raises his price by the same amount; he will increase his market share by 2% if Charley holds his price at the same level; and he will increase his market share by 1% if Charley lowers his price by $1.
a. Construct the payoff matrix for this game.
b. Show that the game is strictly determined.
c. If neither party is willing to lower the price he charges for a haircut, show that both should keep their present price structures.

In Exercises 24–26, determine whether the statement is true or false. If it is true, explain why it is true. If it is false, give an example to show why it is false.

24. In a zero-sum game, the payments made by the players at the end of each play add up to zero.

25. In a strictly determined game, the value of the game is given by the saddle point of the game.

26. If the value of a strictly determined game is not negative, it favors the row player.

SOLUTIONS TO SELF-CHECK EXERCISES 9.4

1. a. Displaying the minimum of each row and the maximum of each column of the payoff matrix A, we obtain

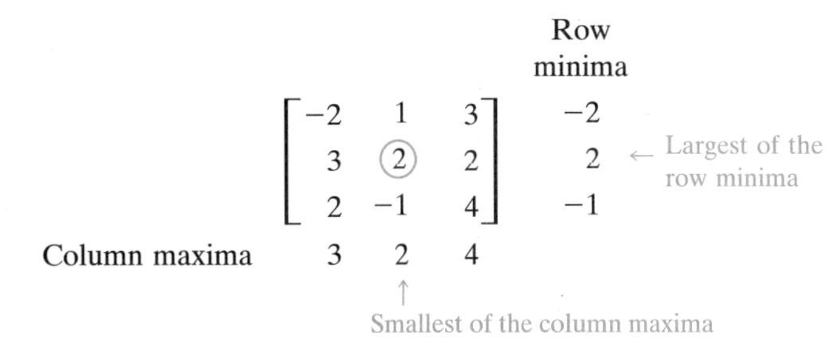

From these results, we see that the circled entry, 2, is simultaneously the smallest entry in its row and the largest entry in its column. Therefore, the game is strictly determined, with the entry $a_{22} = 2$ as its saddle point.

b. From these results, we see that the optimal strategy for the row player is to make the move represented by the second row of the matrix, and the optimal strategy for the column player is to make the move represented by the second column.

c. The value of the game is 2, which implies that if both players adopt their best strategy, the row player will win 2 units in a play. Consequently, the game favors the row player.

2. a. We may view this situation as a game in which the row player is Delta Corporation and the column player is the home mortgage rate. The required payoff matrix is

		Mortgage rate		
		Decrease	Steady	Increase
Delta	Project	750	600	350
Corp.	No project	400	350	250

(All figures are in thousands of dollars.)

b. From part (a), the payoff matrix under consideration is

$$\begin{bmatrix} 750 & 600 & 350 \\ 400 & 350 & 250 \end{bmatrix}$$

Proceeding in the usual manner, we find

				Row minima	
	750	600	(350)	350	← Larger of the row minima
	400	350	250	250	
Column maxima	750	600	350		
			↑ Smallest of the column maxima		

From these results, we see that the entry $a_{13} = 350$ is a saddle point and the game is strictly determined. We can also conclude that the company should go ahead with the project.

9.5 Games with Mixed Strategies

In Section 9.4 we discussed strictly determined games and found that the optimal strategy for the row player is to select the row containing a saddle point for the game, and the optimal strategy for the column player is to select the column containing a saddle point. Furthermore, in repeated plays of the game, each player's optimal strategy consists of making the same move over and over again, since the discovery of the opponent's optimal strategy cannot be used to advantage. Such strategies are called **pure strategies.** In this section we look at games that are not strictly determined and the strategies associated with such games.

MIXED STRATEGIES

As a simple example of a game that is not strictly determined, let's consider the following slightly modified version of the coin-matching game played by Richie and Chuck (see Example 1, Section 9.4). Suppose Richie wins \$3 if both parties choose heads and \$1 if both choose tails and loses \$2 if one chooses heads and the other tails. Then, the payoff matrix for this game is given by

$$
\begin{array}{cc}
 & \begin{array}{cc} & C\text{'s moves} \\ & \begin{array}{cc} C_1 & C_2 \\ \text{(heads)} & \text{(tails)} \end{array} \end{array} \\
R\text{'s moves} \begin{array}{c} R_1 \text{ (heads)} \\ R_2 \text{ (tails)} \end{array} & \begin{bmatrix} 3 & -2 \\ -2 & 1 \end{bmatrix}
\end{array}
$$

A quick examination of this matrix reveals that it contains no entry that is simultaneously the smallest entry in its row and the largest entry in its column; that is, the game has no saddle point and is therefore not strictly determined. What strategy might Richie adopt for the game? Offhand, it would seem that he should consistently select row 1 since he stands to win \$3 by playing this row and only \$1 by playing row 2, at a risk, in either case, of losing \$2. However, if Chuck discovers that Richie is playing row 1 consistently, he would counter this strategy by playing column 2, causing Richie to lose \$2 on each play! In view of this, Richie is led to consider a strategy whereby he chooses row 1 some of the time and row 2 at other times. A similar analysis of the game from Chuck's point of view suggests that he might consider choosing column 1 some of the time and column 2 at other times. Such strategies are called **mixed strategies.**

From a practical point of view, there are many ways in which a player may choose moves in a game with mixed strategies. For example, in the game just mentioned, if Richie decides to play heads half the time and tails the other half of the time, he could toss an unbiased coin before each move and let the outcome of the toss determine which move he should make. Another

FIGURE 9.3

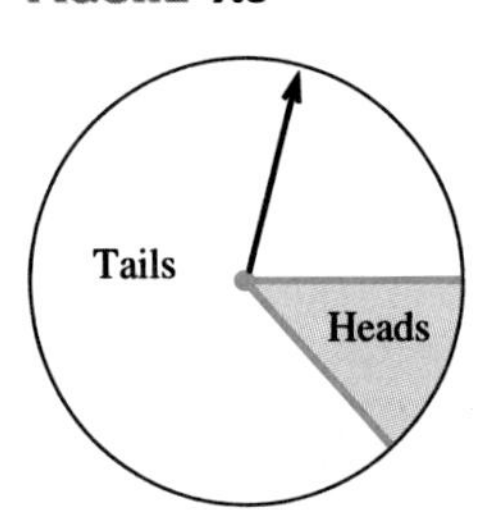

more general but less practical way of deciding on the choice of a move is: Having determined beforehand the proportion of the time row 1 is to be chosen (and therefore the proportion of the time row 2 is to be chosen), Richie might construct a spinner (Figure 9.3) in which the areas of the two sectors reflect these proportions and let the move be decided by the outcome of a spin. These two methods for determining a player's move in a game with mixed strategies guarantee that the strategy will not fall into a pattern that will be discovered by the opponent.

From the mathematical point of view, we may describe the mixed strategy of a row player in terms of a row vector whose dimension coincides with the number of possible moves the player has. For example, if Richie had decided on a strategy in which he chose to play row 1 half the time and row 2 the other half of the time, then this strategy is represented by the row vector

$$[.5 \quad .5]$$

Similarly, the mixed strategy for a column vector may be represented by a column vector of appropriate dimension. For example, returning to our illustration, suppose Chuck has decided that 20% of the time he will choose column 1 and 80% of the time he will choose column 2. This strategy is represented by the column vector

$$\begin{bmatrix} .2 \\ .8 \end{bmatrix}$$

EXPECTED VALUE OF A GAME

For the purpose of comparing the merits of a player's different mixed strategies in a game, it is convenient to introduce a number called the *expected value* of a game. The expected value measures the average payoff to the row player when both players adopt a particular set of mixed strategies. We now explain this notion using a 2×2 matrix game whose payoff matrix has the general form

$$A = \begin{bmatrix} a_{11} & a_{12} \\ a_{21} & a_{22} \end{bmatrix}$$

Suppose that in repeated plays of the game the row player R adopts the mixed strategy

$$P = [p_1 \quad p_2]$$

That is, the player selects row 1 with probability p_1 and row 2 with probability p_2, and the column player C adopts the mixed strategy

$$Q = \begin{bmatrix} q_1 \\ q_2 \end{bmatrix}$$

That is, the column player selects column 1 with probability q_1 and column 2 with probability q_2. Now, in each play of the game, there are four possible outcomes that may be represented by the ordered pairs

(row 1, column 1),
(row 1, column 2),
(row 2, column 1),
(row 2, column 2)

where the first number of each ordered pair represents R's selection and the second number of each ordered pair represents C's selection. Since the choice of moves is made by one player without knowing the other's choice, each pair of events (for example, the events "row 1" and "column 1") constitutes a pair of independent events. Therefore, the probability of R choosing row 1 and C choosing column 1, P(row 1, column 1), is given by

$$\begin{aligned} P(\text{row } 1, \text{column } 1) &= P(\text{row } 1) \cdot P(\text{column } 1) \\ &= p_1 q_1 \end{aligned}$$

In a similar manner, we compute the probability of each of the other three outcomes. These calculations, together with the payoffs associated with each of the four possible outcomes, may be summarized as follows:

Outcome	Probability	Payoff
(row 1, column 1)	p_1q_1	a_{11}
(row 1, column 2)	p_1q_2	a_{12}
(row 2, column 1)	p_2q_1	a_{21}
(row 2, column 2)	p_2q_2	a_{22}

Then, the *expected payoff* E of the game is the sum of the products of the payoffs and the corresponding probabilities (see Section 8.2). Thus,

$$E = p_1q_1a_{11} + p_1q_2a_{12} + p_2q_1a_{21} + p_2q_2a_{22}$$

In terms of the matrices P, A, and Q, we have the following relatively simple expression for E—namely,

$$E = PAQ$$

which you may verify (Exercise 22). This result may be generalized as follows:

Expected Value of a Game

Let

$$P = [p_1 p_2 \cdots p_m] \quad \text{and} \quad Q = \begin{bmatrix} q_1 \\ q_2 \\ \vdots \\ q_n \end{bmatrix}$$

be the vectors representing the mixed strategies for the row player R and the column player C, respectively, in a game with an $m \times n$ payoff matrix

$$A = \begin{bmatrix} a_{11} & a_{12} & \cdots & a_{1n} \\ a_{21} & a_{22} & \cdots & a_{2n} \\ \vdots & \vdots & & \vdots \\ a_{m1} & a_{m2} & \cdots & a_{mn} \end{bmatrix}$$

Then the **expected value of the game** is given by

$$E = PAQ = [p_1 p_2 \cdots p_m] \begin{bmatrix} a_{11} & a_{12} & \cdots & a_{1n} \\ a_{21} & a_{22} & \cdots & a_{2n} \\ \vdots & \vdots & & \vdots \\ a_{m1} & a_{m2} & \cdots & a_{mn} \end{bmatrix} \begin{bmatrix} q_1 \\ q_2 \\ \vdots \\ q_n \end{bmatrix}$$

We now look at several examples involving the computation of the expected value of a game.

EXAMPLE 1

Consider a coin-matching game played by Richie and Chuck with a payoff matrix given by

$$A = \begin{bmatrix} 3 & -2 \\ -2 & 1 \end{bmatrix}$$

Compute the expected payoff of the game if Richie adopts the mixed strategy P and Chuck adopts the mixed strategy Q, where

a. $P = [.5 \quad .5]$ and $Q = \begin{bmatrix} .5 \\ .5 \end{bmatrix}$

b. $P = [.8 \quad .2]$ and $Q = \begin{bmatrix} .1 \\ .9 \end{bmatrix}$

SOLUTION ✔ **a.** We compute

$$E = PAQ = [.5 \quad .5]\begin{bmatrix} 3 & -2 \\ -2 & 1 \end{bmatrix}\begin{bmatrix} .5 \\ .5 \end{bmatrix}$$
$$= [.5 \quad -.5]\begin{bmatrix} .5 \\ .5 \end{bmatrix}$$
$$= 0$$

Thus, in repeated plays of the game, it may be expected in the long run that the game will end in a draw.

b. We compute

$$E = PAQ = [.8 \quad .2]\begin{bmatrix} 3 & -2 \\ -2 & 1 \end{bmatrix}\begin{bmatrix} .1 \\ .9 \end{bmatrix}$$
$$= [2 \quad -1.4]\begin{bmatrix} .1 \\ .9 \end{bmatrix}$$
$$= -1.06.$$

That is, in the long run Richie may be expected to lose $1.06 on the average in each play. ■■■■

EXAMPLE 2 The payoff matrix for a certain game is given by

$$A = \begin{bmatrix} 1 & -2 \\ -1 & 2 \\ 3 & -3 \end{bmatrix}$$

a. Find the expected payoff to the row player if the row player R uses her maximin pure strategy and the column player C uses her minimax pure strategy.

b. Find the expected payoff to the row player if R uses her maximin strategy 50% of the time and chooses each of the other two rows 25% of the time, while C chooses each column 50% of the time.

SOLUTION ✔ **a.** The maximin and minimax strategies for the row and column players, respectively, may be found using the method of the last section. Thus,

			Row minima	
	1	−2	−2	
	−1	2	(−1)	← Largest of the row minima
	3	−3	−3	
Column maxima	3	(2)		
		↑ Smaller of the column maxima		

From these results, we see that R's optimal pure strategy is to choose row 2, whereas C's optimal pure strategy is to choose column 2. Furthermore, if both players use their respective optimal strategy, then the expected payoff to R is 2 units.

b. In this case, R's mixed strategy may be represented by the row vector

$$P = [.25 \quad .50 \quad .25]$$

and C's mixed strategy may be represented by the column vector

$$Q = \begin{bmatrix} .5 \\ .5 \end{bmatrix}$$

The expected payoff to the row player will then be given by

$$\begin{aligned} E = PAQ &= [.25 \quad .50 \quad .25] \begin{bmatrix} 1 & -2 \\ -1 & 2 \\ 3 & -3 \end{bmatrix} \begin{bmatrix} .5 \\ .5 \end{bmatrix} \\ &= [.25 \quad .50 \quad .25] \begin{bmatrix} -.5 \\ .5 \\ 0 \end{bmatrix} \\ &= .125 \end{aligned}$$

■■■■

In the last section we studied optimal strategies associated with strictly determined games and found them to be precisely the maximin and minimax pure strategies adopted by the row and column players. We now look at optimal mixed strategies associated with matrix games that are not strictly determined. In particular, we state, without proof, the optimal mixed strategies to be adopted by the players in a 2×2 matrix game.

As we saw earlier, a player in a nonstrictly determined game should adopt a mixed strategy since a pure strategy will soon be detected by the opponent, who may then use this knowledge to his advantage in devising a counterstrategy. Since there are infinitely many mixed strategies for each player in such a game, the question arises as to how an optimal mixed strategy may be discovered for each player. Recall that an optimal mixed strategy for a player is one in which the row player seeks to maximize his expected payoff and the column player simultaneously seeks to minimize it.

More precisely, the optimal mixed strategy for the row player is arrived at using the following argument: The row player anticipates that any mixed strategy he adopts will be met by a counterstrategy by the column player that will minimize the row player's payoff. Consequently, the row player adopts the mixed strategy for which the expected payoff to the row player (when the column player uses his best counterstrategy) is maximized.

Similarly, the optimal mixed strategy for the column player is arrived at using the following argument: The column player anticipates that the row player will choose a counterstrategy that will maximize the row player's payoff regardless of whatever mixed strategy he (the column player) chooses. Consequently, the column player adopts the mixed strategy for which the expected payoff to the row player (who will use his best counterstrategy) is minimized.

Without going into unnecessary details, let's note that the problem of finding the optimal mixed strategies for the players in a nonstrictly determined game is equivalent to one of solving the related linear programming problem. However, for a 2×2 nonstrictly determined game, the optimal mixed strategies for the players may be found by employing the formulas contained in the following result, which we state without proof.

Optimal Strategies for Nonstrictly Determined Games

Let

$$\begin{bmatrix} a & b \\ c & d \end{bmatrix}$$

be the payoff matrix for a nonstrictly determined game. Then, the **optimal mixed strategy for the row player** is given by

$$P = [p_1 \quad p_2] \tag{2a}$$

where

$$p_1 = \frac{d - c}{a + d - b - c} \quad \text{and} \quad p_2 = 1 - p_1$$

and the **optimal mixed strategy for the column player** is given by

$$Q = \begin{bmatrix} q_1 \\ q_2 \end{bmatrix} \tag{2b}$$

where

$$q_1 = \frac{d - b}{a + d - b - c} \quad \text{and} \quad q_2 = 1 - q_1$$

Furthermore, the **value of the game** is given by the expected value of the game $E = PAQ$, where P and Q are the optimal mixed strategies for the row and column players, respectively. Thus,

$$\begin{aligned} E &= PAQ \\ &= \frac{ad - bc}{a + d - b - c} \end{aligned} \tag{2c}$$

The next example illustrates the use of these formulas in finding the optimal mixed strategies and in finding the value of a 2×2 (nonstrictly determined) game.

EXAMPLE 3 Consider the coin-matching game played by Richie and Chuck with the payoff matrix

$$A = \begin{bmatrix} 3 & -2 \\ -2 & 1 \end{bmatrix} \qquad \text{(See Example 1.)}$$

a. Find the optimal mixed strategies for both Richie and Chuck.
b. Find the value of the game. Does it favor one player over the other?

SOLUTION ✓

a. The game under consideration has no saddle point and is accordingly nonstrictly determined. Using Formula (2a) with $a = 3$, $b = -2$, $c = -2$, and $d = 1$, we find that

$$p_1 = \frac{d - c}{a + d - b - c} = \frac{1 - (-2)}{3 + 1 - (-2) - (-2)} = \frac{3}{8}$$

$$p_2 = 1 - p_1 = 1 - \frac{3}{8} = \frac{5}{8}$$

so Richie's optimal mixed strategy is given by

$$P = [p_1 \quad p_2] = [\tfrac{3}{8} \quad \tfrac{5}{8}]$$

Using (2b), we find that

$$q_1 = \frac{d - b}{a + d - b - c} = \frac{1 - (-2)}{3 + 1 - (-2) - (-2)} = \frac{3}{8}$$

$$q_2 = 1 - q_1 = 1 - \frac{3}{8} = \frac{5}{8}$$

giving Chuck's optimal mixed strategy as

$$Q = \begin{bmatrix} \frac{3}{8} \\ \frac{5}{8} \end{bmatrix}$$

b. The value of the game may be found by computing the matrix product PAQ, where P and Q are the vectors found in part (a). Equivalently, using (2c) we find that

$$E = \frac{ad - bc}{a + d - b - c} = \frac{(3)(1) - (-2)(-2)}{3 + 1 - (-2) - (-2)} = -\frac{1}{8}$$

Since the value of the game is negative, we conclude that the coin-matching game with the particular given payoff matrix favors Chuck (the column player) over Richie. Over the long run, in repeated plays of the game, where each player uses his optimal strategy, Chuck is expected to win 1/8, or 12.5 cents, on the average per play. ■■■■

EXAMPLE 4 As part of their investment strategy, the Carringtons have earmarked $40,000 for short-term investments in the stock market and the money market. The performance of the investments depends on the prime rate (that is, the interest rate that banks charge their best customers). An increase in the prime rate generally favors their investment in the money market, whereas a decrease in the prime rate generally favors their investment in the stock market. Suppose the following payoff matrix gives the expected percentage increase or decrease in the value of each investment for each state of the prime rate:

	Prime rate Up	Prime rate Down
Money-market investment	15	10
Stock market investment	−5	25

a. Determine the optimal investment strategy for the Carringtons' short-term investment of $40,000.

b. What short-term profit can the Carringtons expect to make on their investments?

SOLUTION **a.** We treat the problem as a matrix game in which the Carringtons are the row player. Letting $p = [p_1 \quad p_2]$ denote their optimal strategy, we find

$$p_1 = \frac{d - c}{a + d - b - c} = \frac{25 - (-5)}{15 + 25 - 10 - (-5)} = \frac{30}{35} = \frac{6}{7}$$

$$p_2 = 1 - p_1 = 1 - \frac{6}{7} = \frac{1}{7}$$

Thus, the Carringtons should put (6/7)(40,000), or approximately $34,300, into the money market and (1/7)($40,000), or approximately $5700, into the stock market.

b. The expected value of the game is given by

$$\begin{aligned} E &= \frac{ad - bc}{a + d - b - c} \\ &= \frac{(15)(25) - (10)(-5)}{15 + 25 - 10 - (-5)} = \frac{425}{35} \\ &\approx 12.14 \end{aligned}$$

Thus, the Carringtons can expect to make a short-term profit of 12.14% on their total investment of $40,000—that is, a profit of (0.1214)(40,000), or $4856.

Group Discussion

A two-person, zero-sum game is defined by the payoff matrix

$$A = \begin{bmatrix} x & 1-x \\ 1-x & x \end{bmatrix}$$

1. For what value(s) of x is the game strictly determined? For what value(s) of x is the game not strictly determined?
2. What is the value of the game?

SELF-CHECK EXERCISES 9.5

1. The payoff matrix for a game is given by

$$A = \begin{bmatrix} 2 & 3 & -1 \\ -3 & 2 & -2 \\ 3 & -2 & 2 \end{bmatrix}$$

a. Find the expected payoff to the row player if the row player R uses the maximin pure strategy and the column player C uses the minimax pure strategy.
b. Assuming that the prime rate follows a pattern described by a player using his optimal mixed strategy, find the expected payoff to the row player if R uses the maximin strategy 40% of the time and chooses each of the other two rows 30% of the time, while C uses the minimax strategy 50% of the time and chooses each of the other two columns 25% of the time.
c. Which pair of strategies favors the row player?

2. A farmer has allocated 2000 acres of his farm for planting two crops. Crop A is more susceptible to frost than crop B. If there is no frost in the growing season, then he can expect to make \$40/acre from crop A and \$25/acre from crop B. If there is mild frost, the expected profits are \$20/acre from crop A and \$30/acre from crop B. How many acres of each crop should the farmer cultivate in order to maximize his profits? What profit could he expect to make using this optimal strategy?

Solutions to Self-Check Exercises 9.5 can be found on page 592.

9.5 Exercises

In Exercises 1–6, find the expected payoff E of each of the following games whose payoff matrix and strategies P and Q [for the row and column players, respectively] are given.

1. $\begin{bmatrix} 3 & 1 \\ -4 & 2 \end{bmatrix}, P = [\frac{1}{2} \quad \frac{1}{2}], Q = \begin{bmatrix} \frac{3}{5} \\ \frac{2}{5} \end{bmatrix}$

2. $\begin{bmatrix} -1 & 4 \\ 3 & -2 \end{bmatrix}, P = [.8 \quad .2], Q = \begin{bmatrix} .6 \\ .4 \end{bmatrix}$

3. $\begin{bmatrix} -4 & 3 \\ 2 & 1 \end{bmatrix}, P = [\frac{1}{3} \quad \frac{2}{3}], Q = \begin{bmatrix} \frac{3}{4} \\ \frac{1}{4} \end{bmatrix}$

4. $\begin{bmatrix} 1 & 2 \\ -3 & 1 \end{bmatrix}, P = [\frac{3}{5} \quad \frac{2}{5}], Q = \begin{bmatrix} \frac{1}{3} \\ \frac{2}{3} \end{bmatrix}$

5. $\begin{bmatrix} 2 & 0 & -2 \\ 1 & -1 & 3 \\ 2 & 1 & -4 \end{bmatrix}, P = [.2 \quad .6 \quad .2], Q = \begin{bmatrix} .2 \\ .6 \\ .2 \end{bmatrix}$

6. $\begin{bmatrix} 1 & -4 & 2 \\ 2 & 1 & -1 \\ 2 & -2 & 0 \end{bmatrix}, P = [.2 \quad .3 \quad .5], Q = \begin{bmatrix} .6 \\ .2 \\ .2 \end{bmatrix}$

7. The payoff matrix for a game is given by

$$\begin{bmatrix} 1 & -2 \\ -2 & 3 \end{bmatrix}$$

Compute the expected payoffs of the game for the pairs of strategies given in parts (a–d) Which of these strategies is most advantageous to R?

a. $P = [1 \quad 0], Q = \begin{bmatrix} 1 \\ 0 \end{bmatrix}$

b. $P = [0 \quad 1], Q = \begin{bmatrix} 1 \\ 0 \end{bmatrix}$

c. $P = [\frac{1}{2} \quad \frac{1}{2}], Q = \begin{bmatrix} \frac{1}{2} \\ \frac{1}{2} \end{bmatrix}$

d. $P = [.5 \quad .5], Q = \begin{bmatrix} .8 \\ .2 \end{bmatrix}$

8. The payoff matrix for a game is given by

$$\begin{bmatrix} 3 & 1 & 1 \\ 0 & 2 & 0 \\ -1 & 0 & 2 \end{bmatrix}$$

Compute the expected payoffs of the game for the pairs of strategies given in parts (a–d). Which of these strategies is most advantageous to R?

a. $P = [\frac{1}{3} \quad \frac{1}{3} \quad \frac{1}{3}], Q = \begin{bmatrix} \frac{1}{3} \\ \frac{1}{3} \\ \frac{1}{3} \end{bmatrix}$

b. $P = [\frac{1}{4} \quad \frac{1}{2} \quad \frac{1}{4}], Q = \begin{bmatrix} \frac{1}{8} \\ \frac{3}{8} \\ \frac{1}{2} \end{bmatrix}$

c. $P = [.4 \quad .3 \quad .3], Q = \begin{bmatrix} .6 \\ .2 \\ .2 \end{bmatrix}$

d. $P = [.1 \quad .5 \quad .4], Q = \begin{bmatrix} .3 \\ .3 \\ .4 \end{bmatrix}$

9. The payoff matrix for a game is given by

$$\begin{bmatrix} -3 & 3 & 2 \\ -3 & 1 & 1 \\ 1 & -2 & 1 \end{bmatrix}$$

a. Find the expected payoff to the row player if the row player R uses the maximin pure strategy and the column player C uses the minimax pure strategy.
b. Find the expected payoff to the row player if R uses the maximin strategy 50% of the time and chooses each of the other two rows 25% of the time, while C uses the minimax strategy 60% of the time and chooses each of the other columns 20% of the time.
c. Which of these strategies favors the row player?

10. The payoff matrix for a game is given by

$$\begin{bmatrix} 4 & -3 & 3 \\ -4 & 2 & 1 \\ 3 & -5 & 2 \end{bmatrix}$$

a. Find the expected payoff to the row player if the row player R uses the maximin pure strategy and the column player C uses the minimax pure strategy.
b. Find the expected payoff to the row player if R uses the maximin strategy 40% of the time and chooses each of the other two rows 30% of the time, while C uses the minimax strategy 50% of the time and chooses each of the other columns 25% of the time.
c. Which of these strategies favors the row player?

In Exercises 11–16, find the optimal strategies, P and Q, for the row and column players, respectively. Also compute the expected payoff E of each matrix game and determine which player it favors, if any, if the row and column players use their optimal strategies.

11. $\begin{bmatrix} 4 & 1 \\ 2 & 3 \end{bmatrix}$

12. $\begin{bmatrix} 2 & 5 \\ 3 & -6 \end{bmatrix}$

13. $\begin{bmatrix} -1 & 2 \\ 1 & -3 \end{bmatrix}$

14. $\begin{bmatrix} -1 & 3 \\ 2 & 0 \end{bmatrix}$

15. $\begin{bmatrix} -2 & -6 \\ -8 & -4 \end{bmatrix}$

16. $\begin{bmatrix} 2 & 5 \\ -2 & 4 \end{bmatrix}$

17. Consider the coin-matching game played by Richie and Chuck (see Examples 1 and 3) with the payoff matrix

$$A = \begin{bmatrix} 4 & -2 \\ -2 & 1 \end{bmatrix}$$

a. Find the optimal strategies for Richie and Chuck.
b. Find the value of the game. Does it favor one player over the other?

18. INVESTMENT STRATEGIES As part of their investment strategy, the Carringtons have decided to put $100,000 into stock market investments and also into purchasing precious metals. The performance of the investments depends on the state of the economy in the next year. In an expanding economy it is expected that their stock market investment will outperform their investment in precious metals, whereas an economic recession will have precisely the opposite effect. Suppose the following payoff matrix gives the expected percentage increase or decrease in the value of each investment for each state of the economy:

	Expanding economy	Economic recession
Stock market investment	20	−5
Commodity investment	10	15

a. Determine the optimal investment strategy for the Carringtons' investment of $100,000.
b. What profit can the Carringtons expect to make on their investments over the year if they use their optimal investment strategy?

19. INVESTMENT STRATEGIES The Maxwells have decided to invest $40,000 in the common stocks of two companies listed on the New York Stock Exchange. One of the companies derives its revenue mainly from its worldwide operation of a chain of hotels, whereas the other company is a major brewery in the country. It is expected that if the economy is in a state of growth, then the hotel stock should outperform the brewery stock; however, the brewery stock is expected to hold its own better than the hotel stock in a recessionary period. Suppose the following payoff matrix gives the expected percentage increase or decrease in the value of each investment for each state of the economy:

	Expanding economy	Economic recession
Investment in hotel stock	25	−5
Investment in brewery stock	10	15

a. Determine the optimal investment strategy for the Maxwells' investment of $40,000.
b. What profit can the Maxwells expect to make on their investments if they use their optimal investment strategy?

20. CAMPAIGN STRATEGIES Bella Robinson and Steve Carson are running for a seat in the U.S. Senate. If both candidates campaign only in the major cities of the state, then Robinson is expected to get 60% of the votes; if both candidates campaign only in the rural areas, then Robinson is expected to get 55% of the votes; if Robinson campaigns exclusively in the city and Carson campaigns exclusively in the rural areas, then Robinson is expected to get 40% of the votes; finally, if Robinson campaigns exclusively in the rural areas and Carson campaigns exclusively in the city, then Robinson is expected to get 45% of the votes.
a. Construct the payoff matrix for the game and show that it is not strictly determined.
b. Find the optimal strategy for both Robinson and Carson.

21. ADVERTISEMENTS Two dentists, Lydia Russell and Jerry Carlton, are planning to establish practices in a newly developed community. Both have allocated approximately the same total budget for advertising in the local newspaper and for the distribution of fliers announcing their practices. Because of the location of their offices, Russell is expected to get 48% of the business if both dentists advertise only in the local newspaper; if both dentists advertise through fliers, then Russell is expected to get 45% of the business; if Russell advertises exclusively in the local newspaper and Carlton advertises exclusively through fliers, then Russell is expected to get 65% of the business. Finally, if Russell advertises through fliers exclusively and Carlton advertises exclusively in the local newspaper, then Russell is expected to get 50% of the business.
a. Construct the payoff matrix for the game and show that it is not strictly determined.
b. Find the optimal strategy for both Russell and Carlton.

22. Let

$$\begin{bmatrix} a_{11} & a_{12} \\ a_{21} & a_{22} \end{bmatrix}$$

be the payoff matrix associated with a 2×2 matrix game. Assume that either the row player uses the optimal mixed strategy $P = [p_1 \quad p_2]$, where

$$p_1 = \frac{d-c}{a+d-b-c} \quad \text{and} \quad p_2 = 1 - p_1$$

or the column player uses the optimal mixed strategy

$$Q = \begin{bmatrix} q_1 \\ q_2 \end{bmatrix}$$

where

$$q_1 = \frac{d-b}{a+d-b-c} \quad \text{and} \quad q_2 = 1 - q_1$$

Show by direct computation that the expected value of the game is given by $E = PAQ$.

23. Let

$$\begin{bmatrix} a & b \\ c & d \end{bmatrix}$$

be the payoff matrix associated with a nonstrictly determined 2×2 matrix game. Prove that the expected payoff of the game is given by

$$E = \frac{ad - bc}{a+d-b-c}$$

Hint: Compute $E = PAQ$, where P and Q are the optimal strategies for the row and column players, respectively.

SOLUTIONS TO SELF-CHECK EXERCISES 9.5

1. a. From the following calculations

				Row minima	
	2	3	−1	(−1)	← Largest of the row minima
	−3	2	−2	−3	
	3	−2	2	−2	
Column maxima	3	3	(2) ↑ Smallest of the column maxima		

we see that R's optimal pure strategy is to choose row 1, whereas C's optimal pure strategy is to choose column 3. Furthermore, if both players use their respective optimal strategies, then the expected payoff to R is -1 unit.

b. R's mixed strategy may be represented by the row vector

$$P = [.4 \quad .3 \quad .3]$$

and C's mixed strategy may be represented by the column vector

$$Q = \begin{bmatrix} .25 \\ .25 \\ .50 \end{bmatrix}$$

The expected payoff to the row player will then be given by

$$\begin{aligned} E = PAQ &= [.4 \quad .3 \quad .3] \begin{bmatrix} 2 & 3 & -1 \\ -3 & 2 & -2 \\ 3 & -2 & 2 \end{bmatrix} \begin{bmatrix} .25 \\ .25 \\ .50 \end{bmatrix} \\ &= [.4 \quad .3 \quad .3] \begin{bmatrix} .75 \\ -1.25 \\ 1.25 \end{bmatrix} \\ &= .3 \end{aligned}$$

c. From the results of parts (a) and (b), we see that the mixed strategies of part (b) will be better for R.

2. We may view this problem as a matrix game with the farmer as the row player and the weather as the column player. The payoff matrix for the game is

$$\begin{array}{l} \\ \\ \text{Crop A} \\ \text{Crop B} \end{array} \begin{array}{c} \begin{array}{cc} \text{No} & \text{Mild} \\ \text{frost} & \text{frost} \end{array} \\ \begin{bmatrix} 40 & 20 \\ 25 & 30 \end{bmatrix} \end{array}$$

The game under consideration has no saddle point and is accordingly nonstrictly determined. Letting $p = [p_1 \;\; p_2]$ denote the farmer's optimal strategy and using the formula for determining the optimal mixed strategies for a 2×2 game with $a = 40$, $b = 20$, $c = 25$, and $d = 30$, we find

$$p_1 = \frac{d - c}{a + d - b - c} = \frac{30 - 25}{40 + 30 - 20 - 25} = \frac{5}{25} = \frac{1}{5}$$

$$p_2 = 1 - p_1 = 1 - \frac{1}{5} = \frac{4}{5}$$

Therefore, the farmer should cultivate (1/5)(2000), or 400, acres of crop A and 1600 acres of crop B. By using his optimal strategy, the farmer can expect to realize a profit of

$$\begin{aligned} E &= \frac{ad - bc}{a + d - b - c} \\ &= \frac{(40)(30) - (20)(25)}{40 + 30 - 20 - 25} \\ &= 28 \end{aligned}$$

or \$28/acre—that is, a total profit of (28)(2000), or \$56,000.

CHAPTER 9 Summary of Principal Formulas and Terms

Formulas

1. Steady-state matrix for an absorbing stochastic matrix

If $A = \left[\begin{array}{c|c} I & S \\ \hline O & R \end{array}\right]$

then the steady-state matrix of A is

$$\left[\begin{array}{c|c} I & S(I - R)^{-1} \\ \hline O & O \end{array}\right]$$

2. Expected value of a game

$$E = PAQ = [p_1 p_2 \cdots p_m] \begin{bmatrix} a_{11} & a_{12} & \cdots & a_{1n} \\ a_{21} & a_{22} & \cdots & a_{2n} \\ \vdots & \vdots & & \vdots \\ a_{m1} & a_{m2} & \cdots & a_{mn} \end{bmatrix} \begin{bmatrix} q_1 \\ q_2 \\ \vdots \\ q_n \end{bmatrix}$$

3. Optimal strategy for a nonstrictly determined game

$P = [p_1 \quad p_2]$,

where $p_1 = \dfrac{d - c}{a + d - b - c}$

and $p_2 = 1 - p_1$

and $Q = \begin{bmatrix} q_1 \\ q_2 \end{bmatrix}$

where $q_1 = \dfrac{d - b}{a + d - b - c}$

and $q_2 = 1 - q_1$

The expected value of the game is

$$E = PAQ = \frac{ad - bc}{a + d - b - c}$$

Terms

Markov chain (process)
transition matrix
stochastic matrix
steady-state (limiting) distribution vector
steady-state matrix
regular Markov chain
absorbing state
absorbing stochastic matrix
absorbing Markov chain
zero-sum game
maximin strategy
minimax strategy
optimal strategy
strictly determined game
saddle point
value of a game
fair game
pure strategy
mixed strategy
expected value of a game

CHAPTER 9 REVIEW EXERCISES

In Exercises 1–4, determine which of the following are regular stochastic matrices.

1. $\begin{bmatrix} 1 & -2 \\ 0 & -8 \end{bmatrix}$

2. $\begin{bmatrix} .3 & 1 \\ .7 & 0 \end{bmatrix}$

3. $\begin{bmatrix} \frac{1}{2} & 0 & \frac{1}{3} \\ 0 & 0 & \frac{1}{3} \\ \frac{1}{2} & 1 & \frac{1}{3} \end{bmatrix}$

4. $\begin{bmatrix} .3 & 0 & .5 \\ .2 & 1 & 0 \\ .1 & 0 & .5 \end{bmatrix}$

In Exercises 5 and 6, find X_2 [the probability distribution of the system after two observations] for the given distribution vector X_0 and the given transition matrix T.

5. $X_0 = \begin{bmatrix} \frac{1}{2} \\ \frac{1}{2} \\ 0 \end{bmatrix}, T = \begin{bmatrix} 0 & \frac{1}{4} & \frac{3}{5} \\ \frac{2}{5} & \frac{1}{2} & \frac{1}{5} \\ \frac{3}{5} & \frac{1}{4} & \frac{1}{5} \end{bmatrix}$

6. $X_0 = \begin{bmatrix} .35 \\ .25 \\ .40 \end{bmatrix}, T = \begin{bmatrix} .2 & .1 & .3 \\ .5 & .4 & .4 \\ .3 & .5 & .3 \end{bmatrix}$

In Exercises 7–10, determine whether the given matrix is an absorbing stochastic matrix.

7. $\begin{bmatrix} 1 & .6 & .1 \\ 0 & .2 & .6 \\ 0 & .2 & .3 \end{bmatrix}$

8. $\begin{bmatrix} .3 & .2 & .1 \\ .7 & .5 & .3 \\ 0 & .3 & .6 \end{bmatrix}$

9. $\begin{bmatrix} .32 & .22 & .44 \\ .68 & .78 & .56 \\ 0 & 0 & 0 \end{bmatrix}$

10. $\begin{bmatrix} .31 & .35 & 0 \\ .32 & .40 & 0 \\ .37 & .25 & 1 \end{bmatrix}$

In Exercises 11–14, find the steady-state matrix for the given transition matrix.

11. $\begin{bmatrix} .6 & .3 \\ .4 & .7 \end{bmatrix}$

12. $\begin{bmatrix} .5 & .4 \\ .5 & .6 \end{bmatrix}$

13. $\begin{bmatrix} .6 & .4 & .3 \\ .2 & .2 & .2 \\ .2 & .4 & .5 \end{bmatrix}$

14. $\begin{bmatrix} .1 & .2 & .6 \\ .3 & .4 & .2 \\ .6 & .4 & .2 \end{bmatrix}$

15. A study conducted by the State Department of Agriculture in a Sun-Belt state reveals an increasing trend toward urbanization of the farmland within the state. Ten years ago, 50% of the land within the state was used for agricultural purposes (A), 15% had been urbanized (U), and the remaining 35% was neither agricultural nor urban (N). Since that time, 10% of the agricultural land has been converted to urban land, 5% has been used for other purposes, and the remaining 85% is still agricultural. Of the urban land, 95% has remained urban, whereas 5% of it has been used for nonagricultural purposes. Of the land that was neither agricultural nor urban, 10% has been converted to agricultural land, 5% has been urbanized, and the remaining 85% remains unchanged.
    a. Construct the transition matrix for the Markov chain that describes the shift in land use within the state.
    b. Find the probability vector describing the distribution of land within the state 10 yr ago.
    c. Assuming that this trend continues, find the probability vector describing the distribution of land within the state 10 yr from now.

16. *Auto Trend* magazine conducted a survey among automobile owners in a certain area of the country to determine what type of car they now own and what type of car they expect to own 4 yr from now. For purposes of classification, automobiles mentioned in the survey were placed into three categories: large, intermediate, and small. Results of the survey follow:

		Present car		
		Large	Intermediate	Small
Future car	Large	.3	.1	.1
	Intermediate	.3	.5	.2
	Small	.4	.4	.7

Assuming that these results indicate the long-term buying trend of car owners within the area, what will be the distribution of cars (relative to size) in this area over the long run?

In Exercises 17–20, determine whether each of the games within the given payoff matrix is strictly determined. If so, give the optimal pure strategies for the row player and the column player and also give the value of the game.

17. $\begin{bmatrix} 1 & 2 \\ 3 & 5 \\ 4 & 6 \end{bmatrix}$

18. $\begin{bmatrix} 1 & 0 & 3 \\ 2 & -1 & -2 \end{bmatrix}$

19. $\begin{bmatrix} 1 & 3 & 6 \\ -2 & 4 & 3 \\ -5 & -4 & -2 \end{bmatrix}$

20. $\begin{bmatrix} 4 & 3 & 2 \\ -6 & 3 & -1 \\ 2 & 3 & 4 \end{bmatrix}$

In Exercises 21–24, find the expected payoff *E* of each of the following games whose payoff matrix and strategies *P* and *Q* [for the row and column players, respectively] are given.

21. $\begin{bmatrix} 4 & 8 \\ 6 & -12 \end{bmatrix}, P = [\frac{1}{2} \;\; \frac{1}{2}], Q = \begin{bmatrix} \frac{1}{4} \\ \frac{3}{4} \end{bmatrix}$

22. $\begin{bmatrix} 3 & 0 & -3 \\ 2 & 1 & 2 \end{bmatrix}, P = [\frac{1}{3} \;\; \frac{2}{3}], Q = \begin{bmatrix} \frac{1}{3} \\ \frac{1}{3} \\ \frac{1}{3} \end{bmatrix}$

23. $\begin{bmatrix} 3 & -1 & 2 \\ 1 & 2 & 4 \\ -2 & 3 & 6 \end{bmatrix}, P = [.2 \;\; .4 \;\; .4], Q = \begin{bmatrix} .2 \\ .6 \\ .2 \end{bmatrix}$

24. $\begin{bmatrix} 2 & -2 & 3 \\ 1 & 2 & -1 \\ -1 & 2 & 3 \end{bmatrix}, P = [.2 \;\; .4 \;\; .4], Q = \begin{bmatrix} .3 \\ .3 \\ .4 \end{bmatrix}$

In Exercises 25–28, find the optimal strategies, *P* and *Q*, for the row player and the column player, respectively. Also compute the expected payoff *E* of each matrix game if the row and column players adopt their optimal strategies and determine which player it favors, if any.

25. $\begin{bmatrix} 1 & -2 \\ 0 & 3 \end{bmatrix}$ **26.** $\begin{bmatrix} 4 & -7 \\ -5 & 6 \end{bmatrix}$

27. $\begin{bmatrix} 3 & -6 \\ 1 & 2 \end{bmatrix}$ **28.** $\begin{bmatrix} 12 & 10 \\ 6 & 14 \end{bmatrix}$

29. Two competing music stores, Disco-Mart and Stereo World, each have the option of selling a certain popular compact disc label at a price of either \$7 or \$8/disc. If both sell the label at the same price, they are each expected to get 50% of the business. If Disco-Mart sells the label at \$7/disc and Stereo World sells the label at \$8/disc, Disco-Mart is expected to get 70% of the business; if Disco-Mart sells the label at \$8/disc and Stereo World sells the label at \$7/disc, Disco-Mart is expected to get 40% of the business.

a. Represent this information in the form of a payoff matrix.

b. Determine the optimal price that each company should sell the compact disc label for to ensure that it captures the largest possible expected market share.

30. The management of a division of the National Motor Corporation that produces compact and subcompact cars has estimated that the quantity demanded of their compact models is 1500 units/week if the price of oil increases at a higher than normal rate, whereas the quantity demanded of their subcompact models is 2500 units/week under similar conditions. However, the quantity demanded of their compact models and subcompact models is 3000 units and 2000 units/week, respectively, if the price of oil increases at a normal rate. Determine the percentages of compact and subcompact cars the division should plan to manufacture to maximize the expected number of cars demanded per week.

A INTRODUCTION TO LOGIC

One of the earliest records of man's efforts to understand the reasoning process can be found in Aristotle's work *Organon* (third century B.C.), in which he presented his ideas on logical arguments. Up until the present day, Aristotelian logic has been the basis for traditional logic, the systematic study of valid inferences.

The field of mathematics known as **symbolic logic,** in which symbols are used to replace ordinary language, had its beginnings in the eighteenth and nineteenth centuries with the works of the German mathematician Gottfried Wilhelm Leibniz (1646–1716) and the English mathematician George Boole (1815–1864). Boole introduced algebraic-type operations to the field of logic, thereby providing us with a systematic method of combining statements. Boolean algebra is the basis of modern-day computer technology, as well as being central to the study of pure mathematics.

In this appendix we introduce the fundamental concepts of symbolic logic. Beginning with the definitions of statements and their truth values, we proceed to a discussion of the combination of statements and valid arguments. In Section A.6 we present an application of symbolic logic that is widely used in computer technology—switching networks.

A.1 Propositions and Connectives

We use deductive reasoning in many of the things we do. Whether in a formal debate or in an expository article, we use language to express our thoughts. In English, sentences are used to express assertions, questions, commands, and wishes. In our study of logic we will be concerned with only one type of sentence, the declarative sentence.

Proposition

A **proposition,** or *statement,* is a declarative sentence that can be classified as either true or false, but not both.

Commands, requests, questions, or exclamations are examples of sentences that are not propositions.

EXAMPLE 1 Which of the following are propositions?

a. Toronto is the capital of Ontario.
b. Close the door!
c. There are 100 trillion connections among the neurons in the human brain.
d. Who is the chief justice of the Supreme Court?
e. The new television sit-com is a successful show.

f. The 2000 Summer Olympic Games were held in Montreal.
g. $x + 3 = 8$
h. How wonderful!
i. Seven is either an odd number or it is even.

SOLUTION ✔ Statements (a), (c), (e), (f), and (i) are propositions. Statement (c) would be difficult to verify, but the validity of the statement can at least theoretically be determined. Statement (e) is a proposition if we assume that the meaning of the term *successful* has been defined. For example, one might use the Nielsen ratings to determine the success of a new show. Statement (f) is a proposition that is false.

Statements (b), (d), (g), and (h) are *not* propositions. Statement (b) is a command, (d) is a question, and (h) is an exclamation. Statement (g) is an open sentence that cannot be classified as true or false. For example, if $x = 5$, then $5 + 3 = 8$ and the sentence is true. On the other hand, if $x = 4$, then $4 + 3 \neq 8$ and the sentence is false. ■■■■

Having considered what is meant by a proposition, we now discuss the ways in which propositions may be combined. For example, the two propositions

a. Toronto is the capital of Ontario

and

b. Toronto is the largest city in Canada

may be joined to form the proposition

c. Toronto is the largest city in Canada and is the capital of Ontario.

Propositions (a) and (b) are called **prime,** or *simple,* **propositions** because they are simple statements expressing a single complete thought. In the ensuing discussion, we use the lowercase letters $p, q, r,$ and so on to denote prime propositions.

Propositions that are combinations of two or more propositions, such as proposition (c), are called **compound propositions.** The words used to combine propositions are called **logical connectives.** The connectives that we will consider are given in Table A.1 along with their symbols. We discuss the first three in this section and the remaining two in Section A.3.

Table A.1

Name	Logical Connective	Symbol
Conjunction	and	$\wedge$
Negation	not	$\sim$
Disjunction	or	$\vee$
Conditional	if · · · then	$\rightarrow$
Biconditional	if and only if	$\leftrightarrow$

Conjunction

A **conjunction** is a statement of the form "p and q" and is represented symbolically by

$$p \wedge q$$

The conjunction $p \wedge q$ is true if *both* p and q are true; it is false otherwise.

We have already encountered a conjunction in an earlier example. The two propositions

p: Toronto is the capital of Ontario
q: Toronto is the largest city in Canada

were combined to form the conjunction

$p \wedge q$: Toronto is the capital of Ontario and is the largest city in Canada.

Disjunction

A **disjunction** is a proposition of the form "p or q" and is represented symbolically by

$$p \vee q$$

The disjunction $p \vee q$ is false if *both* statements p and q are false; it is true in all other cases.

The use of the word *or* in this definition is meant to convey the meaning "one or the other, or both." Since this disjunction is true when *both* p and q are true, as well as when either p or q is true, it is sometimes referred to as an **inclusive disjunction.**

EXAMPLE 2 Consider the propositions

p: Dorm residents can purchase meal plans.
q: Dorm residents can purchase à la carte cards.

The disjunction is

$p \vee q$: Dorm residents can purchase meal plans or à la carte cards. ■■■■

The disjunction in Example 2 is true when dorm residents can purchase either meal plans or à la carte cards or both meal plans and à la carte cards.

Exclusive Disjunction

An **exclusive disjunction** is a proposition of the form "p or q" and is denoted by

$$p \veebar q$$

The disjunction $p \veebar q$ is true if *either* p or q is true.

In contrast to an inclusive disjunction, an exclusive disjunction is *false* when both p and q are true. In Example 2 the exclusive disjunction would not be true if dorm residents could buy both meal plans and à la carte cards. The difference between exclusive and inclusive disjunction is further illustrated in the next example.

EXAMPLE 3 Consider the propositions

p: The base price of each condominium unit includes a private deck.

q: The base price of each condominium unit includes a private patio.

Find the exclusive disjunction $p \veebar q$.

SOLUTION ✓ The exclusive disjunction is

$p \veebar q$: The base price of each condominium unit includes either a private deck or a patio.

Observe that this statement is not true when both p and q are true. In other words, the base price of each condominium unit does not include both a private deck and a patio. ■■■■

In our everyday use of language, the meaning of the word *or* is not always clear. In legal documents the two cases are distinguished by the words *and/or* (inclusive) and *either/or* (exclusive). In mathematics we use the word *or* in the inclusive sense, unless otherwise specified.

Negation

A **negation** is a proposition of the form "not p" and is represented symbolically by

$$\sim p$$

The proposition $\sim p$ is true if p is false and vice versa.

EXAMPLE 4 Form the negation of the proposition

p: Prices rose on the New York Stock Exchange today.

SOLUTION ✓ The negation is

$\sim p$: Prices did not rise on the New York Stock Exchange today. ■■■■

EXAMPLE 5 Consider the two propositions

p: The birth rate declined in the United States last year.
q: The population of the United States increased last year.

Write the following statements in symbolic form:

a. Last year the birth rate declined and the population increased in the United States.
b. Either the birth rate declined or the population increased in the United States last year.
c. It is not true that the birth rate declined and the population increased in the United States last year.

The symbolic form of each statement is given by

a. $p \wedge q$ **b.** $p \veebar q$ **c.** $\sim(p \wedge q)$ ■■■■

A.1 Exercises

In Exercises 1–14, determine whether the given statement is a proposition.

1. The defendant was convicted of grand larceny.
2. Who won the 1988 presidential election?
3. The first month of the year is February.
4. The number 2 is odd.
5. $x - 1 \geq 0$
6. Keep off the grass!
7. Coughing may be caused by a lack of water in the air.
8. The first McDonald's fast-food restaurant was opened in California.
9. Exercise protects men and women from sudden heart attacks.
10. If voters do not pass the proposition, then taxes will be increased.
11. Don't swim immediately after eating!
12. What ever happened to Baby Jane?
13. $\dfrac{x^2+1}{x^2+1} = 1$
14. Several major corporations will enter or expand operations into the home-security products market this year.

In Exercises 15–20, identify the logical connective that is used in the given statement.

15. Housing starts in the United States did not increase last month.
16. Mel Bieber's will is valid if and only if he was of sound mind and memory when he made the will.
17. Americans are saving less and spending more this year.
18. If you traveled on your job and weren't reimbursed, then you can deduct these expenses from your taxable income.
19. Both loss of appetite and irritability are symptoms of mental stress.
20. Prices for many imported goods have either stayed flat or dropped slightly this year.

In Exercises 21–26, state the negation of the given proposition.

21. New orders for manufactured goods fell last month.

22. For many men, housecleaning is not a familiar task.

23. Drinking during pregnancy affects the size and weight of babies.

24. Not all patients suffering from influenza lose weight.

25. The commuter airline industry is now undergoing a shakeup.

26. The Dow–Jones industrial average registered its fourth consecutive decline today.

27. Let p and q denote the propositions

p: Domestic car sales increased over the past year.

q: Foreign car sales decreased over the past year.

Express the following compound propositions in words:

a. $p \vee q$ **b.** $p \wedge q$ **c.** $p \veebar q$
d. $\sim p$ **e.** $\sim p \vee q$ **f.** $\sim p \vee \sim q$

28. Let p and q denote the propositions

p: Every employee is required to be fingerprinted.

q: Every employee is required to take an oath of allegiance.

Express the following compound propositions in words:

a. $p \vee q$ **b.** $p \wedge q$ **c.** $\sim p \vee \sim q$
d. $\sim p \wedge \sim q$ **e.** $p \vee \sim q$

29. Let p and q denote the propositions

p: The doctor recommended surgery to treat his hyperthyroidism.

q: The doctor recommended radioactive iodine to treat his hyperthyroidism.

a. State the exclusive disjunction for these propositions in words.
b. State the inclusive disjunction for these propositions in words.

30. Let p and q denote the propositions

p: The investment newsletter recommended buying bond mutual funds.

q: The investment newsletter recommended buying stock mutual funds.

a. State the exclusive disjunction for these propositions in words.
b. State the inclusive disjunction for these propositions in words.

31. Let p and q denote the propositions

p: The SAT verbal scores improved in this school district last year.

q: The SAT math scores improved in this school district last year.

Express each of the following statements symbolically.
a. The SAT verbal scores and the SAT math scores improved in this school district last year.
b. Either the SAT verbal scores or the SAT math scores improved in this school district last year.
c. Neither the SAT verbal scores nor the SAT math scores improved in this school district last year.
d. It is not true that the SAT math scores did not improve in this school district last year.

32. Let p and q denote the propositions

p: Laura purchased a VHS videocassette recorder.

q: Laura did not purchase a DVD player.

Express each of the following statements symbolically.
a. Laura purchased either a VHS or a DVD player.
b. Laura purchased a VHS and a DVD player.
c. It is not true that Laura purchased a DVD player.
d. Laura purchased neither a VHS nor a DVD player.

33. Let p, q, and r denote the propositions

p: The popularity of prime-time soaps increased this year.

q: The popularity of prime-time situation comedies increased this year.

r: The popularity of prime-time detective shows decreased this year.

Express each of the following propositions in words.

a. $\sim p \wedge \sim q$ **b.** $\sim p \vee r$
c. $\sim(\sim r) \vee \sim q$ **d.** $\sim p \veebar \sim q$

A.2 Truth Tables

A primary objective of studying logic is to determine whether a given proposition is true or false. In other words, we wish to determine the **truth value** of a given statement.

Consider, for example, the conjunction $p \wedge q$ formed from the prime propositions p and q. There are four possible truth values for the given conjunction—namely,

1. p is true and q is true.
2. p is true and q is false.
3. p is false and q is true.
4. p is false and q is false.

Table A.2

p	q	$p \wedge q$
T	T	T
T	F	F
F	T	F
F	F	F

Conjunction

By definition the conjunction is true when both p and q are true, and it is false otherwise. We can summarize this information in the form of a truth table, as shown in Table A.2.

In a similar manner we can construct the truth tables for inclusive disjunction, exclusive disjunction, and negation [Tables A.3 (a– c)].

Table A.3

p	q	$p \vee q$
T	T	T
T	F	T
F	T	T
F	F	F

(a) Inclusive disjunction

p	q	$p \veebar q$
T	T	F
T	F	T
F	T	T
F	F	F

(b) Exclusive disjunction

p	$\sim p$
T	F
F	T

(c) Negation

In general, when we are given a compound proposition, we are concerned with the problem of finding every possible combination of truth values associated with the compound proposition. To systematize this procedure, we construct a truth table exhibiting all possible truth values for the given proposition. The next several examples illustrate this method.

EXAMPLE 1 Construct the truth table for the proposition $\sim p \vee q$.

SOLUTION ✓ The truth table is constructed in the following manner.

1. The two prime propositions p and q are placed at the head of the first two columns (Table A.4).
2. The two propositions containing the connectives, $\sim p$ and $\sim p \vee q$, are placed at the head of the next two columns.
3. The possible truth values for p are entered in the column headed by p, and then the possible truth values for q are entered in the column headed by q. Notice that the *possible T values are always exhausted first.*

Table A.4

p	q	$\sim p$	$\sim p \vee q$
T	T	F	T
T	F	F	F
F	T	T	T
F	F	T	T

4. The possible truth values for the negation $\sim p$ are entered in the column headed by $\sim p$, and then the possible truth values for the disjunction $\sim p \vee q$ are entered in the column headed by $\sim p \vee q$. ■■■■

EXAMPLE 2 Construct the truth table for the proposition $\sim p \vee (p \wedge q)$.

SOLUTION ✓ Proceeding as in the previous example, we construct the truth table as shown in Table A.5. Since the given proposition is the disjunction of the propositions $\sim p$ and $(p \wedge q)$, columns are introduced for $\sim p$, $p \wedge q$, and $\sim p \vee (p \wedge q)$.

Table A.5

p	q	$\sim p$	$p \wedge q$	$\sim p \vee (p \wedge q)$
T	T	F	T	T
T	F	F	F	F
F	T	T	F	T
F	F	T	F	T

■■■■

The following is an example involving three prime propositions.

EXAMPLE 3 Construct the truth table for the proposition $(p \vee q) \vee (r \wedge \sim p)$.

SOLUTION ✓ Following the method of the previous examples, we construct the truth table as shown in Table A.6.

Table A.6

p	q	r	$p \vee q$	$\sim p$	$r \wedge \sim p$	$(p \vee q) \vee (r \wedge \sim p)$
T	T	T	T	F	F	T
T	T	F	T	F	F	T
T	F	T	T	F	F	T
T	F	F	T	F	F	T
F	T	T	T	T	T	T
F	T	F	T	T	F	T
F	F	T	F	T	T	T
F	F	F	F	T	F	F

■■■■

Observe that the truth table in Example 3 contains eight rows, whereas the truth tables in Examples 1 and 2 contain four rows. In general, *if a compound proposition $P(p, q, \ldots)$ contains the n prime propositions $p, q, \ldots$, then the corresponding truth table contains 2^n rows.* For example, since the compound proposition in Example 3 contains three prime propositions, its corresponding truth table contains 2^3, or 8, rows.

A.2 Exercises

In Exercises 1–18, construct a truth table for the given compound proposition.

1. $p \vee {\sim}q$

2. ${\sim}p \wedge {\sim}q$

3. ${\sim}({\sim}p)$

4. ${\sim}(p \wedge q)$

5. $p \vee {\sim}p$

6. ${\sim}({\sim}p \vee {\sim}q)$

7. ${\sim}p \wedge (p \vee q)$

8. $(p \vee {\sim}q) \wedge q$

9. $(p \vee q) \wedge (p \wedge {\sim}q)$

10. $(p \vee q) \wedge {\sim}p$

11. $(p \vee q) \wedge {\sim}(p \vee q)$

12. $(p \vee q) \vee ({\sim}p \wedge q)$

13. $(p \vee q) \wedge (p \vee r)$

14. $p \wedge (q \vee r)$

15. $(p \wedge q) \vee {\sim}r$

16. $({\sim}p \vee q) \wedge {\sim}r$

17. $(p \wedge {\sim}q) \vee (p \wedge r)$

18. ${\sim}(p \wedge q) \vee (q \wedge r)$

19. If a compound proposition consists of the prime propositions p, q, r, and s, how many rows does its corresponding truth table contain?

A.3 The Conditional and the Biconditional Connectives

In this section we introduce two other connectives: the conditional and the biconditional. We also discuss three variations of conditional statements: the inverse, the contrapositive, and the converse.

We often use expressions of the form

If it rains, *then* the baseball game will be postponed

to specify the conditions under which a statement will be true. The "if . . . then" statement is the building block on which deductive reasoning is based, and an understanding of its use in forming logical proofs is of fundamental importance.

Conditional Statement

A **conditional statement** is a proposition of the form "if p, then q" and is represented symbolically by

$$p \rightarrow q$$

The connective "if . . . then" is called the **conditional connective**, the proposition p, the **hypothesis**, and the proposition q the **conclusion**. A conditional statement is false if the hypothesis is true and the conclusion is false; it is true in all other cases.

Table A.7

p	q	$p \rightarrow q$
T	T	T
T	F	F
F	T	T
F	F	T

The truth table determined by the conditional $p \rightarrow q$ is shown in Table A.7.

One question that is often asked is: Why is a conditional statement true when its hypothesis is false? We can answer this question by considering the following conditional statement made by a mother to her son.

If you do your homework, then you may watch TV.

Think of the statement consisting of the two prime propositions

p: You do your homework
q: You may watch TV

as a *promise* made by the mother to her son. Four cases arise:

1. The son does his homework, and his mother lets him watch TV.
2. The son does his homework, and his mother does not let him watch TV.
3. The son does not do his homework, and his mother lets him watch TV.
4. The son does not do his homework, and his mother does not let him watch TV.

In case 1, p and q are both true, the promise has been kept and, consequently, $p \rightarrow q$ is true. In case 2, p is true and q is false, and the mother has broken her promise. Therefore, $p \rightarrow q$ is false. In cases 3 and 4, p is not true, and the promise is *not* broken. Thus, $p \rightarrow q$ is regarded as a true statement. In other words, the conditional statement is regarded as false only if the "promise" is broken.

There are several equivalent expressions for the conditional connective "if . . . then." Among them are

1. p implies q
2. p only if q
3. q if p
4. q whenever p
5. Suppose p, then q

Care should be taken not to confuse the conditional "q if p" with the conditional $q \rightarrow p$, because the two statements have quite different meanings. For example, the conditional $p \rightarrow q$ formed from the two propositions

p: There is a fire
q: You call the fire department

is

If there is a fire, then you call the fire department.

The conditional $q \rightarrow p$ is

If you call the fire department, then there is a fire.

Obviously, the two statements have quite different meanings.

We refer to statements that are variations of the conditional $p \to q$ as **logical variants.** We define the three logical variants of the conditional $p \to q$ as follows:

1. The **converse** is a compound statement of the form "if q, then p" and is represented symbolically by

$$q \to p$$

2. The **contrapositive** is a compound statement of the form "if not q, then not p" and is represented symbolically by

$$\sim q \to \sim p$$

3. The **inverse** is a compound statement of the form "if not p, then not q" and is represented symbolically by

$$\sim p \to \sim q$$

EXAMPLE 1 Given the two propositions

p: You vote in the presidential election

q: You are a registered voter

a. State the conditional $p \to q$.
b. State the converse, the contrapositive, and the inverse of $p \to q$.

SOLUTION ✓

a. The conditional is "If you vote in the presidential election, then you are a registered voter."
b. The converse is "If you are a registered voter, then you vote in the presidential election." The contrapositive is "If you are not a registered voter, then you do not vote in the presidential election." The inverse is "If you do not vote in the presidential election, then you are not a registered voter."

■■■■

The truth table for the conditional $p \to q$ and its three logical variants is shown in Table A.8. Notice that the conditional $p \to q$ and its contrapositive $\sim q \to \sim p$ have identical truth tables. In other words, the conditional statement and its contrapositive have the same meaning.

Table A.8

p	q	Conditional $p \to q$	Converse $q \to p$	$\sim p$	$\sim q$	Contrapositive $\sim q \to \sim p$	Inverse $\sim p \to \sim q$
T	T	T	T	F	F	T	T
T	F	F	T	F	T	F	T
F	T	T	F	T	F	T	F
F	F	T	T	T	T	T	T

Logical Equivalence

Two propositions p and q are **logically equivalent**, denoted by

$$p \Leftrightarrow q$$

if they have identical truth tables.

Referring once again to the truth table in Table A.8, we see that $p \to q$ is logically equivalent to its contrapositive. Similarly, the converse of a conditional statement is logically equivalent to the inverse.

It is sometimes easier to prove the contrapositive of a conditional statement than it is to prove the conditional itself. We may use this to our advantage in establishing a proof, as shown in the next example.

EXAMPLE 2 Prove that if n^2 is an odd number, then n is an odd number.

SOLUTION ✓ Let

p: n^2 is odd.

q: n is an odd number.

Since $p \to q$ is logically equivalent to $\sim q \to \sim p$, it suffices to prove $\sim q \to \sim p$. Thus, we wish to prove that if n is not an odd number, then n^2 is an even number. If n is even, then $n = 2k$, where k is an integer. Therefore,

$$n^2 = (2k)(2k) = 2(2k^2)$$

Since $2(2k^2)$ is a multiple of 2, it is an even number and consequently not odd. Thus, we have shown that the contrapositive $\sim q \to \sim p$ is true, and it follows that the conditional $p \to q$ is also true. ■■■■

We now turn our attention to the last of the five basic connectives.

Biconditional Propositions

Statements of the form "p if and only if q" are called **biconditional propositions** and are represented symbolically by

$$p \leftrightarrow q$$

The connective "if and only if" is called the biconditional connective. The biconditional $p \leftrightarrow q$ is true whenever p and q are *both true* or *both false.*

Table A.9

p	q	$p \leftrightarrow q$
T	T	T
T	F	F
F	T	F
F	F	T

The truth table for $p \leftrightarrow q$ is shown in Table A.9.

EXAMPLE 3 Let

p: Mark is going to the senior prom.

q: Linda is going to the senior prom.

State the biconditional $p \leftrightarrow q$.

SOLUTION ✓ The required statement is

Mark is going to the senior prom,
if and only if,
Linda is going to the senior prom. ■■■■

As suggested by its name, the biconditional statement is actually composed of two conditional statements. For instance, an *equivalent expression* for the biconditional statement in Example 3 is given by the two conditional statements

Mark is going to the senior prom
if Linda is going to the senior prom.

Linda is going to the senior prom
if Mark is going to the senior prom.

Thus, "p if q and q if p" is equivalent to "p if and only if q."

The equivalent expressions that are most commonly used in forming conditional and biconditional statements are summarized in Table A.10. The

Table A.10

		Equivalent Form
Conditional Statement	if p, then q	p is sufficient for q p only if q q is necessary for p q, if p
Biconditional Statement	p if and only if q	If p then q, if q then p p is necessary and sufficient for q

words *necessary* and *sufficient* occur frequently in mathematical proofs. When we say that p is **sufficient** for q we mean that *when p is true, q is also true*—that is, $p \to q$. In like manner, when we say that p is **necessary** for q, we mean that *if p is not true, then q is not true*—that is, $\sim p \to \sim q$. But this last expression is logically equivalent to $q \to p$. Thus, when we say that p is necessary and sufficient for q, it follows that $p \to q$ and $q \to p$. Example 4 illustrates the use of these expressions.

EXAMPLE 4 Let

p: The stock market will go up.

q: Interest rates decrease.

Represent the following statements symbolically:

a. If interest rates decrease, then the stock market will go up.
b. If interest rates do not decrease, then the stock market will not go up.
c. The stock market will go up if and only if interest rates decrease.
d. Decreasing interest rates is a sufficient condition for the stock market to go up.

e. Decreasing interest rates is a necessary and sufficient condition for a rising stock market.

SOLUTION ✓

a. $q \to p$ **b.** $\sim q \to \sim p$ or $p \to q$ **c.** $p \leftrightarrow q$

d. $q \to p$ **e.** $q \leftrightarrow p$

Just as there is an order of precedence when we use the arithmetical operations $\times$, $\div$, $+$, and $-$, there is also an order of precedence that must be observed when we use the logical connectives. The following list dictates the order of precedence for logical connectives.

Order of Precedence of Logical Connectives

$$\sim, \ \wedge, \ \vee, \ \to, \ \leftrightarrow$$

Thus, the connective $\sim$ should be applied first, followed by $\wedge$, and so on. For example, using the order of precedence, we see that $\sim p \vee q \to p \wedge r \vee s$ is $[(\sim p) \vee q] \to [(p \wedge r) \vee s]$.

Also, as in the case of arithmetical operations, parentheses take the highest priority in the order of precedence.

A.3 Exercises

In Exercises 1–4, write the converse, the contrapositive, and the inverse of the given conditional statement.

1. $p \to \sim q$
2. $\sim p \to \sim q$
3. $q \to p$
4. $\sim p \to q$

In Exercises 5 and 6, refer to the following propositions p and q:

p: It is snowing.

q: The temperature is below freezing.

5. Express the conditional and the biconditional of p and q in words.
6. Express the converse, contrapositive, and inverse of the conditional $p \to q$ in words.

In Exercises 7 and 8, refer to the following propositions p and q:

p: The company's union and management reach a settlement.

q: The workers will not strike.

7. Express the conditional and biconditional of p and q in words.
8. Express the converse, contrapositive, and inverse of the conditional $p \to q$ in words.

In Exercises 9–12, determine whether the given statement is true or false.

9. A conditional proposition and its converse are logically equivalent.
10. The converse and the inverse of a conditional proposition are logically equivalent.
11. A conditional proposition and its inverse are logically equivalent.
12. The converse and the contrapositive of a conditional proposition are logically equivalent.
13. Consider the conditional statement "If the owner lowers the selling price of the house, then I will buy it." Under what conditions is the conditional statement false?
14. Consider the biconditional statement "I will buy the house if and only if the owner lowers the selling price." Under what conditions is the biconditional statement false?

In Exercises 15–28, construct a truth table for the given compound proposition.

15. $\sim(p \to q)$
16. $\sim(q \to \sim p)$
17. $\sim(p \to q) \wedge p$
18. $(p \to q) \vee (q \to p)$
19. $(p \to \sim q) \veebar \sim p$
20. $(p \to q) \wedge (\sim p \vee q)$
21. $(p \to q) \leftrightarrow (\sim q \to \sim p)$
22. $(p \to q) \leftrightarrow (\sim p \vee q)$
23. $(p \wedge q) \to (p \vee q)$
24. $\sim q \to (\sim p \wedge \sim q)$
25. $(p \vee q) \to \sim r$
26. $(p \leftrightarrow q) \vee r$
27. $p \to (q \vee r)$
28. $[(p \to q) \vee (q \to r)] \to (p \to r)$

In Exercises 29–36, determine whether the given compound propositions are logically equivalent.

29. $\sim p \vee q$; $p \to q$
30. $\sim(p \vee q)$; $\sim p \wedge \sim q$
31. $q \to p$; $\sim p \to \sim q$
32. $\sim p \to q$; $\sim p \vee q$
33. $p \wedge q$; $p \to \sim q$
34. $\sim(p \wedge \sim q)$; $\sim p \vee q$
35. $(p \to q) \to r$; $(p \vee q) \vee r$
36. $p \vee (q \wedge r)$; $(p \vee q) \wedge (p \vee r)$

37. Let p and q denote the following propositions:

    p: Taxes are increased.
    q: The federal deficit increases.

    Represent the following statements symbolically.
    a. If taxes are increased, then the federal deficit will not increase.
    b. If taxes are not increased, then the federal deficit will increase.
    c. The federal deficit will not increase if and only if taxes are increased.
    d. Increased taxation is a sufficient condition for halting the growth of the federal deficit.
    e. Increased taxation is a necessary and sufficient condition for halting the growth of the federal deficit.

38. Let p and q denote the following propositions:

    p: The unemployment rate decreases.
    q: Consumer confidence will improve.

    Represent the following statements symbolically.
    a. If the unemployment rate does not decrease, consumer confidence will not improve.
    b. Consumer confidence will improve if and only if the unemployment rate does not decrease.
    c. A decreasing unemployment rate is a sufficient condition for consumer confidence to improve.
    d. A decreasing unemployment rate is a necessary and sufficient condition for consumer confidence to improve.

A.4 Laws of Logic

Just as the laws of algebra enable us to perform operations with real numbers, the **laws of logic** provide us with a systematic method of connecting statements.

Laws of Logic

Let p, q, and r be any three propositions. Then

1. $p \wedge p \Leftrightarrow p$ — *Idempotent law for conjunction*
2. $p \vee p \Leftrightarrow p$ — *Idempotent law for disjunction*
3. $(p \wedge q) \wedge r \Leftrightarrow p \wedge (q \wedge r)$ — *Associative law for conjunction*
4. $(p \vee q) \vee r \Leftrightarrow p \vee (q \vee r)$ — *Associative law for disjunction*

5. $p \wedge q \Leftrightarrow q \wedge p$	*Commutative law for conjunction*
6. $p \vee q \Leftrightarrow q \vee p$	*Commutative law for disjunction*
7. $p \wedge (q \vee r) \Leftrightarrow (p \wedge q) \vee (p \wedge r)$	*Distributive law for conjunction*
8. $p \vee (q \wedge r) \Leftrightarrow (p \vee q) \wedge (p \vee r)$	*Distributive law for disjunction*
9. $\sim(p \vee q) \Leftrightarrow \sim p \wedge \sim q$	*De Morgan's Law*
10. $\sim(p \wedge q) \Leftrightarrow \sim p \vee \sim q$	*De Morgan's Law*

To verify any of these laws, we need only construct a truth table to show that the given statements are logically equivalent. We illustrate this procedure in the next example.

EXAMPLE 1

Prove the distributive law for conjunction.

SOLUTION ✔

We wish to prove that $p \wedge (q \vee r)$ is logically equivalent to $(p \wedge q) \vee (p \wedge r)$. This is easily done by constructing the associated truth table, as shown in Table A.11. Since the entries in the last two columns are the same, we conclude that $p \wedge (q \vee r) \Leftrightarrow (p \wedge q) \vee (p \wedge r)$.

Table A.11

p	q	r	$q \vee r$	$p \wedge q$	$p \wedge r$	$p \wedge (q \vee r)$	$(p \wedge q) \vee (p \wedge r)$
T	T	T	T	T	T	T	T
T	T	F	T	T	F	T	T
T	F	T	T	F	T	T	T
T	F	F	F	F	F	F	F
F	T	T	T	F	F	F	F
F	T	F	T	F	F	F	F
F	F	T	T	F	F	F	F
F	F	F	F	F	F	F	F

■■■■

The proofs of the other laws are left to you as an exercise.

De Morgan's laws, Laws 9 and 10, are useful in forming the negation of a statement. For example, if we wish to state the negation of the proposition

Steve plans to major in business administration or economics

we can represent the statement symbolically by $p \vee q$, where the prime propositions are

p: Steve plans to major in business administration.

q: Steve plans to major in economics.

Then, the negation of $p \vee q$ is $\sim(p \vee q)$. Using De Morgan's laws, we have

$$\sim(p \vee q) \Leftrightarrow \sim p \wedge \sim q$$

Thus, the required statement is

> Steve does not plan to major in business administration *and* he does not plan to major in economics.

Up to this point we have considered propositions that have both true and false entries in their truth tables. Some statements have the property that they are always true; other propositions have the property that they are always false.

Tautologies and Contradictions

A **tautology** is a statement that is always true.

A **contradiction** is a statement that is always false.

EXAMPLE 2 Show that:

a. $p \vee \sim p$ is a tautology.
b. $p \wedge \sim p$ is a contradiction.

SOLUTION ✓

a. The truth table associated with $p \vee \sim p$ is shown in Table A.12a. Since all the entries in the last column are Ts, the proposition is a tautology.
b. The truth table for $p \wedge \sim p$ is shown in Table A.12b. Since all the entries in the last column are Fs, the proposition is a contradiction.

Table A.12

p	$\sim p$	$p \vee \sim p$
T	F	T
F	T	T

(a)

p	$\sim p$	$p \vee \sim p$
T	F	F
F	T	F

(b)

■■■■

In Section A.3 we defined logically equivalent statements as statements that have identical truth tables. The definition of equivalence may also be restated in terms of a tautology.

Logical Equivalence

Two propositions p and q are **logically equivalent** if the biconditional $p \leftrightarrow q$ is a tautology.

We can see why these two definitions are really the same by looking more closely at the truth table for $p \leftrightarrow q$ (Table A.13a).

Table A.13

(a)

	p	q	$p \leftrightarrow q$
Case 1	T	T	T
Case 2	T	F	F
Case 3	F	T	F
Case 4	F	F	T

(b)

p	q	$p \leftrightarrow q$
T	T	T
F	F	T

If the biconditional is always true, then the second and third cases shown in the truth table are excluded and we are left with the truth table in Table A.13b. Notice that the entries in each row of the p and q columns are identical. In other words, p and q have identical truth values and hence are logically equivalent.

In addition to the ten laws stated earlier, we have the following four laws involving tautologies and contradictions.

Let t be a tautology and c a contradiction. Then,

11. $p \vee {\sim}p \Leftrightarrow t$
12. $p \wedge {\sim}p \Leftrightarrow c$
13. $p \vee t \Leftrightarrow t$
14. $p \wedge t \Leftrightarrow p$

It now remains only to show how these laws are used to simplify proofs.

EXAMPLE 3 Using the laws of logic, show that

$$p \vee ({\sim}p \wedge q) \Leftrightarrow (p \vee q)$$

SOLUTION ✓

$$\begin{aligned} p \vee ({\sim}p \wedge q) &\Leftrightarrow (p \vee {\sim}p) \wedge (p \vee q) && \text{(By Law 8)} \\ &\Leftrightarrow t \wedge (p \vee q) && \text{(By Law 11)} \\ &\Leftrightarrow p \vee q && \text{(By Law 14)} \end{aligned}$$

■■■■

EXAMPLE 4 Using the laws of logic, show that

$${\sim}(p \vee q) \vee ({\sim}p \wedge q) \Leftrightarrow {\sim}p$$

SOLUTION ✓

$$\begin{aligned} &{\sim}(p \vee q) \vee ({\sim}p \wedge q) \\ &\Leftrightarrow ({\sim}p \wedge {\sim}q) \vee ({\sim}p \wedge q) && \text{(By Law 9)} \\ &\Leftrightarrow {\sim}p \wedge ({\sim}q \vee q) && \text{(By Law 7)} \\ &\Leftrightarrow {\sim}p \wedge t && \text{(By Law 11)} \\ &\Leftrightarrow {\sim}p && \text{(By Law 14)} \end{aligned}$$

■■■■

A.4 Exercises

1. Prove the idempotent law for conjunction, $p \wedge p \Leftrightarrow p$.
2. Prove the idempotent law for disjunction, $p \vee p \Leftrightarrow p$.
3. Prove the associative law for conjunction, $(p \wedge q) \wedge r \Leftrightarrow p \wedge (q \wedge r)$.
4. Prove the associative law for disjunction, $(p \vee q) \vee r \Leftrightarrow p \vee (q \vee r)$.
5. Prove the commutative law for conjunction, $p \wedge q \Leftrightarrow q \wedge p$.
6. Prove the commutative law for disjunction, $p \vee q \Leftrightarrow q \vee p$.
7. Prove the distributive law for disjunction, $p \vee (q \wedge r) \Leftrightarrow (p \vee q) \wedge (p \vee r)$.
8. Prove De Morgan's laws
   a. $\sim(p \vee q) \Leftrightarrow \sim p \wedge \sim q$
   b. $\sim(p \wedge q) \Leftrightarrow \sim p \vee \sim q$

In Exercises 9–18, determine whether the given statement is a tautology, a contradiction, or neither.

9. $(p \to q) \leftrightarrow (\sim p \vee q)$
10. $(p \veebar q) \wedge (p \leftrightarrow q)$
11. $p \to (p \vee q)$
12. $(p \to q) \vee (q \to p)$
13. $(p \to q) \leftrightarrow (\sim q \to \sim p)$
14. $p \wedge (p \to q) \to q$
15. $(p \to q) \wedge (\sim q) \to (\sim p)$
16. $[(p \to q) \wedge (q \to r)] \to (p \to r)$
17. $[(p \to q) \vee (q \to r)] \to (p \to r)$
18. $[p \wedge (q \vee r)] \leftrightarrow [(p \wedge q) \vee (p \wedge r)]$
19. Let p and q denote the statements

    p: The candidate opposes changes in the Social Security system.

    q: The candidate supports the ERA.

    Use De Morgan's laws to state the negation of $p \wedge q$ and the negation of $p \vee q$.
20. Let p and q denote the statements

    p: The recycling bill was passed by the voters.

    q: The tax on oil and hazardous materials was not approved by the voters.

    Use De Morgan's laws to state the negation of $p \wedge q$ and the negation of $p \vee q$.

In Exercises 21–26, use the laws of logic to prove the given proposition.

21. $[p \wedge (q \vee \sim q) \vee (p \wedge q)] \Leftrightarrow p \vee (p \wedge q)$
22. $p \vee (\sim p \wedge \sim q) \Leftrightarrow p \vee \sim q$
23. $(p \wedge \sim q) \vee (p \wedge \sim r) \Leftrightarrow p \wedge (\sim q \vee \sim r)$
24. $(p \vee q) \vee \sim q \Leftrightarrow t$
25. $p \wedge [\sim(q \wedge r)] \Leftrightarrow (p \wedge \sim q) \vee (p \wedge \sim r)$
26. $p \vee (q \vee r) \Leftrightarrow r \vee (q \vee p)$

A.5 Arguments

In this section we discuss arguments and the methods used to determine the validity of arguments.

Argument

An **argument** or **proof** consists of a set of propositions $p_1, p_2, \ldots, p_n$, called the *premises,* and a proposition q, called the *conclusion.* An argument is *valid* if and only if the conclusion is true whenever the premises are all true. An argument that is *not* valid is called a *fallacy,* or an *invalid* argument.

The next example illustrates the form in which an argument is presented. Notice that the premises are written separately above the horizontal line and the conclusion is written below the line.

EXAMPLE 1 An argument is presented in the following way:

Premises: If Pam studies diligently, she passes her exams.
Pam studies diligently.
____
Conclusion: Pam passes her exams. ■■■■

To determine the validity of an argument, we first write the argument in symbolic form and then construct the associated truth table containing the prime propositions, the premises, and the conclusion. We then *check the rows in which the premises are all true. If the conclusion in each of these rows is also true, then the argument is valid. Otherwise, it is a fallacy.*

EXAMPLE 2 Determine the validity of the argument in Example 1.

SOLUTION ✓ The symbolic form of the argument is

$$\begin{array}{l} p \to q \\ \underline{p \quad\quad} \\ \therefore q \end{array}$$

Observe that the conclusion is preceded by the symbol $\therefore$, which is used to represent the word *therefore.* The truth table associated with this argument is shown in Table A.14. Observe that only row 1 contains true values for both premises. Since the conclusion is also true in this row, we conclude that the argument is valid.

Table A.14

Propositions		Premises		Conclusion
p	q	$p \to q$	p	q
T	T	T	T	T
T	F	F	T	F
F	T	T	F	T
F	F	T	F	F

■■■■

EXAMPLE 3 Determine the validity of the argument

If Michael is overtired, then he is grumpy.
Michael is grumpy.
____
Therefore, Michael is overtired.

SOLUTION ✓

The argument is written symbolically as follows:

$$\begin{array}{l} p \to q \\ \underline{q \qquad} \\ \therefore p \end{array}$$

Table A.15

p	q	$p \to q$	q	p
T	T	T	T	T
T	F	F	F	T
F	T	(T	T	F)
F	F	T	F	F

The associated truth table is shown in Table A.15. Observe that the entries for the premises in the third row are both true, but the corresponding entry for the conclusion is false. We conclude that the argument is a fallacy.

When we consider the validity of an argument, we are concerned only with the *form* of the argument and not the truth or falsity of the premises. In other words, the conclusion of an argument may follow validly from the premises, but the premises themselves may be false. The next example demonstrates this point.

EXAMPLE 4

Determine the validity of the argument

Reggie is a wealthy man.
Wealthy men are happy.
Therefore, Reggie is happy.

SOLUTION ✓

The symbolic form of the argument is

$$\begin{array}{l} p \to q \\ \underline{q \to r} \\ \therefore p \to r \end{array}$$

The associated truth table is shown in Table A.16. Since the conclusion is true in each of the rows that contain true values for both premises, we conclude that the argument is valid. Note that the validity of the argument is not affected by the truth or falsity of the premise "Wealthy men are happy."

Table A.16

p	q	r	$p \to q$	$q \to r$	$p \to r$	
T	T	T	(T	T	T)	(√)
T	T	F	T	F	F	
T	F	T	F	T	T	
T	F	F	F	T	F	
F	T	T	(T	T	T)	(√)
F	T	F	T	F	T	
F	F	T	(T	T	T)	(√)
F	F	F	(T	T	T)	(√)

A question that may already have arisen in your mind is: How does one determine the validity of an argument if the associated truth table does not contain true values for all the premises? The next example provides the answer to this question.

EXAMPLE 5

Determine the validity of the argument

The door is locked.
The door is unlocked.
Therefore, the door is locked.

SOLUTION ✓

The symbolic form of the argument is

$$\begin{array}{l} p \\ \underline{\sim p} \\ \therefore p \end{array}$$

Table A.17

p	$\sim p$	p
T	F	T
F	T	F

The associated truth table is shown in Table A.17. Observe that no rows contain true values for both premises. Nevertheless, the argument is considered to be valid since the condition that the conclusion is true whenever the premises are all true is not violated. Again, we remind you that the validity of an argument is not determined by the truth or falsity of its premises or conclusion, but rather it is determined only by the form of the argument. ■■■■

The next proposition provides us with an alternative method for determining the validity of an argument.

Propositions

Suppose an argument consists of the premises $p_1, p_2, \ldots, p_n$ and conclusion q. Then, the argument is valid if and only if the proposition $[p_1 \wedge p_2 \wedge \ldots \wedge p_n] \to q$ is a tautology.

To prove this proposition we must show that (a) if an argument is valid, then the given proposition is a tautology, and (b) if the given proposition is a tautology, then the argument is valid.

Proof of (a): Since the argument is valid, it follows that q is true whenever all of the premises $p_1, p_2, \ldots, p_n$ are true. But the conjunction $[p_1 \wedge p_2 \wedge \ldots \wedge p_n]$ is true only when all the premises are true (by the definition of conjunction). Hence, q is true whenever the conjunction is true, and we conclude that the conditional $[p_1 \wedge p_2 \wedge \ldots \wedge p_n] \to q$ is true. (Recall that a conditional statement is always true when its hypotheses are false. Hence, to show that a conditional is a tautology, we need only prove that its conclusion is always true when its hypotheses are true.)

Proof of (b): Since the given proposition is a tautology, q is always true. Therefore, q is true when the conjunction

$$(p_1 \wedge p_2 \wedge \ldots \wedge p_n)$$

is true. But the conjunction is true only when all of the premises $p_1, p_2, \ldots, p_n$ are true. Hence, q is true whenever the premises are all true, and the argument is valid.

Having proved this proposition, we can now use it to determine the validity of an argument. If we are given an argument, we construct a truth table for $(p_1 \wedge p_2 \wedge \ldots \wedge p_n) \to q$. If the truth table contains all the true values in its last column, then the argument is valid; otherwise it is invalid. This method of proof is illustrated in the next example.

EXAMPLE 6 Determine the validity of the argument

You return the book on time, or you will have to pay a fine.
You do not return the book on time.
Therefore, you will have to pay a fine.

SOLUTION The symbolic form of the argument is

$$\begin{array}{r} p \vee q \\ \underline{\sim p} \\ \therefore q \end{array}$$

Following the method described, we construct the truth table for

$$[(p \vee q) \wedge \sim p] \to q$$

as shown in Table A.18. Since the entries in the last column are all Ts, the proposition is a tautology, and we conclude that the argument is valid.

Table A.18

p	q	$p \vee q$	$\sim p$	$(p \vee q) \wedge \sim p$	$[(p \vee q) \wedge \sim p] \to q$
T	T	T	F	F	T
T	F	T	F	F	T
F	T	T	T	T	T
F	F	F	T	F	T

By familiarizing ourselves with a few of the most commonly used argument forms, we can simplify the problem of determining the validity of an argument. Some of the argument forms most commonly used are:

1. **Modus ponens** (a manner of affirming), or rule of detachment

$$\begin{array}{l} p \to q \\ \underline{p \qquad} \\ \therefore q \end{array}$$

2. **Modus tollens** (a manner of denying)

$$\begin{array}{l} p \to q \\ \underline{\sim q} \\ \therefore \sim p \end{array}$$

3. **Law of syllogisms**

$$\begin{array}{l} p \to q \\ \underline{q \to r} \\ \therefore p \to r \end{array}$$

The truth tables verifying modus ponens and the law of syllogisms have already been constructed (see Tables A.14 and A.16, respectively). The verification of modus tollens is left to you as an exercise. The next two examples illustrate the use of these argument forms.

EXAMPLE 7 Determine the validity of the argument

If the battery is dead, the car will not start.
The car starts.
Therefore, the battery is not dead.

SOLUTION As before, we express the argument in symbolic form. Thus,

$$\begin{array}{l} p \to q \\ \underline{\sim q} \\ \therefore \sim p \end{array}$$

Identifying the form of this argument as modus tollens, we conclude that the given argument is valid. ■■■■

EXAMPLE 8 Determine the validity of the argument

If mortgage rates are lowered, housing sales increase.
If housing sales increase, then the prices of houses increase.
Therefore, if mortgage rates are lowered, the prices of houses increase.

SOLUTION The symbolic form of the argument is

$$\begin{array}{l} p \to q \\ \underline{q \to r} \\ \therefore p \to r \end{array}$$

Using the law of syllogisms, we conclude that the argument is valid. ■■■■

A.5 Exercises

In Exercises 1–16, determine whether the given argument is valid.

1. $p \to q$
$q \to r$
$\therefore p \to r$

2. $p \vee q$
$\sim p$
$\therefore q$

3. $p \wedge q$
$\sim p$
$\therefore q$

4. $p \to q$
$\sim q$
$\therefore \sim p$

5. $p \to q$
$\sim p$
$\therefore \sim q$

6. $p \to q$
$q \wedge r$
$\therefore p \vee r$

7. $p \leftrightarrow q$
q
$\therefore p$

8. $p \wedge q$
$\sim p \to \sim q$
$\therefore p \wedge \sim q$

9. $p \to q$
$q \to p$
$\therefore p \leftrightarrow q$

10. $p \to \sim q$
$p \veebar q$
$\therefore \sim q$

11. $p \leftrightarrow q$
$q \leftrightarrow r$
$\therefore p \leftrightarrow r$

12. $p \to q$
$q \leftrightarrow r$
$\therefore p \wedge r$

13. $p \veebar r$
$q \wedge r$
$\therefore p \to r$

14. $p \leftrightarrow q$
$q \vee r$
$\sim r$
$\therefore p \to \sim r$

15. $p \leftrightarrow q$
$q \vee r$
$\sim p$
$\therefore \sim p \to \sim r$

16. $p \to q$
$r \to q$
$p \wedge q$
$\therefore p \vee r$

In Exercises 17–22, represent the given argument symbolically and determine whether it is a valid argument.

17. If Carla studies, then she passes her exams.
Carla did not study.
Therefore, Carla did not pass her exams.

18. If Tony is wealthy, he is either intelligent or a good businessman.
Tony is intelligent and he is not a good businessman.
Therefore, Tony is not wealthy.

19. Steve will attend the matinee and/or the evening show.
If Steve doesn't go to the matinee show, then he will not go to the evening show.
Therefore, Steve will attend the matinee show.

20. If Mary wins the race, then Stacy loses the race.
Neither Mary nor Linda won the race.
Therefore, Stacy won the race.

21. If mortgage rates go up, then housing prices will go up.
If housing prices go up, more people will rent houses.
More people are renting houses.
Therefore, mortgage rates went up.

22. If taxes are cut, then retail sales increase.
If retail sales increase, then the unemployment rate will decrease.
If the unemployment rate decreases, then the incumbent will win the election.
Therefore, if taxes are not cut, the incumbent will not win the election.

23. Given the statement "All good cooks prepare gourmet food," which of the following conclusions follow logically?
a. If George prepares gourmet food, then he is a good cook.
b. If Brenda does not prepare gourmet food, then she is not a good cook.
c. Everyone who prepares gourmet food is a good cook.
d. Some people who prepare gourmet food are not good cooks.

24. What conclusion can be drawn from the following statements?

His date is pretty, or she is tall and skinny.
If his date is tall, then she is a brunette.
His date is not a brunette.

25. Show that modus tollens is a valid form of argument.

A.6 Applications of Logic to Switching Networks

In this section we see how the principles of logic can be used in the design and analysis of switching networks. A **switching network** is an arrangement of wires and switches connecting two terminals. Such networks are used extensively in digital computers.

A switch may be open or closed. If a switch is closed, current will flow through the wire. If it is open, no current will flow through the wire. Because a switch has exactly two states, it can be represented by a proposition p that is true if the switch is closed and false if the switch is open.

Now let's consider a circuit with two switches p and q. If the circuit is connected as shown in Figure A.1, the switches p and q are said to be **in series.**

FIGURE A.1
Two switches connected in series

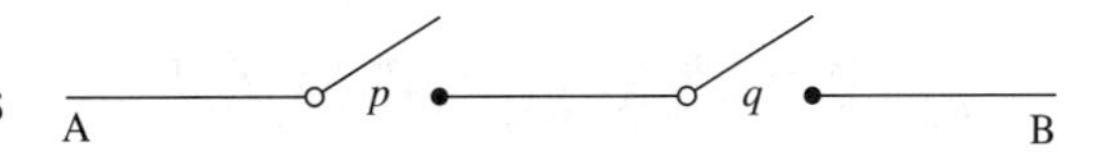

For such a network, current will flow from A to B if and only if both p and q are closed, but no current will flow if one or more of the switches is open. Thinking of p and q as propositions, we have the truth table shown in Table A.19. (Recall that T corresponds to the situation in which the switch is closed.) From the truth table we see that two switches p and q connected in series are analogous to the conjunction $p \wedge q$ of the two propositions p and q.

Table A.19

p	q	
T	T	T
T	F	F
F	T	F
F	F	F

If a circuit is connected as shown in Figure A.2, the switches p and q are said to be connected **in parallel.**

FIGURE A.2
Two switches connected in parallel

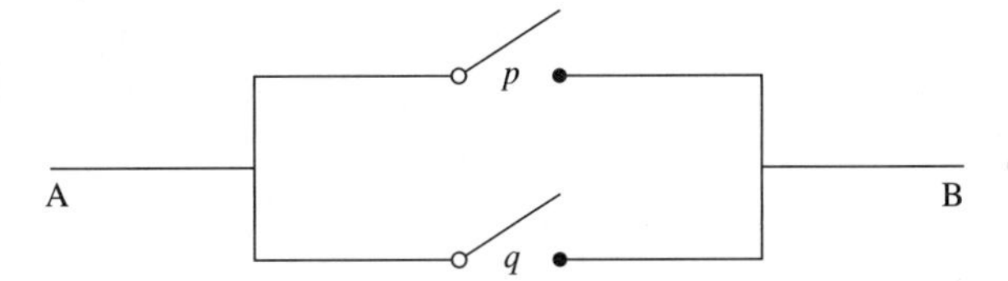

For such a network, current will flow from A to B if and only if one or more of the switches p or q is closed. Once again, thinking of p and q as propositions, we have the truth table shown in Table A.20. From the truth table, we conclude that two switches p and q connected in parallel are analogous to the inclusive disjunction $p \vee q$ of the two propositions p and q.

Table A.20

p	q	
T	T	T
T	F	T
F	T	T
F	F	F

EXAMPLE 1

Find a logic statement that represents the network shown in Figure A.3. By constructing the truth table for this logic statement, determine the conditions under which current will flow from A to B in the network.

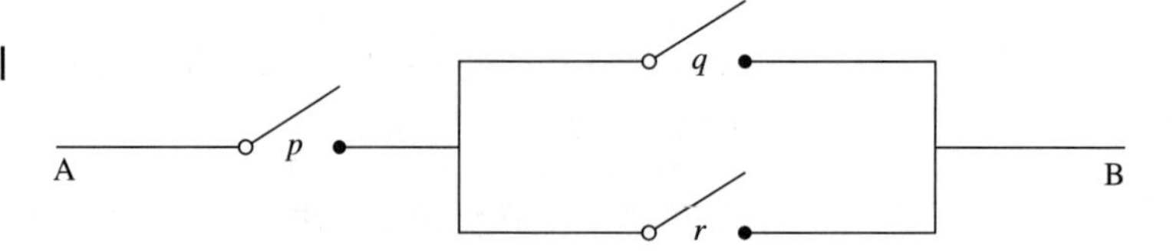

FIGURE A.3
We want to determine when the current will flow from A to B.

SOLUTION ✓

The required logic statement is $p \wedge (q \vee r)$. Next, we construct the truth table for the logic statement $p \wedge (q \vee r)$ (Table A.21).

Table A.21

p	q	r	$q \vee r$	$p \wedge (q \vee r)$
T	T	T	T	T
T	T	F	T	T
T	F	T	T	T
T	F	F	F	F
F	T	T	T	F
F	T	F	T	F
F	F	T	T	F
F	F	F	F	F

From the truth table, we conclude that current will flow from A to B if and only if one of the following conditions is satisfied:

1. p, q, and r are all closed.
2. p and q are closed, but r is open.
3. p and r are closed, but q is open.

In other words, current will flow from A to B if and only if p is closed and either q or r or both q and r are closed. ■■■■

Before looking at some additional examples, let's remark that $\sim p$ represents a switch that is open when p is closed and vice versa. Furthermore, a circuit that is always closed is represented by a tautology, $p \vee \sim p$, whereas a circuit that is always open is represented by a contradiction, $p \wedge \sim p$. (Why?)

EXAMPLE 2

Given the logic statement $(p \vee q) \wedge (r \vee s \vee \sim p)$, draw the corresponding network.

SOLUTION ✓

Recalling that the disjunction $p \vee q$ of the propositions p and q represents two switches p and q connected in parallel and the conjunction $p \wedge q$ represents two switches connected in series, we obtain the following network (Figure A.4).

FIGURE A.4
The network that corresponds to the logic statement $(p \lor q) \land (r \lor s \lor \sim p)$

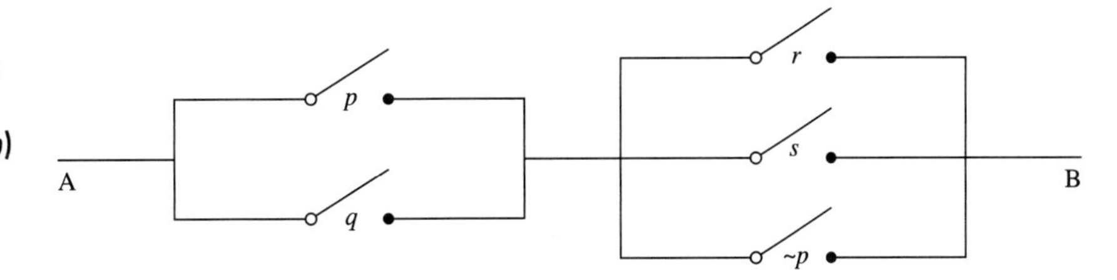

The theory of networks developed so far, when used in conjunction with the laws of logic, is a useful tool in network analysis. In particular, network analysis enables us to find equivalent, and often simpler, networks, as the following example shows.

EXAMPLE 3 Find a logic statement representing the network shown in Figure A.5. Also, find a simpler but equivalent network.

FIGURE A.5
We want to find the logic statement that corresponds to the network shown in the figure.

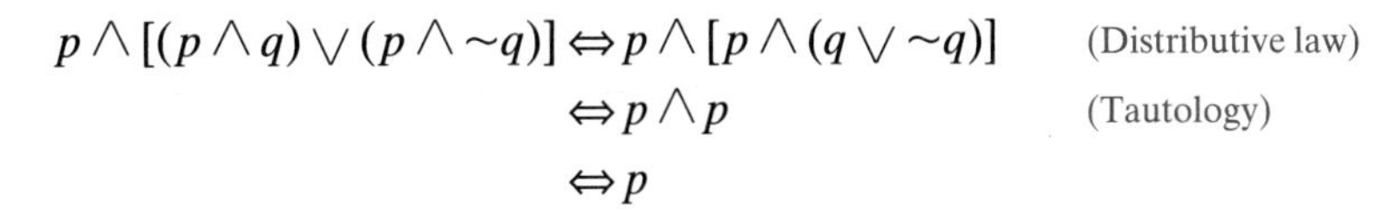

SOLUTION The logic statement corresponding to the given network is $p \land [(p \land q) \lor (p \land \sim q)]$. Next, using the rules of logic to simplify this statement, we obtain

$$\begin{aligned} p \land [(p \land q) \lor (p \land \sim q)] &\Leftrightarrow p \land [p \land (q \lor \sim q)] && \text{(Distributive law)} \\ &\Leftrightarrow p \land p && \text{(Tautology)} \\ &\Leftrightarrow p \end{aligned}$$

FIGURE A.6
This network is equivalent to the one shown in Figure A.5.

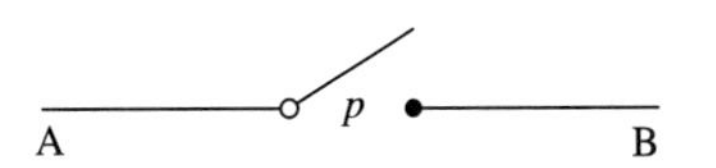

Thus, the given network is equivalent to the one shown in Figure A.6.

A.6 Exercises

In Exercises 1–5, find a logic statement corresponding to the given network. Determine the conditions under which current will flow from A to B.

1.

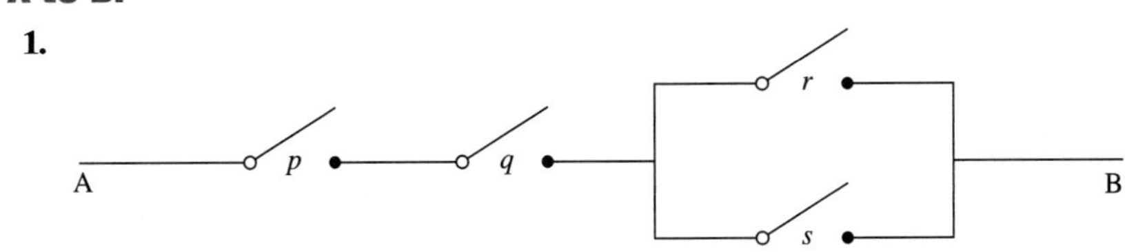

2.

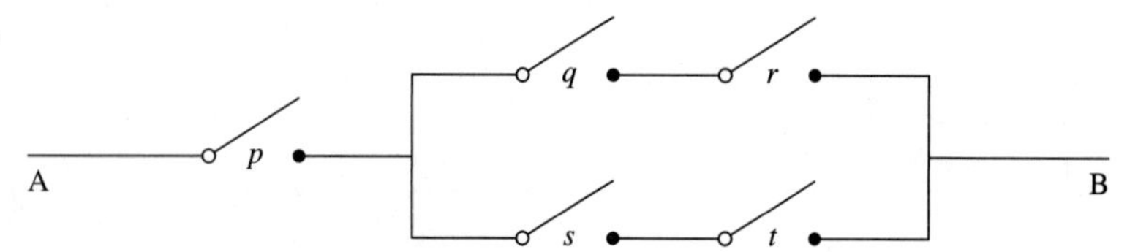

3.

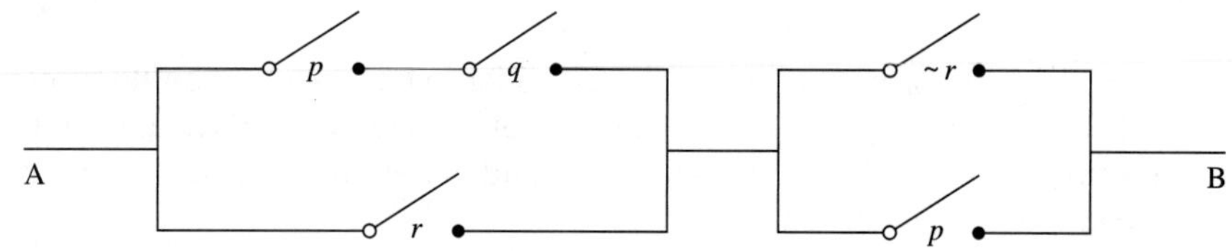

4.

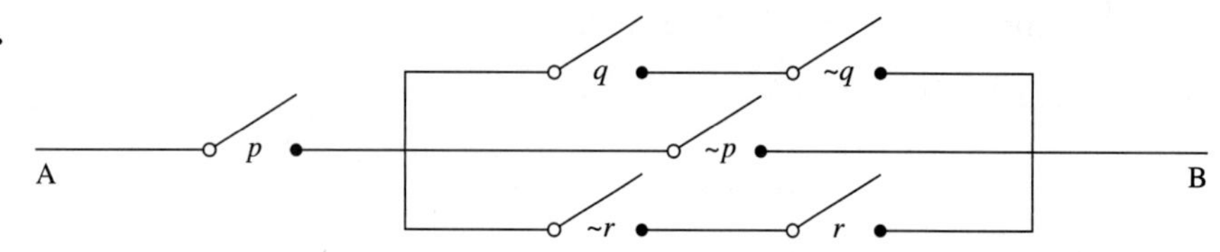

5.

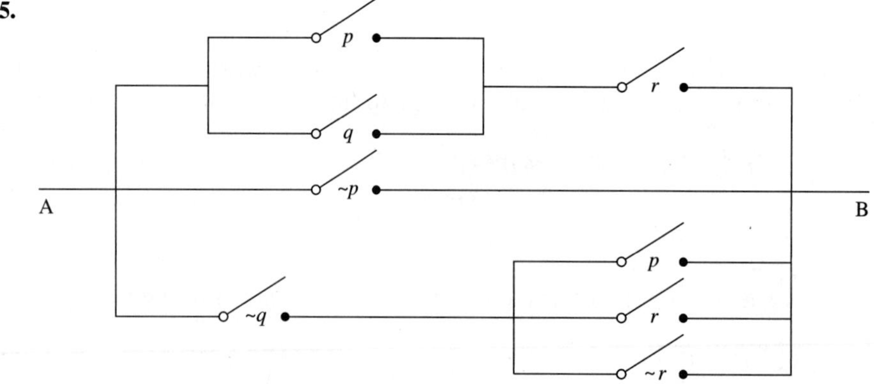

In Exercises 6–11, draw the network corresponding to the given logic statement.

6. $p \vee (q \wedge r)$ **7.** $(p \wedge q) \wedge r$

8. $[p \vee (q \wedge r)] \wedge \sim q$

9. $(p \vee q) \vee [r \wedge (\sim r \vee \sim p)]$

10. $(\sim p \vee q) \wedge (p \vee \sim q)$

11. $(p \wedge q) \vee [(r \vee \sim q) \wedge (s \vee \sim p)]$

In Exercises 12–15, find a logic statement corresponding to the given network. Then find a simpler but equivalent network.

12.

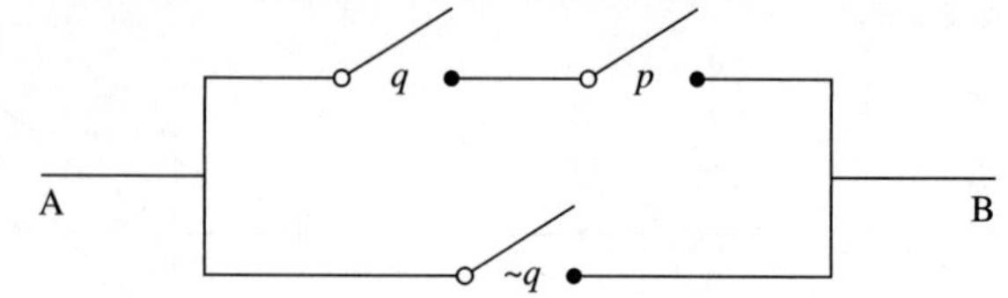

13.

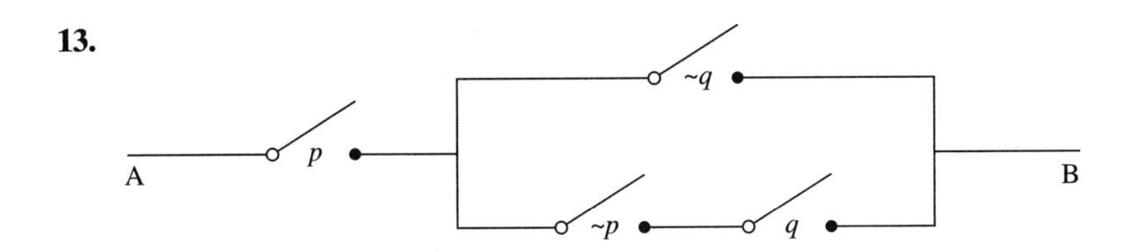

14.

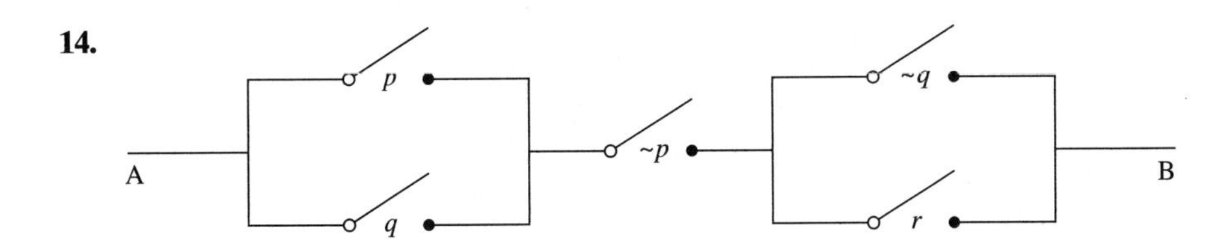

15.

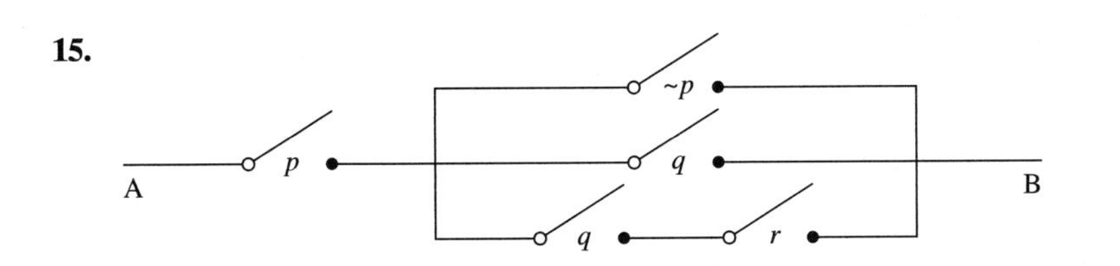

16. A hallway light is to be operated by two switches, one located at the bottom of the staircase and the other located at the top of the staircase. Design a suitable network.

Hint: Let p and q be the switches. Construct a truth table for the associated propositions.

B THE SYSTEM OF REAL NUMBERS

In this appendix we briefly review the **system of real numbers.** This system consists of a set of objects called real numbers together with two operations, addition and multiplication, that enable us to combine two or more real numbers to obtain other real numbers. These operations are subjected to certain rules that we will state after first recalling the set of real numbers.

The set of real numbers may be constructed from the set of **natural** (also called **counting**) **numbers**

$$N = \{1, 2, 3, \ldots\}$$

by adjoining other objects (numbers) to it. Thus, the set

$$W = \{0, 1, 2, 3, \ldots\}$$

obtained by adjoining the single number 0 to N is called the set of **whole numbers.** By adjoining the *negatives* of the numbers 1, 2, 3, . . . to the set W of whole numbers, we obtain the set of **integers**

$$I = \{\ldots, -3, -2, -1, 0, 1, 2, 3, \ldots\}$$

Next, consider the set

$$Q = \left\{ \frac{a}{b} \;\middle|\; a \text{ and } b \text{ are integers with } b \neq 0 \right\}$$

Now, the set I of integers is contained in the set Q of **rational numbers.** To see this, observe that each integer may be written in the form a/b with $b =$ 1, thus qualifying as a member of the set Q. The converse, however, is false, for the rational numbers (fractions) such as

$$\frac{1}{2}, \quad \frac{23}{25}, \quad \text{and so on}$$

are clearly not integers.

The sets N, W, I, and Q constructed thus far have

$$N \subset W \subset I \subset Q$$

That is, N is a proper subset of W, W is a proper subset of I, and so on.

Finally, consider the set Ir of all numbers that cannot be expressed in the form a/b, where a, b are integers ($b \neq 0$). The members of this set, called the set of **irrational numbers,** include $\sqrt{2}$, $\sqrt{3}$, π, and so on. The set

$$R = Q \cup Ir$$

That is, the set composed of all rational numbers as well as irrational numbers, is called the set of **real numbers** (Figure B.1).

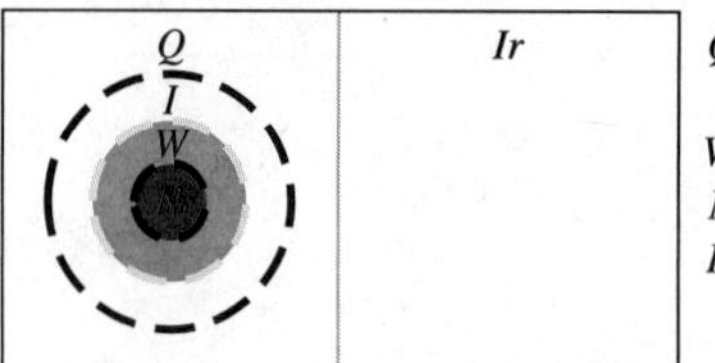

FIGURE B.1
The set of all real numbers consists of the set of rational numbers plus the set of irrational numbers.

Note the following important representation of real numbers: Every real number has a decimal representation; a rational number has a representation in terms of a repeated decimal. For example,

$$\frac{1}{7} = 0.142857142857142857\ldots$$ (Note that the block of integers 142857 repeats.)

On the other hand, the irrational number $\sqrt{2}$ has a representation in terms of a nonrepeating decimal. Thus,

$$\sqrt{2} = 1.41421\ldots$$

As mentioned earlier, any two real numbers may be combined to obtain another real number. The operation of *addition,* written $+$, enables us to combine any two numbers a and b to obtain their sum, denoted by $a + b$. Another operation, called *multiplication,* and written $\cdot$, enables us to combine any two real numbers a and b to form their product, the number $a \cdot b$, or, written more simply, ab. These two operations are subjected to the following rules of operation: Given any three real numbers a, b, and c, we have

I. Under addition

1. $a + b = b + a$ (Commutative law of addition)

2. $a + (b + c) = (a + b) + c$ (Associative law of addition)

3. $a + 0 = a$ (Identity law of addition)

4. $a + (-a) = 0$ (Inverse law of addition)

II. Under multiplication

1. $ab = ba$ (Commutative law of multiplication)

2. $a(bc) = (ab)c$ (Associative law of multiplication)

3. $a \cdot 1 = a$ (Identity law of multiplication)

4. $a(1/a) = 1 \quad (a \neq 0)$ (Inverse law of multiplication)

III. Under addition and multiplication

1. $a(b + c) = ab + ac$ (Distributive law for multiplication with respect to addition)

TABLES C

Table 1 Compound Amount, Present Value, and Annuity

	$i = \frac{1}{4}\%$			$i = \frac{1}{2}\%$		
n	$(1+i)^n$	$s_{\overline{n}\rvert i}$	$a_{\overline{n}\rvert i}$	$(1+i)^n$	$s_{\overline{n}\rvert i}$	$a_{\overline{n}\rvert i}$
1	1.0025 0000	1.0000 0000	0.9975 0623	1.0050 0000	1.0000 0000	0.9950 2488
2	1.0050 0625	2.0025 0000	1.9925 2492	1.0100 2500	2.0050 0000	1.9850 9938
3	1.0075 1877	3.0075 0625	2.9850 6227	1.0150 7513	3.0150 2500	2.9702 4814
4	1.0100 3756	4.0150 2502	3.9751 2446	1.0201 5050	4.0301 0013	3.9504 9566
5	1.0125 6266	5.0250 6258	4.9627 1766	1.0252 5125	5.0502 5063	4.9258 6633
6	1.0150 9406	6.0376 2523	5.9478 4804	1.0303 7751	6.0755 0188	5.8963 8441
7	1.0176 3180	7.0527 1930	6.9305 2174	1.0355 2940	7.1058 7939	6.8620 7404
8	1.0201 7588	8.0703 5110	7.9107 4487	1.0407 0704	8.1414 0879	7.8229 5924
9	1.0227 2632	9.0905 2697	8.8885 2357	1.0459 1058	9.1821 1583	8.7790 6392
10	1.0252 8313	10.1132 5329	9.8638 6391	1.0511 4013	10.2280 2641	9.7304 1186
11	1.0278 4634	11.1385 3642	10.8367 7198	1.0563 9583	11.2791 6654	10.6770 2673
12	1.0304 1596	12.1663 8277	11.8072 5384	1.0616 7781	12.3355 6237	11.6189 3207
13	1.0329 9200	13.1967 9872	12.7753 1555	1.0669 8620	13.3972 4018	12.5561 5131
14	1.0355 7448	14.2297 9072	13.7409 6314	1.0723 2113	14.4642 2639	13.4887 0777
15	1.0381 6341	15.2653 6520	14.7042 0264	1.0776 8274	15.5365 4752	14.4166 2465
16	1.0407 5882	16.3035 2861	15.6650 4004	1.0830 7115	16.6142 3026	15.3399 2502
17	1.0433 6072	17.3442 8743	16.6234 8133	1.0884 8651	17.6973 0141	16.2586 3186
18	1.0459 6912	18.3876 4815	17.5795 3250	1.0939 2894	18.7857 8791	17.1727 6802
19	1.0485 8404	19.4336 1727	18.5331 9950	1.0993 9858	19.8797 1685	18.0823 5624
20	1.0512 0550	20.4822 0131	19.4844 8828	1.1048 9558	20.9791 1544	18.9874 1915
21	1.0538 3352	21.5334 0682	20.4334 0477	1.1104 2006	22.0840 1101	19.8879 7925
22	1.0564 6810	22.5872 4033	21.3799 5488	1.1159 7216	23.1944 3107	20.7840 5896
23	1.0591 0927	23.6437 0843	22.3241 4452	1.1215 5202	24.3104 0322	21.6756 8055
24	1.0617 5704	24.7028 1770	23.2659 7957	1.1271 5978	25.4319 5524	22.5628 6622
25	1.0644 1144	25.7645 7475	24.2054 6591	1.1327 9558	26.5591 1502	23.4456 3803
26	1.0670 7247	26.8289 8619	25.1426 0939	1.1384 5955	27.6919 1059	24.3240 1794
27	1.0697 4015	27.8960 5865	26.0774 1585	1.1441 5185	28.8303 7015	25.1980 2780
28	1.0724 1450	28.9657 9880	27.0098 9112	1.1498 7261	29.9745 2200	26.0676 8936
29	1.0750 9553	30.0382 1330	27.9400 4102	1.1556 2197	31.1243 9461	26.9330 2423
30	1.0777 8327	31.1133 0883	28.8678 7134	1.1614 0008	32.2800 1658	27.7940 5397
31	1.0804 7773	32.1910 9210	29.7933 8787	1.1672 0708	33.4414 1666	28.6507 9997
32	1.0831 7892	33.2715 6983	30.7165 9638	1.1730 4312	34.6086 2375	29.5032 8355
33	1.0858 8687	34.3547 4876	31.6375 0262	1.1789 0833	35.7816 6686	30.3515 2592
34	1.0886 0159	35.4406 3563	32.5561 1234	1.1848 0288	36.9605 7520	31.1955 4818
35	1.0913 2309	36.5292 3722	33.4724 3126	1.1907 2689	38.1453 7807	32.0353 7132
36	1.0940 5140	37.6205 6031	34.3864 6510	1.1966 8052	39.3361 0496	32.8710 1624
37	1.0967 8653	38.7146 1171	35.2982 1955	1.2026 6393	40.5327 8549	33.7025 0372
38	1.0995 2850	39.8113 9824	36.2077 0030	1.2086 7725	41.7354 4942	34.5298 5445
39	1.1022 7732	40.9109 2673	37.1149 1302	1.2147 2063	42.9441 2666	35.3530 8900
40	1.1050 3301	42.0132 0405	38.0198 6336	1.2207 9424	44.1588 4730	36.1722 2786
41	1.1077 9559	43.1182 3706	38.9225 5697	1.2268 9821	45.3796 4153	36.9872 9141
42	1.1105 6508	44.2260 3265	39.8229 9947	1.2330 3270	46.6065 3974	37.7982 9991
43	1.1133 4149	45.3365 9774	40.7211 9648	1.2391 9786	47.8395 7244	38.6052 7354
44	1.1161 2485	46.4499 3923	41.6171 5359	1.2453 9385	49.0787 7030	39.4082 3238
45	1.1189 1516	47.5660 6408	42.5108 7640	1.2516 2082	50.3241 6415	40.2071 9640
46	1.1217 1245	48.6849 7924	43.4023 7047	1.2578 7892	51.5757 8497	41.0021 8547
47	1.1245 1673	49.8066 9169	44.2916 4137	1.2641 6832	52.8336 6390	41.7932 1937
48	1.1273 2802	50.9312 0842	45.1786 9463	1.2704 8916	54.0978 3222	42.5803 1778
49	1.1301 4634	52.0585 3644	46.0635 3580	1.2768 4161	55.3683 2138	43.3635 0028
50	1.1329 7171	53.1886 8278	46.9461 7037	1.2832 2581	56.6451 6299	44.1427 8635

Table 1 *(continued)*

n	$i = \frac{1}{4}\%$ $(1+i)^n$	$s_{\overline{n}\rvert i}$	$a_{\overline{n}\rvert i}$	$i = \frac{1}{2}\%$ $(1+i)^n$	$s_{\overline{n}\rvert i}$	$a_{\overline{n}\rvert i}$
51	1.1358 0414	54.3216 5449	47.8266 0386	1.2896 4194	57.9283 8880	44.9181 9537
52	1.1386 4365	55.4574 5862	48.7048 4176	1.2960 9015	59.2180 3075	45.6897 4664
53	1.1414 9026	56.5961 0227	49.5808 8953	1.3025 7060	60.5141 2090	46.4574 5934
54	1.1443 4398	57.7375 9252	50.4547 5265	1.3090 8346	61.8166 9150	47.2213 5258
55	1.1472 0484	58.8819 3650	51.3264 3656	1.3156 2887	63.1257 7496	47.9814 4535
56	1.1500 7285	60.0291 4135	52.1959 4669	1.3222 0702	64.4414 0384	48.7377 5657
57	1.1529 4804	61.1792 1420	53.0632 8847	1.3288 1805	65.7636 1086	49.4903 0505
58	1.1558 3041	62.3321 6223	53.9284 6730	1.3354 6214	67.0924 2891	50.2391 0950
59	1.1587 1998	63.4879 9264	54.7914 8858	1.3421 3946	68.4278 9105	50.9841 8855
60	1.1616 1678	64.6467 1262	55.6523 5769	1.3488 5015	69.7700 3051	51.7255 6075
61	1.1645 2082	65.8083 2940	56.5110 7999	1.3555 9440	71.1188 8066	52.4632 4453
62	1.1674 3213	66.9728 5023	57.3676 6083	1.3623 7238	72.4744 7507	53.1972 5824
63	1.1703 5071	68.1402 8235	58.2221 0557	1.3691 8424	73.8368 4744	53.9276 2014
64	1.1732 7658	69.3106 3306	59.0744 1952	1.3760 3016	75.2060 3168	54.6543 4839
65	1.1762 0977	70.4839 0964	59.9246 0800	1.3829 1031	76.5820 6184	55.3774 6109
66	1.1791 5030	71.6601 1942	60.7726 7631	1.3898 2486	77.9649 7215	56.0969 7621
67	1.1820 9817	72.8392 6971	61.6186 2974	1.3967 7399	79.3547 9701	56.8129 1165
68	1.1850 5342	74.0213 6789	62.4624 7355	1.4037 5785	80.7515 7099	57.5252 8522
69	1.1880 1605	75.2064 2131	63.3042 1302	1.4107 7664	82.1553 2885	58.2341 1465
70	1.1909 8609	76.3944 3736	64.1438 5339	1.4178 3053	83.5661 0549	58.9394 1756
71	1.1939 6356	77.5854 2345	64.9813 9989	1.4249 1968	84.9839 3602	59.6412 1151
72	1.1969 4847	78.7793 8701	65.8168 5774	1.4320 4428	86.4088 5570	60.3395 1394
73	1.1999 4084	79.9763 3548	66.6502 3216	1.4392 0450	87.8408 9998	61.0343 4222
74	1.2029 4069	81.1762 7632	67.4815 2834	1.4464 0052	89.2801 0448	61.7257 1366
75	1.2059 4804	82.3792 1701	68.3107 5146	1.4536 3252	90.7265 0500	62.4136 4543
76	1.2089 6291	83.5851 6505	69.1379 0670	1.4609 0069	92.1801 3752	63.0981 5466
77	1.2119 8532	84.7941 2797	69.9629 9920	1.4682 0519	93.6410 3821	63.7792 5836
78	1.2150 1528	86.0061 1329	70.7860 3411	1.4755 4622	95.1092 4340	64.4569 7350
79	1.2180 5282	87.2211 2857	71.6070 1657	1.4829 2395	96.5847 8962	65.1313 1691
80	1.2210 9795	88.4391 8139	72.4259 5169	1.4903 3857	98.0677 1357	65.8023 0538
81	1.2241 5070	89.6602 7934	73.2428 4458	1.4977 9026	99.5580 5214	66.4699 5561
82	1.2272 1108	90.8844 3004	74.0577 0033	1.5052 7921	101.0558 4240	67.1342 8419
83	1.2302 7910	92.1116 4112	74.8705 2402	1.5128 0561	102.5611 2161	67.7953 0765
84	1.2333 5480	93.3419 2022	75.6813 2072	1.5203 6964	104.0739 2722	68.4530 4244
85	1.2364 3819	94.5752 7502	76.4900 9548	1.5279 7148	105.5942 9685	69.1075 0491
86	1.2395 2928	95.8117 1321	77.2968 5335	1.5356 1134	107.1222 6834	69.7587 1135
87	1.2426 2811	97.0512 4249	78.1015 9935	1.5432 8940	108.6578 7968	70.4066 7796
88	1.2457 3468	98.2938 7060	78.9043 3850	1.5510 0585	110.2011 6908	71.0514 2086
89	1.2488 4901	99.5396 0527	79.7050 7581	1.5587 6087	111.7521 7492	71.6929 5608
90	1.2519 7114	100.7884 5429	80.5038 1627	1.5665 5468	113.3109 3580	72.3312 9958
91	1.2551 0106	102.0404 2542	81.3005 6486	1.5743 8745	114.8774 9048	72.9664 6725
92	1.2582 3882	103.2955 2649	82.0953 2654	1.5822 5939	116.4518 7793	73.5984 7487
93	1.2613 8441	104.5537 6530	82.8881 0628	1.5901 7069	118.0341 3732	74.2273 3818
94	1.2645 3787	105.8151 4972	83.6789 0900	1.5981 2154	119.6243 0800	74.8530 7282
95	1.2676 9922	107.0796 8759	84.4677 3966	1.6061 1215	121.2224 2954	75.4756 9434
96	1.2708 6847	108.3473 8681	85.2546 0315	1.6141 4271	122.8285 4169	76.0952 1825
97	1.2740 4564	109.6182 5528	86.0395 0439	1.6222 1342	124.4426 8440	76.7116 5995
98	1.2772 3075	110.8923 0091	86.8224 4827	1.6303 2449	126.0648 9782	77.3250 3478
99	1.2804 2383	112.1695 3167	87.6034 3967	1.6384 7611	127.6952 2231	77.9353 5799
100	1.2836 2489	113.4499 5550	88.3824 8346	1.6466 6849	129.3336 9842	78.5426 4477

Table 1 *(continued)*

	$i = \frac{3}{4}\%$			$i = 1\%$		
n	$(1+i)^n$	$s_{\overline{n}\vert i}$	$a_{\overline{n}\vert i}$	$(1+i)^n$	$s_{\overline{n}\vert i}$	$a_{\overline{n}\vert i}$
1	1.0075 0000	1.0000 0000	0.9925 5583	1.0100 0000	1.0000 0000	0.9900 9901
2	1.0150 5625	2.0075 0000	1.9777 2291	1.0201 0000	2.0100 0000	1.9703 9506
3	1.0226 6917	3.0225 5625	2.9555 5624	1.0303 0100	3.0301 0000	2.9409 8521
4	1.0303 3919	4.0452 2542	3.9261 1041	1.0406 0401	4.0604 0100	3.9019 6555
5	1.0380 6673	5.0755 6461	4.8894 3961	1.0510 1005	5.1010 0501	4.8534 3124
6	1.0458 5224	6.1136 3135	5.8455 9763	1.0615 2015	6.1520 1506	5.7954 7647
7	1.0536 9613	7.1594 8358	6.7946 3785	1.0721 3535	7.2135 3521	6.7281 9453
8	1.0615 9885	8.2131 7971	7.7366 1325	1.0828 5671	8.2856 7056	7.6516 7775
9	1.0695 6084	9.2747 7856	8.6715 7642	1.0936 8527	9.3685 2727	8.5660 1758
10	1.0775 8255	10.3443 3940	9.5995 7958	1.1046 2213	10.4622 1254	9.4713 0453
11	1.0856 6441	11.4219 2194	10.5206 7452	1.1156 6835	11.5668 3467	10.3676 2825
12	1.0938 0690	12.5075 8636	11.4349 1267	1.1268 2503	12.6825 0301	11.2550 7747
13	1.1020 1045	13.6013 9325	12.3423 4508	1.1380 9328	13.8093 2804	12.1337 4007
14	1.1102 7553	14.7034 0370	13.2430 2242	1.1494 7421	14.9474 2132	13.0037 0304
15	1.1186 0259	15.8136 7923	14.1369 9495	1.1609 6896	16.0968 9554	13.8650 5252
16	1.1269 9211	16.9322 8183	15.0243 1261	1.1725 7864	17.2578 6449	14.7178 7378
17	1.1354 4455	18.0592 7394	15.9050 2492	1.1843 0443	18.4304 4314	15.5622 5127
18	1.1439 6039	19.1947 1849	16.7791 8107	1.1961 4748	19.6147 4757	16.3982 6858
19	1.1525 4009	20.3386 7888	17.6468 2984	1.2081 0895	20.8108 9504	17.2260 0850
20	1.1611 8414	21.4912 1897	18.5080 1969	1.2201 9004	22.0190 0399	18.0455 5297
21	1.1698 9302	22.6524 0312	19.3627 9870	1.2323 9194	23.2391 9403	18.8569 8313
22	1.1786 6722	23.8222 9614	20.2112 1459	1.2447 1586	24.4715 8598	19.6603 7934
23	1.1875 0723	25.0009 6336	21.0533 1473	1.2571 6302	25.7163 0183	20.4558 2113
24	1.1964 1353	26.1884 7059	21.8891 4614	1.2697 3465	26.9734 6485	21.2433 8726
25	1.2053 8663	27.3848 8412	22.7187 5547	1.2824 3200	28.2431 9950	22.0231 5570
26	1.2144 2703	28.5902 7075	23.5421 8905	1.2952 5631	29.5256 3150	22.7952 0366
27	1.2235 3523	29.8046 9778	24.3594 9286	1.3082 0888	30.8208 8781	23.5596 0759
28	1.2327 1175	31.0282 3301	25.1707 1251	1.3212 9097	32.1290 9669	24.3164 4316
29	1.2419 5709	32.2609 4476	25.9758 9331	1.3345 0388	33.4503 8766	25.0657 8530
30	1.2512 7176	33.5029 0184	26.7750 8021	1.3478 4892	34.7848 9153	25.8077 0822
31	1.2606 5630	34.7541 7361	27.5683 1783	1.3613 2740	36.1327 4045	26.5422 8537
32	1.2701 1122	36.0148 2991	28.3556 5045	1.3749 4068	37.4940 6785	27.2695 8947
33	1.2796 3706	37.2849 4113	29.1371 2203	1.3886 9009	38.8690 0853	27.9896 9255
34	1.2892 3434	38.5645 7819	29.9127 7621	1.4025 7699	40.2576 9862	28.7026 6589
35	1.2989 0359	39.8538 1253	30.6826 5629	1.4166 0276	41.6602 7560	29.4085 8009
36	1.3086 4537	41.1527 1612	31.4468 0525	1.4307 6878	43.0768 7836	30.1075 0504
37	1.3184 6021	42.4613 6149	32.2052 6576	1.4450 7647	44.5076 4714	30.7995 0994
38	1.3283 4866	43.7798 2170	32.9580 8016	1.4595 2724	45.9527 2361	31.4846 6330
39	1.3383 1128	45.1081 7037	33.7052 9048	1.4741 2251	47.4122 5085	32.1630 3298
40	1.3483 4861	46.4464 8164	34.4469 3844	1.4888 6373	48.8863 7336	32.8346 8611
41	1.3584 6123	47.7948 3026	35.1830 6545	1.5037 5237	50.3752 3709	33.4996 8922
42	1.3686 4969	49.1532 9148	35.9137 1260	1.5187 8989	51.8789 8946	34.1581 0814
43	1.3789 1456	50.5219 4117	36.6389 2070	1.5339 7779	53.3977 7936	34.8100 0806
44	1.3892 5642	51.9008 5573	37.3587 3022	1.5493 1757	54.9317 5715	35.4554 5352
45	1.3996 7584	53.2901 1215	38.0731 8136	1.5648 1075	56.4810 7472	36.0945 0844
46	1.4101 7341	54.6897 8799	38.7823 1401	1.5804 5885	58.0458 8547	36.7272 3608
47	1.4207 4971	56.0999 6140	39.4861 6775	1.5962 6344	59.6263 4432	37.3536 9909
48	1.4314 0533	57.5207 1111	40.1847 8189	1.6122 2608	61.2226 0777	37.9739 5949
49	1.4421 4087	58.9521 1644	40.8781 9542	1.6283 4834	62.8348 3385	38.5880 7871
50	1.4529 5693	60.3942 5732	41.5664 4707	1.6446 3182	64.4631 8218	39.1961 1753

Table 1 *(continued)*

	$i = \frac{3}{4}\%$			$i = 1\%$		
n	$(1+i)^n$	$s_{\overline{n}\rvert i}$	$a_{\overline{n}\rvert i}$	$(1+i)^n$	$s_{\overline{n}\rvert i}$	$a_{\overline{n}\rvert i}$
51	1.4638 5411	61.8472 1424	42.2495 7525	1.6610 7814	66.1078 1401	39.7981 3617
52	1.4748 3301	63.3110 6835	42.9276 1812	1.6776 8892	67.7688 9215	40.3941 9423
53	1.4858 9426	64.7859 0136	43.6006 1351	1.6944 6581	69.4465 8107	40.9843 5072
54	1.4970 3847	66.2717 9562	44.2685 9902	1.7114 1047	71.1410 4688	41.5686 6408
55	1.5082 6626	67.7688 3409	44.9316 1193	1.7285 2457	72.8524 5735	42.1471 9216
56	1.5195 7825	69.2771 0035	45.5896 8926	1.7458 0982	74.5809 8192	42.7199 9224
57	1.5309 7509	70.7966 7860	46.2428 6776	1.7632 6792	76.3267 9174	43.2871 2102
58	1.5424 5740	72.3276 5369	46.8911 8388	1.7809 0060	78.0900 5966	43.8486 3468
59	1.5540 2583	73.8701 1109	47.5346 7382	1.7987 0960	79.8709 6025	44.4045 8879
60	1.5656 8103	75.4241 3693	48.1733 7352	1.8166 9670	81.6696 6986	44.9550 3841
61	1.5774 2363	76.9898 1795	48.8073 1863	1.8348 6367	83.4863 6655	45.5000 3803
62	1.5892 5431	78.5672 4159	49.4365 4455	1.8532 1230	85.3212 3022	46.0396 4161
63	1.6011 7372	80.1564 9590	50.0610 8640	1.8717 4443	87.1744 4252	46.5739 0258
64	1.6131 8252	81.7576 6962	50.6809 7906	1.8904 6187	89.0461 8695	47.1028 7385
65	1.6252 8139	83.3708 5214	51.2962 5713	1.9093 6649	90.9366 4882	47.6266 0777
66	1.6374 7100	84.9961 3353	51.9069 5497	1.9284 6015	92.8460 1531	48.1451 5621
67	1.6497 5203	86.6336 0453	52.5131 0667	1.9477 4475	94.7744 7546	48.6585 7050
68	1.6621 2517	88.2833 5657	53.1147 4607	1.9672 2220	96.7222 2021	49.1669 0149
69	1.6745 9111	89.9454 8174	53.7119 0677	1.9868 9442	98.6894 4242	49.6701 9949
70	1.6871 5055	91.6200 7285	54.3046 2210	2.0067 6337	100.6763 3684	50.1685 1435
71	1.6998 0418	93.3072 2340	54.8929 2516	2.0268 3100	102.6831 0021	50.6618 9539
72	1.7125 5271	95.0070 2758	55.4768 4880	2.0470 9931	104.7099 3121	51.1503 9148
73	1.7253 9685	96.7195 8028	56.0564 2561	2.0675 7031	106.7570 3052	51.6340 5097
74	1.7383 3733	98.4449 7714	56.6316 8795	2.0882 4601	108.8246 0083	52.1129 2175
75	1.7513 7486	100.1833 1446	57.2026 6794	2.1091 2847	110.9128 4684	52.5870 5124
76	1.7645 1017	101.9346 8932	57.7693 9746	2.1302 1975	113.0219 7530	53.0564 8638
77	1.7777 4400	103.6991 9949	58.3319 0815	2.1515 2195	115.1521 9506	53.5212 7364
78	1.7910 7708	105.4769 4349	58.8902 3141	2.1730 3717	117.3037 1701	53.9814 5905
79	1.8045 1015	107.2680 2056	59.4443 9842	2.1947 6754	119.4767 5418	54.4370 8817
80	1.8180 4398	109.0725 3072	59.9944 4012	2.2167 1522	121.6715 2172	54.8882 0611
81	1.8316 7931	110.8905 7470	60.5403 8722	2.2388 8237	123.8882 3694	55.3348 5753
82	1.8454 1691	112.7222 5401	61.0822 7019	2.2612 7119	126.1271 1931	55.7770 8666
83	1.8592 5753	114.5676 7091	61.6201 1930	2.2838 8390	128.3883 9050	56.2149 3729
84	1.8732 0196	116.4269 2845	62.1539 6456	2.3067 2274	130.6722 7440	56.6484 5276
85	1.8872 5098	118.3001 3041	62.6838 3579	2.3297 8997	132.9789 9715	57.0776 7600
86	1.9014 0536	120.1873 8139	63.2097 6257	2.3530 8787	135.3087 8712	57.5026 4951
87	1.9156 6590	122.0887 8675	63.7317 7427	2.3766 1875	137.6618 7499	57.9234 1535
88	1.9300 3339	124.0044 5265	64.2499 0002	2.4003 8494	140.0384 9374	58.3400 1520
89	1.9445 0865	125.9344 8604	64.7641 6875	2.4243 8879	142.4388 7868	58.7524 9030
90	1.9590 9246	127.8789 9469	65.2746 0918	2.4486 3267	144.8632 6746	59.1608 8148
91	1.9737 8565	129.8380 8715	65.7812 4981	2.4731 1900	147.3119 0014	59.5652 2919
92	1.9885 8905	131.8118 7280	66.2841 1892	2.4978 5019	149.7850 1914	59.9655 7346
93	2.0035 0346	133.8004 6185	66.7832 4458	2.5228 2869	152.2828 6933	60.3619 5392
94	2.0185 2974	135.8039 6531	67.2786 5467	2.5480 5698	154.8056 9803	60.7544 0982
95	2.0336 6871	137.8224 9505	67.7703 7685	2.5735 3755	157.3537 5501	61.1429 8002
96	2.0489 2123	139.8561 6377	68.2584 3856	2.5992 7293	159.9272 9256	61.5277 0299
97	2.0642 8814	141.9050 8499	68.7428 6705	2.6252 6565	162.5265 6548	61.9086 1682
98	2.0797 7030	143.9693 7313	69.2236 8938	2.6515 1831	165.1518 3114	62.2857 5923
99	2.0953 6858	146.0491 4343	69.7009 3239	2.6780 3349	167.8033 4945	62.6591 6755
100	2.1110 8384	148.1445 1201	70.1746 2272	2.7048 1383	170.4813 8294	63.0288 7877

Table 1 *(continued)*

n	$i = 1\frac{1}{4}\%$ $(1+i)^n$	$s_{\overline{n}\vert i}$	$a_{\overline{n}\vert i}$	$i = 1\frac{1}{2}\%$ $(1+i)^n$	$s_{\overline{n}\vert i}$	$a_{\overline{n}\vert i}$
1	1.0125 0000	1.0000 0000	0.9876 5432	1.0150 0000	1.0000 0000	0.9852 2167
2	1.0251 5625	2.0125 0000	1.9631 1538	1.0302 2500	2.0150 0000	1.9558 8342
3	1.0379 7070	3.0376 5625	2.9265 3371	1.0456 7838	3.0452 2500	2.9122 0042
4	1.0509 4534	4.0756 2695	3.8780 5798	1.0613 6355	4.0909 0338	3.8543 8465
5	1.0640 8215	5.1265 7229	4.8178 3504	1.0772 8400	5.1522 6693	4.7826 4497
6	1.0773 8318	6.1906 5444	5.7460 0992	1.0934 4326	6.2295 5093	5.6971 8717
7	1.0908 5047	7.2680 3762	6.6627 2585	1.1098 4491	7.3229 9419	6.5982 1396
8	1.1044 8610	8.3588 8809	7.5681 2429	1.1264 9259	8.4328 3911	7.4859 2508
9	1.1182 9218	9.4633 7420	8.4623 4498	1.1433 8998	9.5593 3169	8.3605 1732
10	1.1322 7083	10.5816 6637	9.3455 2591	1.1605 4083	10.7027 2167	9.2221 8455
11	1.1464 2422	11.7139 3720	10.2178 0337	1.1779 4894	11.8632 6249	10.0711 1779
12	1.1607 5452	12.8603 6142	11.0793 1197	1.1956 1817	13.0412 1143	10.9075 0521
13	1.1752 6395	14.0211 1594	11.9301 8466	1.2135 5244	14.2368 2960	11.7315 3222
14	1.1899 5475	15.1963 7988	12.7705 5275	1.2317 5573	15.4503 8205	12.5433 8150
15	1.2048 2918	16.3863 3463	13.6005 4592	1.2502 3207	16.6821 3778	13.3432 3301
16	1.2198 8955	17.5911 6382	14.4202 9227	1.2689 8555	17.9323 6984	14.1312 6405
17	1.2351 3817	18.8110 5336	15.2299 1829	1.2880 2033	19.2013 5539	14.9076 4931
18	1.2505 7739	20.0461 9153	16.0295 4893	1.3073 4064	20.4893 7572	15.6725 6089
19	1.2662 0961	21.2967 6893	16.8193 0759	1.3269 5075	21.7967 1636	16.4261 6837
20	1.2820 3723	22.5629 7854	17.5993 1613	1.3468 5501	23.1236 6710	17.1686 3879
21	1.2980 6270	23.8450 1577	18.3696 9495	1.3670 5783	24.4705 2211	17.9001 3673
22	1.3142 8848	25.1430 7847	19.1305 6291	1.3875 6370	25.8375 7994	18.6208 2437
23	1.3307 1709	26.4573 6695	19.8820 3744	1.4083 7715	27.2251 4364	19.3308 6145
24	1.3473 5105	27.7880 8403	20.6242 3451	1.4295 0281	28.6335 2080	20.0304 0537
25	1.3641 9294	29.1354 3508	21.3572 6865	1.4509 4535	30.0630 2361	20.7196 1120
26	1.3812 4535	30.4996 2802	22.0812 5299	1.4727 0953	31.5139 6896	21.3986 3172
27	1.3985 1092	31.8808 7337	22.7962 9925	1.4948 0018	32.9866 7850	22.0676 1746
28	1.4159 9230	33.2793 8429	23.5025 1778	1.5172 2218	34.4814 7867	22.7267 1671
29	1.4336 9221	34.6953 7659	24.2000 1756	1.5399 8051	35.9987 0085	23.3760 7558
30	1.4516 1336	36.1290 6880	24.8889 0623	1.5630 8022	37.5386 8137	24.0158 3801
31	1.4697 5853	37.5806 8216	25.5692 9010	1.5865 2642	39.1017 6159	24.6461 4582
32	1.4881 3051	39.0504 4069	26.2412 7418	1.6103 2432	40.6882 8801	25.2671 3874
33	1.5067 3214	40.5385 7120	26.9049 6215	1.6344 7918	42.2986 1233	25.8789 5442
34	1.5255 6629	42.0453 0334	27.5604 5644	1.6589 9637	43.9330 9152	26.4817 2849
35	1.5446 3587	43.5708 6963	28.2078 5822	1.6838 8132	45.5920 8789	27.0755 9458
36	1.5639 4382	45.1155 0550	28.8472 6737	1.7091 3954	47.2759 6921	27.6606 8431
37	1.5834 9312	46.6794 4932	29.4787 8259	1.7347 7663	48.9851 0874	28.2371 2740
38	1.6032 8678	48.2629 4243	30.1025 0133	1.7607 9828	50.7198 8538	28.8050 5163
39	1.6233 2787	49.8662 2921	30.7185 1983	1.7872 1025	52.4806 8366	29.3645 8288
40	1.6436 1946	51.4895 5708	31.3269 3316	1.8140 1841	54.2678 9391	29.9158 4520
41	1.6641 6471	53.1331 7654	31.9278 3522	1.8412 2868	56.0819 1232	30.4589 6079
42	1.6849 6677	54.7973 4125	32.5213 1874	1.8688 4712	57.9231 4100	30.9940 5004
43	1.7060 2885	56.4823 0801	33.1074 7530	1.8968 7982	59.7919 8812	31.5212 3157
44	1.7273 5421	58.1883 3687	33.6863 9536	1.9253 3302	61.6888 6794	32.0406 2223
45	1.7489 4614	59.9156 9108	34.2581 6825	1.9542 1301	63.6142 0096	32.5523 3718
46	1.7708 0797	61.6646 3721	34.8228 8222	1.9835 2621	65.5684 1398	33.0564 8983
47	1.7929 4306	63.4354 4518	35.3806 2442	2.0132 7910	67.5519 4018	33.5531 9195
48	1.8153 5485	65.2283 8824	35.9314 8091	2.0434 7829	69.5652 1929	34.0425 5365
49	1.8380 4679	67.0437 4310	36.4755 3670	2.0741 3046	71.6086 9758	34.5246 8339
50	1.8610 2237	68.8817 8989	37.0128 7575	2.1052 4242	73.6828 2804	34.9996 8807

Table 1 *(continued)*

n	$i = 1\frac{1}{4}\%$ $(1+i)^n$	$s_{\overline{n}\rceil i}$	$a_{\overline{n}\rceil i}$	$i = 1\frac{1}{2}\%$ $(1+i)^n$	$s_{\overline{n}\rceil i}$	$a_{\overline{n}\rceil i}$
51	1.8842 8515	70.7428 1226	37.5435 8099	2.1368 2106	75.7880 7046	35.4676 7298
52	1.9078 3872	72.6270 9741	38.0677 3431	2.1688 7337	77.9248 9152	35.9287 4185
53	1.9316 8670	74.5349 3613	38.5854 1660	2.2014 0647	80.0937 6489	36.3829 9690
54	1.9558 3279	76.4666 2283	39.0967 0776	2.2344 2757	82.2951 7136	36.8305 3882
55	1.9802 8070	78.4224 5562	39.6016 8667	2.2679 4398	84.5295 9893	37.2714 6681
56	2.0050 3420	80.4027 3631	40.1004 3128	2.3019 6314	86.7975 4292	37.7058 7863
57	2.0300 9713	82.4077 7052	40.5930 1855	2.3364 9259	89.0995 0606	38.1338 7058
58	2.0554 7335	84.4378 6765	41.0795 2449	2.3715 3998	91.4359 9865	38.5555 3751
59	2.0811 6676	86.4933 4099	41.5600 2419	2.4071 1308	93.8075 3863	38.9709 7292
60	2.1071 8135	88.5745 0776	42.0345 9179	2.4432 1978	96.2146 5171	39.3802 6889
61	2.1335 2111	90.6816 8910	42.5033 0054	2.4798 6807	98.6578 7149	39.7835 1614
62	2.1601 9013	92.8152 1022	42.9662 2275	2.5170 6609	101.1377 3956	40.1808 0408
63	2.1871 9250	94.9754 0034	43.4234 2988	2.5548 2208	103.6548 0565	40.5722 2077
64	2.2145 3241	97.1625 9285	43.8749 9247	2.5931 4442	106.2096 2774	40.9578 5298
65	2.2422 1407	99.3771 2526	44.3209 8022	2.6320 4158	108.8027 7215	41.3377 8618
66	2.2702 4174	101.6193 3933	44.7614 6195	2.6715 2221	111.4348 1374	41.7121 0461
67	2.2986 1976	103.8895 8107	45.1965 0563	2.7115 9504	114.1063 3594	42.0808 9125
68	2.3273 5251	106.1882 0083	45.6261 7840	2.7522 6896	116.8179 3098	42.4442 2783
69	2.3564 4442	108.5155 5334	46.0505 4656	2.7935 5300	119.5701 9995	42.8021 9490
70	2.3858 9997	110.8719 9776	46.4696 7562	2.8354 5629	122.3637 5295	43.1548 7183
71	2.4157 2372	113.2578 9773	46.8836 3024	2.8779 8814	125.1992 0924	43.5023 3678
72	2.4459 2027	115.6736 2145	47.2924 7431	2.9211 5796	128.0771 9738	43.8446 6677
73	2.4764 9427	118.1195 4172	47.6962 7093	2.9649 7533	130.9983 5534	44.1819 3771
74	2.5074 5045	120.5960 3599	48.0950 8240	3.0094 4996	133.9633 3067	44.5142 2434
75	2.5387 9358	123.1034 8644	48.4889 7027	3.0545 9171	136.9727 8063	44.8416 0034
76	2.5705 2850	125.6422 8002	48.8779 9533	3.1004 1059	140.0273 7234	45.1641 3826
77	2.6026 6011	128.2128 0852	49.2622 1761	3.1469 1674	143.1277 8292	45.4819 0962
78	2.6351 9336	130.8154 6863	49.6416 9640	3.1941 2050	146.2746 9967	45.7949 8485
79	2.6681 3327	133.4506 6199	50.0164 9027	3.2420 3230	149.4688 2016	46.1034 3335
80	2.7014 8494	136.1187 9526	50.3866 5706	3.2906 6279	152.7108 5247	46.4073 2349
81	2.7352 5350	138.8202 8020	50.7522 5389	3.3400 2273	156.0015 1525	46.7067 2265
82	2.7694 4417	141.5555 3370	51.1133 3717	3.3901 2307	159.3415 3798	47.0016 9720
83	2.8040 6222	144.3249 7787	51.4699 6264	3.4409 7492	162.7316 6105	47.2923 1251
84	2.8391 1300	147.1290 4010	51.8221 8532	3.4925 8954	166.1726 3597	47.5786 3301
85	2.8746 0191	149.9681 5310	52.1700 5958	3.5449 7838	169.6652 2551	47.8607 2218
86	2.9105 3444	152.8427 5501	52.5136 3909	3.5981 5306	173.2102 0389	48.1386 4254
87	2.9469 1612	155.7532 8945	52.8529 7688	3.6521 2535	176.8083 5695	48.4124 5571
88	2.9837 5257	158.7002 0557	53.1881 2531	3.7069 0723	180.4604 8230	48.6822 2237
89	3.0210 4948	161.6839 5814	53.5191 3611	3.7625 1084	184.1673 8954	48.9480 0234
90	3.0588 1260	164.7050 0762	53.8460 6036	3.8189 4851	187.9299 0038	49.2098 5452
91	3.0970 4775	167.7638 2021	54.1689 4850	3.8762 3273	191.7488 4889	49.4678 3696
92	3.1357 6085	170.8608 6796	54.4878 5037	3.9343 7622	195.6250 8162	49.7220 0686
93	3.1749 5786	173.9966 2881	54.8028 1518	3.9933 9187	199.5594 5784	49.9724 2055
94	3.2146 4483	177.1715 8667	55.1138 9154	4.0532 9275	203.5528 4971	50.2191 3355
95	3.2548 2789	180.3862 3151	55.4211 2744	4.1140 9214	207.6061 4246	50.4622 0054
96	3.2955 1324	183.6410 5940	55.7245 7031	4.1758 0352	211.7202 3459	50.7016 7541
97	3.3367 0716	186.9365 7264	56.0242 6698	4.2384 4057	215.8960 3811	50.9376 1124
98	3.3784 1600	190.2732 7980	56.3202 6368	4.3020 1718	220.1344 7868	51.1700 6034
99	3.4206 4620	193.6516 9580	56.6126 0610	4.3665 4744	224.4364 9586	51.3990 7422
100	3.4634 0427	197.0723 4200	56.9013 3936	4.4320 4565	228.8030 4330	51.6247 0367

Table 1 *(continued)*

n	$i = 1\frac{3}{4}\%$ $(1+i)^n$	$s_{\overline{n}\|i}$	$a_{\overline{n}\|i}$	$i = 2\%$ $(1+i)^n$	$s_{\overline{n}\|i}$	$a_{\overline{n}\|i}$
1	1.0175 0000	1.0000 0000	0.9828 0098	1.0200 0000	1.0000 0000	0.9803 9216
2	1.0353 0625	2.0175 0000	1.9486 9875	1.0404 0000	2.0200 0000	1.9415 6094
3	1.0534 2411	3.0528 0625	2.8979 8403	1.0612 0800	3.0604 0000	2.8838 8327
4	1.0718 5903	4.1062 3036	3.8309 4254	1.0824 3216	4.1216 0800	3.8077 2870
5	1.0906 1656	5.1780 8939	4.7478 5508	1.1040 8080	5.2040 4016	4.7134 5951
6	1.1097 0235	6.2687 0596	5.6489 9762	1.1261 6242	6.3081 2096	5.6014 3089
7	1.1291 2215	7.3784 0831	6.5346 4139	1.1486 8567	7.4342 8338	6.4719 9107
8	1.1488 8178	8.5075 3045	7.4050 5297	1.1716 5938	8.5829 6905	7.3254 8144
9	1.1689 8721	9.6564 1224	8.2604 9432	1.1950 9257	9.7546 2843	8.1622 3671
10	1.1894 4449	10.8253 9945	9.1012 2291	1.2189 9442	10.9497 2100	8.9825 8501
11	1.2102 5977	12.0148 4394	9.9274 9181	1.2433 7431	12.1687 1542	9.7868 4805
12	1.2314 3931	13.2251 0371	10.7395 4969	1.2682 4179	13.4120 8973	10.5753 4122
13	1.2529 8950	14.4565 4303	11.5376 4097	1.2936 0663	14.6803 3152	11.3483 7375
14	1.2749 1682	15.7095 3253	12.3220 0587	1.3194 7876	15.9739 3815	12.1062 4877
15	1.2972 2786	16.9844 4935	13.0928 8046	1.3458 6834	17.2934 1692	12.8492 6350
16	1.3199 2935	18.2816 7721	13.8504 9677	1.3727 8571	18.6392 8525	13.5777 0931
17	1.3430 2811	19.6016 0656	14.5950 8282	1.4002 4142	20.0120 7096	14.2918 7188
18	1.3665 3111	20.9446 3468	15.3268 6272	1.4282 4625	21.4123 1238	14.9920 3125
19	1.3904 4540	22.3111 6578	16.0460 5673	1.4568 1117	22.8405 5863	15.6784 6201
20	1.4147 7820	23.7016 1119	16.7528 8130	1.4859 4740	24.2973 6980	16.3514 3334
21	1.4395 3681	25.1163 8938	17.4475 4919	1.5156 6634	25.7833 1719	17.0112 0916
22	1.4647 2871	26.5559 2620	18.1302 6948	1.5459 7967	27.2989 8354	17.6580 4820
23	1.4903 6146	28.0206 5490	18.8012 4764	1.5768 9926	28.8449 6321	18.2922 0412
24	1.5164 4279	29.5110 1637	19.4606 8565	1.6084 3725	30.4218 6247	18.9139 2560
25	1.5429 8054	31.0274 5915	20.1087 8196	1.6406 0599	32.0302 9972	19.5234 5647
26	1.5699 8269	32.5704 3969	20.7457 3166	1.6734 1811	33.6709 0572	20.1210 3576
27	1.5974 5739	34.1404 2238	21.3717 2644	1.7068 8648	35.3443 2383	20.7068 9780
28	1.6254 1290	35.7378 7977	21.9869 5474	1.7410 2421	37.0512 1031	21.2812 7236
29	1.6538 5762	37.3632 9267	22.5916 0171	1.7758 4469	38.7922 3451	21.8443 8466
30	1.6828 0013	39.0171 5029	23.1858 4934	1.8113 6158	40.5680 7921	22.3964 5555
31	1.7122 4913	40.6999 5042	23.7698 7650	1.8475 8882	42.3794 4079	22.9377 0152
32	1.7422 1349	42.4121 9955	24.3438 5897	1.8845 4059	44.2270 2961	23.4683 3482
33	1.7727 0223	44.1544 1305	24.9079 6951	1.9222 3140	46.1115 7020	23.9885 6355
34	1.8037 2452	45.9271 1527	25.4623 7789	1.9606 7603	48.0338 0160	24.4985 9172
35	1.8352 8970	47.7308 3979	26.0072 5100	1.9998 8955	49.9944 7763	24.9986 1933
36	1.8674 0727	49.5661 2949	26.5427 5283	2.0398 8734	51.9943 6719	25.4888 4248
37	1.9000 8689	51.4335 3675	27.0690 4455	2.0806 8509	54.0342 5453	25.9694 5341
38	1.9333 3841	53.3336 2365	27.5862 8457	2.1222 9879	56.1149 3962	26.4406 4060
39	1.9671 7184	55.2669 6206	28.0946 2857	2.1647 4477	58.2372 3841	26.9025 8883
40	2.0015 9734	57.2341 3390	28.5942 2955	2.2080 3966	60.4019 8318	27.3554 7924
41	2.0366 2530	59.2357 3124	29.0852 3789	2.2522 0046	62.6100 2284	27.7994 8945
42	2.0722 6624	61.2723 5654	29.5678 0136	2.2972 4447	64.8622 2330	28.2347 9358
43	2.1085 3090	63.3446 2278	30.0420 6522	2.3431 8936	67.1594 6777	28.6615 6233
44	2.1454 3019	65.4531 5367	30.5081 7221	2.3900 5314	69.5026 5712	29.0799 6307
45	2.1829 7522	67.5985 8386	30.9662 6261	2.4378 5421	71.8927 1027	29.4901 5987
46	2.2211 7728	69.7815 5908	31.4164 7431	2.4866 1129	74.3305 6447	29.8923 1360
47	2.2600 4789	72.0027 3637	31.8589 4281	2.5363 4352	76.8171 7576	30.2865 8196
48	2.2995 9872	74.2627 8425	32.2938 0129	2.5870 7039	79.3535 1927	30.6731 1957
49	2.3398 4170	76.5623 8298	32.7211 8063	2.6388 1179	81.9405 8966	31.0520 7801
50	2.3807 8893	78.9022 2468	33.1412 0946	2.6915 8803	84.5794 0145	31.4236 0589

Table 1 *(continued)*

n	$i = 1\frac{3}{4}\%$ $(1+i)^n$	$s_{\overline{n}\|i}$	$a_{\overline{n}\|i}$	$i = 2\%$ $(1+i)^n$	$s_{\overline{n}\|i}$	$a_{\overline{n}\|i}$
51	2.4224 5274	81.2830 1361	33.5540 1421	2.7454 1979	87.2709 8948	31.7878 4892
52	2.4648 4566	83.7054 6635	33.9597 1913	2.8003 2819	90.0164 0927	32.1449 4992
53	2.5079 8046	86.1703 1201	34.3584 4632	2.8563 3475	92.8167 3746	32.4950 4894
54	2.5518 7012	88.6782 9247	34.7503 1579	2.9134 6144	95.6730 7221	32.8382 8327
55	2.5965 2785	91.2301 6259	35.1354 4550	2.9717 3067	98.5865 3365	33.1747 8752
56	2.6419 6708	93.8266 9043	35.5139 5135	3.0311 6529	101.5582 6432	33.5046 9365
57	2.6882 0151	96.4686 5752	35.8859 4727	3.0917 8859	104.5894 2961	33.8281 3103
58	2.7352 4503	99.1568 5902	36.2515 4523	3.1536 2436	107.6812 1820	34.1452 2650
59	2.7831 1182	101.8921 0405	36.6108 5526	3.2166 9685	110.8348 4257	34.4561 0441
60	2.8318 1628	104.6752 1588	36.9639 8552	3.2810 3079	114.0515 3942	34.7608 8668
61	2.8813 7306	107.5070 3215	37.3110 4228	3.3466 5140	117.3325 7021	35.0596 9282
62	2.9317 9709	110.3884 0522	37.6521 3000	3.4135 8443	120.6792 2161	35.3526 4002
63	2.9831 0354	113.3202 0231	37.9873 5135	3.4818 5612	124.0928 0604	35.6398 4316
64	3.0353 0785	116.3033 0585	38.3168 0723	3.5514 9324	127.5746 6216	35.9214 1486
65	3.0884 2574	119.3386 1370	38.6405 9678	3.6225 2311	131.1261 5541	36.1974 6555
66	3.1424 7319	122.4270 3944	38.9588 1748	3.6949 7357	134.7486 7852	36.4681 0348
67	3.1974 6647	125.5695 1263	39.2715 6509	3.7688 7304	138.4436 5209	36.7334 3478
68	3.2534 2213	128.7669 7910	39.5789 3375	3.8442 5050	142.2125 2513	36.9935 6351
69	3.3103 5702	132.0204 0124	39.8810 1597	3.9211 3551	146.0567 7563	37.2485 9168
70	3.3682 8827	135.3307 5826	40.1779 0267	3.9995 5822	149.9779 1114	37.4986 1929
71	3.4272 3331	138.6990 4653	40.4696 8321	4.0795 4939	153.9774 6937	37.7437 4441
72	3.4872 0990	142.1262 7984	40.7564 4542	4.1611 4038	158.0570 1875	37.9840 6314
73	3.5482 3607	145.6134 8974	41.0382 7560	4.2443 6318	162.2181 5913	38.2196 6975
74	3.6103 3020	149.1617 2581	41.3152 5857	4.3292 5045	166.4625 2231	38.4506 5662
75	3.6735 1098	152.7720 5601	41.5874 7771	4.4158 3546	170.7917 7276	38.6771 1433
76	3.7377 9742	156.4455 6699	41.8550 1495	4.5041 5216	175.2076 0821	38.8991 3170
77	3.8032 0888	160.1833 6441	42.1179 5081	4.5942 3521	179.7117 6038	39.1167 9578
78	3.8697 6503	163.9865 7329	42.3763 6443	4.6861 1991	184.3059 9558	39.3301 9194
79	3.9374 8592	167.8563 3832	42.6303 3359	4.7798 4231	188.9921 1549	39.5394 0386
80	4.0063 9192	171.7938 2424	42.8799 3474	4.8754 3916	193.7719 5780	39.7445 1359
81	4.0765 0378	175.8002 1617	43.1252 4298	4.9729 4794	198.6473 9696	39.9456 0156
82	4.1478 4260	179.8767 1995	43.3663 3217	5.0724 0690	203.6203 4490	40.1427 4663
83	4.2204 2984	184.0245 6255	43.6032 7486	5.1738 5504	208.6927 5180	40.3360 2611
84	4.2942 8737	188.2449 9239	43.8361 4237	5.2773 3214	213.8666 0683	40.5255 1579
85	4.3694 3740	192.5392 7976	44.0650 0479	5.3828 7878	219.1439 3897	40.7112 8999
86	4.4459 0255	196.9087 1716	44.2899 3099	5.4905 3636	224.5268 1775	40.8934 2156
87	4.5237 0584	201.3546 1971	44.5109 8869	5.6003 4708	230.0173 5411	41.0719 8192
88	4.6028 7070	205.8783 2555	44.7282 4441	5.7123 5402	235.6177 0119	41.2470 4110
89	4.6834 2093	210.4811 9625	44.9417 6355	5.8266 0110	241.3300 5521	41.4186 6774
90	4.7653 8080	215.1646 1718	45.1516 1037	5.9431 3313	247.1566 5632	41.5869 2916
91	4.8487 7496	219.9299 9798	45.3578 4803	6.0619 9579	253.0997 8944	41.7518 9133
92	4.9336 2853	224.7787 7295	45.5605 3860	6.1832 3570	259.1617 8523	41.9136 1895
93	5.0199 6703	229.7124 0148	45.7597 4310	6.3069 0042	265.3450 2094	42.0721 7545
94	5.1078 1645	234.7323 6850	45.9555 2147	6.4330 3843	271.6519 2135	42.2276 2299
95	5.1972 0324	239.8401 8495	46.1479 3265	6.5616 9920	278.0849 5978	42.3800 2254
96	5.2881 5429	245.0373 8819	46.3370 3455	6.6929 3318	284.6466 5898	42.5294 3386
97	5.3806 9699	250.3255 4248	46.5228 8408	6.8267 9184	291.3395 9216	42.6759 1555
98	5.4748 5919	255.7062 3947	46.7055 3718	6.9633 2768	298.1663 8400	42.8195 2505
99	5.5706 6923	261.1810 9866	46.8850 4882	7.1025 9423	305.1297 1168	42.9603 1867
100	5.6681 5594	266.7517 6789	47.0614 7304	7.2446 4612	312.2323 0591	43.0983 5164

Table 1 *(continued)*

	$i = 2\frac{1}{4}\%$			$i = 2\frac{1}{2}\%$		
n	$(1 + i)^n$	$s_{\overline{n}\|i}$	$a_{\overline{n}\|i}$	$(1 + i)^n$	$s_{\overline{n}\|i}$	$a_{\overline{n}\|i}$
1	1.0225 0000	1.0000 0000	0.9779 9511	1.0250 0000	1.0000 0000	0.9756 0976
2	1.0455 0625	2.0225 0000	1.9344 6955	1.0506 2500	2.0250 0000	1.9274 2415
3	1.0690 3014	3.0680 0625	2.8698 9687	1.0768 9063	3.0756 2500	2.8560 2356
4	1.0930 8332	4.1370 3639	3.7847 4021	1.1038 1289	4.1525 1563	3.7619 7421
5	1.1176 7769	5.2301 1971	4.6794 5253	1.1314 0821	5.2563 2852	4.6458 2850
6	1.1428 2544	6.3477 9740	5.5544 7680	1.1596 9342	6.3877 3673	5.5081 2536
7	1.1685 3901	7.4906 2284	6.4102 4626	1.1886 8575	7.5474 3015	6.3493 9060
8	1.1948 3114	8.6591 6186	7.2471 8461	1.2184 0290	8.7361 1590	7.1701 3717
9	1.2217 1484	9.8539 9300	8.0657 0622	1.2488 6297	9.9545 1880	7.9708 6553
10	1.2492 0343	11.0757 0784	8.8662 1635	1.2800 8454	11.2033 8177	8.7520 6393
11	1.2773 1050	12.3249 1127	9.6491 1134	1.3120 8666	12.4834 6631	9.5142 0871
12	1.3060 4999	13.6022 2177	10.4147 7882	1.3448 8882	13.7955 5297	10.2577 6460
13	1.3354 3611	14.9082 7176	11.1635 9787	1.3785 1104	15.1404 4179	10.9831 8497
14	1.3654 8343	16.2437 0788	11.8959 3924	1.4129 7382	16.5189 5284	11.6909 1217
15	1.3962 0680	17.6091 9130	12.6121 6551	1.4482 9817	17.9319 2666	12.3813 7773
16	1.4276 2146	19.0053 9811	13.3126 3131	1.4845 0562	19.3802 2483	13.0550 0266
17	1.4597 4294	20.4330 1957	13.9976 8343	1.5216 1826	20.8647 3045	13.7121 9772
18	1.4925 8716	21.8927 6251	14.6676 6106	1.5596 5872	22.3863 4871	14.3533 6363
19	1.5261 7037	23.3853 4966	15.3228 9590	1.5986 5019	23.9460 0743	14.9788 9134
20	1.5605 0920	24.9115 2003	15.9637 1237	1.6386 1644	25.5446 5761	15.5891 6229
21	1.5956 2066	26.4720 2923	16.5904 2775	1.6795 8185	27.1832 7405	16.1845 4857
22	1.6315 2212	28.0676 4989	17.2033 5232	1.7215 7140	28.8628 5590	16.7654 1324
23	1.6682 3137	29.6991 7201	17.8027 8955	1.7646 1068	30.5844 2730	17.3321 1048
24	1.7057 6658	31.3674 0338	18.3890 3624	1.8087 2595	32.3490 3798	17.8849 8583
25	1.7441 4632	33.0731 6996	18.9623 8263	1.8539 4410	34.1577 6393	18.4243 7642
26	1.7833 8962	34.8173 1628	19.5231 1260	1.9002 9270	36.0117 0803	18.9506 1114
27	1.8235 1588	36.6007 0590	20.0715 0376	1.9478 0002	37.9120 0073	19.4640 1087
28	1.8645 4499	38.4242 2178	20.6078 2764	1.9964 9502	39.8598 0075	19.9648 8866
29	1.9064 9725	40.2887 6677	21.1323 4977	2.0464 0739	41.8562 9577	20.4535 4991
30	1.9493 9344	42.1952 6402	21.6453 2985	2.0975 6758	43.9027 0316	20.9302 9259
31	1.9932 5479	44.1446 5746	22.1470 2186	2.1500 0677	46.0002 7074	21.3954 0741
32	2.0381 0303	46.1379 1226	22.6376 7419	2.2037 5694	48.1502 7751	21.8491 7796
33	2.0839 6034	48.1760 1528	23.1175 2977	2.2588 5086	50.3540 3445	22.2918 8094
34	2.1308 4945	50.2599 7563	23.5868 2618	2.3153 2213	52.6128 8531	22.7237 8628
35	2.1787 9356	52.3908 2508	24.0457 9577	2.3732 0519	54.9282 0744	23.1451 5734
36	2.2278 1642	54.5696 1864	24.4946 6579	2.4325 3532	57.3014 1263	23.5562 5107
37	2.2779 4229	56.7974 3506	24.9336 5848	2.4933 4870	59.7339 4794	23.9573 1812
38	2.3291 9599	59.0753 7735	25.3629 9118	2.5556 8242	62.2272 9664	24.3486 0304
39	2.3816 0290	61.4045 7334	25.7828 7646	2.6195 7448	64.7829 7906	24.7303 4443
40	2.4351 8897	63.7861 7624	26.1935 2221	2.6850 6384	67.4025 5354	25.1027 7505
41	2.4899 8072	66.2213 6521	26.5951 3174	2.7521 9043	70.0876 1737	25.4661 2200
42	2.5460 0528	68.7113 4592	26.9879 0390	2.8209 9520	72.8398 0781	25.8206 0683
43	2.6032 9040	71.2573 5121	27.3720 3316	2.8915 2008	75.6608 0300	26.1664 4569
44	2.6618 6444	73.8606 4161	27.7477 0969	2.9638 0808	78.5523 2308	26.5038 4945
45	2.7217 5639	76.5225 0605	28.1151 1950	3.0379 0328	81.5161 3116	26.8330 2386
46	2.7829 9590	79.2442 6243	28.4744 4450	3.1138 5086	84.5540 3443	27.1541 6962
47	2.8456 1331	82.0272 5834	28.8258 6259	3.1916 9713	87.6678 8530	27.4674 8255
48	2.9096 3961	84.8728 7165	29.1695 4777	3.2714 8956	90.8595 8243	27.7731 5371
49	2.9751 0650	87.7825 1126	29.5056 7019	3.3532 7680	94.1310 7199	28.0713 6947
50	3.0420 4640	90.7576 1776	29.8343 9627	3.4371 0872	97.4843 4879	28.3623 1168

Table 1 *(continued)*

	$i = 2\frac{1}{4}\%$			$i = 2\frac{1}{2}\%$		
n	$(1+i)^n$	$s_{\overline{n}\vert i}$	$a_{\overline{n}\vert i}$	$(1+i)^n$	$s_{\overline{n}\vert i}$	$a_{\overline{n}\vert i}$
51	3.1104 9244	93.7996 6416	30.1558 8877	3.5230 3644	100.9214 5751	28.6461 5774
52	3.1804 7852	96.9101 5661	30.4703 0687	3.6111 1235	104.4444 9395	28.9230 8072
53	3.2520 3929	100.0906 3513	30.7778 0623	3.7013 9016	108.0556 0629	29.1932 4948
54	3.3252 1017	103.3426 7442	31.0785 3910	3.7939 2491	111.7569 9645	29.4568 2876
55	3.4000 2740	106.6678 8460	31.3726 5438	3.8887 7303	115.5509 2136	29.7139 7928
56	3.4765 2802	110.0679 1200	31.6602 9768	3.9859 9236	119.4396 9440	29.9648 5784
57	3.5547 4990	113.5444 4002	31.9416 1142	4.0856 4217	123.4256 8676	30.2096 1740
58	3.6347 3177	117.0991 8992	32.2167 3489	4.1877 8322	127.5113 2893	30.4484 0722
59	3.7165 1324	120.7339 2169	32.4858 0429	4.2924 7780	131.6991 1215	30.6813 7290
60	3.8001 3479	124.4504 3493	32.7489 5285	4.3997 8975	135.9915 8995	30.9086 5649
61	3.8856 3782	128.2505 6972	33.0063 1086	4.5097 8449	140.3913 7970	31.1303 9657
62	3.9730 6467	132.1362 0754	33.2580 0573	4.6225 2910	144.9011 6419	31.3467 2836
63	4.0624 5862	136.1092 7221	33.5041 6208	4.7380 9233	149.5236 9330	31.5577 8377
64	4.1538 6394	140.1717 3083	33.7449 0179	4.8565 4464	154.2617 8563	31.7636 9148
65	4.2473 2588	144.3255 9477	33.9803 4405	4.9779 5826	159.1183 3027	31.9645 7705
66	4.3428 9071	148.5729 2066	34.2106 0543	5.1024 0721	164.0962 8853	32.1605 6298
67	4.4406 0576	152.9158 1137	34.4357 9993	5.2299 6739	169.1986 9574	32.3517 6876
68	4.5405 1939	157.3564 1713	34.6560 3905	5.3607 1658	174.4286 6314	32.5383 1099
69	4.6426 8107	161.8969 3651	34.8714 3183	5.4947 3449	179.7893 7971	32.7203 0340
70	4.7471 4140	166.5396 1758	35.0820 8492	5.6321 0286	185.2841 1421	32.8978 5698
71	4.8539 5208	171.2867 5898	35.2881 0261	5.7729 0543	190.9162 1706	33.0710 7998
72	4.9631 6600	176.1407 1106	35.4895 8691	5.9172 2806	196.6891 2249	33.2400 7803
73	5.0748 3723	181.1038 7705	35.6866 3756	6.0651 5876	202.6063 5055	33.4049 5417
74	5.1890 2107	186.1787 1429	35.8793 5214	6.2167 8773	208.6715 0931	33.5658 0895
75	5.3057 7405	191.3677 3536	36.0678 2605	6.3722 0743	214.8882 9705	33.7227 4044
76	5.4251 5396	196.6735 0941	36.2521 5262	6.5315 1261	221.2605 0447	33.8758 4433
77	5.5472 1993	202.0986 6337	36.4324 2310	6.6948 0043	227.7920 1709	34.0252 1398
78	5.6720 3237	207.6458 8329	36.6087 2675	6.8621 7044	234.4868 1751	34.1709 4047
79	5.7996 5310	213.3179 1567	36.7811 5085	7.0337 2470	241.3489 8795	34.3131 1265
80	5.9301 4530	219.1175 6877	36.9497 8079	7.2095 6782	248.3827 1265	34.4518 1722
81	6.0635 7357	225.0477 1407	37.1147 0004	7.3898 0701	255.5922 8047	34.5871 3875
82	6.2000 0397	231.1112 8763	37.2759 9026	7.5745 5219	262.9820 8748	34.7191 5976
83	6.3395 0406	237.3112 9160	37.4337 3130	7.7639 1599	270.5566 3966	34.8479 6074
84	6.4821 4290	243.6507 9567	37.5880 0127	7.9580 1389	278.3205 5566	34.9736 2023
85	6.6279 9112	250.1329 3857	37.7388 7655	8.1569 6424	286.2785 6955	35.0962 1486
86	6.7771 2092	256.7609 2969	37.8864 3183	8.3608 8834	294.4355 3379	35.2158 1938
87	6.9296 0614	263.5380 5060	38.0307 4018	8.5699 1055	302.7964 2213	35.3325 0671
88	7.0855 2228	270.4676 5674	38.1718 7304	8.7841 5832	311.3663 3268	35.4463 4801
89	7.2449 4653	277.5531 7902	38.3099 0028	9.0037 6228	320.1504 9100	35.5574 1269
90	7.4079 5782	284.7981 2555	38.4448 9025	9.2288 5633	329.1542 5328	35.6657 6848
91	7.5746 3688	292.2060 8337	38.5769 0978	9.4595 7774	338.3831 0961	35.7714 8144
92	7.7450 6621	299.7807 2025	38.7060 2423	9.6960 6718	347.8426 8735	35.8746 1604
93	7.9193 3020	307.5257 8645	38.8322 9754	9.9384 6886	357.5387 5453	35.9752 3516
94	8.0975 1512	315.4451 1665	38.9557 9221	10.1869 3058	367.4772 2339	36.0734 0016
95	8.2797 0921	323.5426 3177	39.0765 6940	10.4416 0385	377.6641 5398	36.1691 7089
96	8.4660 0267	331.8223 4099	39.1946 8890	10.7026 4395	388.1057 5783	36.2626 0574
97	8.6564 8773	340.2883 4366	39.3102 0920	10.9702 1004	398.8084 0177	36.3537 6170
98	8.8512 5871	348.9448 3139	39.4231 8748	11.2444 6530	409.7786 1182	36.4426 9434
99	9.0504 1203	357.7960 9010	39.5336 7968	11.5255 7693	421.0230 7711	36.5294 5790
100	9.2540 4630	366.8465 0213	39.6417 4052	11.8137 1635	432.5486 5404	36.6141 0526

Table 1 (continued)

n	$i = 2\frac{3}{4}\%$ $(1+i)^n$	$s_{\overline{n}\vert i}$	$a_{\overline{n}\vert i}$	$i = 3\%$ $(1+i)^n$	$s_{\overline{n}\vert i}$	$a_{\overline{n}\vert i}$
1	1.0275 0000	1.0000 0000	0.9732 3601	1.0300 0000	1.0000 0000	0.9708 7379
2	1.0557 5625	2.0275 0000	1.9204 2434	1.0609 0000	2.0300 0000	1.9134 6970
3	1.0847 8955	3.0832 5625	2.8422 6213	1.0927 2700	3.0909 0000	2.8286 1135
4	1.1146 2126	4.1680 4580	3.7394 2787	1.1255 0881	4.1836 2700	3.7170 9840
5	1.1452 7334	5.2826 6706	4.6125 8186	1.1592 7407	5.3091 3581	4.5797 0719
6	1.1767 6836	6.4279 4040	5.4623 6678	1.1940 5230	6.4684 0988	5.4171 9144
7	1.2091 2949	7.6047 0876	6.2894 0806	1.2298 7387	7.6624 6218	6.2302 8296
8	1.2423 8055	8.8138 3825	7.0943 1441	1.2667 7008	8.8923 3605	7.0196 9219
9	1.2765 4602	10.0562 1880	7.8776 7826	1.3047 7318	10.1591 0613	7.7861 0892
10	1.3116 5103	11.3327 6482	8.6400 7616	1.3439 1638	11.4638 7931	8.5302 0284
11	1.3477 2144	12.6444 1585	9.3820 6926	1.3842 3387	12.8077 9569	9.2526 2411
12	1.3847 8378	13.9921 3729	10.1042 0366	1.4257 6089	14.1920 2956	9.9540 0399
13	1.4228 6533	15.3769 2107	10.8070 1086	1.4685 3371	15.6177 9045	10.6349 5533
14	1.4619 9413	16.7997 8639	11.4910 0814	1.5125 8972	17.0863 2416	11.2960 7314
15	1.5021 9896	18.2617 8052	12.1566 9892	1.5579 6742	18.5989 1389	11.9379 3509
16	1.5435 0944	19.7639 7948	12.8045 7315	1.6047 0644	20.1568 8130	12.5611 0203
17	1.5859 5595	21.3074 8892	13.4351 0769	1.6528 4763	21.7615 8774	13.1661 1847
18	1.6295 6973	22.8934 4487	14.0487 6661	1.7024 3306	23.4144 3537	13.7535 1308
19	1.6743 8290	24.5230 1460	14.6460 0157	1.7535 0605	25.1168 6844	14.3237 9911
20	1.7204 2843	26.1973 9750	15.2272 5213	1.8061 1123	26.8703 7449	14.8774 7486
21	1.7677 4021	27.9178 2593	15.7929 4612	1.8602 9457	28.6764 8572	15.4150 2414
22	1.8163 5307	29.6855 6615	16.3434 9987	1.9161 0341	30.5367 8030	15.9369 1664
23	1.8663 0278	31.5019 1921	16.8793 1861	1.9735 8651	32.4528 8370	16.4436 0839
24	1.9176 2610	33.3682 2199	17.4007 9670	2.0327 9411	34.4264 7022	16.9355 4212
25	1.9703 6082	35.2858 4810	17.9083 1795	2.0937 7793	36.4592 6432	17.4131 4769
26	2.0245 4575	37.2562 0892	18.4022 5592	2.1565 9127	38.5530 4225	17.8768 4242
27	2.0802 2075	39.2807 5467	18.8829 7413	2.2212 8901	40.7096 3352	18.3270 3147
28	2.1374 2682	41.3609 7542	19.3508 2640	2.2879 2768	42.9309 2252	18.7641 0823
29	2.1962 0606	43.4984 0224	19.8061 5708	2.3565 6551	45.2188 5020	19.1884 5459
30	2.2566 0173	45.6946 0831	20.2493 0130	2.4272 6247	47.5754 1571	19.6004 4135
31	2.3186 5828	47.9512 1003	20.6805 8520	2.5000 8035	50.0026 7818	20.0004 2849
32	2.3824 2138	50.2698 6831	21.1003 2623	2.5750 8276	52.5027 5852	20.3887 6553
33	2.4479 3797	52.6522 8969	21.5088 3332	2.6523 3524	55.0778 4128	20.7657 9178
34	2.5152 5626	55.1002 2765	21.9064 0712	2.7319 0530	57.7301 7652	21.1318 3668
35	2.5844 2581	57.6154 8391	22.2933 4026	2.8138 6245	60.4620 8181	21.4872 2007
36	2.6554 9752	60.1999 0972	22.6699 1753	2.8982 7833	63.2759 4427	21.8322 5250
37	2.7285 2370	62.8554 0724	23.0364 1609	2.9852 2668	66.1742 2259	22.1672 3544
38	2.8035 5810	65.5839 3094	23.3931 0568	3.0747 8348	69.1594 4927	22.4924 6159
39	2.8806 5595	68.3874 8904	23.7402 4884	3.1670 2698	72.2342 3275	22.8082 1513
40	2.9598 7399	71.2681 4499	24.0781 0106	3.2620 3779	75.4012 5973	23.1147 7197
41	3.0412 7052	74.2280 1898	24.4069 1101	3.3598 9893	78.6632 9753	23.4123 9997
42	3.1249 0546	77.2692 8950	24.7269 2069	3.4606 9589	82.0231 9645	23.7013 5920
43	3.2108 4036	80.3941 9496	25.0383 6563	3.5645 1677	85.4838 9234	23.9819 0213
44	3.2991 3847	83.6050 3532	25.3414 7507	3.6714 5227	89.0484 0911	24.2542 7392
45	3.3898 6478	86.9041 7379	25.6364 7209	3.7815 9584	92.7198 6139	24.5187 1254
46	3.4830 8606	90.2940 3857	25.9235 7381	3.8950 4372	96.5014 5723	24.7754 4907
47	3.5788 7093	93.7771 2463	26.2029 9154	4.0118 9503	100.3965 0095	25.0247 0783
48	3.6772 8988	97.3559 9556	26.4749 3094	4.1322 5188	104.4083 9598	25.2667 0664
49	3.7784 1535	101.0332 8544	26.7395 9215	4.2562 1944	108.5406 4785	25.5016 5693
50	3.8823 2177	104.8117 0079	26.9971 6998	4.3839 0602	112.7968 6729	25.7297 6401

Table 1 *(continued)*

	$i = 2\frac{3}{4}\%$			$i = 3\%$		
n	$(1+i)^n$	$s_{\overline{n}\vert i}$	$a_{\overline{n}\vert i}$	$(1+i)^n$	$s_{\overline{n}\vert i}$	$a_{\overline{n}\vert i}$
51	3.9890 8562	108.6940 2256	27.2478 5400	4.5154 2320	117.1807 7331	25.9512 2719
52	4.0987 8547	112.6831 0818	27.4918 2871	4.6508 8590	121.6961 9651	26.1662 3999
53	4.2115 0208	116.7818 9365	27.7292 7368	4.7904 1247	126.3470 8240	26.3749 9028
54	4.3273 1838	120.9933 9573	27.9603 6368	4.9341 2485	131.1374 9488	26.5776 6047
55	4.4463 1964	125.3207 1411	28.1852 6879	5.0821 4859	136.0716 1972	26.7744 2764
56	4.5685 9343	129.7670 3375	28.4041 5454	5.2346 1305	141.1537 6831	26.9654 6373
57	4.6942 2975	134.3356 2718	28.6171 8203	5.3916 5144	146.3883 8136	27.1509 3566
58	4.8233 2107	139.0298 5692	28.8245 0806	5.5534 0098	151.7800 3280	27.3310 0549
59	4.9559 6239	143.8531 7799	29.0262 8522	5.7200 0301	157.3334 3379	27.5058 3058
60	5.0922 5136	148.8091 4038	29.2226 6201	5.8916 0310	163.0534 3680	27.6755 6367
61	5.2322 8827	153.9013 9174	29.4137 8298	6.0683 5120	168.9450 3991	27.8403 5307
62	5.3761 7620	159.1336 8002	29.5997 8879	6.2504 0173	175.0133 9110	28.0003 4279
63	5.5240 2105	164.5098 5622	29.7808 1634	6.4379 1379	181.2637 9284	28.1556 7261
64	5.6759 3162	170.0338 7726	29.9569 9887	6.6310 5120	187.7017 0662	28.3064 7826
65	5.8320 1974	175.7098 0889	30.1284 6605	6.8299 8273	194.3327 5782	28.4528 9152
66	5.9924 0029	181.5418 2863	30.2953 4409	7.0348 8222	201.1627 4055	28.5950 4031
67	6.1571 9130	187.5342 2892	30.4577 5581	7.2459 2868	208.1976 2277	28.7330 4884
68	6.3265 1406	193.6914 2022	30.6158 2074	7.4633 0654	215.4435 5145	28.8670 3771
69	6.5004 9319	200.0179 3427	30.7696 5522	7.6872 0574	222.9068 5800	28.9971 2399
70	6.6792 5676	206.5184 2746	30.9193 7247	7.9178 2191	230.5940 6374	29.1234 2135
71	6.8629 3632	213.1976 8422	31.0650 8270	8.1553 5657	238.5118 8565	29.2460 4015
72	7.0516 6706	220.0606 2054	31.2068 9314	8.4000 1727	246.6672 4222	29.3650 8752
73	7.2455 8791	227.1122 8760	31.3449 0816	8.6520 1778	255.0672 5949	29.4806 6750
74	7.4448 4158	234.3578 7551	31.4792 2936	8.9115 7832	263.7192 7727	29.5928 8107
75	7.6495 7472	241.8027 1709	31.6099 5558	9.1789 2567	272.6308 5559	29.7018 2628
76	7.8599 3802	249.4522 9181	31.7371 8304	9.4542 9344	281.8097 8126	29.8075 9833
77	8.0760 8632	257.3122 2983	31.8610 0540	9.7379 2224	291.2640 7469	29.9102 8964
78	8.2981 7869	265.3883 1615	31.9815 1377	10.0300 5991	301.0019 9693	30.0099 8994
79	8.5263 7861	273.6864 9485	32.0987 9685	10.3309 6171	311.0320 5684	30.1067 8635
80	8.7608 5402	282.2128 7345	32.2129 4098	10.6408 9056	321.3630 1855	30.2007 6345
81	9.0017 7751	290.9737 2747	32.3240 3015	10.9601 1727	332.0039 0910	30.2920 0335
82	9.2493 2639	299.9755 0498	32.4321 4613	11.2889 2079	342.9640 2638	30.3805 8577
83	9.5036 8286	309.2248 3137	32.5373 6850	11.6275 8842	354.2529 4717	30.4665 8813
84	9.7650 3414	318.7285 1423	32.6397 7469	11.9764 1607	365.8805 3558	30.5500 8556
85	10.0335 7258	328.4935 4837	32.7394 4009	12.3357 0855	377.8569 5165	30.6311 5103
86	10.3094 9583	338.5271 2095	32.8364 3804	12.7057 7981	390.1926 6020	30.7098 5537
87	10.5930 0696	348.8366 1678	32.9308 3994	13.0869 5320	402.8984 4001	30.7862 6735
88	10.8843 1465	359.4296 2374	33.0227 1527	13.4795 6180	415.9853 9321	30.8604 5374
89	11.1836 3331	370.3139 3839	33.1121 3165	13.8839 4865	429.4649 5500	30.9324 7936
90	11.4911 8322	381.4975 7170	33.1991 5489	14.3004 6711	443.3489 0365	31.0024 0714
91	11.8071 9076	392.9887 5492	33.2838 4905	14.7294 8112	457.6493 7076	31.0702 9820
92	12.1318 8851	404.7959 4568	33.3662 7644	15.1713 6556	472.3788 5189	31.1362 1184
93	12.4655 1544	416.9278 3418	33.4464 9776	15.6265 0652	487.5502 1744	31.2002 0567
94	12.8083 1711	429.3933 4962	33.5245 7202	16.0953 0172	503.1767 2397	31.2623 3560
95	13.1605 4584	442.2016 6674	33.6005 5671	16.5781 6077	519.2720 2568	31.3226 5592
96	13.5224 6085	455.3622 1257	33.6745 0775	17.0755 0559	535.8501 8645	31.3812 1934
97	13.8943 2852	468.8846 7342	33.7464 7956	17.5877 7076	552.9256 9205	31.4380 7703
98	14.2764 2255	482.7790 0194	33.8165 2512	18.1154 0388	570.5134 6281	31.4932 7867
99	14.6690 2417	497.0554 2449	33.8846 9598	18.6588 6600	588.6288 6669	31.5468 7250
100	15.0724 2234	511.7244 4867	33.9510 4232	19.2186 3198	607.2877 3270	31.5989 0534

Table 1 *(continued)*

	$i = 3\frac{1}{2}\%$			$i = 4\%$		
n	$(1+i)^n$	$s_{\overline{n}\rvert i}$	$a_{\overline{n}\rvert i}$	$(1+i)^n$	$s_{\overline{n}\rvert i}$	$a_{\overline{n}\rvert i}$
1	1.0350 0000	1.0000 0000	0.9661 8357	1.0400 0000	1.0000 0000	0.9615 3846
2	1.0712 2500	2.0350 0000	1.8996 9428	1.0816 0000	2.0400 0000	1.8860 9467
3	1.1087 1788	3.1062 2500	2.8016 3698	1.1248 6400	3.1216 0000	2.7750 9103
4	1.1475 2300	4.2149 4288	3.6730 7921	1.1698 5856	4.2464 6400	3.6298 9522
5	1.1876 8631	5.3624 6588	4.5150 5238	1.2166 5290	5.4163 2256	4.4518 2233
6	1.2292 5533	6.5501 5218	5.3285 5302	1.2653 1902	6.6329 7546	5.2421 3686
7	1.2722 7926	7.7794 0751	6.1145 4398	1.3159 3178	7.8982 9448	6.0020 5467
8	1.3168 0904	9.0516 8677	6.8739 5554	1.3685 6905	9.2142 2626	6.7327 4487
9	1.3628 9735	10.3684 9581	7.6076 8651	1.4233 1181	10.5827 9531	7.4353 3161
10	1.4105 9876	11.7313 9316	8.3166 0532	1.4802 4428	12.0061 0712	8.1108 9578
11	1.4599 6972	13.1419 9192	9.0015 5104	1.5394 5406	13.4863 5141	8.7604 7671
12	1.5110 6866	14.6019 6164	9.6633 3433	1.6010 3222	15.0258 0546	9.3850 7376
13	1.5639 5606	16.1130 3030	10.3027 3849	1.6650 7351	16.6268 3768	9.9856 4785
14	1.6186 9452	17.6769 8636	10.9205 2028	1.7316 7645	18.2919 1119	10.5631 2293
15	1.6753 4883	19.2956 8088	11.5174 1090	1.8009 4351	20.0235 8764	11.1183 8743
16	1.7339 8604	20.9710 2971	12.0941 1681	1.8729 8125	21.8245 3114	11.6522 9561
17	1.7946 7555	22.7050 1575	12.6513 2059	1.9479 0050	23.6975 1239	12.1656 6885
18	1.8574 8920	24.4996 9130	13.1896 8173	2.0258 1652	25.6454 1288	12.6592 9697
19	1.9225 0132	26.3571 8050	13.7098 3742	2.1068 4918	27.6712 2940	13.1339 3940
20	1.9897 8886	28.2796 8181	14.2124 0330	2.1911 2314	29.7780 7858	13.5903 2634
21	2.0594 3147	30.2694 7068	14.6979 7420	2.2787 6807	31.9692 0172	14.0291 5995
22	2.1315 1158	32.3289 0215	15.1671 2484	2.3699 1879	34.2479 6979	14.4511 1533
23	2.2061 1448	34.4604 1373	15.6204 1047	2.4647 1554	36.6178 8858	14.8568 4167
24	2.2833 2849	36.6665 2821	16.0583 6760	2.5633 0416	39.0826 0412	15.2469 6314
25	2.3632 4498	38.9498 5669	16.4815 1459	2.6658 3633	41.6459 0829	15.6220 7994
26	2.4459 5856	41.3131 0168	16.8903 5226	2.7724 6978	44.3117 4462	15.9827 6918
27	2.5315 6711	43.7590 6024	17.2853 6451	2.8833 6858	47.0842 1440	16.3295 8575
28	2.6201 7196	46.2906 2734	17.6670 1885	2.9987 0332	49.9675 8298	16.6630 6322
29	2.7118 7798	48.9107 9930	18.0357 6700	3.1186 5145	52.9662 8630	16.9837 1463
30	2.8067 9370	51.6226 7728	18.3920 4541	3.2433 9751	56.0849 3775	17.2920 3330
31	2.9050 3148	54.4294 7098	18.7362 7576	3.3731 3341	59.3283 3526	17.5884 9356
32	3.0067 0759	57.3345 0247	19.0688 6547	3.5080 5875	62.7014 6867	17.8735 5150
33	3.1119 4235	60.3412 1005	19.3902 0818	3.6483 8110	66.2095 2742	18.1476 4567
34	3.2208 6033	63.4531 5240	19.7006 8423	3.7943 1634	69.8579 0851	18.4111 9776
35	3.3335 9045	66.6740 1274	20.0006 6110	3.9460 8899	73.6522 2486	18.6646 1323
36	3.4502 6611	70.0076 0318	20.2904 9381	4.1039 3255	77.5983 1385	18.9082 8195
37	3.5710 2543	73.4578 6930	20.5705 2542	4.2680 8986	81.7022 4640	19.1425 7880
38	3.6960 1132	77.0288 9472	20.8410 8736	4.4388 1345	85.9703 3626	19.3678 6423
39	3.8253 7171	80.7249 0604	21.1024 9987	4.6163 6599	90.4091 4971	19.5844 8484
40	3.9592 5972	84.5502 7775	21.3550 7234	4.8010 2063	95.0255 1570	19.7927 7388
41	4.0978 3381	88.5095 3747	21.5991 0371	4.9930 6145	99.8265 3633	19.9930 5181
42	4.2412 5799	92.6073 7128	21.8348 8281	5.1927 8391	104.8195 9778	20.1856 2674
43	4.3897 0202	96.8486 2928	22.0626 8870	5.4004 9527	110.0123 8169	20.3707 9494
44	4.5433 4160	101.2383 3130	22.2827 9102	5.6165 1508	115.4128 7696	20.5488 4129
45	4.7023 5855	105.7816 7290	22.4954 5026	5.8411 7568	121.0293 9204	20.7200 3970
46	4.8669 4110	110.4840 3145	22.7009 1813	6.0748 2271	126.8705 6772	20.8846 5356
47	5.0372 8404	115.3509 7255	22.8994 3780	6.3178 1562	132.9453 9043	21.0429 3612
48	5.2135 8898	120.3882 5659	23.0912 4425	6.5705 2824	139.2632 0604	21.1951 3088
49	5.3960 6459	125.6018 4557	23.2765 6450	6.8333 4937	145.8337 3429	21.3414 7200
50	5.5849 2686	130.9979 1016	23.4556 1787	7.1066 8335	152.6670 8366	21.4821 8462

Table 1 *(continued)*

n	$i = 3\frac{1}{2}\%$ $(1+i)^n$	$s_{\overline{n}\vert i}$	$a_{\overline{n}\vert i}$	$i = 4\%$ $(1+i)^n$	$s_{\overline{n}\vert i}$	$a_{\overline{n}\vert i}$
51	5.7803 9930	136.5828 3702	23.6286 1630	7.3909 5068	159.7737 6700	21.6174 8521
52	5.9827 1327	142.3632 3631	23.7957 6454	7.6865 8871	167.1647 1768	21.7475 8193
53	6.1921 0824	148.3459 4958	23.9572 6043	7.9940 5226	174.8513 0639	21.8726 7493
54	6.4088 3202	154.5380 5782	24.1132 9510	8.3138 1435	182.8453 5865	21.9929 5667
55	6.6331 4114	160.9468 8984	24.2640 5323	8.6463 6692	191.1591 7299	22.1086 1218
56	6.8653 0108	167.5800 3099	24.4097 1327	8.9922 2160	199.8055 3991	22.2198 1940
57	7.1055 8662	174.4453 3207	24.5504 4760	9.3519 1046	208.7977 6151	22.3267 4943
58	7.3542 8215	181.5509 1869	24.6864 2281	9.7259 8688	218.1496 7197	22.4295 6676
59	7.6116 8203	188.9052 0085	24.8177 9981	10.1150 2635	227.8756 5885	22.5284 2957
60	7.8780 9090	196.5168 8288	24.9447 3412	10.5196 2741	237.9906 8520	22.6234 8997
61	8.1538 2408	204.3949 7378	25.0673 7596	10.9404 1250	248.5103 1261	22.7148 9421
62	8.4392 0793	212.5487 9786	25.1858 7049	11.3780 2900	259.4507 2511	22.8027 8289
63	8.7345 8020	220.9880 0579	25.3003 5796	11.8331 5016	270.8287 5412	22.8872 9124
64	9.0402 9051	229.7225 8599	25.4109 7388	12.3064 7617	282.6619 0428	22.9685 4927
65	9.3567 0068	238.7628 7650	25.5178 4916	12.7987 3522	294.9683 8045	23.0466 8199
66	9.6841 8520	248.1195 7718	25.6211 1030	13.3106 8463	307.7671 1567	23.1218 0961
67	10.0231 3168	257.8037 6238	25.7208 7951	13.8431 1201	321.0778 0030	23.1940 4770
68	10.3739 4129	267.8268 9406	25.8172 7489	14.3968 3649	334.9209 1231	23.2635 0740
69	10.7370 2924	278.2008 3535	25.9104 1052	14.9727 0995	349.3177 4880	23.3302 9558
70	11.1128 2526	288.9378 6459	26.0003 9664	15.5716 1835	364.2904 5876	23.3945 1498
71	11.5017 7414	300.0506 8985	26.0873 3975	16.1944 8308	379.8620 7711	23.4562 6440
72	11.9043 3624	311.5524 6400	26.1713 4275	16.8422 6241	396.0565 6019	23.5156 3885
73	12.3209 8801	323.4568 0024	26.2525 0508	17.5159 5290	412.8988 2260	23.5727 2966
74	12.7522 2259	335.7777 8824	26.3309 2278	18.2165 9102	430.4147 7550	23.6276 2468
75	13.1985 5038	348.5300 1083	26.4066 8868	18.9452 5466	448.6313 6652	23.6804 0834
76	13.6604 9964	361.7285 6121	26.4798 9244	19.7030 6485	467.5766 2118	23.7311 6187
77	14.1386 1713	375.3890 6085	26.5506 2072	20.4911 8744	487.2796 8603	23.7799 6333
78	14.6334 6873	389.5276 7798	26.6189 5721	21.3108 3494	507.7708 7347	23.8268 8782
79	15.1456 4013	404.1611 4671	26.6849 8281	22.1632 6834	529.0817 0841	23.8720 0752
80	15.6757 3754	419.3067 8685	26.7487 7567	23.0497 9907	551.2449 7675	23.9153 9185
81	16.2243 8835	434.9825 2439	26.8104 1127	23.9717 9103	574.2947 7582	23.9571 0755
82	16.7922 4195	451.2069 1274	26.8699 6258	24.9306 6267	598.2665 6685	23.9972 1879
83	17.3799 7041	467.9991 5469	26.9275 0008	25.9278 8918	623.1972 2952	24.0357 8730
84	17.9882 6938	485.3791 2510	26.9830 9186	26.9650 0475	649.1251 1870	24.0728 7241
85	18.6178 5881	503.3673 9448	27.0368 0373	28.0436 0494	676.0901 2345	24.1085 3116
86	19.2694 8387	521.9852 5329	27.0886 9926	29.1653 4914	704.1337 2839	24.1428 1842
87	19.9439 1580	541.2547 3715	27.1388 3986	30.3319 6310	733.2990 7753	24.1757 8694
88	20.6419 5285	561.1986 5295	27.1872 8489	31.5452 4163	763.6310 4063	24.2074 8745
89	21.3644 2120	581.8406 0581	27.2340 9168	32.8070 5129	795.1762 8225	24.2379 6870
90	22.1121 7595	603.2050 2701	27.2793 1564	34.1193 3334	827.9833 3354	24.2672 7759
91	22.8861 0210	625.3172 0295	27.3230 1028	35.4841 0668	862.1026 6688	24.2954 5923
92	23.6871 1568	648.2033 0506	27.3652 2732	36.9034 7094	897.5867 7356	24.3225 5695
93	24.5161 6473	671.8904 2074	27.4060 1673	38.3796 0978	934.4902 4450	24.3486 1245
94	25.3742 3049	696.4065 8546	27.4454 2680	39.9147 9417	972.8698 5428	24.3736 6582
95	26.2623 2856	721.7808 1595	27.4835 0415	41.5113 8594	1012.7846 4845	24.3977 5559
96	27.1815 1006	748.0431 4451	27.5202 9387	43.1718 4138	1054.2960 3439	24.4209 1884
97	28.1328 6291	775.2246 5457	27.5558 3948	44.8987 1503	1097.4678 7577	24.4431 9119
98	29.1175 1311	803.3575 1748	27.5901 8308	46.6946 6363	1142.3665 9080	24.4646 0692
99	30.1366 2607	832.4750 3059	27.6233 6529	48.5624 5018	1189.0612 5443	24.4851 9896
100	31.1914 0798	862.6116 5666	27.6554 2540	50.5049 4818	1237.6237 0461	24.5049 9900

Table 1 *(continued)*

	$i = 5\%$			$i = 6\%$		
n	$(1+i)^n$	$s_{\overline{n}\mid i}$	$a_{\overline{n}\mid i}$	$(1+i)^n$	$s_{\overline{n}\mid i}$	$a_{\overline{n}\mid i}$
1	1.0500 0000	1.0000 0000	0.9523 8095	1.0600 0000	1.0000 0000	0.9433 9623
2	1.1025 0000	2.0500 0000	1.8594 1043	1.1236 0000	2.0600 0000	1.8333 9267
3	1.1576 2500	3.1525 0000	2.7232 4803	1.1910 1600	3.1836 0000	2.6730 1195
4	1.2155 0625	4.3101 2500	3.5459 5050	1.2624 7696	4.3746 1600	3.4651 0561
5	1.2762 8156	5.5256 3125	4.3294 7667	1.3382 2558	5.6370 9296	4.2123 6379
6	1.3400 9564	6.8019 1281	5.0756 9207	1.4185 1911	6.9753 1854	4.9173 2433
7	1.4071 0042	8.1420 0845	5.7863 7340	1.5036 3026	8.3938 3765	5.5823 8144
8	1.4774 5544	9.5491 0888	6.4632 1276	1.5938 4807	9.8974 6791	6.2097 9381
9	1.5513 2822	11.0265 6432	7.1078 2168	1.6894 7896	11.4913 1598	6.8016 9227
10	1.6288 9463	12.5778 9254	7.7217 3493	1.7908 4770	13.1807 9494	7.3600 8705
11	1.7103 3936	14.2067 8716	8.3064 1422	1.8982 9856	14.9716 4264	7.8868 7458
12	1.7958 5633	15.9171 2652	8.8632 5164	2.0121 9647	16.8699 4120	8.3838 4394
13	1.8856 4914	17.7129 8285	9.3935 7299	2.1329 2826	18.8821 3767	8.8526 8296
14	1.9799 3160	19.5986 3199	9.8986 4094	2.2609 0396	21.0150 6593	9.2949 8393
15	2.0789 2818	21.5785 6359	10.3796 5804	2.3965 5819	23.2759 6988	9.7122 4899
16	2.1828 7459	23.6574 9177	10.8377 6956	2.5403 5168	25.6725 2808	10.1058 9527
17	2.2920 1832	25.8403 6636	11.2740 6625	2.6927 7279	28.2128 7976	10.4772 5969
18	2.4066 1923	28.1323 8467	11.6895 8690	2.8543 3915	30.9056 5255	10.8276 0348
19	2.5269 5020	30.5390 0391	12.0853 2086	3.0255 9950	33.7599 9170	11.1581 1649
20	2.6532 9771	33.0659 5410	12.4622 1034	3.2071 3547	36.7855 9120	11.4699 2122
21	2.7859 6259	35.7192 5181	12.8211 5271	3.3995 6360	39.9927 2668	11.7640 7662
22	2.9252 6072	38.5052 1440	13.1630 0258	3.6035 3742	43.3922 9028	12.0415 8172
23	3.0715 2376	41.4304 7512	13.4885 7388	3.8197 4966	46.9958 2769	12.3033 7898
24	3.2250 9994	44.5019 9887	13.7986 4179	4.0489 3464	50.8155 7735	12.5503 5753
25	3.3863 5494	47.7270 9882	14.0939 4457	4.2918 7072	54.8645 1200	12.7833 5616
26	3.5556 7269	51.1134 5376	14.3751 8530	4.5493 8296	59.1563 8272	13.0031 6619
27	3.7334 5632	54.6691 2645	14.6430 3362	4.8223 4594	63.7057 6568	13.2105 3414
28	3.9201 2914	58.4025 8277	14.8981 2726	5.1116 8670	68.5281 1162	13.4061 6428
29	4.1161 3560	62.3227 1191	15.1410 7358	5.4183 8790	73.6397 9832	13.5907 2102
30	4.3219 4238	66.4388 4750	15.3724 5103	5.7434 9117	79.0581 8622	13.7648 3115
31	4.5380 3949	70.7607 8988	15.5928 1050	6.0881 0064	84.8016 7739	13.9290 8599
32	4.7649 4147	75.2988 2937	15.8026 7667	6.4533 8668	90.8897 7803	14.0840 4339
33	5.0031 8854	80.0637 7084	16.0025 4921	6.8405 8988	97.3431 6471	14.2302 2961
34	5.2533 4797	85.0669 5938	16.1929 0401	7.2510 2528	104.1837 5460	14.3681 4114
35	5.5160 1537	90.3203 0735	16.3741 9429	7.6860 8679	111.4347 7987	14.4982 4636
36	5.7918 1614	95.8363 2272	16.5468 5171	8.1472 5200	119.1208 6666	14.6209 8713
37	6.0814 0694	101.6281 3886	16.7112 8734	8.6360 8712	127.2681 1866	14.7367 8031
38	6.3854 7729	107.7095 4580	16.8678 9271	9.1542 5235	135.9042 0578	14.8460 1916
39	6.7047 5115	114.0950 2309	17.0170 4067	9.7035 0749	145.0584 5813	14.9490 7468
40	7.0399 8871	120.7997 7424	17.1590 8635	10.2857 1794	154.7619 6562	15.0462 9687
41	7.3919 8815	127.8397 6295	17.2943 6796	10.9028 6101	165.0476 8356	15.1380 1592
42	7.7615 8756	135.2317 5110	17.4232 0758	11.5570 3267	175.9505 4457	15.2245 4332
43	8.1496 6693	142.9933 3866	17.5459 1198	12.2504 5463	187.5075 7724	15.3061 7294
44	8.5571 5028	151.1430 0559	17.6627 7331	12.9854 8191	199.7580 3188	15.3831 8202
45	8.9850 0779	159.7001 5587	17.7740 6982	13.7646 1083	212.7435 1379	15.4558 3209
46	9.4342 5818	168.6851 6366	17.8800 6650	14.5904 8748	226.5081 2462	15.5243 6990
47	9.9059 7109	178.1194 2185	17.9810 1571	15.4659 1673	241.0986 1210	15.5890 2821
48	10.4012 6965	188.0253 9294	18.0771 5782	16.3938 7173	256.5645 2882	15.6500 2661
49	10.9213 3313	198.4266 6259	18.1687 2173	17.3775 0403	272.9584 0055	15.7075 7227
50	11.4673 9979	209.3479 9572	18.2559 2546	18.4201 5427	290.3359 0458	15.7618 6064

Table 1 *(continued)*

	$i = 7\%$			$i = 8\%$		
n	$(1 + i)^n$	$s_{\overline{n}\vert i}$	$a_{\overline{n}\vert i}$	$(1 + i)^n$	$s_{\overline{n}\vert i}$	$a_{\overline{n}\vert i}$
1	1.0700 0000	1.0000 0000	0.9345 7944	1.0800 0000	1.0000 0000	0.9259 2593
2	1.1449 0000	2.0700 0000	1.8080 1817	1.1664 0000	2.0800 0000	1.7832 6475
3	1.2250 4300	3.2149 0000	2.6243 1604	1.2597 1200	3.2464 0000	2.5770 9699
4	1.3107 9601	4.4399 4300	3.3872 1126	1.3604 8896	4.5061 1200	3.3121 2684
5	1.4025 5173	5.7507 3901	4.1001 9744	1.4693 2808	5.8666 0096	3.9927 1004
6	1.5007 3035	7.1532 9074	4.7665 3966	1.5868 7432	7.3359 2904	4.6228 7966
7	1.6057 8148	8.6540 2109	5.3892 8940	1.7138 2427	8.9228 0336	5.2063 7006
8	1.7181 8618	10.2598 0257	5.9712 9851	1.8509 3021	10.6366 2763	5.7466 3894
9	1.8384 5921	11.9779 8875	6.5152 3225	1.9990 0463	12.4875 5784	6.2468 8791
10	1.9671 5136	13.8164 4796	7.0235 8154	2.1589 2500	14.4865 6247	6.7100 8140
11	2.1048 5195	15.7835 9932	7.4986 7434	2.3316 3900	16.6454 8746	7.1389 6426
12	2.2521 9159	17.8884 5127	7.9426 8630	2.5181 7012	18.9771 2646	7.5360 7802
13	2.4098 4500	20.1406 4286	8.3576 5074	2.7196 2373	21.4952 9658	7.9037 7594
14	2.5785 3415	22.5504 8786	8.7454 6799	2.9371 9362	24.2149 2030	8.2442 3698
15	2.7590 3154	25.1290 2201	9.1079 1401	3.1721 6911	27.1521 1393	8.5594 7869
16	2.9521 6375	27.8880 5355	9.4466 4860	3.4259 4264	30.3242 8304	8.8513 6916
17	3.1588 1521	30.8402 1730	9.7632 2299	3.7000 1805	33.7502 2569	9.1216 3811
18	3.3799 3228	33.9990 3251	10.0590 8691	3.9960 1950	37.4502 4374	9.3718 8714
19	3.6165 2754	37.3789 6479	10.3355 9524	4.3157 0106	41.4462 6324	9.6035 9920
20	3.8696 8446	40.9954 9232	10.5940 1425	4.6609 5714	45.7619 6430	9.8181 4741
21	4.1405 6237	44.8651 7678	10.8355 2733	5.0338 3372	50.4229 2144	10.0168 0316
22	4.4304 0174	49.0057 3916	11.0612 4050	5.4365 4041	55.4567 5516	10.2007 4366
23	4.7405 2986	53.4361 4090	11.2721 8738	5.8714 6365	60.8932 9557	10.3710 5895
24	5.0723 6695	58.1766 7076	11.4693 3400	6.3411 8074	66.7647 5922	10.5287 5828
25	5.4274 3264	63.2490 3772	11.6535 8318	6.8484 7520	73.1059 3995	10.6747 7619
26	5.8073 5292	68.6764 7036	11.8257 7867	7.3963 5321	79.9544 1515	10.8099 7795
27	6.2138 6763	74.4838 2328	11.9867 0904	7.9880 6147	87.3507 6836	10.9351 6477
28	6.6488 3836	80.6976 9091	12.1371 1125	8.6271 0639	95.3388 2983	11.0510 7849
29	7.1142 5705	87.3465 2927	12.2776 7407	9.3172 7490	103.9659 3622	11.1584 0601
30	7.6122 5504	94.4607 8632	12.4090 4118	10.0626 5689	113.2832 1111	11.2577 8334
31	8.1451 1290	102.0730 4137	12.5318 1419	10.8676 6944	123.3458 6800	11.3497 9939
32	8.7152 7080	110.2181 5426	12.6465 5532	11.7370 8300	134.2135 3744	11.4349 9944
33	9.3253 3975	118.9334 2506	12.7537 9002	12.6760 4964	145.9506 2044	11.5138 8837
34	9.9781 1354	128.2587 6481	12.8540 0936	13.6901 3361	158.6266 7007	11.5869 3367
35	10.6765 8148	138.2368 7835	12.9476 7230	14.7853 4429	172.3168 0368	11.6545 6822
36	11.4239 4219	148.9134 5984	13.0352 0776	15.9681 7184	187.1021 4797	11.7171 9279
37	12.2236 1814	160.3374 0202	13.1170 1660	17.2456 2558	203.0703 1981	11.7751 7851
38	13.0792 7141	172.5610 2017	13.1934 7345	18.6252 7563	220.3159 4540	11.8288 6899
39	13.9948 2041	185.6402 9158	13.2649 2846	20.1152 9768	238.9412 2103	11.8785 8240
40	14.9744 5784	199.6351 1199	13.3317 0884	21.7245 2150	259.0565 1871	11.9246 1333
41	16.0226 6989	214.6095 6983	13.3941 2041	23.4624 8322	280.7810 4021	11.9672 3457
42	17.1442 5678	230.6322 3972	13.4524 4898	25.3394 8187	304.2435 2342	12.0066 9867
43	18.3443 5475	247.7764 9650	13.5069 6167	27.3666 4042	329.5830 0530	12.0432 3951
44	19.6284 5959	266.1208 5125	13.5579 0810	29.5559 7166	356.9496 4572	12.0770 7362
45	21.0024 5176	285.7493 1084	13.6055 2159	31.9204 4939	386.5056 1738	12.1084 0150
46	22.4726 2338	306.7517 6260	13.6500 2018	34.4740 8534	418.4260 6677	12.1374 0880
47	24.0457 0702	329.2243 8598	13.6916 0764	37.2320 1217	452.9001 5211	12.1642 6741
48	25.7289 0651	353.2700 9300	13.7304 7443	40.2105 7314	490.1321 6428	12.1891 3649
49	27.5299 2997	378.9989 9951	13.7667 9853	43.4274 1899	530.3427 3742	12.2121 6341
50	29.4570 2506	406.5289 2947	13.8007 4629	46.9016 1251	573.7701 5642	12.2334 8464

Table 2 Binomial Probabilities

		p										
n	*x*	0.05	0.1	0.2	0.3	0.4	0.5	0.6	0.7	0.8	0.9	0.95
2	0	0.902	0.810	0.640	0.490	0.360	0.250	0.160	0.090	0.040	0.010	0.002
	1	0.095	0.180	0.320	0.420	0.480	0.500	0.480	0.420	0.320	0.180	0.095
	2	0.002	0.010	0.040	0.090	0.160	0.250	0.360	0.490	0.640	0.810	0.902
3	0	0.857	0.729	0.512	0.343	0.216	0.125	0.064	0.027	0.008	0.001	
	1	0.135	0.243	0.384	0.441	0.432	0.375	0.288	0.189	0.096	0.027	0.007
	2	0.007	0.027	0.096	0.189	0.288	0.375	0.432	0.441	0.384	0.243	0.135
	3		0.001	0.008	0.027	0.064	0.125	0.216	0.343	0.512	0.729	0.857
4	0	0.815	0.656	0.410	0.240	0.130	0.062	0.026	0.008	0.002		
	1	0.171	0.292	0.410	0.412	0.346	0.250	0.154	0.076	0.026	0.004	
	2	0.014	0.049	0.154	0.265	0.346	0.375	0.346	0.265	0.154	0.049	0.014
	3		0.004	0.026	0.076	0.154	0.250	0.346	0.412	0.410	0.292	0.171
	4			0.002	0.008	0.026	0.062	0.130	0.240	0.410	0.656	0.815
5	0	0.774	0.590	0.328	0.168	0.078	0.031	0.010	0.002			
	1	0.204	0.328	0.410	0.360	0.259	0.156	0.077	0.028	0.006		
	2	0.021	0.073	0.205	0.309	0.346	0.312	0.230	0.132	0.051	0.008	0.001
	3	0.001	0.008	0.051	0.132	0.230	0.312	0.346	0.309	0.205	0.073	0.021
	4			0.006	0.028	0.077	0.156	0.259	0.360	0.410	0.328	0.204
	5				0.002	0.010	0.031	0.078	0.168	0.328	0.590	0.774
6	0	0.735	0.531	0.262	0.118	0.047	0.016	0.004	0.001			
	1	0.232	0.354	0.393	0.303	0.187	0.094	0.037	0.010	0.002		
	2	0.031	0.098	0.246	0.324	0.311	0.234	0.138	0.060	0.015	0.001	
	3	0.002	0.015	0.082	0.185	0.276	0.312	0.276	0.185	0.082	0.015	0.002
	4		0.001	0.015	0.060	0.138	0.234	0.311	0.324	0.246	0.098	0.031
	5			0.002	0.010	0.037	0.094	0.187	0.303	0.393	0.354	0.232
	6				0.001	0.004	0.016	0.047	0.118	0.262	0.531	0.735
7	0	0.698	0.478	0.210	0.082	0.028	0.008	0.002				
	1	0.257	0.372	0.367	0.247	0.131	0.055	0.017	0.004			
	2	0.041	0.124	0.275	0.318	0.261	0.164	0.077	0.025	0.004		
	3	0.004	0.023	0.115	0.227	0.290	0.273	0.194	0.097	0.029	0.003	
	4		0.003	0.029	0.097	0.194	0.273	0.290	0.227	0.115	0.023	0.004
	5			0.004	0.025	0.077	0.164	0.261	0.318	0.275	0.124	0.041
	6				0.004	0.017	0.055	0.131	0.247	0.367	0.372	0.257
	7					0.002	0.008	0.028	0.082	0.210	0.478	0.698
8	0	0.663	0.430	0.168	0.058	0.017	0.004	0.001				
	1	0.279	0.383	0.336	0.198	0.090	0.031	0.008	0.001			
	2	0.051	0.149	0.294	0.296	0.209	0.109	0.041	0.010	0.001		
	3	0.005	0.033	0.147	0.254	0.279	0.219	0.124	0.047	0.009		
	4		0.005	0.046	0.136	0.232	0.273	0.232	0.136	0.046	0.005	
	5			0.009	0.047	0.124	0.219	0.279	0.254	0.147	0.033	0.005
	6			0.001	0.010	0.041	0.109	0.209	0.296	0.294	0.149	0.051
	7				0.001	0.008	0.031	0.090	0.198	0.336	0.383	0.279
	8					0.001	0.004	0.017	0.058	0.168	0.430	0.663

Table 2 *(continued)*

		p										
n	*x*	0.05	0.1	0.2	0.3	0.4	0.5	0.6	0.7	0.8	0.9	0.95
9	0	0.630	0.387	0.134	0.040	0.010	0.002					
	1	0.299	0.387	0.302	0.156	0.060	0.018	0.004				
	2	0.063	0.172	0.302	0.267	0.161	0.070	0.021	0.004			
	3	0.008	0.045	0.176	0.267	0.251	0.164	0.074	0.021	0.003		
	4	0.001	0.007	0.066	0.172	0.251	0.246	0.167	0.074	0.017	0.001	
	5		0.001	0.017	0.074	0.167	0.246	0.251	0.172	0.066	0.007	0.001
	6			0.003	0.021	0.074	0.164	0.251	0.267	0.176	0.045	0.008
	7				0.004	0.021	0.070	0.161	0.267	0.302	0.172	0.063
	8					0.004	0.018	0.060	0.156	0.302	0.387	0.299
	9						0.002	0.010	0.040	0.134	0.387	0.630
10	0	0.599	0.349	0.107	0.028	0.006	0.001					
	1	0.315	0.387	0.268	0.121	0.040	0.010	0.002				
	2	0.075	0.194	0.302	0.233	0.121	0.044	0.011	0.001			
	3	0.010	0.057	0.201	0.267	0.215	0.117	0.042	0.009	0.001		
	4	0.001	0.011	0.088	0.200	0.251	0.205	0.111	0.037	0.006		
	5		0.001	0.026	0.103	0.201	0.246	0.201	0.103	0.026	0.001	
	6			0.006	0.037	0.111	0.205	0.251	0.200	0.088	0.011	0.001
	7			0.001	0.009	0.042	0.117	0.215	0.267	0.201	0.057	0.010
	8				0.001	0.011	0.044	0.121	0.233	0.302	0.194	0.075
	9					0.002	0.010	0.040	0.121	0.268	0.387	0.315
	10						0.001	0.006	0.028	0.107	0.349	0.599
11	0	0.569	0.314	0.086	0.020	0.004						
	1	0.329	0.384	0.236	0.093	0.027	0.005	0.001				
	2	0.087	0.213	0.295	0.200	0.089	0.027	0.005	0.001			
	3	0.014	0.071	0.221	0.257	0.177	0.081	0.023	0.004			
	4	0.001	0.016	0.111	0.220	0.236	0.161	0.070	0.017	0.002		
	5		0.002	0.039	0.132	0.221	0.226	0.147	0.057	0.010		
	6			0.010	0.057	0.147	0.226	0.221	0.132	0.039	0.002	
	7			0.002	0.017	0.070	0.161	0.236	0.220	0.111	0.016	0.001
	8				0.004	0.023	0.081	0.177	0.257	0.221	0.071	0.014
	9				0.001	0.005	0.027	0.089	0.200	0.295	0.213	0.087
	10					0.001	0.005	0.027	0.093	0.236	0.384	0.329
	11							0.004	0.020	0.086	0.314	0.569
12	0	0.540	0.282	0.069	0.014	0.002						
	1	0.341	0.377	0.206	0.071	0.017	0.003					
	2	0.099	0.230	0.283	0.168	0.064	0.016	0.002				
	3	0.017	0.085	0.236	0.240	0.142	0.054	0.012	0.001			
	4	0.002	0.021	0.133	0.231	0.213	0.121	0.042	0.008	0.001		
	5		0.004	0.053	0.158	0.227	0.193	0.101	0.029	0.003		
	6			0.016	0.079	0.177	0.226	0.177	0.079	0.016		
	7			0.003	0.029	0.101	0.193	0.227	0.158	0.053	0.004	
	8			0.001	0.008	0.042	0.121	0.213	0.231	0.133	0.021	0.002
	9				0.001	0.012	0.054	0.142	0.240	0.236	0.085	0.017
	10					0.002	0.016	0.064	0.168	0.283	0.230	0.099
	11						0.003	0.017	0.071	0.206	0.377	0.341
	12							0.002	0.014	0.069	0.282	0.540

Table 2 *(continued)*

n	x	0.05	0.1	0.2	0.3	0.4	0.5	0.6	0.7	0.8	0.9	0.95
13	0	0.513	0.254	0.055	0.010	0.001						
	1	0.351	0.367	0.179	0.054	0.011	0.002					
	2	0.111	0.245	0.268	0.139	0.045	0.010	0.001				
	3	0.021	0.100	0.246	0.218	0.111	0.035	0.006	0.001			
	4	0.003	0.028	0.154	0.234	0.184	0.087	0.024	0.003			
	5		0.006	0.069	0.180	0.221	0.157	0.066	0.014	0.001		
	6		0.001	0.023	0.103	0.197	0.209	0.131	0.044	0.006		
	7			0.006	0.044	0.131	0.209	0.197	0.103	0.023	0.001	
	8			0.001	0.014	0.066	0.157	0.221	0.180	0.069	0.006	
	9				0.003	0.024	0.087	0.184	0.234	0.154	0.028	0.003
	10				0.001	0.006	0.035	0.111	0.218	0.246	0.100	0.021
	11					0.001	0.010	0.045	0.139	0.268	0.245	0.111
	12						0.002	0.011	0.054	0.179	0.367	0.351
	13							0.001	0.010	0.055	0.254	0.513
14	0	0.488	0.229	0.044	0.007	0.001						
	1	0.359	0.356	0.154	0.041	0.007	0.001					
	2	0.123	0.257	0.250	0.113	0.032	0.006	0.001				
	3	0.026	0.114	0.250	0.194	0.085	0.022	0.003				
	4	0.004	0.035	0.172	0.229	0.155	0.061	0.014	0.001			
	5		0.008	0.086	0.196	0.207	0.122	0.041	0.007			
	6		0.001	0.032	0.126	0.207	0.183	0.092	0.023	0.002		
	7			0.009	0.062	0.157	0.209	0.157	0.062	0.009		
	8			0.002	0.023	0.092	0.183	0.207	0.126	0.032	0.001	
	9				0.007	0.041	0.122	0.207	0.196	0.086	0.008	
	10				0.001	0.014	0.061	0.155	0.229	0.172	0.035	0.004
	11					0.003	0.022	0.085	0.194	0.250	0.114	0.026
	12					0.001	0.006	0.032	0.113	0.250	0.257	0.123
	13						0.001	0.007	0.041	0.154	0.356	0.359
	14							0.001	0.007	0.044	0.229	0.488
15	0	0.463	0.206	0.035	0.005							
	1	0.366	0.343	0.132	0.031	0.005						
	2	0.135	0.267	0.231	0.092	0.022	0.003					
	3	0.031	0.129	0.250	0.170	0.063	0.014	0.002				
	4	0.005	0.043	0.188	0.219	0.127	0.042	0.007	0.001			
	5	0.001	0.010	0.103	0.206	0.186	0.092	0.024	0.003			
	6		0.002	0.043	0.147	0.207	0.153	0.061	0.012	0.001		
	7			0.014	0.081	0.177	0.196	0.118	0.035	0.003		
	8			0.003	0.035	0.118	0.196	0.177	0.081	0.014		
	9			0.001	0.012	0.061	0.153	0.207	0.147	0.043	0.002	
	10				0.003	0.024	0.092	0.186	0.206	0.103	0.010	0.001
	11				0.001	0.007	0.042	0.127	0.219	0.188	0.043	0.005
	12					0.002	0.014	0.063	0.170	0.250	0.129	0.031
	13						0.003	0.022	0.092	0.231	0.267	0.135
	14							0.005	0.031	0.132	0.343	0.366
	15								0.005	0.035	0.206	0.463

Table 3 The Standard Normal Distribution

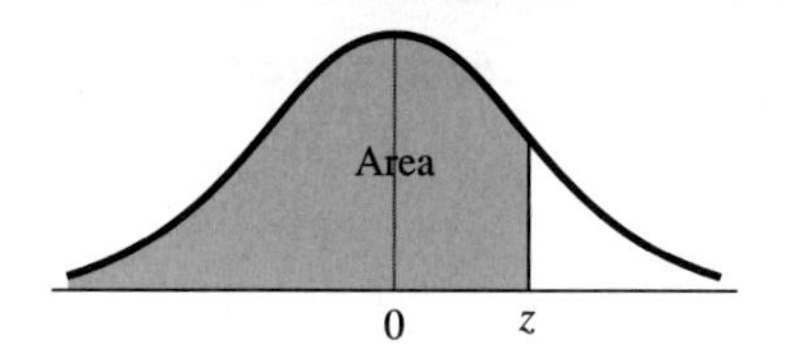

$$F_z(z) = P[Z \le z]$$

z	0.00	0.01	0.02	0.03	0.04	0.05	0.06	0.07	0.08	0.09
−3.4	0.0003	0.0003	0.0003	0.0003	0.0003	0.0003	0.0003	0.0003	0.0003	0.0002
−3.3	0.0005	0.0005	0.0005	0.0004	0.0004	0.0004	0.0004	0.0004	0.0004	0.0003
−3.2	0.0007	0.0007	0.0006	0.0006	0.0006	0.0006	0.0006	0.0005	0.0005	0.0005
−3.1	0.0010	0.0009	0.0009	0.0009	0.0008	0.0008	0.0008	0.0008	0.0007	0.0007
−3.0	0.0013	0.0013	0.0013	0.0012	0.0012	0.0011	0.0011	0.0011	0.0010	0.0010
−2.9	0.0019	0.0018	0.0017	0.0017	0.0016	0.0016	0.0015	0.0015	0.0014	0.0014
−2.8	0.0026	0.0025	0.0024	0.0023	0.0023	0.0022	0.0021	0.0021	0.0020	0.0019
−2.7	0.0035	0.0034	0.0033	0.0032	0.0031	0.0030	0.0029	0.0028	0.0027	0.0026
−2.6	0.0047	0.0045	0.0044	0.0043	0.0041	0.0040	0.0039	0.0038	0.0037	0.0036
−2.5	0.0062	0.0060	0.0059	0.0057	0.0055	0.0054	0.0052	0.0051	0.0049	0.0048
−2.4	0.0082	0.0080	0.0078	0.0075	0.0073	0.0071	0.0069	0.0068	0.0066	0.0064
−2.3	0.0107	0.0104	0.0102	0.0099	0.0096	0.0094	0.0091	0.0089	0.0087	0.0084
−2.2	0.0139	0.0136	0.0132	0.0129	0.0125	0.0122	0.0119	0.0116	0.0113	0.0110
−2.1	0.0179	0.0174	0.0170	0.0166	0.0162	0.0158	0.0154	0.0150	0.0146	0.0143
−2.0	0.0228	0.0222	0.0217	0.0212	0.0207	0.0202	0.0197	0.0192	0.0188	0.0183
−1.9	0.0287	0.0281	0.0274	0.0268	0.0262	0.0256	0.0250	0.0244	0.0239	0.0233
−1.8	0.0359	0.0352	0.0344	0.0336	0.0329	0.0322	0.0314	0.0307	0.0301	0.0294
−1.7	0.0446	0.0436	0.0427	0.0418	0.0409	0.0401	0.0392	0.0384	0.0375	0.0367
−1.6	0.0548	0.0537	0.0526	0.0516	0.0505	0.0495	0.0485	0.0475	0.0465	0.0455
−1.5	0.0668	0.0655	0.0643	0.0630	0.0618	0.0606	0.0594	0.0582	0.0571	0.0559
−1.4	0.0808	0.0793	0.0778	0.0764	0.0749	0.0735	0.0722	0.0708	0.0694	0.0681
−1.3	0.0968	0.0951	0.0934	0.0918	0.0901	0.0885	0.0869	0.0853	0.0838	0.0823
−1.2	0.1151	0.1131	0.1112	0.1093	0.1075	0.1056	0.1038	0.1020	0.1003	0.0985
−1.1	0.1357	0.1335	0.1314	0.1292	0.1271	0.1251	0.1230	0.1210	0.1190	0.1170
−1.0	0.1587	0.1562	0.1539	0.1515	0.1492	0.1469	0.1446	0.1423	0.1401	0.1379
−0.9	0.1841	0.1814	0.1788	0.1762	0.1736	0.1711	0.1685	0.1660	0.1635	0.1611
−0.8	0.2119	0.2090	0.2061	0.2033	0.2005	0.1977	0.1949	0.1922	0.1894	0.1867
−0.7	0.2420	0.2389	0.2358	0.2327	0.2296	0.2266	0.2236	0.2206	0.2177	0.2148
−0.6	0.2743	0.2709	0.2676	0.2643	0.2611	0.2578	0.2546	0.2514	0.2483	0.2451
−0.5	0.3085	0.3050	0.3015	0.2981	0.2946	0.2912	0.2877	0.2843	0.2810	0.2776

Table 3 *(continued)*

	$F_Z(z) = P[Z \leq z]$									
z	**0.00**	**0.01**	**0.02**	**0.03**	**0.04**	**0.05**	**0.06**	**0.07**	**0.08**	**0.09**
−0.4	0.3446	0.3409	0.3372	0.3336	0.3300	0.3264	0.3228	0.3192	0.3156	0.3121
−0.3	0.3821	0.3783	0.3745	0.3707	0.3669	0.3632	0.3594	0.3557	0.3520	0.3483
−0.2	0.4207	0.4168	0.4129	0.4090	0.4052	0.4013	0.3974	0.3936	0.3897	0.3859
−0.1	0.4602	0.4562	0.4522	0.4483	0.4443	0.4404	0.4364	0.4325	0.4286	0.4247
−0.0	0.5000	0.4960	0.4920	0.4880	0.4840	0.4801	0.4761	0.4721	0.4681	0.4641
0.0	0.5000	0.5040	0.5080	0.5120	0.5160	0.5199	0.5239	0.5279	0.5319	0.5359
0.1	0.5398	0.5438	0.5478	0.5517	0.5557	0.5596	0.5636	0.5675	0.5714	0.5753
0.2	0.5793	0.5832	0.5871	0.5910	0.5948	0.5987	0.6026	0.6064	0.6103	0.6141
0.3	0.6179	0.6217	0.6255	0.6293	0.6331	0.6368	0.6406	0.6443	0.6480	0.6517
0.4	0.6554	0.6591	0.6628	0.6664	0.6700	0.6736	0.6772	0.6808	0.6844	0.6879
0.5	0.6915	0.6950	0.6985	0.7019	0.7054	0.7088	0.7123	0.7157	0.7190	0.7224
0.6	0.7257	0.7291	0.7324	0.7357	0.7389	0.7422	0.7454	0.7486	0.7517	0.7549
0.7	0.7580	0.7611	0.7642	0.7673	0.7704	0.7734	0.7764	0.7794	0.7823	0.7852
0.8	0.7881	0.7910	0.7939	0.7967	0.7995	0.8023	0.8051	0.8078	0.8106	0.8133
0.9	0.8159	0.8186	0.8212	0.8238	0.8264	0.8289	0.8315	0.8340	0.8365	0.8389
1.0	0.8413	0.8438	0.8461	0.8485	0.8508	0.8531	0.8554	0.8577	0.8599	0.8621
1.1	0.8643	0.8665	0.8686	0.8708	0.8729	0.8749	0.8770	0.8790	0.8810	0.8830
1.2	0.8849	0.8869	0.8888	0.8907	0.8925	0.8944	0.8962	0.8980	0.8997	0.9015
1.3	0.9032	0.9049	0.9066	0.9082	0.9099	0.9115	0.9131	0.9147	0.9162	0.9177
1.4	0.9192	0.9207	0.9222	0.9236	0.9251	0.9265	0.9278	0.9292	0.9306	0.9319
1.5	0.9332	0.9345	0.9357	0.9370	0.9382	0.9394	0.9406	0.9418	0.9429	0.9441
1.6	0.9452	0.9463	0.9474	0.9484	0.9495	0.9505	0.9515	0.9525	0.9535	0.9545
1.7	0.9554	0.9564	0.9573	0.9582	0.9591	0.9599	0.9608	0.9616	0.9625	0.9633
1.8	0.9641	0.9649	0.9656	0.9664	0.9671	0.9678	0.9686	0.9693	0.9699	0.9706
1.9	0.9713	0.9719	0.9726	0.9732	0.9738	0.9744	0.9750	0.9756	0.9761	0.9767
2.0	0.9772	0.9778	0.9783	0.9788	0.9793	0.9798	0.9803	0.9808	0.9812	0.9817
2.1	0.9821	0.9826	0.9830	0.9834	0.9838	0.9842	0.9846	0.9850	0.9854	0.9857
2.2	0.9861	0.9864	0.9868	0.9871	0.9875	0.9878	0.9881	0.9884	0.9887	0.9890
2.3	0.9893	0.9896	0.9898	0.9901	0.9904	0.9906	0.9909	0.9911	0.9913	0.9916
2.4	0.9918	0.9920	0.9922	0.9925	0.9927	0.9929	0.9931	0.9932	0.9934	0.9936
2.5	0.9938	0.9940	0.9951	0.9943	0.9945	0.9946	0.9948	0.9949	0.9951	0.9952
2.6	0.9953	0.9955	0.9956	0.9957	0.9959	0.9960	0.9961	0.9962	0.9963	0.9964
2.7	0.9965	0.9966	0.9967	0.9968	0.9969	0.9970	0.9971	0.9972	0.9973	0.9974
2.8	0.9974	0.9975	0.9976	0.9977	0.9977	0.9978	0.9979	0.9979	0.9980	0.9981
2.9	0.9981	0.9982	0.9982	0.9983	0.9984	0.9984	0.9985	0.9985	0.9986	0.9986
3.0	0.9987	0.9987	0.9987	0.9988	0.9988	0.9989	0.9989	0.9989	0.9990	0.9990
3.1	0.9990	0.9991	0.9991	0.9991	0.9992	0.9992	0.9992	0.9992	0.9993	0.9993
3.2	0.9993	0.9993	0.9994	0.9994	0.9994	0.9994	0.9994	0.9995	0.9995	0.9995
3.3	0.9995	0.9995	0.9995	0.9996	0.9996	0.9996	0.9996	0.9996	0.9996	0.9997
3.4	0.9997	0.9997	0.9997	0.9997	0.9997	0.9997	0.9997	0.9997	0.9997	0.9998

Answers to Odd-Numbered Exercises

Chapter 1

Exercises 1.1, page 10

1. (3, 3); Quadrant I **3.** (2, −2); Quadrant IV

5. (−4, −6); Quadrant III **7.** A **9.** E, F, and G

11. F **13.–19.** See the following figure.

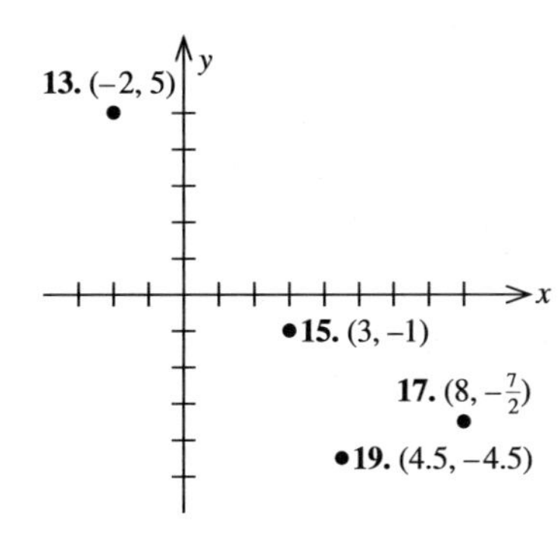

21. 5 **23.** $\sqrt{61}$ **25.** (−8, −6) and (8, −6)

29. $(x-2)^2 + (y+3)^2 = 25$ **31.** $x^2 + y^2 = 25$

33. $(x-2)^2 + (y+3)^2 = 34$ **35.** 3400 miles

37. Route 1 **39.** Model C

41. **a.** $d = \sqrt{1300}\,t$ **b.** 72.11 miles **43.** True

Exercises 1.2, page 25

1. 1/2 **3.** Not defined **5.** 5 **7.** 5/6

9. $\dfrac{d-b}{c-a}$ $(a \neq c)$ **11.** **a.** 4 **b.** −8 **13.** Parallel

15. Perpendicular **17.** $a = -5$ **19.** $y = -3$

21. e **23.** a **25.** f **27.** $y = 2x - 10$

29. $y = 2$ **31.** $y = 3x - 2$ **33.** $y = x + 1$

35. $y = 3x + 4$ **37.** $y = 5$

39. $y = \frac{1}{2}x$; $m = \frac{1}{2}$; $b = 0$

41. $y = \frac{2}{3}x - 3$; $m = \frac{2}{3}$; $b = -3$

43. $y = -\frac{1}{2}x + \frac{7}{2}$; $m = -\frac{1}{2}$; $b = \frac{7}{2}$

45. $y = \frac{1}{2}x + 3$ **47.** $y = -6$ **49.** $y = b$

51. $y = \frac{2}{3}x - \frac{2}{3}$ **53.** $k = 8$

55. **57.**

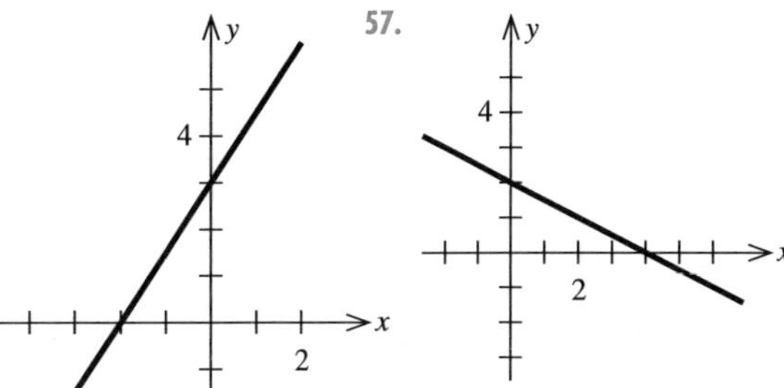

59.

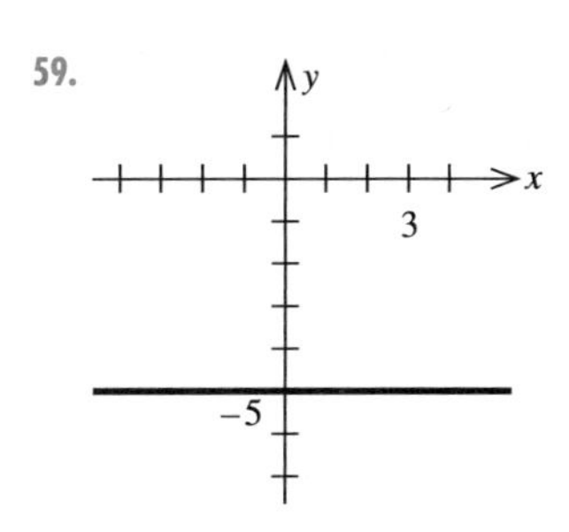

63. $y = -2x - 4$ **65.** $y = \frac{1}{8}x - \frac{1}{2}$ **67.** Yes

69. **a.** $y = 0.55x$ **b.** 2000 **71.** 82.4%

73. **a.** and **b.**

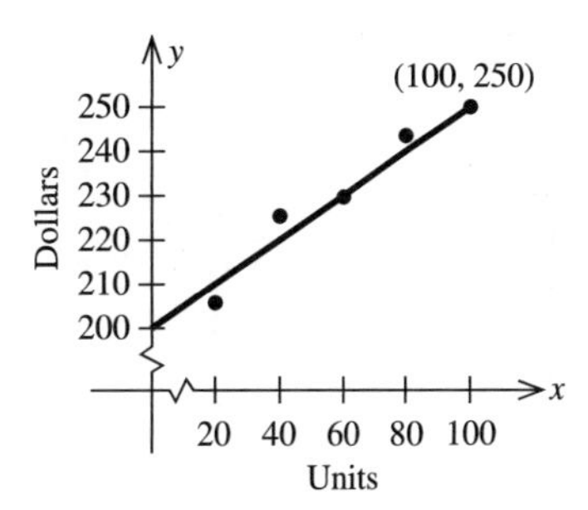

c. $y = \frac{1}{2}x + 200$ **d.** $227

75. **a.** and **b.**

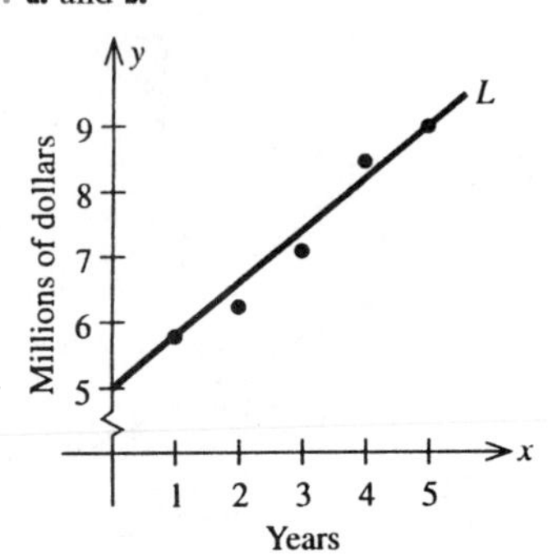

c. $y = 0.8x + 5$ **d.** \$12.2 million

77. **a.** A family of parallel lines having slope m

b. A family of straight lines that pass through the point $(0, b)$

79. True 81. True

Using Technology Exercises 1.2, page 31

1.

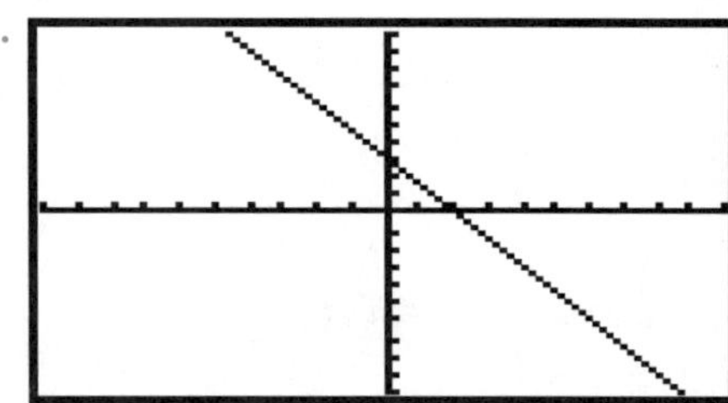

3.

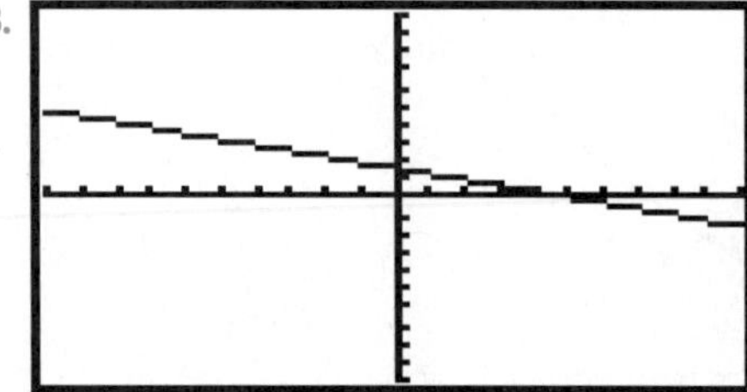

5.

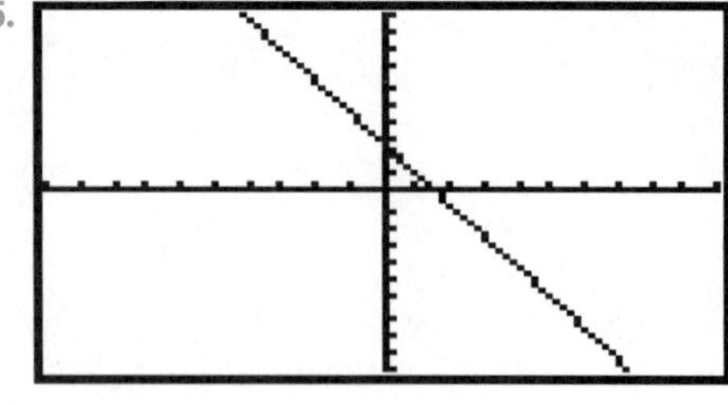

7. **a.**

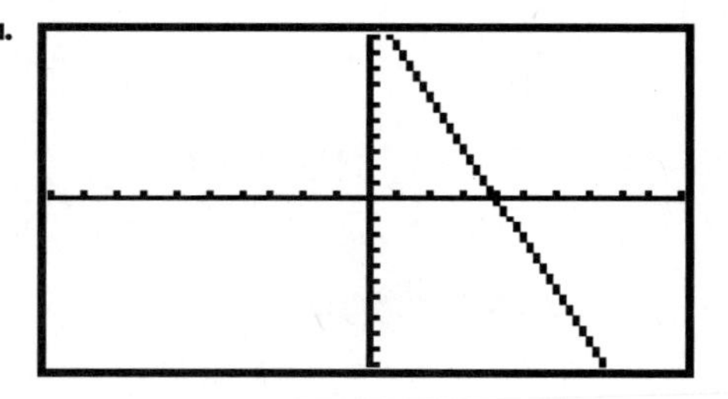

b.

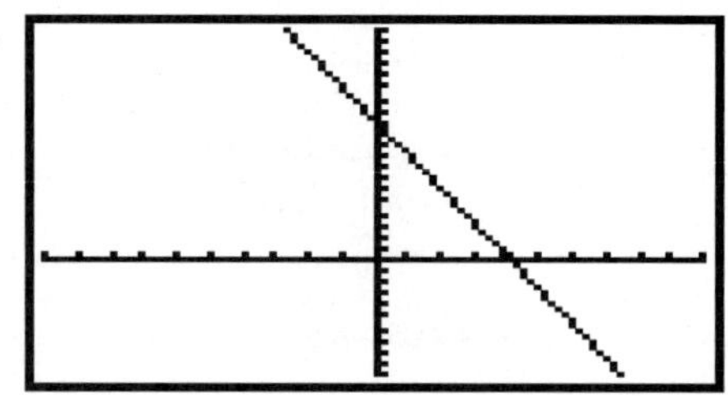

9. **a.**

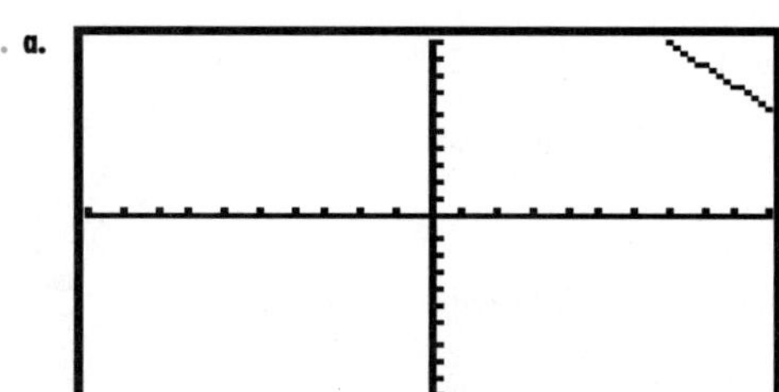

b.

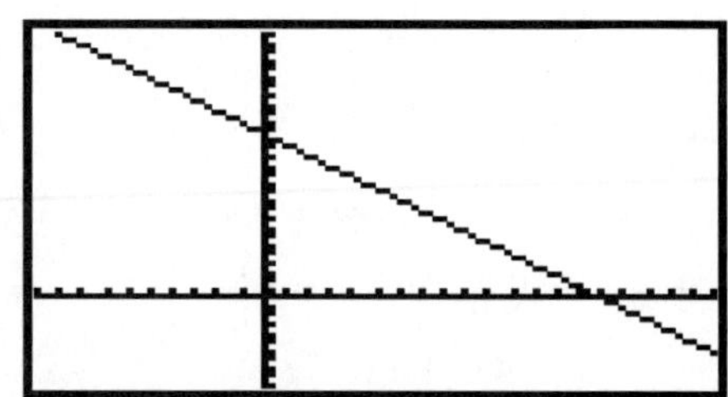

11.

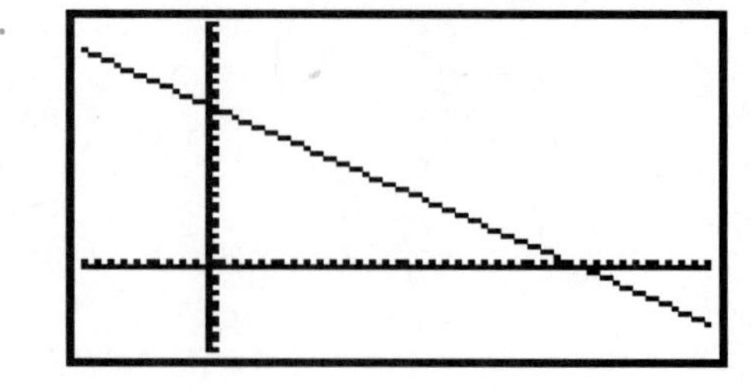

13.

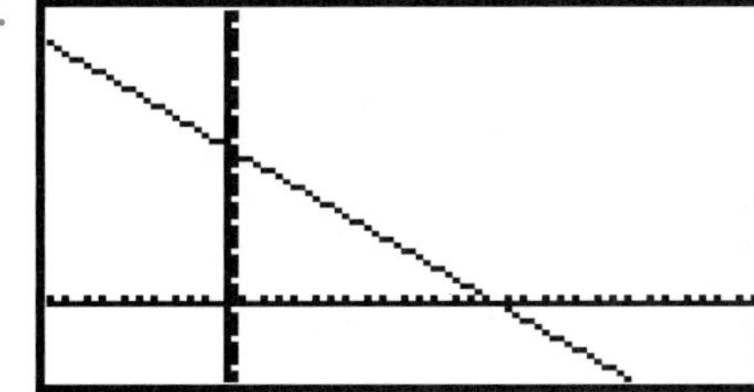

15.

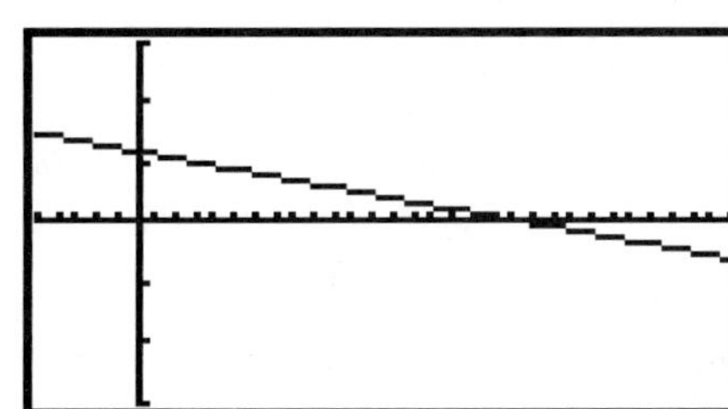

17.

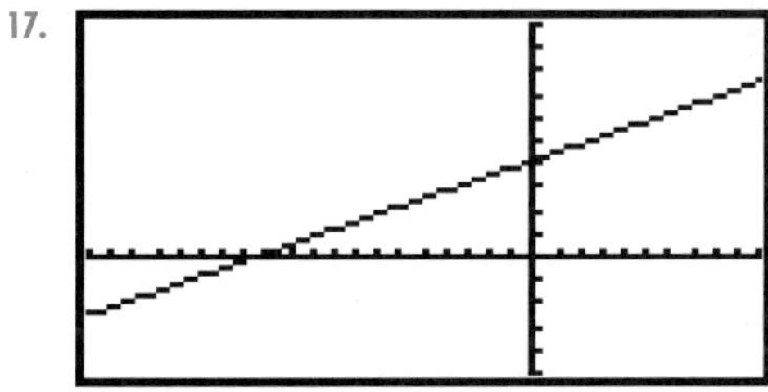

Exercises 1.3, page 41

1. Yes; $y = -\frac{2}{3}x + 2$ **3.** Yes; $y = \frac{1}{2}x + 2$

5. Yes; $y = \frac{1}{2}x + \frac{9}{4}$ **7.** No **9.** No

11. a. $C(x) = 8x + 40{,}000$ **b.** $R(x) = 12x$
c. $P(x) = 4x - 40{,}000$
d. Loss of \$8000; profit of \$8000

13. $m = -1, b = 2$ **15.** \$900,000; \$800,000

17. \$6 billion; \$43.5 billion; \$81 billion

19. a. $y = 1.053x$ **b.** \$652.86

21. $C(x) = 0.6x + 12{,}100$; $R(x) = 1.15x$; $P(x) = 0.55x - 12{,}100$

23. a. \$12,000/yr
b. $V = 60{,}000 - 12{,}000t$
c.

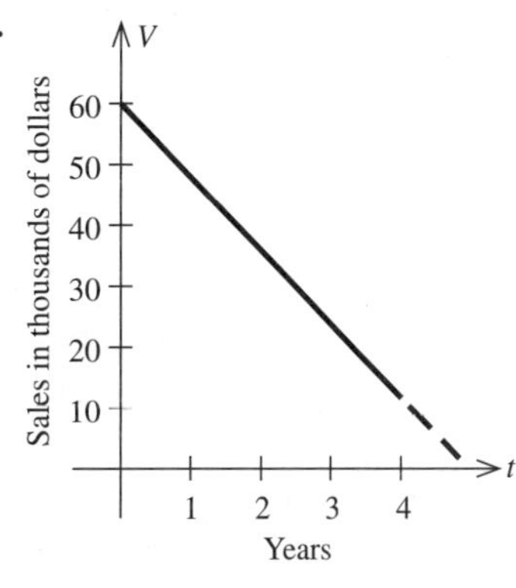

d. \$24,000

25. \$900,000; \$800,000

27. a. $m = a/1.7$; $b = 0$ **b.** 117.65 mg

29. a. $F = \frac{9}{5}C + 32$ **b.** 68°F **c.** 21.1°C **31.** L_2

33. a.

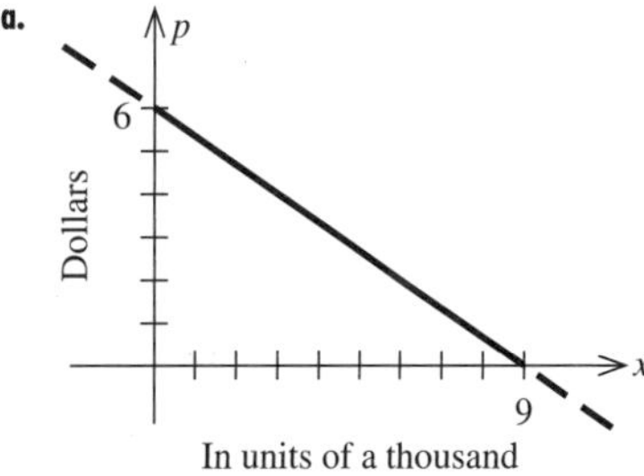

b. 3000

35. a.

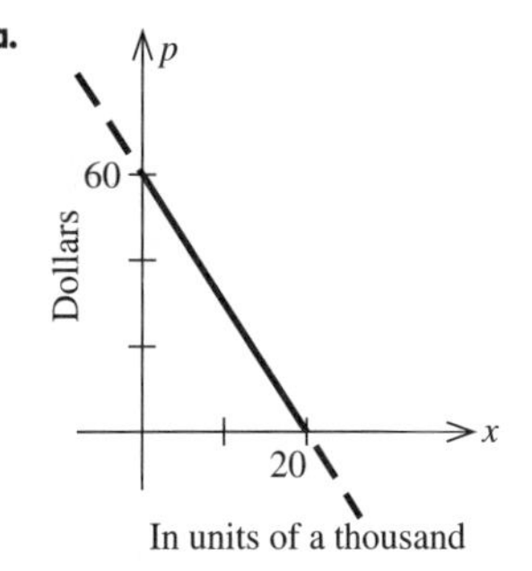

b. 10,000

37. $p = -\frac{3}{40}x + 130$; \$130; 1733

39. $p = -0.001x + 10$; 2500 units

41. a.

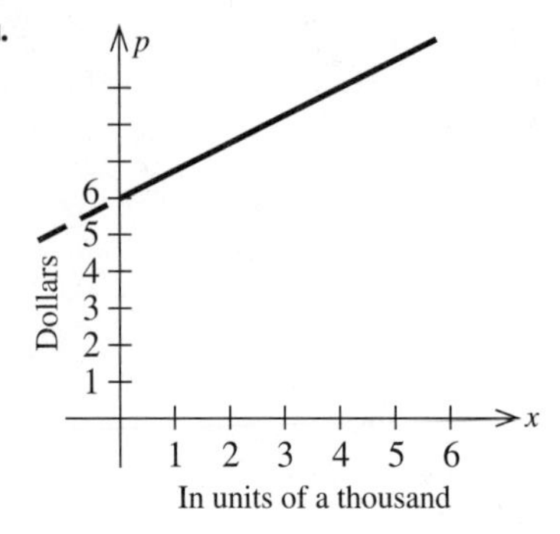

b. 2667 units

43. a.

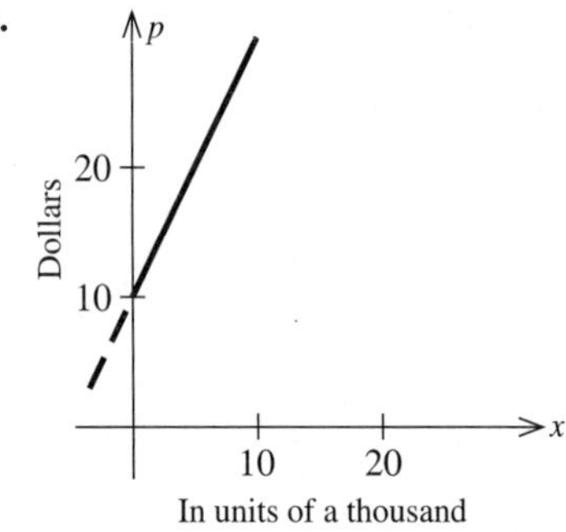

b. 2000 units

45. $p = \frac{1}{2}x + 40$ (x is measured in units of a thousand)

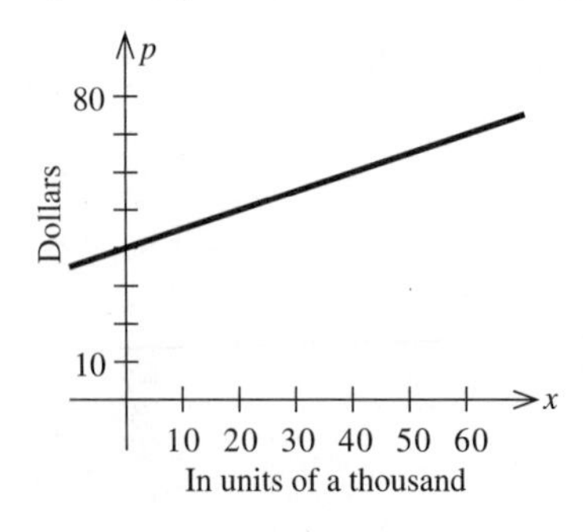

60,000 units

47. False

Using Technology Exercises 1.3, page 45

1. 2.2875 **3.** 2.880952381 **5.** 7.2851648352

7. 2.4680851064

Exercises 1.4, page 55

1. (2, 10) **3.** (4, 2/3) **5.** (−4, −6)

7. 1000 units; $15,000 **9.** 600 units; $240

11. a.

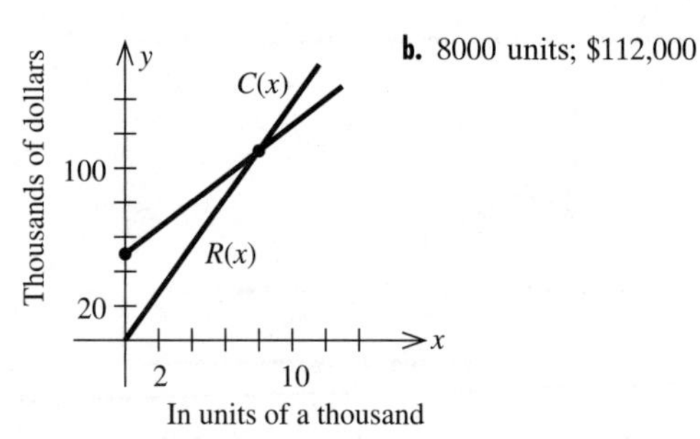

b. 8000 units; $112,000

c.

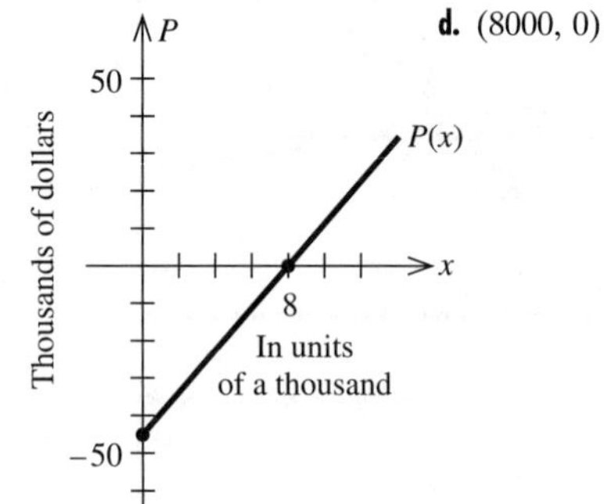

d. (8000, 0)

13. 9259 units; $83,331

15. a. $C_1(x) = 18{,}000 + 15x$
$C_2(x) = 15{,}000 + 20x$

b.

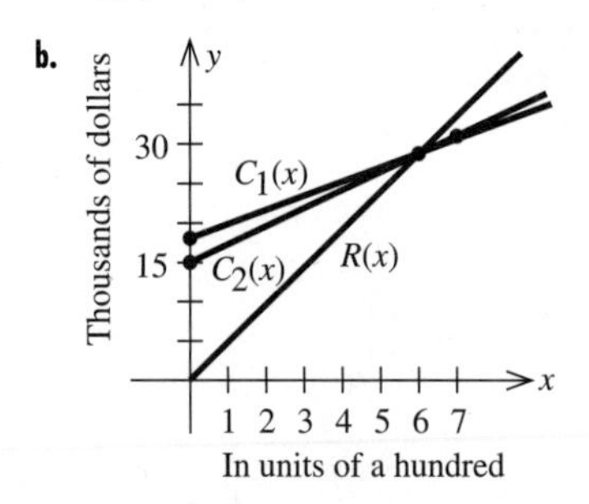

c. Machine II; machine II; machine I

d. ($1500); $1500; $4750

17. 8000 units; $9 **19.** 2000 units; $18

21. a. $p = -0.08x + 725$ **b.** $p = 0.09x + 300$
c. 2500 VCRs; $525

23. 300 fax machines; $600

25. a. $\dfrac{b-d}{c-a}$; $\dfrac{bc-ad}{c-a}$

b. If c is increased, x gets smaller and p gets larger.
c. If b is decreased, x decreases and p decreases.

27. True

29. a. $m_1 = m_2$ and $b_2 \neq b_1$
 b. $m_1 \neq m_2$
 c. $m_1 = m_2$ and $b_1 = b_2$

Using Technology Exercises 1.4, page 59

1. (0.6, 6.2) 3. (3.8261, 0.1304)

5. (386.9091, 145.3939)

Exercises 1.5, page 65

1. a. $y = 2.3x + 1.5$

 b.

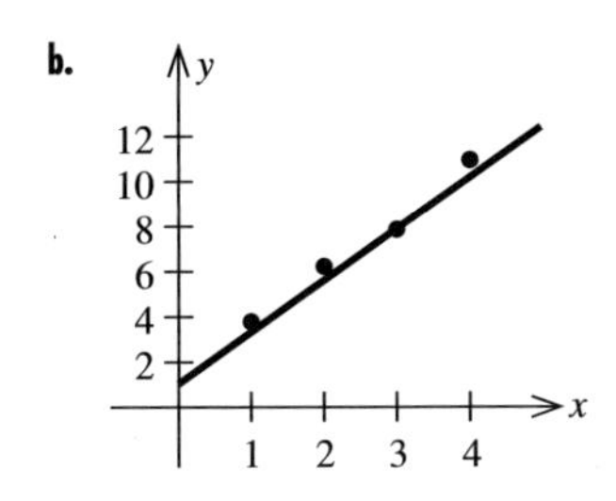

3. a. $y = -0.77x + 5.74$

 b.

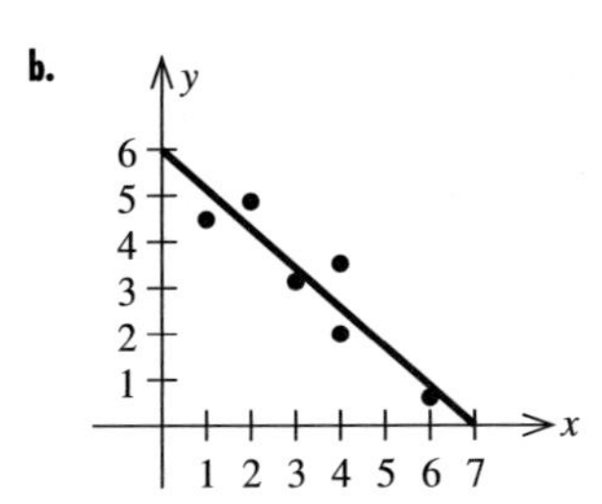

5. a. $y = 1.2x + 2$

 b.

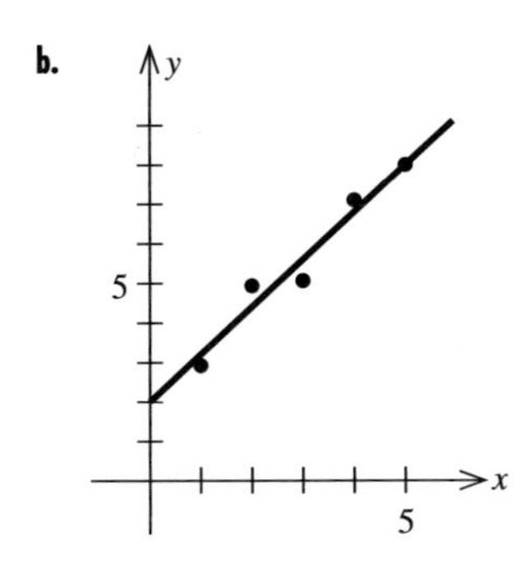

7. a. $y = 0.34x - 0.9$

 b.

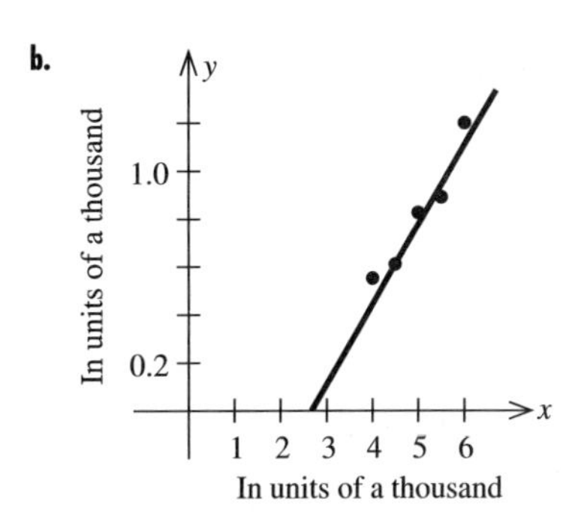

 c. 1276 applications

9. a. $y = -2.8x + 440$

 b.

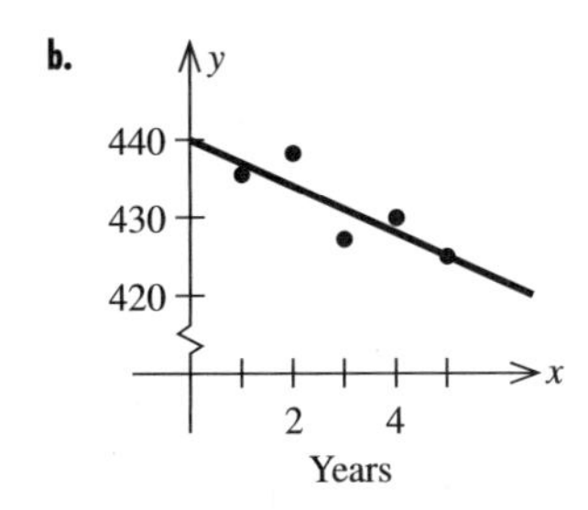

 c. 420

11. a. $y = 5.70x + 176$ b. 535 acres

13. a. $y = 2.8x + 17.6$ b. $40,000,000

15. a. $y = 4.842x + 11.842$ b. 103.8 billion cans

17. a. $y = 12.2x + 20.9$ b. 81.9 million

19. a. $y = 98.176x - 231.7$ b. $1732/capita

21. True

Using Technology Exercises 1.5, page 69

1. $y = 3.8639 + 2.3596x$ 3. $y = 3.5525 - 1.1948x$

5. a. $y = 13.321x + 72.571$ b. 192 million tons

Chapter 1 Review Exercises, page 73

1. 5 3. 5 5. $x = -2$ 7. $x + 10y - 38 = 0$

9. $5x - 2y + 18 = 0$ 11. $y = -\frac{1}{2}x - 3$

13. $3x + 4y - 18 = 0$ 15. $3x + 2y + 14 = 0$

17.

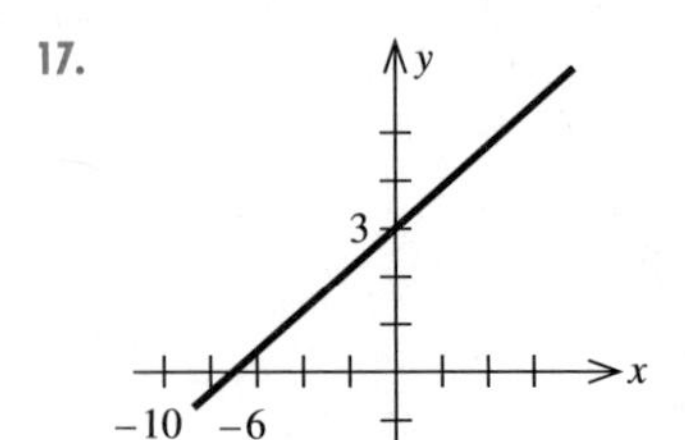

19. a. $f(x) = x + 2.4$ **b.** \$5.4 million

21. b. ≈117 mg

23. a. \$22,500/yr **b.** $V = -22{,}500t + 300{,}000$

25. $p = -0.05x + 200$

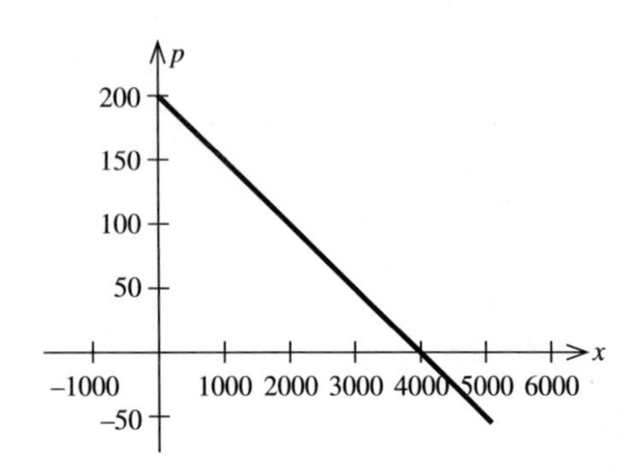

27. $(2, -3)$ **29.** $(2500, 50{,}000)$

31. a. $y = 0.25x$ **b.** 1600

CHAPTER 2

Exercises 2.1, page 84

1. Unique solution; $(2, 1)$ **3.** No solution

5. Unique solution; $(3, 2)$

7. Infinitely many solutions; $(t, \frac{2}{5}t - 2)$; t, a parameter

9. Unique solution; $(1, -2)$

11. No solution **13.** $k = -2$

15. $\begin{aligned} x + y &= 500 \\ 42x + 30y &= 18{,}600 \end{aligned}$ **17.** $\begin{aligned} x + y &= 100 \\ 2.5x + 3y &= 280 \end{aligned}$

19. $\begin{aligned} x + y &= 1000 \\ .25x + .75y &= 650 \end{aligned}$

21. $\begin{aligned} .06x + .08y + .12z &= 21{,}600 \\ z &= 2x \\ .12z &= .08y \end{aligned}$

23. $\begin{aligned} 8000x + 12{,}000y + 16{,}000z &= 1{,}000{,}000 \\ x &= 2y \\ x + y + z &= 100 \end{aligned}$

25. $\begin{aligned} 10x + 6y + 8z &= 100 \\ 10x + 12y + 6z &= 100 \\ 5x + 4y + 12z &= 100 \end{aligned}$

27. True

Exercises 2.2, page 98

1. $\left[\begin{array}{cc|c} 2 & -3 & 7 \\ 3 & 1 & 4 \end{array}\right]$

3. $\left[\begin{array}{ccc|c} 0 & -1 & 2 & 6 \\ 2 & 2 & -8 & 7 \\ 0 & 3 & 4 & 0 \end{array}\right]$

5. $\begin{aligned} 3x + 2y &= -4 \\ x - y &= 5 \end{aligned}$ **7.** $\begin{aligned} x + 3y + 2z &= 4 \\ 2x &= 5 \\ 3x - 3y + 2z &= 6 \end{aligned}$

9. Yes **11.** No **13.** Yes **15.** No **17.** No

19. $\left[\begin{array}{cc|c} 1 & 2 & 4 \\ 0 & -5 & -10 \end{array}\right]$ **21.** $\left[\begin{array}{cc|c} 1 & -2 & -3 \\ 0 & 16 & 20 \end{array}\right]$

23. $\left[\begin{array}{ccc|c} 1 & 2 & 3 & 6 \\ 0 & -1 & -5 & -7 \\ 0 & -7 & -7 & -14 \end{array}\right]$

25. $\left[\begin{array}{ccc|c} -6 & -11 & 0 & -5 \\ 2 & 4 & 1 & 3 \\ 1 & -2 & 0 & -10 \end{array}\right]$

27. $\left[\begin{array}{cc|c} 3 & 9 & 6 \\ 2 & 1 & 4 \end{array}\right] \xrightarrow{\frac{1}{3}R_1} \left[\begin{array}{cc|c} 1 & 3 & 2 \\ 2 & 1 & 4 \end{array}\right]$

$\xrightarrow{R_2 - 2R_1} \left[\begin{array}{cc|c} 1 & 3 & 2 \\ 0 & -5 & 0 \end{array}\right] \xrightarrow{-\frac{1}{5}R_2}$

$\left[\begin{array}{cc|c} 1 & 3 & 2 \\ 0 & 1 & 0 \end{array}\right] \xrightarrow{R_1 - 3R_2} \left[\begin{array}{cc|c} 1 & 0 & 2 \\ 0 & 1 & 0 \end{array}\right]$

29. $\left[\begin{array}{ccc|c} 1 & 3 & 1 & 3 \\ 3 & 8 & 3 & 7 \\ 2 & -3 & 1 & -10 \end{array}\right] \xrightarrow[R_3 - 2R_1]{R_2 - 3R_1}$

$\left[\begin{array}{ccc|c} 1 & 3 & 1 & 3 \\ 0 & -1 & 0 & -2 \\ 0 & -9 & -1 & -16 \end{array}\right] \xrightarrow{-R_2}$

$\left[\begin{array}{ccc|c} 1 & 3 & 1 & 3 \\ 0 & 1 & 0 & 2 \\ 0 & -9 & -1 & -16 \end{array}\right] \xrightarrow[R_3 + 9R_2]{R_1 - 3R_2}$

$\left[\begin{array}{ccc|c} 1 & 0 & 1 & -3 \\ 0 & 1 & 0 & 2 \\ 0 & 0 & -1 & 2 \end{array}\right] \xrightarrow[-R_3]{R_1 + R_3}$

$\left[\begin{array}{ccc|c} 1 & 0 & 0 & -1 \\ 0 & 1 & 0 & 2 \\ 0 & 0 & 1 & -2 \end{array}\right]$

31. (2, 0) **33.** (−1, 2, −2)

35. (4, −2) **37.** (−1, 2)

39. $(\frac{7}{9}, -\frac{1}{9}\ -\frac{2}{3})$ **41.** (19, −7, −15)

43. (3, 0, 2) **45.** (1, −2, 1)

47. (−20, −28, 13) **49.** (4, −1, 3)

51. 300 acres of corn, 200 acres of wheat

53. In 100 lb of blended coffee, use 40 lb of the \$2.50/lb coffee and 60 lb of the \$3/lb coffee.

55. 200 children and 800 adults

57. \$40,000 in a savings account, \$120,000 in mutual funds, \$80,000 in bonds

59. 60 compact, 30 intermediate, and 10 full-size cars

61. 4 oz of food I, 2 oz of food II, 6 oz of food III

63. 240 front orchestra seats, 560 rear orchestra seats, 200 front balcony seats

65. False

Using Technology Exercises 2.2, page 104

1. $x_1 = 3$, $x_2 = 1$, $x_3 = -1$, and $x_4 = 2$

3. $x_1 = 5$, $x_2 = 4$, $x_3 = -3$, and $x_4 = -4$

5. $x_1 = 1$, $x_2 = -1$, $x_3 = 2$, $x_4 = 0$, and $x_5 = 3$

Exercises 2.3, page 113

1. **a.** One solution **b.** (3, −1, 2)

3. **a.** One solution **b.** (2, 4)

5. **a.** Infinitely many solutions
b. $(4 - t, -2, t)$; t, a parameter

7. **a.** No solution

9. **a.** Infinitely many solutions
b. $(2, -1, 2 - t, t)$; t, a parameter

11. **a.** Infinitely many solutions
b. $(2 - 3s, 1 + s, s, t)$; s, t, parameters

13. (2, 1) **15.** No solution **17.** (1, −1)

19. $(2 + 2t, t)$; t, a parameter

21. $(\frac{4}{3} - \frac{2}{3}t, t)$; t, a parameter

23. $(-2 + \frac{1}{2}s - \frac{1}{2}t, s, t)$; s, t, parameters

25. $(-1, \frac{17}{7}, \frac{23}{7})$

27. $(1 - \frac{1}{4}s + \frac{1}{4}t, s, t)$; s, t, parameters

29. No solution **31.** (2, −1, 4)

33. $x = 20 + z$, $y = 40 - 2z$; 25 compact cars, 30 mid-sized cars, and 5 full-sized cars; 30 compact cars, 20 mid-sized cars, and 10 full-sized cars

37. **a.**
$$\begin{aligned} x_1 - x_2 \qquad\qquad\qquad\qquad &= 200 \\ x_1 \qquad\qquad - x_5 \qquad &= -100 \\ - x_2 + x_3 \qquad\qquad + x_6 &= 600 \\ - x_3 + x_4 \qquad\qquad &= 200 \\ x_4 - x_5 + x_6 &= 700 \end{aligned}$$

b.
$$\begin{aligned} x_1 &= s + 100 \\ x_2 &= s - 100 \\ x_3 &= s - t + 500 \\ x_4 &= s - t + 700 \\ x_5 &= s \\ x_6 &= t \end{aligned}$$
(250, 50, 600, 800, 150, 50)
(300, 100, 600, 800, 200, 100)

c. (300, 100, 700, 900, 200, 0)

39. $k = 6$; (−2, 2) **41.** False

Using Technology Exercises 2.3, page 117

1. $(1 + t, 2 + t, t)$; t, a parameter

3. $(-\frac{17}{7} + \frac{6}{7}t, 3 - t, -\frac{18}{7} + \frac{1}{7}t, t)$; t, a parameter

5. No solution

Exercises 2.4, page 127

1. 4×4; 4×3; 1×5; 4×1 **3.** 2; 3; 8

5. D; $D^T = [1 \quad 3 \quad -2 \quad 0]$

7. 3×2; 3×2; 3×3; 3×3

9. $\begin{bmatrix} 1 & 6 \\ 6 & -1 \\ 2 & 2 \end{bmatrix}$ **11.** $\begin{bmatrix} 1 & 1 & -4 \\ -1 & -8 & 1 \\ 6 & 3 & 1 \end{bmatrix}$

13. $\begin{bmatrix} 3 & 5 & 9 \\ 4 & 10 & 13 \end{bmatrix}$ **15.** $\begin{bmatrix} 3 & -4 & -16 \\ 17 & -4 & 16 \end{bmatrix}$

17. $\begin{bmatrix} -1.9 & 3.0 & -0.6 \\ 6.0 & 9.6 & 1.2 \end{bmatrix}$

19. $\begin{bmatrix} \frac{7}{2} & 3 & -1 & \frac{10}{3} \\ -\frac{19}{6} & \frac{2}{3} & -\frac{17}{2} & \frac{23}{3} \\ \frac{29}{3} & \frac{17}{6} & -1 & -2 \end{bmatrix}$

21. $x = \frac{5}{2}$, $y = 7$, $z = 2$, and $u = 3$

23. $x = 2$, $y = 2$, $z = -\frac{7}{3}$, and $u = 15$

31. $\begin{bmatrix} 3 \\ 2 \\ -1 \\ 5 \end{bmatrix}$ **33.** $\begin{bmatrix} 1 & 3 & 0 \\ -1 & 4 & 1 \\ 2 & 2 & 0 \end{bmatrix}$

35. $\begin{bmatrix} 220 & 215 & 210 & 205 \\ 220 & 210 & 200 & 195 \\ 215 & 205 & 195 & 190 \end{bmatrix}$

37. a. $D = \begin{bmatrix} 2960 & 1510 & 1150 \\ 1100 & 550 & 490 \\ 1230 & 590 & 470 \end{bmatrix}$

b. $E = \begin{bmatrix} 3256 & 1661 & 1265 \\ 1210 & 605 & 539 \\ 1353 & 649 & 517 \end{bmatrix}$

39. False

Using Technology Exercises 2.4, page 131

1. $\begin{bmatrix} 15 & 38.75 & -67.5 & 33.75 \\ 51.25 & 40 & 52.5 & -38.75 \\ 21.25 & 35 & -65 & 105 \end{bmatrix}$

3. $\begin{bmatrix} -5 & 6.3 & -6.8 & 3.9 \\ 1 & 0.5 & 5.4 & -4.8 \\ 0.5 & 4.2 & -3.5 & 5.6 \end{bmatrix}$

5. $\begin{bmatrix} 16.44 & -3.65 & -3.66 & 0.63 \\ 12.77 & 10.64 & 2.58 & 0.05 \\ 5.09 & 0.28 & -10.84 & 17.64 \end{bmatrix}$

7. $\begin{bmatrix} 7.4 & 7.2 & 2.9 \\ -0.1 & 5.9 & 1.4 \\ -4 & 3 & -6.9 \\ 1.5 & -1.4 & 11.2 \end{bmatrix}$

Exercises 2.5, page 140

1. 2×5; not defined **3.** 1×1; 7×7

5. $n = s$ and $m = t$ **7.** $\begin{bmatrix} -1 \\ 3 \end{bmatrix}$ **9.** $\begin{bmatrix} 9 \\ -10 \end{bmatrix}$

11. $\begin{bmatrix} 4 & -2 \\ 9 & 13 \end{bmatrix}$ **13.** $\begin{bmatrix} 2 & 9 \\ 5 & 16 \end{bmatrix}$ **15.** $\begin{bmatrix} 0.57 & 1.93 \\ 0.64 & 1.76 \end{bmatrix}$

17. $\begin{bmatrix} 6 & -3 & 0 \\ -2 & 1 & -8 \\ 4 & -4 & 9 \end{bmatrix}$ **19.** $\begin{bmatrix} 5 & 1 & -3 \\ 1 & 7 & -3 \end{bmatrix}$

21. $\begin{bmatrix} -4 & -20 & 4 \\ 4 & 12 & 0 \\ 12 & 32 & 20 \end{bmatrix}$ **23.** $\begin{bmatrix} 4 & -3 & 2 \\ 7 & 1 & -5 \end{bmatrix}$

27. $AB = \begin{bmatrix} 10 & 7 \\ 22 & 15 \end{bmatrix}$, $BA = \begin{bmatrix} 5 & 8 \\ 13 & 20 \end{bmatrix}$

31. $A = \begin{bmatrix} -2 & -1 \\ 5 & 2 \end{bmatrix}$ **33. a.** $A^T = \begin{bmatrix} 2 & 5 \\ 4 & -6 \end{bmatrix}$

35. $AX = B$, where $A = \begin{bmatrix} 2 & -3 \\ 3 & -4 \end{bmatrix}$, $X = \begin{bmatrix} x \\ y \end{bmatrix}$, and $B = \begin{bmatrix} 7 \\ 8 \end{bmatrix}$

37. $AX = B$, where $A = \begin{bmatrix} 2 & -3 & 4 \\ 0 & 2 & -3 \\ 1 & -1 & 2 \end{bmatrix}$, $X = \begin{bmatrix} x \\ y \\ z \end{bmatrix}$, and $B = \begin{bmatrix} 6 \\ 7 \\ 4 \end{bmatrix}$

39. $AX = B$, where $A = \begin{bmatrix} -1 & 1 & 1 \\ 2 & -1 & -1 \\ -3 & 2 & 4 \end{bmatrix}$, $X = \begin{bmatrix} x_1 \\ x_2 \\ x_3 \end{bmatrix}$, and $B = \begin{bmatrix} 0 \\ 2 \\ 4 \end{bmatrix}$

41. a. $AB = \begin{bmatrix} 51{,}400 \\ 54{,}200 \end{bmatrix}$

b. The first entry shows that William's total stockholdings are \$51,400; the second shows that Michael's stockholdings are \$54,200.

43. a. $B = \begin{bmatrix} 20{,}000 \\ 22{,}000 \\ 25{,}000 \\ 30{,}000 \end{bmatrix}$

b. \$7,160,000 in New York, \$2,860,000 in Connecticut, and \$1,430,000 in Massachusetts; the total profit is \$11,450,000.

45.
$$\begin{matrix} & \text{D} & \text{R} & \text{I} \\ BA = [41{,}000 & 35{,}000 & 14{,}000] \end{matrix}$$

47. $AB = \begin{bmatrix} 1575 & 1590 & 1560 & 975 \\ 410 & 405 & 415 & 270 \\ 215 & 205 & 225 & 155 \end{bmatrix}$

49. a. $AC = \begin{bmatrix} 348{,}400 \\ 273{,}200 \\ 632{,}400 \end{bmatrix}$ **b.** $AD = \begin{bmatrix} 478{,}200 \\ 369{,}800 \\ 877{,}400 \end{bmatrix}$

c. $BC = \begin{bmatrix} 233{,}400 \\ 239{,}000 \\ 517{,}600 \end{bmatrix}$ **d.** $BD = \begin{bmatrix} 320{,}100 \\ 324{,}500 \\ 718{,}200 \end{bmatrix}$

e. $(A + B)C = \begin{bmatrix} 581{,}800 \\ 512{,}200 \\ 1{,}150{,}000 \end{bmatrix}$

f. $(A + B)D = \begin{bmatrix} 798{,}300 \\ 694{,}300 \\ 1{,}595{,}600 \end{bmatrix}$

g. $A(D - C) = \begin{bmatrix} 129{,}800 \\ 96{,}600 \\ 245{,}000 \end{bmatrix}$

h. $B(D - C) = \begin{bmatrix} 86{,}700 \\ 85{,}500 \\ 200{,}600 \end{bmatrix}$

i. $(A + B)(D - C) = \begin{bmatrix} 216{,}500 \\ 182{,}100 \\ 445{,}600 \end{bmatrix}$

51. False

Using Technology Exercises 2.5, page 145

1. $\begin{bmatrix} 18.66 & 15.2 & -12 \\ 24.48 & 41.88 & 89.82 \\ 15.39 & 7.16 & -1.25 \end{bmatrix}$

3. $\begin{bmatrix} 20.09 & 20.61 & -1.3 \\ 44.42 & 71.6 & 64.89 \\ 20.97 & 7.17 & -60.65 \end{bmatrix}$

5. $\begin{bmatrix} 32.89 & 13.63 & -57.17 \\ -12.85 & -8.37 & 256.92 \\ 13.48 & 14.29 & 181.64 \end{bmatrix}$

7. $\begin{bmatrix} 18.66 & 24.48 & 15.39 \\ 15.2 & 41.88 & 7.16 \\ -12 & 89.82 & -1.25 \end{bmatrix}$

9. $\begin{bmatrix} 87 & 68 & 110 & 82 \\ 119 & 176 & 221 & 143 \\ 51 & 128 & 142 & 94 \\ 28 & 174 & 174 & 112 \end{bmatrix}$; $\begin{bmatrix} 113 & 117 & 72 & 101 & 90 \\ 72 & 85 & 36 & 72 & 76 \\ 81 & 69 & 76 & 87 & 30 \\ 133 & 157 & 56 & 121 & 146 \\ 154 & 157 & 94 & 127 & 122 \end{bmatrix}$

11. $\begin{bmatrix} 170 & 18.1 & 133.1 & -106.3 & 341.3 \\ 349 & 226.5 & 324.1 & 164 & 506.4 \\ 245.2 & 157.7 & 231.5 & 125.5 & 312.9 \\ 310 & 245.2 & 291 & 274.3 & 354.2 \end{bmatrix}$

Exercises 2.6, page 158

5. $\begin{bmatrix} 3 & -5 \\ -1 & 2 \end{bmatrix}$ **7.** Does not exist

9. $\begin{bmatrix} 2 & -11 & -3 \\ 1 & -6 & -2 \\ 0 & -1 & 0 \end{bmatrix}$ **11.** Does not exist

13. $\begin{bmatrix} -\frac{13}{10} & \frac{7}{5} & \frac{1}{2} \\ \frac{2}{5} & -\frac{1}{5} & 0 \\ -\frac{7}{10} & \frac{3}{5} & \frac{1}{2} \end{bmatrix}$

15. $\begin{bmatrix} 3 & 4 & -6 & 1 \\ -2 & -3 & 5 & -1 \\ -4 & -4 & 7 & -1 \\ -4 & -5 & 8 & -1 \end{bmatrix}$

17. a. $A = \begin{bmatrix} 2 & 5 \\ 1 & 3 \end{bmatrix}$, $X = \begin{bmatrix} x \\ y \end{bmatrix}$, $B = \begin{bmatrix} 3 \\ 2 \end{bmatrix}$

b. $x = -1, y = 1$

19. a. $A = \begin{bmatrix} 2 & -3 & -4 \\ 0 & 0 & -1 \\ 1 & -2 & 1 \end{bmatrix}$, $X = \begin{bmatrix} x \\ y \\ z \end{bmatrix}$, $B = \begin{bmatrix} 4 \\ 3 \\ -8 \end{bmatrix}$

b. $x = -1, y = 2$, and $z = -3$

21. a. $A = \begin{bmatrix} 1 & 4 & -1 \\ 2 & 3 & -2 \\ -1 & 2 & 3 \end{bmatrix}$, $X = \begin{bmatrix} x \\ y \\ z \end{bmatrix}$, $B = \begin{bmatrix} 3 \\ 1 \\ 7 \end{bmatrix}$

b. $x = 1, y = 1$, and $z = 2$

23. a. $A = \begin{bmatrix} 1 & 1 & -1 & 1 \\ 2 & 1 & 1 & 0 \\ 2 & 1 & 0 & 1 \\ 2 & -1 & -1 & 3 \end{bmatrix}$, $X = \begin{bmatrix} x_1 \\ x_2 \\ x_3 \\ x_4 \end{bmatrix}$,

$B = \begin{bmatrix} 6 \\ 4 \\ 7 \\ 9 \end{bmatrix}$

b. $x_1 = 1, x_2 = 2, x_3 = 0$, and $x_4 = 3$

25. b.(i) $x = 24/5, y = 23/5$ **(ii)** $x = 2/5, y = 9/5$

27. b.(i) $x = -1, y = 3, z = 2$
(ii) $x = 1, y = 8, z = -12$

29. b.(i) $x = -2/17, y = -10/17, z = -60/17$
(ii) $x = 1, y = 0, z = -5$

31. b.(i) $x_1 = 1, x_2 = -4, x_3 = 5, x_4 = -1$
(ii) $x_1 = 12, x_2 = -24, x_3 = 21, x_4 = -7$

33. a. $A^{-1} = \begin{bmatrix} -\frac{5}{2} & -\frac{3}{2} \\ 2 & 1 \end{bmatrix}$

35. a. $ABC = \begin{bmatrix} 4 & 10 \\ 2 & 3 \end{bmatrix}$, $A^{-1} = \begin{bmatrix} 3 & -5 \\ 1 & -2 \end{bmatrix}$,

$B^{-1} = \begin{bmatrix} 1 & -3 \\ -1 & 4 \end{bmatrix}$, $C^{-1} = \begin{bmatrix} \frac{1}{8} & -\frac{3}{8} \\ \frac{1}{4} & \frac{1}{4} \end{bmatrix}$

37. a. 3214; 3929 **b.** 4286; 3571 **c.** 3929; 5357

39. a. $6 million to Organization I, $10 million to Organization II, and $8 million to Organization III
b. $8 million to Organization I, $6 million to Organization II, and $5 million to Organization III

41. False

43. a–b. If $a \neq 0$, then $A^{-1} = \begin{bmatrix} \dfrac{d}{ad - bc} & -\dfrac{b}{ad - bc} \\ \dfrac{-c}{ad - bc} & \dfrac{a}{ad - bc} \end{bmatrix}$,

provided $ad - bc \neq 0$.

If $a = 0$, then $A^{-1} = \begin{bmatrix} -\dfrac{d}{bc} & \dfrac{1}{c} \\ \dfrac{1}{b} & 0 \end{bmatrix}$, provided $bc \neq 0$.

Using Technology Exercises 2.6, page 161

1. $\begin{bmatrix} 0.36 & 0.04 & -0.36 \\ 0.06 & 0.05 & 0.20 \\ -0.19 & 0.10 & 0.09 \end{bmatrix}$

3. $\begin{bmatrix} 0.01 & -0.09 & 0.31 & -0.11 \\ -0.25 & 0.58 & -0.15 & -0.02 \\ 0.86 & -0.42 & 0.07 & -0.37 \\ -0.27 & 0.01 & -0.05 & 0.31 \end{bmatrix}$

5. $\begin{bmatrix} 0.30 & 0.85 & -0.10 & -0.77 & -0.11 \\ -0.21 & 0.10 & 0.01 & -0.26 & 0.21 \\ 0.03 & -0.16 & 0.12 & -0.01 & 0.03 \\ -0.14 & -0.46 & 0.13 & 0.71 & -0.05 \\ 0.10 & -0.05 & -0.10 & -0.03 & 0.11 \end{bmatrix}$

Exercises 2.7, page 171

1. a. $10 million **b.** $160 million
c. Agricultural; manufacturing and transportation

3. $x = 23.75$ and $y = 21.25$

5. $x = 42.85$ and $y = 57.14$

9. a. $318.2 million worth of agricultural goods and $336.4 million worth of manufactured products
b. $198.2 million worth of agricultural products and $196.4 million worth of manufactured goods

11. a. $443.75 million, $381.25 million, and $281.25 million worth of agricultural products, manufactured goods, and transportation, respectively
b. $243.75 million, $281.25 million, and $221.25 million worth of agricultural products, manufactured goods, and transportation, respectively

13. $45 million and $75 million

15. $34.4 million, $33 million, and $21.6 million

Using Technology Exercises 2.7, page 173

1. The final outputs of the first, second, third, and fourth industries are 602.62, 502.30, 572.57, and 523.46 units, respectively.

3. The final outputs of the first, second, third, and fourth industries are 143.06, 132.98, 188.59, and 125.53 units, respectively.

Chapter 2 Review Exercises, page 178

1. $\begin{bmatrix} 2 & 2 \\ -1 & 4 \\ 3 & 3 \end{bmatrix}$ **3.** $[-6 \quad -2]$

5. $x = 2, y = 3, z = 1$, and $w = 3$

7. $a = 3, b = 4, c = -2, d = 2$, and $e = -3$

9. $\begin{bmatrix} 8 & 9 & 11 \\ -10 & -1 & 3 \\ 11 & 12 & 10 \end{bmatrix}$ **11.** $\begin{bmatrix} 6 & 18 & 6 \\ -12 & 6 & 18 \\ 24 & 0 & 12 \end{bmatrix}$

13. $\begin{bmatrix} -11 & -16 & -15 \\ -4 & -2 & -10 \\ -6 & 14 & 2 \end{bmatrix}$ **15.** $\begin{bmatrix} -3 & 17 & 8 \\ -2 & 56 & 27 \\ 74 & 78 & 116 \end{bmatrix}$

17. $x = 1, y = -1$ **19.** $x = 1, y = 2$, and $z = 3$

21. No solution

23. $x = 1, y = 0$, and $z = 1$

25. $\begin{bmatrix} \frac{2}{5} & -\frac{1}{5} \\ -\frac{1}{5} & \frac{3}{5} \end{bmatrix}$ **27.** $\begin{bmatrix} -1 & 2 \\ 1 & -\frac{3}{2} \end{bmatrix}$

29. $\begin{bmatrix} \frac{5}{4} & \frac{1}{4} & -\frac{7}{4} \\ -\frac{1}{4} & -\frac{1}{4} & \frac{3}{4} \\ -\frac{3}{4} & \frac{1}{4} & \frac{5}{4} \end{bmatrix}$ **31.** $\begin{bmatrix} -\frac{1}{5} & \frac{2}{5} & 0 \\ \frac{2}{3} & -\frac{1}{3} & \frac{1}{3} \\ -\frac{1}{30} & \frac{1}{15} & -\frac{1}{6} \end{bmatrix}$

33. $\begin{bmatrix} \frac{3}{2} & 1 \\ -\frac{7}{2} & -1 \end{bmatrix}$ **35.** $\begin{bmatrix} \frac{2}{5} & -\frac{3}{5} \\ \frac{1}{5} & \frac{1}{5} \end{bmatrix}$

37. $A^{-1} = \begin{bmatrix} \frac{2}{7} & \frac{3}{7} \\ \frac{1}{7} & -\frac{2}{7} \end{bmatrix}$; $x = -1, y = -2$

39. $A^{-1} = \begin{bmatrix} 1 & -\frac{2}{5} & \frac{4}{5} \\ -1 & 1 & -1 \\ -\frac{1}{2} & \frac{3}{5} & -\frac{7}{10} \end{bmatrix}$; $x = 1, y = 2$, and $z = 4$

41. \$5180, \$4540, and \$5550

43. 30 of each type

Chapter 3

Exercises 3.1, page 188

1.

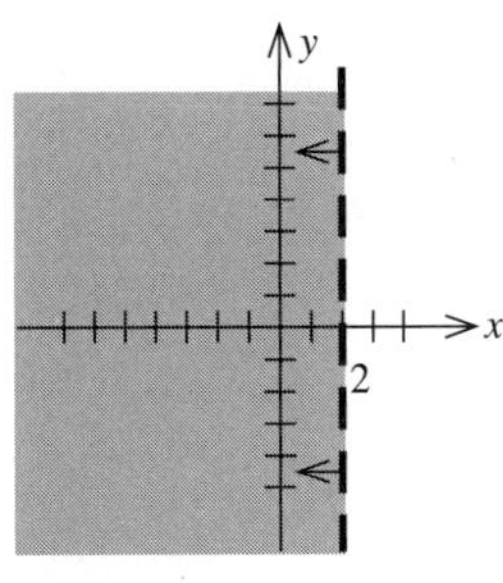

3.

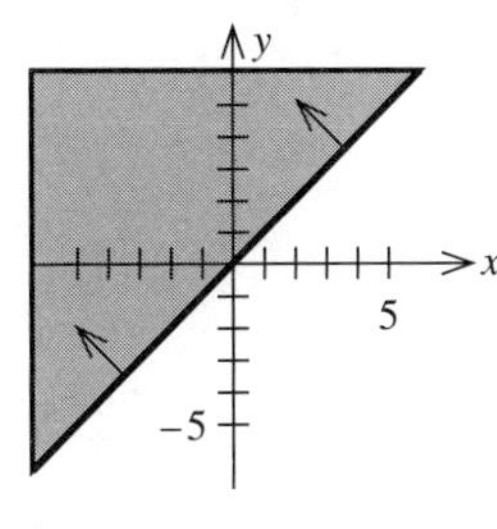

5.

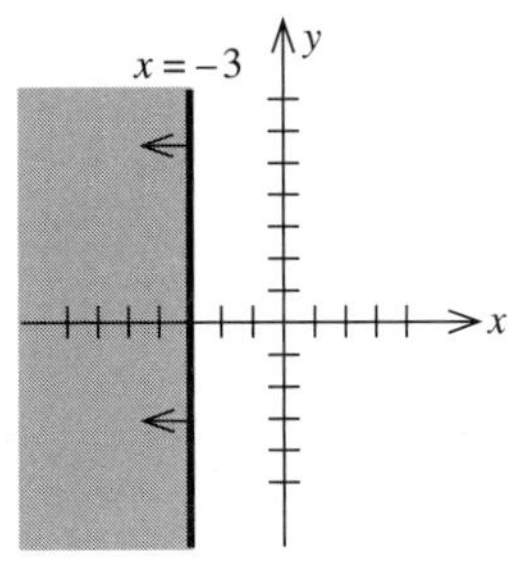

7.

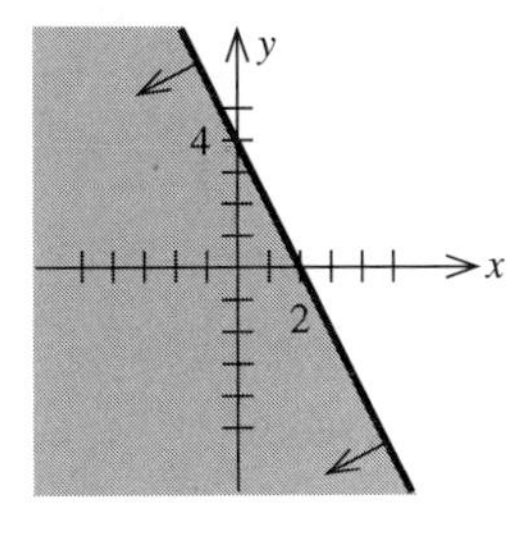

9.

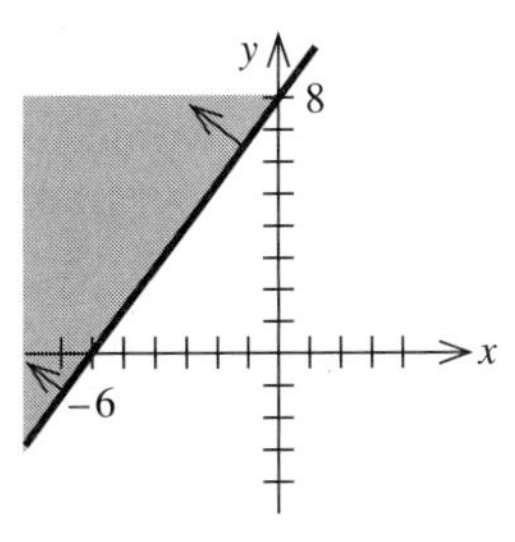

11. $x \geq 1$, $x \leq 5$, $y \geq 2$, and $y \leq 4$

13. $2x - y \geq 2$, $5x + 7y \geq 35$, and $x \leq 4$

15. $x - y \geq -10$, $7x + 4y \leq 140$, and $x + 3y \geq 30$

17. $x + y \geq 7$, $x \geq 2$, $y \geq 3$, and $y \leq 7$

19.

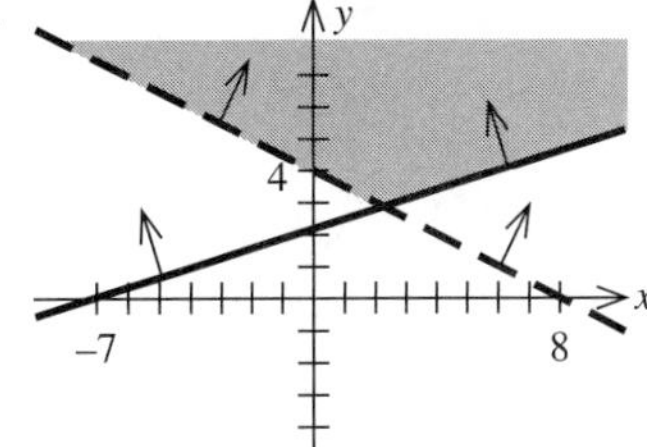

21.

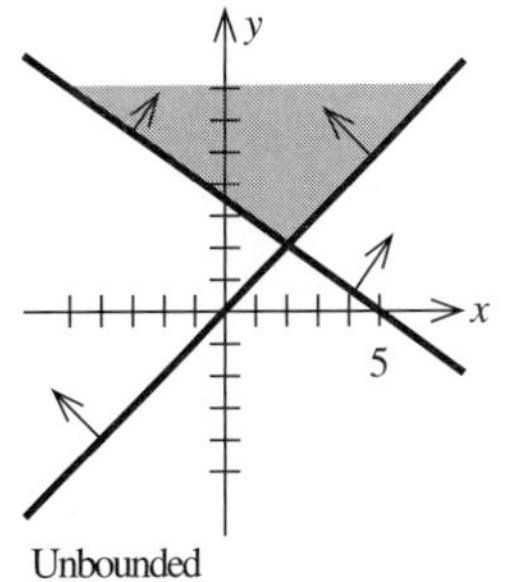

23.

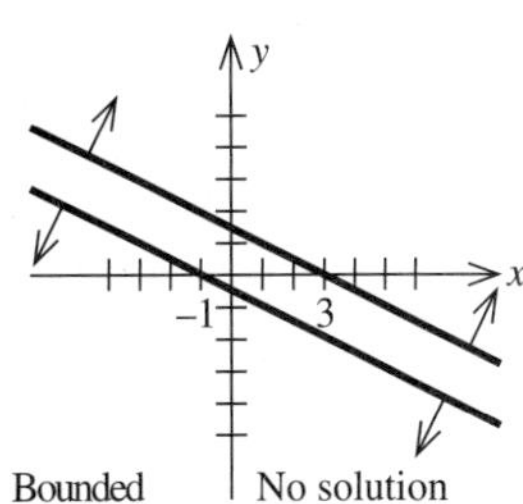

25.

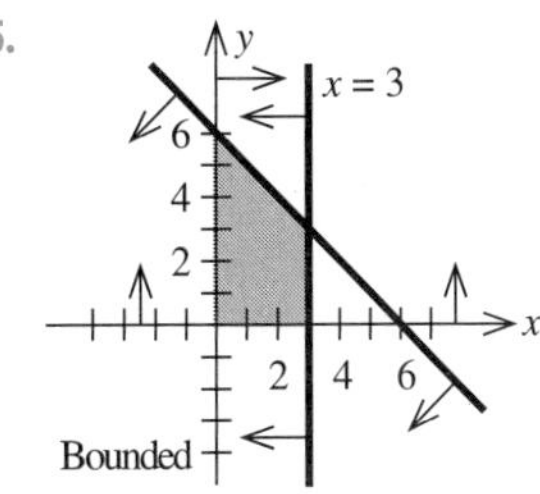

27.

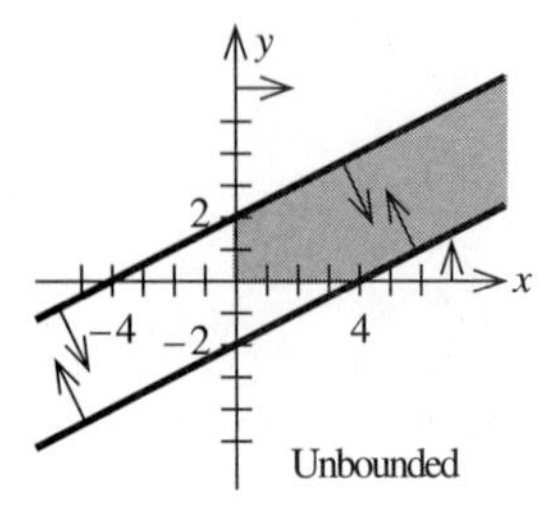

29.

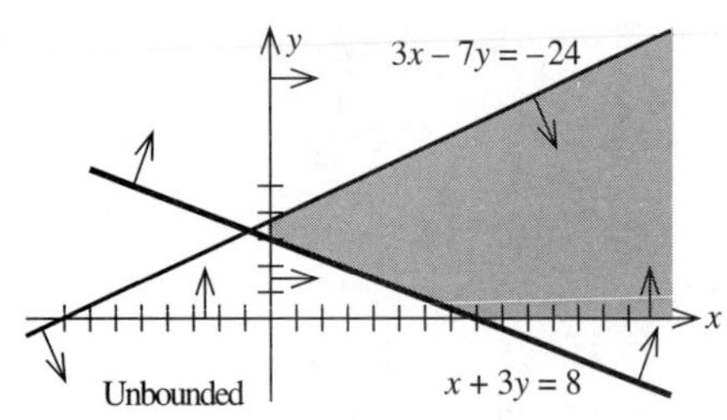

31.

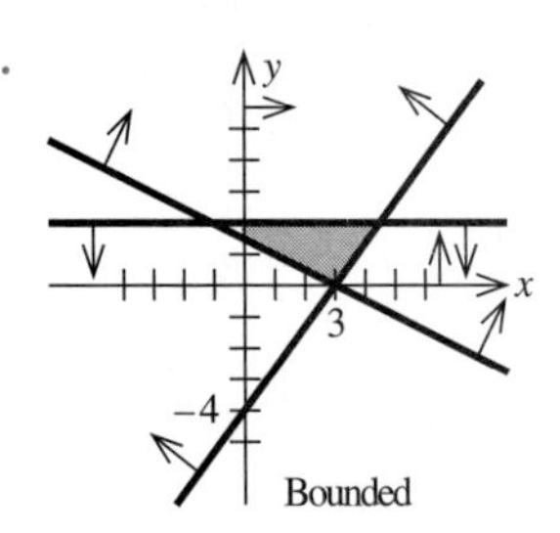

33.

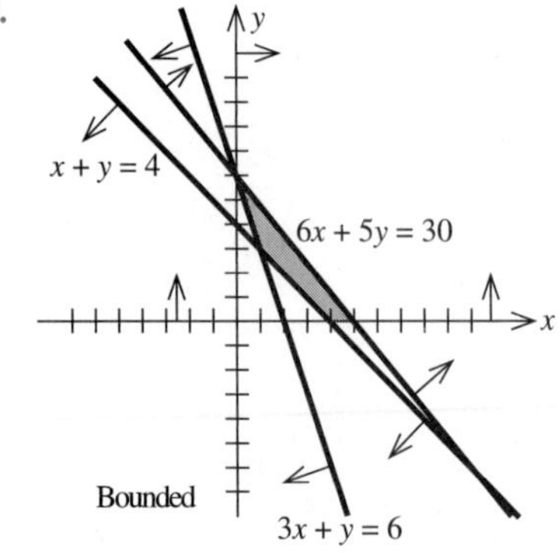

35.

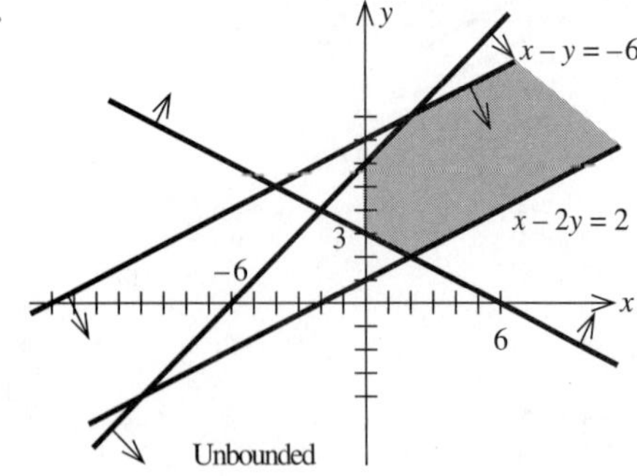

37. False **39.** True

Exercises 3.2, page 197

1. Maximize $P = 3x + 4y$
subject to $6x + 9y \leq 300$
$5x + 4y \leq 180$
$x \geq 0, y \geq 0$

3. Maximize $P = 2x + (3/2)y$
subject to $3x + 4y \leq 1000$
$6x + 3y \leq 1200$
$x \geq 150, y \geq 0$

5. Maximize $P = 50x + 40y$
subject to $\frac{1}{200}x + \frac{1}{200}y \leq 1$
$\frac{1}{100}x + \frac{1}{300}y \leq 1$
$x \geq 0, y \geq 0$

7. Minimize $C = 2x + 5y$
subject to $30x + 25y \geq 400$
$x + 0.5y \geq 10$
$2x + 5y \geq 40$
$x \geq 0, y \geq 0$

9. Minimize $C = 32{,}000 - x - 3y$
subject to $x + y \leq 6000$
$x + y \geq 2000$
$x \leq 3000$
$y \leq 4000$
$x \geq 0, y \geq 0$

11. Maximize $P = 18x + 12y + 15z$
subject to $2x + y + 2z \leq 900$
$3x + y + 2z \leq 1080$
$2x + 2y + z \leq 840$
$x \geq 0, y \geq 0, z \geq 0$

13. Maximize $P = 26x + 28y + 24z$
subject to $\frac{5}{4}x + \frac{3}{2}y + \frac{3}{2}z \leq 310$
$x + y + \frac{3}{4}z \leq 205$
$x + y + \frac{1}{2}z \leq 190$
$x \geq 0, y \geq 0, z \geq 0$

15. Minimize $C = 60x_1 + 60x_2 + 80x_3 + 80x_4 + 70x_5 + 50x_6$
subject to $x_1 + x_2 + x_3 \leq 300$
$x_4 + x_5 + x_6 \leq 250$
$x_1 + x_4 \geq 200$
$x_2 + x_5 \geq 150$
$x_3 + x_6 \geq 200$
$x_1 \geq 0, x_2 \geq 0, \ldots, x_6 \geq 0$

17. False

Exercises 3.3, page 210

1. Max: 35; min: 5 **3.** No max. value; min: 27

5. Max: 44; min: 15 **7.** $x = 0$, $y = 6$, $P = 18$

9. $x = 14$, $y = 3$, $C = 58$

11. Any point (x, y) lying on the line segment joining $(5, 20)$ to $(12, 6)$, $C = 90$

13. $x = 3$, $y = 3$, $C = 75$

15. $x = 15$, $y = 17.5$, $P = 115$

17. $x = 10$, $y = 38$, $P = 134$

19. Max: $x = 6$, $y = 33/2$, $P = 258$
Min: $x = 15$, $y = 3$, $P = 186$

21. Max: $x = 5$, $y = 15$, $P = 70$
Min: $x = 0$, $y = 5$, $P = 20$

23. No model A, 2000 model B; $P = \$80{,}000$

25. 120 model A, 160 model B; $P = \$480$

27. \$16 million in homeowner loans, \$4 million in auto loans; $P = \$2.08$ million

29. 65 acres of crop A, 80 acres of crop B; $P = \$25{,}750$

31. 80 from I to A, 20 from I to B, 0 from II to A, 50 from II to B; $C = \$12{,}700$

33. 2000 from I to A, 4000 from I to B, 1000 from II to A, 0 from II to B; $C = \$18{,}000$

35. \$22,500 in growth stocks and \$7500 in speculative stocks; maximum return: \$5250

37. True   39. False

43. **a.**

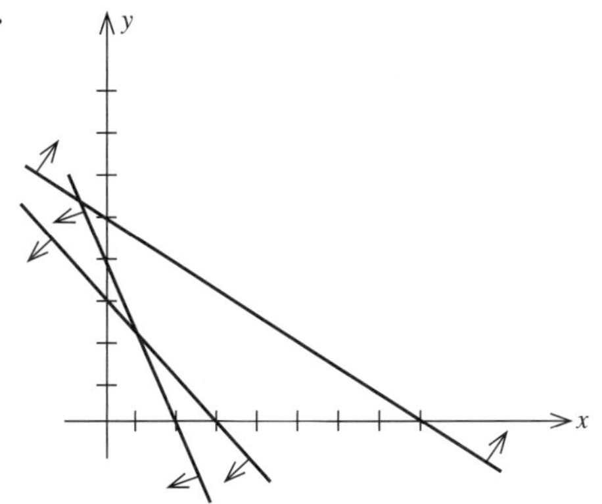

b. No solution

Chapter 3 Review Exercises, page 216

1. Max: 18; any point (x, y) lying on the line segment joining $(0, 6)$ to $(3, 4)$; min: 0

3. $x = 0$, $y = 4$, $P = 20$ 5. $x = 0$, $y = 0$, $C = 0$

7. $x = 3$, $y = 10$, $P = 29$

9. $x = 20$, $y = 0$, $C = 40$

11. Max: $x = 22$, $y = 0$, $Q = 22$
Min: $x = 3$, $y = 5/2$, $Q = 11/2$

13. \$40,000 in each company; $P = \$13{,}600$

15. 93 model A, 180 model B

CHAPTER 4

Exercises 4.1, page 238

1. In final form; $x = \frac{30}{7}$, $y = \frac{20}{7}$, $u = 0$, $v = 0$, $P = \frac{220}{7}$

3. Not in final form; pivot element is $\frac{1}{2}$, lying in the first row, second column.

5. In final form: $x = \frac{1}{3}$, $y = 0$, $z = \frac{13}{3}$, $u = 0$, $v = 6$, $w = 0$, $P = 17$

7. Not in final form; pivot element is 1, lying in the third row, second column.

9. In final form; $x = 30$, $y = 0$, $z = 0$, $u = 10$, $v = 0$, $P = 60$ and $x = 0$, $y = 30$, $z = 0$, $u = 10$, $v = 0$, $P = 60$

11. $x = 6$, $y = 3$, $u = 0$, $v = 0$, $P = 96$

13. $x = 6$, $y = 6$, $u = 0$, $v = 0$, $w = 0$, $P = 60$

15. $x = 0$, $y = 4$, $z = 4$, $u = 0$, $v = 0$, $P = 36$

17. $x = 0$, $y = 3$, $z = 0$, $u = 90$, $v = 0$, $w = 75$, $P = 12$

19. $x = 15$, $y = 3$, $z = 0$, $u = 2$, $v = 0$, $w = 0$, $P = 78$

21. $x = \frac{5}{4}$, $y = \frac{15}{2}$, $z = 0$, $u = 0$, $v = \frac{15}{4}$, $w = 0$, $P = 90$

23. $x = 2$, $y = 1$, $z = 1$, $u = 0$, $v = 0$, $w = 0$, $P = 87$

27. No model A, 2000 model B; $P = \$80{,}000$

29. 65 acres of crop A, 80 acres of crop B; $P = \$25{,}750$

31. 22 min of morning advertising time, 3 min of evening advertising time; maximum exposure: 6,200,000 viewers

33. 80 units of model A, 80 units of model B, and 60 units of model C; maximum profit: \$5760

35. 9000 bottles of formula I, 7833 bottles of formula II, 6000 bottles of formula III; maximum profit: \$4986.67

37. True

Using Technology Exercises 4.1, page 243

1. $x = 1.2$, $y = 0$, $z = 1.6$, $w = 0$, and $P = 8.8$

3. $x = 1.6$, $y = 0$, $z = 0$, $w = 3.6$, and $P = 12.4$

Exercises 4.2, page 256

1. $x = 4, y = 0, C = -8$

3. $x = 4, y = 3, C = -18$

5. $x = 0, y = 13, z = 18, w = 14, C = -111$

7. $x = \frac{5}{4}, y = \frac{1}{4}, u = 2, v = 3$, and $C = P = 13$

9. $x = 5, y = 10, z = 0, u = 1, v = 2$, and $C = P = 80$

11. Maximize $P = 4u + 6v$ subject to
$$\begin{aligned} u + 3v &\le 2 \\ 2u + 2v &\le 5; \quad x = 4, y = 0, C = 8 \\ u \ge 0, v &\ge 0 \end{aligned}$$

13. Maximize $P = 60u + 40v + 30w$ subject to
$$\begin{aligned} 6u + 2v + w &\le 6 \\ u + v + w &\le 4; \quad x = 10, y = 20, C = 140 \\ u \ge 0, v \ge 0, w &\ge 0 \end{aligned}$$

15. Maximize $P = 10u + 20v$ subject to
$$\begin{aligned} 20u + v &\le 200 \\ 10u + v &\le 150; \quad x = 0, y = 0, z = 10, C = 1200 \\ u + 2v &\le 120 \\ u \ge 0, v &\ge 0 \end{aligned}$$

17. Maximize $P = 10u + 24v + 16w$ subject to
$$\begin{aligned} u + 2v + w &\le 6 \\ 2u + v + w &\le 8; \quad x = 8, y = 0, z = 8, C = 80 \\ 2u + v + w &\le 4 \\ u \ge 0, v \ge 0, w &\ge 0 \end{aligned}$$

19. Maximize $P = 6u + 2v + 4w$ subject to
$$\begin{aligned} 2u + 6v \qquad &\le 30 \\ 4u \qquad + 6w &\le 12; \quad x = \tfrac{1}{3}, y = \tfrac{4}{3}, z = 0, C = 26 \\ 3u + v + 2w &\le 20 \\ u \ge 0, v \ge 0, w &\ge 0 \end{aligned}$$

21. Loc. I: 500 to warehouse A, 200 to warehouse B; Loc. II: 200 to warehouse B, 400 to warehouse C; shipping costs: \$20,800

23. 8 oz. of orange juice; 6 oz. of pink grapefruit juice; 178 calories

25. True

Using Technology Exercises 4.2, page 260

1. $x = \frac{4}{3}, y = \frac{10}{3}, z = 0, C = \frac{14}{3}$

3. $x = 0.9524, y = 4.2857, z = 0, C = 6.0952$

Exercises 4.3, page 275

1. Maximize $C = -P = -2x + 3y$ subject to
$$\begin{aligned} -3x - 5y &\le -20 \\ 3x + y &\le 16 \\ -2x + y &\le 1 \\ x \ge 0, y &\ge 0 \end{aligned}$$

3. Maximize $P = -C = -5x - 10y - z$ subject to
$$\begin{aligned} -2x - y - z &\le -4 \\ -x - 2y - 2z &\le -2 \\ 2x + 4y + 3z &\le 12 \\ x \ge 0, y \ge 0, z &\ge 0 \end{aligned}$$

5. $x = 5, y = 2, P = 9$

7. $x = 4, y = 0, C = -8$

9. $x = 4, y = \frac{2}{3}, P = \frac{20}{3}$

11. $x = 3, y = 2, P = 7$

13. $x = 24, y = 0, z = 0, P = 120$

15. $x = 0, y = 17, z = 1, C = -33$

17. $x = \frac{46}{7}, y = 0, z = \frac{50}{7}, P = \frac{142}{7}$

19. $x = 0, y = 0, z = 10, P = 30$

21. 80 acres of crop A, 68 acres of crop B; $P = \$25,600$

23. \$50 million worth of home loans, \$10 million worth of commercial-development loans; maximum return: \$4.6 million

25. 0 units of product A, 280 units of product B, 280 units of product C; $P = \$7560$

27. 10 oz of food A, 4 oz of food B, 40 mg of cholesterol; infinitely many solutions

Chapter 4 Review Exercises, page 280

1. $x = 3, y = 4, u = 0, v = 0, P = 25$

3. $x = 56/5, y = 2/5, z = 0, u = 0, v = 0, P = 23\frac{3}{5}$

5. $x = 3/2, y = 1, C = 13/2$

7. $x = 3/4, y = 0, z = 7/4, C = 60$

9. $x = 45, y = 0, P = 135$

11. $x = 5, y = 2, P = 16$

13. 30 units product B; $P = \$180$

15. \$50,000 in stocks, \$100,000 in bonds, \$50,000 in money-market funds; maximum return: \$21,500/yr

CHAPTER 5

Exercises 5.1, page 293

1. \$80; \$580 3. \$836 5. \$1000 7. 146 days

9. 10%/yr 11. \$1718.19 13. \$4974.47

15. \$27,566.93 17. \$261,751.04 19. \$214,986.69

21. $10\frac{1}{4}$%/yr 23. 8.3%/yr 25. \$29,277.61

27. \$30,255.95 29. 24%/yr 31. \$558.34

33. $182,326 **35.** $3.8 million **37.** $26,267.49

39. a. $34,626.88 **b.** $33,886.16 **c.** $33,506.76

41. $23,329.48 **43.** 4.2% **45.** 8.5%

47. a. A family of straight lines with varying slope and P-intercept
b. A family of straight lines emanating from the point $(0, P)$ with varying slope

49. False **51.** False

Using Technology Exercises 5.1, page 297

1. $5872.78 **3.** $475.49 **5.** 8.95%/yr

7. 10.20%/yr

9.
```
:PROGRAM: PREVAL
:Disp "A"
:Input A
:Disp "r"
:Input r
:Disp "t"
:Input t
:Disp "m"
:Input m
:A(1+r/m)^(−m*t)→P
:Disp "PRESENT VALUE IS"
:Disp P
```

11. $94,038.74

13. $62,244.96

Exercises 5.2, page 306

1. $15,937.42 **3.** $54,759.35 **5.** $37,965.57

7. $28,733.19 **9.** $15,558.61 **11.** $15,011.29

13. $109,658.91 **15.** $455.70 **17.** $44,526.45

19. $9850.12 **21.** $608.54

23. Between $120,141 and $143,927

25. Between $103,875 and $123,593 **27.** True

Using Technology Exercises 5.2, page 309

1. $59,622.15 **3.** $8453.59

5.
```
:PROGRAM: PVAN
:Disp "R"
:Input R
:Disp "i"
:Input i
:Disp "N"
:Input N
:(R/i)(1−(1+i)^(−N))→P
:Disp "AMOUNT IS"
:Disp P
```

7. $45,983.53

9. $18,344.08

Exercises 5.3, page 317

1. $14,902.95 **3.** $444.24 **5.** $622.13

7. $731.79 **9.** $1491.19 **11.** $516.76

13. $172.95 **15.** $16,274.54

17. a. $212.27 **b.** $1316.36; $438.79

19. a. $387.21; $304.35 **b.** $1939.56; $2608.80

21. $1174.02; $27,888; $39,641; $83,236

23. $3135.48 **25.** $1449.74

27. $2090.41; $4280.21 **29.** $33,835.20

31. a. $1048.82; $1369.29
b. $131,103

Using Technology Exercises 5.3, page 321

1. $3645.40 **3.** $18,443.75

5.
```
:PROGRAM: SINKFD
:Disp "S"
:Input S
:Disp "i"
:Input i
:Disp "N"
:Input N
:S*i/((1+i)^N−1)→R
:Disp "R is"
:Disp R
```

7. $916.26

9. $809.31 **11.** $45,069.31

Exercises 5.4, page 328

1. 30 **3.** −4.5 **5.** −3, 8, 19, 30, 41 **7.** $x + 6y$

9. 795 **11.** 792 **13.** 550

15. a. 275 **b.** −280 **17.** 37 wk **19.** $7.90

21. b. $800 **23.** GP; 256; 508 **25.** Not a GP

27. GP; 1/3; $364\frac{1}{3}$ **29.** 3; 0 **31.** 293,866

33. $41,149.12 **35.** Annual raise of 8%/yr

37. a. $20,113.57 **b.** $87,537.38 **39.** $25,165.82

41. $39,321.60; $110,678.40 **43.** True

Chapter 5 Review Exercises, page 332

1. a. $7320.50 **b.** $7387.28 **c.** $7422.53
d. $7446.77

3. a. 12% **b.** 12.36% **c.** 12.5509%
d. 12.6825%

5. $30,000.29 **7.** $5557.68 **9.** $7861.70

11. \$694.49 **13.** \$332.73 **15.** 7.179%

17. \$2,592,702; \$8,612,002 **19.** \$15,000

21. \$218.64 **23.** \$13,026.89

25. a. \$965.55 **b.** \$227,598 **c.** \$42,684

27. \$19,573.56 **29.** \$205.09

Chapter 6

Exercises 6.1, page 344

1. $\{x \mid x$ is a gold medalist in the 2000 Summer Olympic Games$\}$

3. $\{x \mid x$ is an integer greater than 2 and less than 8$\}$

5. $\{2, 3, 4, 5, 6\}$ **7.** $\{-2\}$

9. a. True **b.** False **11. a.** False **b.** False

13. True **15. a.** True **b.** False

17. a. and **b.**

19. a. $\varnothing, \{1\}, \{2\}, \{1, 2\}$
b. $\varnothing, \{1\}, \{2\}, \{3\}, \{1, 2\}, \{1, 3\}, \{2, 3\}, \{1, 2, 3\}$
c. $\varnothing, \{1\}, \{2\}, \{3\}, \{4\}, \{1, 2\}, \{1, 3\}, \{1, 4\}, \{2, 3\}, \{2, 4\}, \{3, 4\}, \{1, 2, 3\}, \{1, 2, 4\}, \{1, 3, 4\}, \{2, 3, 4\}, \{1, 2, 3, 4\}$

21. $\{1, 2, 3, 4, 6, 8, 10\}$

23. {Jill, John, Jack, Susan, Sharon}

25. a.

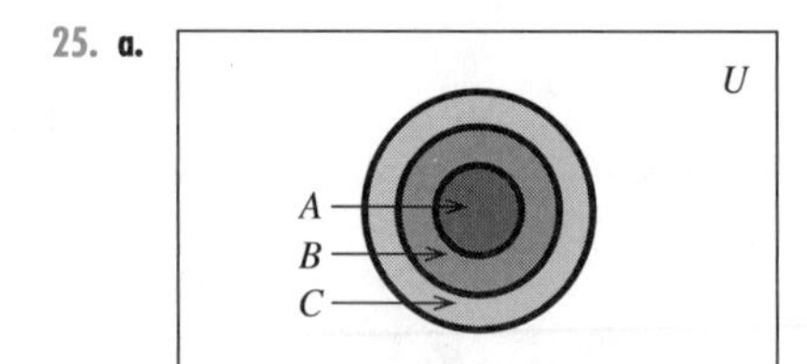

b.

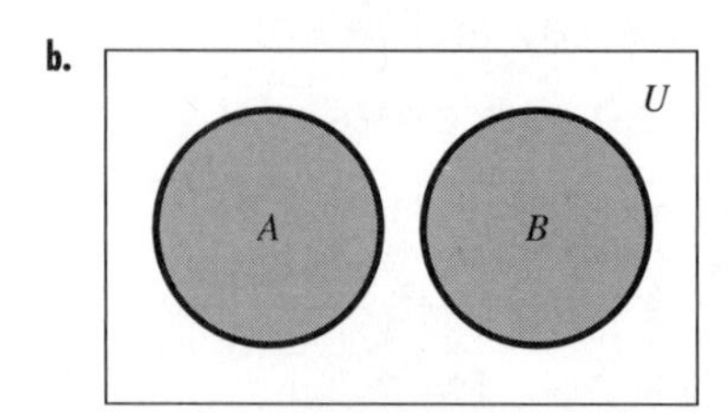

c.

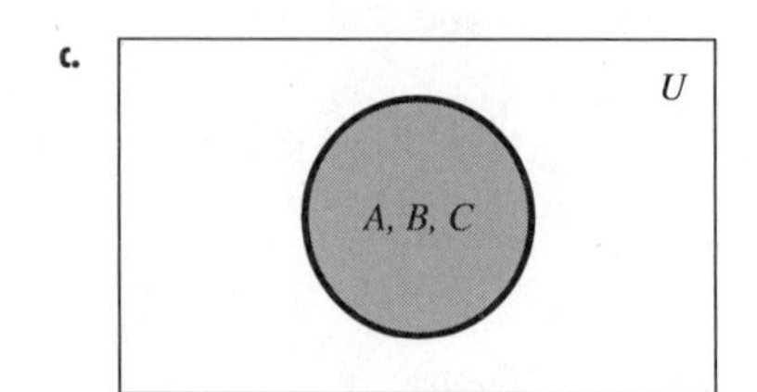

27. a.

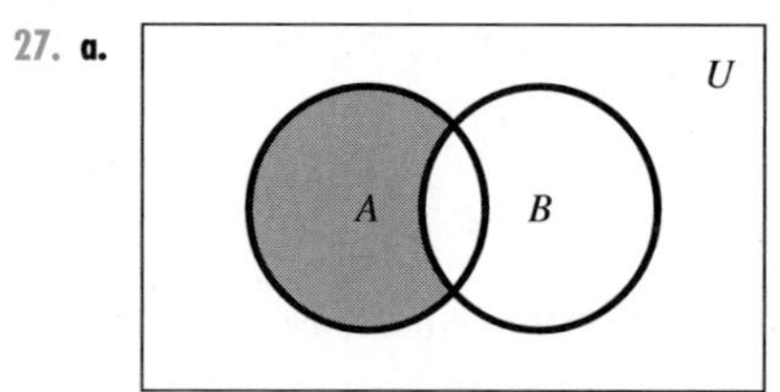

b.

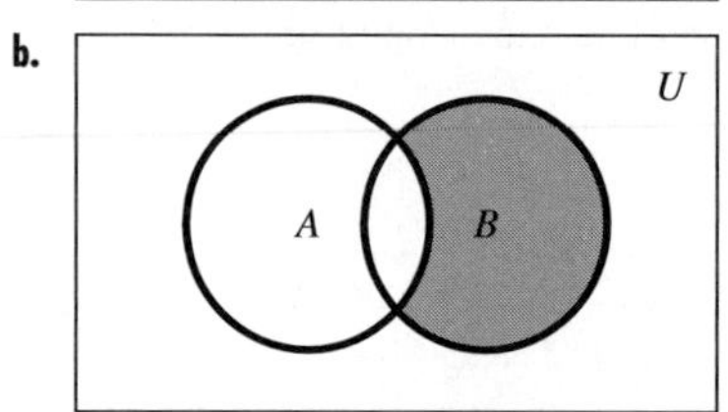

29. a.

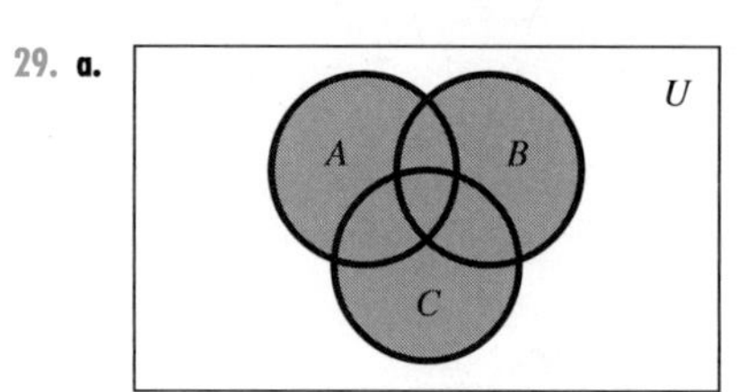

b.

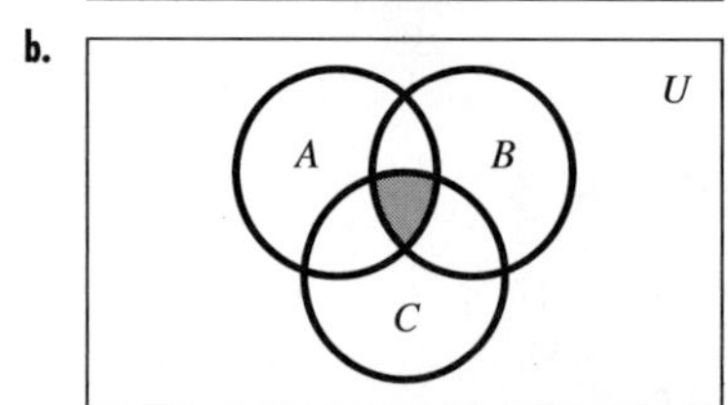

31. a. **b.**

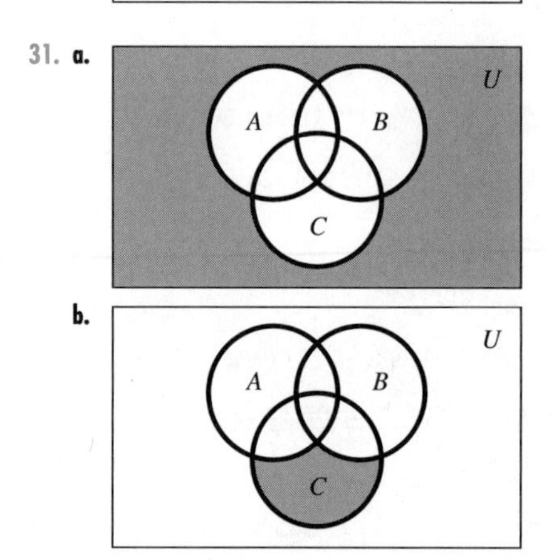

33. a. $\{2, 4, 6, 8, 10\}$ **b.** $\{1, 2, 4, 5, 6, 8, 9, 10\}$
c. U

35. a. $C = \{1, 2, 4, 5, 8, 9\}$ **b.** $\varnothing$ **c.** U

37. a. Not disjoint **b.** Disjoint

39. a. The set of all employees at the Universal Life Insurance Company who do not drink tea
b. The set of all employees at the Universal Life Insurance Company who do not drink coffee

41. **a.** The set of all employees at the Universal Life Insurance Company who drink tea but not coffee
b. The set of all employees at the Universal Life Insurance Company who drink coffee but not tea

43. **a.** The set of all employees in a hospital who are not doctors
b. The set of all employees in a hospital who are not nurses

45. **a.** The set of all employees in a hospital who are female doctors
b. The set of all employees in a hospital who are both doctors and administrators

47. **a.** $D \cap F$ **b.** $R \cap F^C \cap L^C$

49. **a.** B^C **b.** $A \cap B$ **c.** $A \cap B \cap C^C$

51. **a.** $A \cap B \cap C$; the set of tourists who have taken the underground, a cab, and a bus over a 1-wk period in London
b. $A \cap C$; the set of tourists who have taken the underground and a bus over a 1-wk period in London
c. B^c; the set of tourists who have not taken a cab over a 1-wk period in London

53. **a.**

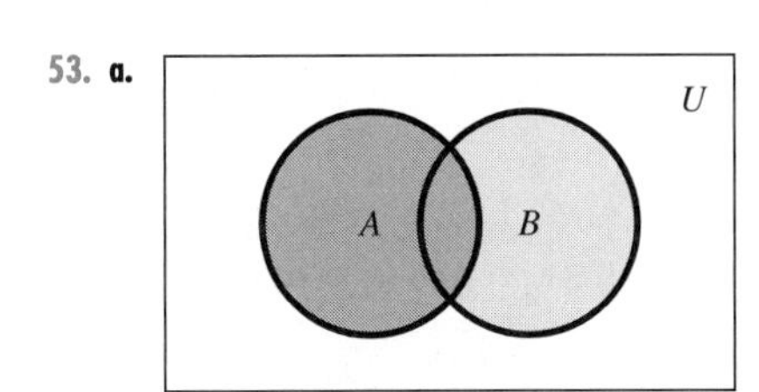

b.

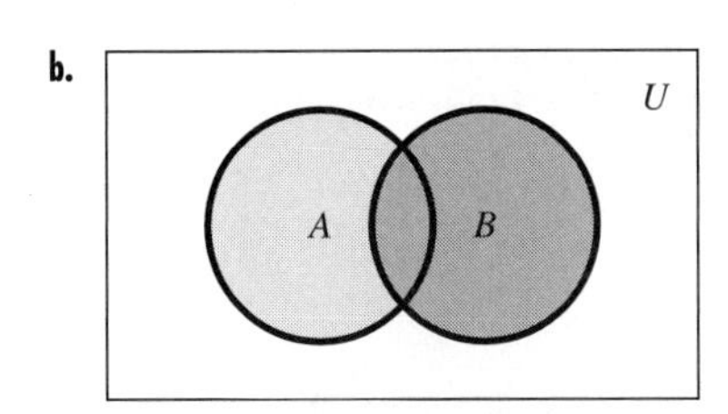

55.

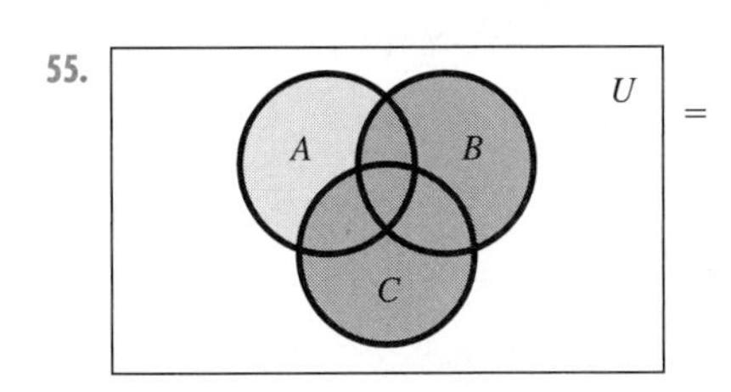

=

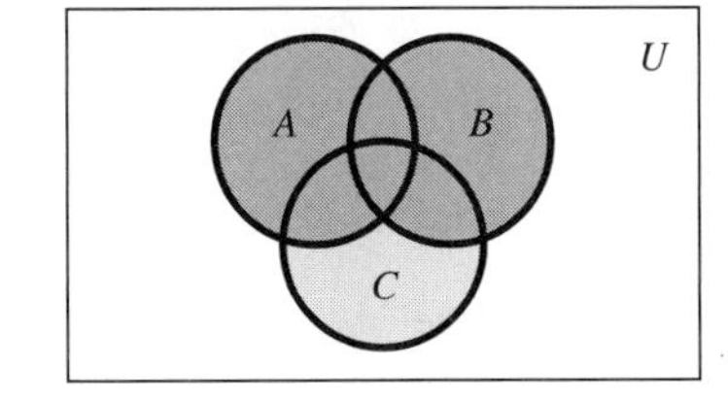

57.

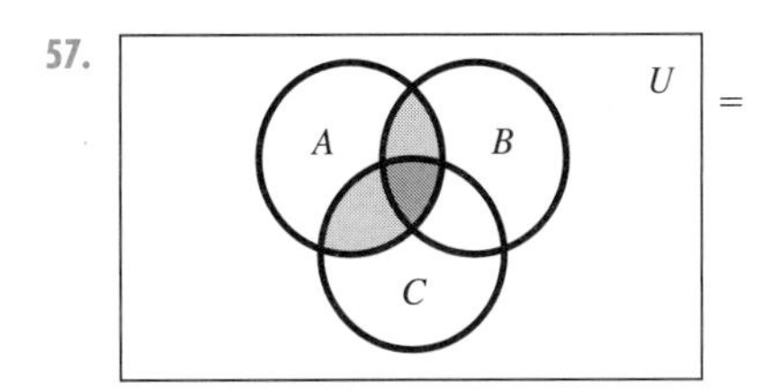

=

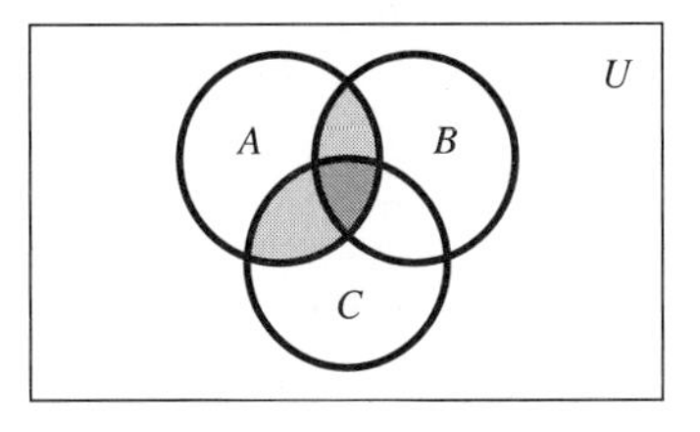

61. **a.** x, y, v, r, w, u **b.** v, r

63. **a.** s, t, y **b.** t, z, w, x, s 65. $A \subset C$

67. False 69. True 71. True

Exercises 6.2, page 352

3. **a.** 4 **b.** 5 **c.** 7 **d.** 2

7. 3

9. **a.** 40 **b.** 120 **c.** 60

11. 11 13. 2 15. 600; 100

17. 61

19. **a.** 106 **b.** 64 **c.** 38 **d.** 14

21. **a.** 182 **b.** 118 **c.** 56 **d.** 18

23. **a.** 13 **b.** 15 25. **a.** 38 **b.** 64

27. 5

29. **a.** 62 **b.** 33 **c.** 25 **d.** 38

31. **a.** 108 **b.** 15 **c.** 45 **d.** 12

33. True

Exercises 6.3, page 359

1. 12 3. 64 5. 24

7. 24 9. 60 11. 1 billion 13. 5^{50}

15. 30 17. 9990

19. **a.** 17,576,000 **b.** 17,576,000

21. 1024; 59,049 **23.** 2730

25. 217 **27.** False

Exercises 6.4, page 372

1. 360 **3.** 10 **5.** 120

7. 20 **9.** n **11.** 1

13. 35 **15.** 1 **17.** 84

19. $\frac{n(n-1)}{2}$ **21.** $\frac{n!}{2}$

23. Permutation **25.** Combination

27. Permutation **29.** Combination

31. $P(4, 4) = 24$ **33.** $P(4, 4) = 24$

35. $P(9, 9) = 362{,}880$ **37.** $C(12, 3) = 220$

39. 151,200 **41.** $C(12, 3) = 220$

43. $C(100, 3) = 161{,}700$ **45.** $P(6, 6) = 720$

47. $P(12, 6) = 665{,}280$

49. **a.** $P(10, 10) = 3{,}628{,}800$
b. $P(3, 3) \cdot P(4, 4) \cdot P(3, 3) \cdot P(3, 3) = 5184$

51. **a.** $P(20, 20) = 20!$
b. $P(5, 5) \cdot P[(4, 4)]^5 = 5!(4!)^5 = 955{,}514{,}880$

53. $P(2, 1) \cdot P(3, 1) = 6$

55. $C(3, 3) \cdot [C(8, 6) + C(8, 7) + C(8, 8)] = 37$

57. **a.** $C(12, 3) = 220$ **b.** $C(11, 2) = 55$
c. $C(5, 1) \cdot C(7, 2) + C(5, 2) \cdot C(7, 1) + C(5, 3) = 185$

59. $P(7, 3) + C(7, 2) \cdot P(3, 2) = 336$

61. $[C(5, 1) \cdot C(3, 1) \cdot C(6, 2)][C(4, 1) + C(3, 1)] = 1575$

63. $C(10, 8) + C(10, 9) + C(10, 10) = 56$

65. $10 \cdot C(4, 1) = 40$

67. $4 \cdot C(13, 5) - 40 = 5108$

69. $13C(4, 3) \cdot 12C(4, 2) = 3744$

71. $C(6, 2) = 15$

73. $C(12, 6) + C(12, 7) + C(12, 8) + C(12, 9) + C(12, 10) + C(12, 11) + C(12, 12) = 2510$

75. $4! = 24$ **79.** True **81.** True

Using Technology Exercises 6.4, page 377

1. $1.307674368 \times 10^{12}$ **3.** $2.56094948229 \times 10^{16}$

5. 674,274,182,400 **7.** 133,784,560 **9.** 4,656,960

11. 658,337,004,000

Chapter 6 Review Exercises, page 379

1. {3} **3.** {4, 6, 8, 10} **5.** Yes **7.** Yes

9.

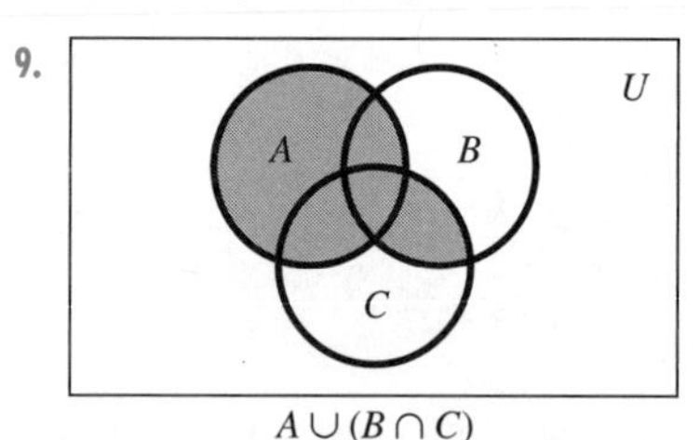

$A \cup (B \cap C)$

11.

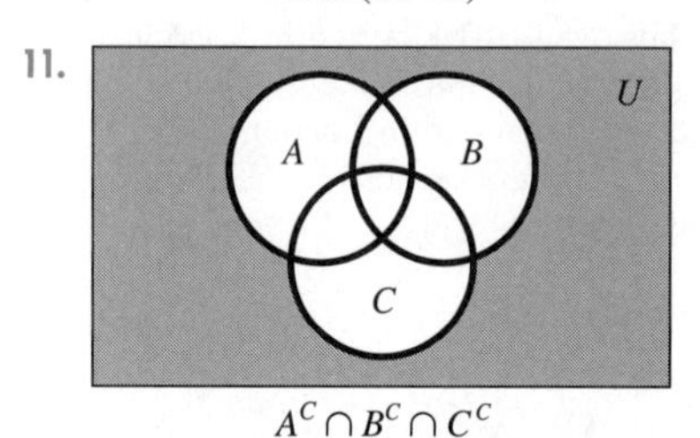

$A^C \cap B^C \cap C^C$

17. The set of all participants in a consumer-behavior survey who both avoided buying a product because it is not recyclable and boycotted a company's products because of its record on the environment

19. The set of all participants in a consumer-behavior survey who both did not use cloth diapers rather than disposable diapers and voluntarily recycled their garbage

21. 150 **23.** 270 **25.** 70 **27.** 190 **29.** 120

31. None **33.** 720 **35.** 720

37. **a.** 50,400 **b.** 5040 **39.** **a.** 5040 **b.** 3600

41. **a.** $C(15, 4) = 1365$ **b.** $C(15, 4) - C(10, 4) = 1155$

CHAPTER 7

Exercises 7.1, page 390

1. $\{a, b, d, f\}$; $\{a\}$ **3.** $\{b, c, e\}$; $\{a\}$ **5.** No **7.** S

9. $\varnothing$ **11.** Yes **13.** $E \cup F$ **15.** G^C

17. $(E \cup F \cup G)^C$

19. $\varnothing$, $\{a\}$, $\{b\}$, $\{c\}$, $\{a, b\}$, $\{a, c\}$, $\{b, c\}$, S

21. **a.** $S = \{B, R\}$ **b.** $\varnothing$, $\{B\}$, $\{R\}$, S

23. a. $S = \{(H, 1), (H, 2), (H, 3), (H, 4), (H, 5), (H, 6), (T, 1), (T, 2), (T, 3), (T, 4), (T, 5), (T, 6)\}$
b. $\{(H, 2), (H, 4), (H, 6)\}$

25. $S = \{(d, d, d), (d, d, n), (d, n, d), (n, d, d), (d, n, n), (n, d, n), (n, n, d), (n, n, n)\}$

27. a. {ABC, ABD, ABE, ACD, ACE, ADE, BCD, BCE, BDE, CDE}
b. 6 c. 3 d. 6

29. a. E^C b. $E^C \cap F^C$ c. $E \cup F$
d. $(E \cap F^C) \cup (E^C \cap F)$

31. a. $\{x \mid x > 0\}$ b. $\{x \mid 0 < x \leq 2\}$
c. $\{x \mid x > 2\}$

33. a. $S = \{0, 1, 2, 3, \ldots, 10\}$ b. $E = \{0, 1, 2, 3\}$
c. $F = \{5, 6, 7, 8, 9, 10\}$

35. a. $S = \{0, 1, 2, \ldots, 20\}$
b. $E = \{0, 1, 2, \ldots, 9\}$ c. $F = \{20\}$

39. 2^n 41. True

Exercises 7.2, page 400

1. {(H, H)}, {(H, T)}, {(T, H)}, {(T, T)}

3. $\{(D, m)\}, \{(D, f)\}, \{(R, m)\}, \{(R, f)\}, \{(I, m)\}, \{(I, f)\}$

5. $\{(1, i)\}, \{(1, d)\}, \{(1, s)\}, \{(2, i\}, \{(2, d)\}, \{(2, s)\}, \ldots, \{(5, i)\}, \{(5, d)\}, \{(5, s)\}$

7. $\{(A, Rh^+)\}, \{(A, Rh^-)\}, \{(B, Rh^+)\}, \{(B, Rh^-)\}, \{(AB, Rh^+)\}, \{(AB, Rh^-)\}, \{(O, Rh^+)\}, \{(O, Rh^-)\}$

9.

Grade	A	B	C	D	F
Probability	.10	.25	.45	.15	.05

11. a. $S = \{(0 < x \leq 200), (200 < x \leq 400), (400 < x \leq 600), (600 < x \leq 800), (800 < x \leq 1000), (x > 1000)\}$

b.

Number of Cars (x)	Probability
$0 < x \leq 200$	.075
$200 < x \leq 400$	.1
$400 < x \leq 600$	.175
$600 < x \leq 800$	.35
$800 < x \leq 1000$	.225
$x > 1000$	.075

13.

Event	A	B	C	D	E
Probability of an Event	.026	.199	.570	.193	.012

15.

Number of Figures Produced (in dozens)	30	31	32
Probability	.125	0	.1875

Number of Figures Produced (in dozens)	33	34	35	36
Probability	.25	.1875	.125	.125

17. .469 19. a. .856 b. .144 21. .46

23. a. 1/4 b. 1/2 c. 1/13 25. 3/8 27. .95

29. a. .633 b. .276

31. There are two ways of obtaining a sum of 7.

33. No 35. No 37. a. 1/6 b. 5/6 c. 1

39. a. 3/8 b. 1/2 c. 1/4 41. .782 43. True

Exercises 7.3, page 411

1. 1/2 3. 1/36 5. 1/9 7. 1/52

9. 3/13 11. 12/13 13. .002; .998

15. $P(a) + P(b) + P(c) \neq 1$

17. Since the five events are not mutually exclusive, Property (3) cannot be used; that is, he could win more than one purse.

19. The two events are not mutually exclusive; hence, the probability of the given event is $\frac{1}{6} + \frac{1}{6} - \frac{1}{36} = \frac{11}{36}$.

21. $E^C \cap F^C = \{e\} \neq \varnothing$

23. $P(G \cup C)^C \neq 1 - P(G) - P(C)$; he has not considered the case in which a customer buys both glasses and contact lenses.

25. a. 0 b. .7 c. .8 d. .3

27. a. 1/2, 3/8 b. 1/2, 5/8 c. 1/8 d. 3/4 29. .33

31. a. .16 b. .38 c. .22

33. a. .90 b. .40 c. .40

35. a. .6 b. .332 c. .232 d. .6

39. True 41. False

Exercises 7.4, page 420

1. 1/32 **3.** 31/32

5. $P(A) = 13 \cdot C(4, 2)/C(52, 2) = .0588$

7. $C(26, 2)/C(52, 2) = .245$

9. $[C(3, 2) \cdot C(5, 2)]/C(8, 4) = 3/7$

11. $[C(5, 3) \cdot C(3, 1)]/C(8, 4) = 3/7$

13. $[C(3, 2) \cdot C(1, 1)]/8 = 3/8$

15. 1/8 **17.** $C(10, 6)/2^{10} = .205$

19. a. $C(4, 2)/C(24, 2) \approx .022$
b. $1 - C(20, 2)/C(24, 2) \approx .312$

21. a. $C(6, 2)/C(80, 2) \approx .005$
b. $1 - C(74, 2)/C(80, 2) \approx .145$

23. a. .12; $C(98, 10)/C(100, 12) \approx .013$
b. .15; .015

25. $[C(12, 8) \cdot C(8, 2) + C(12, 9) \cdot C(8, 1) + C(12, 10)]/C(20, 10) \approx .085$

27. a. 3/5 **b.** $C(3, 1)/C(5, 3) = .3$
c. $1 - C(3, 3)/C(5, 3) = .9$

29. 1/729 **31.** .0001 **33.** .10

35. $40/C(52, 5) \approx .0000154$

37. $[4C(13, 5) - 40]/C(52, 5) \approx .00197$

39. $[13C(4, 3) \cdot 12C(4, 2)]/C(52, 5) \approx .00144$

41. a. .618 **b.** .059 **43.** .03

Exercises 7.5, page 436

1. a. .4 **b.** .33 **3.** .3 **5.** Independent

7. Independent **9. a.** .24 **b.** .76

11. a. .5 **b.** .4 **c.** .2 **d.** .35
e. No **f.** No

13. a. .4 **b.** .3 **c.** .12 **d.** .30
e. Yes **f.** Yes

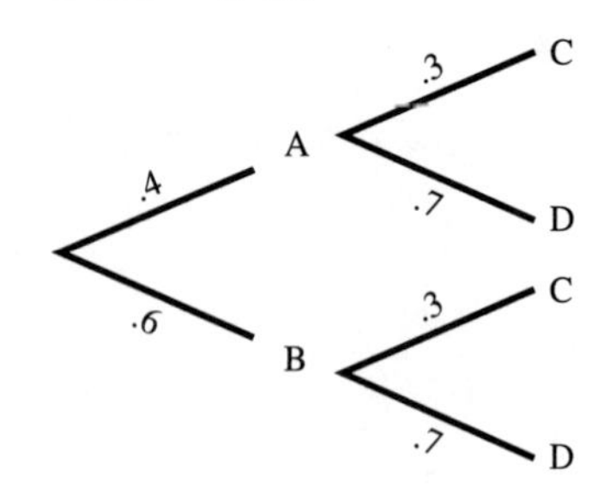

15. $\frac{4}{11}$ **17.** Independent **19.** .1875

21. a. $\frac{4}{9}$ **b.** $\frac{4}{9}$ **23. a.** $\frac{1}{21}$ **b.** $\frac{1}{3}$ **25.** .25

27. $\frac{1}{7}$

29. a.

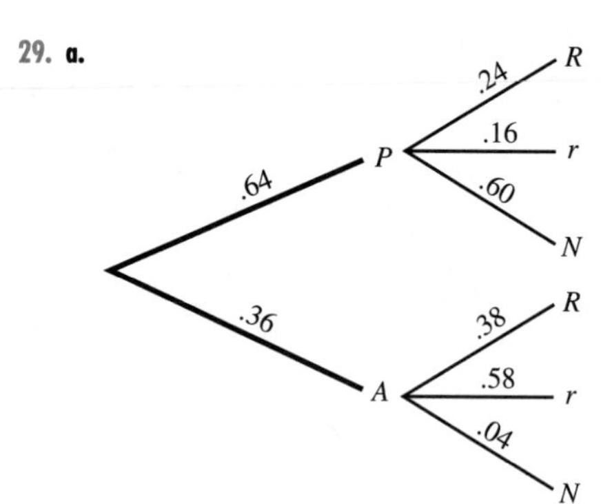

P = Professional
A = Amateur
R = Recovered within 48 hr
r = Recovered after 48 hr
N = Never recovered

b. .24 **c.** .40

31. a. .16 **b.** .424 **c.** .1696

33. a. .092 **b.** ≈.008

35. a. .280; .390; .180; .643; .292 **b.** Not independent

37. Not independent **39.** .0000068 **41. a.** $\frac{7}{10}$ **b.** $\frac{1}{5}$

43. 3 **47.** True **49.** True

Exercises 7.6, page 446

1.

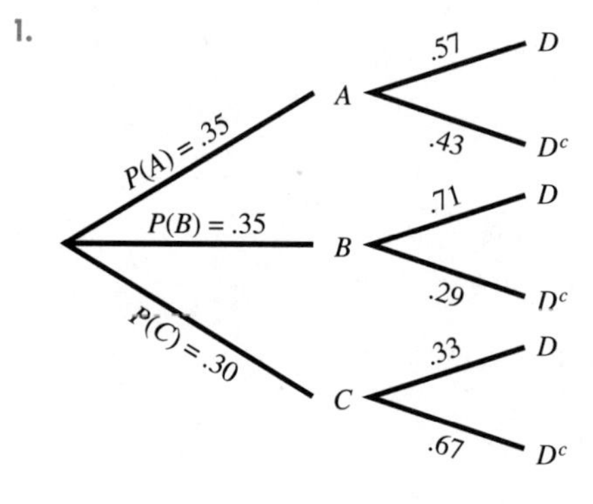

3. a. .45 **b.** .22 **5. a.** .48 **b.** .33

7. a. .08 **b.** .15 **c.** .35

9. **a.** 1/12 **b.** 1/4 **c.** 1/18 **d.** 3/14

11. 4/17 13. .0784

15.

2/5 W, 4/9 W, 5/9 B; 3/5 B, 1/3 W, 2/3 B

17. .53 19. .422

21. **a.** 3/4 **b.** 2/9 23. .856

25. **a.** .03 **b.** .29 27. .35

29. **a.** .30 **b.** .10

31. .62 33. .65 35. .93 37. **a.** .763 **b.** .276 **c.** .724

Chapter 7 Review Exercises, page 452

1. **a.** 0 **b.** .6 **c.** .6 **d.** .4 **e.** 1

3. **a.** .49 **b.** .39 **c.** .48

5. **a.** .019 **b.** .981

7. .18 9. .06 11. .49

13. **a.** .284 **b.** .984 15. 2/15

17. .01 19. .510 21. .457

CHAPTER 8

Exercises 8.1, page 463

1. **a.** See part (b)

b.

Outcome	GGG	GGR	GRG	RGG
Value	3	2	2	2

Outcome	GRR	RGR	RRG	RRR
Value	1	1	1	0

c. {GGG}

3. Any positive integer 5. 1/6

7. Any positive integer; infinite discrete

9. $0 \le x < \infty$, continuous

11. Any positive integer; infinite discrete

13. **a.** .20 **b.** .60 **c.** .30 **d.** 1

15.

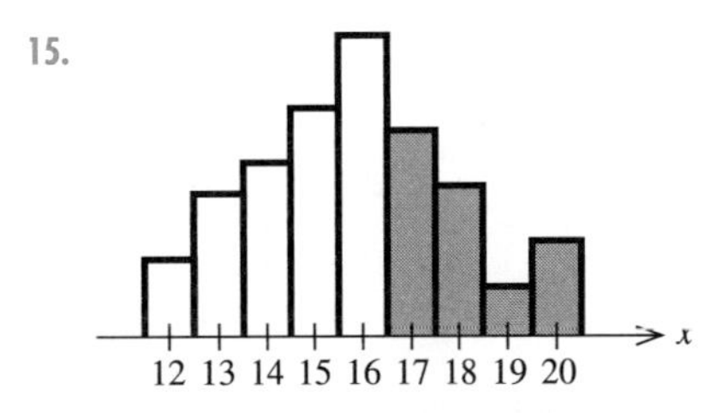

17. **a.**

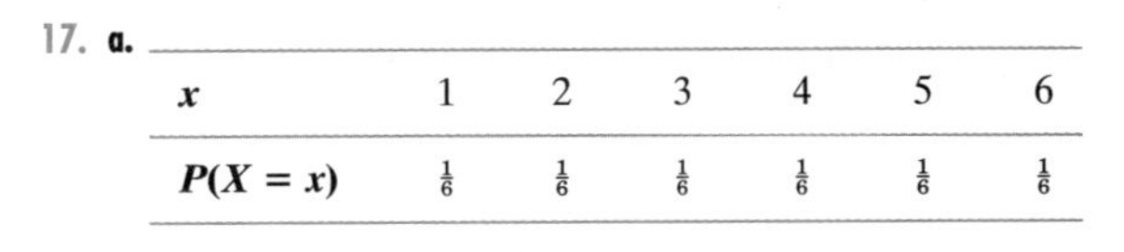

x	1	2	3	4	5	6
$P(X = x)$	$\frac{1}{6}$	$\frac{1}{6}$	$\frac{1}{6}$	$\frac{1}{6}$	$\frac{1}{6}$	$\frac{1}{6}$

y	1	2	3	4	5	6
$P(Y = y)$	$\frac{1}{6}$	$\frac{1}{6}$	$\frac{1}{6}$	$\frac{1}{6}$	$\frac{1}{6}$	$\frac{1}{6}$

b.

$x + y$	2	3	4	5	6	7
$P(X + Y = x + y)$	$\frac{1}{36}$	$\frac{2}{36}$	$\frac{3}{36}$	$\frac{4}{36}$	$\frac{5}{36}$	$\frac{6}{36}$

$x + y$	8	9	10	11	12
$P(X + Y = x + y)$	$\frac{5}{36}$	$\frac{4}{36}$	$\frac{3}{36}$	$\frac{2}{36}$	$\frac{1}{36}$

19. **a.**

x	0	1	2	3	4
$P(X = x)$	.017	.067	.033	.117	.233

x	5	6	7	8	9	10
$P(X = x)$	.133	.167	.1	.05	.067	.017

b.

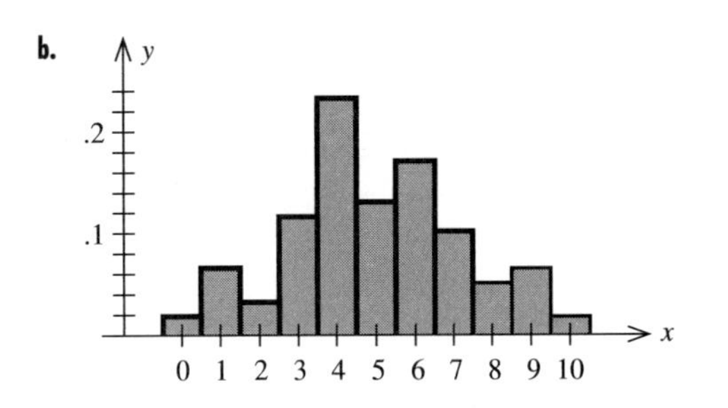

21.

x	1	2	3	4	5
$P(X = x)$	.007	.029	.021	.079	.164

x	6	7	8	9	10
$P(X = x)$	.15	.20	.207	.114	.029

23. False

Using Technology Exercises 8.1, page 465

1.

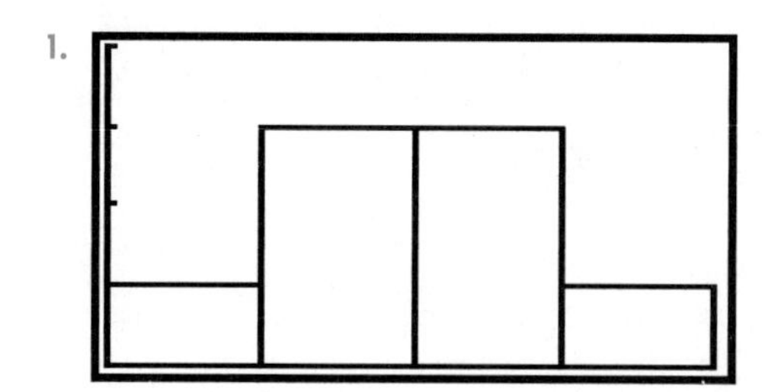

3.

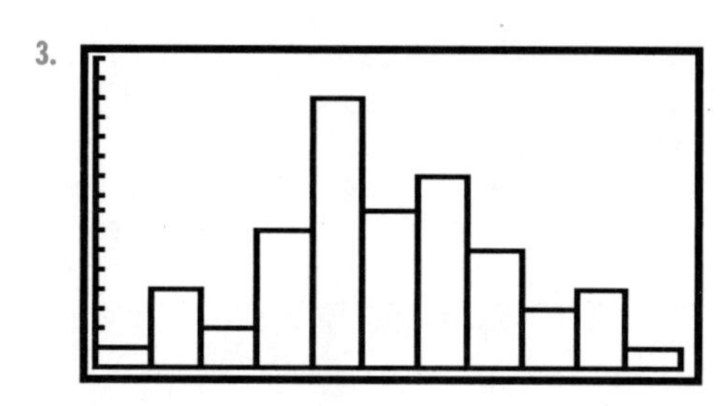

Exercises 8.2, page 478

1. **a.** 2.6
b.

x	0	1	2	3	4
$P(X = x)$	0	.1	.4	.3	.2

; 2.6

3. 0.86 5. $78.50 7. 0.91

9. 0.12 11. 1.73 13. $800

15. $162.50 17. First project

19. **a.** Dahl: 8.52; Farthington: 7.25
b. Farthington Auto Sales

21. **a.** 3 to 7 **b.** 7 to 3

23. −10.5¢ 25. 4 to 1; 1 to 4 27. .5625

29. .6154 31. .5714 33. **b.** −$15.79

35. Mean: $10.94; mode: $11.00; median: $10.95

37. True

Exercises 8.3, page 490

1. $\mu = 2$, $\text{Var}(X) = 1$, $\sigma = 1$

3. $\mu = 0$, $\text{Var}(X) = 1$, $\sigma = 1$

5. $\mu = 518$, $\text{Var}(X) = 1891$, $\sigma = 43.5$

7. Figure (a) 9. 1.56

11. $\mu = 4.5$, $\text{Var}(X) = 5.25$

13. **a.** Let X = The annual birth rate during the years 1981–1990
b.

x	15.5	15.6	15.7	15.9	16.2	16.7
$P(X = x)$	.2	.1	.3	.2	.1	.1

c. $\mu = 15.84$, $\text{Var}(X) = 0.1224$, $\sigma = 0.350$

15. **a.** Mutual fund A: μ = \$620, $\text{Var}(X)$ = \$267,600; Mutual fund B: μ = \$520, $\text{Var}(X)$ = \$137,600
b. Mutual fund A **c.** Mutual fund B

17. 1

19. μ = \$239,600; $\text{Var}(X)$ = \$1,443,840,000; σ = \$37,998

21. **a.** At least .75 **b.** At least .96

23. At least 75% 25. At least 7/16 27. $c \geq 0.1$ 29. True

Using Technology Exercises 8.3, page 495

1. **a.**

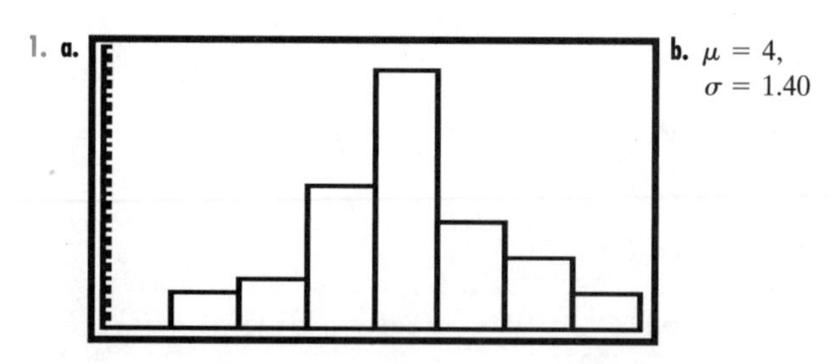

b. $\mu = 4$, $\sigma = 1.40$

3. **a.**

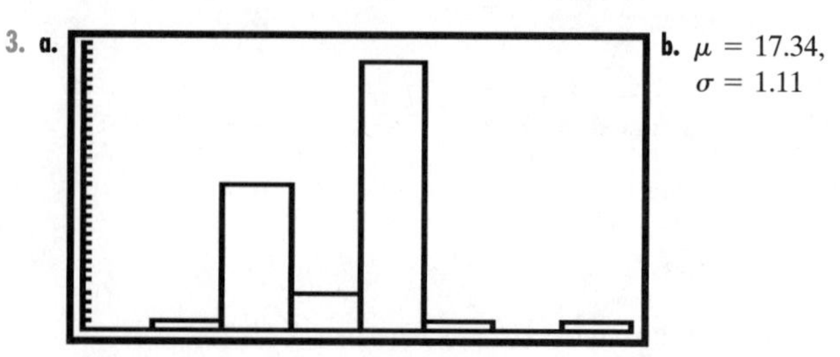

b. $\mu = 17.34$, $\sigma = 1.11$

5. **a.** Let X denote the random variable that gives the weight of a carton of sugar.

b.

x	4.96	4.97	4.98	4.99	5.00	5.01
$P(X = x)$	$\frac{3}{30}$	$\frac{4}{30}$	$\frac{4}{30}$	$\frac{1}{30}$	$\frac{1}{30}$	$\frac{5}{30}$

x	5.02	5.03	5.04	5.05	5.06
$P(X = x)$	$\frac{3}{30}$	$\frac{3}{30}$	$\frac{4}{30}$	$\frac{1}{30}$	$\frac{1}{30}$

c. $\mu \approx 5.00$; $\sigma \approx 0.03$

7. a.

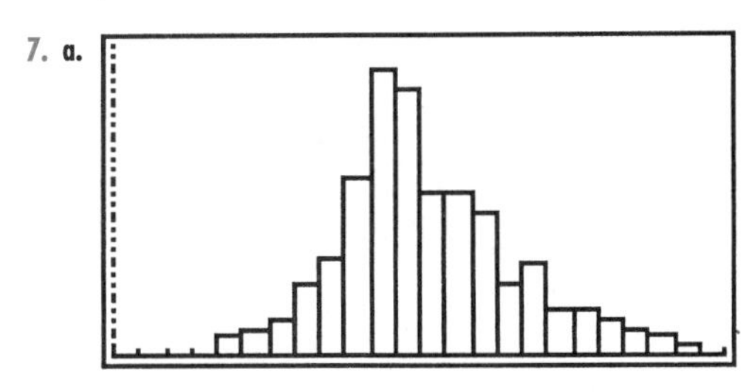

b. 65.875; 1.73

Exercises 8.4, page 504

1. Yes

3. No. There are more than two outcomes to the experiment.

5. No. The probability of an accident on a clear day is not the same as the probability of an accident on a rainy day.

7. .296 **9.** .0512 **11.** .132

13. 21/32 **15.** .0041 **17.** $\approx$ .116

19. a. $P(X = 0) \approx .08$, $P(X = 1) \approx .26$, $P(X = 2) \approx .35$, $P(X = 3) \approx .23$, $P(X = 4) \approx .08$, $P(X = 5) \approx .01$

b.

x	0	1	2	3	4	5
$P(X = x)$	.08	.26	.35	.23	.08	.01

.4
.3
.2
.1
0 1 2 3 4 5

c. $\mu = 2$, $\sigma \approx 1.1$

21. No. The probability that at most 1 is defective is $P(X = 0) + P(X = 1) = .74$.

23. $\approx$. 0002

25. a. $\approx$ .633 **b.** $\approx$ .367

27. a. $\approx$ .0329 **b.** $\approx$ .3292

29. a. $\approx$ .273 **b.** $\approx$ .650

31. .3125 **33.** .9133

35. a. $\approx$ .003988 **b.** .000006 **c.** $\approx 3.997 \times 10^{-9}$

37. $\approx$ .392 **39.** 7 times

41. a. 1200 **b.** $\approx$ 21.91 **43.** True

Exercises 8.5, page 515

1. .9265 **3.** .0401 **5.** .8657

7. a.

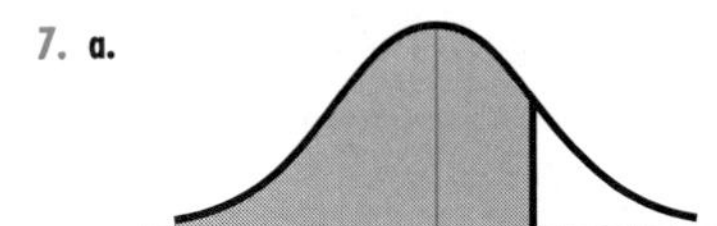

b. .9147

9. a.

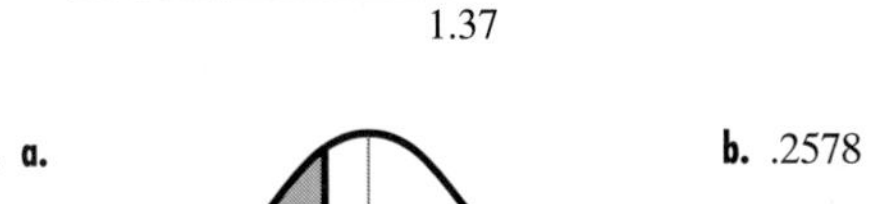

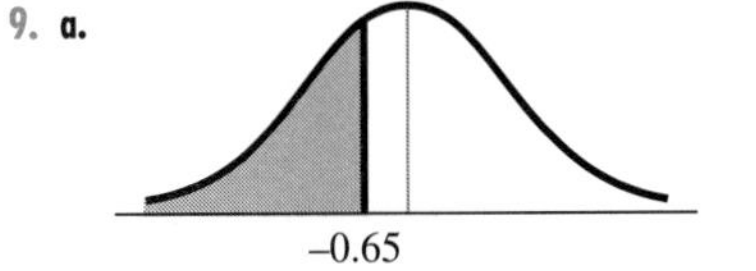

b. .2578

11. a.

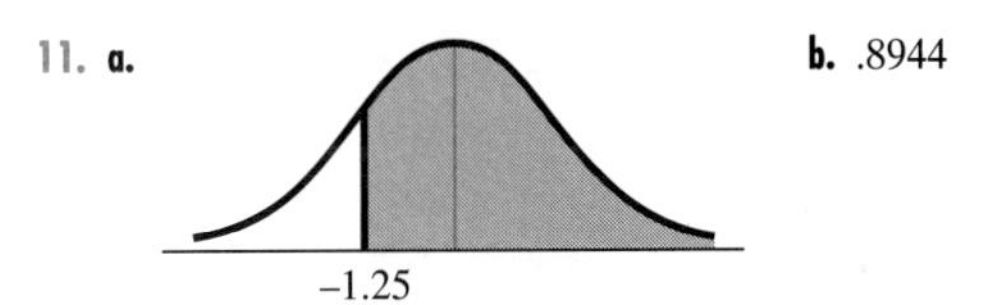

b. .8944

13. a.

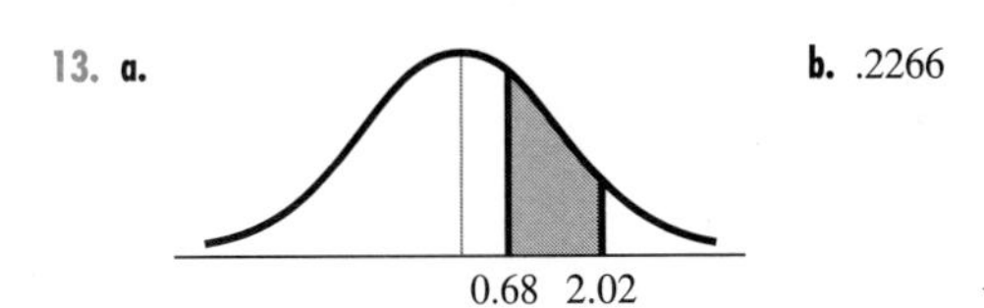

b. .2266

15. a. 1.23 **b.** −0.81 **17. a.** 1.9 **b.** −1.9

19. a. .9772 **b.** .9192 **c.** .7333

Exercises 8.6, page 525

1. a. .2206 **b.** .2206 **c.** .3034

3. a. .0228 **b.** .0228 **c.** .4772 **d.** .7258

5. a. .0038 **b.** .0918 **c.** .4082 **d.** .2514

7. .6247 **9.** 0.62%

11. A:80; B:73; C:62; D:54

13. a. .4207 **b.** .4254 **c.** .0122

15. a. .2877 **b.** .0008 **c.** .7287

17. .9265

19. a. .0037 **b.** The drug is very effective.

21. 2142

Chapter 8 Review Exercises, page 528

1. a. {WWW, BWW, WBW, WWB, BBW, BWB, WBB, BBB}

b.

Outcome	WWW	BWW	WBW	WWB
Value of X	0	1	1	1

Outcome	BBW	BWB	WBB	BBB
Value of X	2	2	2	3

c.

x	0	1	2	3
$P(X = x)$	$\frac{1}{35}$	$\frac{12}{35}$	$\frac{18}{35}$	$\frac{4}{35}$

d.

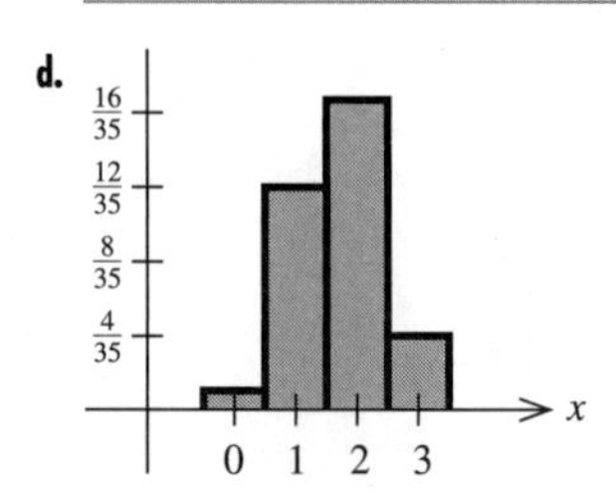

3. a. .8 **b.** $\mu = 2.7$; $\sigma = 1.42$

5. .6915

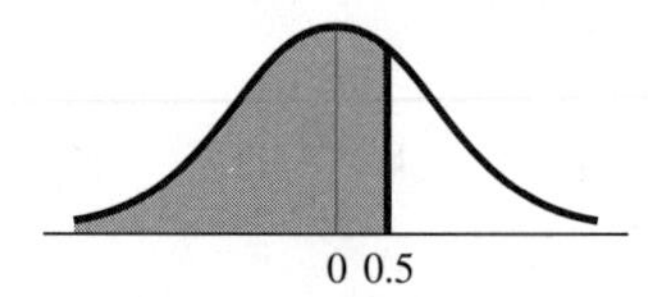

7. .4649

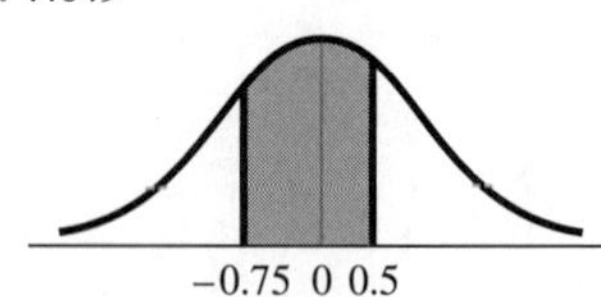

9. 2.42 **11.** −2.03 **13.** .6915 **15.** .2417

17. .2646; .9163 **19.** At least .75

21. $\mu = 120$, $\sigma = 10.1$ **23.** 0.9738

Chapter 9

Exercises 9.1, page 539

1. Yes **3.** Yes **5.** No **7.** Yes **9.** No

11. a. Given that the outcome state 1 has occurred, the conditional probability that the outcome state 1 will occur is .3.

b. .7 **c.** $\begin{bmatrix} .48 \\ .52 \end{bmatrix}$

13. $TX_0 = \begin{bmatrix} .4 \\ .6 \end{bmatrix}$;

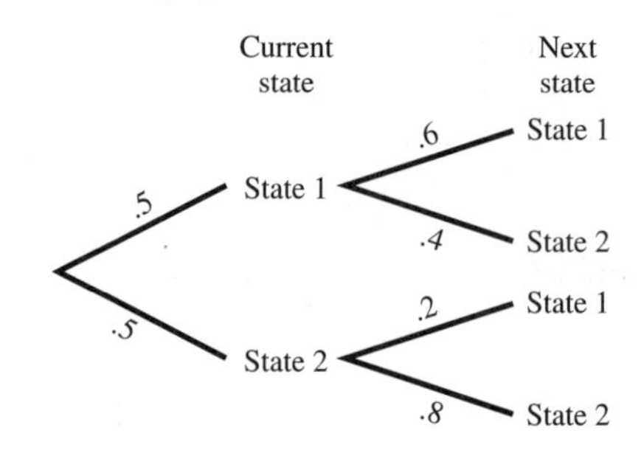

15. $X_2 = \begin{bmatrix} .576 \\ .424 \end{bmatrix}$ **17.** $X_2 = \begin{bmatrix} \frac{5}{16} \\ \frac{27}{64} \\ \frac{17}{64} \end{bmatrix}$

19. a.

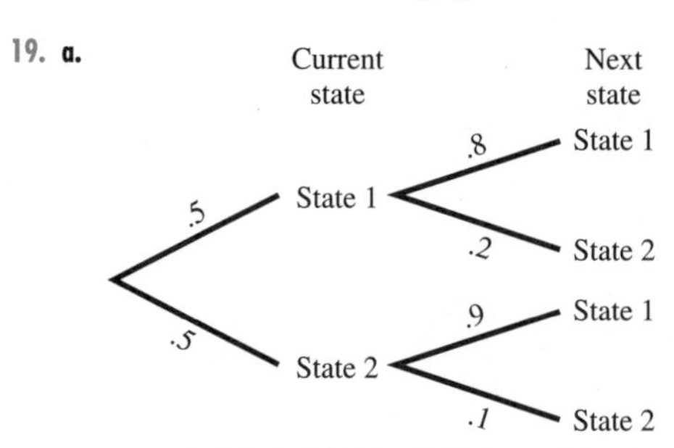

b. $T = \begin{array}{c} \\ L \\ R \end{array} \begin{array}{c} \begin{array}{cc} L & R \end{array} \\ \begin{bmatrix} .8 & .9 \\ .2 & .1 \end{bmatrix} \end{array}$ **c.** $X_0 = \begin{array}{c} L \\ R \end{array} \begin{bmatrix} .5 \\ .5 \end{bmatrix}$ **d.** .85

21. a. Vote is evenly split. **b.** Democrat

23. 50% in zone I, 30% in zone II, 20% in zone III

25. University: 37%, Campus: 35%, Book Mart: 28%; University: 34.5%, Campus: 31.35%, Book Mart: 34.15%

27. Business: 36%, Humanities: 23.8%, Education: 15%, Natural sciences and others: 25.1%

29. False

Using Technology Exercises 9.1, page 543

1. $X_5 = \begin{bmatrix} .204489 \\ .131869 \\ .261028 \\ .186814 \\ .2158 \end{bmatrix}$

3. Manufacturer A will have 23.95% of the market, Manufacturer B will have 49.71% of the market share, and Manufacturer C will have 26.34% of the market share.

Exercises 9.2, page 552

1. Regular 3. Not regular 5. Regular

7. Not regular 9. $\begin{bmatrix} \frac{3}{11} \\ \frac{8}{11} \end{bmatrix}$ 11. $\begin{bmatrix} \frac{2}{7} \\ \frac{5}{7} \end{bmatrix}$

13. $\begin{bmatrix} \frac{3}{13} \\ \frac{8}{13} \\ \frac{2}{13} \end{bmatrix}$ 15. $\begin{bmatrix} \frac{3}{19} \\ \frac{8}{19} \\ \frac{8}{19} \end{bmatrix}$ 17. 81.8%

19. **a.** 40.8% one wage earner and 59.2% two wage earners
b. 30% one wage earner and 70% two wage earners

21. **a.** 72.5% in single-family homes and 27.5% in condominiums
b. 70% in single-family homes and 30% in condominiums

23. **a.** 31.7% ABC, 37.35% CBS, 30.95% NBC
b. $33\frac{1}{3}$% ABC, $33\frac{1}{3}$% CBS, $33\frac{1}{3}$% NBC

25. 25% red, 50% pink, 25% white 27. False

Using Technology Exercises 9.2, page 557

1. $X = \begin{bmatrix} .2045 \\ .1319 \\ .2610 \\ .1868 \\ .2158 \end{bmatrix}$

Exercises 9.3, page 565

1. Yes 3. Yes 5. Yes 7. Yes

9. $\left[\begin{array}{c|c} 1 & .4 \\ \hline 0 & .6 \end{array}\right]$, $R = [.6]$, $S = [.4]$

11. $\left[\begin{array}{c|cc} 1 & .4 & .5 \\ \hline 0 & .4 & .5 \\ 0 & .2 & 0 \end{array}\right]$, $R = \begin{bmatrix} .4 & .5 \\ .2 & 0 \end{bmatrix}$, and $S = [.4 \quad .5]$ or

$\left[\begin{array}{c|cc} 1 & .5 & .4 \\ \hline 0 & 0 & .2 \\ 0 & .5 & .4 \end{array}\right]$, $R = \begin{bmatrix} 0 & .2 \\ .5 & .4 \end{bmatrix}$, and $S = [.5 \quad .4]$

13. $\left[\begin{array}{cc|cc} 1 & 0 & .2 & .4 \\ 0 & 1 & .3 & 0 \\ \hline 0 & 0 & .3 & .2 \\ 0 & 0 & .2 & .4 \end{array}\right]$,

$R = \begin{bmatrix} .3 & .2 \\ .2 & .4 \end{bmatrix}$, and $S = \begin{bmatrix} .2 & .4 \\ .3 & 0 \end{bmatrix}$, or

$\left[\begin{array}{cc|cc} 1 & 0 & .4 & .2 \\ 0 & 1 & 0 & .3 \\ \hline 0 & 0 & .4 & .2 \\ 0 & 0 & .2 & .3 \end{array}\right]$,

$R = \begin{bmatrix} .4 & .2 \\ .2 & .3 \end{bmatrix}$, $S = \begin{bmatrix} .4 & .2 \\ 0 & .3 \end{bmatrix}$, and so forth

15. $\left[\begin{array}{c|c} 1 & 1 \\ \hline 0 & 0 \end{array}\right]$ 17. $\left[\begin{array}{c|cc} 1 & 1 & 1 \\ \hline 0 & 0 & 0 \\ 0 & 0 & 0 \end{array}\right]$

19. $\left[\begin{array}{cc|cc} 1 & 0 & 1 & 1 \\ 0 & 1 & 0 & 0 \\ \hline 0 & 0 & 0 & 0 \\ 0 & 0 & 0 & 0 \end{array}\right]$ 21. $\left[\begin{array}{cc|cc} 1 & 0 & \frac{1}{2} & \frac{1}{2} \\ 0 & 1 & \frac{1}{2} & \frac{1}{2} \\ \hline 0 & 0 & 0 & 0 \\ 0 & 0 & 0 & 0 \end{array}\right]$

23. $\left[\begin{array}{ccc|cc} 1 & 0 & 0 & \frac{7}{22} & \frac{3}{11} \\ 0 & 1 & 0 & \frac{5}{22} & \frac{9}{22} \\ 0 & 0 & 1 & \frac{5}{11} & \frac{7}{22} \\ \hline 0 & 0 & 0 & 0 & 0 \\ 0 & 0 & 0 & 0 & 0 \end{array}\right]$

25. **a.**
$$\begin{array}{c} \\ \text{UL} \\ \text{L} \end{array} \begin{array}{c} \begin{array}{cc} \text{UL} & \text{L} \end{array} \\ \left[\begin{array}{c|c} 1 & .2 \\ \hline 0 & .8 \end{array}\right] \end{array}, \quad R = [.8], \quad S = [.2]$$

b. $\left[\begin{array}{c|c} 1 & 1 \\ \hline 0 & 0 \end{array}\right]$; eventually, only unleaded fuel will be used.

27. .25; .50; .75

29. **a.**
$$\begin{array}{c} \\ D \\ G \\ 1 \\ 2 \end{array} \begin{array}{c} \begin{array}{cccc} D & G & 1 & 2 \end{array} \\ \left[\begin{array}{cc|cc} 1 & 0 & .25 & .1 \\ 0 & 1 & 0 & .9 \\ \hline 0 & 0 & 0 & 0 \\ 0 & 0 & .75 & 0 \end{array}\right] \end{array}$$

b. $\left[\begin{array}{cc|cc} 1 & 0 & .325 & .1 \\ 0 & 1 & .675 & .9 \\ \hline 0 & 0 & 0 & 0 \\ 0 & 0 & 0 & 0 \end{array}\right]$

c. .675

31. False

Exercises 9.4, page 577

1. R: row 1; C: column 2 **3.** R: row 1; C: column 1

5. R: row 1 or row 3; C: column 3

7. R: row 1 or row 3; C: column 2

9. Strictly determined;
a. 2 **b.** R: row 1; C: column 1
c. 2 **d.** Favors row player.

11. Strictly determined;
a. 1 **b.** R: row 1; C: column 1
c. 1 **d.** Favors row player.

13. Strictly determined;
a. 1 **b.** R: row 1; C: column 1
c. 1 **d.** Favors row player.

15. Not strictly determined

17. Not strictly determined

19. a. $\begin{bmatrix} 2 & -3 & 4 \\ -3 & 4 & -5 \\ 4 & -5 & 6 \end{bmatrix}$

b. Robin; row 1; Cathy: column 1 or column 2
c. Not strictly determined
d. Not strictly determined

21. a.

		Economy	
		Good	Recess.
Mgmt.	Expand	200,000	120,000
	Not exp.	50,000	150,000

b. Yes

23. a.

		Charley		
		Raises	Holds	Lowers
Roland	Raises	3	−1	−3
	Holds	2	0	−2
	Lowers	5	2	1

25. True

Exercises 9.5, page 589

1. 3/10 **3.** 5/12 **5.** 0.16

7. a. 1 **b.** −2 **c.** 0
d. −3/10; (a) is most advantageous

9. a. 1 **b.** −7/20 **c.** The first strategy

11. $P = [\frac{1}{4} \;\; \frac{3}{4}]$, $Q = \begin{bmatrix} \frac{1}{2} \\ \frac{1}{2} \end{bmatrix}$;
$E = 2.5$; favors row player

13. $P = [\frac{4}{7} \;\; \frac{3}{7}]$, $Q = \begin{bmatrix} \frac{5}{7} \\ \frac{2}{7} \end{bmatrix}$;
$E = -\frac{1}{7}$; favors column player

15. $P = [\frac{1}{2} \;\; \frac{1}{2}]$, $Q = \begin{bmatrix} \frac{1}{4} \\ \frac{3}{4} \end{bmatrix}$;
$E = -5$; favors column player

17. a. $P = [\frac{1}{3} \;\; \frac{2}{3}]$, $Q = \begin{bmatrix} \frac{1}{3} \\ \frac{2}{3} \end{bmatrix}$
b. $E = 0$; no

19. a. \$5714 in hotel stocks; \$34,286 in brewery stock
b. \$4857

21. a.

		C	
		N	F
R	N	.48	.65
	F	.50	.45

C = Carlton; R = Russell
N = local newspaper; F = flyer

b. Russell's strategy: $P = [.23 \;\; .77]$

Carlton's strategy: $Q = \begin{bmatrix} .91 \\ .09 \end{bmatrix}$

Chapter 9 Review Exercises, page 595

1. Not regular **3.** Regular

5. $\begin{bmatrix} .3675 \\ .36 \\ .2725 \end{bmatrix}$ **7.** Yes

9. No **11.** $\begin{bmatrix} \frac{3}{7} & \frac{3}{7} \\ \frac{4}{7} & \frac{4}{7} \end{bmatrix}$

13. $\begin{bmatrix} .457 & .457 & .457 \\ .200 & .200 & .200 \\ .343 & .343 & .343 \end{bmatrix}$

15. a.

$$\begin{array}{c} \\ A \\ U \\ N \end{array}\begin{array}{c} \begin{array}{ccc} A & U & N \end{array} \\ \begin{bmatrix} .85 & 0 & .10 \\ .10 & .95 & .05 \\ .05 & .05 & .85 \end{bmatrix} \end{array}$$

b. $\begin{array}{c} A \\ U \\ N \end{array}\begin{bmatrix} .50 \\ .15 \\ .35 \end{bmatrix}$ **c.** $\begin{array}{c} A \\ U \\ N \end{array}\begin{bmatrix} .424 \\ .262 \\ .314 \end{bmatrix}$

17. Favors row player; strictly determined; R: row 3; C: column 1; value is 4.

19. Favors row player; strictly determined; R: row 1; C: column 1; value is 1.

21. $-1/4$ **23.** 2

25. $P = [\frac{1}{2} \ \frac{1}{2}]$, $Q = \begin{bmatrix} \frac{5}{6} \\ \frac{1}{6} \end{bmatrix}$;

$E = \frac{1}{2}$; favors row player

27. $P = [\frac{1}{10} \ \frac{9}{10}]$, $Q = \begin{bmatrix} \frac{4}{5} \\ \frac{1}{5} \end{bmatrix}$;

$E = 1.2$; favors row player

29. a. $\begin{bmatrix} .5 & .7 \\ .4 & .5 \end{bmatrix}$ **b.** \$7

APPENDIX A

Exercises A.1, page 602

1. Yes **3.** Yes **5.** No

7. Yes **9.** Yes **11.** No

13. No **15.** Negation **17.** Conjunction

19. Conjunction

21. New orders for manufactured goods did not fall last month.

23. Drinking during pregnancy does not affect both the size and weight of babies.

25. The commuter airline industry is not now undergoing a shakeup.

27. a. Domestic car sales increased over the past year, and/or foreign car sales decreased over the past year.
b. Domestic car sales increased over the past year, and foreign car sales decreased over the past year.
c. Either domestic car sales increased over the past year or foreign car sales decreased over the past year.
d. Domestic car sales did not increase over the past year.
e. Domestic car sales did not increase over the past year, or foreign car sales decreased over the past year.
f. Domestic car sales did not increase over the past year, or foreign car sales did not decrease over the past year.

29. a. Either the doctor recommended surgery to treat his hyperthyroidism or the doctor recommended radioactive iodine to treat his hyperthyroidism.
b. The doctor recommended surgery to treat his hyperthyroidism, and/or the doctor recommended radioactive iodine to treat his hyperthyroidism.

31. a. $p \wedge q$ **b.** $p \veebar q$ **c.** $\sim(p \wedge q)$ **d.** $\sim(\sim q)$

33. a. Both the popularity of prime-time soaps and prime-time situation comedies did not increase this year.
b. The popularity of prime-time soaps did not increase this year, and/or the popularity of prime-time detective shows decreased this year.
c. The popularity of prime-time soaps increased this year, and the popularity of prime-time situation comedies did not increase this year.
d. Either the popularity of prime-time soaps did not increase this year or the popularity of prime-time detective shows did not decrease this year.

Exercises A.2, page 606

1.

p	q	$\sim q$	$p \vee \sim q$
T	T	F	T
T	F	T	T
F	T	F	F
F	F	T	T

3.

p	$\sim p$	$\sim(\sim p)$
T	F	T
F	T	F

5.

p	$\sim p$	$p \vee \sim p$
T	F	T
F	T	T

7.

p	$\sim p$	q	$p \vee q$	$\sim p \wedge (p \vee q)$
T	F	T	T	F
T	F	F	T	F
F	T	T	T	T
F	T	F	F	F

9.

p	q	$\sim q$	$p \vee q$	$p \wedge \sim q$	$(p \vee q) \wedge (p \wedge \sim q)$
T	T	F	T	F	F
T	F	T	T	T	T
F	T	F	T	F	F
F	F	T	F	F	F

11.

p	q	$p \vee q$	$\sim(p \vee q)$	$(p \vee q) \wedge \sim(p \vee q)$
T	T	T	F	F
T	F	T	F	F
F	T	T	F	F
F	F	F	T	F

13.

p	q	r	$p \vee q$	$p \vee r$	$(p \vee q) \wedge (p \vee r)$
T	T	T	T	T	T
T	T	F	T	T	T
T	F	T	T	T	T
T	F	F	T	T	T
F	T	T	T	T	T
F	T	F	T	F	F
F	F	T	F	T	F
F	F	F	F	F	F

15.

p	q	r	$p \wedge q$	$\sim r$	$(p \wedge q) \vee \sim r$
T	T	T	T	F	T
T	T	F	T	T	T
T	F	T	F	F	F
T	F	F	F	T	T
F	T	T	F	F	F
F	T	F	F	T	T
F	F	T	F	F	F
F	F	F	F	T	T

17.

p	q	r	$\sim q$	$p \wedge \sim q$	$p \wedge r$	$(p \wedge \sim q) \vee (p \wedge r)$
T	T	T	F	F	T	T
T	T	F	F	F	F	F
T	F	T	T	T	T	T
T	F	F	T	T	F	T
F	T	T	F	F	F	F
F	T	F	F	F	F	F
F	F	T	T	F	F	F
F	F	F	T	F	F	F

19. 16 rows

Exercises A.3, page 611

1. $\sim q \to p$; $q \to \sim p$; $\sim p \to q$

3. $p \to q$; $\sim p \to \sim q$; $\sim q \to \sim p$

5. Conditional: If it is snowing, then the temperature is below freezing.
   Biconditional: It is snowing if and only if the temperature is below freezing.

7. Conditional: If the company's union and management reach a settlement, then the workers will not strike.
   Biconditional: The company's union and management will reach a settlement if and only if the workers do not strike.

9. False 11. False

13. It is false when
   a. I buy the house and the owner does not lower the selling price and
   b. I do not buy the house and the owner lowers the selling price.

15.

p	q	$p \to q$	$\sim(p \to q)$
T	T	T	F
T	F	F	T
F	T	T	F
F	F	T	F

17.

p	q	$p \to q$	$\sim(p \to q)$	$\sim(p \to q) \wedge p$
T	T	T	F	F
T	F	F	T	T
F	T	T	F	F
F	F	T	F	F

19.

p	q	$\sim p$	$\sim q$	$p \to \sim q$	$(p \to \sim q) \veebar \sim p$
T	T	F	F	F	F
T	F	F	T	T	T
F	T	T	F	T	F
F	F	T	T	T	F

21.

p	q	$\sim p$	$\sim q$	$p \to q$	$\sim q \to \sim p$	$(p \to q) \leftrightarrow (\sim q \to \sim p)$
T	T	F	F	T	T	T
T	F	F	T	F	F	F
F	T	T	F	T	T	F
F	F	T	T	T	T	T

23.

p	q	$p \wedge q$	$p \vee q$	$(p \wedge q) \rightarrow (p \vee q)$
T	T	T	T	T
T	F	F	T	T
F	T	F	T	T
F	F	F	F	T

25.

p	q	r	$p \vee q$	r	$(p \vee q) \rightarrow r$
T	T	T	T	F	F
T	T	F	T	T	T
T	F	T	T	F	F
T	F	F	T	T	T
F	T	T	T	F	F
F	T	F	T	T	T
F	F	T	F	F	T
F	F	F	F	T	T

27.

p	q	r	$p \vee r$	$p \rightarrow (q \vee r)$
T	T	T	T	T
T	T	F	T	T
T	F	T	T	T
T	F	F	F	F
F	T	T	T	T
F	T	F	T	T
F	F	T	T	T
F	F	F	F	T

29. Logically equivalent

31. Not logically equivalent

33. Not logically equivalent

35. Not logically equivalent

37. **a.** $p \rightarrow \sim q$ **b.** $\sim p \rightarrow q$ **c.** $\sim q \leftrightarrow p$
d. $p \rightarrow \sim q$ **e.** $p \leftrightarrow \sim q$

Exercises A.4, page 616

1.

p	p	$p \wedge p$
T	T	T
F	F	F

3.

p	q	r	$p \wedge q$
T	T	T	T
T	T	F	T
T	F	T	F
T	F	F	F
F	T	T	F
F	T	F	F
F	F	T	F
F	F	F	F

$(p \wedge q) \wedge r$	$q \wedge r$	$p \wedge (q \wedge r)$
T	T	T
F	F	F
F	F	F
F	F	F
F	T	F
F	F	F
F	F	F
F	F	F

5.

p	q	$p \wedge q$	$q \wedge p$
T	T	T	T
T	F	F	F
F	T	F	F
F	F	F	F

7.

p	q	r	$q \wedge r$	$p \vee (q \wedge r)$
T	T	T	T	T
T	T	F	F	T
T	F	T	F	T
T	F	F	F	T
F	T	T	T	T
F	T	F	F	F
F	F	T	F	F
F	F	F	F	F

$p \vee q$	$p \vee r$	$(p \vee q) \wedge (p \vee r)$
T	T	T
T	T	T
T	T	T
T	T	T
T	T	T
T	F	F
F	T	F
F	F	F

9. Tautology 11. Tautology

13. Tautology

15. Tautology 17. Neither

19. $\sim(p \wedge q)$: The candidate does not oppose changes in the Social Security system, or the candidate does not support the ERA.
$\sim(p \vee q)$: The candidate does not oppose changes in the Social Security system, and the candidate does not support the ERA.

21. $[p \wedge (q \vee \sim q) \vee (p \wedge q)]$
$\Leftrightarrow p \wedge t \vee (p \wedge q)]$ (By Law 11)
$\Leftrightarrow p \vee (p \vee q)$ (By Law 14)

23. $(p \wedge \sim q) \vee (p \wedge \sim r)$
$\Leftrightarrow p \wedge (\sim q \vee \sim r)$ (By Law 7)

25. $(p \wedge [\sim(q \wedge r)]$
$\Leftrightarrow p \wedge (\sim q \vee \sim r)$ (By Law 10)
$\Leftrightarrow (p \vee \sim q) \vee (p \wedge \sim r)$ (By Law 7)

Exercises A.5, page 622

1. Valid 3. Valid 5. Invalid 7. Valid

9. Valid 11. Valid 13. Valid 15. Invalid

17. $p \to q$; invalid
$\underline{\sim p}$
$\therefore \sim q$

19. $p \vee q$; valid
$\underline{\sim p \to \sim q}$
$\therefore p$

21. $p \to q$; invalid
$q \to r$
$\underline{r}$
$\therefore p$

23. b

25.

p	q	$p \to q$	$\sim q$	$\sim p$
T	T	T	F	F
T	F	F	T	F
F	T	T	F	T
F	F	T	T	T

Exercises A.6, page 625

1. $p \wedge q \wedge (r \vee s)$

3. $[(p \wedge q) \vee r] \wedge (\sim r \vee p)$

5. $[(p \vee q) \wedge r] \vee (\sim p) \vee [\sim q \wedge (p \vee r \vee \sim r)]$

7.

9.

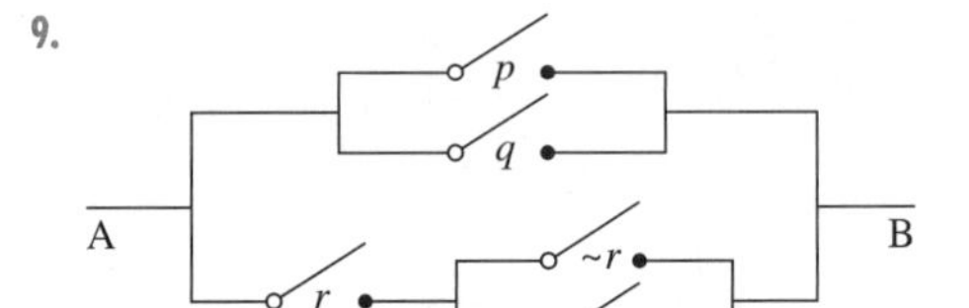

11.

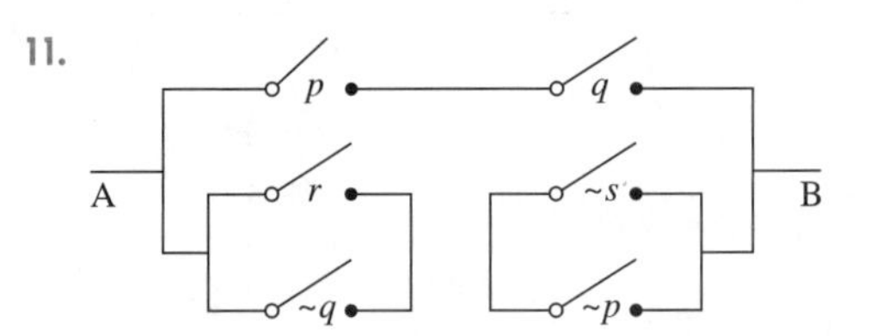

13. $p \wedge [\sim q \vee (\sim p \wedge q)]$; $p \wedge \sim q$

15. $p \wedge [\sim p \vee q \vee (q \wedge r)]$; $p \wedge q$

Index

FORMULAS

EQUATION OF A STRAIGHT LINE

a. point-slope form $y - y_1 = m(x - x_1)$
b. slope-intercept form $y = mx + b$
c. general form $Ax + By + C = 0$

EQUATION OF THE LEAST-SQUARES LINE

$$y = mx + b$$

where m and b satisfy the **normal equations**

$$nb + (x_1 + x_2 + \cdots + x_n)m = y_1 + y_2 + \cdots + y_n$$

$$(x_1 + x_2 + \cdots + x_n)b + (x_1^2 + x_2^2 + \cdots + x_n^2)m = x_1y_1 + x_2y_2 + \cdots + x_ny_n$$

COMPOUND INTEREST

$$A = P(1 + i)^n \qquad (i = r/m,\ n = mt)$$

where A is the accumulated amount at the end of n conversion periods, P is the principal, r is the interest rate per year, m is the number of conversion periods per year, and t is the number of years.

EFFECTIVE RATE OF INTEREST

$$r_{\text{eff}} = \left(1 + \frac{r}{m}\right)^m - 1$$

where r_{eff} is the effective rate of interest, r is the nominal interest rate per year, and m is the number of conversion periods per year.

FUTURE VALUE OF AN ANNUITY

$$S = R\left[\frac{(1 + i)^n - 1}{i}\right]$$

PRESENT VALUE OF AN ANNUITY

$$P = R\left[\frac{(1 - (1 + i)^{-n}}{i}\right]$$

AMORTIZATION FORMULA

$$R = \frac{Pi}{1 - (1 + i)^{-n}}$$